Mathematical Modeling of Complex Reaction Systems in the Oil and Gas Industry

Mathematical Modeling of Complex Reaction Systems in the Oil and Gas Industry

Edited by

Jorge Ancheyta
Instituto Mexicano del Petróleo
Mexico City, Mexico

Andrey Zagoruiko
Boreskov Institute of Catalysis
Novosibirsk, Russia

Andrey Elyshev
University of Tyumen
Tyumen, Russia

WILEY

This edition first published 2024

Registered Offices
John Wiley & Sons, Inc., 111 River Street, Hoboken, NJ 07030, USA
John Wiley & Sons Ltd, The Atrium, Southern Gate, Chichester, West Sussex, PO19 8SQ, UK

For details of our global editorial offices, customer services, and more information about Wiley products visit us at www.wiley.com.

Wiley also publishes its books in a variety of electronic formats and by print-on-demand. Some content that appears in standard print versions of this book may not be available in other formats.

Library of Congress Cataloging-in-Publication Data is Applied for
Hardback ISBN: 9781394220021

Cover Design: Wiley
Cover Image: © Created and Designed by Darya Obanina together with the Shedevrum neural network

Set in 9.5/12.5pt STIXTwoText by Straive, Pondicherry, India
Printed and bound by CPI Group (UK) Ltd, Croydon, CR0 4YY

C9781394220021_030724

Contents

List of Contributors

Osman Abdulla
Tyumen State University, Tyumen, Russia

Jorge Ancheyta
Department of Petroleum Engineering
Kazan Federal University, Kazan, Russia;
Instituto Mexicano del Petróleo
Mexico City, Mexico; and Escuela Superior de
Ingeniería Química e Industrias Extractivas
Instituto Politécnico Nacional, UPALM
Zacatenco, Mexico City Mexico

Cristian Calderón
Tecnológico de Monterrey, Mexico State
Mexico

Andrew Elyshev
Tyumen State University, Tyumen, Russia

Guillermo Félix
Department of Petroleum Engineering
Kazan Federal University, Kazan
Russia and Tecnológico Nacional de
México/IT de Los Mochis, Los Mochis
Sinaloa, Mexico

Nikita Glazov
Boreskov Institute of Catalysis SB RAS
Prospekt Akademika Lavrent'yeva
5, Novosibirsk, Novosibirsk Oblast
Russian Federation

Javier Jurado
Instituto Politécnico Nacional, Centro de
Investigación en Ciencia Aplicada y Tecnología
Avanzada, Unidad Legaria, Legaria 694
Col. Irrigación, Ciudad de México, Mexico

Sergey Lopatin
Boreskov Institute of Catalysis, Novosibirsk
Russia and Tyumen State University
Tyumen, Russia

César I. Méndez
Centro de Investigación en Ciencia Aplicada
y Tecnología Avanzada del Instituto Politécnico
Nacional, Mexico City, Mexico

Alexis Tirado
Department of Petroleum Engineering
Kazan Federal University, Kazan, Russia

Fernando Trejo
Instituto Politécnico Nacional
CICATA-Legaria, Ciudad de México, Mexico

Andrey Zagoruiko
Boreskov Institute of Catalysis, Novosibirsk
Russia and Tyumen State University
Tyumen, Russia

Sergey Zazhigalov
Boreskov Institute of Catalysis, Novosibirsk
Russia and Tyumen State University
Tyumen, Russia

Vicente Samano
Instituto Politécnico Nacional
Centro de Investigación en Ciencia
Aplicada y Tecnología Avanzada
Unidad Legaria, Legaria 694
Col. Irrigación
Ciudad de México, Mexico

Preface

After more than 30 years of experience working in the kinetic and reactor modeling of different processes for the petroleum and gas industry as well as for environmental protection applications, we have identified the need to have a collection of the most relevant cases that would have great impact on the industry. The general areas that are described in this book correspond to reaction kinetics, catalyst deactivation, reactor model, computational fluid dynamics, and steady-state and dynamics simulations using own experimental data or information reported in the literature.

Mathematical Modeling of Complex Reaction Systems in the Oil and Gas Industry is focused on the step-by-step description of kinetics and reactor modeling of various important processes used in the oil and gas industry. The topics covered in this book are hydrocracking of heavy oils, catalyst deactivation, oxidative regeneration of catalyst, adsorption–desorption catalytic processes, molecular reconstruction, production of green diesel via catalytic hydrotreating, slurry-phase hydrocracking reactor, Fischer–Tropsch synthesis and structured beds of micro-fibrous catalyst. Nowadays, all these topics are relevant to the oil and gas industry, particularly in this era of energy transition and decarbonization, which makes the book original and unique among previous reactor modeling books.

Each chapter describes the development of kinetic and reactor models for steady-state and dynamic simulations. To develop the kinetic model for each reaction or set of reaction for further reactor modeling, exhaustive experimental data either reported in the literature or generated in own laboratories are used. The developed models are validated with laboratory experimental data and simulation are reported to predict the commercial performance of the involved reactors. In addition, such a modeling in some cases allows for a deeper understanding of the qualitative essence of the phenomena observed in experiments, that is, to apply the augmented reality approach and make visible what cannot be directly observed in experiments, or at least to reproduce the most plausible picture of the physical events taking place.

This book is expected to be a reference guide for researchers, PhD students, postdoctoral researchers, catalyst manufacturers, process designers, and professors, to help them develop kinetic and reactor models at different reaction scales. It is also anticipated that this textbook can be used to cover part of the content of courses of different carriers at undergraduate and postgraduate levels.

Jorge Ancheyta, National Polytechnic Institute and Mexican Institute of Petroleum, Mexico
Andrey Zagoruiko, Boreskov Institute of Catalysis, Novosibirsk, Russia
Andrey Elyshev, Tyumen State University, Russia

1

Modeling the Kinetics of Hydrocracking of Heavy Oil with Mineral Catalyst

Guillermo Félix[1,2], Fernando Trejo[3], and Jorge Ancheyta[4,5]

[1] *Department of Petroleum Engineering, Kazan Federal University, Kazan, Russia*
[2] *Tecnológico Nacional de México/IT de Los Mochis, Los Mochis, Sinaloa, Mexico*
[3] *Instituto Politécnico Nacional, CICATA-Legaria. Legaria 694, Col. Irrigación, Ciudad de México, Mexico*
[4] *Instituto Mexicano del Petróleo, Eje Central Lázaro Cárdenas Norte 152, San Bartolo Atepehuacan, Mexico City, Mexico*
[5] *Escuela Superior de Ingeniería Química e Industrias Extractivas, Instituto Politécnico Nacional, UPALM, Zacatenco, Mexico City, Mexico*

1.1 Introduction

The continuously increasing global demand for energy and the decrease of light crude oil reserves have caused researchers to look toward heavy crude oil reserves. Despite the importance of renewable sources is growing, fossil energy sources will continue to meet most of the global energy requirements (Bellussi et al. 2013; Félix and Ancheyta 2019a; Martínez-Grimaldo et al. 2014; Rana et al. 2007; Santos et al. 2014). The driving forces of the oil industries to improve heavy crude oils and vacuum residues (VRs) are as follows: (i) obtain high-quality transportation fuels, (ii) middle distillates, (iii) reduction of conventional crude oil supply, and (iv) price increase. Heavy crude oils or VRs are considered suitable alternative sources for transportation fuels, energy, and petrochemicals to meet the demands of industries. However, because of the low quality and difficulties in transporting heavy crude oils, it is necessary to develop technologies to upgrade their properties (Castañeda et al. 2014; Martínez-Grimaldo et al. 2014; Sahu et al. 2015).

In recent decades, researchers have prioritized refining techniques that use low-cost feedstock, such as coal, low-grade oil or wax, heavy crude oil, VRs, and natural gases, which are aimed to improve the cost-effectiveness of the refining process. The goal of these techniques is to transform such feedstock into transportation fuels and other low-boiling point liquid products. Refining heavy crude oils and VRs decreases their viscosity, initial boiling point, metal contents, sulfur, and other impurities, while increasing the H/C ratio to meet the commercial standards (Rodríguez et al. 2018; Sahu et al. 2015; Speight 2004).

1.1.1 Reserves and Production of Heavy Crude Oils

The assessment of production capacity of oil field reveals that conventional light and medium oil reserves peaked in the 1960s. Since then, these reserves have steadily declined. An important focus of the global oil industry is to develop new technologies to be feasible: the use of unconventional resources as heavy and extra heavy crude oils (Heidary et al. 2017; Martínez et al. 2010; Santos et al. 2014). At the end of 2021, more than 1730 billion barrels of proven reserves are reported worldwide,

Mathematical Modeling of Complex Reaction Systems in the Oil and Gas Industry, First Edition.
Edited by Jorge Ancheyta, Andrey Zagoruiko, and Andrey Elyshev.

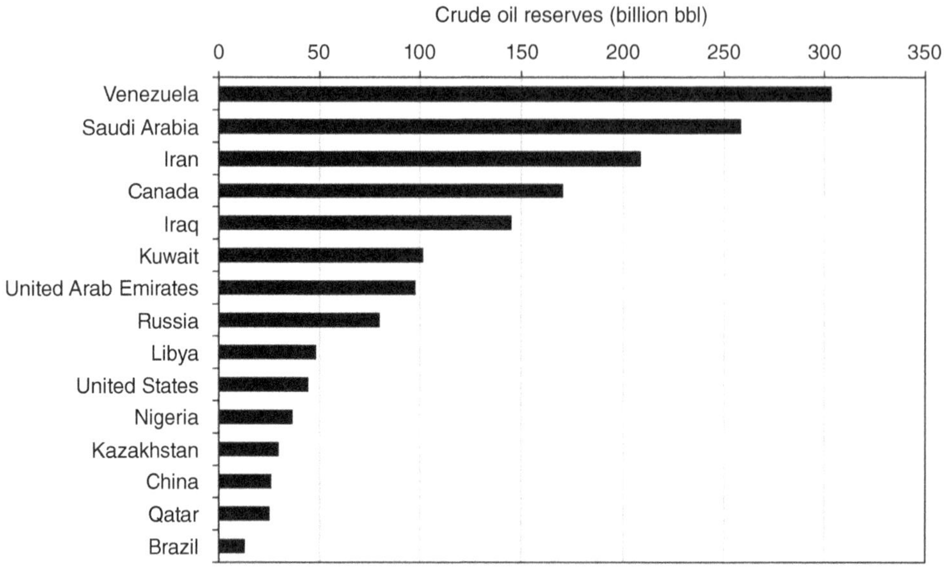

Figure 1.1 Proven reserves of crude oil by country.

as shown in Figure 1.1 (*International—U.S. Energy Information Administration (EIA)* 2021; *Oil Gas Journal* 2022; *Organization of the Petroleum Exporting Countries (OPEC)* 2020). Venezuela, Saudi Arabia, and Iran possess the largest crude oil reserves; however, the quality of their crudes varies significantly. Currently, these reserves contain 434.3 billion barrels of heavy and extra-heavy crude oil, while 650.7 billion barrels correspond to bitumen, representing over 60% of the total volume (Bata et al. 2019; Pratama and Babadagli 2022).

1.1.2 Heavy Crude Oil Upgrading Processes

The processing of the heavy crude oil presents difficulties due to the complex nature of its heavier fractions, which commonly possess low API gravity (high density), high viscosity, elevated initial boiling point, higher contents of heteroatoms (sulfur, nitrogen, and oxygen), metals (Ni and V), and asphaltenes. In general, most of sulfur and nitrogen species present in crude oils are found in heavier fractions, while metals are found in porphyrins and as concentrated, exclusively in residue fractions (Ancheyta 2011; Memon et al. 2010; Quitian and Ancheyta 2016b). Therefore, the addition of hydrogen or rejection of carbon is the main process focused to increase the hydrogen-to-carbon (H/C) ratio in heavy crudes. These two routes can be catalytic and noncatalytic (Ancheyta 2013; Félix et al. 2017).

Over the years, several technologies based on carbon rejection, the addition of hydrogen, and the combination of both have been used to upgrade heavy crude oils (Figure 1.2). The noncatalytic (deasphalting, gasification, coking, and visbreaking) and catalytic (catalytic cracking of VR) carbon rejection technologies account for 56.6% of the global processing technology due to its cost-effective operation. The noncatalytic (hydrovisbreaking) and catalytic (hydrotreating and hydrocracking) hydrogen addition technologies display advantages and disadvantages when applied to heavy crude oil processing as well as carbon rejection technologies (Castañeda et al. 2014; Rana et al. 2007; Sahu et al. 2015; Speight 2004).

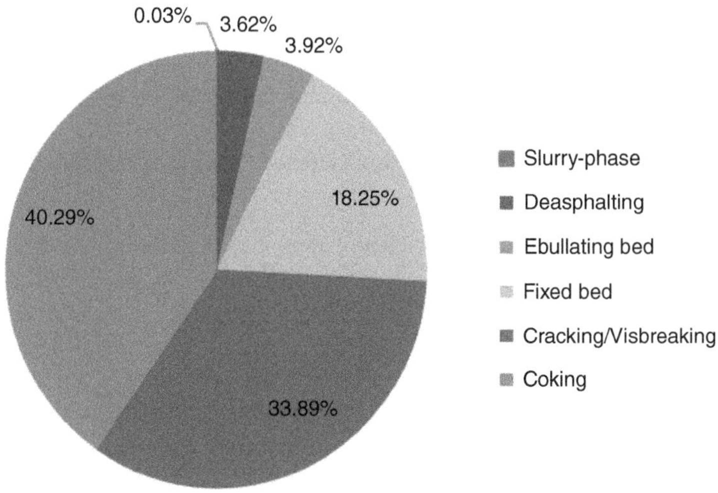

Figure 1.2 Processing capacity for worldwide heavy crude oil upgrading technologies.

For heavy and extra-heavy crude oils, the choice between carbon rejection and hydrogen addition technologies depends on technical and economic assessments, as well as the quality of products from downstream refining processes. Therefore, deciding on which upgrading technology is appropriate for certain heavy crude oil is not an easy task and several factors must be taken into consideration, such as the following (Ancheyta 2013; Ancheyta and Speight 2007; Quitian and Ancheyta 2016b):

✓ Price of heavy crude oil
✓ Amount of impurities in feedstock
✓ Level of quality of the upgraded oil
✓ Process outline of the refinery

In the past, the carbon rejection technologies were preferred for feed with high metal content, such as heavy crude oils. Nonetheless, these technologies have the following disadvantages (Ancheyta 2013; Sahu et al. 2015):

✓ Low yield of upgraded oil
✓ Enhanced formation of coke
✓ High aromatic content in the upgraded oil
✓ Lower yield and quality of gasoline and diesel fuels
✓ Low value of upgraded oil

Technologies based on the hydrogen addition produce crude oils that possess superior quality and greater added value than carbon rejection methods do; hence, these approaches are becoming increasingly attractive. Highlights of the hydrogen addition technologies are the following:

✓ Minimizes the coke formation
✓ Enhances the yield of liquid products obtained by hydrocracking or hydrogenolysis mechanisms
✓ Increases the H/C ratio of the products
✓ Reduction of impurities, such as sulfur, nitrogen, metals, and asphaltenes
✓ Increases hydrogen consumption, giving higher liquid yields

These technologies also present the following disadvantages (Ancheyta 2013; Quitian and Ancheyta 2016b; Rana et al. 2007):

✓ Large volumes of hydrogen are necessary to carry out the hydrogenation of carbon-rich crude oil
✓ Difficulties arise from the carbon and metal deposition on the catalyst surface
✓ Requires well-designed catalysts that are able to process high levels of metals and asphaltenes within the feed

There are also some emerging technologies at different levels of development or even close to be commercialized, which have been developed due to the growing production of heavy and extra-heavy crude oils (Castañeda et al. 2014; Speight 2004). Hydrocracking is a well-known refining process where heavy crude oil in hydrogen atmosphere is transformed into light products with better quality. There are different types of reactors to carry out this process and the choice depends mainly on the level of conversion required and the number of impurities (metals and asphaltenes) in the feed. Three-phase catalytic reactors are highly valued due to their broad applications in various reaction systems and being notably used in the petroleum industry, including hydrocracking processes. Fixed-bed reactors have stable and reliable performance, but strong limitations are feed properties due to fast catalyst deactivation by coking and metal depositions. Ebullated-bed reactors perform better than fixed-bed reactors since they can process feed containing higher quantities of metals and coke at the expense to be conversion-limited and reaching values up to 80%. This is because of the rapid formation of sediments, by which commonly aromatic solvents are used to keep the soluble in the reacting phases. Slurry phase reactors achieve higher conversions performing superior levels of upgradation of feed rich in sulfur, metals, and asphaltenes (Ancheyta 2011; Jarullah et al. 2011; Martínez-Grimaldo et al. 2014; Sahu et al. 2015).

The degree of conversion is affected by the properties and nature of the feed, thus lighter crude oil possesses lower number of impurities, while heavy fractions contain higher content of asphaltenes (Alonso et al. 2019; Marafi et al. 2005; Ortega-García et al. 2012; Stanislaus et al. 2005; Stratiev et al. 2014, 2019; Tirado and Ancheyta 2018). Asphaltene is the most polar fraction found in heavy crudes and residues that is soluble in aromatic solvents but not in alkanes. This fraction comprises of both aromatic and aliphatic carbons, which contribute to its high molecular weight. The complex structure (Figure 1.3) makes asphaltenes one of the most studied fractions as they are coke precursors. The asphaltene fraction is formed by large aromatic sheets, side alkyl chains, heteroatoms, and metalloporphyrins. These latter components exhibit a remarkable influence of p-electronic interactions. The alteration in solubility during the processing of heavy crudes or residues is linked to solid formation. This disruption in solubility equilibrium relies on the chemical characteristics of the feed and the processing conditions. Based on the colloidal model (Figure 1.4), asphaltenes are surrounded by resin-building micelles keeping the asphaltenes dispersed in crude oil that avoids the contact with saturated compounds, since their nonpolar nature cause the instability of asphaltene micelles. Sediment formation is the result of a disturbance in the balance between asphaltenes and those compounds that keep them dispersed in solution during hydrotreating/hydrocracking; in other words, resins lose their peptization properties causing aggregation and precipitation of asphaltenes (Ashoori et al. 2017; Félix and Ancheyta 2019a; Ortega-García et al. 2012; Rana et al. 2007; Rogel et al. 2013; Stratiev et al. 2014, 2019; Tirado and Ancheyta 2018).

In addition to the loss of solubility of asphaltenes by oil and resin fractions, hydrocracking leads to an increase in the aromaticity and condensation of unconverted asphaltene cores. Reports indicate that the asphaltenes found in upgraded products have higher aromaticity and a lower H/C ratio in comparison with original asphaltenes in the feed. The results of structural analysis of the hexane-insoluble fraction indicate a higher degree of polarity due to the presence of a significant number of

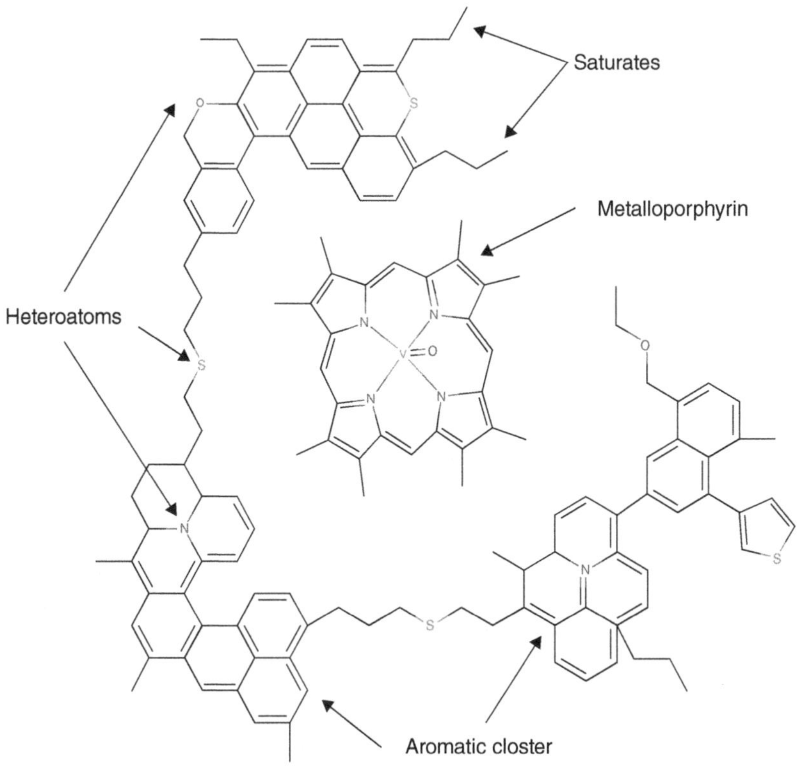

Figure 1.3 Hypothetical asphaltene molecule.

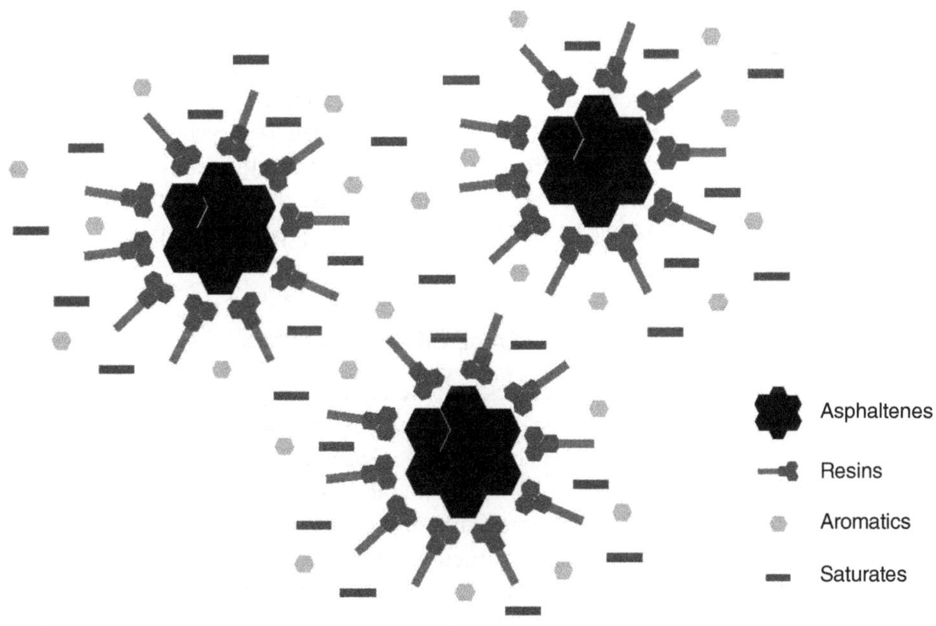

Figure 1.4 Heavy crude oil composition based on SARA fractions.

heteroatoms with shorter alkyl chains because of the dealkylation reactions, which causes asphaltenes to become more aromatic and precipitate. The precipitation of sediment has also been attributed to the possible liquid–liquid phase separation of the material at high temperatures of processing (Mochida et al. 1989; Rogel et al. 2013; Tirado and Ancheyta 2018).

Sediment formation typically has a maximum limit of 0.8–1.0 wt.% in refineries. Higher values may cause commercial plants to shut down due to plugging issues. For this reason, sediment formation is a critical process parameter during hydrotreating and hydrocracking of heavy oils and thus dictates the maximum allowable residue conversion. Furthermore, a more extensive comprehension of sediment formation during upgrading processes is essential for enhancing both performance and control strategies. Improving the predictability of sediment deposition is highly desirable as it would enable the preliminary screening of feed and the optimization of processing (Alonso et al. 2019; Rogel et al. 2013; Tirado and Ancheyta 2018).

1.1.3 Reactions During Slurry Phase Hydrocracking

For hydrocracking reactions in slurry phase, the operating conditions can be classified as follows (Quitian and Ancheyta 2016a; Speight 2013):

- Mild conditions can be attained if the pressure and temperature do not exceed 80–100 kg/cm^2 and 410 °C, respectively, while the hydrocracking of heavy crude oil or residua leads to a VR of less than 50%.
- Severe conditions are reached at pressure and temperature exceeding 80–100 kg/cm^2 and 410 °C, respectively, and the VR conversion is greater than 50% during the hydrocracking of heavy crudes or residua.

Hydrocracking and hydrogenolysis are exothermic reactions and may occur at different extents depending on the operating conditions. At mild temperatures, hydrogenation is the dominant reaction unlike hydrocracking and thermal cracking reactions that take place at high temperatures (Quitian and Ancheyta 2016a; Raşeev 2003). The polyaromatic compounds break the weak bonds, such as C–S, C–O, and so on, and transform into free radicals. Meanwhile, the H_2 molecule dissociates to form Mo–H or S–H bonds, which in turn release hydrogen radicals to cap unstable compounds. Consequently, the formation and saturation of free radicals is a cyclic process. It has been reported that thermal and catalytic hydrocracking occur in series, the residue is converted to vacuum gas oil (VGO), medium distillate, naphtha, and gases, and the combination of thermal cracking and hydrogenation promotes the formation of the requested products (Martínez-Grimaldo et al. 2014; Nguyen et al. 2016; Quitian et al. 2015).

1.1.4 Catalysts for Hydrocracking of Heavy Crude Oils in Slurry Phase

The types of catalysts used in hydrocracking process are the solid heterogeneous powders and dispersed catalysts. The first type contains at least one metal (cobalt, molybdenum, nickel, iron) supported on alumina, silica–alumina, etc., by which the catalytic system is heterogeneous, containing the feed in liquid phase, the hydrogen in gaseous phase, and the catalyst in solid phase. As the reaction takes place, heavier molecules block the pore mouths and reduce the catalytic activity, thereby, reducing the reaction rate (Nguyen et al. 2016; Rodríguez et al. 2018). To solve this issue, unsupported catalysts were developed. The dispersed catalysts are particulate materials that enhance the catalytic hydrotreating at high reaction rates due to their large specific surface area, which makes the kinetics in the slurry phase faster than in fixed-, moving-, or ebullated-bed reactors. The faster the mass and energy transfer, the lower the over-hydrocracking activity, and thereby, the lower the coke and gas formation (Liu et al. 2009; Quitian and Ancheyta 2016a).

Dispersed catalysts are highly stable by which optimized contact between oil and hydrogen is possible. As a result of high metal contents in dispersed catalysts, the heavy hydrocarbons reach the active sites very quickly and avoid plugging the pores as in supported catalysts. For these reasons, dispersed catalysts can be used directly in presence of heavier feedstock making slurry processes more suitable than other hydroconversion technologies when processing heavy feedstock (Angeles et al. 2014; Bellussi et al. 2013). Dispersed catalysts consist in different types: (i) soluble in water, (ii) soluble in oil (in this case heavy crude oil), (iii) mineral, and (iv) ionic liquid. The dispersed oil-soluble catalysts are generally more expensive than the water-soluble catalysts, but the water can evaporate during the reaction and reduce the catalytic activity, resulting in particle agglomeration. Meanwhile, the oil-soluble catalyst can be uniformly dispersed in the feed enhancing hydrogen uptake and suppressing the coke formation. The conversion as well as the quality of liquid product increases (Galarraga et al. 2012; Kim et al. 2017; Liu et al. 2009; Nguyen et al. 2016).

The mineral catalysts are natural ores with a high content of one or more transition metals, usually the Fe-based and Mo-based minerals are the most employed for slurry-phase hydrocracking. Different works have been employed magnetite, limonite, hematite, ferrite, laterite, pyrite, and molybdenite, among others. This type of catalyst does not require any special preparation, instead drying, grinding, and sieving are the usual operations required before using, making them cheaper than other catalysts (Al-Attas et al. 2019; Fukuyama and Terai 2007; Quitian and Ancheyta 2017). The ionic liquid catalysts are a relatively new alternative to dispersed catalysts. These compounds are organic salts that can be liquefied at near room temperature. Their catalytic activity depends primarily on the cations and anions used during synthesis and usually contain molybdenum, cobalt, iron, or nickel (Cai et al. 2022; Ma et al. 2022). It is preferred to use molybdenum compounds as dispersed catalysts due to their greater hydrogenation activity (Calderon and Ancheyta 2016; Du et al. 2015; Panariti et al. 2000). It has been shown that metal sulfide is the active species in hydrocracking process, and it is formed by reacting the metal precursors with sulfur compounds in the feed or additional sources of sulfur (Nguyen et al. 2016; Panariti et al. 2000). The sulfidation (activation of catalysts) is done before the hydrocracking reaction and starts by two routes: (i) *ex situ* and (ii) *in situ* sulfidation. In the case of *ex situ* sulfidation, the sulfur sources are often found in carbonaceous compounds (as heavy crude oils) or using hydrogen sulfide (Nguyen et al. 2016; Quitian and Ancheyta 2016a). Small concentration of low-cost, finely dispersed catalyst makes recovery impractical after reaction. In addition, slurry catalysts are deactivated after reaction, which occurs by coke and metal deposition hindering its reuse (Quitian and Ancheyta 2016a; Rezaei 2013).

1.2 Kinetic Models

For adequate modeling of a new hydrocracking process is imperative to take into account the feed properties, design, performance, process control, and cost to meet the specification of the products. The main issue when modeling hydrocracking process is the complex nature of feedstock compared with model compounds by which it is necessary to understand the changes in the rate constants, conversion, and API gravity at different reaction scales. This will improve the methodologies for upscaling reaction parameters from the laboratory to industrial scale (Ancheyta et al. 2005; Becker et al. 2016; Celse et al. 2015; Elizalde et al. 2009; Hassanzadeh and Abedi 2010; Marafi et al. 2010; Rashidzadeh et al. 2011).

1.2.1 General Types of Kinetic Models

Discrete and continuous lumping models have been largely used to represent the reaction kinetics. Recently, the use of more complex and fundamental microkinetic models to simulate hydrocracking has increased the interest of the research community (Becker et al. 2016; Martínez and Ancheyta 2012; Sánchez et al. 2005a). The yield of gases, naphtha, middle distillates, VGO, and residue based on boiling point intervals is adequately predicted by kinetic models. In addition, the so-called SARA fractions, that is, saturates, aromatics, resins, and asphaltenes, are also included in the modeling. Kinetic studies for the transformation of asphaltenes into maltenes fractions (saturates, aromatics, and resins) have received less attention than those involving the pseudocomponents based on boiling point ranges (Asaee et al. 2014; Félix and Ancheyta 2019a; Moustafa and Froment 2003).

1.2.1.1 Lumping Kinetic Models

Traditional models by pseudocomponents take macroscopic properties (for example, measurable properties in the product) and derive the equation system to be solved. These models can be specifically grouped into compounds in the feed or products in a few cuts, often characterized by ranges of boiling points that undergo different reactions in series or parallel (Ancheyta 2013; Ancheyta et al. 2005; Balasubramanian and Pushpavanam 2008; Elizalde et al. 2009; Galarraga et al. 2012; Hassanzadeh and Abedi 2010; Moghadassi et al. 2011).

Models based on pseudocomponents (up to 10) are commonly reported in the literature to simplify the large number of hydrocarbons present in heavy crude oil fractions for lumping purposes during hydrocracking. These models are used to simulate the yield of fractions of interest for refiners or process designers in spite of the reduced amount of information obtained from pseudocomponent-based models. When a distillation curve is required, a small number of pseudocomponents must be interpolated, while the number of kinetic parameters to be calculated increases exponentially, which clearly is not feasible when modeling (Asaee et al. 2014; Becker et al. 2016; Quitian and Ancheyta 2016a; Sadighi et al. 2010).

1.2.1.2 Continuous Lumping Kinetic Models

Continuous models overcome limitations in lumping models by considering that petroleum is formed by an infinite number of components, allowing crude to be considered as a continuous mixture instead of pseudocomponents. Continuous mixing takes into account a large number of undefined species contained in the oil fraction, and the difference between adjacent species is relatively smaller (Ancheyta 2013; Elizalde et al. 2009).

A continuous function describes the reactivity, while the distillation curve is considered to solve the model keeping the number of parameters constant. In addition, properties, such as the carbon number or the true boiling points in distillation, are used to characterize the feedstock and products in the continuous kinetic modeling. The combination of continuous models and pseudocomponents allows more than one continuous distribution to be obtained. However, this model is highly depending on feed properties, and its advantages are as follows (Ancheyta 2013; Becker et al. 2016):

- ✓ Prediction of the distillation curve with few parameters to be estimated
- ✓ Track the chemistry of the process accurately
- ✓ Individual reaction order has physical meaning
- ✓ The properties of fractions between the initial and final boiling points are easily obtained
- ✓ The heteroatom distribution curve is well predicted
- ✓ Easy to adapt to heavy, extra-heavy crudes and residue kinetics

The model involves the dimensionless temperature (θ) and weight fraction, which is calculated as follows:

$$\theta = \frac{\text{TBP} - \text{TBP(l)}}{\text{TBP(h)} - \text{TBP(l)}} \tag{1.1}$$

where TBP is the true boiling point temperature (in kelvin) of any component, TBP(h) and TBP(l) are the highest and lowest boiling points in the mixture, respectively. Because the TBP of the distillation curve is a function of the reactivity (λ), a coordinate transformation between discrete and continuous forms (from the TBP to reactivity for any mixture) is expressed as follows:

$$\frac{\lambda}{\lambda_{\text{max}}} = \theta^{1/\alpha} \tag{1.2}$$

where λ_{max} and α are positive constants and model parameters; λ_{max} is the reactivity for species with the highest dimensionless temperature. The distribution function of species type ($D(\lambda)$) is necessary to transform the coordinates adequately from θ to λ, and it is defined as follows:

$$D(\lambda) = \frac{n_s\alpha}{\lambda_{\text{max}}^{\alpha}}\lambda^{\alpha-1} \tag{1.3}$$

where n_s is the total number of species in the mixture. The equation of mass balance as a function of the reactivity and residence time (t or LHSV^{-1}) after changing the coordinates, becomes

$$\frac{dc(\lambda,t)}{dt} = -\lambda c(\lambda,t) + \int_{\lambda}^{\lambda_{\text{max}}} p(\lambda,\Lambda)\Lambda c(\Lambda,t)D(\Lambda)d\Lambda \tag{1.4}$$

where $c(\lambda, t)$ is the concentration of the component with reactivity λ, $c(\Lambda, t)$ is the concentration of the component with reactivity Λ, $p(\lambda, \Lambda)$ is the probability density function of species with reactivity Λ that can form species with reactivity λ, Λ is the maximum reactivity of the same component with reactivity λ, and $D(\Lambda)$ is the Jacobian that transforms the discrete coordinate i into a continuous coordinate Λ. The form of the function $p(\lambda, \Lambda)$ and its parameters are given by:

$$p(\lambda,\Lambda) = \frac{1}{S_0\sqrt{2\pi}}\left[e^{-\left(\left\{\left[\frac{\lambda}{\Lambda}\right]^{a_0}-0.5\right\}/a_1\right)^2} - A_1 + B_1\right] \tag{1.5}$$

$$A_1 = e^{-\left(0.5/a_1\right)^2} \tag{1.6}$$

$$B_1 = \sigma\left[1 - \left(\frac{\lambda}{\Lambda}\right)\right] \tag{1.7}$$

where a_0, a_1, and σ are parameters of the model, and S_0 is obtained from the following mass balance criteria:

$$\int_0^{\Lambda} p(\lambda,\Lambda)D(\lambda)d\lambda = 1 \tag{1.8}$$

where $D(\lambda)$ is the Jacobian that transforms the discrete coordinate i into the continuous coordinate λ. Using this equation and solving Eq. (1.5) to S_0, the following solution is reached:

$$S_0 = \int_0^{\Lambda} \frac{1}{\sqrt{2\pi}}\left[e^{-\left(\left\{\left[\frac{\lambda}{\Lambda}\right]^{a_0}-0.5\right\}/a_1\right)^2} - A_1 + B_1\right]D(\lambda)d\lambda \tag{1.9}$$

The hydrocracking phenomena occurring within the reactor are described by the differential equation, which illustrates the formation of compounds resulting from the cracking of components with higher boiling points (second term on the right side of the equation) and the cracking of lower boiling products (first term of the equation). This equation is solved for any residence time in terms of $c(\lambda, t)$, and the desired cuts in mass fraction depending on its dimensionless temperature as follows:

$$C_{\theta_1,\theta_2}(t) = \int_{\lambda_1}^{\lambda_2} c(\lambda,t)D(\lambda)d\lambda \tag{1.10}$$

where $C_{\theta_1,\theta_2}(t)$ is the mass fraction with dimensionless temperature ranging from θ_1 to θ_2 with its corresponding reactivity λ_1 and λ_2 as shown in Eq. (1.2) (Ancheyta 2013; Becker et al. 2016; Elizalde et al. 2009, 2016; Elizalde and Ancheyta 2014; Laxminarasimhan et al. 1996; Martínez-Grimaldo et al. 2011).

1.2.1.3 Single-Event Kinetic Models

Microkinetic models are developed at a molecular scale considering the chemistry of the reaction system by deriving the kinetic equations for individual molecules. The solution adequately converges in reacting systems that contain a limited number of chemical species. However, the set of equations suggested for intricate systems, such as those represented by heavy crude fractions, presents a challenging task to accomplish. Even modern computational tools are unsatisfactory because the size of the reaction network increases exponentially with the number of carbons (Ancheyta et al. 2005; Becker et al. 2016).

The advancement of these types of models for hydrocracking of heavy crude oils is propelled by the rising availability of experimental data gathered using sophisticated analytical techniques and increasing computational capacities. The microkinetic modeling method (single events) was initially proposed by Froment et al. (Baltanas et al. 1989; Baltanas and Froment 1985; Froment 1987; Hillewaert et al. 1988), and the reaction coefficients were calculated for each individual reaction. Computational regrouping algorithms reduce the size of reaction networks without losing information. These can be applied in various reactions, such as hydrocracking, catalytic reformation, isomerization and alkylation; however, the original regrouping method remains as unsuitable for implementation in huge network of reactions, notwithstanding the current computational capabilities. To find an approach, two alternative methods have been proposed (Becker et al. 2016; Guillaume et al. 2011; Mitsios et al. 2009; Valéry et al. 2007): (i) a method based on decomposition of side chains that was developed by Valéry et al. (2007) and (ii) a method based on structural classes, which was developed by Martens and Marin (2001).

1.2.2 Kinetic Models Reported in the Literature for Hydrocracking of Heavy Crude Oils Using Dispersed Catalysts

1.2.2.1 Kinetic Models Based on Distillation Curves

Hassanzadeh and Abedi (2010) as well as Galarraga et al. (2012) carried out experiments for the catalytic heavy crude oil upgrading with dispersed NiWMo catalyst and Athabasca bitumen (9.5 °API) as feed in a batch reactor. The tests were performed at temperature ranges of 320–380 °C, 3–70 h of reaction time, total hydrogen pressure of 3.45 MPa, and stirring rate of 500 rpm. Both works used the same kinetic model (Figure 1.5a) that was first reported by Sánchez et al. (2005a) based on boiling point intervals. The reaction scheme takes into account five lumps: Residue (R, 538 °C+), VGO (343–538 °C), distillates (D, 204–343 °C), naphtha (N, IBP-204 °C), and

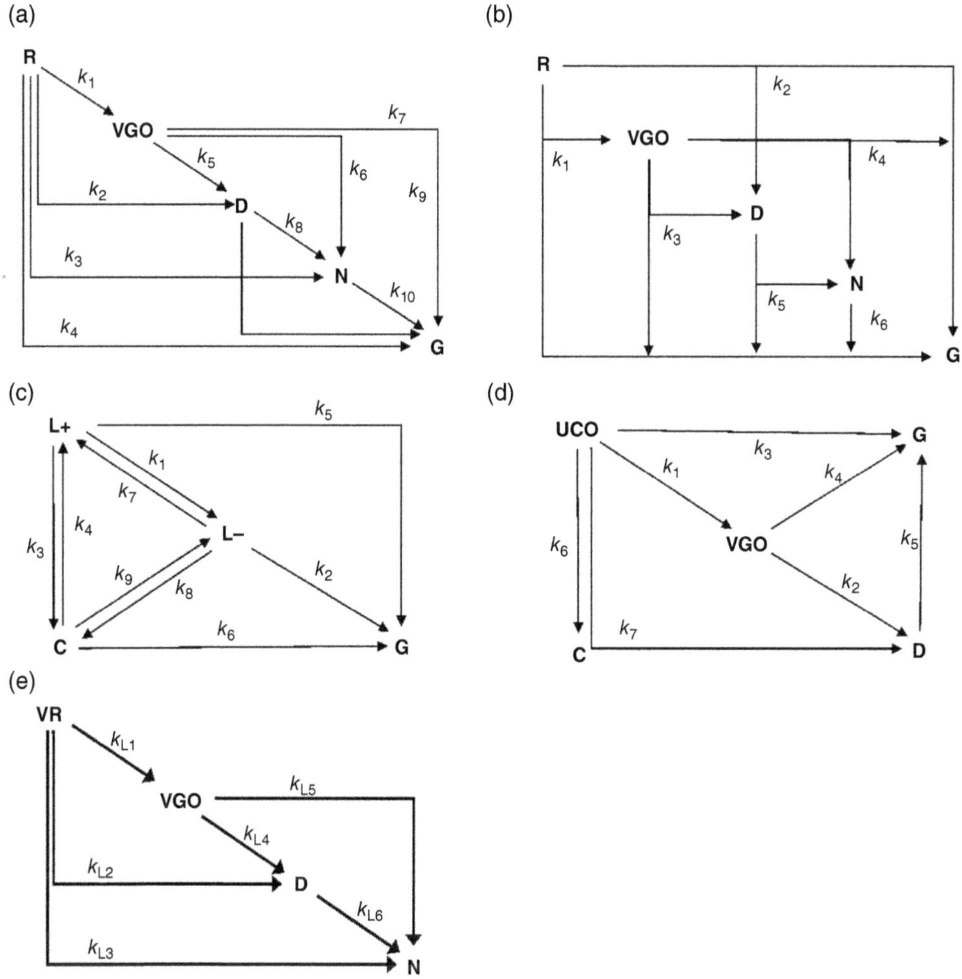

Figure 1.5 Four- and five-lump kinetic models based on distillation curves reported in the literature for heavy oil hydrocracking using dispersed catalyst.

gases (G). Nonetheless, some pseudocomponents proposed by Galarraga et al. (2012) have different boiling point ranges: residue (545 °C+), VGO (343–545 °C), D (216–343 °C), N (IBP–216 °C). The mass balance equations are the same in both cases, and a first-order reaction was considered giving the following expressions:

$$r_R = -(k_1 + k_2 + k_3 + k_4)y_R \tag{1.11}$$

$$r_{VGO} = k_1 y_R - (k_5 + k_6 + k_7)y_{VGO} \tag{1.12}$$

$$r_D = k_2 y_R + k_5 y_{VGO} - (k_8 + k_9)y_D \tag{1.13}$$

$$r_N = k_3 y_R + k_6 y_{VGO} + k_8 y_D - k_{10} y_N \tag{1.14}$$

$$r_G = k_4 y_R + k_7 y_{VGO} + k_9 y_D + k_{10} y_N \tag{1.15}$$

where r_i is the reaction rate of the component i, y_i the mass fractions of component i, k_j the reaction rate coefficients for each reaction pathway j. The only difference is that Hassanzadeh and Abedi

(2010) used the weight fraction, while Galarraga et al. (2012) employed a dimensionless concentration (the concentration at any time divided by the initial concentration). Furthermore, the latter authors performed a general reaction for the residue conversion into light products assuming first-order reaction with k_0 as the only reaction rate coefficient as follows:

$$\text{Residue} \xrightarrow{k_0} \text{Products} \tag{1.16}$$

After the catalytic hydrocracking of Athabasca bitumen with ultradispersed catalyst at bench-scale, kinetic studies on a large-scale reactor were developed by Loria et al. (2011) using the same model as Galarraga et al. (2012). The experimental results were obtained in a pilot plant tubular reactor at reaction temperatures of 320–380 °C, residence times (τ, for continuous reactors is assumed to be the ratio of the reactor volume to the volumetric flow rate) of 9–51 h, keeping the total pressure constant (2.76 MPa), and a hydrogen-to-oil ratio of 3509 (scf/bbl). Although the reaction scheme is the same, some boiling point intervals are different: VGO (343–550 °C) and R (550 °C+). Because an ultradispersed catalyst is employed in the study, the internal and external diffusion and the axial dispersion were negligible in a continuous gas-oil hydrocracking reactor, which allowed to set the reaction equations as follows:

$$\int_{y(\tau_0)}^{y(\tau_f)} dy_R = -k_1\int_{\tau_0}^{\tau_f} y_R d\tau - k_2\int_{\tau_0}^{\tau_f} y_R d\tau - k_3\int_{\tau_0}^{\tau_f} y_R d\tau - k_4\int_{\tau_0}^{\tau_f} y_R d\tau \tag{1.17}$$

$$\int_{y(\tau_0)}^{y(\tau_f)} dy_{VGO} = k_1\int_{\tau_0}^{\tau_f} y_R d\tau - k_5\int_{\tau_0}^{\tau_f} y d\tau - k_6\int_{\tau_0}^{\tau_f} y_{VGO} d\tau - k_7\int_{\tau_0}^{\tau_f} y_{VGO} d\tau \tag{1.18}$$

$$\int_{y(\tau_0)}^{y(\tau_f)} dy_D = k_2\int_{\tau_0}^{\tau_f} y_R d\tau + k_5\int_{\tau_0}^{\tau_f} y_{VGO} d\tau - k_8\int_{\tau_0}^{\tau_f} y_D d\tau - k_9\int_{\tau_0}^{\tau_f} y_D d\tau \tag{1.19}$$

$$\int_{y(\tau_0)}^{y(\tau_f)} dy_N = k_3\int_{\tau_0}^{\tau_f} y_R d\tau + k_6\int_{\tau_0}^{\tau_f} y_{VGO} d\tau + k_8\int_{\tau_0}^{\tau_f} y_D d\tau - k_{10}\int_{\tau_0}^{\tau_f} y_N d\tau \tag{1.20}$$

$$\int_{y(\tau_0)}^{y(\tau_f)} dy_G = k_4\int_{\tau_0}^{\tau_f} y_R d\tau + k_7\int_{\tau_0}^{\tau_f} y_{VGO} d\tau + k_9\int_{\tau_0}^{\tau_f} y_D d\tau + k_{10}\int_{\tau_0}^{\tau_f} y_N d\tau \tag{1.21}$$

where τ_0 and τ_f are the initial and final residence times, respectively, and $y(\tau_0)$ and $y(\tau_f)$ are the initial and final weight percentages, respectively, as a function of the residence time.

Nguyen et al. (2013) carried out a kinetic study for the hydrocracking of the Arabian light atmospheric residue in a batch reactor using Mo-dispersed catalyst (molybdenum naphthenate). The experiments were carried out at temperatures of 420–430 °C, residence times of 0.5–1 h, and hydrogen pressure of 15.2 MPa. The proposed reaction scheme (Figure 1.5b) considered the following boiling point intervals: R (510 °C+), VGO (350–510 °C), D (180–350 °C), and N (IBP–180 °C). A first order was assumed for all the reactions and the reaction from R to N was neglected;

furthermore, the stoichiometric coefficients of the reactions were considered, giving the following equations:

$$r_{\mathrm{R}} = v_{\mathrm{R}}^{1} k_1 C_{\mathrm{R}}^{\mathrm{L}} C_{\mathrm{H_2}}^{\mathrm{L}} + v_{\mathrm{R}}^{2} k_2 C_{\mathrm{R}}^{\mathrm{L}} C_{\mathrm{H_2}}^{\mathrm{L}} \tag{1.22}$$

$$r_{\mathrm{VGO}} = v_{\mathrm{VGO}}^{1} k_1 C_{\mathrm{R}}^{\mathrm{L}} C_{\mathrm{H_2}}^{\mathrm{L}} + v_{\mathrm{VGO}}^{3} k_3 C_{\mathrm{VGO}}^{\mathrm{L}} C_{\mathrm{H_2}}^{\mathrm{L}} + v_{\mathrm{VGO}}^{4} k_4 C_{\mathrm{VGO}}^{\mathrm{L}} C_{\mathrm{H_2}}^{\mathrm{L}} \tag{1.23}$$

$$r_{\mathrm{D}} = v_{\mathrm{D}}^{2} k_2 C_{\mathrm{R}}^{\mathrm{L}} C_{\mathrm{H_2}}^{\mathrm{L}} + v_{\mathrm{D}}^{3} k_3 C_{\mathrm{VGO}}^{\mathrm{L}} C_{\mathrm{H_2}}^{\mathrm{L}} + v_{\mathrm{D}}^{5} k_5 C_{\mathrm{D}}^{\mathrm{L}} C_{\mathrm{H_2}}^{\mathrm{L}} \tag{1.24}$$

$$r_{\mathrm{N}} = v_{\mathrm{N}}^{4} k_4 C_{\mathrm{VGO}}^{\mathrm{L}} C_{\mathrm{H_2}}^{\mathrm{L}} + v_{\mathrm{N}}^{5} k_5 C_{\mathrm{D}}^{\mathrm{L}} C_{\mathrm{H_2}}^{\mathrm{L}} + v_{\mathrm{N}}^{6} k_6 C_{\mathrm{N}}^{\mathrm{L}} C_{\mathrm{H_2}}^{\mathrm{L}} \tag{1.25}$$

$$\begin{aligned} r_{\mathrm{G}} = {} & v_{\mathrm{G}}^{1} k_1 C_{\mathrm{R}}^{\mathrm{L}} C_{\mathrm{H_2}}^{\mathrm{L}} + v_{\mathrm{G}}^{2} k_2 C_{\mathrm{R}}^{\mathrm{L}} C_{\mathrm{H_2}}^{\mathrm{L}} + v_{\mathrm{G}}^{3} k_3 C_{\mathrm{VGO}}^{\mathrm{L}} C_{\mathrm{H_2}}^{\mathrm{L}} + v_{\mathrm{G}}^{4} k_4 C_{\mathrm{VGO}}^{\mathrm{L}} C_{\mathrm{H_2}}^{\mathrm{L}} + v_{\mathrm{G}}^{5} k_5 C_{\mathrm{D}}^{\mathrm{L}} C_{\mathrm{H_2}}^{\mathrm{L}} \\ & + v_{\mathrm{G}}^{6} k_6 C_{\mathrm{N}}^{\mathrm{L}} C_{\mathrm{H_2}}^{\mathrm{L}} \end{aligned} \tag{1.26}$$

$$\begin{aligned} r_{\mathrm{H_2}} = {} & v_{\mathrm{H_2}}^{1} k_1 C_{\mathrm{R}}^{\mathrm{L}} C_{\mathrm{H_2}}^{\mathrm{L}} + v_{\mathrm{H_2}}^{2} k_2 C_{\mathrm{R}}^{\mathrm{L}} C_{\mathrm{H_2}}^{\mathrm{L}} + v_{\mathrm{H_2}}^{3} k_3 C_{\mathrm{VGO}}^{\mathrm{L}} C_{\mathrm{H_2}}^{\mathrm{L}} + v_{\mathrm{H_2}}^{4} k_4 C_{\mathrm{VGO}}^{\mathrm{L}} C_{\mathrm{H_2}}^{\mathrm{L}} + v_{\mathrm{H_2}}^{5} k_5 C_{\mathrm{D}}^{\mathrm{L}} C_{\mathrm{H_2}}^{\mathrm{L}} \\ & + v_{\mathrm{H_2}}^{6} k_6 C_{\mathrm{N}}^{\mathrm{L}} C_{\mathrm{H_2}}^{\mathrm{L}} \end{aligned} \tag{1.27}$$

where v_i^j is the stoichiometric coefficients of species i for reaction pathway j and C_i^{L} the molar concentration of species i in the liquid.

A kinetic model of four pseudocomponents (Figure 1.5c) for VR hydrocracking with Maya crude and powder NiMo/Al_2O_3 catalyst was developed by Puron et al. (2014). The experimental data were obtained from a batch reactor at three temperatures (400, 425, and 450 °C), 18.5 MPa of H_2 pressure and four reaction times (10, 30, 60, and 90 min). The model was divided on four fractions: (i) boiling point products >450 °C (L+), which are formed by asphaltenes and heavy maltenes, (ii) boiling point products <450 °C (L−) that include light maltenes, (iii) gases (G), and (iv) coke (C). The reaction rate equations were set as follows:

Boiling point products (L+):

$$r_{\mathrm{L+}} = -(k_1 + k_3 + k_5) y_{\mathrm{L+}} + k_4 y_{\mathrm{C}} + k_7 y_{\mathrm{L-}} \tag{1.28}$$

Boiling point products (L−):

$$r_{\mathrm{L-}} = k_1 y_{\mathrm{L+}} + k_9 y_{\mathrm{C}} - (k_2 + k_7 + k_8) y_{\mathrm{L-}} \tag{1.29}$$

Gases (G):

$$r_{\mathrm{G}} = k_2 y_{\mathrm{L-}} + k_5 y_{\mathrm{L+}} + k_6 y_{\mathrm{C}} \tag{1.30}$$

Coke (C):

$$r_{\mathrm{C}} = k_3 y_{\mathrm{L+}} + k_8 y_{\mathrm{L-}} - (k_4 + k_6 + k_9) y_{\mathrm{C}} \tag{1.31}$$

da Silva De Andrade (2014) established a kinetic model for *in situ* upgrading of Athabasca bitumen VR and pitch (asphaltene-rich phase) with ultradispersed catalyst. All of the tests were conducted in a downflow reactor at temperatures between 320 and 395 °C, residence times of 20–50 h, hydrogen pressure of 3.45 MPa, and hydrogen-to-oil ratio of 90 standard cm^3/cm^3. The same lumped kinetic model reported by Hassanzadeh and Abedi (2010) was used by the authors.

A six-lump kinetic model was developed by Asaee et al. (2014) using the experimental information reported by Song et al. (2004) for the Chinese Gudao residue upgrading with ammonium phosphomolybdate catalyst. The reaction scheme (Figure 1.6a) includes R (480 °C+), VGO (350–480 °C),

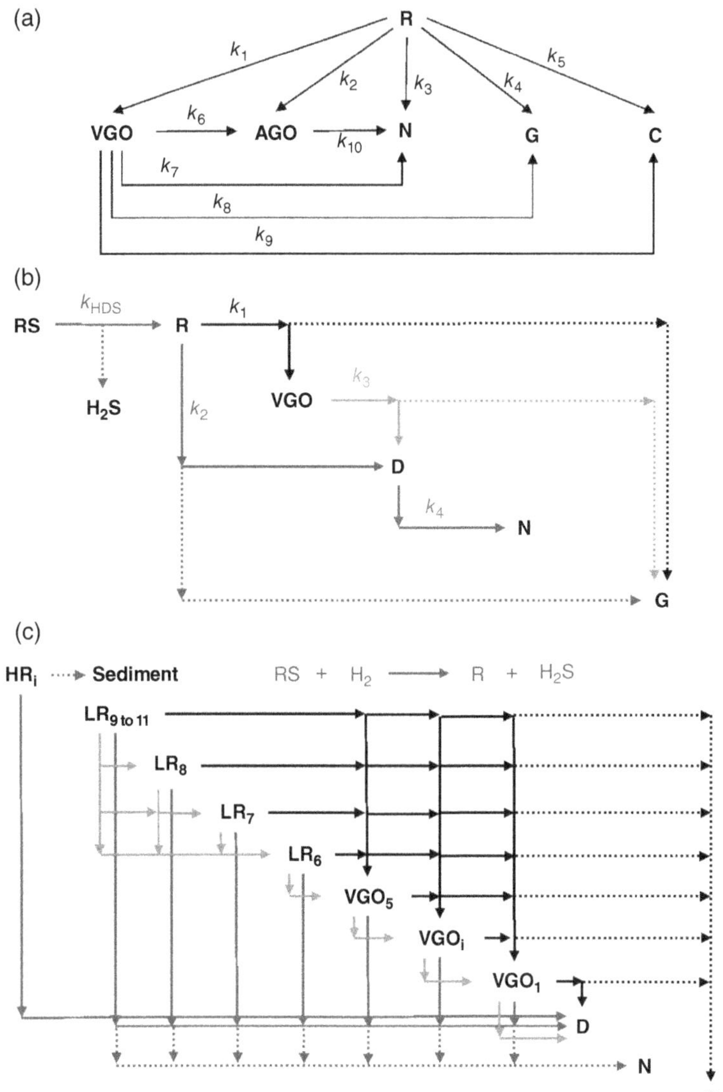

Figure 1.6 Detailed-lumping kinetic models based on distillation curves reported in the literature for heavy oil hydrocracking using dispersed catalyst.

atmospheric gas oil (AGO, 180–350 °C), N (IBP-180 °C), G, and C. The heavier fraction cracks into lower molecular weight fractions, and G and C are generated only from heavy fractions (R and VGO), obtaining the following equations:

$$r_R = -(k_1 + k_2 + k_3 + k_4 + k_5)y_R \tag{1.32}$$

$$r_{VGO} = k_1 y_R - (k_6 + k_7 + k_8)y_{VGO} \tag{1.33}$$

$$r_{AGO} = k_2 y_R + k_6 y_{VGO} - (k_9 + k_{10})y_{AGO} \tag{1.34}$$

$$r_N = k_3 y_R + k_7 y_{VGO} + k_9 y_{AGO} \tag{1.35}$$

$$r_G = k_4 y_R + k_8 y_{VGO} + k_{10} y_{AGO} \tag{1.36}$$

$$r_C = k_5 y_R \tag{1.37}$$

The kinetic studies for *in situ* upgrading process of VR and heavy oil at Mexican carbonate reservoir conditions were developed by Orozco Castillo (2016) and were performed in a plug flow reactor employing a Ni–W–Mo catalyst (the same as in the aforementioned studies) at temperatures of 320–360 °C, residence times of 24–96 h (heavy oil) and 48–120 h (VR), and hydrogen pressure of 10.34 MPa. The kinetic model is the same as used by Loria et al. (2011).

Kim et al. (2017) used an oil-soluble (MoS_2) catalyst for the hydrocracking of a residue fraction at temperatures of 380–400 °C, reaction times of 1–16 h, and hydrogen pressure of 9.5 MPa in a batch reactor. The five-lump reaction scheme (Figure 1.5d) includes unconverted oil (UCO, 560 °C+), VGO (320–560 °C), D (IBP-320 °C), G, and C. The cracking of heavier fractions produces lighter fractions, except when UCO forms C by condensation reactions. A second-order of reaction was considered for UCO, a first-order of reaction for VGO, and zero-order of reaction for D and G, obtaining the following equations:

$$r_{UCO} = -(k_1 + k_3 + k_6 + k_7) y_{UCO}^2 \tag{1.38}$$

$$r_{VGO} = k_1 y_{UCO}^2 - (k_2 + k_4) y_{VGO} \tag{1.39}$$

$$r_D = k_2 y_{VGO} + k_7 y_{UCO}^2 - k_5 \tag{1.40}$$

$$r_G = k_3 y_{UCO}^2 + k_4 y_{VGO} + k_5 \tag{1.41}$$

$$r_C = k_6 y_{UCO}^2 \tag{1.42}$$

A kinetic model for the hydrocracking of Iranian heavy crude oil with an oil-soluble dispersed catalyst in a batch reactor was developed by Huang et al. (2017). The experimental data were carried out at temperatures of 405–435 °C, reaction times of 1–10 h, and hydrogen pressure of 9 MPa. The kinetic model was the same as reported by Hassanzadeh and Abedi (2010), but the boiling point ranges were similar to those reported by Nguyen et al. (2013), with different R (524 °C+) and VGO (350–524 °C) boiling points.

Ortega-García et al. (2017) conducted a kinetic study for heavy oil (12°API) upgrading in a continuous stirred tank reactor (CSTR) reactor using a liquid-acid catalyst at moderate conditions: temperatures of 350–370 °C, residence times of 12–180 h, hydrogen pressure of 9.8 MPa, and hydrogen-to-oil ratio of 62.9 standard cm^3/cm^3. The five-lump kinetic model is similar to that reported by Hassanzadeh and Abedi (2010) (with different names, VGO is named HGO and D is termed LGO) based on the following boiling point ranges: R (540 °C+), heavy gas oil (HGO, 343–540 °C), light gas oil (LGO, 221–343 °C), N (IBP-221 °C).

Elahi et al. (2019) performed a kinetic study for *in situ* upgrading inside a carbonate rock (porous media) by injecting VR from Mexican heavy oil, hydrogen, and NiMo-ultradispersed nanocatalyst. The experimental part was carried out in a continuous reactor where a mixture of VR and nanocatalysts (0.22 wt.% of the carbonate cores) is submitted at reservoir conditions (temperatures of 320–360 °C, residence times of 12–72 h, hydrogen pressure of 10 MPa, and hydrogen-to-oil ratio of 150 standard cm^3/cm^3). The kinetic model scheme is the same as employed by Hassanzadeh and Abedi (2010).

Félix and Ancheyta (2019b) proposed a four-lump kinetic model (Figure 1.5e) for the hydrocracking of heavy crude oil using molybdenite ore catalyst in a batch reactor. The tests were conducted at

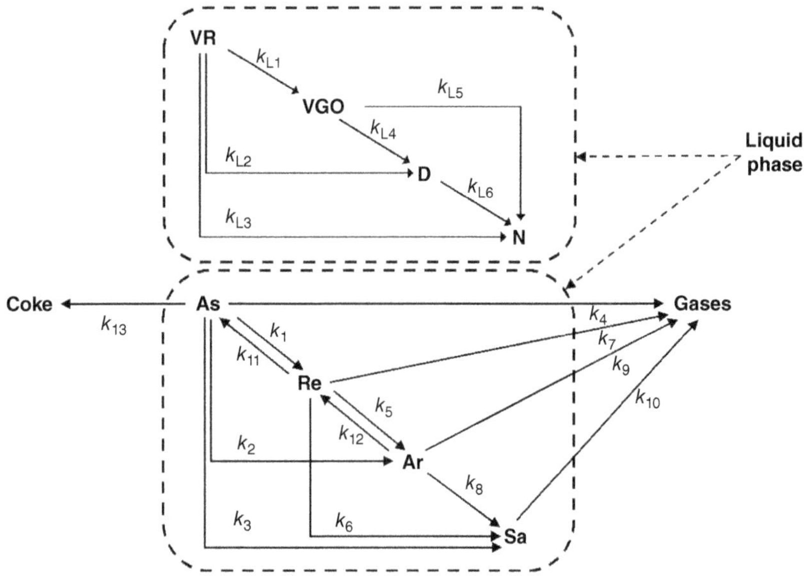

Figure 1.7 Global kinetic model to predict liquid, gas, and coke yields combining kinetic models based on SARA fraction and distillation curves.

mild conditions (temperatures of 360–400 °C, reaction times of 2–5 h, and hydrogen pressure of 3.9 MPa). The boiling point ranges for R, VGO, D, and N were similar to those reported by Sánchez et al. (2005a, b): R, VGO, D, and N. The reaction scheme is based on the conversion from heavier fractions to lighter fractions, but the G and solid (Coke, C) fractions were calculated (Figure 1.7) following a complementary kinetic model (Félix and Ancheyta 2019a), which is presented later, giving the following equations:

$$r_R = -(k_1 + k_2 + k_3)y_R \tag{1.43}$$

$$r_{VGO} = k_1 y_R - (k_4 + k_5)y_{VGO} \tag{1.44}$$

$$r_D = k_2 y_R + k_4 y_{VGO} - k_6 y_D \tag{1.45}$$

$$r_N = k_3 y_R + k_5 y_{VGO} + k_6 y_D \tag{1.46}$$

Complementary reactions rate equations:

$$r_G = k_4 y_{As} + k_7 y_{Re} + k_9 y_{Ar} + k_{10} y_{Sa} \tag{1.47}$$

$$r_C = k_{13} y_{As} \tag{1.48}$$

A detailed (seven-lump) kinetic model was developed by Álvarez et al. (2019) based on the residua upgrading (Safaniya residue and Arabian light atmospheric residue) in a semibatch reactor with an oil-soluble Mo precursor catalyst (molybdenum octoate). The boiling point range is similar to those reported by Nguyen et al. (2013), and the pseudocomponents are involved in one hydrodesulfurization (HDS) and four hydrocracking reactions (Figure 1.6b). All reactions are performed in series, except for the parallel cracking of R into D. The cracking of R and VGO to produce lighter fractions also forms G, while D, which comes from heavy fractions (R and VGO), only produces *N*.

The stoichiometric coefficients are taken into account in hydrocracking reaction rates, and the reaction orders for each lump are estimated, leading to the resulting kinetic equation:

$$r_i = \sum_{j=1}^{4} v_i^j A_{0,j} e^{-\frac{E_{A,j}}{RT}} \left(C_i^L\right)^{n_i} \tag{1.49}$$

where $A_{0,j}$ is the pre-exponential or collision factor for the reaction j, $E_{A,j}$ is the activation energy of the reaction j, n_i the order of reaction for component i, R is the universal gas constant, and T is the reaction temperature. Considering the sulfur removal from the residue fraction (RS) to produce H_2S, the stoichiometry of this reaction is based on the addition of hydrogen to 4,6-DMDBT to obtain the following rate equation:

$$r_{\mathrm{HDS}} = v_{\mathrm{HDS}} A_{0,\mathrm{HDS}} e^{-\frac{E_{A,\mathrm{HDS}}}{RT}} \left(C_{\mathrm{RS}}^{\mathrm{L}}\right)^{n_{RS}} C_{\mathrm{H_2}}^{\mathrm{L}} \tag{1.50}$$

Another detailed kinetic model (Figure 1.6c) was reported by Browning et al. (2019) for the hydrocracking of VR in a semibatch reactor using the same catalyst as reported by Álvarez et al. (2019) at similar operating conditions. The main fractions were divided in different boiling point ranges: R (525 °C+), VGO (350–525 °C), D (160–350 °C), N (IBP-160 °C), and G. Additionally, the VGO and R fractions were separated into intervals of 35 °C in detailed lumps: five lumps for VGO, six lumps for light residue (LR, 525–770 °C), and seven lumps for heavy residue (HR, 770–980 °C). One HDS reaction and four hydrocracking reactions are assumed as reported by Álvarez et al. (2019), and the conversion from LR sublumps to VGO sublumps produced G. All reactions are believed to consume hydrogen and the cracking of LR and VGO sublumps to produce D also forms N. Furthermore, the HR lump yields D and sediments, giving the following general equation:

$$r_i = v_i^j k_j C_i^{n_i} \tag{1.51}$$

where C_i is the concentration of the species i. Only the reaction in series from LR sublumps to VGO sublumps to D is assessed without the formation of any by-products. The first-order reaction for hydrogen is assumed in the latter and HDS reactions obtaining the following overall equation:

$$r_i = v_i^j k_j C_i^{n_i} C_{\mathrm{H_2}} \tag{1.52}$$

A kinetic model for *in situ* upgrading of residue fraction from Aguacate heavy oil using ultradispersed nanocatalyst (Ni–Mo) in four dolomite cores was developed by Duran Armas (2021) at near-reservoir conditions: temperatures of 335–365 °C, residence times of 12–72 h, and hydrogen pressure of 10 MPa. The five-lump kinetic model is similar to that used by Elahi et al. (2019) and Hassanzadeh and Abedi (2010).

Pham et al. (2021) reported a kinetic model for the slurry-phase hydrocracking of VR using an oil-soluble catalyst (Mo-octoate) in a CSTR. Severe operating conditions were employed to obtain the experimental data: temperatures of 400–450 °C, residence times of 1–4 h, hydrogen pressure of 16 MPa, and hydrogen-to-oil ratio of 1500 standard cm^3/cm^3. The study utilized a five-lump kinetic model resembling that of Galarraga et al. (2012), encompassing the following cuts: R (524 °C+), VGO (343–524 °C), D (177–343 °C), N (IBP-177 °C), and G. A second-order of reaction for residue is supposed, giving the following equations:

$$r_{\mathrm{R}} = -(k_1 + k_2 + k_3 + k_4) y_{\mathrm{R}}^2 \tag{1.53}$$

$$r_{\mathrm{VGO}} = k_1 y_{\mathrm{R}}^2 - (k_5 + k_6 + k_7) y_{\mathrm{VGO}} \tag{1.54}$$

$$r_{\mathrm{D}} = k_2 y_{\mathrm{R}}^2 + k_5 y_{\mathrm{VGO}} - (k_8 + k_9) y_{\mathrm{D}} \tag{1.55}$$

$$r_N = k_3 y_R^2 + k_6 y_{VGO} + k_8 y_D - k_{10} y_N \tag{1.56}$$

$$r_G = k_4 y_R^2 + k_7 y_{VGO} + k_9 y_D + k_{10} y_N \tag{1.57}$$

A fine-dispersed catalyst (nickel nanoparticles in polyethylene glycol, PEG300) was employed by Coronel-García et al. (2021) for the heavy oil upgrading in a batch reactor at mild conditions (temperatures of 310–370 °C, reaction times of 4–72, and hydrogen pressure of 7 MPa). The kinetic model is the same as reported by Hassanzadeh and Abedi (2010), with a reaction order for residue conversion of 2.5 used to establish the subsequent system of differential equations:

$$r_R = -(k_1 + k_2 + k_3 + k_4) y_R^{2.5} \tag{1.58}$$

$$r_{VGO} = k_1 y_R^{2.5} - (k_5 + k_6 + k_7) y_{VGO} \tag{1.59}$$

$$r_D = k_2 y_R^{2.5} + k_5 y_{VGO} - (k_8 + k_9) y_D \tag{1.60}$$

$$r_N = k_3 y_R^{2.5} + k_6 y_{VGO} + k_8 y_D - k_{10} y_N \tag{1.61}$$

$$r_G = k_4 y_R^{2.5} + k_7 y_{VGO} + k_9 y_D + k_{10} y_N \tag{1.62}$$

Cai et al. (2022) reported a kinetic model for the hydrocracking of Karamay atmospheric residue using Mo-based ionic liquid (1-alkyl-3-methylimidazolium cations with molybdate anion, $[C_n mim]_2[MoO_4]$) as dispersed catalyst in a batch reactor at temperatures of 390–430 °C, reaction times of 1–8 h, and hydrogen pressure of 12 MPa. A similar kinetic model to that reported by Kim et al. (2017) was employed with different boiling points: R (524 °C+) instead of UCO, VGO (350–524 °C), and D (<350 °C). The set of differential equations is analogous in both cases because the UCO and R followed a second-order kinetics.

The trialkylmethylammonium molybdate ionic liquid was used for the slurry-phase hydrocracking of Karamay atmospheric residue in a batch reactor by Ma et al. (2022). The experimental data were obtained at temperatures of 380–430 °C, reaction times of 0.5–6 h, and hydrogen pressure of 12.3 MPa, and the kinetic model utilized was in line with that of Cai et al. (2022).

1.2.2.2 Kinetic Models Based on SARA Fractions

A combined kinetic model (based on SARA fractions and distillation curves) was reported by Fukuyama and Terai (2007) for the hydrocracking of VR using two different catalysts (pyrite and pyrite-active carbon) in a semibatch reactor at temperatures of 415–445 °C, residence times of 1–4 h, and a hydrogen pressure of 10 MPa. The products were classified as gas (G), naphtha (N, IBP-171 °C), kerosene (K, 171–232 °C), gas oil (GO, 232–343 °C), VGO (343–525 °C), vacuum residue (VR, 525 °C+), and coke (C). Additionally, the VR was further separated before and after reaction into SARA fractions: Asphaltenes (As), resins (Re), aromatics (Ar), and saturates (Sa). The authors supposed a reaction scheme where As in VR is the main reactant, which only produces C and VGO, meanwhile VGO generates G, N, K, and GO lumped into light oil (LO), which is also produced by Sa. Concerning the SARA fractions, Ar cracks into Sa, LO, and VGO, while Re yields VGO. Furthermore, other condensation reactions are performed from Ar to Re and Re to As. All of these hydrocracking reactions (Figure 1.8) are assumed to be first-order of reaction, resulting in the subsequent equations:

$$\frac{dy_{As}}{dt} = -(k_8 + k_9) y_{As} + k_3 y_{Re} \tag{1.63}$$

$$\frac{dy_{Re}}{dt} = -(k_3 + k_7) y_{Re} + k_2 y_{Ar} \tag{1.64}$$

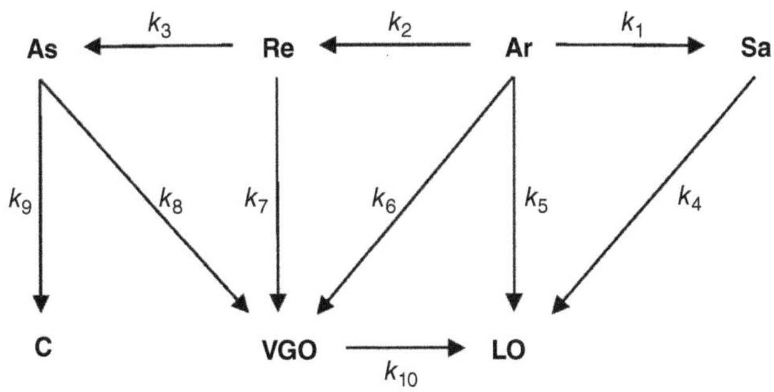

Figure 1.8 SARA- and distillation curves-based kinetic model for heavy oil hydrocracking using mineral catalysts.

$$\frac{dy_{\mathrm{Ar}}}{dt} = -(k_1 + k_2 + k_5 + k_6)y_{\mathrm{Ar}} \tag{1.65}$$

$$\frac{dy_{\mathrm{Sa}}}{dt} = k_1 y_{\mathrm{Ar}} - k_4 y_{\mathrm{Sa}} \tag{1.66}$$

$$\frac{dy_{\mathrm{LO}}}{dt} = k_4 y_{\mathrm{Sa}} + k_5 y_{\mathrm{Ar}} + k_{10} y_{\mathrm{VGO}} \tag{1.67}$$

$$\frac{dy_{\mathrm{VGO}}}{dt} = -k_{10} y_{\mathrm{VGO}} + k_6 y_{\mathrm{Ar}} + k_7 y_{\mathrm{Re}} + k_8 y_{\mathrm{As}} \tag{1.68}$$

$$\frac{dy_{\mathrm{C}}}{dt} = k_9 y_{\mathrm{As}} \tag{1.69}$$

Sheng et al. (2017) developed a kinetic model for asphaltenes liquefaction using a small volume of industrial distillate as hydrogen donor and dialkyldithiocarbamate molybdenum as a dispersed oil-soluble catalyst. A four-lump model (asphaltenes, maltenes, gas, and coke) describes the reaction scheme (Figure 1.9a). The reactions were carried out in a 250-mL batch reactor at 7 MPa of

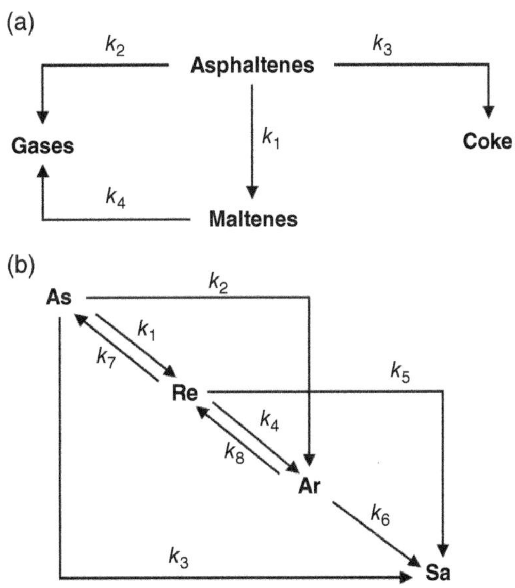

Figure 1.9 Four-lump kinetic models based on SARA fractions reported in the literature for heavy oil hydrocracking using dispersed catalyst.

hydrogen pressure, temperature of 380–400 °C, reaction time of 3–9 h, 5:1 asphaltenes-to-hydrogen donor mass ratio, and 500 μg Mo/g asphaltenes. A first-order reaction was assumed and the mathematical equations of the mentioned kinetic model are as follows:

Asphaltene (y_1):

$$\frac{dy_1}{dt} = -(k_{12} + k_{13} + k_{14})y_1 \quad (1.70)$$

Maltene (y_2):

$$\frac{dy_2}{dt} = k_{12}y_1 - k_{23}y_2 \quad (1.71)$$

Gases (y_3):

$$\frac{dy_3}{dt} = k_{13}y_1 + k_{23}y_2 \quad (1.72)$$

Coke (y_4):

$$\frac{dy_4}{dt} = k_{14}y_1 \quad (1.73)$$

Based on the data reported by Félix and Ancheyta (2019b), these authors developed four kinetic models for the hydrocracking of heavy crude oil using dispersed Mo-based catalyst based on SARA fractions (Félix and Ancheyta 2019a), where the heavy fractions produce lighter fractions by parallel and in series reactions. The first model (Figure 1.9b) is a four-lump kinetic model that takes into account only SARA fractions in the liquid phase; furthermore, Ar and Re fractions can undergo condensation reactions to generate Re and As fractions, respectively. A first-order reaction is assumed to obtaining the following equations:

$$r_{As} = -(k_1 + k_2 + k_3)y_{As} + k_7y_{Re} \quad (1.74)$$

$$r_{Re} = k_1y_{As} + k_8y_{Ar} - (k_4 + k_5 + k_7)y_{Re} \quad (1.75)$$

$$r_{Ar} = k_2y_{As} + k_4y_{Re} - (k_6 + k_8)y_{Ar} \quad (1.76)$$

$$r_{Sa} = k_3y_{As} + k_5y_{Re} + k_6y_{Ar} \quad (1.77)$$

The second kinetic model (Figure 1.10a) considers the same reactions between SARA fractions carried out in the four-lump model. Additionally, this reaction mechanism (five-lump) also considers the production of G from all SARA fractions, giving the following equations:

$$r_{As} = -(k_1 + k_2 + k_3 + k_4)y_{As} + k_{11}y_{Re} \quad (1.78)$$

$$r_{Re} = k_1y_{As} + k_{12}y_{Ar} - (k_5 + k_6 + k_7 + k_{11})y_{Re} \quad (1.79)$$

$$r_{Ar} = k_2y_{As} + k_5y_{Re} - (k_8 + k_9 + k_{12})y_{Ar} \quad (1.80)$$

$$r_{Sa} = k_3y_{As} + k_6y_{Re} + k_8y_{Ar} - k_{10}y_{Sa} \quad (1.81)$$

$$r_G = k_4y_{As} + k_7y_{Re} + k_9y_{Ar} + k_{10}y_{Sa} \quad (1.82)$$

Another five-lump kinetic model (Figure 1.10b) was reported by Félix and Ancheyta (2019b), which considers the same reaction mechanism as in the four-lump kinetic model. Furthermore, the condensation reaction of As to generate C is included in this model. Therefore, in Eq. (1.74), the term $-k_9y_{As}$ is added as well as in the reaction rate equation for C:

$$r_C = k_9y_{As} \quad (1.83)$$

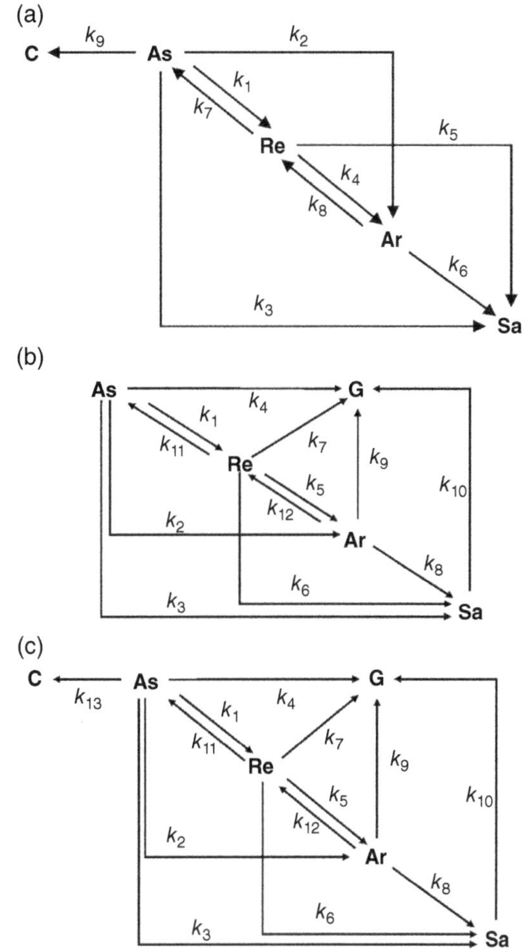

Figure 1.10 Five- and six-lump kinetic models based on SARA fractions reported in the literature for heavy oil hydrocracking using dispersed catalyst.

A six-lump kinetic model (Figure 1.10c) was also reported by Félix and Ancheyta (2019b) considering the five-lump reaction mechanism mentioned before that includes the production of G. In addition, the condensation reaction of As to produce C is considered and the term $-k_{13}y_{As}$ is included in Eq. (1.78), likewise the reaction rate equation for the C fraction. This kinetic model is employed to calculate the G and C fractions in the four-lump model based on distillation curves that were previously discussed.

$$r_C = k_{13}y_{As} \tag{1.84}$$

1.2.3 Kinetic Models Based on Continuous Lumping

A continuous kinetic model was developed by Martínez-Grimaldo et al. (2011) for the Maya heavy oil upgrading in a batch reactor using a dispersed catalyst (powder ammonium heptamolybdate) at 400 °C, reaction times of 1–24 h, and hydrogen pressure of 5.5 MPa. The continuous kinetic model forecasted the composition based on distillation curves of liquid phase during the hydrocracking reaction using Eqs. (1.1)–(1.10). Since the continuous kinetic model solely predicts the composition of the liquid phase, the quantities of gas (G), liquid (L), and solid (S) are computed as follows:

$$\frac{dL}{dt} = \frac{-k_L(L - L_L)}{(1 + K_L L + K_S S + K_G G)^{\gamma}} \tag{1.85}$$

$$\frac{dS}{dt} = P_S \frac{k_L(L - L_L)}{(1 + K_L L + K_S S + K_G G)^{\gamma}} \tag{1.86}$$

$$\frac{dG}{dt} = P_G \frac{k_L(L - L_L)}{(1 + K_L L + K_S S + K_G G)^{\gamma}} \tag{1.87}$$

$$P_S + P_G = 1 \tag{1.88}$$

where k_L, K_L, L_L, K_S, P_S, K_G, P_G, and γ involves the liquid, solid, and gas phases to be used as parameters in the continuous lumping equations.

1.2.4 Thermodynamic Model to Predict the Asphaltenes Flocculation and Sediments Formation

Asphaltene fraction is in equilibrium surrounded by resins, aromatics, and saturates in crude oil, with the resins acting as peptizing agents. Therefore, a regular solution solubility model can be applied (Akbarzadeh et al. 2004), contemplating the liquid–solid equilibrium. The equilibrium ratios of each component are used to determine the quantity of precipitated asphaltene. The activity of the solid phase for asphaltenes approaches 1 when this fraction is in solid–liquid equilibrium. Furthermore, by disregarding the heat of fusion ($\Delta H/RT$), the equilibrium ratio (K_i) can be obtained according to the following:

$$K_i = \frac{x_i^S}{x_i^L} = \exp\left[1 - \frac{V_i^L}{V_{mix}} + \ln\left(\frac{V_i^L}{V_{mix}}\right) + \frac{V_i^L}{RT}(\delta_i - \delta_{mix})^2\right] \tag{1.89}$$

where x_i^S and x_i^L are the mass fractions of the component i in the solid and liquid phase, respectively, V_i^L is the liquid molar volume of the component i, V_{mix} is the average molar volume, δ_i is the solubility parameters of the component i, and δ_{mix} represents the solubility parameters of the mixture. The solubility parameter of the liquid (δ^L) correlates with the internal energy of vaporization (ΔU^{LV}) and the molar volume of the liquid phase (V^L) according to the equation:

$$\delta^L = \left(\frac{\Delta U^{LV}}{V^L}\right)^{1/2} \tag{1.90}$$

The energy of vaporization and the molar volume are calculated using a cubic equation, which enables the determination of the solubility parameter. It has been reported that the SRK EOS with a Peneloux correction (c) gives the best approaches of liquid volumes whereby accurate values of the solubility parameter are obtained (Akbarzadeh et al. 2004). The SRK EOS has the following form:

$$P = \frac{RT}{V^{SRK} - b} - \frac{a(T)}{V^{SRK}(V^{SRK} + b)} \tag{1.91}$$

where P is the pressure, V^{SRK} is the molar volume calculated with the SRK EOS, and $a(T)$ and b are EOS parameters. The corrected liquid molar volume is obtained as

$$V^L = V^{SRK} - c \tag{1.92}$$

Substituting this term in Eq. (1.90) gives

$$P = \frac{RT}{V^{L} - (b - c)} - \frac{a(T)}{\left(V^{L} + c\right)\left(V^{L} + c + b\right)} \tag{1.93}$$

The Peneloux correction term (c) can be estimated by

$$c = \frac{0.40768 T_c}{P_c}(0.29441 - Z_{RA}) \tag{1.94}$$

where T_c is the critical temperature, P_c is the critical pressure, and Z_{RA} is the Racket compressibility factor. If the value of the Z_{RA} is not available, it is calculated using the acentric factor (ω) with the following equation:

$$Z_{RA} = 0.29056 - 0.08775\omega \tag{1.95}$$

The internal energy of vaporization is related with the EOS parameters (a and b) by

$$\Delta U^{LV} = \left(\frac{a}{b} - \frac{T}{b}\frac{da}{dT}\right)\ln\left(1 + \frac{b}{V^{SRK}}\right) \tag{1.96}$$

Therefore, at any temperature and pressure, Eq. (1.92) can be solved for the molar volume, and the solubility parameter is calculated by

$$\delta^{L} = \left[\frac{a - T\dfrac{da}{dT}}{bV^{L}}\ln\left(1 + \frac{b}{V^{L} + c}\right)\right]^{1/2} \tag{1.97}$$

The SRK EOS parameters (a and b) for a given component are calculated from the critical properties and the acentric factor as follows:

$$a(T) = \frac{0.42748 R^2 T_c^2}{P_c}\left[1 + \left(0.48 + 1.574\omega - 0.176\omega^2\right)\left(1 - \left(\frac{T}{T_c}\right)^{1/2}\right)\right]^2 \tag{1.98}$$

$$b = \frac{0.08664 R T_c}{P_c} \tag{1.99}$$

The critical properties and the acentric factor of each SARA fraction to calculate the EOS parameters can be estimated with some correlations reported in literature (Akbarzadeh et al. 2004; Riazi 2005). For the saturate fraction, the correlation depends on some constants shown in Table 1.1 and molar mass as can be seen in the following equation:

$$\varepsilon = \varepsilon_{\infty} - \exp\left(A - EM^{C}\right) \tag{1.100}$$

Table 1.1 Constants and their correlations for Eq. (1.100).

ε	ε_{∞}	A	E	C
T_b	1070	6.98291	0.02013	0.67
T_b/T_c	1.15	−0.41966	0.02436	0.58
$-P_c$	0	4.65757	0.13423	0.50
$-w$	0.3	−3.06826	−1.04987	0.20

where ε, ε_{∞}, A, E, and C are constants, and M is the molar mass. For aromatics, resins, and asphaltenes, the correlations are solely based on the molar mass. Due to the fact that these fractions consist of polynuclear aromatics with slight structural variations, a correction factor (c_f, Table 1.6) was implemented for adjusting the acentric factor and account for such variations, even in the saturate fraction. These correlations are presented as follows:

$$T_c = 77.856M^{0.4708} \tag{1.101}$$

$$P_c = 1891.4M^{-0.7975} \tag{1.102}$$

$$w = c_f[0.5837\ln(M) - 2.5389] \tag{1.103}$$

The application of mixing rules allows for the computation of crude oil properties, based on the SARA fractions:

$$M_{\text{mix}} = \sum_{i=\text{SARA}} x_i M_i \tag{1.104}$$

$$V_{\text{mix}} = \sum_{i=\text{SARA}} x_i V_i \tag{1.105}$$

$$\delta_{\text{mix}} = \sum_{i=\text{SARA}} \frac{x_i V_i \delta_i}{V_{\text{mix}}} \tag{1.106}$$

where M_{mix} is the molar mass of crude oil, x_i is the mass fraction of component i, and M_i is the molar mass of component i. After estimating the equilibrium ratio, the subsequent task is to solve the phase equilibrium for asphaltene precipitation. When flocculation starts, the fraction of asphaltene that is dissolved is equivalent to the highest fraction that can dissolve in crude oil. The solid phase fraction is 1 when asphaltenes are precipitated, enabling the maximum soluble fraction to be obtained as described below:

$$x_{\text{max}}^L = \frac{1}{K} \tag{1.107}$$

where x_{max}^L is the maximum fraction soluble in the crude oil. Once the maximum amount of soluble asphaltenes within crude oil has been determined, this value is compared with the calculated asphaltene amount through the SARA-based kinetic model to establish the sediment fraction as follows:

$$\text{Sediment fraction} = y_{\text{As}}^{\text{cal}} - x_{\text{max}}^L \tag{1.108}$$

where $y_{\text{As}}^{\text{cal}}$ is the calculated composition of asphaltenes. Finally, the sediment formation is determined through asphaltene precipitation, and this methodology is depicted in Figure 1.11.

1.3 Kinetic Parameters Estimation

Once experimental data have been collected, it is essential to devise a reaction scheme capable of accurately predicting the product composition. The reaction rate equations are obtained by deriving them from the proposed reaction scheme, while the mass balance equations are obtained by making some assumptions. For estimating parameters in kinetic modeling of complex reaction networks, a nonlinear optimization process is typically employed. However, when the number of parameters to be estimated is significant, convergence issues may arise. Additionally, the initialization

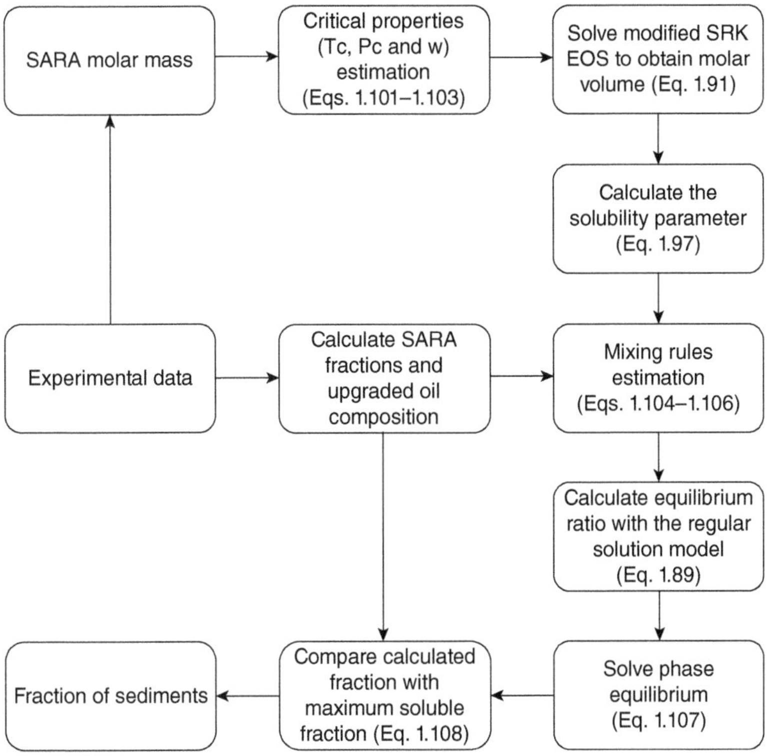

Figure 1.11 Methodology to calculate the fraction of precipitated asphaltenes to form sediments.

of the kinetic parameter values is another problem frequently found in nonlinear estimation that may converge to a local minimum of the objective function instead the global minimum (Ancheyta and Sotelo-Boyás 2000; Quitian and Ancheyta 2016a).

After optimizing the reaction rate coefficients, the activation energies and collision factors are calculated using the Arrhenius equation.

$$k_j = A_{0,j} e^{\left(\frac{-E_{A,j}}{RT}\right)} \tag{1.109}$$

MATLAB software is widely used for estimating kinetic parameters (Álvarez et al. 2019; Browning et al. 2019; Félix and Ancheyta 2019a, b; Huang et al. 2017; Loria et al. 2011; Ortega-García et al. 2017; Pham et al. 2021; Puron et al. 2014; Sheng et al. 2017); however, other commercial packages, such as Microsoft Excel (da Silva De Andrade 2014; Duran Armas 2021; Elahi et al. 2019; Galarraga et al. 2012), R (Coronel-García et al. 2021), 1st opt (Cai et al. 2022; Ma et al. 2022), and PROII, have been used to calculate kinetic parameters (Nguyen et al. 2013). Regarding the optimization algorithms, the majority of them use the Levenberg–Marquardt methodology (Coronel-García et al. 2021; Félix and Ancheyta 2019a, b; Huang et al. 2017; Kim et al. 2017; Nguyen et al. 2013; Ortega-García et al. 2017; Pham et al. 2021). In addition, the trust-region-reflective (Álvarez et al. 2019; Browning et al. 2019) and the Universal Global Optimization are also reported (Cai et al. 2022; Ma et al. 2022). The mass balance equations are preferred to be solved numerically by integral factor technique (Hassanzadeh and Abedi 2010), the trapezoidal rule (Galarraga et al. 2012), the Petzold–Gear backward differentiation formulae (Nguyen et al. 2013), Runge–Kutta

(Félix and Ancheyta 2019a, b; Huang et al. 2017; Ortega-García et al. 2017; Pham et al. 2021; Puron et al. 2014; Sheng et al. 2017), and the Runge–Kutta–Fehlberg method (Cai et al. 2022; Ma et al. 2022) instead of obtaining analytical solutions (Asaee et al. 2014; Coronel-García et al. 2021). Regarding the objective function, the most used is the SSE (sum of the squared error) (Álvarez et al. 2019; Browning et al. 2019; Cai et al. 2022; Félix and Ancheyta 2019a, b; Galarraga et al. 2012; Huang et al. 2017; Kim et al. 2017; Loria et al. 2011; Ma et al. 2022; Ortega-García et al. 2017; Pham et al. 2021; Puron et al. 2014). Nonetheless, the SSECC (correlation coefficient) (da Silva De Andrade 2014; Duran Armas 2021; Elahi et al. 2019) and the SSEWF (weighting factor) (Asaee et al. 2014; Coronel-García et al. 2021) are also employed.

1.3.1 Assumptions

The mass balance equations for kinetic models during the hydrocracking reactions carried out in batch reactors using slurry phase catalysts assume the following considerations (Matsumura et al. 2005; Quitian and Ancheyta 2016a; Rezaei et al. 2010):

✓ Liquid and gas phases are solely present when using dispersed catalysts
✓ Each phase is considered as pseudohomogeneous
✓ Reactions are carried out a constant temperature
✓ The volume is kept as a constant
✓ The mass transfer between phases is invariant with the time
✓ Reactions take place in liquid phase

Kinetic studies can be conducted under various experimental conditions, depending on whether kinetics is intrinsic or not. However, data from reaction kinetics acquired under conditions of mass transfer limitations cannot be directly applied to develop kinetic models, as these limitations can conceal results and cause misinterpretations. There are specific prerequisites for conducting kinetic experiments in circumstances where transport resistances are insignificant. If this is possible, the intrinsic kinetics and its parameters are obtained (rate constant, activation energies, and collision factors). Therefore, it is important to perform kinetic experiments under conditions where transport resistances are negligible (Bej 2002; Mederos et al. 2009; Perego and Peratello 1999).

Gradients of concentration and temperature may arise at the phase boundary within the reactor. In the case of solid catalysts, these gradients can occur between the internal solid surface and the bulk fluid (interphase), as well as within the catalyst particles themselves (intraphase). These gradients are also referred to as external or internal, respectively, and their presence complicates data analysis (Fogler 1992; Perego and Peratello 1999). For batch reactors, the agitation or external diffusion test (Figure 1.12a) is useful to avoid interphase gradients. In these experiments, the influence of changing the stirring rate on conversion at constant operating conditions is analyzed. The external transport-free operation regime allows the conversion to remain constant as the stirring speed is increased, whereas the internal diffusion test (Figure 1.12b) is employed to find the particle size of the dispersed catalyst at which kinetic control is carried out. Intraphase gradients are excluded during reaction if catalyst is finely crushed into progressively smaller particles until the reactant conversion is not changed (Perego and Peratello 1999).

A significant number of researchers determine the reaction order from the experimental findings or adopt values documented in the literature. The most inferred reaction order in the kinetic model for the hydrocracking of heavy oil using dispersed catalyst is the first order. Only few works corroborate the assumed order by linearizing the reaction rate equation with the supposed reaction order and plotting the yield (y_i) or conversion of the main reactant (X_i) term

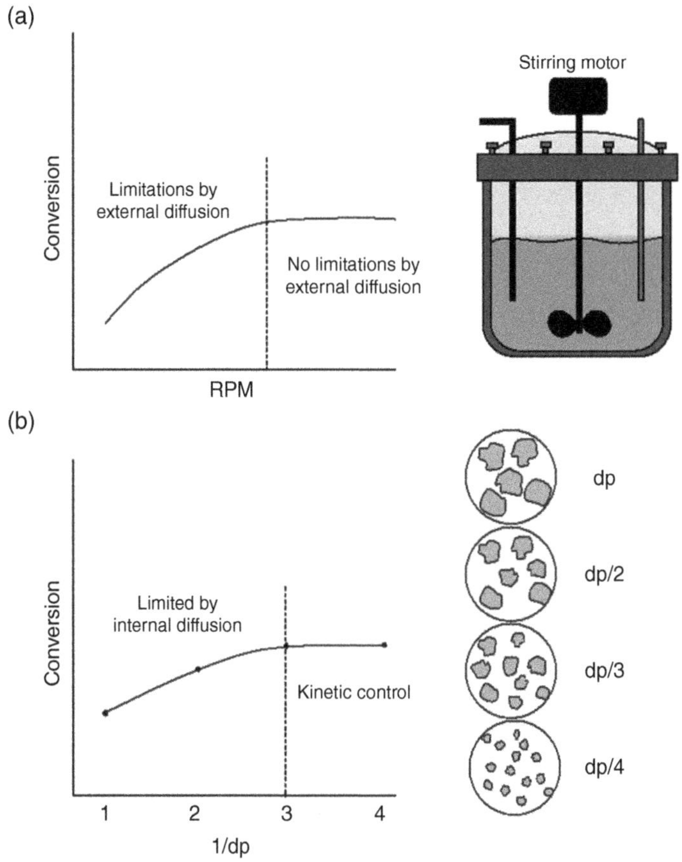

Figure 1.12 External diffusion test (a) and internal test (b) to evaluate the influence of catalyst particle size on the conversion at constant operating conditions.

against the reaction time. For reaction order different than 1 ($n \neq 1$), "$y_i^{1-n} - y_{i,0}^{1-n}$" or "$(1 - X_i)^{1-n} - 1$" is plotted against t, being the slope "$(n-1)k$" or "$(n-1)y_{i,0}^{n-1}k$", respectively, while for $n=1$, "$\ln(y_{i,0}/y_i)$" or "$\ln(1/(1-X_i))$" is plotted against t and the slope is k in both cases.

1.3.2 Initialization of Parameters

The parameter initialization is the foremost step of the optimization process as the estimation of kinetic parameters heavily relies on the initial guesses. Commonly, the initial guess is obtained from literature or calculated using the Monte Carlo algorithm, as documented earlier with satisfactory outcomes (Asaee et al. 2014; Félix et al. 2019; Félix and Ancheyta 2019a, b; Hassanzadeh and Abedi 2010; Huang et al. 2017; Ortega-García et al. 2017; Pham et al. 2021). The Monte Carlo method assigns random numbers to the kinetic parameters and the objective function is calculated. This calculation is repeated, and a plot with different kinetic parameters and their corresponding objective function is made. The kinetic parameters that minimize the objective function are selected as initial values for nonlinear optimization.

1.3.3 Nonlinear Optimization

Some models may contain numerous parameters requiring a precise estimation due to their high nonlinearity. In such cases, when calculating kinetic parameter values, multiple solutions of the objective function can arise during the optimization process by which optimal values could not be reached. The optimal solution depends mostly on initial guesses of kinetic parameters (Alcázar and Ancheyta 2007; Félix et al. 2019). A general method for estimating kinetic parameters (Figure 1.13) that considers various steps to calculate and validate the best parameter set minimizing the objective function has been found to yield favorable results in the literature (Alcázar and Ancheyta 2007).

Once the mass balance of reactants and products is obtained, the kinetic parameters are calculated by minimizing the difference between the experimental and calculated values. The set of ordinary differential equations (ODEs) is solved analytically (integrating factor for example) or using numerical methods (fourth-order Runge–Kutta method). The nonlinear optimization problem can be solved using optimization algorithms, such as Levenberg–Marquardt, interior-point, trust-region-reflective, among others, to find the minimum value of the objective function (Alcázar and Ancheyta 2007; Angeles et al. 2014; Ortiz Moreno et al. 2014; Quitian and Ancheyta 2016a).

1.3.4 Objective Function

The most used objective function for the kinetic modeling during the hydrocracking of heavy oil is the SSE. Nevertheless, other objective functions, such as the SSEWF and the SSECC, also can be used together with SSE. Similar objective functions, such as the average absolute difference (AAD), and more recently, the average absolute error (AAE) have been utilized to compare calculated and experimental data (Alcázar and Ancheyta 2007; Ancheyta and Sotelo-Boyás 2000; Asaee et al. 2014; da Silva De Andrade 2014; Stratiev et al. 2021). The AAE objective function has recently gained attention (Félix et al. 2022a,b; Stratiev et al. 2021; Tirado et al. 2022) due to its ability to provide more balanced weightings to all yields, even when their differences in order of magnitude are high, which contributes to have an improved level of accuracy.

$$SSE = \sum_{i=1}^{nc} \left(y_i^{exp} - y_i^{cal}\right)^2 \tag{1.110}$$

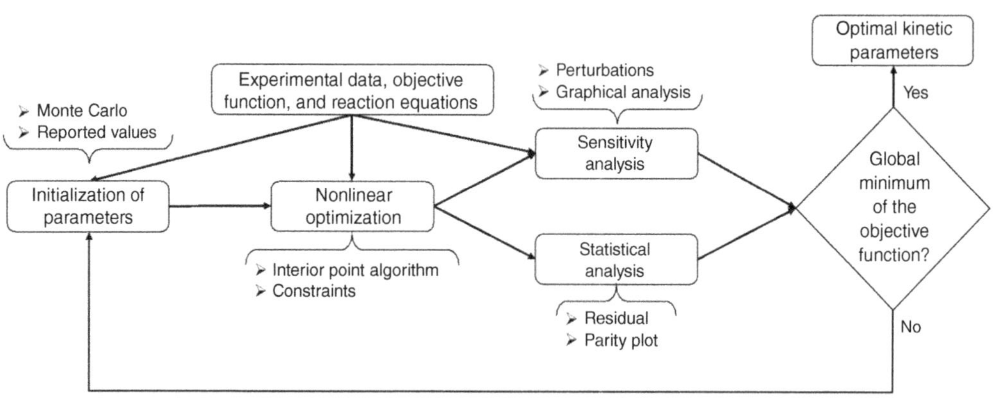

Figure 1.13 Methodology for optimal estimation of kinetic parameters.

$$\text{SSEWF} = \sum_{i=1}^{nc} w_i \left(y_i^{\text{exp}} - y_i^{\text{cal}}\right)^2 \tag{1.111}$$

$$\text{SSECC} = w_i \text{SSE} + W_R \left(nr - \sum_{j=1}^{nr} \phi_j^2\right) \tag{1.112}$$

$$\text{AAD} = \sum_{i=1}^{nc} \left|y_i^{\text{exp}} - y_i^{\text{cal}}\right| \tag{1.113}$$

$$\text{AAE} = \frac{\sum_{i=1}^{nc} \left(\left|y_i^{\text{exp}} - y_i^{\text{cal}}\right| / y_i^{\text{exp}}\right) \times 100}{N} \tag{1.114}$$

where w_i is the weighting factor, W_R is the Arrhenius fitting index weight factor, and ϕ_j^2 is the correlation coefficient of the reaction j. The most used weighting factors are $1/y_{\text{exp}}^2$ and the ratio of the arithmetic mean yield of the main reactant to the arithmetic mean yield of each pseudocomponent (Alcázar and Ancheyta 2007; Asaee et al. 2014; Félix et al. 2022a,b; Gauthier et al. 2006; Rana et al. 2021; Saleh 2018; Stratiev et al. 2021).

1.3.5 Sensitivity and Statistical Analyses

1.3.5.1 Perturbations

After estimating the kinetic parameters, the sensitivity analysis is applied through perturbations by changing one parameter at a time while keeping the others constant. This allows to re-evaluate the objective function for each perturbation and plot the corresponding percentage of perturbation against the value of the objective function. If the original parameters (0% perturbation) give the minimum value for the objective function after applying all perturbations, it can be concluded that the global minimum has been achieved. Conversely, if one or more parameters fail to achieve the global minimum at 0% perturbation, it suggests that the nonlinear optimization was inadequate and the kinetic parameters should be recalculated.

1.3.5.2 Parity Plots

The parity plot is another tool to evaluate the adjustment between experimental and calculated values. A satisfactory model fit is observed when values align with a 45° line (Alcázar and Ancheyta 2007; Félix et al. 2019).

1.3.5.3 Residuals

It is common to use a plot of residual values against the number of experimental observations to verify the adequacy of the proposed model and calculated parameters. The plot should demonstrate a regular distribution of residuals with no prediction bias. In the case of a pattern or tendency, it can be concluded that the parameters were estimated incorrectly. Another useful method to observe the fit quality is the parity plot, in which experimental values are plotted against calculated values (Alcázar and Ancheyta 2007; Félix et al. 2019; Félix et al. 2022a,b; Sámano et al. 2020).

$$\text{Residual} = y_i^{\text{exp}} - y_i^{\text{cal}} \tag{1.115}$$

1.3.5.4 AIC and BIC

To compare the kinetic models based on SARA fractions, both the Akaike and Bayesian information criteria (AIC and BIC, respectively) were used (Sánchez et al. 2007). The AIC takes into account the intricacy of a model and how well it fits the data. BIC is more rigorous in penalizing free parameters in comparison with AIC. Both approaches strive to choose the model with the best fit and minimum number of free parameters, as illustrated below:

$$\mathrm{AIC} = 2h + Nln\left(\frac{\mathrm{SSE}}{N}\right) \tag{1.116}$$

$$\mathrm{BIC} = hln(N) + Nln\left(\frac{\mathrm{SSE}}{N}\right) \tag{1.117}$$

A penalization term ($2h$) is incorporated in the AIC that increases with the number of parameters (h), enabling effective comparison of models with different parameters. In both cases, the preferred model is that with the lowest value of AIC and BIC.

1.4 Results and Discussion

Among all the reported kinetic models for the hydrocracking of heavy oil using dispersed catalyst, only the kinetic models developed by Félix and Ancheyta (2019a, b), and Fukuyama and Terai (2007) employed mineral catalyst (molybdenite and pyrite, respectively), thus this section is focused to their results.

1.4.1 Kinetic Parameters

1.4.1.1 Assumptions

In the assumptions of both kinetic models, the fist-order of reaction for asphaltenes and residue conversion is supposed, as well as external and internal diffusion phenomena are neglected. The plot to corroborate the first-order of reaction for asphaltenes conversion reported by Félix and Ancheyta (2019a, b) during the hydrocracking of heavy oil using mineral catalysts is shown in Figure 1.14. Here, t versus $\ln(1/(1 - X_{\mathrm{As}}))$ is plotted at the different studied temperatures, where the experimental data follow a straight-line relationship, which suggests that the reaction order is indeed one. In addition to the hydrocracking of asphaltenes, Figure 1.14 shows a plot of t versus $\ln(1/(1 - X_{\mathrm{R}}))$ to confirm first-order reaction in hydrocracking of residue at different temperatures, as reported by the same authors. The experimental data follow a straight line, which confirms the first-order of reaction. These results are in accordance with most of the reported kinetic models for residua (Asaee et al. 2014; da Silva De Andrade 2014; Elahi et al. 2019; Galarraga et al. 2012; Hassanzadeh and Abedi 2010; Huang et al. 2017; Loria et al. 2011; Nguyen et al. 2013; Orozco Castillo 2016; Ortega-García et al. 2017; Puron et al. 2014) and asphaltenes (Sheng et al. 2017) conversion during the hydrocracking of heavy oils using dispersed catalysts. Nonetheless, this supposition not always is corroborated as in the case of Fukuyama and Terai (2007), and only few studies have corroborated the supposed first-order reaction (Duran Armas 2021; Elahi et al. 2019; Orozco Castillo 2016).

Additional experiments were carried out by Félix and Ancheyta (2019a, b) to find the conditions at which the external and internal diffusions are disregarded and the kinetic control predominates. The external diffusion experiments, where different stirred rates versus the residue conversion are plotted, can be noticed in Figure 1.15. The higher the stirring rate, the higher the residue

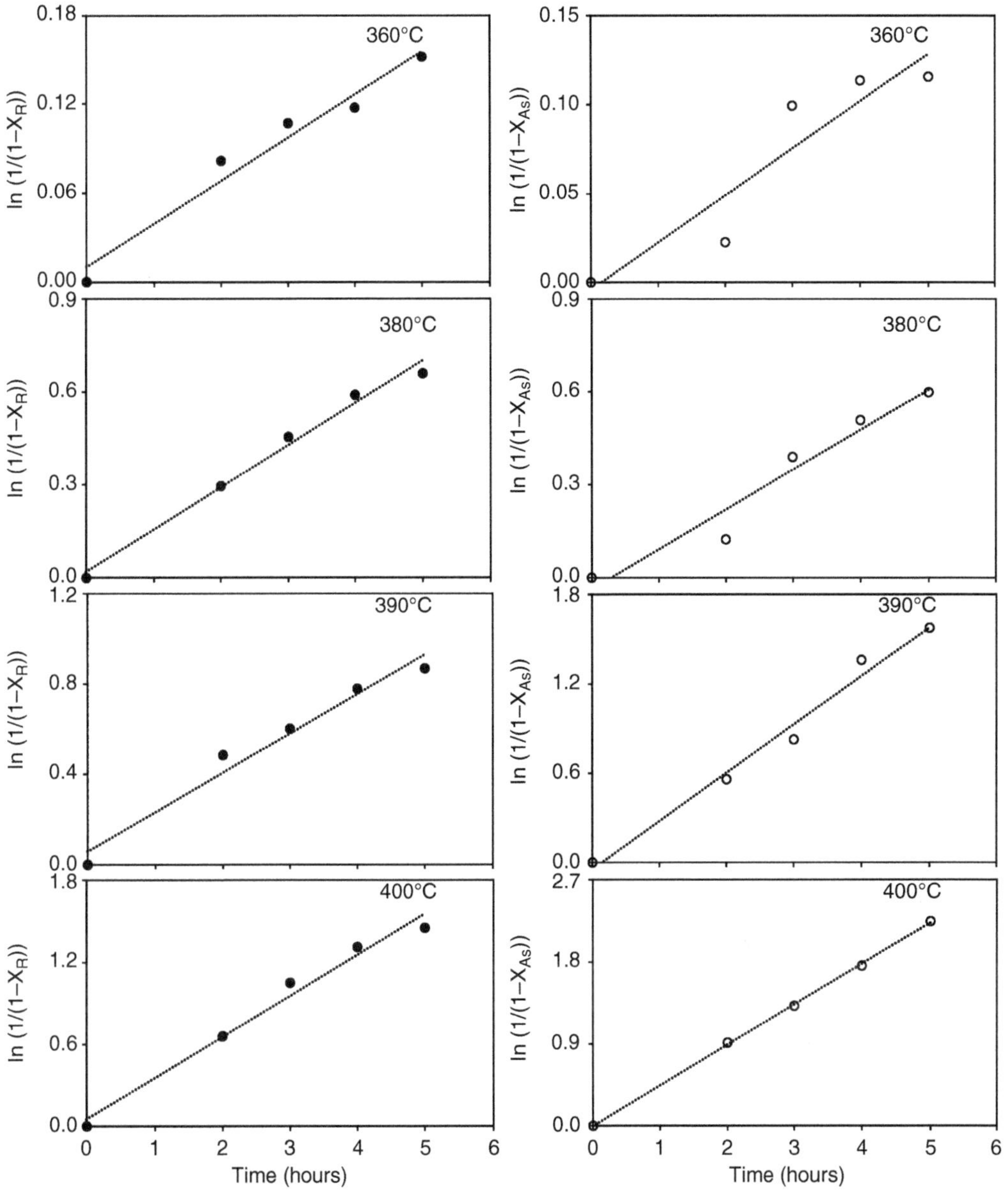

Figure 1.14 First-order reaction corroboration for R and As conversion during the hydrocracking of heavy oil using molybdenite.

conversion. This trend is observed up to 800 rpm, beyond which the conversion remains constant and unaffected by diffusion. Figure 1.15 also shows the internal diffusion experiments to find the catalyst particle size at which the intrinsic kinetic is performed. The residue conversion against the inverse of the particle size was plotted, and it can be noticed that the conversion is increased as the particle size is reduced (the inverse of size is increased). Nevertheless, when the inverse of particle size achieves the value 0.01571/μm (63.5 μm), the residue conversion is constant. Hence, with 800 rpm of stirring rate and 63.5 μm of molybdenite particle size, the intrinsic kinetics can be obtained.

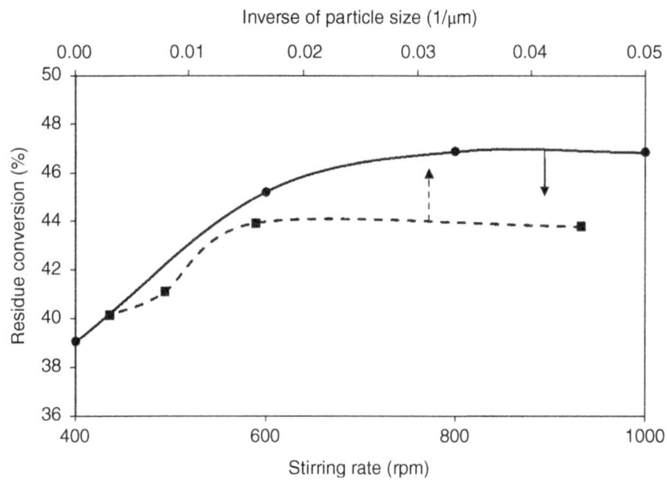

Figure 1.15 Experiments for internal and external diffusion tests for the hydrocracking of heavy oil using molybdenite.

1.4.1.2 Reaction Rate Coefficients

SARA-Based Model The reaction rate coefficients for the kinetic models based on SARA fractions reported by Félix and Ancheyta (2019a, b) during the hydrocracking heavy oils using mineral catalyst are summarized in Table 1.2. The kinetic parameter values of the four-lump kinetic model (Figure 1.9b, Model 1) suggest that specific reactions do not occur at particular temperatures. The reaction pathways from asphaltenes to resins (k_1) and resins to saturates (k_5) are not carried out at 360 °C. Furthermore, the agglomeration of aromatics to produce resins (k_8) is only carried out at 400 °C because the thermal cracking is enough to break small molecules from heavy fractions (asphaltenes and resins) but unable to produce heavy fractions with lower molecular weight, such as resins. In addition, the thermal cracking produces free radicals from resins that form asphaltenes by agglomeration. However, aromatics do not form resins since thermal cracking requires higher temperature. Some of these kinetic parameters are not affected by temperature changes (k_2, k_3, k_7) under the mild conditions used. Accordingly, changes are noticeable only at high temperature (400 °C).

The reaction rate coefficients for the five-lump kinetic model considering the SARA fractions and coke (Figure 1.10a, Model 2) are observed in this table. The conversion of resins to saturates cannot be performed, as the constant k_5, obtained by optimizing the reaction rate coefficients, is zero at all studied temperatures. Moreover, the reactions that involve the coefficients k_1 and k_8 are not carried out at 360 °C as in the previous model, as well as the same reaction pathways (k_2, k_3, k_7) are not influenced by the temperature increase.

The kinetic parameters obtained for five-lump SARA-based Model 3 (considering all SARA and gas fractions, Figure 1.10b) are observed in Table 1.2. The reaction from aromatics to gas (k_9) is not performed in this model. The reaction rate coefficients k_1 (asphaltenes to resins), k_6 (resins to saturates), and k_{10} (saturates to gases) are not carried out at 360 °C, while k_4 (asphaltenes to gases) and k_{12} (aromatics to resins) are performed at higher temperatures (380 and 400 °C, respectively). As stated earlier, breaking these molecules at low temperatures is difficult, and thermal cracking requires temperatures exceeding 400 °C to occur. Additionally, most of the reaction rate coefficients are not sensitive to the temperature change until high temperatures are reached.

The kinetic parameters obtained for the 6-lump kinetic model based on SARA fractions (Figure 1.10c, Model 4) are shown in the same table. In this model as in Model 3, the reaction from

Table 1.2 Kinetic parameters reported for kinetic models based on SARA fractions during the heavy oil hydrocracking using molybdenite catalyst.

Temperature (°C)	As→Re	As→Ar	As→Sa	As→G	As→C	Re→Ar	Re→Sa	Re→G	Re→As	Ar→Sa	Ar→G	Ar→Re	Sa→G	SSE	AAE (%)
Model 1 (4 lumps, 8 constants)															
360	0	0.09499	0.02088			0.01666	0		0.07782	0.01936		0		4.87×10^{-4}	2.07
380	0.11916	0.09499	0.02088			0.01666	0.00321		0.07782	0.05004		0		2.07×10^{-3}	4.09
390	0.35913	0.09499	0.02088			0.08568	0.00815		0.07782	0.05004		0		2.03×10^{-3}	6.20
400	0.35913	0.23637	0.05205			1.58695	0.00815		0.07782	0.05004		1.17453		4.03×10^{-3}	9.28
E_A (kJ/mol)	202.64	63.34	63.47			367.34	170.97		—	87.15		—			
A_0 (h^{-1})	2.29×0^{15}	1.33×10^{04}	3.01×10^{03}			1.49×10^{28}	1.76×10^{11}		—	3.49×10^{05}		—			
Model 2 (5 lumps, 9 constants)															
360	0	0.09305	0.02058		0.00248	0.01737			0.07760	0.01932		0		4.86×10^{-4}	3.18
380	0.10049	0.09305	0.02058		0.03636	0.01737			0.08391	0.04959		0.00668		2.50×10^{-3}	25.70
390	0.26486	0.09305	0.02058		0.12208	0.01737			0.08391	0.04959		0.00668		3.74×10^{-3}	18.38
400	0.26497	0.13053	0.03286		0.25659	2.74437			0.08391	0.04959		2.03092		4.39×10^{-3}	11.75
E_A (kJ/mol)	178.10	23.52	32.50		419.72	396.63			7.18	86.52		1039.55			
A_0 (h^{-1})	2.06×10^{13}	$7.59 \text{x} 10^{00}$	9.01×10^{00}		1.17×10^{32}	3.64×10^{30}			3.07×10^{-01}	3.08×10^{05}		3.52×10^{80}			
Model 3 (5 lumps, 12 constants)															
360	0	0.09562	0.01892	0		0.01365	0	0.00752	0.07514	0.01967		0	0	4.79×10^{-4}	2.71
380	0.12056	0.09562	0.01892	0		0.01365	0.00703	0.00752	0.07514	0.05295		0	0.01484	2.03×10^{-3}	5.50
390	0.33659	0.09562	0.01892	0.02374		0.07866	0.00942	0.00752	0.07514	0.05295		0	0.01484	1.99×10^{-3}	6.36
400	0.33659	0.21383	0.07895	0.02374		1.43887	0.00942	0.00752	0.07514	0.05295		1.06501	0.03190	3.40×10^{-3}	9.67
E_A (kJ/mol)	188.59	55.91	99.27	—		377.86	53.66	—	—	90.87		—	139.18		
A_0 (h^{-1})	$1.72 \text{x} 10^{14}$	3.35×10^{03}	2.21×10^{06}	—		8.90×10^{28}	1.44×10^{02}	—	—	7.24×10^{05}		—	$1.76 \text{x} 10^{09}$		
Model 4 (6 lumps, 13 constants)															
360	0	0.09381	0.01858		0.00248	0.01428		0.00753	0.07499	0.01965		0	0	4.77×10^{-4}	3.46
380	0.09647	0.09381	0.01858		0.03586	0.01428		0.00753	0.07772	0.05547		0.00361	0.01489	2.41×10^{-3}	23.26
390	0.25858	0.09381	0.01858		0.11950	0.06722		0.01393	0.07772	0.05547		0.00361	0.01991	3.49×10^{-3}	17.33
400	0.25858	0.11424	0.05680		0.24924	1.58861		0.01393	0.07772	0.05547		1.17341	0.03459	3.77×10^{-3}	11.53
E_A (kJ/mol)	181.11	13.69	77.62		417.01	375.34		61.83	3.28	95.26		1051.84	153.85		
A_0 (h^{-1})	3.45×10^{13}	1.22×10^{00}	3.78×10^{04}		7.00×10^{31}	5.65×10^{28}		8.68×10^{02}	1.41×10^{-01}	1.68×10^{06}		1.81×10^{81}	2.86×10^{10}		

aromatics to gas is not carried out at the temperatures studied, furthermore, k_4 and k_6 corresponding to the reactions from asphaltenes to gas and resins to saturates, respectively, are not accomplished. This is due to the fact that, under these conditions, the reactions occur primarily in series rather than in parallel. Furthermore, parameters k_1 (asphaltenes to resins), k_{10} (saturates to gases), and k_{13} (aromatics to resins) exhibit a value of zero at 360 °C. Some reactions are not affected by temperature as the reaction rate coefficients (k_2, k_3, and k_{11}) remain invariant despite the increase in temperature.

The sensitivity analysis performed to the obtained reaction rate coefficients for the SARA-based kinetic models at 400 °C is depicted in Figure 1.16. Almost all the obtained reaction rate coefficients achieved the minimum value of the objective function (SSE) at 0% of perturbation. Thus, these parameter values are optimal and have been accurately calculated. For Model 1, some of the obtained reaction rate coefficients (k_2, k_5, k_6, and k_7) at 400 °C give a lower value of the objective function when the perturbation is negative (lower than 0%). For Model 2, the acquired kinetic parameters k_2, k_3, k_6, and k_7 at 390 and 400 °C have a lower value of the SSE in a perturbation lower than the original value, whereas for Model 3, some of the obtained reaction rate coefficients (k_2, k_3, k_4, k_6, k_7, k_{10} and k_{11}) have a lower value of the objective function when the perturbation is negative at high temperatures (390 and 400 °C). Concerning Model 4, the optimized kinetic parameters for the reactions from asphaltenes to aromatics, aromatics to saturates, resins to gases, aromatics to gases, and resins to asphaltenes at all temperatures except for 360 °C exhibit a lower objective function value with a perturbation of less than 0%. This suggests that the parameter value ought to be lower, but it cannot be the case, since to satisfy the Arrhenius equation, the reaction rate coefficient has to rise with temperature.

Distillation Curves-Based Models The reaction rate coefficients for combined kinetic model obtained by Fukuyama and Terai (2007) are depicted in Table 1.3. All the reaction pathways are carried out at the conditions studied. Nevertheless, the reaction from Ar to Re for both catalysts, and the production of VGO from Ar and As for the pyrite-activated carbon catalyst are not sensitive to the temperature because the condensation reactions of Ar to produce heavier fraction, such as Re and VGO, require higher temperature to be performed, meanwhile for As being a high polar and polycondensed fraction is easier to agglomerate to produce C than crack into VGO. The highest reaction rate coefficient was k_4 (Sa to LO) for pyrite and k_5 (Ar to LO) for pyrite-activated carbon catalysts, due to the similarity of these light fractions, such as Sa with N, and Ar with K, which enhance the reaction. Therefore, the lowest values are for k_2 (Ar to Re) and k_1 (Ar to Sa) for the pyrite and pyrite-activated carbon catalysts, respectively.

The estimated kinetic parameters for the distillation curves-based kinetic model reported by Félix and Ancheyta (2019a, b) are presented in Table 1.4. Some reaction pathways (VR to D, VR to N, and D to N) do not occur at 360 °C because their kinetic parameters (k_{L2}, k_{L3} and k_{L6}) were zero. This trend is also shown at 380 °C for the reaction from D to N (k_{L6}) owing to the mild reaction conditions used in the experiments, which cannot achieve severe hydrocracking to produce lighter distillates. Among the three reaction pathways of VR hydrocracking (k_{L1}, k_{L2}, and k_{L3}), k_{L1} (VR to VGO) gave the lowest value, while k_{L2} (VR to D) exhibited higher values, indicating that VR hydrocracking is more selective toward light distillates. k_{L4}, k_{L5}, and k_{L6} had even lower values indicating that over cracking of VGO and D proceeded at low rates.

As expected, all the obtained reaction rate coefficients are in accordance with Arrhenius equation, contrary to some reaction rate coefficients reported in the literature (Coronel-García et al. 2021; Galarraga et al. 2012; Hassanzadeh and Abedi 2010; Kim et al. 2017; Loria et al. 2011; Orozco Castillo 2016) that displayed higher values at lower temperatures. The sensitivity analysis on the obtained reaction rate coefficients for distillation curves-based kinetic model is shown in Figure 1.17. With the original value of reaction rate coefficients (0% of perturbation), the global

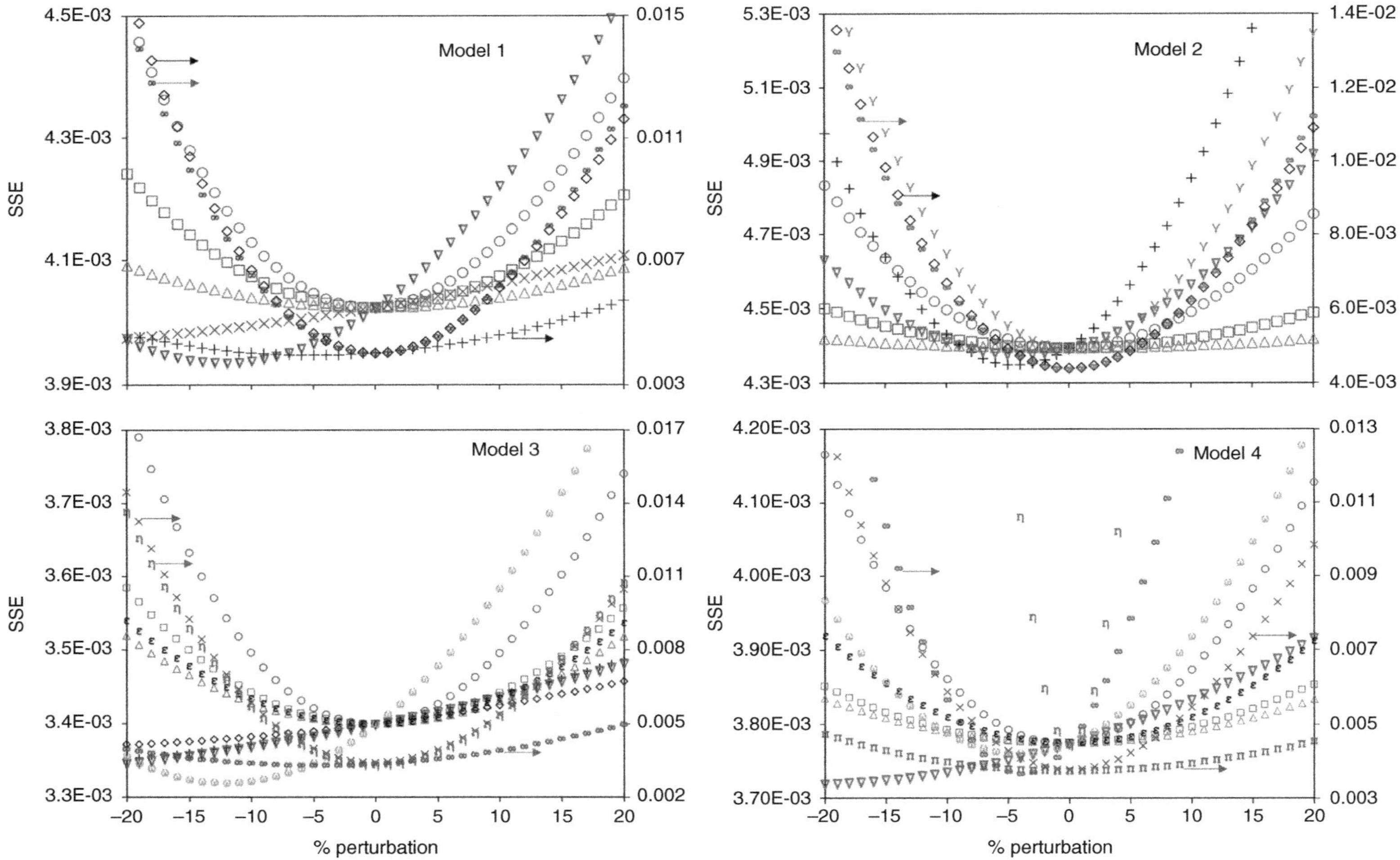

Figure 1.16 Sensitivity analysis of the obtained reaction rate coefficients (k_j) for the SARA-based kinetic models at 400 °C. k_1 (○), k_2 (□), k_3 (△), k_4 (⬦), k_5 (×), k_6 (+), k_7 (▽), k_8 (∞), k_9 (γ), k_{10} (ε), k_{11} (ω), k_{12} (η), and k_{13} (Π).

Table 1.3 Kinetic parameters obtained for the kinetic model based on combined fractions (SARA and distillation curves) in the hydrocracking of heavy oil using mineral catalysts.

	Temperature (°C)				
Parameter	**415**	**425**	**435**	E_A **(kJ/mol)**	A_0 **(h^{-1})**
Pyrite					
k_1	4.76×10^{-3}	1.66×10^{-2}	3.03×10^{-2}	253.00	5.17×10^{18}
k_2	1.35×10^{-5}	1.41×10^{-5}	1.46×10^{-5}	10.00	4.94×10^{-3}
k_3	3.54×10^{-6}	6.34×10^{-4}	7.62×10^{-3}	1050.00	1.29×10^{76}
k_4	1.82×10^{-2}	4.65×10^{-2}	7.27×10^{-2}	189.00	2.65×10^{14}
k_5	7.96×10^{-4}	5.67×10^{-3}	1.45×10^{-2}	397.00	7.45×10^{28}
k_6	2.44×10^{-5}	3.45×10^{-3}	3.71×10^{-3}	1002.00	2.14×10^{73}
k_7	5.42×10^{-3}	1.66×10^{-2}	2.83×10^{-2}	226.00	4.90×10^{16}
k_8	4.84×10^{-6}	1.30×10^{-4}	6.32×10^{-4}	667.00	1.28×10^{47}
k_9	3.56×10^{-5}	4.23×10^{-3}	4.18×10^{-2}	967.00	6.17×10^{70}
k_{10}	7.09×10^{-4}	7.11×10^{-3}	2.15×10^{-2}	467.00	1.19×10^{34}
Pyrite-activated carbon					
k_1	8.84×10^{-6}	1.18×10^{-5}	1.25×10^{-5}	47.00	2.18×10^{0}
k_2	1.49×10^{-5}	1.54×10^{-5}	1.57×10^{-5}	7.00	3.02×10^{-3}
k_3	1.22×10^{-5}	1.37×10^{-5}	1.45×10^{-5}	24.00	4.74×10^{-2}
k_4	1.41×10^{-5}	1.27×10^{-3}	1.09×10^{-2}	910.00	1.14×10^{66}
k_5	4.68×10^{-3}	1.95×10^{-2}	3.87×10^{-2}	289.00	2.65×10^{21}
k_6	1.40×10^{-5}	1.49×10^{-5}	1.54×10^{-5}	13.00	8.07×10^{-3}
k_7	4.65×10^{-3}	1.36×10^{-2}	2.28×10^{-2}	217.00	9.31×10^{15}
k_8	1.24×10^{-5}	1.29×10^{-5}	1.31×10^{-5}	7.00	2.66×10^{-3}
k_9	1.57×10^{-5}	9.12×10^{-4}	6.39×10^{-3}	822.00	2.61×10^{59}
k_{10}	1.29×10^{-3}	4.28×10^{-3}	7.61×10^{-3}	243.00	2.38×10^{17}

Table 1.4 Kinetic parameters obtained for the kinetic model based on distillation curves during the heavy oil hydrocracking using molybdenite catalyst.

	Temperature (°C)					
Parameter	**360**	**380**	**390**	**400**	E_A **(kJ/mol)**	A_0 **(h^{-1})**
k_{L1}	0.03203	0.03443	0.03615	0.07771	64.69	6.01×10^{3}
k_{L2}	0	0.06499	0.09512	0.12896	125.33	6.92×10^{8}
k_{L3}	0	0.04349	0.06515	0.11827	182.66	1.70×10^{13}
k_{L4}	0.00219	0.00230	0.00543	0.06503	263.40	6.42×10^{18}
k_{L5}	0.00339	0.00355	0.01119	0.01175	123.08	4.08×10^{7}
k_{L6}	0	0	0.01088	0.02711	338.81	5.29×10^{24}
SSE	3.96E−4	4.97E−4	1.61E−3	7.04E−4		
AAE (%)	2.25	2.43	3.24	2.84		

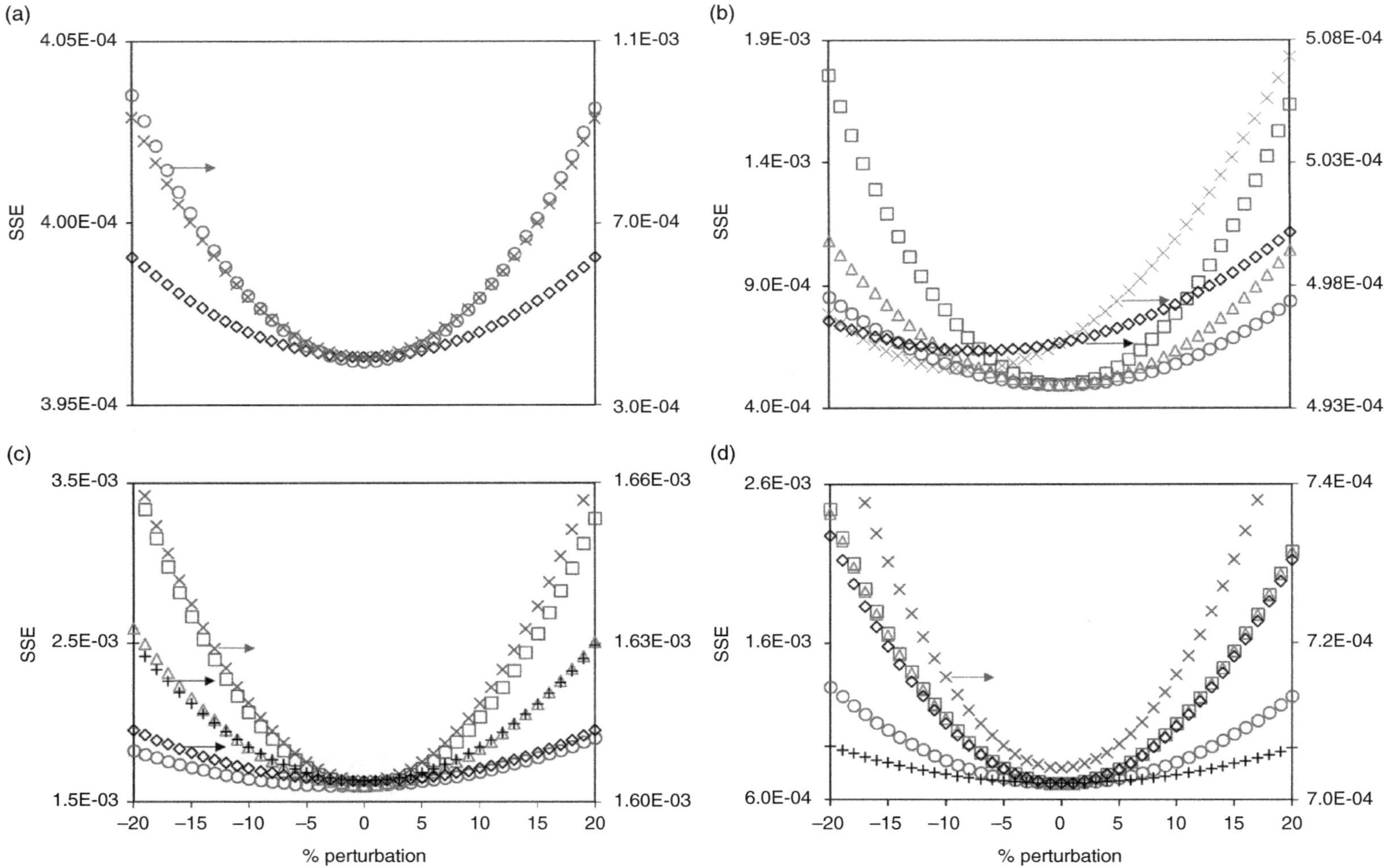

Figure 1.17 Sensitivity analysis of the reaction rate coefficients (k_j) for the kinetic model based on distillation curves at 360 (a), 380 (b), 390 (c), and 400 °C (d). k_1 (□), k_2 (△), k_3 (○), k_4 (◇), k_5 (+), and k_6 (×).

minimum of the objective function (SSE) is achieved confirming that they are the optimal values. The same results were observed at different temperatures demonstrating accurate estimation of optimal values even at other temperatures.

1.4.1.3 Activation Energies

SARA-Based Models As can be seen in Table 1.2, certain activation energies were not calculated because several reactions did not proceed as Re to Sa (Models 2 and 4.), As to G, and Ar to G (Models 3 and 4) at any temperature, and Ar to Re (Models 1 and 3) is only carried out at 400 °C. Additionally, other reactions were unchanged with variations of the temperature, such as Re to As (all models), and Re to G (Model 3), whereas the range of the activation energies values is lower for Models 1 (63.34–367.34 kJ/mol) and 3 (53.66–377.86 kJ/mol) than Models 2 (7.18–1039.55 kJ/mol) and 4 (3.28–1051.84 kJ/mol). This is owed to the reaction from Ar to Re is carried out only at 400 °C for the former models, while for the latter ones the reaction is performed at 380–400 °C with low-rate values at low temperature (380 and 390 °C) and considerably higher at 400 °C, resulting in a high activation energy. Based on the activation energies in all models, the cracking of the heavy fractions (As and Re) into lighter fractions (Ar and Sa) through parallel reactions are more feasible under the studied conditions, while higher values of activation energies were calculated for in-series reactions (As to Re to Ar to Sa).

Distillation Curves-Based Models The activation energies for the kinetic model based on distillation curves using pyrite and pyrite-activated carbon (pyrite-AC) are noticed in Table 1.3. Some of the activation energies for both catalysts showed low values because of the reactions are insensitive to temperature, such as Ar to Re. The values of the activation energies are in the range of 10–1050 kJ/mol (pyrite) and 7–910 kJ/mol (pyrite-AC). All the activation energies present lower values using pyrite–AC catalyst, indicating that this catalyst enhances all the reactions. The greater activation energies observed when using pyrite as a catalyst pertain to condensation reactions (Re to As, Ar to VGO, and As to C), while employing pyrite–AC as catalyst the largest values corresponded to the cracking of *Sa* and condensation of As. The lower activation energies using pyrite corresponded to the cracking of light fractions (Ar and Sa), meanwhile utilizing pyrite–AC, lower values were observed for condensation reactions (Ar to Re, Ar to VGO, and Re to As) because these reactions were not influenced by temperature. Thus, the addition of activated carbon to pyrite restricts the secondary cracking of light fractions, such as Sa.

The activation energies (64.69–338.81 kJ/mol) reported by Félix and Ancheyta (2019a) for the distillation curves-based kinetic model are depicted in Table 1.4. The lower values of activation energies pertained to the cracking of VR, and the activation energy increased as the molecular weight of products decreased (VR to VGO < VR to D < VR to N). This behavior suggests that the conversion of heavy fractions was performed with ease, while the conversion of D demands high activation energy. Similar results were obtained by other authors (da Silva De Andrade 2014; Galarraga et al. 2012; Loria et al. 2011) using dispersed catalysts since the lighter the products, the higher the activation energy for the cracking of heavy fractions.

1.4.2 Accuracy of the Kinetic Models

1.4.2.1 SARA-Based Models

The calculated and experimental mass fractions varying the reaction times at 390 and 400 °C for the kinetic models based on SARA fractions are observed in Figure 1.18. In all models, the mass fraction of As decreases while Sa and G (only considered in Models 3 and 4) fractions increases as reaction

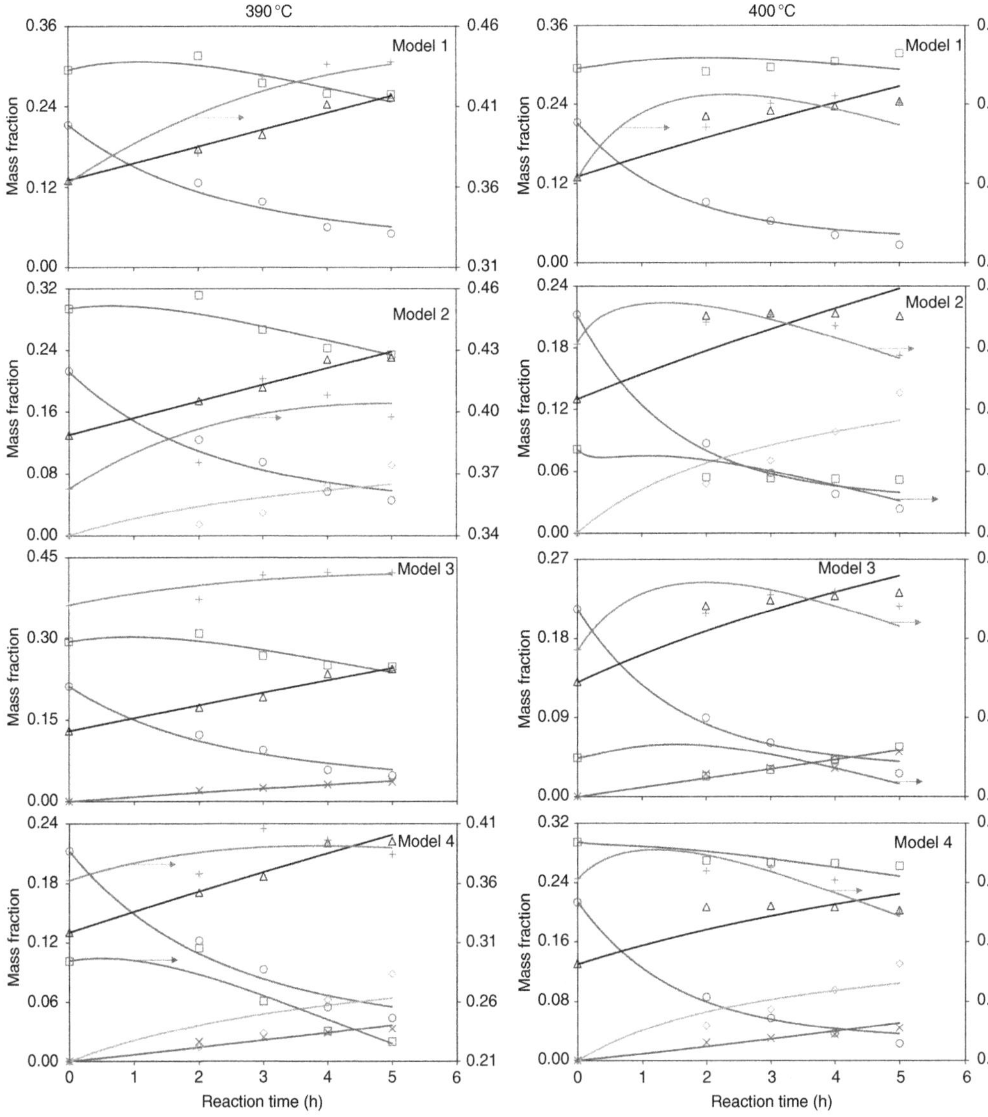

Figure 1.18 Calculated (lines) and experimental (symbols) mass fractions for the kinetic models based on SARA fractions at 390 and 400 °C. As (○), Re (□), Ar (+), Sa (△), G (×), and C (◇).

time and temperature raise. This is due to the thermal cracking that produces lighter fractions (Sa and G), and these products do not undergo further cracking. In addition, as temperature and reaction time increase, C (only considered in Models 2 and 4) fraction increases by the formation of free radicals in thermal cracking. All the experimental data are accurately predicted in the four SARA-based models, with the C fraction being the most difficult to forecast.

The residual plots of the kinetic models based on SARA fractions are shown in Figure 1.19. The residual plot confirms the correct estimation of kinetic parameters, as they neither overestimate nor underestimate the experimental data. This is evidenced by the number of positive and negative residuals in the statistical analysis (Table 1.5). The residual values showed that the prediction is better

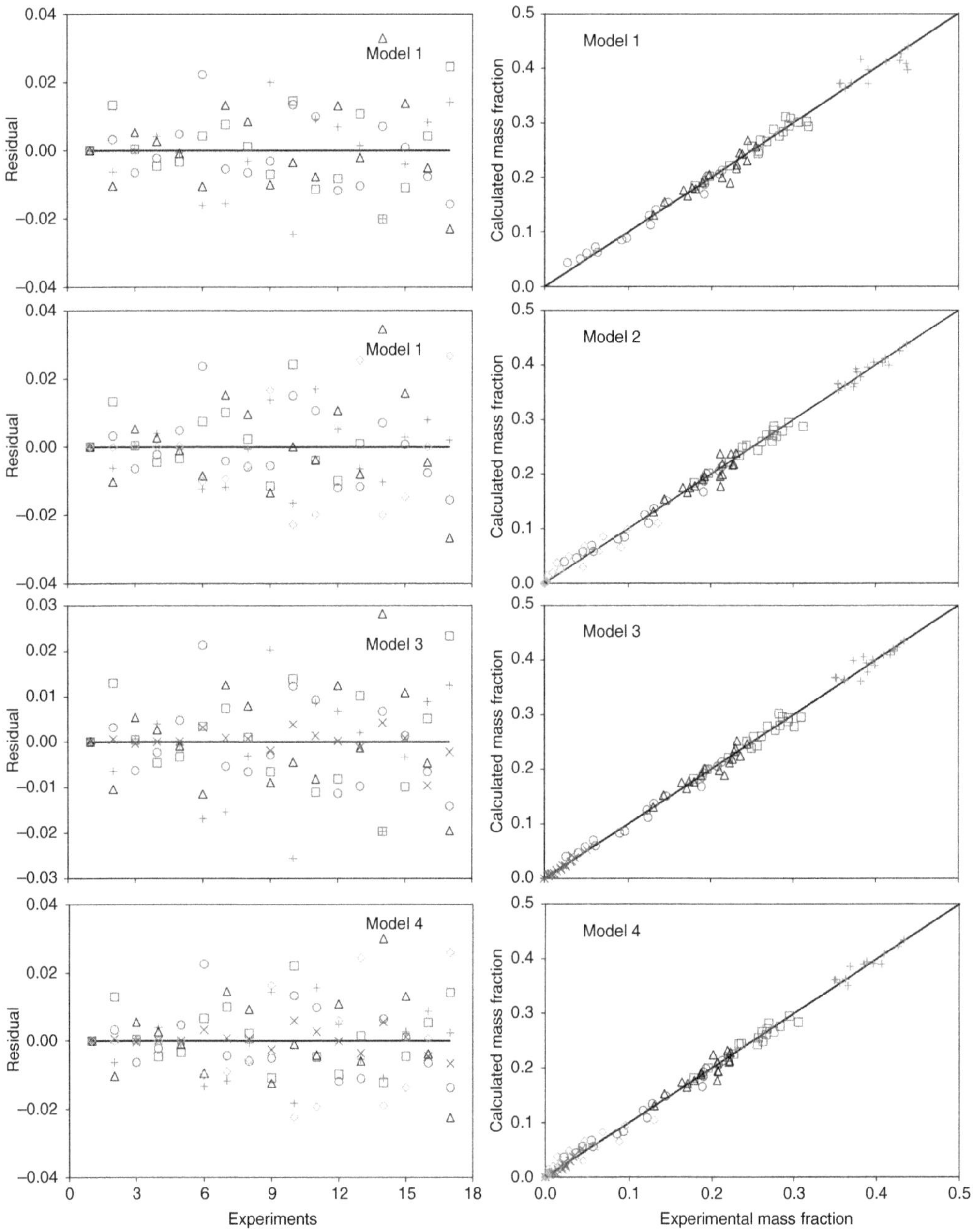

Figure 1.19 Residual and parity plots for the kinetic models based on SARA fractions. As (○), Re (□), Ar (+), Sa (△), G (×), and C (◇).

at low temperatures, especially for coke fraction because the values increase as temperature rises due to the constraint that the reaction rate coefficients must increase with temperature restraining their values. The lowest residuals corresponded to the gas fraction since its mass fraction is smaller, thus the residuals were closer to zero. In Figure 1.19, the parity plots of the kinetic models based on

Table 1.5 Statistical analysis for the kinetic models during the heavy oil hydrocracking using molybdenite catalyst.

Model	Residuals (+)	Residuals (−)	Residual range	Intercept	Slope	AIC	BIC
SARA-based							
1	31	33	0.0575	0.0023	0.9909	−23.94	−27.07
2	40	40	0.0613	0.0031	0.9887	−20.84	−24.35
3	42	38	0.0537	0.0009	0.9960	−16.23	−20.92
4	49	47	0.0525	0.0028	0.9945	−13.19	−18.27
Distillation curves-based							
—	26	38	0.0428	0.0021	1.0084	—	—

SARA fractions is also shown. The experimental and calculated data follow a straight line presenting good fit. Furthermore, in the statistical analysis (Table 1.5), the slope and the intercept to "*y*" axis are close to 1 and 0, respectively, confirming the good prediction.

The AIC and BIC values for the kinetic models based on SARA fractions are shown in Table 1.5. Model 1 is deemed the most appropriate model due to its lower AIC and BIC values. Nevertheless, all models presented similar values for both criteria and accurately forecasted the experimental data, regardless of the number of parameters present.

1.4.2.2 Distillation Curves-Based Models

The experimental and calculated yields at different reaction times and temperatures using a kinetic model combining distillation curves and SARA fractions is noticed in Figure 1.20. The yields of VR, Re, and Ar are diminished as the reaction time and temperature are increased, indicating that the main reactants to produce light and middle fractions in VR are Re and Ar using both catalysts, whereas the yields of LO and C rise as the reaction time and temperature increase, due to the cracking of Ar and Re, as well as the in-series condensation reactions of the latter to produce As and then C. The yield of VGO increases with the reaction time at 415 °C; however, at 425 °C, it increased and then decreased, because the production of this fraction is mainly from the cracking of Re, while at higher temperature (425 °C) the rate of secondary cracking is higher and results in a slight decrement.

The yield of As does not show any noticeable change except a slight decrement at 425 °C owed to the higher rate of free radicals production to generate C, which agrees with the increment of this latter fraction. It can be noticed that the addition of activated carbon to pyrite restricts the condensation reactions, enhancing the generation of lighter fractions, such as LO. A reasonably good fit of the kinetic model to the experimental data is observed. The sensitivity or statistical analyses are not provided for this model. Hence, some of the experimental fractions are underestimated or overestimated, due to the lack of optimization procedure.

The calculated and experimental mass fractions at different reaction times and temperatures of 390 and 400 °C for the kinetic model based on distillation curves are observed in Figure 1.21. The mass fraction of VR is decreased as the reaction time and temperature are raised, while the opposite behavior is observed for the mass fractions of D, N, G, and C. This is because the cracking and

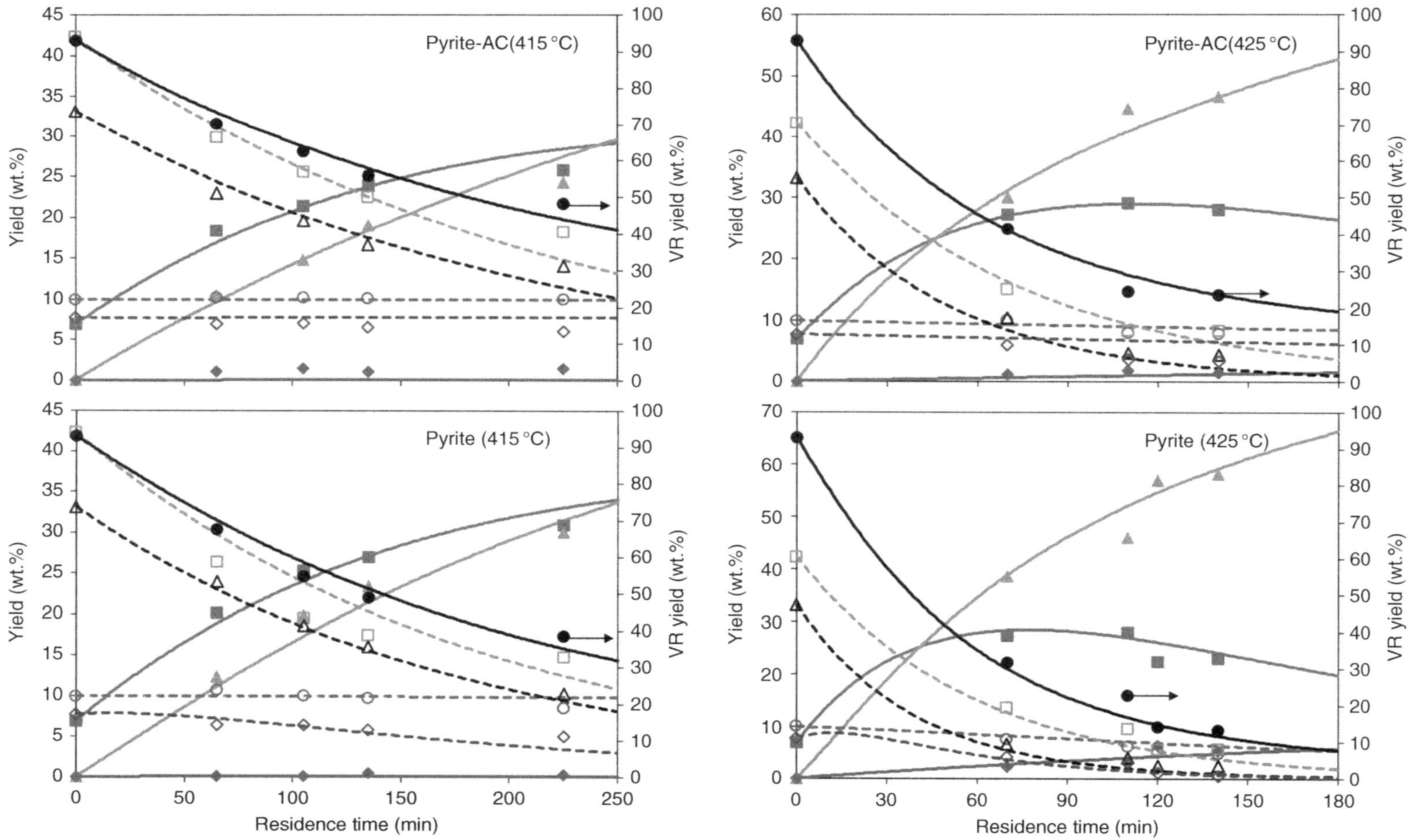

Figure 1.20 Calculated (lines) and experimental (symbols) yields for the combined kinetic model based on SARA fractions and distillation curves. VR (●), VGO (■), LO (▲), C (◆), As (○), Re (□), Ar (△), and Sa (◇).

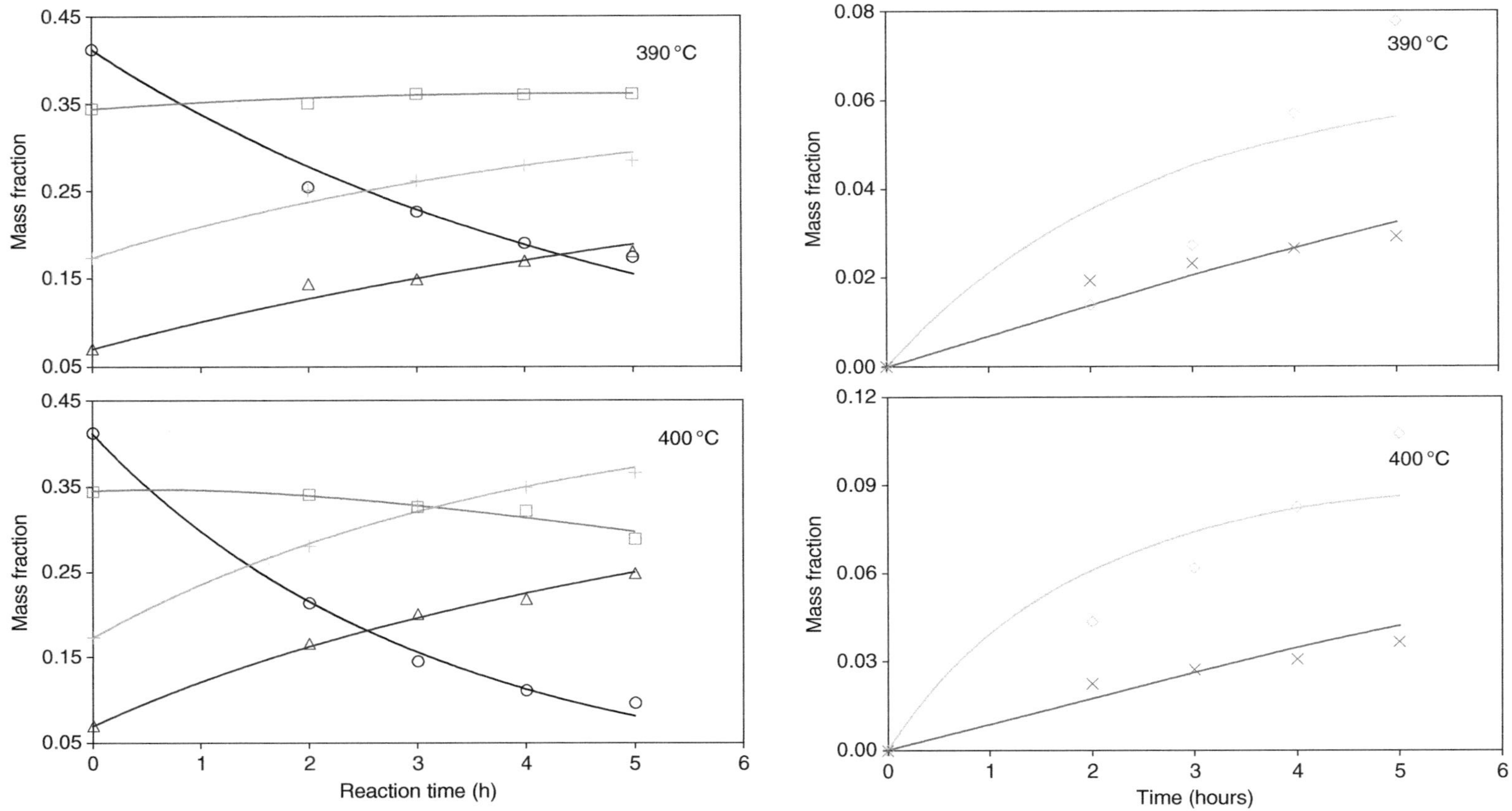

Figure 1.21 Calculated (lines) and experimental (symbols) mass fractions for the kinetic model based on distillation curves at 390 and 400 °C. VR (○), VGO (□), D (+), N (△), G (×), and C (◇).

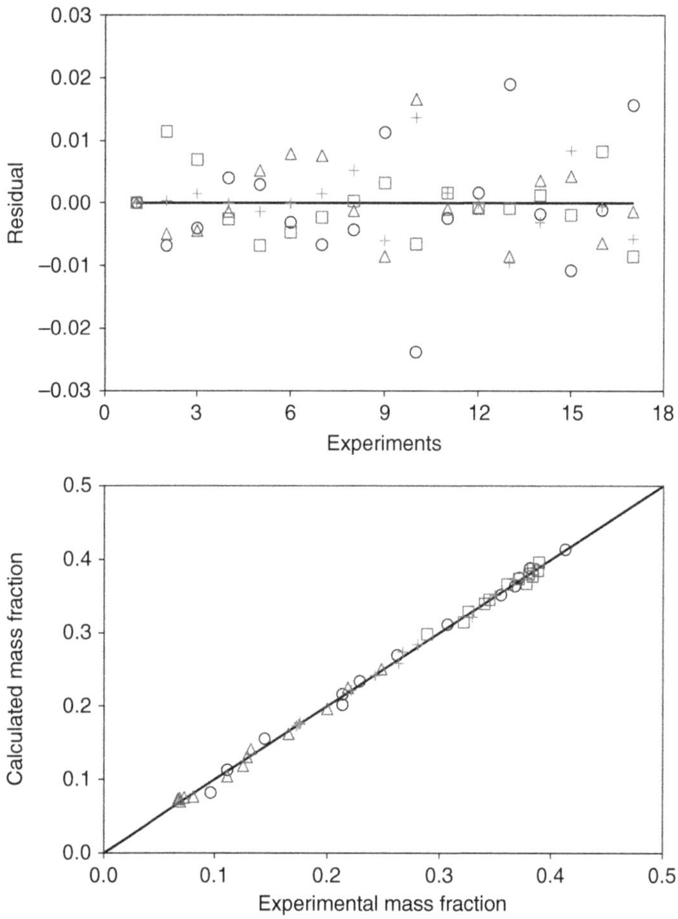

Figure 1.22 Residual and parity plots for the kinetic model based on distillation curves. VR (○), VGO (□), D (+), and N (△).

condensation reactions of VR generates lighter fractions, such as D, N, and G, as well as highly polycondensed molecules, such as C. Concerning the VGO mass fraction, at 390 °C, it slightly increases with the reaction time because of its small production from VR and secondary cracking into lighter fractions is presented. Nevertheless, at 400 °C the secondary cracking rate is higher, causing a reduction in the mass fraction of VGO as reaction time increases. Almost all mass fractions are well predicted, being the C fraction the most difficult to forecast since the free radical production that leads to C is faster than the hydrogenation reactions at higher temperatures, hindering to predict its behavior at high temperatures.

Figure 1.22 shows the residual and parity plots for the kinetic model based on distillation curves. The small residual values confirm that the kinetic parameters were assessed appropriately. Additionally, the parity plot demonstrates that the experimental and calculated mass fractions exhibit similarity, as evidenced by a 45°-straight line in the data, thereby, validating the fitting of the kinetic model.

1.4.3 Reactions in Parallel and in Series

The results obtained from the kinetic models based on SARA fractions indicate that some in-series (As to Re to Ar) and in-parallel reactions (As to Ar) are benefited using molybdenite as catalyst.

Regarding the production of G, the reactions of As to G and Ar to G are hard to be performed at the conditions used, where the formation of G fraction is mainly carried out from Sa fraction. Furthermore, the condensation reaction of middle (Ar to Re) and heavy (Re to As and As to C) fractions are intricate to be done as higher temperatures are required.

In the combined kinetic model, the kinetic parameters suggest that the in-series reactions from Ar to Sa to LO and the parallel reaction from Re to VGO are easily carried out using pyrite as catalyst. Meanwhile, the addition of activated carbon to pyrite induces in-series reaction from Re to VGO to LO. Furthermore, both catalysts (in higher degree for pyrite–AC) hinder the condensation in-series reactions from Ar to Re to As to C. The results of distillation curves-based kinetic model infer that the cracking of heavy fraction (VR and VGO) by parallel reactions (VR to D, VR to N, and VGO to N) are favored. Additionally, the cracking of D is hindered at these conditions.

The experimental and calculated As and VR conversions of kinetic models based on distillation curves and SARA fractions for hydrocracking of heavy crude oil using molybdenite as catalyst is observed in Figure 1.23. The similarity between the experimental As and VR conversions suggests that a majority of the As fraction is present in VR. Furthermore, the similarity between the calculated and experimental As and VR conversions confirms the reported kinetic models have a good fit.

1.4.4 Thermodynamic Model

Figure 1.24 illustrates the impact of temperature on the molar mass of SARA fractions in the heavy oil hydrocracking. It is evident that the molar mass of all fractions (As, Re, Ar, and Sa) decreases with an increase in temperature. This trend can be attributed to the hydrogenation of aromatic rings and dealkylation reactions of side aliphatic chains in the polyaromatic fractions. Linear correlations are observed among all molar masses, and the equations showed below are obtained accordingly:

$$M_{\mathrm{As}} = -16.517T + 8816.521 \tag{1.118}$$

$$M_{\mathrm{Re}} = -8.326T + 4446.031 \tag{1.119}$$

$$M_{\mathrm{Ar}} = -11.304T + 5282.505 \tag{1.120}$$

$$M_{\mathrm{Sa}} = -10.686T + 4768.261 \tag{1.121}$$

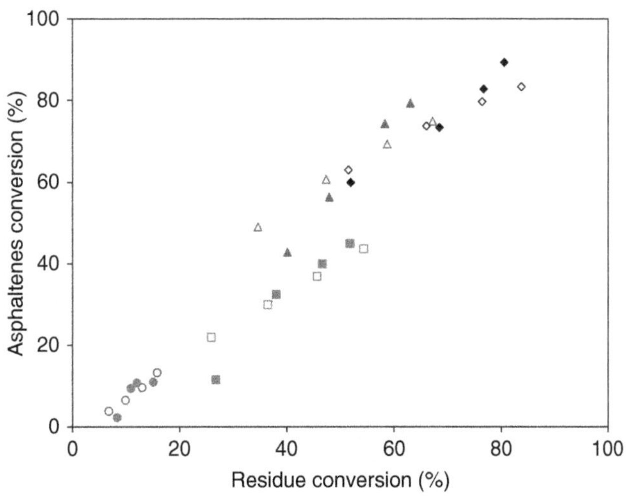

Figure 1.23 Experimental (full symbols) and calculated (empty symbols) residue and asphaltenes conversions at 360 (circle), 380 (square), 390 (triangle), and 400 °C (diamond).

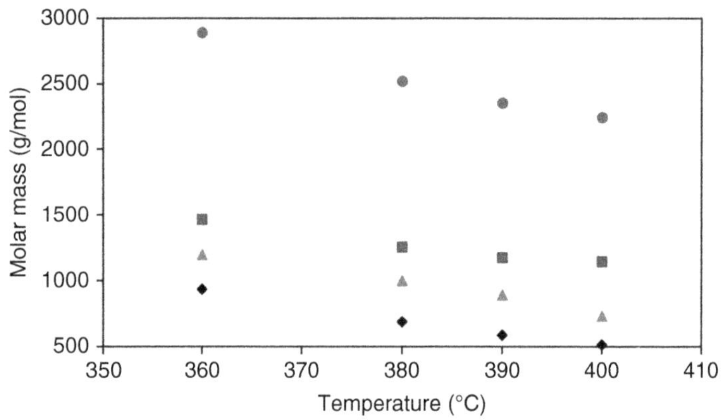

Figure 1.24 Effect of temperature on the molar mass of As (●), Re (■), Ar (▲), and Sa (◆).

The sediment content resulting from the experiments at varying reaction times and temperatures is plotted in Figure 1.25a. As depicted, increasing both temperature and reaction time results in elevated sediment content. When conducting reactions at low temperature (360 °C), sediment formation is insignificant for all reaction durations, whereas at high temperature, sediment formation occurs even during brief reaction intervals. Based on the experimental findings, it was observed that the Sa fraction increases with an increase in reaction time and temperature. As the solubility of As in crude oil decreases, the agglomeration and flocculation of this fraction intensify and form sediments.

The kinetic model based on SARA fractions (6-lump) was employed as well as the regular solution model to calculate the molar volumes and solubility parameters of SARA fractions, and then estimate the content of sediments during the hydrocracking of heavy oil using molybdenite as catalyst. Molar volumes and solubility parameters for SARA fractions at various temperatures are displayed in Table 1.6. The molar volume increases with increasing molecular weight of the fractions. Furthermore, the molar volume of asphaltene, resin, and aromatic fractions decreases, while the saturate fraction increases as the reaction temperature rises. At low temperatures, the solubility parameter of the four fractions is similar and diminishes as the temperature increases. This behavior was more noticeable in the Sa fraction as it is responsible for altering the solubility of asphaltenes in crude oil.

The experimental and calculated sediment contents using the regular solution model for different residue conversions and temperatures are shown in Figure 1.25b. The fit of the proposed thermodynamic model is moderate, since the molar masses of SARA fractions are calculated at different temperatures using the obtained correlations. Nevertheless, these molar masses also vary as the reaction time increases, but in this case, they are assumed to be constant because there are no such correlations reported. The formation of sediments is minimal at temperatures below 380 °C, corresponding to a residue conversion lower than 40–50%. However, when residue conversion is higher than this value, the sediment formation increases substantially. This behavior is also reported in literature due to insolubility of heavy fractions as aliphatic chains are converted (Marroquín 2007). Similar results are attained for the experimental and calculated sediments content obtained for different asphaltene conversions and temperatures (Figure 1.25c), since the formation of sediment increases rapidly when asphaltene conversion is near 50%. The higher the temperature, the higher the conversion of asphaltene and residue. Furthermore, the formation of sediments is observed. The calculated and experimental sediment contents using the thermodynamic model for different Sa/As ratios and temperatures are depicted in Figure 1.25d. It is

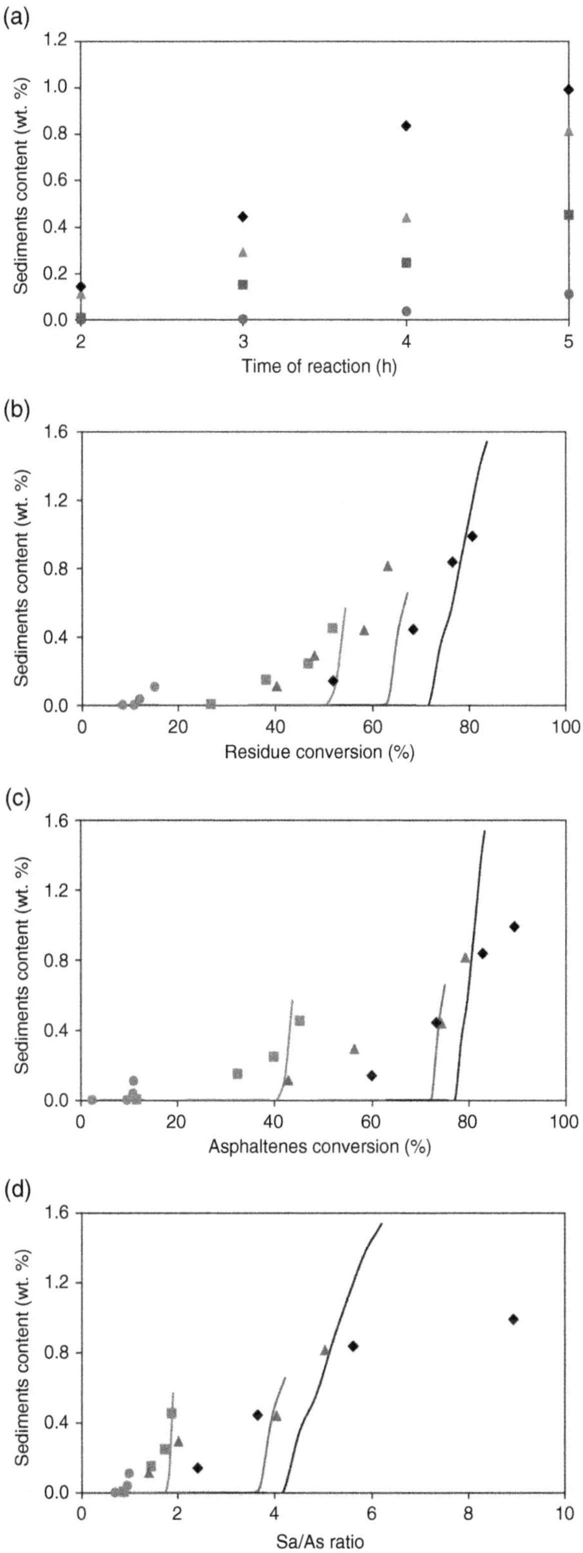

Figure 1.25 Experimental (symbols) and calculated (lines) sediment content for different reaction times (a), residue conversion (b), asphaltenes conversion (c), and Sa/As ratio (d) at 360 °C (●), 380 °C (■), 390 °C (▲), and 400 °C (◆).

Table 1.6 Correction factor for acentric factor, molar volume, and solubility parameter obtained for the different fractions during the heavy oil hydrocracking using molybdenite catalyst.

Fraction	Correction factor (c_f)	Molar volume (cm^3/mol)	Solubility parameter ($MPa^{0.5}$)	Temperature (°C)
Saturates	0.9024	447.98	19.36	360
Aromatics	0.8098	1131.76	18.78	
Resins	0.7910	1368.47	18.54	
Asphaltenes	0.7940	2188.98	19.58	
Saturates	0.9024	833.32	12.30	380
Aromatics	0.8098	984.85	18.45	
Resins	0.7910	1210.03	18.28	
Asphaltenes	0.7940	2014.36	19.22	
Saturates	0.9024	826.31	11.40	390
Aromatics	0.8098	898.99	18.25	
Resins	0.7910	1148.58	18.16	
Asphaltenes	0.7940	1925.81	19.05	
Saturates	0.9024	787.71	10.87	400
Aromatics	0.8098	762.82	17.92	
Resins	0.7910	1129.45	18.07	
Asphaltenes	0.7940	1866.94	18.93	

noticed that the regular solution model predicts the behavior since the sediment content increases as Sa/As ratio increases. This is because when the Sa fraction increases, the insolubility of As also increases, leading to the agglomeration and flocculation of this fraction, resulting in sediment formation.

1.4.5 General Comments

It was corroborated that the hydrocracking reactions at mild conditions using molybdenite indeed follow a first-order reaction. Furthermore, the optimal particle size and stirring rate were determined to obtain the intrinsic kinetics for hydrocracking reactions using mineral catalysts.

According to the AIC and BIC criteria, Model 1 is the most accurate among all the kinetic models based on SARA fractions. Nonetheless, the analysis solely focuses on the liquid phase, which involves the smallest number of reaction pathways. Additionally, the AAE is marginally elevated. Model 2 considers the liquid and solid phases, but it displays the highest AAE due to the difficulty to predict C fraction. Model 3 exhibit similar AAE values to Model 1, even considering two phases (liquid and gas) because of the good prediction of gas phase. However, it was not possible to calculate some activation energies because most of the kinetic parameters do not change with temperature. Finally, Model 4 considers the greatest quantity of pseudocomponents, reaction pathways, reaction rate coefficients, and phases (liquid, solid, and gas). Although it presents high AAE values, it enables the calculation of all activation energies. All these features turn into the most complete model and explain better As hydrocracking under the reaction conditions studied. The results for these SARA-based kinetic models indicate that some hydrocracking in-series (As to Re to Ar) and in-parallel (As to Ar) reactions are promoted, differing from the hydrocracking of lighter fractions (Ar and Sa). However, the formation of G fraction is caused by the breaking of Sa molecules. The activation energies of condensation reactions (Ar to Re to As) are not calculated or their values are significantly high due to their insensitivity to temperature. These reactions require more energy to be carried out.

Concerning the distillation curves- and SARA fractions-based kinetic models, the results indicate that some condensation reactions need higher temperatures to be performed. Moreover, the hydrocracking of middle fractions, such as Ar and Re, to produce LO are preferred at the conditions studied. Both catalysts promote the in-series reactions following different pathways: with pyrite as catalyst, the reactions from Ar to Sa to LO are enhanced as well as from Re to VGO, while for pyrite–AC catalyst, the reactions from Re to VGO to LO are promoted. The activation energies for this kinetic model exhibit higher values using pyrite as catalyst than the pyrite–AC catalyst. In addition, a synergic effect between pyrite and activated carbon hinders the condensation reactions (Ar to Re to As to C), increasing the yield of LO. The results for the kinetic model based on distillation curves imply that the hydrocracking of heavy fractions (VR and VGO) is promoted by parallel reactions (VR to D, VR to N, and VGO to N), while the secondary cracking of D is difficult to be performed at the studied conditions.

According to the statistical analysis, the kinetic parameters for the models based on SARA fractions and distillation curves were accurately estimated. Furthermore, the sensitivity analysis for all these kinetic parameters denotes that all these values are optimized for the experimental data. Nevertheless, some of the reaction rate coefficients did not reach the minimum value of the objective function (especially at high temperatures), since the established constraints (the reaction rate coefficient must increase with temperature) limited the range of permissible values. These analyses are not performed to the estimated parameters for the combined (distillation curves and SARA fractions) kinetic model. Therefore, these kinetic parameters might not be the optimal set as certain fractions are not accurately predicted.

The molar masses of SARA fractions display a linear relationship with the temperature since they are reduced as the temperature increased. Hence, a correlation for each SARA fraction was obtained. The solubility parameters for SARA fractions exhibit a slightly reduction with the increment of temperature, excluding those for Sa fractions since the drop is considerable. With these data, a regular solution model and a kinetic model based on SARA fractions were applied to predict the As flocculation to form sediments, which increases as the reaction time and temperature rise. The experimental and calculated sediment amounts with the thermodynamic model were compared with As and VR conversion, obtaining a remarkable increase in the sediment contents after reaching a value of 50% and 40–50%, respectively. Additionally, it was confirmed that increasing the amount of Sa in heavy oil rises the flocculation of As to produce sediments, because more sediments are obtained when the Sa/As ratio is incremented. Likewise, it was proved that almost all As are present in the VR fraction because similar As and VR conversions are acquired at the different conditions studied.

Comparing the results obtained for the hydrocracking of heavy crude oil using mineral catalyst with other processes (supported and dispersed catalysts), it is noticed that different levels of VR or As conversion are achieved. Higher conversions are obtained employing commercial or synthesized catalysts at high reaction temperatures (above 420 °C) or at high hydrogen pressures than using mineral catalysts. The conversion of VR or As using mineral catalysts reaches 80% at high temperature (400 °C) with formation of sediments. One advantage of utilizing this kind of catalysts is that the mineral catalysts are cheaper compared with commercial catalyst used in other technologies. In addition, if mild reaction conditions are used in these works, the technical and operational costs are reduced.

1.5 Conclusion

From the results obtained from the developed kinetic models for the catalytic hydrocracking of heavy oil using mineral catalysts, the following can be concluded:

- The reaction conditions to estimate the intrinsic kinetics using molybdenite as catalysts were obtained
- The first-order of reaction for the As and VR hydrocracking was confirmed
- The 6-lump model proved to be the most accurate one (largest number of parameters and pseudocomponents) in comparison with other SARA-based kinetic models
- The reactions from Re to Sa, Ar to G, and As to G did not occur at the reaction conditions studied using molybdenite as a catalyst
- The reactions in-series (As to Re to Ar) and in-parallel (As to Ar) of heavy fractions are promoted using molybdenite as a catalyst
- For the combined-based (distillation curves and SARA fractions) kinetic model, the pyrite (Re to VGO and Ar to Sa to LO) and the pyrite–AC (Re to VGO to LO) catalysts promoted in-series reactions
- The addition of activated carbon to pyrite suppresses the condensation reactions (Ar to Re to As to C) and enhances the yield of LO
- For the kinetic model based on distillation curves, the use of molybdenite enhances the hydrocracking of VR and VGO by parallel reactions (VR to D, VR to N, and VGO to N)
- Based on the statistical and sensitivity analyses, the kinetic parameters for the models based on SARA fractions and distillation curves were accurately estimated and corresponded to optimal values
- All the kinetic models fit the experimental data well, although predicting the C fraction proved to be the most difficult to forecast
- Linear correlations were developed for each SARA fraction because their molar masses reduced as the temperature increased
- The sediment content enlarged with the increase of the reaction time, temperature, and Sa/As ratio during hydrocracking of heavy oils using molybdenite
- The sediment formation increases promptly when As and VR conversions reached 40–50% and 50%, respectively
- Most of As fraction in heavy oil is included on VR fraction

References

Akbarzadeh, K., Ayatollahi, S., Moshfeghian, M. et al. (2004). Estimation of SARA fraction properties with the SRK EOS. *Journal of Canadian Petroleum Technology* 43: 31–39.

Al-Attas, T.A., Ali, S.A., Zahir, M.H. et al. (2019). Recent advances in heavy oil upgrading using dispersed catalysts. *Energy Fuels* 33: 7917–7949.

Alcázar, L.A. and Ancheyta, J. (2007). Sensitivity analysis based methodology to estimate the best set of parameters for heterogeneous kinetic models. *Chemical Engineering Journal* 128: 85–93.

Alonso, F., Ancheyta, J., Centeno, G. et al. (2019). Effect of reactor configuration on the hydrotreating of atmospheric residue. *Energy Fuels* 33: 1649–1658.

Álvarez, P., Browning, B., Jansen, T. et al. (2019). Modeling of atmospheric and vacuum petroleum residue hydroconversion in a slurry semi-batch reactor: study of hydrogen consumption. *Fuel Processing Technology* 185: 68–78.

Ancheyta, J. (2011). *Modelling and Simulation of Catalytic Reactors for Petroleum Refining*, 1ste. Wiley.

Ancheyta, J. (2013). *Modeling of Processes and Reactors for Upgrading of Heavy Petroleum*, 1ste. CRC Press.

Ancheyta, J. and Sotelo-Boyás, R. (2000). Estimation of kinetic constants of a five-lump model for fluid catalytic cracking process using simpler sub-models. *Energy Fuels* 14: 1226–1231.

Ancheyta, J. and Speight, J.G. (2007). *Hydroprocessing of Heavy Oils and Residual*, 1ste. CRC Press.

Ancheyta, J., Sánchez, S., and Rodríguez, M.A. (2005). Kinetic modeling of hydrocracking of heavy oil fractions: a review. *Catalysis Today* 109: 76–92.

Angeles, M.J., Leyva, C., Ancheyta, J., and Ramírez, S. (2014). A review of experimental procedures for heavy oil hydrocracking with dispersed catalyst. *Catalysis Today* 220–222: 274–294.

Asaee, S.D.S., Vafajoo, L., and Khorasheh, F. (2014). A new approach to estimate parameters of a lumped kinetic model for hydroconversion of heavy residue. *Fuel* 134: 343–353.

Ashoori, S., Sharifi, M., Masoumi, M., and Mohammad Salehi, M. (2017). The relationship between SARA fractions and crude oil stability. *Egyptian Journal of Petroleum* 26: 209–213.

Balasubramanian, P. and Pushpavanam, S. (2008). Model discrimination in hydrocracking of vacuum gas oil using discrete lumped kinetics. *Fuel* 87: 1660–1672.

Baltanas, M.A. and Froment, G.F. (1985). Computer generation of reaction networks and calculation of product distributions in the hydroisomerization and hydrocracking of paraffins on Pt-containing bifunctional catalysts. *Computers Chemical Engineering* 9: 71–81.

Baltanas, M.A., Van Raemdonck, K.K., Froment, G.F., and Mohedas, S.R. (1989). Fundamental kinetic modeling of hydroisomerization and hydrocracking on noble metal-loaded faujasites. 1. Rate parameters for hydroisomerization. *Industrial Engineering Chemistry Research* 28: 899–910.

Bata, T., Schamel, S., Fustic, M., and Ibatulin, R. (2019). AAPG Energy Minerals Division Bitumen and Heavy Oil Committee Annual Commodity Report. In *American Association of Petroleum Geologists (AAPG)*.

Becker, P.J., Serrand, N., Celse, B. et al. (2016). Comparing hydrocracking models: continuous lumping vs single events. *Fuel* 165: 306–315.

Bej, S.K. (2002). Performance evaluation of hydroprocessing catalysts. A review of experimental techniques. *Energy Fuels* 16: 774–784.

Bellussi, G., Rispoli, G., Landoni, A. et al. (2013). Hydroconversion of heavy residues in slurry reactors: developments and perspectives. *Journal of Catalysis* 308: 189–200.

Browning, B., Pitault, I., Couenne, F. et al. (2019). Distributed lump kinetic modeling for slurry phase vacuum residue hydroconversion. *Chemical Engineering Journal* 377: 119811.

Cai, Z., Ma, Y., Zhang, J. et al. (2022). Tunable ionic liquids as oil-soluble precursors of dispersed catalysts for suspended-bed hydrocracking of heavy residues. *Fuel* 313: 122664.

Calderon, C.J. and Ancheyta, J. (2016). Modeling of slurry-phase reactors for hydrocracking of heavy oils. *Energy Fuels* 30: 2525–2543.

Castañeda, L.C., Muñoz Arroyo, J.A.D., and Ancheyta, J. (2014). Current situation of emerging technologies for upgrading of heavy oils. *Catalysis Today* 220–222: 248–273.

Celse, B., Costa, V., Wahl, F., and Verstraete, J.J. (2015). Dealing with uncertainties: sensitivity analysis of vacuum gas oil hydrotreatment. *Chemical Engineering Journal* 278: 469–478.

Coronel-García, M.A., Reyes de la Torre, A.I., Domínguez-Esquivel, J.M. et al. (2021). Heavy oil hydrocracking kinetics with nano-nickel dispersed in PEG300 as slurry phase catalyst using batch reactor. *Fuel* 283: 118930.

Du, H., Liu, D., Li, M. et al. (2015). Effects of the temperature and initial hydrogen pressure on the isomerization reaction in heavy oil slurry-phase hydrocracking. *Energy Fuels* 29: 626–633.

Duran Armas, J. L. (2021). *Application of In-situ Upgrading in Naturally Fractured Reservoirs* [PhD. Thesis, University of Calgary].

Elahi, S.M., Scott, C.E., Chen, Z., and Pereira-Almao, P. (2019). In-situ upgrading and enhanced recovery of heavy oil from carbonate reservoirs using nano-catalysts: upgrading reactions analysis. *Fuel* 252: 262–271.

Elizalde, I. and Ancheyta, J. (2014). Modeling the deactivation by metal deposition of heavy oil hydrotreating catalyst. *Catalysis Today* 220–222: 221–227.

Elizalde, I., Rodríguez, M.A., and Ancheyta, J. (2009). Application of continuous kinetic lumping modeling to moderate hydrocracking of heavy oil. *Applied Catalysis A: General* 365: 237–242.

Elizalde, I., Trejo, F., Muñoz, J.A.D. et al. (2016). Dynamic modeling and simulation of a bench-scale reactor for the hydrocracking of heavy oil by using the continuous kinetic lumping approach. *Reaction Kinetics, Mechanisms and Catalysis* 118: 299–311.

Félix, G. and Ancheyta, J. (2019a). Comparison of hydrocracking kinetic models based on SARA fractions obtained in slurry-phase reactor. *Fuel* 241: 495–505.

Félix, G. and Ancheyta, J. (2019b). Using separate kinetic models to predict liquid, gas, and coke yields in heavy oil hydrocracking. *Industrial Engineering Chemistry Research* 58: 7973–7979.

Félix, G., Quitian, A., Rodríguez, E. et al. (2017). Methods to calculate hydrogen consumption during hydrocracking experiments in batch reactors. *Energy Fuels* 31: 11690–11697.

Félix, G., Ancheyta, J., and Trejo, F. (2019). Sensitivity analysis of kinetic parameters for heavy oil hydrocracking. *Fuel* 241: 836–844.

Félix, G., Ríos, J.J., Tirado, A. et al. (2022a). Monte Carlo and sensitivity analysis methods for kinetic parameters optimization: application to heavy oil slurry-phase hydrocracking. *Energy Fuels* 36: 9251–9260.

Félix, G., Tirado, A., Al-Muntaser, A. et al. (2022b). SARA-based kinetic model for non-catalytic aquathermolysis of heavy crude oil. *Journal of Petroleum Science and Engineering* 216: 110845.

Fogler, H.S. (1992). *Elements of Chemical Reaction Engineering*, 5the. Prentice-Hall.

Froment, G.F. (1987). Kinetics of the hydroisomerization and hydrocracking of paraffins on a platinum containing bifunctional Y-zeolite. *Catalysis Today* 1: 455–473.

Fukuyama, H. and Terai, S. (2007). Kinetic study on the hydrocracking reaction of vacuum residue using a lumping model. *Petroleum Science and Technology* 25: 277–287.

Galarraga, C.E., Scott, C., Loria, H., and Pereira-Almao, P. (2012). Kinetic models for upgrading athabasca bitumen using unsupported NiWMo catalysts at low severity conditions. *Industrial and Engineering Chemistry Research* 51: 140–146.

Gauthier, S., Keane, C.J., Niemela, J.J. et al. (2006). Optimization algorithms and weighting factors for analysis of dynamic PET studies. *Physics in Medicine Biology* 51: 4217.

Guillaume, D., Valéry, E., Verstraete, J. et al. (2011). Single event kinetic modelling without explicit generation of large networks: application to hydrocracking of long paraffins. *Oil Gas Science and Technology – Revue d'IFP Energies Nouvelles* 66: 399–422.

Hassanzadeh, H. and Abedi, J. (2010). Modelling and parameter estimation of ultra-dispersed in situ catalytic upgrading experiments in a batch reactor. *Fuel* 89: 2822–2828.

Heidary, S., Dehghan, A.A., and Mahdavi, S. (2017). Feasibility study on application of the recent enhanced heavy oil recovery methods (VAPEX, SAGD, CAGD and THAI) in an Iranian heavy oil reservoir. *Petroleum Science and Technology* 35: 2059–2065.

Hillewaert, L.P., Dierickx, J.L., and Froment, G.F. (1988). Computer generation of reaction schemes and rate equations for thermal cracking. *AIChE Journal* 34: 17–24.

Huang, T., Liu, B., Wang, Z., and Guo, X. (2017). Kinetic model for hydrocracking of Iranian heavy crude with dispersed catalysts in slurry-phase. *Petroleum Science and Technology* 35: 1846–1851.

International – U.S. Energy Information Administration (EIA). (2021). https://www.eia.gov/

Jarullah, A.T., Mujtaba, I.M., and Wood, A.S. (2011). Kinetic parameter estimation and simulation of trickle-bed reactor for hydrodesulfurization of crude oil. *Chemical Engineering Science* 66: 859–871.

Kim, S.-H., Kim, K.-D., and Lee, Y.-K. (2017). Effects of dispersed MoS2 catalysts and reaction conditions on slurry phase hydrocracking of vacuum residue. *Journal of Catalysis* 347: 127–137.

Laxminarasimhan, C.S., Verma, R.P., and Ramachandran, P.A. (1996). Continuous lumping model for simulation of hydrocracking. *AIChE Journal* 42: 2645–2653.

Liu, D., Kong, X., Li, M., and Que, G. (2009). Study on a water-soluble catalyst for slurry-phase hydrocracking of an atmospheric residue. *Energy Fuels* 23: 958–961.

Loria, H., Trujillo-Ferrer, G., Sosa-Stull, C., and Pereira-Almao, P. (2011). Kinetic modeling of bitumen hydroprocessing at in-reservoir conditions employing ultradispersed catalysts. *Energy Fuels* 25: 1364–1372.

Ma, Y., Zhang, J., Wu, W. et al. (2022). Trialkylmethylammonium molybdate ionic liquids as novel oil-soluble precursors of dispersed metal catalysts for slurry-phase hydrocracking of heavy oils. *Chemical Engineering Science* 253: 117516.

Marafi, M., Al-Barood, A., and Stanislaus, A. (2005). Effect of diluents in controlling sediment formation during catalytic hydrocracking of Kuwait vacuum residue. *Petroleum Science and Technology* 23: 899–908.

Marafi, A., Stanislaus, A., and Furimsky, E. (2010). Kinetics and modeling of petroleum residues hydroprocessing. *Catalysis Reviews – Science and Engineering* 52: 204–324.

Marroquín, G. (2007). *Sediment formation during the catalytic hydrotreatment of heavy crude oils* [PhD. Thesis, Instituto Politécnico Nacional].

Martens, G.G. and Marin, G.B. (2001). Kinetics for hydrocracking based on structural classes: model development and application. *AIChE Journal* 47: 1607–1622.

Martínez, J. and Ancheyta, J. (2012). Kinetic model for hydrocracking of heavy oil in a CSTR involving short term catalyst deactivation. *Fuel* 100: 193–199.

Martínez, J., Sánchez, J.L., Ancheyta, J., and Ruiz, R.S. (2010). A review of process aspects and modeling of ebullated bed reactors for hydrocracking of heavy oils. *Catalysis Reviews* 52: 60–105.

Martínez-Grimaldo, H.J., Chavarria-Hernandez, J.C., Ramírez, J. et al. (2011). Prediction of sulfur content, API gravity, and viscosity using a continuous mixture kinetic model for maya crude oil hydrocracking in a slurry-phase reactor. *Energy Fuels* 25: 3605–3614.

Martínez-Grimaldo, H.J., Ortiz Moreno, H., Sánchez Minero, F. et al. (2014). Hydrocracking of Maya crude oil in a slurry-phase reactor. I. Effect of reaction temperature. *Catalysis Today* 220–222: 295–300.

Matsumura, A., Sato, S., Kondo, T. et al. (2005). Hydrocracking Marlim vacuum residue with natural limonite. Part 2: Experimental cracking in a slurry-type continuous reactor. *Fuel* 84: 417–421.

Mederos, F.S., Ancheyta, J., and Chen, J. (2009). Review on criteria to ensure ideal behaviors in trickle-bed reactors. *Applied Catalysis A: General* 355: 1–19.

Memon, A. I., Gao, J., Taylor, S. D., Engel, T. L., and Jia, N. (2010). A systematic workflow process for heavy oil characterization: experimental techniques and challenges. *Canadian Unconventional Resources and International Petroleum Conference.*

Mitsios, M., Guillaume, D., Galtier, P., and Schweich, D. (2009). Single-event microkinetic model for long-chain paraffin hydrocracking and hydroisomerization on an amorphous Pt/SiO2·Al2O3 catalyst. *Industrial Engineering Chemistry Research* 48: 3284–3292.

Mochida, I., Zhao, X., Sakanishi, K. et al. (1989). Structure and properties of sludges produced in the catalytic hydrocracking of vacuum residue. *Industrial Engineering Chemistry Research* 28: 418–421.

Moghadassi, A., Amini, N., Fadavi, O., and Bahmani, M. (2011). Hydrocracking lumped kinetic model with catalyst deactivation in Arak refinery hydrocracker unit. *Journal of Petroleum Science and Technology* 1: 31–37.

Moustafa, T.M. and Froment, G.F. (2003). Kinetic modeling of coke formation and deactivation in the catalytic cracking of vacuum gas oil. *Industrial Engineering Chemistry Research* 42: 14–25.

Nguyen, T.S., Tayakout-Fayolle, M., Ropars, M., and Geantet, C. (2013). Hydroconversion of an atmospheric residue with a dispersed catalyst in a batch reactor: kinetic modeling including vapor-liquid equilibrium. *Chemical Engineering Science* 94: 214–223.

Nguyen, M.T., Nguyen, N.T., Cho, J. et al. (2016). A review on the oil-soluble dispersed catalyst for slurry-phase hydrocracking of heavy oil. *Journal of Industrial and Engineering Chemistry* 43: 1–12.

Oil Gas Journal. (2022). https://www.ogj.com/

Organization of the Petroleum Exporting Countries (OPEC). (2020). https://www.opec.org/opec_web/en/index.htm

Orozco Castillo, C. (2016). In situ heavy oil upgrading through ultra-disperse nano-catalyst injection in naturally fracture reservoirs [PhD. Thesis, University of Calgary].

Ortega-García, F.J., Mar Juárez, E., and Schacht Hernández, P. (2012). Controlling sediments in the ebullated bed hydrocracking process. *Energy Fuels* 26: 2948–2952.

Ortega-García, F.J., Muñoz Arroyo, J.A.D., Flores Sánchez, P. et al. (2017). Hydrocracking kinetics of a heavy crude oil on a liquid catalyst. *Energy Fuels* 31: 6794–6799.

Ortiz Moreno, H., Ramírez, J., Sanchez-Minero, F. et al. (2014). Hydrocracking of Maya crude oil in a slurry-phase batch reactor. II. Effect of catalyst load. *Fuel* 130: 263–272.

Panariti, N., Del Bianco, A., and Marchionna, M. (2000). Petroleum residue upgrading with dispersed catalysts Part 1. Catalysts activity and selectivity. *Applied Catalysis A: General* 204: 203–213.

Perego, C. and Peratello, S. (1999). Experimental methods in catalytic kinetics. *Catalysis Today* 52: 133–145.

Pham, H.H., Kim, K.H., Go, K.S. et al. (2021). Hydrocracking and hydrotreating reaction kinetics of heavy oil in CSTR using a dispersed catalyst. *Journal of Petroleum Science and Engineering* 107997.

Pratama, R.A. and Babadagli, T. (2022). A review of the mechanics of heavy-oil recovery by steam injection with chemical additives. *Journal of Petroleum Science and Engineering* 208: 109717.

Puron, H., Chin, K.K., Pinilla, J.L. et al. (2014). Kinetic analysis of vacuum residue hydrocracking in early reaction stages. *Fuel* 117: 408–414.

Quitian, A. and Ancheyta, J. (2016a). Experimental methods for developing kinetic models for hydrocracking reactions with slurry-phase catalyst using batch reactors. *Energy Fuels* 30: 4419–4437.

Quitian, A. and Ancheyta, J. (2016b). Partial upgrading of heavy crude oil by slurry-phase hydrocracking with analytical grade and ore catalysts. *Energy Fuels* 30: 10117–10125.

Quitian, A. and Ancheyta, J. (2017). Characterization of upgraded oil fractions obtained by slurry-phase hydrocracking at low-severity conditions using analytical and ore catalysts. *Energy Fuels* 31: 9162–9178.

Quitian, A., Leyva, C., Ramírez, S., and Ancheyta, J. (2015). Exploratory study for the upgrading of transport properties of heavy oil by slurry-phase hydrocracking. *Energy Fuels* 29: 9–15.

Rana, M.S., Sámano, V., Ancheyta, J., and Diaz, J.A.I. (2007). A review of recent advances on process technologies for upgrading of heavy oils and residua. *Fuel* 86: 1216–1231.

Rana, A., Khan, I., and Saleh, T.A. (2021). Advances in carbon nanostructures and nanocellulose as additives for efficient drilling fluids: trends and future perspective – a review. *Energy Fuels* 35: 7319–7339.

Raşeev, S.D. (2003). *Thermal and Catalytic Processes in Petroleum Refining*, 1ste. Marcel Dekker.

Rashidzadeh, M., Ahmad, A., and Sadighi, S. (2011). Studying of catalyst deactivation in a commercial hydrocracking process (ISOMAX). *Petroleum Science and Technologyetroleum* 1: 46–54.

Rezaei, H. (2013). *Catalyst Recycle in Slurry-phase Residue Upgrading* [PhD. Thesis, The University of British Columbia].

Rezaei, H., Liu, X., Ardakani, S.J. et al. (2010). A study of Cold Lake vacuum residue hydroconversion in batch and semi-batch reactors using unsupported MoS2 catalysts. *Catalysis Today* 150: 244–254.

Riazi, M.R. (2005). *Characterization and Properties of Petroleum*, 1ste, vol. 407. ASTM International.

Rodríguez, E., Félix, G., Ancheyta, J., and Trejo, F. (2018). Modeling of hydrotreating catalyst deactivation for heavy oil hydrocarbons. *Fuel* 225: 118–133.

Rogel, E., Ovalles, C., Pradhan, A. et al. (2013). Sediment formation in residue hydroconversion processes and its correlation to asphaltene behavior. *Energy Fuels* 27: 6587–6593.

Sadighi, S., Ahmad, A., and Seif Mohaddecy, S.R. (2010). 6-Lump kinetic model for a commercial vacuum gas oil hydrocracker. *International Journal of Chemical Reactor Engineering* 8.

Sahu, R., Song, B.J., Im, J.S. et al. (2015). A review of recent advances in catalytic hydrocracking of heavy residues. *Journal of Industrial and Engineering Chemistry* 27: 12–24.

Saleh, T.A. (2018). *Nanotechnology in Oil and Gas Industries*, 1ste. Springer.

Sámano, V., Tirado, A., Félix, G., and Ancheyta, J. (2020). Revisiting the importance of appropriate parameter estimation based on sensitivity analysis for developing kinetic models. *Fuel* 267: 117113.

Sánchez, S., Rodríguez, M.A., and Ancheyta, J. (2005a). Kinetic model for moderate hydrocracking of heavy oils. *Industrial Engineering Chemistry Research* 44: 9409–9413.

Sánchez, S., Ancheyta, J., and McCaffrey, W.C. (2007). Comparison of probability distribution functions for fitting distillation curves of petroleum. *Energy Fuels* 21: 2955–2963.

Santos, R.G., Loh, W., Bannwart, A.C., and Trevisan, O.v. (2014). An overview of heavy oil properties and its recovery and transportation methods. *Brazilian Journal of Chemical Engineering* 31: 571–590.

Sheng, Q., Wang, G., Zhang, Q. et al. (2017). Kinetic model for liquid-phase liquefaction of asphaltene by hydrogenation with industrial distillate narrow fraction as hydrogen donor. *Fuel* 209: 54–61.

da Silva De Andrade, F. J. (2014). *Kinetic modeling of catalytic in situ upgrading for Athabasca bitumen, deasphalting pitch and vacuum residue* [PhD. Thesis, University of Calgary].

Song, F.M., Liu, C.G., Zhou, G.S., and Yang, G.J. (2004). Thermal hydrocracking kinetics of Chinese gudao vacuum residue. *Petroleum Science and Technology* 22: 689–708.

Speight, J.G. (2004). New approaches to hydroprocessing. *Catalysis Today* 98: 55–60.

Speight, J.G. (2013). *Heavy and Extra-Heavy Oil Upgrading Technologies*, 1ste. Elsevier Science.

Stanislaus, A., Hauser, A., and Marafi, M. (2005). Investigation of the mechanism of sediment formation in residual oil hydrocracking process through characterization of sediment deposits. *Catalysis Today* 109: 167–177.

Stratiev, D.S., Russell, C.A., Sharpe, R. et al. (2014). Investigation on sediment formation in residue thermal conversion based processes. *Fuel Processing Technology* 128: 509–518.

Stratiev, D.S., Shishkova, I.K., Nikolaychuk, E. et al. (2019). Effect of catalyst condition on sedimentation and conversion in the ebullated bed vacuum residue H-Oil hydrocracking. *Petroleum Science and Technology* 37: 1463–1470.

Stratiev, D., Nenov, S., Nedanovski, D. et al. (2021). Different nonlinear regression techniques and sensitivity analysis as tools to optimize oil viscosity modeling. *Resources* 10: 99.

Tirado, A. and Ancheyta, J. (2018). A batch reactor study of the effect of aromatic diluents to reduce sediment formation during hydrotreating of heavy oil. *Energy Fuels* 32: 60–66.

Tirado, A., Félix, G., Kwofie, M. et al. (2022). Kinetics of heavy oil non-catalytic aquathermolysis with and without stoichiometric coefficients. *Fuel* 323: 124365.

Valéry, E., Guillaume, D., Surla, K. et al. (2007). Kinetic modeling of acid catalyzed hydrocracking of heavy molecules: application to squalane. *Industrial Engineering Chemistry Research* 46: 4755–4763.

2

Modeling Catalyst Deactivation of Hydrotreating of Heavy Oils

Javier Jurado[1], Vicente Samano[1], and Jorge Ancheyta[2]

[1] *Instituto Politécnico Nacional, Centro de Investigación en Ciencia Aplicada y Tecnología Avanzada, Unidad Legaria, Legaria 694, Col. Irrigación, Ciudad de México, Mexico*
[2] *Instituto Mexicano del Petróleo, Eje Central Lázaro Cárdenas Norte 152, San Bartolo Atepehuacan, Ciudad de México, Mexico*

2.1 Introduction

The demand for high-value petroleum products, such as middle distillate, gasoline, and lube oil, is increasing, while the demand for low-value products, such as fuel oil and residua-based products, is decreasing. Therefore, maximizing liquid products' yield from various processes and valorization of residues are of immediate attention to refiners. Also, the changing prices of crude oils as well as increasing production of heavy and extra-heavy crude oils has motivated more research and development aiming at upgrading such heavy materials. These trends have emphasized the importance of processes that convert the heavier oil fractions into lighter and more valuable clean products, such as hydrotreating (HDT) and hydrocracking (Castañeda et al. 2014; Rana et al. 2007).

Catalytic HDT is a process to hydrogenate and cleave the C—C bonds of hydrocarbons using hydrogen gas and a catalyst, and thereby converting them into compounds of lower molecular weight (Miki et al. 1983). Although it is often mentioned that HDT is a mature technology and there is not too much to investigate, the increasing availability of heavy crude oils and the need to upgrade them have motivated the researchers to continue to study the conversion of these heavy hydrocarbons (Rana et al. 2007). The nonmetallic (sulfur, oxygen, and nitrogen) and metallic (nickel and vanadium) compounds mainly present in heavy crude oils are undesirable even before refinery processing because they are poisons for many catalysts. Thus, reducing the content of heteroatoms and asphaltenes of heavy crude oils improves their processability considerably in downstream conversion processes (Gary et al. 2007).

In industrial HDT, trickle-bed reactor (TBR) is the most used for heavy fractions because of the advantages it offers compared with other types of reactors (Mederos et al. 2009), such as low catalyst loss, capability to operate at high pressures and temperatures, lower investment and operating costs, among others, as compared with other technologies, for example, ebullated-bed reactor, slurry-phase reactor (SPR) (Gianetto and Specchia 1992). SPR offers several advantages over other types of reactors, such as low pressure drop, nearly isothermal operations, capacity to achieve high conversion rates, operational flexibility, low catalyst concentration and online catalyst addition. However, reduction of conversion due to high back-mixing, separation of solid particles, among others make SPR to operate at high temperature and pressure with the consequent high investment

Mathematical Modeling of Complex Reaction Systems in the Oil and Gas Industry, First Edition.
Edited by Jorge Ancheyta, Andrey Zagoruiko, and Andrey Elyshev.

and operating costs. That is the main reason why SPR-based technologies for conversion of residues have shown poor commercial experience (Calderón and Ancheyta 2017). However, the main disadvantage during HDT of heavy crude oils is catalyst deactivation due to the metals and coke deposition, which is more dramatic at long time-on-stream (TOS) and high reaction severity (Mohammed et al. 2006). Coke is formed very rapidly during the first hours of stream, meanwhile deposition of metals takes place at longer period. During HDT reactions at high temperature, asphaltene molecules are mostly converted into coke and together with metals deposited on the catalyst in due course of the reaction (Ancheyta et al. 2003; Elizalde and Ancheyta 2014a).

The study of catalyst deactivation during HDT of heavy oil fractions is one of the most important aspects to improve the catalytic performance in petroleum refining processes (Ancheyta et al. 2002). A good commercial catalyst is known through three characteristics, namely activity, selectivity, and stability (Moghadassi et al. 2011; Ward 1993).

The loss of catalytic activity and/or selectivity over time is a problem of great and continuing concern in the practice of industrial catalytic processes. The degree of catalyst deactivation depends mainly on the properties of the feed and operating conditions (Kohli et al. 2016). The performance of catalysts decreases with time due to the already commented reasons, thus to maintain constant product yields and/or quality, the loss of catalytic activity must be compensated by periodic increases of reaction temperature (Ancheyta et al. 2003; Moghadassi et al. 2011). Costs to industry for catalyst replacement and process shutdown are of billions of dollars per year. However, it is inevitable that all catalysts activity will decay (Bartholomew 2001).

Kinetic studies provide valuable information that is part of the database used for comparison and selection of catalysts as well as matching the type of catalyst with reactor and feed. The parameters determined from such studies form the basis for the development of kinetic and deactivation models, which can be used for assessing the reactivity of heavy feeds, selection of catalysts and optimizing the operation of catalytic systems. Kinetic and deactivation parameters are part of the models used for predicting catalyst life. In fact, the determination of such parameters for a particular catalyst may be viewed as the modeling of a catalyst activity scale (Marafi et al. 2010a,b).

2.2 Mechanisms of Deactivation

Catalyst deactivation during HDT of heavy oils comprises three stages: (i) a rapid loss of activity due to coke deposition on the catalyst surface, usually within the first hundred hours on stream, (ii) a slower stage where metal covers the catalyst surface, and finally (iii) a rapid loss of activity due to metal and coke deposited on the catalyst (Callejas et al. 2001; Furimsky and Massoth 1999). Deactivation of HDT catalysts occurs by single or multiple causes, such as active site blockage, active site coverage by deposits of coke and/or metals, pore mouth blockage, and sintering of the active phase (Dautzenberg et al. 1978). The degree and type of deactivation will depend on the feed properties and reaction conditions and will be evidenced by an S-shape curve of temperature vs TOS (Ancheyta 2016) as shown in Figure 2.1. Stage 1 is characterized by rapid formation of initial coke with significant loss in catalyst activity; Stage 2 continues with gradual buildup or accumulation of metal deposit, causing further loss in the activity of the catalyst; and Stage 3 finishes with the constriction or blockage of pore by continuous deposition of metals with complete loss of catalytic activity due to diffusion limitation (Gray et al. 1999). Initially, coke deposition causes a rapid deactivation, which in several hours attains a pseudo-steady state (Stage 1). Then, deactivation by metal

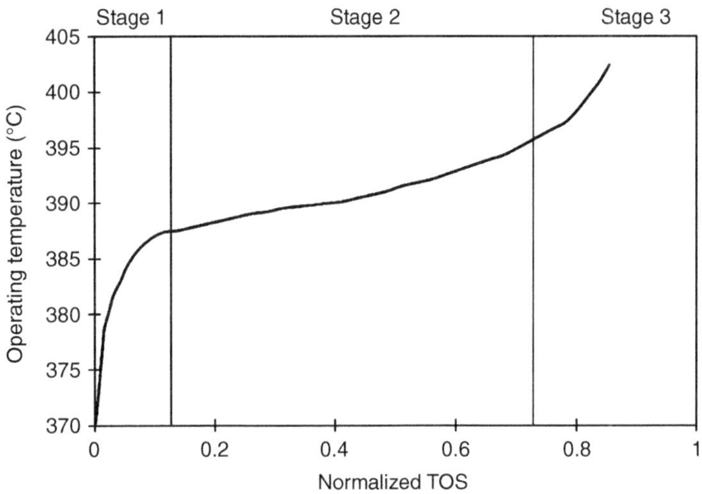

Figure 2.1 Typical profile of catalyst deactivation. Stage 1: deactivation by coke, Stage 2: deactivation by metals, and Stage 3: pore blockage (Ancheyta 2016 / John Wiley & Sons).

deposits is observed during a longer period of time until a drastic deactivation caused by pore restriction and blockage is observed (Ancheyta 2016).

In catalytic HDT of heavy oil, asphaltenes are the main cause for coke formation via different reaction mechanisms. They also contribute to metal deposition as nickel and vanadium are mostly concentrated in asphaltenes, and when they are hydrocracked, metals are released and deposited on the catalyst surface. Asphaltenes are widely known to be extremely large and complex molecules, and their diffusion into catalyst pores is difficult. Therefore, when designing catalysts for HDT of heavy oil, it is important to study the properties of such large molecules (not only bulk properties but also structural and crystalline properties) and textural properties of the catalyst (pore size distribution, average pore size). For instance, various authors have reported that the activity of the catalyst is strictly related to the properties of the catalyst, such as size and volume of the pores (Galiasso et al. 1983; Oyekunle et al. 2005; Shimura et al. 1986; Song et al. 1991).

In the case of heavy feeds, resins play the role of stabilizer. If there is not a proper ratio of asphaltenes, oil, and resins, the colloid stability of the feed cannot be maintained provoking asphaltene precipitation (Mochida et al. 1989). During HDT, it is important to remove the resins at about the same rate as asphaltenes, otherwise, precipitation of asphaltenes from the feed can be a major cause of coke formation (Beuther et al. 1980a). Hydrotreated asphaltenes become more aromatic than the original asphaltenes. Condensation of asphaltenes on the catalysts surface may produce coke containing small anisotropic regions consistent with mesophase development (Eser et al. 1986; Nagaishi et al. 1997).

Coke deposition increases with the boiling range of the feed. However, for feeds with similar boiling range, the one with higher concentration of coke precursors, such as aromatics and heterocyclic compound, will cause more severe poisoning. About one half of metals (Ni, V) present in heavy feeds are present as porphyrins, the other part of metals are in less-characterized forms that include bonds with nitrogen, oxygen, and sulfur in the defect centers of asphaltene sheets. Due to the bulkiness of the porphyrins and asphaltene molecules, the demetallization reaction is affected by pore diffusion limitations. Consequently, metals are generally deposited on the outer zone of the catalyst particle (Ancheyta 2016; Beaton and Bertolacini 1991; Mitchell 1990).

2.2.1 Coking Deposition (Fouling)

Fouling is the physical deposition of species from the fluid phase onto the catalyst surface, which results in activity loss due to blockage of site and/or pores mainly due to deposits of coke in porous catalysts. The deactivating influence of coke on catalyst depends on the nature of coke, its structure and morphology, and the exact location of its deposition on the catalyst surface. The chemical structures of cokes formed in catalytic processes vary with reaction type, catalyst type, and reaction condition. When coke deposition occurs mainly on the carrier surface and not directly on the active sites, the catalyst can tolerate even 6–10 wt.% coke without much loss in activity (Bartholomew 1982; Menon 1990; Nielsen and Trimm 1977; Trimm 1977, 1983).

The coke content of the catalyst typically reaches its maximum steady-state value at about 15 to 25% in a very short period (Thakur and Thomas 1985). About one-third of the total pore volume of fresh catalyst is filled by coke in the first hours of operation of the process. The formation of coke is due to polymerization and polycondensation of high molecular weight compounds of the feedstock (Furimsky and Massoth 1999). Analyses made by Tamm et al. (1981) have shown that the coke deposition starts on the external surface of the catalyst and its concentration in the interior of the catalyst increases with process time. In residue HDT, the rapid decline in catalyst activity during the early hours-on-stream is due to buildup of strongly adsorbed surface coke (Hauser et al. 2005), which forms a monolayer on the catalyst pellet (Richardson et al. 1996).

Coke deposition increases with the boiling range of the feed. However, for feeds with similar boiling range, the one with higher concentration of coke precursors, such as aromatics and heterocyclic compounds, will cause a more severe deactivation (Ancheyta 2016). Beuther et al. (1980b) classified coke deposits formed during hydrodesulfurization (HDS) of residue into three types: Type I deposits are the normal aromatic background of non-asphaltenic structures deposited during the first part of the cycle at low temperature. Type II deposits are structures of asphaltenes deposited early in the coking process. Type III deposits are materials laid after long reaction time and high temperature, resulting from condensation of aromatic concentrates into clusters and then crystals which constitute a mesophase.

2.2.2 Metal Deposition (Poisoning)

Poisoning is the strong chemisorption of reactants, products, and impurities, which may affect catalytic activity, changing the geometry of the catalytic surface. Poisoning has an operational meaning: whether a species acts as a poison, its effect on catalyst activity depends on the adsorption strength relative to the other species competing for the catalytic sites (Bartholomew 2001; Hegedus and McCabe 1994; Maxted 1951).

In heavy feeds, porphyrin forms of V and Ni are the main organometallic compounds, they are the main cause of the metal deposits formation on catalyst surface, and most of these metals are associated with the asphaltene entities. In this type of feed, the amount of V-porphyrins is greater than that of Ni-porphyrins. Therefore, the metal deposition patterns are influenced by the former to a much greater extent (Marafi et al. 2010a,b).

The most commonly used catalytic materials for removing both vanadium and nickel present in heavy feeds are Ni/Mo and Co/Mo sulfide catalysts supported on alumina (Speight 1999). The specific surface area of the catalyst is usually large (200–300 m^2/g), but almost the entire surface is contained within the pore space of the alumina. The Co, Mo, and Ni metals are dispersed in a thin layer within the pore system of the alumina to form active sites. These types of catalysts are frequently deactivated and the main cause of deactivation is the metal sulfide deposits (Oyekunle et al. 2005).

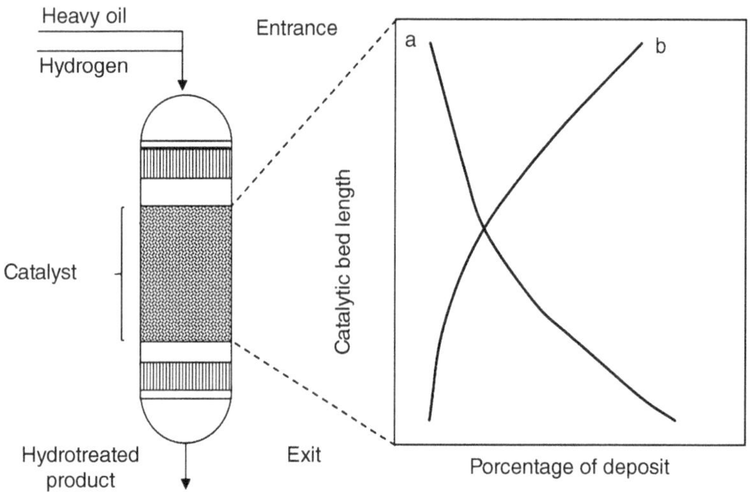

Figure 2.2 Schematic representation of position-dependent foulant deposition profiles in a fixed-bed reactor: (a) coke and (b) metals.

The deposits tend to grow with TOS blocking the pore mouths, reducing the pore radius to the same magnitude as the size of the organometallic molecules, and thus the effect of restricted diffusivities is presented at the pore mouth of the catalyst (Kam et al. 2005; Rajagopalan 1979). In a fixed-bed reactor coke content increases from the entrance to the exit, while metal content decreases from the entrance to the exit of the reactor as shown in Figure 2.2 (Thakur and Thomas 1985).

2.3 Deactivation Models

From technical point of view, studying the deactivation phenomena due to HDT reactions is a very complex task, because it requires high costs of experiments and time-consuming, for estimating the performance and life of the catalysts (Elizalde and Ancheyta 2014a). Mathematical modeling is a powerful tool to minimize the costs and time required in experimental work to achieve such research targets (Kam et al. 2005). For this reason, it is desirable to account for a model that simulates the deactivation patterns during catalytic HDT reactions. There are various reactor modeling works reported in the literature (Alvarez and Ancheyta 2008; Ferreira et al. 2010; Korsten and Hoffmann 1996; Kumar et al. 1997; Mederos et al. 2006, 2012; Sau et al. 1997; Verstraete et al. 2007) that do not take into consideration the catalytic deactivation to describe long time operations, that is, middle-of-run (MOR) and end-of-run (EOR) periods. Various models have been reported in the literature to explain the catalysts deactivation based on different mechanisms of coke and metal deposition. The most relevant deactivation models are summarized in the following sections.

2.3.1 Deactivation Models by Coke Deposition

It is known that coke is deposited on HDT catalyst surfaces very quickly in the initial stage of run (Kam et al. 2005). Newson (1975) observed that this behavior is due to a period of uncontrolled hydrocracking in a short time, about 40 h; subsequently, coke content on catalyst remains constant with increasing process time. Gualda and Kasztelan (1996) reported that carbon deposits accumulated on the catalyst surface at a level of 11 wt.% after 6 h and reach 14 wt.% after 140 h on stream. This initial deposition of coke was termed "young coke" and had a relative high hydrogen-to-carbon ratio (H/C), ranging from 0.8 to 1.2 (Kam et al. 2005).

One of the most used coking model to describe coke deposition was proposed by Froment and Bischoff (1962), which involves a deactivation function with coke content on the catalyst throughout the reactor, relating the carbon content to the reaction rate coefficient as follows:

$$k = k_{\eta}\varphi(c) \tag{2.1}$$

where k_{η} is the reaction rate coefficient when no carbon has been deposited, k is the reaction rate coefficient when carbon has been deposited, and φ is an activity function. An exponential form of φ could be used as follows:

$$\varphi = \exp(-\propto C_0) \tag{2.2}$$

where $\propto$ is the deactivation constant and C_0 the carbon content on catalyst. This is based on the observation that after a strong initial decrease of conversion, the product of interest's yield decreases slowly tending asymptotically to a limiting value.

Another form of the activity function was also suggested (Eq. 2.3), which may arise from Langmuir–Hinshelwood concepts. For mathematical simplicity, it neglects the adsorption of the components of the main reaction, so that the constant α was then a type of adsorption equilibrium constant that expresses the proportionality between the amount of deposited carbonaceous compound, measured as carbon, and the number of active sites fouled.

$$\varphi = \frac{1}{(1 + \alpha\, C_0)^{\beta}} \tag{2.3}$$

Most of the authors (Jacob et al. 1976; Lee et al. 1989; Martínez and Ancheyta 2012; Van et al. 1996) represent the catalytic activity decline by the exponential law or the hyperbolic function given by Eqs. (2.2) and (2.3), respectively, to simplify the overall kinetic model and parameter estimation.

Voorhies (1945) proposed a model to describe coke deposition on a cracking catalyst in a fixed-bed reactor, considering only the catalytic carbon produced in the reaction and not including the removable carbon formed by the absorbed and interleaved gaseous and liquid hydrocarbon remaining in the spent catalyst after the cracking. This model considers that the weight percent of carbon formed on the catalyst is approximately a logarithmic function of the length of the time elapsed since the catalyst is freshly regenerated, by means of the following equation:

$$C_0 = A(\tau/60)^{V} \tag{2.4}$$

for $0.5 \leq V \leq 1$

where C_0 is the amount of coke on the catalyst at time τ, and the correlation coefficients A and V are obtained experimentally. This equation is not recommended for hydroprocessing catalyst where the effect of coke on the activity loss is more complicated than that of cracking catalysts.

Rashidzadeh (2010) proposed a 5-lump kinetic model for hydrocracking of vacuum gas oil (VGO) in a commercial plant. The model considers VGO, diesel, kerosene, naphtha, and gas as products and that deactivation of catalyst is caused by its surface coke. Catalyst decay function is based on its life, an exponential form formulated as follows:

$$\varphi_{ij} = e^{\left(-\alpha_{ij}\ \text{Life}\right)} \tag{2.5}$$

where $\propto_{ij}$ shows the deactivation constant for converting lump i to lump j and life value can be expressed in days on stream.

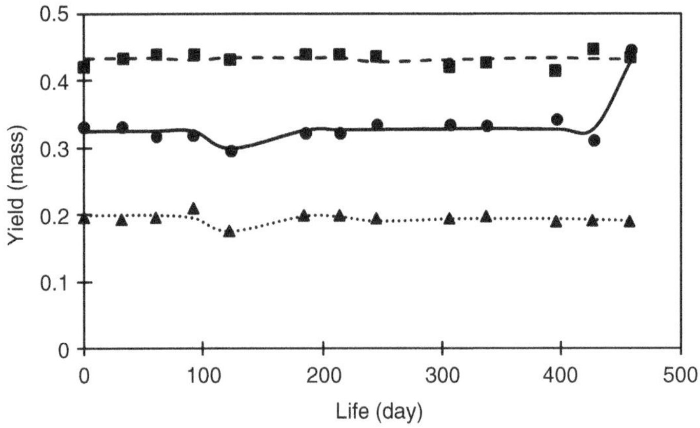

Figure 2.3 Comparison of predicted yield of model with measured data. Diesel, predicted (---) and measured (■). Kerosene, predicted (—) and measured (●). Naphtha, predicted (···) and measured (▲) (Adapted from Rashidzadeh et al. 2010).

The catalyst used was an amorphous type with the following properties: specific surface area of 176 m^2/g, pore volume of 0.51 cm^3/g, and bulk density of 0.75 g/cm^3. The feed was a mixture of fresh VGO and unconverted oil. The reactor had 4 beds with a total amount of loaded catalyst in the reactor of about 69,000 kg. Feed flow rate and reactor operating condition are the following: fresh feed and recycle feed of 48,545.5 kg/h and 32,634.3 kg/h, respectively, temperature of 400 °C, and total pressure of 153 kg/cm^2.

Figure 2.3 shows comparisons between measured product yields from the industrial hydrocracking reactor and the model predictions. The prediction of the model for 1.5 years is in good agreement with the actual commercial data. The authors reported deviations of predictions for naphtha, kerosene, and diesel yields of about 1.78%, 1.98%, and 1.97%, respectively.

Richardson et al. (1996) reported a model to describe the initial coke buildup and to predict coke formation during processing of heavy residues in the first hours of operation, using bitumen and commercial NiMo/γ-Al_2O_3 catalyst in batch and continuous stirred tank reactor (CSTR) reactors, represented by the following equation:

$$C_0 = C_{\max}[1 - \exp(-Kw)] \tag{2.6}$$

where w is the cumulative feed-to-catalyst ratio, K an adsorption constant, and $C_{\max}$ is the maximum carbon deposition that can be accumulated on the surface equal to monolayer coverage. If the substance adsorbed on the surface is similar in carbon content, molecular weight (M_W), and density of the residue fraction of the oil (ρ), the radius for a typical residue molecule (r_M) would be given by the following equation:

$$r_M = \left(\frac{3M_W}{N\rho 4\pi}\right)^{1/3} \tag{2.7}$$

where N is the Avogadro's number. To obtain the maximum carbon deposition, $C_{\max}$ is given as follows:

$$C_{\max} = \frac{M_W X_c}{N\pi\, r_M^2} \tag{2.8}$$

where X_c is the mass fraction of carbon in the residue.

Ancheyta and Lopez (2000) analyzed five deactivation models based on TOS theory for studying the deactivation of catalyst for catalytic cracking of VGO, in which the main cause of catalyst deactivation is coke deposition. The models were derived from a general m-order expression: Model 1, exponential law; Model 2, a second-order model; Model 3, power law; Model 4, hyperbolic function; and Model 5, an empirical model. The following deactivation rate was considered:

$$-\frac{d\varphi}{dt} = k_d \varphi^m \tag{2.9}$$

Integration of Eq. (2.9) for $m \neq 1$ gives,

$$\varphi = [1 + k_d(m-1)\,t]^{-1/m-1} \tag{2.10}$$

Using Eqs. (2.9) and (2.10), the following reported models can be deduced:
Model 1: for $m = 1$, integrating Eq. (2.9) gives the exponential law,

$$\varphi = e^{-k_d t} \tag{2.11}$$

Model 2: for $m = 2$, a second-order model is obtained from Eq. (2.10):

$$\varphi = \frac{1}{1 + k_d t} \tag{2.12}$$

Model 3: for $m > 2$, a power law model is obtained from Eq. (2.10):

$$\varphi = A\, t^{-B} \tag{2.13}$$

where

$$A = [k_d(m-1)]^{-1/(m-1)} \tag{2.14}$$

$$B = 1/(m-1) \tag{2.15}$$

Model 4: the hyperbolic function is obtained from Eq. (2.10):

$$\varphi = (1 + Gt)^{-B} \tag{2.16}$$

where

$$G = k_d\,(m-1) \tag{2.17}$$

Model 5: Empirical deactivation model, reported elsewhere (Jacob et al. 1976; Oliveira and Biscaia 1989), $g = 1$ and $d = k_d$

$$\varphi = \frac{a}{p^q(1 + d\,t^g)} \tag{2.18}$$

To test the models, experiments were carried out with industrial catalyst and feedstock using the 3-lump kinetic scheme proposed by Weekman (1968) in a micro-activity unit with a fixed-bed reactor at a reaction temperature of 500 °C and constant cat-to-oil ratio of 5.

Figure 2.4 shows the calculated and experimental conversion, gasoline yield, and gas plus coke yield for each deactivation model. The authors found that Models 4 and 5 were the best for representing the deactivation of catalyst by coke, and Models 1 and 3 were slightly inferior to Models 4 and 5; however, they represent the catalyst deactivation sufficiently well. Model 2 was highly inaccurate.

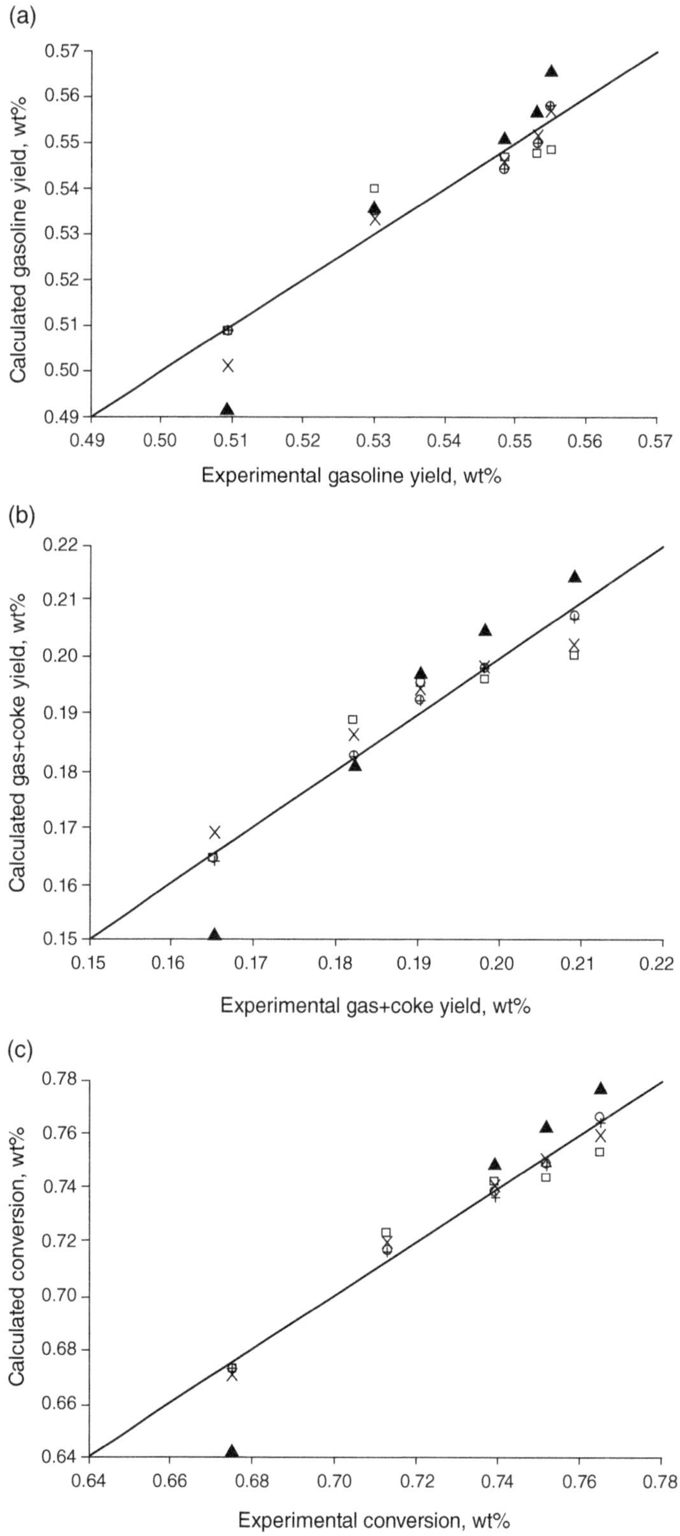

Figure 2.4 Experimental vs calculated values: (a) Gasoline yield, wt.%; (b) Gas + coke yield, wt.%; (c) Conversion, wt.%. Model 1 (€), Model 2 (▲), Model 3 (×), Model 4 (o), and Model 5 (+) (Ancheyta and Lopez 2000 / Scientific Electronic Library Online (scielo)).

2.3.2 Deactivation Models by Metal Deposition

Oyekunle and Ikpekri (2004) and Oyekunle et al. (2005) developed a model to describe catalyst deactivation by metal deposition during residuum HDS. They used three types of catalysts with macro-, micro-, and random pore structures. It is assumed that the HDS reaction occurs in macro- and micropores, and there are interconnections between them. An idealized pore model of Wheeler (1951) was applied making use of an average value for tortuosity factor for interconnecting pores. The model considers activity decays with time both linearly and nonlinearly. To determine the catalyst activity decline with TOS, the following equations had to be solved:

$$\frac{\partial^2 \alpha_D}{\partial {\beta_D}^2} = \Phi^2 {\alpha_D}^2 + \frac{\partial \alpha_D}{\partial \tau_D} \tag{2.19}$$

$$\frac{d}{dt} = \frac{\epsilon k_{SA}(C_F) M_{MS} {\gamma_D}^2}{\rho_m r} \tag{2.20}$$

Oyekunle and Ikpekri (2004) tested the model using data of hydroprocessing of Maya heavy crude oil published by Ancheyta et al. (2002, 2003). The catalyst life was estimated, and the total activity loss was predicted to occur after 462, 316, and 150 days for macro-, random, and micropore systems.

Bourseau et al. (1985) and Toulhoat et al. (1990) presented a deactivation model for hydrodemetallization (HDM) catalysts, which considers the catalyst's textural properties, its initial activity, the catalyst size, the feed reactivity, and includes the concept of ultimate storage capacity (USC), which is defined as the mass of metal accumulated in the catalyst particle, relative to the unit mass of fresh catalyst, until the catalyst activity became zero. The USC can be expressed as follows:

$$\text{USC} = \frac{\Omega \rho_m \epsilon_0 C_m}{\rho_{cat}} \tag{2.21}$$

where ϵ_0 is the the catalyst's initial porosity, ρ_{cat} is its initial grain density, ρ_m is the deposited metal sulfides' true density, C_m is the metal content on the catalyst, and Ω is the fraction of initial porosity occupied by the solid deposited on catalyst.

Pereira (1990) proposed a model that describes the influence of feed and catalyst properties on the demetallation performance of a HDT catalyst. This model includes the metal distribution factor, θ_M (Tamm et al. 1981).

$$\theta_M = \left\{ \int_0^1 C_m \, \xi \, d\xi \right\} \Big/ \left\{ C_m^{max} \int_0^1 \xi \, d\xi \right\} \tag{2.22}$$

where C_m is the local concentration of metal deposit, C_m^{max} is the maximum concentration of metals, and ξ is the fractional distance from the center of the pellet. For demetalation reactions, θ_M is small when the intraparticle metal deposit profile is not uniform, and approaches unity when metals are uniformly deposited throughout the catalyst particle. A set of mathematical equations was developed, and their solution was obtained for first-order HDM kinetics. A series of correlations was developed which involves the change of parameters, such as distribution factor, pore radius, the Thiele modulus, reaction rate, and so on, with TOS (Pereira 1990).

Dautzenberg et al. (1978) developed a model for describing the deactivation behavior of residue desulfurization catalysts. The deposition of metal compounds was the principal parameter to describe the deactivation phenomena. They assumed that in residue hydrodesulfurization (RDS) two reactions take place in parallel: one-and-a-half order HDS; and pseudo-first-order vanadium removal. Sulfur is removed as H_2S and ends up in the gas stream, while the metal removed is deposited on the walls of the catalyst pores as sulfide, diminishing the original desulfurization activity, due to the reduction of the pore radius (r_A) with time (t) according to the following:

$$\frac{\partial r_A}{\partial t} = k_v\, C_v\, V_{\text{dep}} \tag{2.23}$$

where k_v, C_v, and V_{dep} are the reaction rate constant for vanadium removal, molar concentration of vanadium, and deposit volume per mole of vanadium, respectively.

The degree of deactivation (θ) is related to the volume deposited per unit surface area per unit time represented by the following equation:

$$\theta = \frac{k_v\, V_{\text{dep}}}{r_{p0}} \int_0^t c_{v0}\, dt \tag{2.24}$$

where r_{p0} and c_{v0} are the pore radius of fresh catalyst and molar vanadium concentration outside the catalyst particle, respectively. This model does not consider configurational diffusion effects and the corresponding changes in diffusivity of metal-bearing molecules with pore diameter.

The experiments were performed at fixed standard temperature and pressure in bench-scale equipment. Two feedstock were used: Caribbean long residue and Middle-East long residue. Catalysts of Ni/Mo and Co/Mo type with different particle size, shape bulk density, and support were tested. The decline in metal-removal activity of catalyst used in stirred-tank reactor experiments was evaluated using a vanadium weight balance. Figure 2.5 shows the metal content of the product (vanadium) as a function of TOS.

Arbabi and Sahimi (1991a) proposed a percolation model catalyst deactivation, where the porous catalyst is represented by a network of interconnected pores. Because of continuous deposition of solid products on the surface of the pores, the morphology of the catalyst's pore space changes with time. The effective radius of the pores is distributed according to a probability density function represented by the following expression:

$$f(r) = \frac{r_{p0} - r_{\min}}{r_a - r_{\min}} \exp\left[\frac{-1}{2}\left(\frac{r_{p0} - r_{\min}}{r_a - r_{\min}}\right)^2\right] \tag{2.25}$$

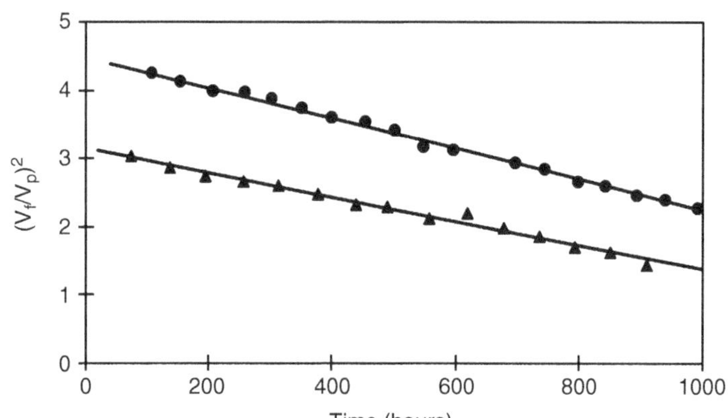

Figure 2.5 Vanadium removal during hydrodesulfurization of Caribbean residue with $Co/Mo/Al_2O_3$ (●) and Middle-East long residue with $Ni/Mo/Al_2O_3$ (▲) (Adapted from Dautzenberg et al. 1978).

where r_{min} is the minimum pore radius and r_a is the average pore radius. The model was applied to hydrodemetalation reaction. The results for the deposition profile were found to be in good agreement with the experimental data (Arbabi and Sahimi 1991b).

Rajagopalan (1979) developed a model for predicting the total life of catalyst during first-order demetallation reaction considering only the period following the initial fast coking. In this model, the demetallation activity decays with time is an almost linear function, due to the influence of the restricted diffusion of the large reactant species in the pores. The model considers the demetallation in a single cylindrical pore, where the rate of metal deposit per unit length of the pore is as follows:

$$\frac{dCm}{dt} = \zeta 2\pi r_{p0} c k_{fm} M_{MS} \tag{2.26}$$

They found that the catalyst becomes inactive when the pore month is reduced to the size of the organometallic molecule, this behavior can be represented by the following equation:

$$t_m = \frac{\rho_{md} r_m}{\zeta k_{fm} C_F M_{MS}} \frac{1-\lambda_0}{\lambda_0} \tag{2.27}$$

where ρ_{md} and r_m are the metal deposit density and the radius of molecule, respectively, M_{MS} is the molecular weight of metal deposit, k_{fm} is the demetallation reaction rate constant, ζ is the number of metal atoms per reactant molecule, C_F is the concentration of reactant, r_{p0} is the pore radius of fresh catalyst, t_m is the maximum lifetime of the catalyst, t is TOS (>50 hours), C_m is the concentration of metal, and λ_0 is the ratio of molecular to pore radius.

The model was used for predicting pore sizes, which yield the optimal initial activity and optimal lifetime activity. Figure 2.6 describes the demetallation rate per unit volume of a slab catalyst as a function of pore radius for several ages of the catalyst. The graph indicates that the pore radius corresponding to the optimum initial activity is 3.1×10^{-9} m. For pores smaller than the optimal radius, increasing the pore radius improves the catalyst performance as this increases the initial activity; however, for pores larger than the optimal size, an increase in pore size reduces the initial activity but increases the total life.

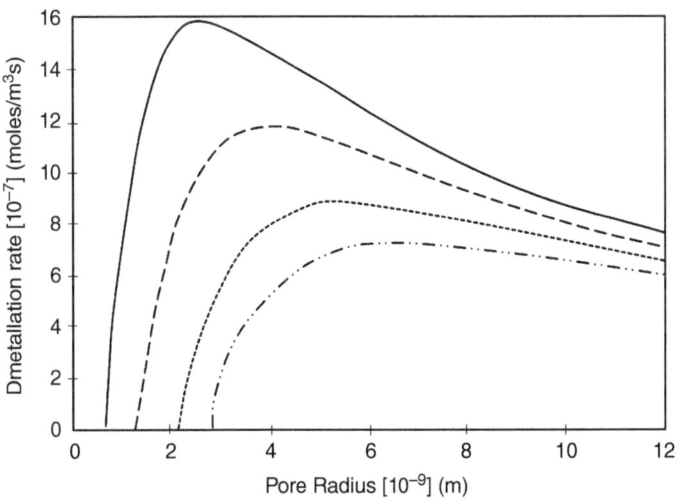

Figure 2.6 Dependence of demetallation rate on catalyst pore size and time-on-stream: 0 days (–), 8.22 days (---), 18.5 days (⋯), and 26.72 days (-⋯-) (Adapted from Rajagopalan 1979).

Khang and Mosby (1986) proposed a model to describe deactivation of hydroprocessing catalyst due to slow and uniform deposition of reaction products in their microporous, mostly metal compounds. The model does not include mass-transfer restriction in the main channels of the pores and considers the creation of active sites by metal compounds in the deposited layers. The remaining pore diameter (R_{pd}) can be represented using the fraction of the volume occupied by metal deposited,

$$R_{pd} = \left(1 - {M_d}/{M_0}\right)^{1/2} \tag{2.28}$$

where M_d is the actual amount of metal deposited and M_0 is the maximum amount of metal deposited when the pore is filled completely. From Eq. (2.28), the term ${M_d}/{M_0}$ as a function of time is

$${M_d}/{M_0} = 1 - \left(1 - {t_d}/{h}\right)^2 \tag{2.29}$$

$$h_z = (k_z/D_z)^2 (r_{p0}/2) \tag{2.30}$$

$$h_0 = (k_0/D_z)^2 (r_{p0}/2) \tag{2.31}$$

$$h = h_0^2/h_z \tag{2.32}$$

where h_z, h_0, and h are dimensionless factors, D_z is the diffusivity of metal, C_m is the metal concentration, r_{p0} is the pore radius of fresh catalyst, k_0 and k_z is the initial demetallation constant and the demetallation rate constant for the additional active sites created by metal deposits, and t_d is the dimensionless time defined as follows:

$$t_d = \frac{k_0 C_m t}{2\rho_L} \tag{2.33}$$

The model is appropriate for predicting deactivation curves of first-order HDM and second-order HDS reactions for less than 50% of pores filled with metals.

Ahn and Smith (1984) formulated a model, which considers two causes of deactivation in HDS catalysts, partial poisoning of the interior pore surface for HDS and pore mouth plugging for HDM first-order reactions; these two mechanisms of deposition are described in terms of two parameters, the Thiele modulus $\emptyset_m$ for metal deposition and a partial poisoning parameter, Θ. These parameters are defined as follows:

$$\emptyset_m = r_0 \left(\frac{\rho_{cat} k_{ms}}{D_{m0}}\right)^{1/2} \tag{2.34}$$

$$\Theta = Q_0 M_p \rho_{cat} / 2\rho_{md} \epsilon_p \tag{2.35}$$

where D_{m0} is the effective diffusivity of the organometallic compounds at zero time, r_0 is the catalyst particle radius, k_m is the rate constant for demetallizaion, ρ_{cat} is the apparent density of catalyst particle, Q_0 is the concentration of deposited metals corresponding to a monolayer, ϵ_p is the porosity of catalyst particle, and M_p and ρ_{md} are the molecular weight and density of metal deposited, respectively.

The rates of demetallization (η_m) and desulfurization (η_s) are represented by effectiveness factor η, which is related to the rate of organometallic entering to the pore with intraparticle diffusion resistance and without intraparticle diffusion resistance.

$$\eta_m(t) = \frac{\left[r_0^2\left(\frac{r_A}{r_{p0}}\right)^2 D_{m0}\left(\frac{dC_{mp}}{dr_p}\right)\right]}{\left(\frac{1}{3}\right) r_0^3 \rho_{\text{cat}}\, k_{ms}\, C_{m0}} \tag{2.36}$$

$$\eta_s(t) = \frac{\int_0^{r_p} r_p{}^2\rho(bk_{ss})C_s dr_p + \int_{ri}^{r0} \rho r_p{}^2 k_{ss}\left(\frac{r_A}{r_{p0}}\right)^2 C_s dr_p}{\left(\frac{1}{3}\right) r_0^3 \rho_{\text{cat}}\, k_{ss}\, C_{s0}} \tag{2.37}$$

A restricted effective diffusivity was employed for metal compounds on spherical pellet by the next equation:

$$D_{\text{eff}} = \frac{\varepsilon D}{\tau}\left(1 - \frac{r_m}{r_A}\right)^2 \tag{2.38}$$

Haynes and Leung (1983) formulated a catalyst deactivation model for HDT processes of heavy fractions. The model considers the deposition of poison on the catalyst and the effects of this deposition on catalyst activity; furthermore, it incorporates the combined effects of pore plugging and active side poisoning. Catalyst deactivation is represented by effectiveness factor, which is defined as the ratio of the observed rate at any TOS to the rate at time zero in the absence of significant resistance to diffusion:

$$\eta = \frac{D_p(X_{0,t})\frac{\partial C}{\partial X}(X_0, t)}{X_0 \rho_{\text{cat}} k C_{AS}} \tag{2.39}$$

where the effective diffusivity is represented by the relation used by Spry and Sawyer (1975):

$$D_P = D_A n_p \pi r_A{}^2 / \tau_f \tag{2.40}$$

The effect of pore plugging in catalytic deactivation is represented by the following equation:

$$\alpha_p = 2W_s / \rho_p\, r_0 \tag{2.41}$$

where the value of W_s corresponds to the mass of adsorbed poison at complete site coverage, ρ_p is the adsorbed poison density, and r_0 is the initial pore radius. This approach may represent a diffusion limited coke type of deactivation.

Galiasso (2007) developed a model based on the effect of recycling the unconverted bottom on catalyst deactivation as a way to improve the hydrocracking conversion. The author assumed that the six-month test represents the catalyst deactivation by metal loading and neglects the coke contribution in the inner part of the particle because its content was approximately constant after 10 days on stream. The model considered that the recycle introduces an additional deactivation through the insoluble material deposition on catalyst in each pass that reduces the incoming molecules' diffusivity. Two sites are considered for catalyst deactivation. A simplified type of the Levenspiel model is used (HCK on Site 1, HDS and HDM reactions on Sites 2):

$$\frac{da_1}{dtc} = k_{d1} C_m a_1 \tag{2.42}$$

$$\frac{da_2}{dtc} = k_{d2} C_m a_2 \tag{2.43}$$

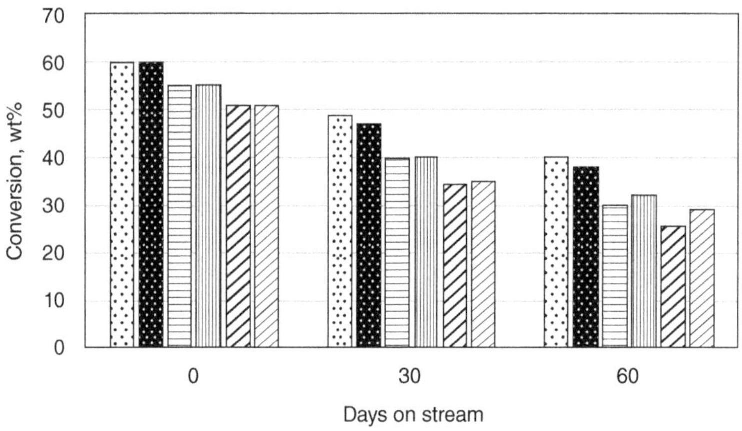

Figure 2.7 Activity as a function of time-on-stream (Galiasso 2007). For (▫) HCK experimental and (■) HCK model, (▤) HDS experimental and (▥) HDS model, (▨) HDM experimental, and (▨) HDM model.

where k_{d1} and k_{d2} are the deactivation constants for Sites 1 and 2, C_m is the metal loading, and tc is the solid residence time.

The experiments were performed at ebullated-bed reactor with a $NiMo/Al_2O_3$ catalyst at the following operating conditions: temperature of 420 °C, pressure of 140 kg/cm^2, and liquid-hourly space-velocity (LHSV) of 0.5 h^{-1} during two-month of TOS. The feedstock was a 500 °C$^+$ residue from extra-heavy crude oil. The residue has 5.02, 0.93, 22, and 21 wt.% of sulfur, nitrogen, Conradson carbon, and C_7 asphaltene content, respectively, and 714 ppm of vanadium. Figure 2.7 shows the results of the model prediction for the aforementioned operating conditions. The model was able to accurately predict the initial activities, with approximately 5% deviation of the 60-day run lengths.

Shah et al. (1976) developed empirical relationships for studying the life cycle of catalyst in a residual oil trickle-bed HDS reactor. Catalyst activity functions for desulfurization and demetallization were developed from pilot plant data. The catalyst activity for sulfur and metal removal reactions decreases with time due to both coke and metal deposition (Newson 1975), however, the authors found that coke deposition is important only after a short TOS, so that most of the activity loss is a result of continued metals deposition, for this reason activity functions, φ_{sf} and φ_m, are dependent only on the metals concentration (c_m) of the catalyst represented by the following equations:

$$\varphi_{sf} = \left(1 - y_{a\sqrt{C_m}}\right)^{y_m} \tag{2.44}$$

$$\varphi_m = \left(1 - y_{b\sqrt{C_m}}\right)^{y_n} \tag{2.45}$$

where y_a, y_b, y_m, and y_n are fitting parameters.

2.3.3 Deactivation Models by Coke and Metal Deposition

Kodama et al. (1980) proposed a deactivation model in which the deposits of coke and metals (mainly V) cause both active site poisoning and pore plugging on catalyst. Second-order kinetics was used to describe both HDM and HDS reaction rates. The amounts of coke and vanadium deposited were expressed as follows:

$$Q_v = 100\int_0^t d_v\,dt/\rho_{\text{cat}} \tag{2.46}$$

$$Q_c = 100\int_0^t d_c\, dt/\rho_{cat} \tag{2.47}$$

In dimensionless forms, these two equations can be rearranged as follows:

$$\varphi_v = Q_v/(100 \cdot P_v \cdot \rho_v) \tag{2.48}$$

$$\varphi_c = Q_c/(100 \cdot P_v \cdot \rho_c) \tag{2.49}$$

The results showed that coke deposition on the catalyst never increased, but if decreased abruptly, then coke deposition may be reversible (Eq. 2.50), indicating that reducing the area of the active surface (A_0C) gives a negative value of coking rate, causing a decrease in the content of coke on the catalyst. Thus vanadium deposition increases, ejecting the coke deposited previously (Eq. 2.51).

$$r_c = \rho_{cat}\, A_{0c}\, k_{c1} - \rho_{cat}\, A_0\, k_{c2}\, C_h\, Q_c \tag{2.50}$$

$$r_V = \rho_{cat}\, A_{0V}\, \eta_V\, k_V\, C_h\, C_V^2 \tag{2.51}$$

Isothermal HDS test was made under constant reaction conditions in an up-flow reactor with the following characteristics: inner diameter of 25.4 mm, thermowell diameter in the center of the reactor of 6.35 mm, and catalyst bed volume of 400 cm^3. An Iranian heavy vacuum residue was used as feedstock at the following operating conditions: pressure of 140 kg/cm^2, LHSV of 0.46 h^{-1}, H_2/oil ratio of 5614 scf/bbl, and temperature of 385 °C and 400 °C. The catalyst used was of Co-Mo type with 0.601 cm^3/g and 155 m^2/g of pore volume and specific surface area, respectively. Figures 2.8 and 2.9 show the calculated results of HDS and HDM respectively, compared with the experimental data.

Skala et al. (1991) suggested the following model of catalyst activity for sulfur and metal removals, which decreases over time due to both coke and metal deposition. Coke deposition was important only during the first part of the run, so that most of the activity loss is the result of continued metal deposition. The model is based on the three reactions (HDS, HDO, and HDM), each of them being pseudo-first-order. The catalyst activity is dependent upon both coke and metals depositions on the

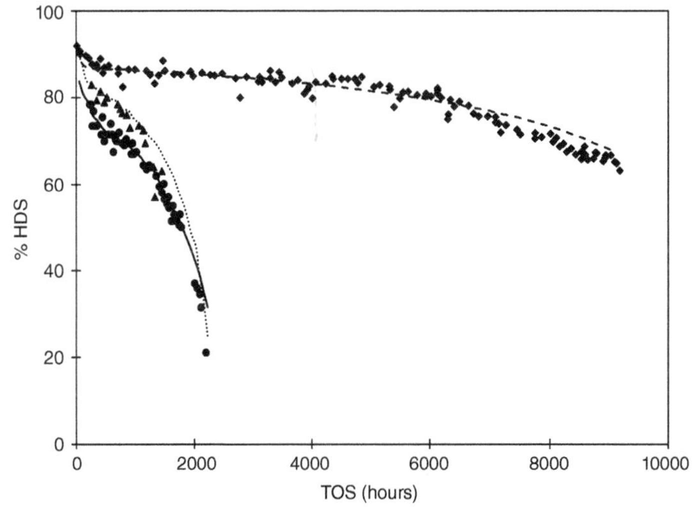

Figure 2.8 HDS results reported by Kodama et al. (1980), experimental data: (●) 385 °C and (▲) 400 °C, simulated data (–) 385 °C and (·) 400 °C. Kam et al. (2005) experimental (– –) and simulated (♦) data.

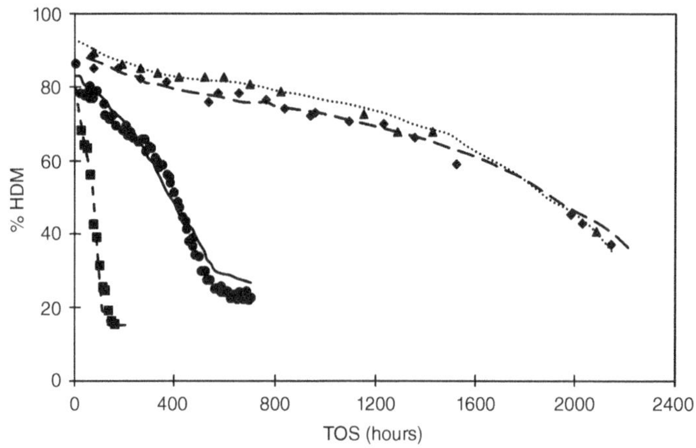

Figure 2.9 HDM results reported by Toulhoat et al. (2005), Catalyst A: experimental (●) and simulated (-) data. Catalyst B: experimental (■) and simulated (---) data. Kodama et al. (1980), experimental data: (◆) 385 °C and (▲) 400 °C, simulated data (---) 385 °C and (·) 400 °C.

catalyst surface. The catalyst activity decline was calculated with the equation proposed by Shah et al. (1976).

$$\varphi = \left(1 - 1.58\, Q^{0.5}\right)^{0.6} \tag{2.52}$$

The concentration of coke and metal (Q) was calculated with the following equation:

$$Q = [0.185\, C_{\text{coke},0} + C_{\text{ash},0}]Lt/w \tag{2.53}$$

where $C_{\text{ash},0}$ is the initial concentration of metals determined as ash of the used oil.

Since the catalyst activity decline is dependent on the concentration of deposits, it was considered obvious that catalyst deactivation is related to the decrease in bed porosity. Skala et al. (1991) assumed that the change in catalyst bed porosity (ϵ) is related to the catalyst activity by an equation similar to Eq. (2.52):

$$\varphi = \left[1 - 1.58(\epsilon_0 - \epsilon_t)^{0.5}\right]^{0.6} \tag{2.54}$$

for the calculation of the decrease in the bed porosity for time different to zero,

$$\epsilon_t = \epsilon_0 - \left[\left(1 - \varphi^{1.67}\right)/1.58\right]^2 \tag{2.55}$$

Newson (1975) proposed a pore-plugging model for residuum hydroprocessing in an axial-flow TBR to describe catalyst deactivation. The model has three stages for determining the loss in porosity of the catalyst: The first stage (θ_{C1}) by fast coke deposition, which takes up one-third of the total porosity; second stage (θ_{C2}) by slow coke deposition; and third stage (θ_{MS}) by metal sulfides plugging. These three stages are represented by the next equation:

$$\theta_T = \theta_{C1} + \theta_{C2} + \theta_{MS} \tag{2.56}$$

The rate of metal sulfides deposition, which depends on the properties of the feed and catalyst, and the operating conditions used are as follows:

$$r_{MS} = \frac{L}{\Delta l} MW_{MS}\, \text{LHSV} \frac{\rho}{\rho_{\text{cat}}} C_F\, e^{-k_1 \tau \left(\frac{l}{L}\right)} \left[1 - e^{-k_1 \tau \left(\frac{\Delta l}{L}\right)}\right] \tag{2.57}$$

Coke contribution to the pore plugging rate is represented by Eq. (2.58), which represents the difference between fast and slow coke, and neglects metal deposited in the first 50 h of operation. The slow coke is assumed to remain approximately constant after the first hours-on-stream:

$$C_0 = \frac{C_{sc} + W_f R_{c2}(t-50)}{1 + W_f R_{c2}(t-50) + W_f R_{MS}(t-50)} \tag{2.58}$$

The rate of deposition of coke plus metal sulfides is estimated from the kinetics of demetallation and desulfurization first-order reactions.

Alvarez et al. (2011) developed the following relationship for describing the catalyst activity decline during the hydroprocessing of heavy oils, caused by coking reaction at start-of-run (SOR) conditions and metal deposition during the whole cycle, the contribution of these two processes to the overall catalyst deactivation was expressed as follows:

$$\Psi(z,t) = \Psi_{\text{coke}}(t) + \Psi_{\text{metal}}(z,t) \tag{2.59}$$

where Ψ_{coke} and Ψ_{metal} stand for the deactivation functions for coking at time $t \leq 100$ h and local metals buildup (Ni + V) at time t, respectively. To represent the effect of these two processes for the overall catalyst deactivation, the authors proposed the following empirical expression:

$$\varphi(z,t) = \frac{1}{(1+\delta t)^{\beta}} - (X_{\text{MOC}}(z,t))^{\gamma} \tag{2.60}$$

The initial activity decline Ψ_{coke} during the first 100 h is represented by the first term of Eq. (2.60), as function of time. At $t = 100$ h, it is considered that the initial period is terminated and the value of Ψ_{coke} is kept constant for the rest of the operation cycle. The author proposes that the best way to describe this process is to link this function to a coking agent instead of time; unfortunately, there was no experimental information available that provided insight into this mechanism. For this reason, the application of such functionality is restricted to the experimental operating conditions. Nevertheless, SOR data obtained with different heavy crudes at different operating conditions (Centeno 2008) indicated that the HDS and HDM activity decay in this period interestingly follows the same pattern even in such a wide range of conditions. This behavior is represented in Figure 2.10 as the change in the deactivation functions of HDM and HDS with respect to the change in time ($\Delta\varphi_{\text{HDM}}/\Delta t$, $\Delta\varphi_{\text{HDS}}/\Delta t$).

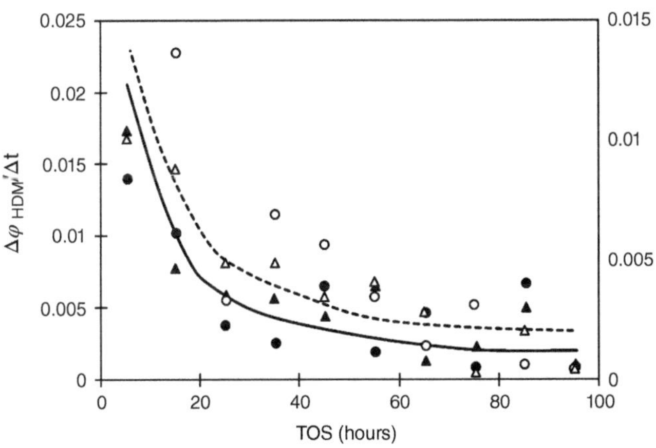

Figure 2.10 HDS and HDM deactivation curves at SOR, experimental data for HDS at (○) 385 C and (Δ) 400 °C, and simulated data (---). HDM experimental data at (●) 385 °C and (▲) 400 °C, and simulated data (-) (Alvarez et al. 2011).

The second term represents the deactivation function due to metals deposit. The amount of metal deposited is expressed in terms of X_{MOC}, defined as the ratio between the local concentration of metal-on-catalyst (MOC) at time t and the specific maximum metals uptake of each catalyst.

MOC is expressed in wt.% and obtained from the following metals mass balance equation:

$$\frac{dMOC(z,t)}{dt} = U_{MOC}\, r_{HDM}(z,t) \tag{2.61}$$

Where r_{HDM} is the local rate of HDM at time t, and U_{MOC} is a unit conversion factor. This equation allows for determining a deactivation profile with respect to time t and axial position z. Experimental data reported elsewhere (Centeno 2008) were used to represent the deactivation of the catalyst caused by these two mechanisms. Kinetics and catalyst aging data were obtained from bench-scale tests using a heavy crude oil (13 °API) at the following operating conditions: LHSV of 0.5 h^{-1}, pressure of 70 kg/cm^2, H_2/oil ratio of 5000 scf/bbl, and temperature of 380 and 400 °C.

Elizalde and Ancheyta (2014a) applied a deactivation model that considers the three stages of deactivation, SOR, MOR, and EOR during HDT of heavy oil in an isothermal bench-scale plant. The model is based on that reported by Idei et al. (1995, 1998, 2003) and Elizalde and Ancheyta (2014b).

$$\varphi = \varphi_A + \varphi_D \tag{2.62}$$

The first term of Eq. (2.62) represents the catalyst decay during SOR with absence of diffusional resistances due to deposits, in which it is considered with different active sites in the catalyst, sites Type I and sites Type II. The former correspond to the sites which are deactivated at fast rate, and the latter correspond to those that are deactivated slower than sites Type I.

After a series of deductions and substituting initial conditions, Elizalde and Ancheyta (2014a) represented the first term of Eq. (2.62) as follows. Detailed deduction can be found elsewhere (Ancheyta 2016).

$$\ln(\varphi_A) = \ln\left(1 + \frac{k_1}{k_2}\, e^{(\alpha_1 - \alpha_2)t}\right) - \left[\ln\left(\frac{k_0}{k_1}\right) + \alpha_1 t\right] \tag{2.63}$$

The loss of catalyst pore volume and pore diameter reduction due to the metal sulfide and the aging coke at MOR provokes diffusional resistance, this fact is represented by the second term of Eq. (2.62) with the following expression:

$$\ln(\varphi_D) = \frac{1}{2}\ln\left(\frac{1}{n_0^2 + (1 - n_0^2)/\left(1 - \frac{t}{t_\infty}\right)}\right) \tag{2.64}$$

Eqs. (2.63) and (2.64) are substituted in Eq. (2.62) and the following final expression is obtained:

$$\ln(\varphi) = \ln\left(1 + \frac{k_1}{k_2}\, e^{(\alpha_1 - \alpha_2)t}\right) - \left[\ln\left(\frac{k_0}{k_1}\right) + \alpha_1 t\right] + \frac{1}{2}\ln\left(\frac{1}{n_0^2 + (1 - n_0^2)/\left(1 - \frac{t}{t_\infty}\right)}\right) \tag{2.65}$$

Experimental data were obtained from a fixed-bed catalytic reactor with internal diameter of 2.54 cm. Details of the bench-scale plant were given elsewhere (Ancheyta et al. 2001). Commercial $NiMo/Al_2O_3$ catalyst was used, its properties are the following: specific surface area of 175 m^2/g, pore volume of 0.56 cm^3/g, and average pore diameter of 127 Å. The heavy oil was hydrotreated under the following reaction conditions: temperature of 400 °C, pressure of 70 kg/cm^2, LHSV of 1 h^{-1}, and H_2/oil ratio of 5000 scf/bbl. Figure 2.11 shows a comparison of experimental data of deactivation for hydrodevanadization (HDV) together with simulated profiles.

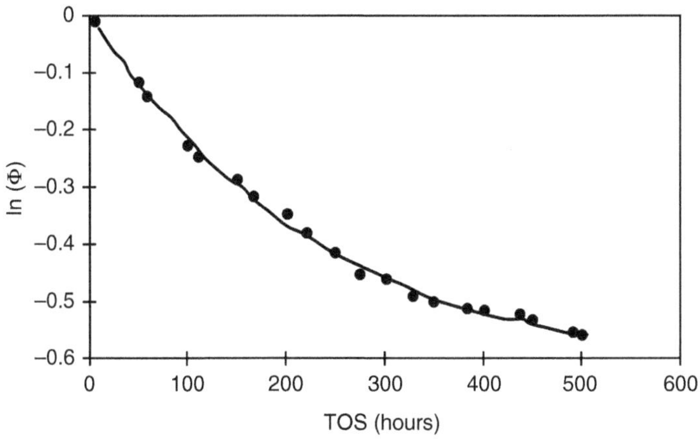

Figure 2.11 Simulation profiles of deactivation for HDV for the range of experimental TOS: (symbols) experimental, (line) simulation (Elizalde and Ancheyta 2014a).

Elizalde and Ancheyta (2014b) used a continuous lumping kinetic model to represent both the reaction kinetics and catalyst deactivation of hydrocracking of heavy oil. The model involves site coverage and pore blockage as the main causes of catalyst deactivation, these functions were represented by power law, considering first-order dependence on activity (Corella et al. 1985). According to Monzón et al. (2003), taking into account the considerations mentioned earlier, modeling the deactivation as TOS function can be expressed as follows:

$$\varphi = \varphi_s + (1 - \varphi_s)e^{-\alpha t} \tag{2.66}$$

where φ is the catalyst activity, α is the parameter decay, and φ_s is the steady-state catalyst activity. In previous experiments (Ancheyta et al. 2002, 2003), a change in activity has been observed in two intervals of TOS. In the first interval, the change in activity is during the first hours of TOS by site coverage due to coke precursors. After certain time, simultaneous sites coverage and pore blockage provoke loss of activity. According to this, the following function defined by intervals is proposed:

$$\varphi = \begin{cases} \varphi_1 & 0 \le t \le t_b \\ \varphi_1 \cdot \varphi_2 & t_b < t \end{cases} \tag{2.67}$$

where φ_1 was considered to be equal to Eq. (2.66) and φ_2 was defined as follows:

$$\varphi_2 = \varphi_s + (1 - \varphi_s)e^{-\omega(t - t_b)} \tag{2.68}$$

In Eq. (2.68), t_b is the time that marks the change in activity during TOS. Experimental information was obtained in a CSTR operated under the following conditions: total initial pressure of 100 kg/cm^2, LHSV of 0.75 hours^{-1}, hydrogen-to-oil ratio of 5000 scf/bbl, temperature of 380 °C and 400 °C, and 1000 rpm of stirring rate. An extruded catalyst was employed and its properties were the following: nominal size of 1/18 in., Ni content of 0.58 wt.%, Mo content of 2.18 wt.%, specific surface area of 197.2 m^2/g, pore volume of 0.85 cm^3/g, and average pore diameter of 172.6 Å. The first part of curve in Figure 2.12 represents the loss of catalyst activity by site coverage, and the sharp changes on reactivity are attributed to some catalyst pore constrictions together with deactivation by coke precursors.

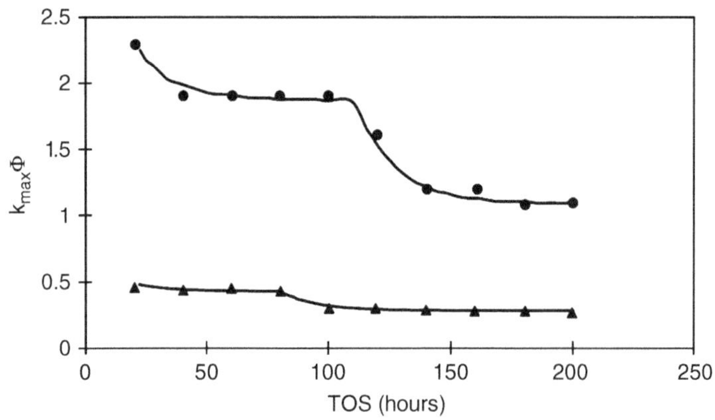

Figure 2.12 Effect of time-on-stream on k_{max} model parameter at (●) 380 °C and (▲) 400 °C (Elizalde and Ancheyta 2014b).

Toulhoat et al. (2005) and Kressmann et al. (1998) developed a model for residue upgrading using several fixed-bed reactors in series. In this model, catalyst deactivation is represented by two mechanisms: coke and metal depositions. They represented the local concentration of metals (Ni, V) in solid phase with the following:

$$\frac{dC_m(r,z,t)}{dt} = \frac{M_m}{\rho_{cat}} \sum_j \rho_m(r,z,t) \tag{2.69}$$

The local coke deposition caused by the reversible dehydrogenation of the asphaltenic material and its first-order reaction rate with respect to a driving force equal to the difference between actual and equilibrium local coke concentrations on catalyst surface are represented by the following equation:

$$\frac{dC_0(r,z,t)}{dt} = k_{coke,k}(r,z,t)S_k(r,z,t)\exp\left(\frac{E^{\pm}_{coke,k}}{RT}\right) \times \left[C_{coke}(r,z,t) - C_0(r,z,t)\right] \tag{2.70}$$

To express the rate of hydroprocessing reactions, a pseudo-Langmuir–Hinshelwood law was used, the next equation is a simplified form of this equation for HDS:

$$MO(r,z,t) = k(r,z,t)S(r,z,t)\exp\left(E^{\pm}/RT\right) \frac{C^{1}_{asph}(r,z,t)}{\left[1 + K_{ads}(T[z,t])\, C^{*}_{asph}(r,z,t)\right]} \tag{2.71}$$

Catalyst tests were run on a bench-scale unit, which has an up-flow trickle-bed tubular reactor. Two catalysts with various textural parameters were used: A is a $NiMo/Al_2O_3$ HDM catalyst, which is characterized by a continuous pore distribution between macropores and mesopores, and B is a $Co/Mo/Al_2O_3$ HDS catalyst that is characterized by a mesoporous monomodal pore distribution. A feedstock (Ni and V content of 196 and 1150 ppm, respectively, and 5.28, 0.67, 13, 12 wt.% of sulfur, nitrogen, asphaltenes, and CCR contents, respectively) was used to evaluate the evolution of HDM and HDS performances vs TOS. This feed was injected until complete deactivation of the catalyst for evaluating metal retention capacity at the following operating conditions: total pressure of 153 kg/cm^2, catalyst volume of 40 cm^3, LHSV of 1 hour^{-1}; H_2/oil ratio of 6737 scf/bbl, temperature of 390 °C, and the TOS was variable.

Figure 2.9 shows the comparisons of simulated and experimental deactivation curves for HDM for catalysts A and B.

Chao et al. (1991) proposed a mathematical model that describes catalyst deactivation in a commercial RDS unit. The model takes into account metal and coke deposition, and the decrease of effective diffusivities during HDS and HDM reactions. The reaction rate coefficients of desulfurization and demetallization reactions, varying according to deposited metal amounts, can be represented as follows:

$$k_{fs} = k_s/\Lambda \quad \text{when} \quad \Delta r_{pm} \geq \Delta r_{pc} \tag{2.72}$$

$$k_{fs} = k_s\left[1 + \left(\frac{1}{\alpha} - 1\right)\left(\frac{\Delta r_{pm}}{\Delta r_{pc}}\right)\right] \quad \text{when} \quad \Delta r_{pm} < \Delta r_{pc} \tag{2.73}$$

$$k_{fm} = \frac{k_m}{\Gamma} \quad \text{when} \quad \Delta r_{pm} \geq \Delta r_{pc} \tag{2.74}$$

$$k_{fm} = k_m\left[1 + \left(\frac{1}{\alpha} - 1\right)\left(\frac{\Delta r_{pm}}{\Delta r_{pc}}\right)\right] \quad \text{when} \quad \Delta r_{pm} < \Delta r_{pc} \tag{2.75}$$

$$\Delta r_{pm} = r_{p0} - r_{pm} \tag{2.76}$$

where Λ and Γ are the relative reaction rate coefficients of fresh catalyst and deposited metal for desulfurization and demetallization, r_{p0} is the pore radius of fresh catalyst, r_{pm} is catalyst pore radius which neglects coke thickness, and Δr_{pc} is the thickness of monolayer deposited metal.

The coke deposited covers the active sites causing them to become inactive, and this fraction of surface is assumed to be proportional to the relative volumes of carbonaceous compounds and catalytic pore volume. In this model, catalyst decay by coke deposition is represented as follows:

$$1 - x_s = \alpha_c Q_c \tag{2.77}$$

$$Q_c = \left[\left(\frac{\Delta r_{pm}}{r_{p0}}\right)^2 - \left(\frac{r_p}{r_{p0}}\right)^2\right] \tag{2.78}$$

where Q_c is the volume of carbonaceous compounds, x_s is the fraction of specific surface area that is left uncovered by coke, and α_c is a proportional constant. The decrease of catalyst effective diffusivity is represented by a second-order equation with respect to pore radius:

$$D_{eff} = D_0\left(\frac{r_p}{r_{p0}}\right)^2 \tag{2.79}$$

Kam et al. (2005) developed a kinetic model of rapid initial catalyst deactivation for HDS, HDM (by nickel and vanadium), and hydrodeasphaltenization reactions. An exponential decay function was more appropriate, and moreover, as the HDS catalyst is protected by the HDM catalyst, the deactivation is only by coke. Therefore, they proposed an empirical decay function (Ψ_i) where the changes in the reaction rate coefficient of the main reactions can be represented as follows:

$$k_{Ai} = k_i + 0.5\Delta k_i \Psi_i \tag{2.80}$$

$$\Psi_i = -e^{\left(\frac{c_{p,i(\text{Target})}}{c_{p,i(\text{Product from previous reactor})}}\right)^n} \tag{2.81}$$

$$\Delta k_{0,i} = k_i|_{t=12\,h} - k_i \tag{2.82}$$

$$k_{0,i}|_{t=12\,h} = \frac{\int\left(\frac{dc}{c_i^n}\right)}{\int dt} \tag{2.83}$$

For $i = 1$ to 4

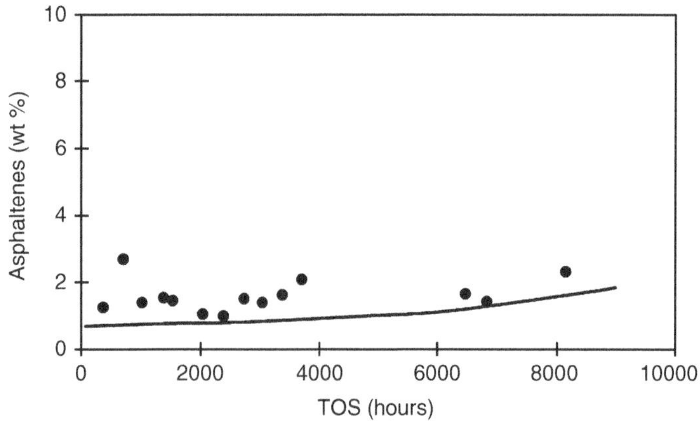

Figure 2.13 Comparison of model predictions (-) and experimental (●) data of asphaltenes with processing time (Adapted from Kam et al. 2005).

where k_{Ai} and k_i are apparent reaction rate coefficients in the initial rapid deactivation period for species i estimated by Eq. (2.80) and determined experimentally, respectively; i is the species index (1 for sulfur, 2 for vanadium, 3 for nickel, and 4 for asphaltenes); C_p and C_f are concentrations of the product and the feed, respectively; and n is the reaction order. The initial rapid deactivation period is completed when the decay function for sulfur becomes $\Psi_i \leq 0.01$. They found that the amount of coke deposited on the catalysts increased rapidly and reached about 15 wt.% within 12 hours. After this period, the rate of coke formation slowed down reaching around 20 wt.% after 240 h.

The model was used to simulate the pilot plant performance and predict the catalyst deactivation beyond the 4000 hours of pilot plant test at the following operating conditions: temperature of 375 °C, pressure of 122 kg/cm^2, LHSV of 1 hours^{-1}, H_2/oil ratio of 3200 scf/bbl, and TOS of 9500 hours. The predicted sulfur and asphaltene concentrations in the product oil beyond the 4100 hours are illustrated in Figures 2.8 and 2.13, respectively. The pilot-plant sulfur and asphaltenes data clearly follow the model predictions.

2.4 Development of Models for HDT Catalyst Deactivation

2.4.1 Important Issues

It has been found from the literature review that deactivation models for HDT catalysts have been developed mainly for heavy oil applications, which is certainly the case that exhibits the most severe loss of catalyst activity. The deactivation models take into consideration the activity decay as a function of coke deposition, metals deposition, and both, which are the main agents (coke, metals) that deactivate the catalyst. Other models use catalyst activity as a function of TOS as the experiments are easier and cheaper to carry out.

Conducting experiments for developing models based on coke-on-catalyst (COC) or MOC requires a number of tests to characterize the spent catalyst, which is developing one test at certain value of TOS, another test at longer TOS, and so on. In each test, the different stages of evaluation must be performed, that is, catalyst loading, catalyst activation, operation at defined conditions, and catalyst unloading. It means that for obtaining one point (content of coke and metals on spent catalyst), a complete experimental run must be done. The following run must be conducted at longer time, which for having enough experimental information for developing a robust deactivation model can be as long as several weeks. All these experiments will be costly and time-consuming, that is why researchers think twice when intending to do them and decide either to develop models based on TOS or to conduct experiments at shorter time to develop models based on COC and early stage of MOC.

The deactivation models reported in the literature have different forms, terms, and parameters for representing activity loss. Usually, they include parameters that are calculated by fitting experimental data; however the applicability of these models is only valid for the range of operating conditions at which they were derived. Tables 2.1–2.3 summarize different deactivation

Table 2.1 Deactivation parameters for deactivation models based on coke deposition and TOS.

Authors	Model	Reactor	Feedstock	Parameter
Froment and Bischoff (1962)	$k = k_{\eta}\varphi(C_0)$ $\varphi = \exp(-\propto C_0)$ $\varphi = \dfrac{1}{(1+\alpha C_0)^{\beta}}$			
Voorhies (1945)	$C_0 = A(\tau/60)^V$	Fixed-bed	East Texas gas oil	$A = 0.52$ $V = 0.38$
		Fluid catalyst cracking	Tinsley gas oil	$A = 0.24$ $V = 0.53$
Rashidzadeh et al. (2010)	$\varphi_{ij} = e^{\left(-\alpha_{ij}\ \mathrm{Life}\right)}$	Trickle-bed	Vacuum gas oil	$\alpha_{DK} = 4.13E{-}05$ $\alpha_{FG} = 8.43E{-}05$ $\alpha_{FN} = 8.28E{-}05$ $\alpha_{NG} = 9.12E{-}05$ $\alpha_{KN} = 1.59E{-}04$
Richardson et al. (1996)	$C_0 = C_{max}[1 - \exp(-Kw)]$	Batch CSTR	Athabasca bitumen feed	
Ancheyta and Lopez-Izunza (2000)	Model 1 $\varphi = e^{-k_d t}$ Model 2 $\varphi = \dfrac{1}{1 + k_d t}$ Model 3 $\varphi = A\,t^{-B}$ Model 4 $\varphi = (1 + Gt)^{-B}$ Model 5 $\varphi = \dfrac{a}{p^q(1 + d\,t^g)}$	Fixed-bed	Vacuum gas oil	Model 1 $k_d = 0.0875$ Model 2 $k_d = 1.2864$ Model 3 $B = 0.7888$ Model 4 $B = 1.5741$ $G = 0.5889$ Model 5 $a = 0.9335$ $g = 1.3576$

Table 2.2 Deactivation parameters for deactivation models based on metal deposition.

	Model	Reaction	Reactor	Feedstock	Parameter
Oyekunle et al. (2004, 2005)	$\frac{\partial^2 \alpha_D}{\partial {\beta_D}^2} = \Phi^2 {\alpha_D}^2 + \frac{\partial \alpha_D}{\partial \tau_D}$	HDM	Fixed-bed	Heavy Maya crude	$\eta = 0.2$ $0 < \beta_D < 0.6$ $k_m = 4.5 \times 10^{-12}$ kg/s m
Toulhoalt et al. (1990)	$\text{USC} = \frac{\Omega \rho_m \epsilon_0 C_m}{\rho_{cat}}$	First-order HDM	CSTR	Deasphalted Boscan crude oil	$\Omega = 4.26$
Shah and Mhaskar (1976)	$\varphi_{sf} = \left(1 - y_{a\sqrt{C_m}}\right)^{y_m}$ $\varphi_m = \left(1 - y_{b\sqrt{C_m}}\right)^{y_n}$	First-order HDM-HDS	Trickle-bed	Petroleum residue	$y_a = 1.58$ $y_b = 1.58$ $y_m = 0.8$ $y_n = 0.6$
Dautzenberg et al. (1978)	$\theta = \frac{k_v V_{dep}}{r_{p0}} \int_0^t c_{v0}\, dt$	First-order HDM	Fixed-bed	Caribbean long residue Middle East long residue	$\eta_m = 0.5$
Rajagopalan (1979)	$\frac{dm}{dt} = \zeta 2\pi r_{p0} k_{fm} M_{MS}$	First-order HDM $k_m = 5 \times 10^{-13}$ m s^{-1}			$\text{Ø}_m = 3$ $\epsilon_p = 0.5$
Khang and Mosby (1986)	$R_{pd} = \left(1 - {M_d}/{M_0}\right)^{1/2}$	First-order HDM	Trickle-bed	Vacuum residues	
Ahn and Smith (1984)	$\text{Ø}_m = r_0 \left(\frac{\rho_{cat} k_{ms}}{D_{m0}}\right)^{1/2}$ $\Theta = Q_0 M_p \rho_{cat} / 2\rho_{md} \epsilon_p$	First-order HDM First-order HDS			$\text{Ø}_m = 3–9$ $\Theta = 0.025–0.075$ $\text{Ø}_s = 1$ $b = 1, 4, 9$
Haynes and Leung (1983)	$\alpha_p = 2W_s/\rho_p\, r_0$				$\tau_f = 3$ $\eta = 0.7, 0.2, 0.1, 0.05$
Galiasso (2007)	$\frac{da_1}{dtc} = k_{d1} C_m a_1$ $\frac{da_2}{dtc} = k_{d2} C_m a_2$		Ebullated-bed	500 °C$^+$ residue	$\eta = 0.83–0.91$ $k_{d1} = 1\text{E} + 15$ $k_{d2} = 1\text{E} + 17$

models based on coke deposition, metal deposition, and coke/metals deposition, respectively. Information about the values of parameters, type of reactor, type of reaction, and type of feedstock obtained by different authors is also included. More details about experimental reaction conditions and additional values of parameters can be found in the corresponding references, as well as shown in the tables.

Metals deposition is present preferentially near the entrance of the pores of the catalyst (Kodama et al. 1980) resulting in the blocking of the access to the active sites, the level of blocking depends, of course, on several factors among them type of support, type of active metals, size of the pores, pore size distribution, amount and size of asphaltenes, amount of asphaltenic and non-asphaltenic metals, etc. For these reasons, deactivation models based on metal deposits take into consideration some properties of the catalyst pore (volume, structure, and size distribution) and intraparticle

Authors	Model	Reaction	Reactor	Feedstock	Parameter
Kodama et al. (1980)	$r_c = \rho_{cat} A_{0c} k_{c1} - \rho_{cat} A_0 k_{c2} C_h Q_c$ $r_V = \rho_{cat} A_{0V} Ef_V k_V C_h C_V^2$	Second-order HDM-HDS	Up-flow	Iranian heavy vacuum residue	$k_s = 1.22 \times 10^{10} e^{-31.3 \times 10^3/RT}$ $k_v = 1.28 \times 10^{10} e^{-24.2 \times 10^3/RT}$ $k_{c1} = 3 \times 10^{-2} e^{-20 \times 10^3/RT}$ $k_{c2} = 5.35 \times 10^{-5} e^{-10 \times \frac{10^3}{RT}}$
Skala et al. (1991)	$\varphi = [1 - 1.58(\epsilon_0 - \epsilon_t)^{0.5}]^{0.6}$	First-order HDM-HDS	Trickle-bed		$y_n = 0.6$
Newson (1975)	$\theta_T = \theta_{C1} + \theta_{C2} + \theta_{MS}$	First-order HDS	Trickle-bed	Residuum	$E_m = 30, E_s = 40$ $\eta_m = 0.2, \eta_s = 0.6$ $m = 1, 2$
Alvarez et al. (2011)	$\Psi(z, t) = \Psi_{\text{coke}}(t) + \Psi_{\text{metal}}(z, t)$	HDM HDS	Fixed-bed	Atmospheric residue	$\delta_{HDS} = 3.81$ $\beta_{HDS} = 0.06$ $\gamma_{HDS} = 0.34$ $\eta_0 = 0.7$
Elizalde and Ancheyta (2014a)	$\varphi = \varphi_A + \varphi_D$	HDM HDS	Fixed-bed	Heavy crude oil (13 °API)	$\alpha_1 = 9.0 \times 10^{-3}$ $\alpha_2 = 4.0 \times 10^{-4}$ $\eta_0 = 0.73$
Elizalde and Ancheyta (2014b)	$\varphi_2 = \varphi_s + (1 - \varphi_s)e^{-\alpha(t - t_b)}$	Kinetic lumping	Stirred tank	Residue	$\alpha_{380\,^\circ C} = 0.070$ $\varphi_s = 0.660$ $t_b = 80$
Toulhoat et al. (2005)	$\frac{dC_m(r,z,t)}{dt} = \frac{M_m}{\rho_{\text{cat}}} \sum_j \rho_m(r,z,t)$	HDM HDS	Fixed-bed	Boscan crude petroleum	$k_s = 16.5 \times 10^3$ $k_m = 3.9 \times 10^4$ $E_m = 40, E_s = 40$
Chao et al. (1991)	$1 - x_s = \alpha_c Q_c$	HDM HDS	Fixed-bed	Iranian heavy A.R.	$k_{c1} = 3 \times 10^{-2} e^{-20 \times 10^3/RT}$ $k_m = 1.28 \times 10^{10} e^{-24.2 \times 10^3/RT}$ $k_s = 1.22 \times 10^{10} e^{31.3 \times 10^3/RT}$ $\alpha_c = 1.8$
Kam et al. (2005)	$\Psi_i = -e^{\left(\frac{C_{p,i(\text{Target})}}{C_{p,i(\text{Product from previous reactor})}}\right)^n}$	HDM HDS	Fixed-bed	Atmospheric residue	$E_s = 26.1$ $k_s = 0.492$ $k_v = 0.021$ $n_{HDS} = 2$ $n_{HDV} = 2$

mass-transfer phenomena, however, they do not consider the effect of coke deposition on activity loss as coke buildup occurs only during the first hour-on-stream.

Most of the authors use first-order kinetics for HDM and HDS reactions for evaluating catalysts performance. Others utilized different values of global effectiveness factor for determining catalyst life (Dautzenberg et al. 1978; Weekman 1968; Wheeler 1951). Idei et al. (1998), Skala et al. (1991), and Arbabi and Sahimi (1991a) calculated deactivation parameters from experimental data and proposed two different deactivation mechanisms: one involves coke and metals deposits and another one considers only metals deposits.

Table 2.3 shows the parameters reported by various authors for deactivation models considering both mechanisms of deactivation. Chao et al. (1991) and Kodama et al. (1980) proposed models based on different properties of the catalyst and obtained good agreement with experimental data. Toulhoat et al. (2005) and Newson (1975) obtained similar activation energies for HDM and HDS reactions. Despite on using different catalysts, these two models were tested in a commercial plant, the model proposed by Newson (1975) showed better agreement with experimental data.

For the development of deactivation models, it is then important to consider several factors, such as chemical properties of the feed (particularly metals and asphaltenes contents) and physical and chemical properties of the catalyst. If the experiments are conducted in tubular reactor, the changes in catalyst properties should be determined as a function of its position along the reactor length (Isaza et al. 2000); of course, this depends on the length of the catalytic bed. For small length, the behavior of the bed may be assumed to behave as integral reactor and the changes in reactor position can be neglected.

Additionally, it is necessary to take into account the different stages of catalyst deactivation since it is well-known that at long TOS, various mechanisms are involved, such as coke deposition during SOR causing loss of catalytic surface area and decrease in mean pore diameter and volume; at MOR, deactivation of catalyst is mainly due to loss of sites by poisoning and pore plugging by metal sulfides deposits; and finally at EOR, catalyst presents severe diffusional resistance due to almost total pore plugging.

Some models reported in the literature do not take into consideration the effective diffusivity, which is affected by coke and metal deposits. Deposits cause decrease in pore diameter and loss of catalytic specific surface area preventing the access to the active sites of the catalyst; consequently in some cases, a wrong value of effectiveness factor is utilized, resulting in falsified data by apparent kinetics.

Most models based on coke deposits are a simple function of time, nevertheless, to develop a better deactivation model, it is important to consider the concentration of reacting species, which provoke coke deposition on catalyst. Some deactivation models are based on TOS and do not consider the concentration of heteroatoms in the feedstock. It should be highlighted that the reason for catalyst deactivation is not TOS, but metals and coke deposition. For heavy oils with high content of impurities (metal and asphaltenes) at long TOS, metal buildup is more relevant; then each feed will exhibit different deactivation patterns, and for this reason, the proper way for describing the catalyst deactivation is with a model that includes the agents present in the feedstock, which provoke activity decay of the catalyst, instead of a function based on TOS.

2.4.2 Final Remarks

The formation of coke and metals deposits during HDT reactions of heavy oil is a real and complex problem that affects the performance of the process and the quality of products due to catalyst deactivation. The main reasons for catalyst deactivation are attributed to loss of active sites due to poisoning and pore mouth plugging. It has been found that most of the coke deposition on the catalyst occurs during the first stage of the process and after that deposited coke remains constant, and catalyst decay continues but mostly affected by metals deposition. For this reason, the models based on both stages of deactivation fit better to experimental data.

The models that consider the deactivation of the catalyst due to metals accumulation, generally take into account the size of the pores, since it is necessary to have a suitable pore size that allows the metal molecules to enter into the pores. Some authors have developed models considering both metal and coke deposits. These types of models are more appropriate for predicting catalyst life because they cover the three stages of deactivation in the HDT process.

It should be recognized that modeling catalyst deactivation during HDT of heavy oils is not an easy task. More efforts need to be put to develop robust models that consider the different stages (SOR, MOR, and EOR) of deactivation, and the effect of the agents that are indeed affecting the activity of the catalyst (i.e., coke and metals).

2.5 Development of a Reactor Model for Heavy Oil Hydrotreating with Catalyst Deactivation Based on Vanadium and Coke Deposition

A quasi-dynamic one-dimensional pseudohomogeneous TBR model for HDT of heavy vacuum residue was developed. The model includes catalyst deactivation based on the concentration of impurities in heavy oil, as well as coke and vanadium deposition on catalyst surface during TOS. Literature experimental data of heavy vacuum residue HDT were used for model validation. The proposed model well-predicted the typical "S" shape profiles during TOS due to catalyst deactivation for heavy oil HDT.

The proposed model was developed considering a previous work reported in the literature (Kodama et al. 1980), which was used to describe experimental data obtained for HDT of a vacuum residue with sulfur and vanadium contents of 3.67 wt.% and 270 ppm, respectively. Experimental data include the concentrations of vanadium and sulfur in products along TOS at 140 kg/cm^2, 1100 m^3/m^3 H_2/oil ratio, and 0.46 hours^{-1} LHSV at three temperatures of 385 °C, 400 °C, and 415 °C. At 1360 hours for the temperature of 400 °C, the amount of vanadium deposits on catalyst along reactor axial length was measured. From the profiles of impurities concentration of products along TOS, the first stage, second stage, and the beginning of the third stage of catalyst deactivation can be clearly distinguished. According to the classification of Froment et al. (2010), the proposed model would be a dynamic one-dimensional pseudohomogeneous model. However, since there is not an accumulation term, the model was classified as a quasi-dynamic. The model may be considered of medium complexity. It can simulate the operation of a TBR at isothermal conditions. By mass balance, the reactor model can predict the concentration profiles of impurities with respect to TOS. Effective diffusivities have the strongest influence on catalytic deactivation, so that the limitation of the access of heteroatom compounds to the catalytic active sites due to pore plugging is included. This consideration is based on literature reports. For instance, Kobayashi et al. (1987) conducted a study of heavy oil HDT considering the catalyst pore size. They found that for catalysts with 3% molybdenum on alumina and on silica alumina, the optimum pore size was from 100 to 150 Å at 400 °C and the optimum pore diameter increases as the temperature increases. Vanadium and nickel impurities deposited on catalyst surface are high during HDT reactions, while coke is deposited in lower amount. Active surface area is affected as pore plugging progresses, reducing conversion of impurities. The base model only includes vanadium deposition; however, it can be easily adapted to consider nickel content separately from vanadium or as a sum of them.

2.5.1 The Model

2.5.1.1 Description

The mass balance equations for the quasi-dynamic one-dimensional pseudohomogeneous reactor model are as follows:

$$u_l \frac{dC_S}{dz} + \varepsilon_S r_S = 0 \tag{2.84}$$

$$u_l \frac{dC_V}{dz} + \varepsilon_S r_V = 0 \tag{2.85}$$

where u_l is the mass flux velocity of liquid phase, C_S is the sulfur concentration in liquid phase, C_V is the vanadium concentration in liquid phase, ε_s is the catalyst fraction in the reactor, and z is the reactor length.

Eqs. (2.84) and (2.85) were integrated by the Runge–Kutta method. The exit values of the reactor at certain TOS can be compared with the respective experimental value. Catalyst fraction (ε_s) was calculated by the relationship between void and solid fractions:

$$\varepsilon_s = 1 - \epsilon_B \tag{2.86}$$

Bed void fraction was calculated by a correlation of particle size and diameter of the reactor proposed by Froment et al. (2010).

$$\epsilon_B = 0.38 + 0.073\left[1 + \frac{\left(d_R/d_{pe} - 2\right)^2}{\left(d_R/d_{pe}\right)^2}\right] \tag{2.87}$$

Reaction rates (r_i) for vanadium or sulfur depend on the remanent catalyst surface area, effectiveness factor, and concentration of heteroatoms in the liquid phase. All these parameters change along the length of the reactor and during TOS. The equations to obtain the reaction rates for S and V are as follows:

$$r_i = \rho_{cat} A_{ri} \eta_i k_i C_h C_i^2 \tag{2.88}$$

where ρ_{cat} is the packed bed density of catalyst, A_{ri} is the remanent surface area, η_i is the effectiveness factor, k_i is the reaction rate coefficient, C_h is the concentration of hydrogen in the liquid phase, C_i is the concentration of heteroatom i in process stream, and i represents sulfur and vanadium. It is well-known that kinetics follows first-order of reaction when model compounds are used, however, when mixture of model compounds or more complex petroleum fractions are studied, the reactor order tends to increase. For the case of HDS and HDV reactions when residues are hydrotreated, the most common reaction order reported in the literature is two. Second-order of reaction was also used in the literature model (Kodama et al. 1980). The concentration of hydrogen in the liquid phase was calculated with

$$C_h = \left(8.91 \times 10^{-6}\right)P + \left(4.16 \times 10^{-6}\right)(T - 273) - 1.40 \times 10^{-3} \tag{2.89}$$

where P is the reactor pressure and T the temperature.

Rate coefficient for each HDT reaction is based on the Arrhenius equation:

$$k_i = A_0 e^{-\frac{E_A}{R_g T}} \tag{2.90}$$

where A_0 is pre-exponential factor, E_A the activation energy, and R_g the ideal gas constant.

Effectiveness factor (η_i) is calculated from the Thiele modulus (ϕ_i):

$$\phi_i = r\sqrt{\frac{\rho_{cat}A_{ri}k_iC_hC_i}{\rho_{oil}D_i}} \tag{2.91}$$

$$\eta_i = \frac{3}{\phi_i}\left(\frac{1}{\tan h\phi_i} - \frac{1}{\phi_i}\right) \tag{2.92}$$

where r is the catalytic particle equivalent radius, ρ_{oil} is the density of the oil, and D_i is the diffusivity of component i inside the pores of the catalyst. Reaction rate equations include catalytic activity in the term of remanent surface area of catalyst active sites. Remanent surface area of catalyst is reduced as the amount of impurities deposited on catalyst pores increases and can be calculated as follows:

$$A_{rs} = A_o(1 - \alpha_V\psi_V - \alpha_c\psi_c) \tag{2.93}$$

$$A_{rV} = A_o(1 - \beta_V\psi_V - \beta_c\psi_c) \tag{2.94}$$

$$A_{rc} = A_o(1 - \gamma_V\psi_V) \tag{2.95}$$

where A_{rs}, A_{rV}, and A_{rc} are the remanent surface areas for HDS, HDV, and for the deposition of coke, respectively; A_o is the fresh surface area; α_V, β_V, and γ_V are the parameters for the effect of deposition of vanadium on HDS, HDV, and coking active sites including the effect of nickel; α_c and β_c are the parameters for the effect of deposition of coke on HDS and HDV active sites. Remanent surface area of active sites for HDS and HDV are affected by both vanadium and coke deposition. Meanwhile remanent surface area for coke deposition is affected by vanadium deposits. ψ is the volumetric amount of impurities deposited on catalyst calculated according to the following:

$$\psi_i = \frac{Q_i}{100P_V\rho_i} \tag{2.96}$$

where Q_i is the dimensionless mass of impurity i deposited on catalyst, P_V is the catalyst pore volume, and ρ_i is the impurity density. The mass amounts of heteroatom deposits on the catalyst in a time interval generated by HDV reaction and coke deposition are obtained by as follows:

$$Q_i = \frac{100}{\rho_{cat}}\int_0^t r_i dt \tag{2.97}$$

where t is TOS.

The diffusivity of heteroatoms in the liquid phase is limited by the deposited metals and coke. The parameter ψ_{max} refers to the maximum retention capacity. From this parameter, the diffusivity is affected as the deactivation of the catalyst progresses.

$$D_s = D_{so}\left(1 - \delta_V\frac{\psi_V}{\psi_{max}}\right) \tag{2.98}$$

$$D_V = D_{Vo}\left(1 - \frac{\psi_V}{\psi_{max}}\right) \tag{2.99}$$

Effective diffusivities depend on the volumetric amount of impurity deposit on catalyst. δ_V is the parameter of the effect of vanadium deposits on sulfur diffusivity. Coke deposition reaction is reversible, and the rate equation has a different expression than those of the vanadium or sulfur:

$$r_C = \rho_{cat}A_{rc}k_{C1} - \rho_{cat}A_ok_{C2}C_hq_c \tag{2.100}$$

where k_{C1} is the reaction rate coefficient of deposition of coke, k_{C2} is the reaction rate coefficient of removal of coke, and q_c is the mass of impurity deposit per volumetric unity of catalyst.

Coke and vanadium deposits generated affect reaction rates by covering surface area and diffusivity of sulfur and vanadium by plugging catalyst pores.

2.5.1.2 Solution of the Model

The sequence of calculus starts with integration of mass balance equation for vanadium and sulfur with input data about properties of feedstock and catalyst and characteristics of the reactor. The deposits of vanadium and coke for fresh catalyst are zero, and the surface area of active sites is the initial value in the fresh catalyst. The deposition rate of sulfur, vanadium, and coke along the catalyst bed determines the amount of deposits at each coordinate according to TOS and axial length steps. Then, diffusivity of sulfur and vanadium, remanent surface areas, and volumetric quantity of impurities on catalyst are determined. The sequence is repeated for different TOS, and in each step calculated and experimental values are used in the objective function for optimization of parameters.

To obtain a more exact simulation, intervals of experimental TOS were divided into forty steps for the proposed model. For literature model (Kodama et al. 1980), it was reported that intervals were divided into 20 steps.

Operating conditions of HDT, and feedstock and catalyst properties are essential input data for better application of the proposed model. It requires at least both concentration of impurities in products and deposits on catalysts to perform a reliable parameter estimation. Pore volume, remanent surface area, and other heteroatoms concentration values are considered additional information that can contribute to more precise modeling.

The structure of the algorithm was coded in Python, whereby the concentration of sulfur and metals in process stream profiles were obtained with respect to TOS. The calculation of model parameters was done by minimizing an objective function based on the squared differences between calculated and experimental values including their covariances. Levenberg–Marquardt's method of optimization was applied at last to rectify the best set of parameter values by assuring the global minimum of the objective function.

2.5.1.3 Advantages of the Model

The following are the main advantages of the proposed model compared with the literature model (Kodama et al. 1980). Some of them are related to unreported data and inconsistencies of the literature model and others more important for simplification of equations, optimization of methods, and verification of model parameters.

- Maximum capacity of vanadium deposit. An inconsistency in deposited vanadium profile with respect to reactor length was observed in the literature model (Kodama et al. 1980). While in a plot a maximum capacity of vanadium deposition of 34.1% was reported, in the text the corresponding value for this parameter is 31.4%. This difference causes a mathematical error because the value of effective diffusivity (Eq. 2.99) becomes negative and then the value of the Thiele modulus (Eq. 2.91) results in erroneous values. In any case, using 31.4% or 34.1% of maximum capacity of vanadium deposition gives erroneous values for the prediction of concentration profiles, which differ from the values reported by the literature model (Kodama et al. 1980). This discrepancy was corrected in the proposed model.
- Calculation of effectiveness factor. The literature model (Kodama et al. 1980) determines the Thiele modulus and effectiveness factor by solving the intraparticle dimensionless concentration

differential equations by numerical methods. In the proposed model, general Eqs. (2.91) and (2.92) were used to calculate effectiveness factor from the Thiele modulus. This change simplifies the calculation and improves optimization of model parameters without losing accuracy. From the results discussed later on, a reduction of effectiveness factor value for HDS and HDV reactions was observed along TOS. The rate and distribution of coke and vanadium deposition on catalyst active sites in residue HDT has a particular behavior compared with HDT of light fractions and catalyst poisoning in other processes. The shape of the catalyst pores in residue HDT changes to a bottle ink as metals and coke depositions occur (Ancheyta et al. 2005). Moreover, residue conversion is reduced as the covering of active sites progresses, being the reduction of effective diffusivity due to pore plugging what contributes considerably to conversion decay. Effective diffusivity in conventional equations is a function of tortuosity, porosity, and constriction factor. In the proposed model, Eqs. (2.16) and (2.98) describe the effect of vanadium deposits on effective diffusivity of sulfur and vanadium compounds, which was accurate due to the impossibility to model the specific effect of bottle ink shape of the catalyst pores for the calculation of effective diffusivity.

- Density of vanadium deposits. The density of vanadium deposited on catalyst was calculated with Eq. (2.96) with maximum vanadium volume and mass capacities of 0.261 m^3/m^3 and 0.314 kg/kg, respectively. The calculated density of vanadium deposits was 2001.77 kg/m^3.
- Density of coke deposits. Coke density in reforming catalyst has been reported to be 966 kg/m^3 (Baghalha et al. 2010). The density of bitumen sample with 82% carbon was calculated to be 1050 kg/m^3 (Richardson et al. 1996). Density ranges for different types of coals were from 1290 kg/m^3 to 1470 kg/m^3 (Flores 2014). Based on these references, for this work, coke density was assumed to be 990 kg/m^3. In the literature model (Kodama et al. 1980), neither vanadium density deposits nor coke density was reported.
- Solid phase fraction volume. In the proposed model, it was calculated through Eqs. (2.86) and (2.87), while literature model (Kodama et al. 1980) does not report this value.
- Method of optimization. In the literature model (Kodama et al. 1980), complex method was used to estimate reaction rate coefficients for HDS and HDV, least squares to estimate the Arrhenius parameters, while coke deposition rate coefficients were estimated by trial and error with α_V, α_c, β_V, β_c, and γ_V. For the proposed model, sequential quadratic programming (SQP) and Powell methods were used to determine reaction rate coefficient by minimization of the following objective function. The use of more accurate optimization methods and different statistical approaches for parameter estimation is a new aspect added to the literature model (Kodama et al. 1980).

$$SSQQ = \sum_{j=1}^{N} \frac{1}{cov_{i,j}} \left(y_{i,j\,\mathrm{exp}} - y_{i,j\mathrm{calc}} \right)^2 \tag{2.101}$$

where i is sulfur or vanadium, j is the number of experimental data at different TOS, and y is the concentration of heteroatom in products or on catalyst surface.

The optimization methods used in the proposed model are much more accurate than those used in the literature model (Kodama et al. 1980), this latter being developed more than 40 years ago.

A sensitivity analysis on the estimated parameters was carried out to ensure their optimal values, and that the objective function corresponds to the global minimum and not local minima (Félix et al. 2019). This type of analysis was not performed in the literature model (Kodama et al. 1980).

2.5.1.4 Procedure for Parameter Estimation

Parameter estimation for the proposed model was done in three steps to obtain better agreement with experimental data. In the first step, the reaction rate coefficients for HDS and HDV are estimated for the middle operation temperature. Initial values were taken from the literature (Kodama et al. 1980). Then, preexponential factors and activation energies are calculated for the SOR condition (300 hours TOS). The other parameters are kept constant in value of 1 and for α_V, α_c, β_V, β_c, γ_V, and ψ_{max} the value was 0.4. The reaction rate coefficients for coke deposition and desorption are estimated based on the content of coke on catalyst at the middle temperature. A simulated curve approaching to experimental data can be observed after this step. The optimization method used to estimate preexponential factor and activation energy for coke deposition and desorption is SQP and Powell for reaction rate coefficients.

The second step of parameter estimation consists of re-estimating α_V, α_c, β_V, β_c, γ_V, δ_V, and ψ_{max} by SQP. This is a key step for having a good agreement with experimental data. So that, the typical "S" shape of the curve can be reproduced. The value of the objective function is drastically reduced since these parameters describe the effect of the deposition of impurities on catalytic deactivation during TOS. For this step, all the experimental values were used.

The last step of parameter estimation consists of carrying out the minimization of the objective function by SQP method changing all parameter values. Parameters in this point considerably changed from the initial guesses, indicating that SQP method only approaches to the optimum set of parameter values. Then, Powell method is used to continue the optimization parameter values. The curve does not change substantially in this step, but minimization by Powell method yields a better set of parameter values. A rectification by Levenberg–Marquardt is done at the end of the parameter estimation. Table 2.4 summarizes the values of parameters in each stage of

Table 2.4 Values of parameters (developed model).

	Values in each step of estimation				
Symbol	**Initial**	**First**	**Second**	**Third**	**Unit**
k_S	1.512	1.212	1.212	1.156	kg/m^{3}h
E_A	20,000	23,679	23,679	23,127.45	cal/mol
k_V	238.0	225.42	225.42	220.98	kg/m^{3}h
E_A	25,000	29,784.2	29,784.2	30,989.38	cal/mol
k_C	1×10^{-9}	0.98×10^{-7}	0.98×10^{-7}	1.04×10^{-8}	kg/m^{3}h
E_A	10,000	11,058.2	11,058.2	11,083.01	cal/mol
k_{DC}	1×10^{-9}	3.89×10^{-8}	3.89×10^{-8}	3.97×10^{-8}	kg/m^{3}h
E_A	10,000	7723.8	7723.8	7612.83	cal/mol
α_V	1.0	1.0	1.8405	1.884	Dimensionless
α_c	1.0	1.0	1.7632	1.7759	Dimensionless
β_V	1.0	1.0	1.7379	1.819	Dimensionless
β_c	1.0	1.0	1.7753	1.85	Dimensionless
δ_V	1.0	1.0	0.1870	1.0697	Dimensionless
γ_V	1.0	1.0	0.6038	0.5768	Dimensionless
ψ_{max}	0.4	0.4	0.3424	0.3248	m_V^3/m_{cat}^3
D_{S0}	2.10×10^{-6}	2.10×10^{-6}	2.10×10^{-6}	8.22×10^{-7}	m^2/h
D_{V0}	1.57×10^{-6}	1.57×10^{-6}	1.57×10^{-6}	7.22×10^{-7}	m^2/h

parameter estimation. To be sure about the optimal parameter values, a sensitivity analysis method reported in literature (Félix et al. 2019) is used.

2.5.2 Results and Discussion

2.5.2.1 Profiles of Sulfur and Vanadium Concentration in Products

The concentration profiles of sulfur and vanadium in products along TOS at 385 °C, 400 °C, and 415 °C obtained with the proposed model are shown in Figure 2.14. The expected trend of profiles with respect to operation temperature is observed, that is the higher the temperature, the higher the removal of vanadium and sulfur. The first stage (SOR) of catalyst deactivation can be observed until approximately 500 hours of TOS, the second stage (MOR) until 1100 hours of TOS, and the beginning of the third stage (EOR) in the few available last points after MOR. The proposed model is able to reproduce the "S" shape profile of sulfur and vanadium concentration in products at constant temperature, which means that the prediction of the first, second, and the start of the third stages of catalyst deactivation is accurate with good agreement with experimental data.

A comparison of predictions with both proposed and literature models (Kodama et al. 1980) is shown in Figure 2.15 for the concentration of sulfur and vanadium profiles at 400 °C and

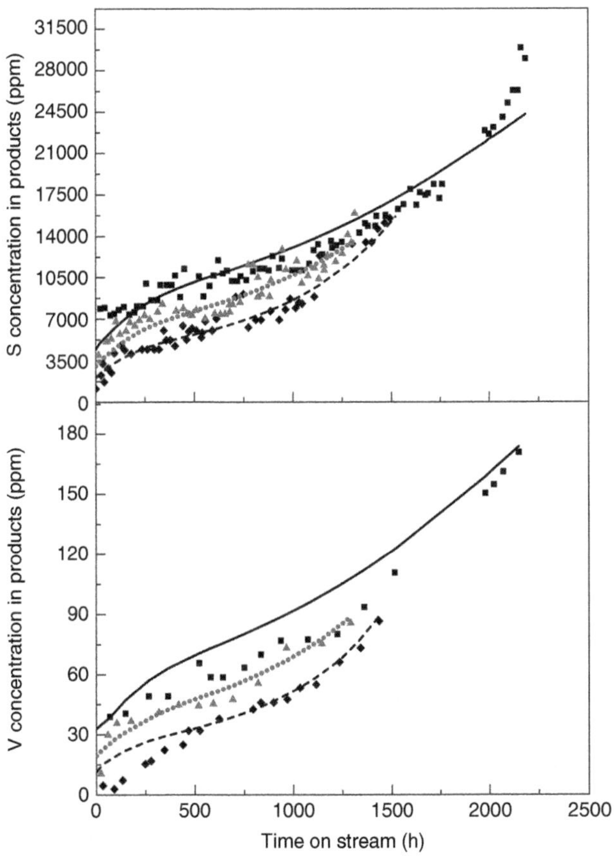

Figure 2.14 Profiles of vanadium and sulfur content in products vs TOS. Calculated with the proposed model: (—) 385 °C, (···) 400 °C, and (---) 415 °C. Experimental data (Kodama et al. 1980): (■) 385 °C, (▲) 400 °C, and (♦) 415 °C.

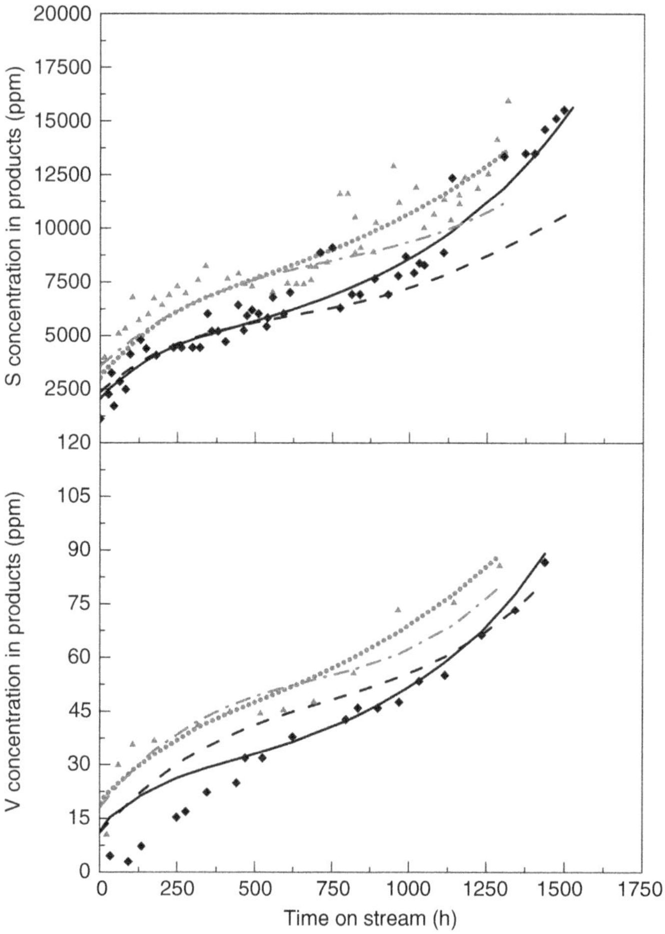

Figure 2.15 Comparison of profiles of vanadium and sulfur content in products vs TOS. Calculated with the proposed model: (···) 400 °C, and (—) 415 °C. Calculated with the literature model (Kodama et al. 1980): (-·-) 400 °C, and (---) 415°C. Experimental data (Adapted from Kodama et al. 1980): (▲) 400 °C, and (♦) 415 °C.

415 °C. The curve corresponding to the first stage of catalyst deactivation for sulfur profiles is the same for both models till 500 hours of TOS. Calculated and experimental vanadium concentrations at short TOS (500 hours) show higher deviations at 415 °C as compared with 385 °C and 400 °C, although the deviations are lower than the literature model (Kodama et al. 1980). For TOS longer than 500 hours, the literature model (Kodama et al. 1980) tends to underpredict the experimental data, which is more notorious for sulfur than for vanadium concentration. The proposed model corrects all these deviations and matches with more accuracy with experimental data.

2.5.2.2 Comparison of Predictions with Literature and Proposed Model

The differences of predictions with the literature (Kodama et al. 1980) and proposed models can be clearly seen in Figure 2.15 for sulfur and vanadium concentrations, respectively. Both models overpredict the smallest values of concentrations corresponding to short TOS (SOR conditions), particularly for vanadium, although the overprediction is more notorious for the literature model (Kodama et al. 1980). For longer TOS, that is, MOR conditions, both models perform similarly.

However, for EOR conditions, the literature model (Kodama et al. 1980) drastically fails in the predictions giving very low concentrations values of sulfur, while the proposed model reproduced the experimental values with high accuracy. The calculated coefficient of determination for proposed model was equal to 0.9119, which confirms a good description of experimental data. The coefficient of determination was also determined for the literature model (Kodama et al. 1980) resulting in a value of 0.879. In other words, the proposed model gives a better description of the experimental data especially at higher temperature of operation and offers the advantage of simplicity. Statistical analysis was carried out for the proposed and literature models (Kodama et al. 1980). Residual balance for the proposed model was 10; meanwhile for the literature model (Kodama et al. 1980), it was −36, which indicates that the proposed model profiles have better distribution of predictions without tendency to over- or sub-estimation of experimental data, which is not the case for the literature model (Kodama et al. 1980). Relative standard error for the proposed model was 0.4471 and for the literature model was 0.4634 (Kodama et al. 1980).

2.5.2.3 Profiles of Coke and Vanadium on Catalyst

Figure 2.16 shows a comparison of predicted and experimental profiles of vanadium and coke deposits along the reactor length. It is observed that vanadium deposits on catalyst decreases from the inlet to the outlet of the reactor. The same behavior is observed for the vanadium concentration profile in liquid phase. In other words, the higher the concentration of vanadium in liquid phase (at reactor inlet), the more abundant the vanadium deposits on catalyst. Even the predictions of experimental data with the proposed model have some deviations, and the literature model (Kodama et al. 1980) also presented high deviations. In spite of these differences, predictions with the proposed model follow the tendencies of coke and vanadium contents along reactor axial coordinate. One aspect that should be taken into consideration is that an inconsistency was found in the

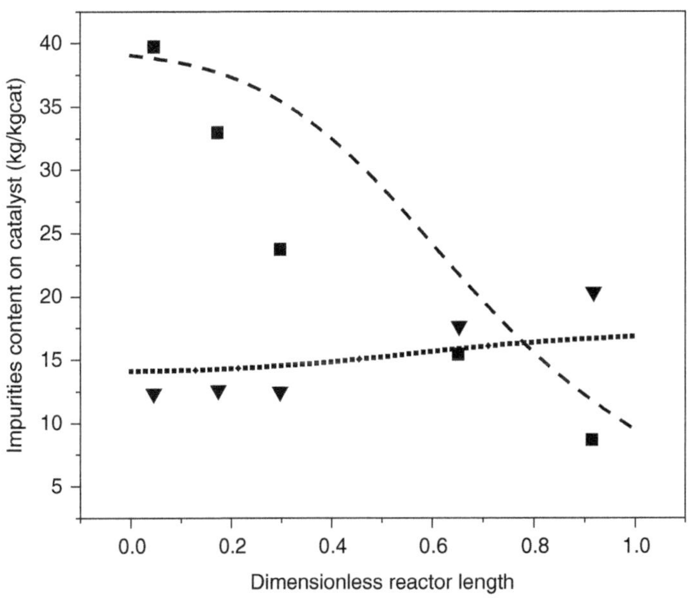

Figure 2.16 Deposits along the axial length of catalyst bed at 400 °C and 1360 h. (---) Calculated content of V, (···) calculated content of coke, (■) experimental content of vanadium, (▼) experimental content of coke (Adapted from Kodama et al. 1980).

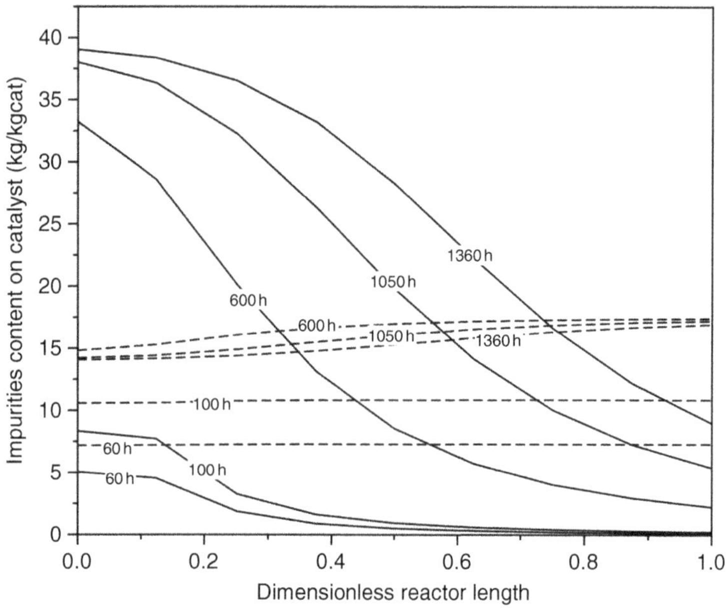

Figure 2.17 Evolution of deposit profiles along TOS. Calculated content of V (—), calculated content of coke (---) at different TOS values.

calculated profiles by authors (Kodama et al. (1980), regarding the profile of vanadium deposits on catalyst, since the maximum quantity of vanadium deposits on catalyst does not agree with the initial calculated profile. As mentioned in Section 5.3, 34.1% is the highest value of the vanadium deposits on catalyst profile, but the reported value is 31.4%. When using this value, an error is obtained in the Thiele modulus calculation (Eq. 2.91), causing the vanadium diffusivity to be negative at the beginning of the profile of vanadium deposits on catalyst.

As shown in Figure 2.17, the concentration of coke on catalyst at SOR depicts a linear profile due to the more uniform distribution of catalyst active sites along reactor axial length. As vanadium deposits occupy catalyst active sites for long TOS, the coke deposits profile tends to increase. This is because near the reactor input catalyst active sites are occupied by more vanadium deposits and more catalyst active sites are free near the exit of the reactor. As remanent surface area of catalyst active sites is reduced as the quantity of deposits increases on catalyst surface, catalyst deactivation is dependent on the poisoning of active sites. Meanwhile, pore plugging metals deposits reduce diffusivity of components to internal remanent active sites. External hindrances are not included in the equations of the proposed model. The increased coke content on catalyst may be also due to the low hydrogen availability at the bottom of the reactor caused by hydrogen consumption. In commercial reactors, high temperatures at the exit of reactor are another reason for high coke content of catalyst; however, in this case simulated reactor is isothermal. The earliest profile of coke deposits on catalyst is uniform since vanadium deposition at SOR is minimal along the catalyst bed. Coke content on catalyst increases until approximately 600 hours of TOS. This TOS value is near the end of the first stage of catalyst deactivation as shown in Figure 2.17. It would be expected that coke content profile on catalyst at 1360 hours had higher values with respect to those at shorter TOS, which is not the case. This is because the coke deposit reaches a maximum at the end of SOR. Then, coke deposition reversibility uncovers active sites surface area and metals are deposited instead. On the contrary, vanadium deposits increase along the catalyst bed and desorption is not occurring.

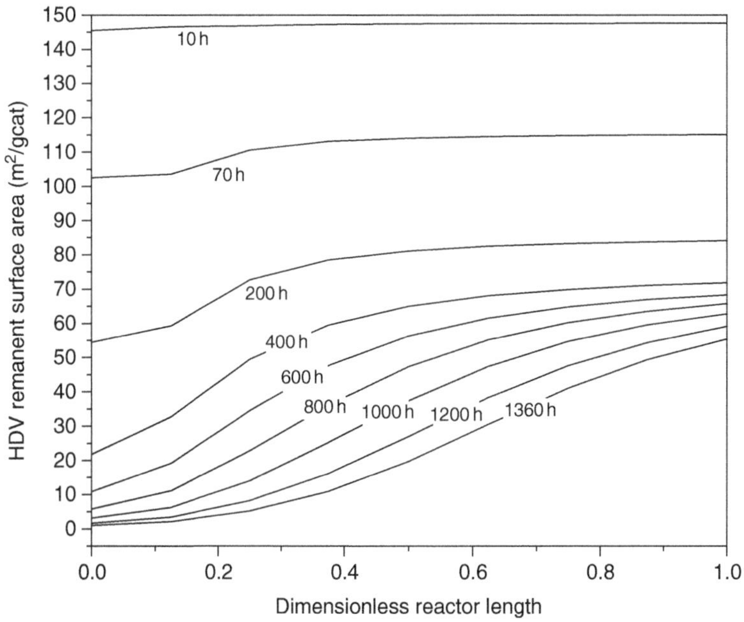

Figure 2.18 Evolution of surface area of HDV active sites along TOS.

The covering of active sites by vanadium deposits reduces the surface area where HDS, HDV, and coking reactions take place. Reduction of surface area by coking is diminished as the content of vanadium deposits increases. The quantity of coke deposits not only is reduced due to its reversibility, but vanadium deposition on catalyst surface reduces the quantity of coke on catalyst until the end of operation. Something important to clarify is that vanadium compounds do not remove coke from catalyst. Figure 2.18 shows the evolution of remanent surface area of HDV active sites along the catalyst bed. As expected, vanadium deposition is higher near the inlet of the reactor and remanent surface area of HDV active sites is lower at the inlet of the reactor. The loss of active surface area during TOS has significant effect on the concentration of sulfur and vanadium in products at SOR and part of MOR. The last stage of catalyst deactivation (EOR) regarding the concentration of sulfur and vanadium in products along TOS is dominated by Eqs. (2.98) and (2.99) due to the considerable fulfilling of catalyst pores, which provokes the nearly zero value of diffusivity and the consequent low value of effectiveness factor.

The proposed model can describe the behavior of catalytic deactivation in heavy oil HDT. It reproduces well the first, second, and the start of the third stages of catalyst deactivation in vacuum residue HDT, that is, SOR, MOR, and EOR conditions.

2.5.2.4 Final Remarks

The proposed model could be described as of intermediate complexity. As shown in Table 2.5, it considers oil and catalyst properties in the model equations, which give the opportunity to simulate atmospheric or vacuum residues. As seen in Table 2.4, final values of HDS and HDV reaction rate coefficients are similar to their initial values, since it is convenient to maintain the intercept near the first experimental data points. Coking formation and desorption reaction rate coefficients change according to the experimental quantity of coke deposits on catalyst. The values of parameters, α_V, α_c, β_V, β_c, γ_V, δ_V, and ψ_{max}, are determined by SQP method according to the performance of a specific HDT catalyst for a given feedstock, that is, these parameter values will be unique for

Table 2.5 Values of input data (developed model).

Symbol	Value	Unit
ρ_{cat}	660	kg/m^3
ρ_{coke}	990	kg/m^3
$\rho_{vanadium}$	2001.77	kg/m^3
$\rho_{oil@T_{avg}}$	766.3	kg/m^3
ε_S	0.511	m^3_{cat}/m^3_R
C_{S0}	0.0367	kg_S/kg_T
C_{V0}	0.00027	kg_S/kg_T
r_{cat}	1.34	m
P_V	0.601	m^3_{pore}/g_{cat}
Q_V	0	$kg_V/kg_{cat}\%$
Q_c	0	$kg_C/kg_{cat}\%$
A_{rs}	155	m^2/g_{cat}
A_{rC}	155	m^2/g_{cat}
A_{rV}	155	m^2/g_{cat}

each combination of catalyst and feedstock. The proposed model can represent more accurately the experimental data at intermediate and high temperatures, which are more common when processing heavy residues. Figures 2.15 and 2.16 help demonstrate the capability of the proposed model to predict the TOS in which coking formation reaches equilibrium and coke desorption begins, as well as dynamic coke and vanadium deposits on the catalyst. Figure 2.18 shows the evolution of remanent surface area for HDV at 10 hours of TOS, which depicts a horizontal line that indicates that coke deposition is predominant and vanadium deposition is small. However, for longer TOS (MOR and EOR), the profile of remanent surface area achieves a hybrid shape, which is a result of both coke and vanadium covering. In general, the proposed model offers a simplified manner to describe residue HDT and allows for determining the evolution of the main parameters, that is, heteroatom concentrations in products, impurity contents on catalyst, and catalyst deactivation. From the results given in Table 2.4, it is observed that at SOR, coke deposition has the highest contribution to the catalyst deactivation. From MOR to EOR, vanadium deposition dominates catalyst deactivation, thus decreasing significantly the rate of HDS and HDV reactions.

The experimental data show a rapid loss of catalyst activity, that is, HDS and HDV reduce with TOS. This behavior is typical in laboratory experiments that are conducted at constant reaction temperature, which is necessary to develop kinetic and deactivation models. In commercial application, this temperature is constantly increased to compensate for catalyst deactivation, so that the quality of the product is maintained constant during TOS. The most important aspect for commercial application of TBR for HDT of petroleum residues is the metal retention capacity of the catalyst, and its life to guarantee operation cycles that can give a positive cost benefit.

Even the number of parameters is large; we used a three-step parameter estimation procedure and verification of their optimal values was done by sensitivity analysis that assures proper calculations. Then the set of estimated parameter values that fits the model to experimental data is unique.

The proposed model gives a better description of experimental data than literature model (Kodama et al. 1980). One important characteristic of the proposed model compared with others is the simpler manner to calculate the accumulation of impurities on catalyst to describe the catalyst deactivation.

According to the results, it is deduced that concentrations of impurities in products and on catalyst at temperatures from 385 °C to 415 °C can be predicted for the same reactor, hydrotreated feedstock, and used catalyst. This can be applied to find the conditions in which vanadium and coke deposition are lower to minimize catalyst deactivation. The proposed model can be adapted to describe experimental data of other residue HDT reactors and reaction conditions. The proposed model can predict the performance of residue HDT reactor with the same feedstock and catalyst at operation temperatures between 385 °C and 415 °C. It should be commented that experimental data of the three stages of catalyst deactivation are necessary to tune the parameters to be able to accurately predict the whole run length, typically observed in commercial operation.

Using the information of short TOS (SOR) for the prediction of EOR conditions is risky. It is necessary to have some experimental data corresponding to EOR for model validation. Although the available experimental data used to develop the proposed model was obtained at SOR, MOR, and the beginning of EOR conditions, and not during the complete EOR, in which catalyst has been totally deactivated, the predicted values are in good agreement with the typical "S" shape profiles of impurities content during TOS. This means that the proposed model is validated at these conditions, and it requires additional experimental values to evaluate its application for long-term operations.

2.5.3 Usefulness of the Model

The proposed model represents a useful tool to make predictions for the HDT of petroleum residues. The model parameters were estimated using a sequential method that assure their optimal values and accurately represent the typical concentration profiles during TOS. It means that the model can predict the effect of temperature and TOS on catalyst deactivation by proper calculation of metals and coke deposition. Changes in feed properties (sulfur and vanadium content) can also be evaluated so that different feed's nature can be correlated with catalyst deactivation. In summary, with some properly designed experiments for new catalyst and feed, the proposed model can be useful for the following:

- Screening of HDT catalysts for heavy oil feeds. Typical experiments for HDT catalyst screening only consider data at SOR conditions, that is, short TOS. The performance of an HDT catalyst can be better than others at SOR, but deficient at long TOS. This situation cannot be anticipated at SOR operation. The model parameters can be tuned with experimental data for new catalyst and feed, so that it can predict the long-term operation of the catalyst.
- Deactivation of catalyst. With short-term experiments, only early catalyst deactivation can be observed. Appropriate evaluation of loss of catalyst activity needs to be carried out at long TOS. The model can predict long-term catalyst deactivation and the extent of metals and coke deposition, as well as losses in specific surface area. This is of great importance to determine the catalyst life.
- Selection of operating conditions. Having the concentration profiles of impurities in the products as function of TOS is useful to define suitable operating conditions to achieve the desired quality of the products. Increases in reactor temperature to mitigate catalyst deactivation can be properly predicted by the model. In addition, proper selection of SOR conditions can be done to optimize the catalyst utilization and life.
- Design of new reactors. Although at present some reactor designs are still done based on experiences collected during decades, and correlations derived from huge number of experimental data, the use of reactor model with catalyst deactivation is also important for design, simulation, and

optimization of new reactors. For HDT of residue, it is well-known that more than one catalyst is needed, for example, HDM, hydrocracking, and HDS catalysts. The amount of each catalyst and operating conditions in a series of HDT reactor depends on the feed properties and desired product quality. The proposed model can be used to help define the reactor configuration, since it properly describes the concentration profiles along the reactor length.

- Economy. The use of reactor models during the development of new processes or optimization of actual processes is a common task for technology developers. The main advantage of a model reactor for HDT of residue is that it substantially reduces the experimental work (pilot plant testing, product characterization, analysis of catalyst) and time; therefore, important savings of time and money are obtained.

2.5.4 Conclusion

The following conclusions can be pointed out from the results obtained with the proposed model:

Catalyst deactivation is modeled through vanadium and coke deposition as a function of TOS. The proposed model assumes that catalyst deactivation is caused by poisoning mechanism through deposition of vanadium and coke on catalyst active sites affecting HDS and HDV conversions. Diffusivity of sulfur and vanadium to internal catalyst active sites is also limited by vanadium and coke deposition due to pore plugging, that is, internal limitation of diffusivity. The proposed model predicts the performance of residue HDT reactor with the same feedstock and catalyst at operation temperatures between 385 °C and 415 °C. It describes the evolution of vanadium and sulfur concentration in products and deposits on catalyst surface in the three stages of catalyst deactivation. For estimation of model parameters, it is highly recommended to have experimental data of the three stages of catalyst deactivation. The predictions with the proposed model are in good agreement with experimental data. By using more accurate methods and simplifying the calculation of some parameters, the proposed model can describe the profiles of heteroatoms content in products and on catalyst. Detailed sequence of parameter estimation and input data were described to help researchers with the use of the proposed model.

2.6 Application of the Deactivation Model for Hydrotreating of Heavy Crude Oil in Bench-Scale Reactor

Experimental data of a heavy crude oil HDT carried out in a bench-scale reactor at 380 °C, 98 bar, 0.5 $hours^{-1}$, and 890 m^3/m^3 H_2/oil ratio for 500 hours were modeled by a quasi-dynamic one-dimensional pseudohomogeneous TBR model.

2.6.1 Properties of Heavy Oil

The characterization of the heavy crude oil was performed according to ASTM standard methods and is reported in Table 2.6.

2.6.2 Properties of the Catalyst

A tetra lobular catalyst was loaded to the TBR of the bench-scale pilot plant. A commercial HDT catalyst was used for experiments, which is composed of NiMo supported on aluminum oxide ($NiMo/Al_2O_3$). The main chemical and physical properties of the catalyst are presented in Table 2.7. The catalyst pore distribution is concentrated in the ranges of 100–200 Å (49.44 vol.%)

Table 2.6 Properties of heavy crude oil.

Property	Value
Specific weight @20/4 °C (ASTM D-1298)	0.9810
API gravity	12.74
Sulfur content (wt.%) (ASTM D-287)	4.997
Asphaltenes content (wt.%) (ASTM D-4294)	21.8
Viscosity at 37.8 °C (cSt) (ASTM D-445)	11,549
Elemental analysis (ASTM D-7042)	
C (wt.%)	83.28
H (wt.%)	10.91
Metals (by atomic absorption)	
Ni (ppm)	85
V (ppm)	456

Table 2.7 Properties of catalyst and dimensions of reactor.

Property	Value
Nominal size (in)	1/18
Average pore diameter (Å)	172.6
Particle diameter (mm)	9.95
Specific surface area (m^2/g)	197.2
Average pore volume (cm^3/g)	0.85
Average pore radio (Å)	86.3
Bulk density (kg/m^3)	533.9
Chemical composition (wt.%)	
NiO/Ni	0.73/0.58
MoO_3/Mo	3.27/2.18
Pore size distribution (vol.%)	
<50 Å	2.32
50–100 Å	14.86
100–200 Å	49.44
200–500 Å	30.34
500–1000 Å	2.26
>1000 Å	0.78
Reactor bed length (cm)	44.5
Reactor diameter (cm)	2.54

and 200–500 Å (30.34 vol.%), which makes it suitable for processing heavy oils with high content of metals and asphaltenes for long time of operation due to more available volume to accumulate deposits without causing severe deactivation.

2.6.3 Bench-Scale Reactor

HDT experiments were carried out in an isothermal TBR. A general scheme of both the bench-scale plant and the reactor are shown in Figures 2.19 and 2.20, respectively. The plant has four sections:

1) Feed section. A stainless-steel container within a heating system and temperature controller over a mass quantifier scale, a controlled velocity gear pump, a feeding hydrogen module, and a mass flow meter.
2) Reaction section. A stainless-steel tubular reactor heated by three electric resistances located along the reactor length. The temperature profile is controlled through three axial thermocouples installed at the bottom, middle, and top of the catalyst bed.
3) Separation section. Two high pressure separators with control temperature system and two low pressure separators, which condensate liquefiable products, are further added to the hydrotreated product stream. A venting line for safe draining of gases also added.
4) Measurement and analysis of gases section. A flow meter to register the flow of reaction gases at the outlet of the plant and an online gas chromatograph.

2.6.4 Catalyst Activation

The catalyst was activated by the following three stages:

1) Drying. Water in catalyst was eliminated under nitrogen gas at 120 °C. The common content of water on alumina-supported catalysts is from 1 to 3 wt.% due to its high hygroscopy.
2) Soaking. Typically, naphtha, kerosene, or light straight-run gas oil (SRGO) is used to soak the external surface area of the catalyst particles in the bed as much as possible. In this case, SRGO was used. The presence of non-wetted areas on catalyst bed reduces the effectiveness of the catalyst.

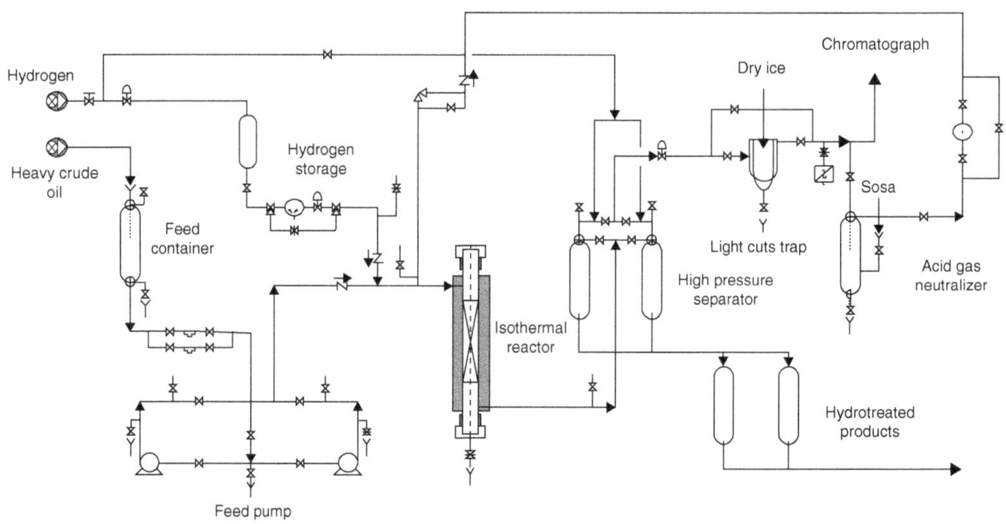

Figure 2.19 Heavy crude oil HDT bench-scale reactor system.

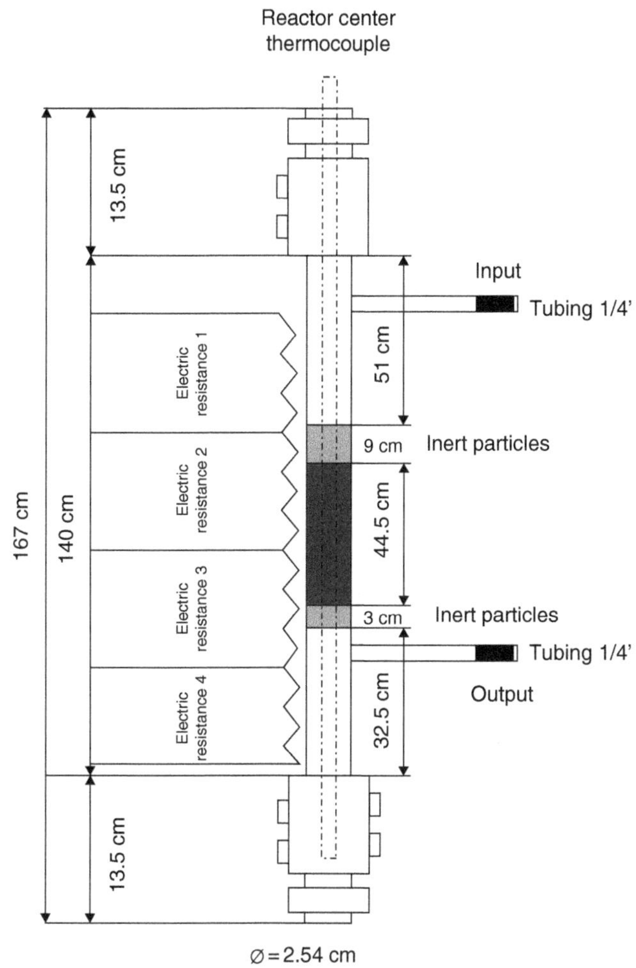

Figure 2.20 Trickle-bed reactor for heavy crude oil HDT.

3) Sulfiding. A sulfiding stream consisting of 98.7 vol.% SRGO and 1.3 vol.% dimethyl disulfide (DMDS) was employed to convert metal oxides into metal sulfides.

The conditions for drying, soaking, and sulfiding steps are detailed in Table 2.8.

2.6.5 Operating Conditions

Experiments were carried out at 380 °C, 98 bar, 0.5 hours^{-1} LHSV and 890 m^3/m^3 H_2/oil ratio for 480 hours. The volume of catalyst bed was 200 mL. The diameter and length of the reactor are 2.54 and 44.5 cm, respectively. Product samples were recovered every 24 hours.

2.6.6 Characterization Methods

For sulfur content determination, a model SLFA 2100 Horiba X ray fluorescence equipment was used according to ASTM D-4294. The quantity of asphaltenes was determined according to ASTM D3279 method. The elemental analysis of feed and products streams was determined based on

Table 2.8 Conditions of catalyst activation.

Variables	Drying	Soaking	Activation
Pressure (bar)	1.013	1.013	27.45
Temperature (°C)	120	150	260, 320
Hydrogen flow (m^3/h)	0.380	0.070	0.1535
Diesel flow (mL/h)	–	500	–
SRGO + DMDS (mL/h)	–	–	200
Time (h)	2	2	3 (@260 °C) 10 (@320 °C)

ASTM D5291 method with a 2400 Series II model Perkin Elmer elemental analyzer. The determination of textural properties through nitrogen physisorption was based on BET method from adsorption isotherm in Nova 2000 model Quantachrome equipment. Obtained values of properties were input data for modeling.

2.6.7 Parameter Estimation

The accuracy of the model has been demonstrated with literature data for HDT of residue. The input data of the model are shown in Table 2.9, which consider the properties of the oil and the catalyst. Parameter estimation was carried out in three steps. Table 2.10 shows the variation of parameters values in each step. The model parameters were initialized with similar order of magnitude of previous parameter estimation (Jurado and Ancheyta 2022). Initial values of reaction rate coefficients for HDS and HDM were obtained through considering the experimental data in the first

Table 2.9 Values of input data for (application of the model).

Symbol	Value	Unit
ρ_{cat}	553	kg/m^3
ρ_{coke}	990	kg/m^3
$\rho_{vanadium}$	2001.77	kg/m^3
$\rho_{oil@T_{avg}}$	831.3	kg/m^3
ε_S	0.484	m^3_{cat}/m^3_R
C_{S0}	0.0499	kg_S/kg_T
C_{V0}	541	ppm
d_p	9.95	m
P_V	0.85	cm^3_{pore}/g_{cat}
Q_V	0	$kg_V/kg_{cat}\%$
Q_c	0	$kg_C/kg_{cat}\%$
A_{rS}	197	m^2/g_{cat}
A_{rC}	197	m^2/g_{cat}
A_{rV}	197	m^2/g_{cat}

Table 2.10 Parameter values (application of the model).

Symbol	Values in each step of estimation				Unit
	Initial	First	Second	Third	
k_S	1.512	0.3021	0.3021	0.3348	kg/m^{3}h
k_V	100.0	82.321	82.321	80.81	kg/m^{3}h
k_C	1×10^{-8}	2.98×10^{-8}	2.98×10^{-8}	1.041×10^{-8}	kg/m^{3}h
k_{DC}	1×10^{-8}	9.23×10^{-7}	9.23×1010^{-7}	8.034×10^{-7}	kg/m^{3}h
α_M	1.0	1.0	1.385	1.4766	Dimensionless
α_c	1.0	1.0	1.256	1.2640	Dimensionless
β_M	1.0	1.0	0.278	0.2556	Dimensionless
β_c	1.0	1.0	1.463	1.4965	Dimensionless
δ_V	1.0	1.0	1.942	1.9641	Dimensionless
γ_V	1.0	1.0	0.502	0.5265	Dimensionless
ψ_{max}	0.4	0.4	1.638	1.7999	m_V^3/m_{cat}^3
D_{S0}	1.0×10^{-7}	1.0×10^{-7}	1.0×10^{-7}	3.51×10^{-7}	m^2/h
D_{V0}	1.0×10^{-6}	1.0×10^{-6}	1.0×10^{-6}	1.73×10^{-6}	m^2/h

100 hours of operation so that the first step of parameter estimation was to obtain the values of reaction rate coefficients based on the set of experimental points by SQP method. The second step of parameter estimation was to obtain the values of Greek letter parameters through SQP method. In the last step, all the parameters were re-estimated through SQP, Powel, and Levenberg–Marquardt methods.

2.6.8 Results and Discussion

2.6.8.1 Evolution of Sulfur and Metals Concentration in Products

Figure 2.21 clearly shows that the model predicted in an accurate manner the experimental data of sulfur and metals concentrations in products during heavy crude oil HDT. The first shorter TOS (<100 hours) and second longer TOS (>100 hours) stages of catalyst deactivation are also well described with respect to experimental data. The rapid deposition of coke is present in the first 100 hours of TOS, which corresponds to the first stage of catalyst deactivation. Then, the second stage can be observed, which is of more metal deposition dominant. The rate of coke deposition is depleted at the end of the first stage of catalyst deactivation (first curvature). As the amount of coke increases, the desorption of coke increases. Hence, there is a kinetic limited quantity of coke in the catalyst. Once the coke content on catalyst reaches a maximum value, it starts to decrease, and metals deposition governs the behavior of the profiles of sulfur and metals concentrations in the product. The deposition of metals on catalyst is carried out in the remaining time of operation. The horizontal tendency in both sulfur and metals concentration profiles is due to the balance between coke formation and desorption. For the sulfur concentration in products, the trend is different than that for metals due to the different impact of metals deposits on HDS and HDM active sites. It should be commented that the catalyst used is the first catalyst in a series of three HDT catalysts for upgrading heavy oils, so that it is designed for metals removal. That is why its metal retention capacity is uniform (constant HDM), while sulfur removal exhibits an increasing

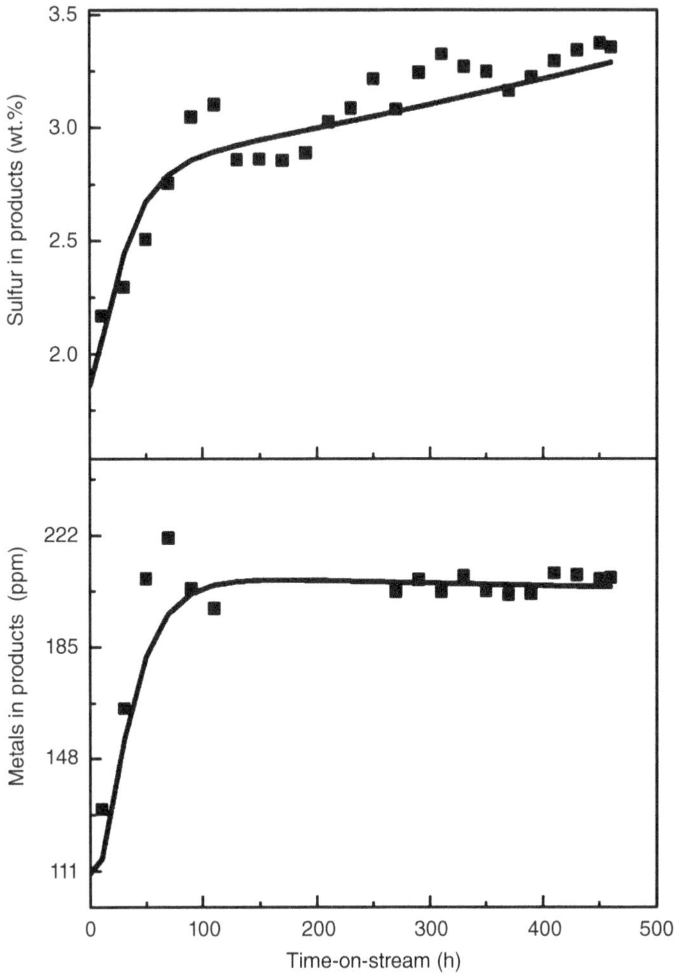

Figure 2.21 Profiles of sulfur and metals content in products vs TOS. (lines) Calculated with the model, (symbols) experimental data.

tendency. The third stage of catalyst deactivation is not observed because the maximum TOS is 480 hours. However, there would be the increasing tendency in both sulfur and metals concentration profiles.

Although typical experimental data of heavy crude oil HDT unclearly indicates the change of regime from coke formation to metals deposits, the understanding of the behavior may be improved with the observation of calculated parameters, for example, impurities content on catalyst and the sulfur and metals concentration in products along TOS.

2.6.8.2 Coke and Metals on Catalyst

The simulated profiles of COC and MOC on catalyst at different TOS along the reactor length can be observed in Figures 2.22 and 2.23, respectively. Coke is formed uniformly along the catalyst bed axial length in the first ten hours of the operation when metals deposition has no notorious effect. The coke content profile along the reactor length increases rapidly during the first 100 hours of TOS

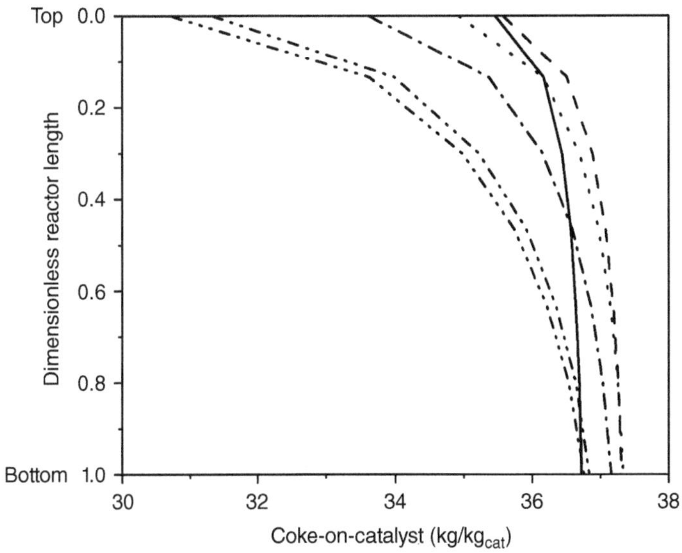

Figure 2.22 Evolution of calculated COC profiles along TOS. 50 hours (—), 100 hours (---), 150 hours (···), 200 hours (—·—), 300 hours (—··—), and 500 hours (—···—).

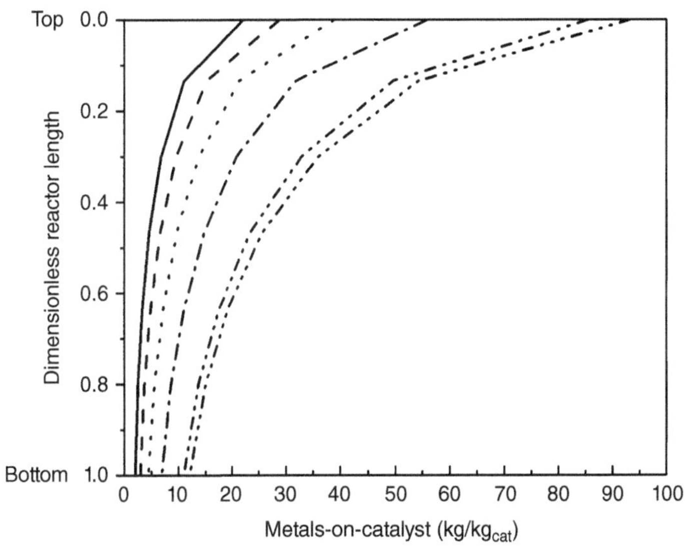

Figure 2.23 Evolution of calculated metals-on-catalyst profiles along TOS. 50 hours (—), 100 hours (---), 150 hours (···), 200 hours (—·—), 300 hours (—··—), and 500 hours (—···—).

until the coke desorption is favored to pursue the chemical equilibrium. The solid line profile indicates the quantity of coke on catalyst bed at 50 hours of TOS. Later at 100 hours (dashed line), the highest values of coke deposits on catalyst are reached. The values of coke deposits are depleted after 100 hours of TOS and do not increase anymore with respect to TOS. A portion of the released catalyst sites by coke desorption may be occupied by metals deposits. This can be observed at the beginning of the second stage of catalyst deactivation in the concentration of sulfur and metals profiles since they coincide with the curve at 100 hours of TOS in Figure 2.21. An increasing tendency in coke profile along the reactor length is because of the effect of metals deposits and the consumption of hydrogen by HDS and HDM reactions. The coke on catalyst increases along the reactor length since near the outlet hydrogen partial pressure is lower due to hydrogen consumption. It is clearly observed that the COC is higher near the 100 hours of TOS than the end of operation. As time passes, the increasing tendency of coke along reactor length is contrasted due to the accumulation of metals deposits majorly near the inlet of the catalyst bed.

The high quantities of both COC and MOC are due to the high pore volume per catalyst particle. Most of metals deposits are at the inlet of the catalyst bed since a high amount of metals in liquid phase is removed in this section. The profile of MOC increases continuously during TOS.

Figure 2.24 shows that the profile of the conversion of HDS decreases continuously as TOS passes because of the effect of COC and MOC. In the first 100 hours of operation, the major reduction of HDS conversion is observed due to the rapid formation of coke. Afterward, the effect of metals deposition can be observed. The decreasing behavior of calculated HDS conversion profile is due to the poisoning of catalytic HDS sites by coke and metals during TOS. At 500 hours, HDS conversion of lower than 5% is obtained in the first 20% of the reactor length due to the limitation by COC and MOC near the inlet of the reactor. The experimental outlet values of HDS conversion are plotted and compared with the outlet value of the calculated profiles at the corresponding TOS.

The highest part of the HDM conversion takes place in the first third part of the reactor length as shown in Figure 2.25. The profile of HDM conversion along the reactor length decreases until 100 hours of TOS. Later, the HDM conversion profile maintains its values since the metals deposit accumulation effect is not considerable on HDM reactions and the depletion of coke formation on catalyst is slowed. This latter observation can also be proved with the low value of β_M. The depletion of

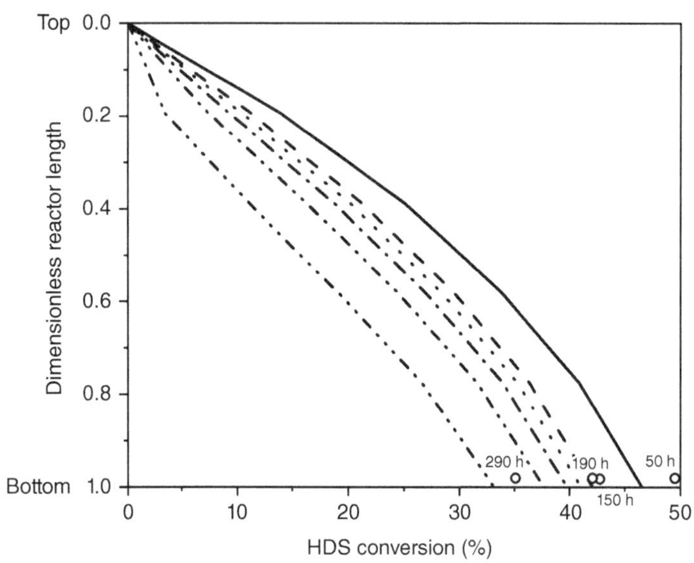

Figure 2.24 Evolution of HDS conversion profiles along TOS. Calculated HDS conversion at 50 hours (—), 100 hours (---), 150 hours (···), 200 hours (—·—), 300 hours (—··—), and 500 hours (—···—) and outlet experimental HDS conversion (O).

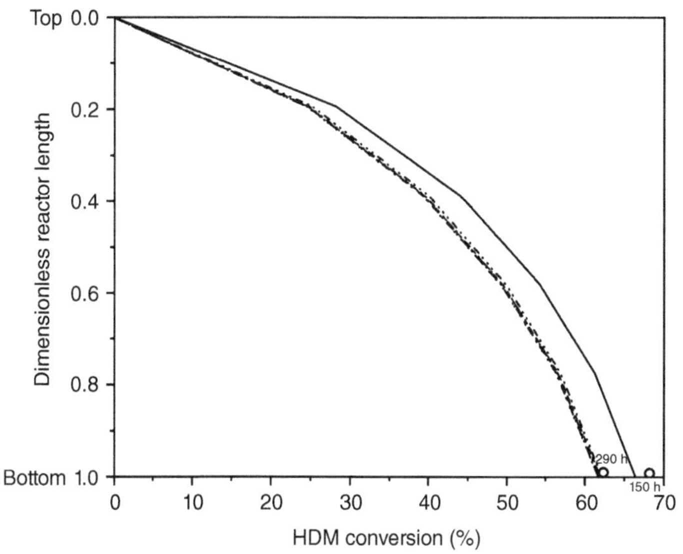

Figure 2.25 Evolution of HDM conversion profiles along TOS. Calculated HDM conversion at 50 hours (—), 100 hours (---), 150 hours (···), 200 hours (—·—), 300 hours (—··—), and 500 hours (—···—) and outlet experimental HDM conversion (○).

the calculated outlet HDM conversion along TOS agrees with the behavior of the experimental values. In the first 20% of reactor length, the highest slope in HDM conversion agrees with the lowest slope of HDS conversion at 500 hours of operation.

2.6.9 Conclusion

Experimental data of the HDT of a heavy crude oil in a bench-scale reactor were generated and modeled at 380 °C, 98 bar, 0.5 $hours^{-1}$ and 890 m^3/m^3 H_2/oil ratio during 500 hours of TOS. The typical behavior of the first and second stages of catalyst deactivation in heavy oils was observed for the profiles of sulfur and metals concentrations in products. The first stage of catalyst deactivation due to coke deposition was accurately described with the deactivation model. The amounts of COC and MOC and their effect on the concentration of sulfur and metals in products along TOS were included in the model. The model shows good agreement with experimental data and offers the advantage of a better understanding of the mechanism of the first and second stages of catalyst deactivation.

Nomenclature

Symbols

A	constant, depending on catalyst, feed stock, and temperature
A_0	pre-exponential factor of reaction rate coefficient, kg/(m^2 h)
A_o	active sites fresh surface area, $m_{area}{}^2/kg_{cat}$
A_{0C}	effective surface area of catalyst under use for coke deposition expressed, m^2/kg
A_{0S}	effective surface area of catalyst under use for desulfurization deposition expressed, m^2/kg

A_{rc} remanent surface area for coke deposition, $m_{area}{}^2/kg_{cat}$
A_{ri} remanent active superficial area for remotion of heteroatom i, $m_{area}{}^2/kg_{cat}$
A_{rs} fresh surface area of HDS active sites, $m_{area}{}^2/kg_{cat}$
A_{rV} hydrodevanadization active sites remanent surface area, $m_{area}{}^2/kg_{cat}$
a constant
a_1 deactivation rate for hydrocracking
a_2 deactivation rate for desulfurization and demetallation
b ratio of desulfurization rate constant of the fresh core to that of the fouled shell
C local reactant concentration, $kmol/m^3$
C_0 carbon content of the catalyst, kg carbon/kg catalyst
C_{AS} surface reactant concentration, $kmol/m^3$
$C_{ash,0}$ initial concentration of metal determined as ash, mass% $(kg/kg_{oil}) \times 100$
C_{coke} equilibrium local coke concentration
$C_{coke,0}$ initial concentration in used oil, mass%, $(kg/kg_{oil}) \times 100$
C_F reactant concentration entering reactor, ppm
C_h concentration of hydrogen, kg/kg
C_m concentration of metals
C_m^{max} maximum concentration of metals
C_{m0} concentration of metals at particle surface
C_{mp} concentration of organometallic compounds in the liquid in the pores, mol/m^3
$C_{p,i}$ product concentration, wt.% or wt-ppt
C_s concentration of sulfur-containing molecules in the liquid in the pores, mol/m^3
C_{s0} concentration of sulfur at particle surface
C_S sulfur concentration in products, kg_S/kg_l
C_{s0} carbon content during first 50 hours of time-on-stream
C_v molar vanadium concentration, inside the pore, $kmol/m^3$
C_V vanadium concentration in products, kg_V/kg_l
C_{vo} molar vanadium concentration, outside the catalyst particle, $kmol/m^3$
D_0 fresh catalyst effective diffusivity
D_A diffusivity in single capillary, m^2/s
D_{m0} effective diffusivity of the organometallic compounds at zero time, m^2/s
D_P particle effective diffusivity, m^2/s
D_z diffusivity of metal-bearing species in the deposited layer, m^2/s
d constant
d_c distance from pellet center for coke, m
d_v distance from pellet center for vanadium, m
d_{pe} equivalent particle diameter, cm
d_R inner diameter of reactor, cm
E_A activation energy for reaction rate coefficient, cal/mol
E_m activation energy for HDM, kcal/mol
E_s activation energy for HDS, kcal/mol
E_v activation energy for HDV, kcal/mol
$f(r)$ pore size distribution function
g constant
K adsorption rate constant, g of catalyst/g of feed

k_1	first-order observed reaction rate constant, $hour^{-1}$
k_0	demetallation rate constant for original active sites, $second^{-1}$
k_{Ai}	apparent rate constant in the initial rapid deactivation period
K_{ads}	adsorption constants of asphaltenes
k_{c1}	intrinsic rate constant for coke formation, kg/m^2 h
k_{c2}	intrinsic rate constant for coke hydrogenation, kg/m^2 h
k_d	deactivation constant, $second^{-1}$
k_{d1}	deactivation constants for hydrocracking
k_{d2}	deactivation constants for desulfurization and demetallation
k_{fm}	rate constant of demetallation, kg/m^2 h
k_{fs}	rate constant of desulfurization, kg/m^2 h
k_i	reaction rate constant
k_m	rate constant of fresh catalyst for demetallation, kg/m^2 h
k_{ms}	rate constant for demetallization in the shell, m^3/kg s
k_s	rate constant of fresh catalyst for desulfurization, kg/m^2 h
k_{SA}	reaction rate constant per unit surface area
k_{ss}	rate constant for desulfurization in the shell, m^3/kg s
k_v	first-order surface reaction rate constant for vanadium removal, m s^{-1}
k_z	demetallation rate constant in the deposited layer, $second^{-1}$
k_n	rate coefficient at η 0, $h^{-1}atm^{-1}$
L	total bed length in plug-flow reactor, cm
$\frac{l}{L}$	fractional bed length in plug-flow reactor
$LHSV$	liquid-hourly space-velocity, $hour^{-1}$
M	molecular weight ratio, metal sulfides to metal
M_0	maximum amount of metal deposits when the pore is completely filled, kg of metal/kg of catalyst
M_d	present amount of metal deposits, kg of metal/kg of catalyst
M_m	average molar mass of metals
M_{MS}	molecular weight of metal sulfide deposit
M_p	molecular weight of metal deposit, kg/kmol
MO	moles of reactant per unit volume of catalyst
MOC	concentration of metals-on-catalyst, wt.%
M_w	molecular weight of residue
m	deactivation order
N	Avogadro's number, molecules/Kmol
n	correlation coefficient or reaction order
n_p	number of pores per unit surface area, m/s
P	reactor pressure, kPa
P_v	pore volume of fresh catalyst, m^3/kg
q_c	mass quantity of coke by volumetric unity, $kg_{coke}/m_{cat}{}^3$
Q_i	dimensionless mass quantity of heteroatom i on catalyst, kg_i/kg_{cat}
Q	quantity of contaminant on catalyst, wt.%
Q_0	concentration of deposited metals corresponding to a monolayer, kmol/kg
Q_c	dimensionless quantity of coke on catalyst
r	mean random pore radius, m

r_0	equivalent radius of catalyst pellet, m
r_A	actual pore radius, m
r_a	average pore radius
R_{c2}	slow coke deposition rate, h^{-1}
R_g	ideal gases constant, cal/mol K
r_{HDM}	rate of hydrodemetallization reaction, mol cm^{-3} s^{-1}
r_i	reaction rate of remotion of compound of heteroatom i from process stream, kg_i/m^3 h
r_M	radius of residue molecule, m
r_m	radius of molecule
r_{min}	minimum pore radius
r_{MS}	metal sulfides deposition rate, h^{-1}
r_p	radial position in catalyst pellet, m
r_{p0}	pore radius of fresh catalyst, m
R_{PD}	remaining pore diameter
r_{pm}	catalyst pore radius which neglects coke thickness, m
r_s	rate of desulfurization, kg/m^3 h
r_v	rate of vanadium removal, kg/m^3 h
S	specific surface area of catalyst
t	any time-on-stream (>50 h)
t_b	time in which it is presented φ_1 and φ_2
t_d	dimensionless time
t_m	maximal life of pore
t_∞	catalyst life
T	temperature, K
u_l	mass flux velocity, kg_{liq}/m^2 s
U_{MOC}	unit conversion factor of Eq. (2.61)
V	a dimensionless time-measuring variable
V_{dep}	deposition volume per mole vanadium, m^3 kmol
W	local mass of adsorbed poison per unit poison-free area, kg/m^2
w	cumulative feed-to-catalyst ratio, a pseudo-time coordinate, g of catalyst/g of feed
W_f	fresh catalyst weight, g
W_{MS}	metal sulfides weight, g
W_s	value of W corresponding to complete active site poisoning
X	distance from particle center, m
X_0	slab half-thickness, m
X_c	mass fraction of carbon in the residue
X_{MOC}	fractional concentration of metals-on-catalyst
x_s	fraction of surface area not covered with coke
y_a	constant
y_b	constant
y_m	constant
y_n	constant
z	axial reactor coordinate, cm

Greek letters

α	deactivation constant
α_1	rate constant of diminution on sites concentration Type I
α_2	rate constant of diminution on sites concentration Type II
α_c	proportional constant
a_D	dimensionless reactant concentration
α_p	relative pore plugging/active site poisoning tendency
α_{DK}	deactivation parameter for diesel to kerosene, $hour^{-1}$
α_{FG}	deactivation parameter for diesel to kerosene, $hour^{-1}$
α_{FN}	deactivation parameter for feed to naphtha, $hour^{-1}$
α_{NG}	deactivation parameter for naphtha to gas, $hour^{-1}$
α_{KN}	deactivation parameter for kerosene to naphtha, $hour^{-1}$
α_V,α_c,β_V,β_c, and γ_V	effective deposition of vanadium and nickel parameters
β	fitting parameter of Eq. (2.60)
β_D	dimensionless distance along the pore
γ	fitting parameter of Eq. (2.60)
Γ	relative rate constant of desulfurization
γ_D	ratio of the instantaneous pore radius to the initial pore radius
δ	fitting parameter of Eq. (2.60)
Δr_{pc}	thickness of monolayer deposited metal, m
ϵ	number of metal sulfide molecules per molecule of reactant
$\in_0$	initial catalyst bed porosity
$\in_B$	catalyst bed void fraction, $cm^3_{(G+L)}/cm^3_r$
ϵ_p	porosity of catalyst particle
$\in_t$	catalyst bed porosity at time t
εS	solid fraction in reactor, $m_S{}^3/m_R{}^3$
Θ	partial surface poisoning, dimensionless
ζ	number of metal atoms per reactant molecule
η	effectiveness factor
η_0	initial effectiveness factor
$\eta_m(t)$	relative activity of demetallation
$\eta_s(t)$	relative activity of desulfurization
θ	relative catalyst age
θ_{C1}	loss in catalyst porosity due to fast coke
θ_{C2}	loss in catalyst porosity due to slow coke
θ_M	metal distribution factor
θ_{MS}	loss in catalyst porosity due to metal sulfides
θ_T	total loss in catalyst porosity due to pore-plugging
Λ	relative rate constant of demetallation
λ_0	ratio of molecular to pore radius defined
ξ	fractional distance from the center of the pellet
ρ	density oil, kg/m^3
ρ_c	density coke, kg/m^3
ρ_{cat}	density catalyst, kg/m^3
ρ_f	density of phase f at process conditions, g_f/cm^3_f

ρ_i	density of deposit on catalyst, kg_i/m_i^3
ρ_L	density of deposited layer, kg/m^3
ρ_m	density of metal sulfides
ρ_{md}	density of metal deposited
ρ_{oil}	feedstock density, kg_{oil}/m_{oil}^3
ρ_p	adsorbed poison density, kg/m^3
ρ_v	density vanadium, kg/m^3
τ	space-time, hour
τ_D	dimensionless time for the restricted-diffusivity
τ_f	tortuosity factor
φ	activity function
φ_1	loss of activity due to site coverage
φ_2	loss of activity due to sites coverage and pore blockage
φ_c	dimensionless concentration of coke
φ_s	steady-state catalyst activity
φ_{sf}	activity function for sulfur removal
φ_m	activity function for metal removal
φ_v	dimensionless concentration of vanadium
Φ	The Thiele modulus for system with restricted diffusivity
$\emptyset_m$	The Thiele modulus for demetallation
ψ	dimensionless volumetric quantity of deposit on catalyst, m_i^3/m_{cat}^3
ψ_{max}	dimensionless volumetric maximum quantity of deposit on catalyst, m_i^3/m_{cat}^3
Ω	fraction of initial porosity occupied by the solid deposited

Subscripts and Superscripts

H_2	molecular hydrogen
i	component designation number
j	number of experimental data point
R	reactor
S	sulfur
V	vanadium

Abbreviation

CSTR, continuous stirred tank reactor
CCR, Conradson carbon residue
COC, coke–on-catalyst
DMDS, dimethyl disulfide
EOR, end-of-run
HDM, hydrodemetallization
HDO, hydrodeoxygenation
HDS, hydrodesulfurization
HDT, hydrotreating

HDV, hydrodevanadization
LHSV, liquid-hourly space-velocity
MOC, metal-on-catalyst
MOR, middle-of-run
RDS, residue hydrodesulfurization
SOR, start-of-run
SQP, sequential quadratic programming
SRGO, straight-run gas oil
TBR, trickle-bed reactor
TOS, time-on-stream
USC, ultimate storage capacity
VGO, vacuum gas oil

References

Ahn, B.J. and Smith, J.M. (1984). Deactivation of hydrodesulfurization catalysts by metals deposition. *AIChE Journal* 30: 739–746.

Alvarez, A. and Ancheyta, J. (2008). Modeling residue hydroprocessing in a multi-fixed-bed reactor system. *Applied Catalysis A: General* 351: 148–158.

Alvarez, A., Ancheyta, J., Centeno, G., and Marroquín, G. (2011). A modeling study of the effect of reactor configuration on the cycle length of heavy oil fixed-bed hydroprocessing. *Fuel* 90: 3551–3560.

Ancheyta, J. (2016). *Deactivation of Heavy Oil Hydroprocessing Catalysts: Fundamentals and Modeling.* New Jersey: Wiley.

Ancheyta, J. and Lopez, F. (2000). *Analysis of deactivation models based on time-on-stream (TOS). Theory for fluid catalytic cracking process. Revista de La Sociedad Química de México* 44: 183–187.

Ancheyta, J., Betancourt, G., Marroquin, G. et al. (2001). *An exploratory study for obtaining synthetic crudes from heavy crude oils via hydrotreating. Energy & Fuels* 15: 120–127.

Ancheyta, J., Betancourt, G., Centeno, G. et al. (2002). *Catalyst deactivation during hydro processing of maya heavy crude oil: evaluation at constant operating conditions. Energy & Fuels* 16: 1438–1443.

Ancheyta, J., Betancourt, G., Centeno, G., and Marroquín, G. (2003). *Catalyst deactivation during hydroprocessing of Maya heavy crude oil. (II) Effect of temperature during time-on-stream. Energy & Fuels* 17: 462–467.

Ancheyta, J., Rana, M., and Furimsky, E. (2005). Hydroprocessing of heavy petroleum feeds: tutorial. *Catalysis Today* 109: 3–15.

Arbabi, S. and Sahimi, M. (1991a). Computer simulations of catalyst deactivation-I. Model formulation and validation. *Chemical Engineering Science* 46: 1739–1747.

Arbabi, S. and Sahimi, M. (1991b). The effect of morphological, transport and kinetic parameters on the performance. *Chemical Engineering Science* 46: 1749–1755.

Baghalha, M., Mohammadi, M., and Ghorbanpour, A. (2010). Coke deposition mechanism on the pores of a commercial Pt–Re/γ-Al2O3 naphtha reforming catalyst. *Fuel Processing Technology* 91: 714–722.

Bartholomew, C.H. (1982). Catalysis reviews: science and reforming and methanation carbon deposition in steam reforming and methanation. *Catalysis Reviews: Science and Engineering* 24: 67–112.

Bartholomew, C.H. (2001). Mechanisms of catalyst deactivation. *Applied Catalysis A: General* 212: 17–60.

Beaton, W.I. and Bertolacini, R.J. (1991). Resid hydroprocessing at Amoco. *Catalysis Reviews* 33: 281–317.

Beuther, H., Larson, O.A., and Perrotta, A.J. (1980a). The mechanism of coke formation on catalysts. In: *Studies in Surface Science and Catalysis*, vol. 6. Issue C.

Bourseau, P., Muratet, G., Saillour, C., and Toulhoat, H. (1985). Evolution au cours du temps d'un grain de catalyseur d'hydrodémétallisation: module et simulation numérique de résultats expérimentaux. *Revue de l'Institut Français Du Pétrole* 40: 595–607.

Calderón, C.J. and Ancheyta, J. (2017). Dynamic modeling and simulation of a slurry-phase reactor for hydrotreating of oil fractions. *Energy & Fuels* 31: 5691–5700.

Callejas, M.A., Martínez, M.T., Fierro, J.L.G. et al. (2001). Structural and morphological study of metal deposition on an aged hydrotreating catalyst. *Applied Catalysis A: General* 220: 93–104.

Castañeda, L.C., Muñoz, J.A., and Ancheyta, J. (2014). Current situation of emerging technologies for upgrading of heavy oils. *Catalysis Today* 220–222: 248–273.

Centeno, G. (2008). Catalyst deactivation during hydroprocessing of heavy oils. *PhD thesis (In Spanish).* Instituto Tecnológico de Ciudad Madero, México

Chao, Y.C., Liaw, H.J., and Huang, H.P. (1991). *A mathematical model for the catalyst deactivation in a commercial residue hydrodesulfurization reactor system. Chemical Engineering Communications* 104: 267–290.

Corella, J., Bilbao, R., Molina, J.A., and Artigas, A. (1985). Variation with time of the mechanism, observable order, and activation energy of the catalyst deactivation by coke in the FCC process. *Industrial and Engineering Chemistry Process Design and Development* 24: 625–636.

Dautzenberg, F., Klinken, J., Pronk, K. et al. (1978). Catalyst deactivation through pore mouth plugging during residue desulfurization. In: *Chemica Reaction Engineering*, vol. 65, 254–267.

Elizalde, I. and Ancheyta, J. (2014a). Application of a three-stage approach for modeling the complete period of catalyst deactivation during hydrotreating of heavy oil. *Fuel* 138: 45–51.

Elizalde, I. and Ancheyta, J. (2014b). Modeling catalyst deactivation during hydrocracking of atmospheric residue by using the continuous kinetic lumping model. *Fuel Processing Technology* 123: 114–121.

Eser, S., Jenkins, R.G., Derbyshire, F.J., and Malladi, M. (1986). Carbonization of coker feedstocks and their fractions. *Carbon* 24: 77–82.

Félix, G., Ancheyta, J., and Trejo, F. (2019). Sensitivity analysis of kinetic parameters for heavy oil hydrocracking. *Fuel* 241: 836–844.

Ferreira, C., Marques, J., Tayakout, M. et al. (2010). Modeling residue hydrotreating. *Chemical Engineering Science* 65: 322–329.

Flores, R.M. (2014). Chapter 1 – introduction and principles. In: *Coal and Coalbed Gas* (ed. R.M. Flores), 1–40. Boston.

Froment, G.F. and Bischoff, K.B. (1962). Kinetic data and product distributions from fixed bed catalytic reactors subject to catalyst fouling. *Chemical Engineering Science* 17: 105–114.

Froment, G.F., Bischoff, K.B., and Wilde, J.D. (2010). *Chemical Reactor Analysis and Design*, 3e. USA: Wiley.

Furimsky, E. and Massoth, F.E. (1999). Deactivation of hydroprocessing catalysts. *Catalysis Today* 52: 381–495.

Galiasso, R. (2007). Effect of recycling the unconverted residue on a hydrocracking catalyst operating in an ebullated bed reactor. *Fuel Processing Technology* 88: 779–785.

Galiasso, R., Blanco, R., Gonzalez, C., and Quinteros, N. (1983). Deactivation of hydrodemetallization catalyst by pore plugging. *Fuel* 62: 817–822.

Gary, J.H., Handwerk, G.E., and Kaiser, M.J. (2007). *Petroleum Refining: Technology and Economics*, 4e. New York: Marcel Dekker, Inc.

Gianetto, A. and Specchia, V. (1992). Trickle-bed reactors: state of art and perspectives. *Chemical Engineering Science* 47: 3197–3213.

Gray, M., Zhao, Y., McKnight, C. et al. (1999). Coking of hydroprocessing catalyst by residue fractions of bitumen. *Energy & Fuels* 13: 1037–1045.

Gualda, G.A. and Kasztelan, S.A. (1996). Initial deactivation of residue hydrodemetallization catalysts. *Journal of Catalysis* 161: 319–337.

Hauser, A., Marafi, A., Stanislaus, A., and Al-Adwani, A. (2005). Relation between feed quality and coke formation in a three-stage atmospheric residue desulfurization (ARDS) process. *Energy & Fuels* 19: 544–553.

Haynes, H. and Leung, K. (1983). Catalyst deactivation by pore plugging and active site poisoning mechanisms. *Chemical Engineering Communications* 23: 161–179.

Hegedus, L. and McCabe, R. (1994). Catalyst poisoning. *Catalysis Reviews* 36: i–ii.

Idei, K., Yamamoto, Y., and Yamazaki, H. (1995). Kinetic model and analysis method of catalysts deactivation in hydrodesulfization of light and heavy oils. *Kagaku Kōgaku Ronbunshū* 21: 972–983.

Idei, K., Yamamoto, Y., and Takehara, S. (1998). Generalized deactivation equation of catalyst in hydrodesulfurization of light and heavy oils. *Kagaku Kōgaku Ronbunshū* 24: 653–659.

Idei, K., Takahashi, T., and Kai, T. (2003). Estimation of coke and metal deposition distribution within hydrodesulfurization catalyst pore at the last stage of operation. *Journal of the Japan Petroleum Institute* 46: 45–52.

Isaza, M., Pachon, Z., Kafarov, V., and Resasco, D. (2000). Deactivation of Ni–Mo/Al2O3 catalysts aged in a commercial reactor during the hydrotreating of deasphalted vacuum residuum. *Applied Catalysis A: General* 199: 263–273.

Jacob, S.M., Gross, B., Voltz, S.E., and Weekman, V.W. (1976). A lumping and reaction scheme for catalytic cracking. *AIChE Journal* 22: 701–713.

Jurado, J. and Ancheyta, J. (2022). Reactor model for heavy oil hydrotreating with catalyst deactivation based on vanadium and coke deposition. *Energy & Fuels* 36: 11132–11141.

Kam, E.K.T., Al-Shamali, M., Juraidan, M., and Qabazard, H. (2005). A hydroprocessing multicatalyst deactivation and reactor performance model-pilot-plant life test applications. *Energy & Fuels* 19: 753–764.

Khang, S. and Mosby, J. (1986). Catalyst deactivation due to deposition of reaction products in macropores during hydroprocessing of petroleum residuals. *Industrial & Engineering Chemistry Process Design and Development* 25: 437–442.

Kobayashi, S., Kushiyama, S., Aizawa, R. et al. (1987). Kinetic study on the hydrotreating of heavy oil. 2. Effect of catalyst pore size. *Industrial & Engineering Chemistry Research* 26: 2245–2250.

Kodama, S., Nitta, H., Takatsuka, T., and Yokoyama, T. (1980). Simulation of residue hydrodesulfurization reaction based on catalyst deactivation model. *Journal of The Japan Petroleum Institute* 23: 310–320.

Kohli, K., Prajapati, R., Maity, S.K. et al. (2016). Deactivation of hydrotreating catalyst by metals in resin and asphaltene parts of heavy oil and residues. *Fuel* 175: 264–273.

Korsten, H. and Hoffmann, U. (1996). Three-phase reactor model for hydrotreating in pilot trickle-bed reactors. *AIChE Journal* 42: 1350–1360.

Kressmann, S., Morel, F., Harl, V., and Kasztelan, S. (1998). Recent developments in fixed-bed catalytic residue upgrading. *Catalysis Today* 43: 203–215.

Kumar, V.R., Balaraman, K.S., Rao, V.S.R., and Ananth, M.S. (1997). Modelling of hydrotreating process in a trickle-bed reactor. *Petroleum Science and Technology* 15: 283–295.

Lee, L.-S., Chen, Y.-W., Huang, T.-N., and Pan, W.-Y. (1989). Four-lump kinetic model for fluid catalytic cracking process. *The Canadian Journal of Chemical Engineering* 67: 615–619.

Marafi, A., Stanislaus, A., and Furimsky, E. (2010a). Kinetics and modeling of petroleum residues hydroprocessing. *Catalysis Reviews – Science and Engineering* 52: 204–324.

Marafi, M., Stanislaus, A., and Furimsky, E. (2010b). *Handbook of spent hydroprocessing catalysts*, 1e.

Martínez, J. and Ancheyta, J. (2012). Kinetic model for hydrocracking of heavy oil in a CSTR involving short term catalyst deactivation. *Fuel* 100: 193–199.

Maxted, E.B. (1951). The poisoning of metallic catalysts. *Advances in Catalysis* 3: 129–178.

Mederos, F.S., Rodríguez, M.A., Ancheyta, J., and Arce, E. (2006). Dynamic modeling and simulation of catalytic hydrotreating reactors. *Energy & Fuels* 20: 936–945.

Mederos, F.S., Elizalde, I., and Ancheyta, J. (2009). Steady-state and dynamic reactor models for hydrotreatment of oil fractions: a review. *Catalysis Reviews* 51: 485–607.

Mederos, F.S., Ancheyta, J., and Elizalde, I. (2012). Dynamic modeling and simulation of hydrotreating of gas oil obtained from heavy crude oil. *Applied Catalysis A: General* 425–426: 13–27.

Menon, P.G. (1990). Coke on catalysts-harmful, harmless, invisible and beneficial types. *Journal of Molecular Catalysis* 59: 207–220.

Miki, Y., Yamadaya, S., Oba, M., and Sugimoto, Y. (1983). Role of catalyst in hydrocracking of heavy oil. *Journal of Catalysis* 83: 371–383.

Mitchell, P.C.H. (1990). Hydrodemetallisation of crude petroleum: fundamental studies. *Catalysis Today* 7: 439–445.

Mochida, I., Zhao, X., Sakanishi, K. et al. (1989). Structure and properties of sludges produced in the catalytic hydrocracking of vacuum residue. *Industrial & Engineering Chemistry Research* 28: 418–421.

Moghadassi, A., Amini, N., Fadavi, O., and Bahmani, M. (2011). Hydrocracking lumped kinetic model with catalyst deactivation in arak refinery hydrocracker unit. *Journal of Petroleum Science and Technology Petroleum Science and Technology* 1: 31–37.

Mohammed, J., Mahmoud, A., Hasan, Q., and Kam, K.T. (2006). A refined hydroprocessing multicatalyst deactivation and reactor performance model pilot-plant accelerated test applications. *Energy & Fuels* 20: 1354–1364.

Monzón, A., Romeo, E., and Borgna, A. (2003). Relationship between the kinetic parameters of different catalyst deactivation models. *Chemical Engineering Journal* 94: 19–28.

Nagaishi, H., Chan, E.W., Sanford, E.C., and Gray, M.R. (1997). Kinetics of high-conversion hydrocracking of bitumen. *Energy & Fuels* 11: 402–410.

Newson, E. (1975). Catalyst deactivation due to pore-plugging by reaction products. *Industrial and Engineering Chemistry Process Design and Development* 14: 27–33.

Nielsen, J.R. and Trimm, D.L. (1977). Mechanisms of carbon formation on nickel-containing catalysts. *Journal of Catalysis* 48: 155–165.

Oliveira, L. and Biscaia, E. (1989). Catalytic cracking kinetic models. Parameter estimation and model evaluation. *Industrial & Engineering Chemistry Research* 28: 264–271.

Oyekunle, L. and Ikpekri, O. (2004). Modeling of hydrodesulfurization catalysts. I. Influence of catalyst pore structures on the rate of demetallization. *Industrial & Engineering Chemistry Research* 43: 6647–6653.

Oyekunle, L.O., Ikpekri, O.B., and Jaiyeola, A. (2005). Modelling of hydrodesulfurization catalysts: II. Effects of catalyst pore structures on deactivation by metal deposits. *Catalysis Today* 109: 128–134.

Pereira, C.J. (1990). Metal deposition in hydrotreating catalysts. 1. A regular perturbation solution approach. *Industrial and Engineering Chemistry Research* 29: 512–519.

Rajagopalan, K. (1979). Influence of catalyst pore size on demetallation rate. *Industrial & Engineering Chemistry Process Design and Development* 18: 459–465.

Rana, M.S., Sámano, V., Ancheyta, J., and Diaz, J.A.I. (2007). A review of recent advances on process technologies for upgrading of heavy oils and residua. *Fuel* 86: 1216–1231.

Rashidzadeh, M., Ahmad, A., and Sadighi, S. (2010). Studying of catalyst deactivation in a commercial hydrocracking process (ISOMAX). *Journal of Petroleum Science and Technology* 1: 46–54.

Richardson, S.M., Nagaishi, H., and Gray, M.R. (1996). Initial coke deposition on a NiMo/γ-Al2O3 bitumen hydroprocessing catalyst. *Industrial & Engineering Chemistry Research* 35: 3940–3950.

Sau, M., Narasimhan, C.S.L., and Verma, R.P. (1997). A kinetic model for hydrodesulfurisation. In: *Studies in Surface Science and Catalysis*, vol. 106 (ed. G.F. Froment, B. Delmon, and P. Grange), 421–435.

Shah, Y.T. and Mhaskar, R.D. (1976). Optimum quench location for a hydrodesulfurization reactor with time-varying catalyst activity. *Industrial & Engineering Chemistry Process Design and Development* 15: 400–406.

Shah, Y.T., Mhaskar, R.D., and Paraskos, J.A. (1976). Optimum quench location for a hydrodesulfurization reactor with time varying catalyst activity. *Industrial & Engineering Chemistry Process Design and Development* 15: 400–406.

Shimura, M., Shiroto, Y., and Takeuchi, C. (1986). Effect of catalyst pore structure on hydrotreating of heavy oil. *Industrial & Engineering Chemistry Fundamentals* 25: 330–337.

Skala, D., Saban, M., Orlovic, M. et al. (1991). Hydrotreating of used oil: prediction of industrial trickle-bed operation from pilot-plant data. *Industrial & Engineering Chemistry Research* 30: 2059–2065.

Song, C., Nihonmatsu, T., and Nomura, M. (1991). Effect of pore structure of nickel-molybdenum/alumina catalysts in hydrocracking of coal-derived and oil sand derived asphaltenes. *Industrial & Engineering Chemistry Research* 30: 1726–1734.

Speight, J.G. (1999). *The Desulfurization of Heavy Oils and Residua*. New York: CRC Press.

Spry, J.C. and Sawyer, W.H. (1975). Configurational diffusion effects in catalytic demetallization of petroleum fractions. *Chemical Engineering Science* 24: 16–20.

Tamm, P.W., Harnsberger, H.F., and Bridge, A.G. (1981). Effects of feed metals on catalyst aging in hydroprocessing residuum. *Industrial and Engineering Chemistry Process Design and Development* 20: 262–273.

Thakur, D. and Thomas, M. (1985). Catalyst deactivation in heavy petroleum and synthetic crude processing: a review. *Applied Catalysis* 15: 197–225.

Toulhoat, H., Szymanski, R., and Plumail, J. (1990). Interrelations between initial pore structure, morphology and distribution of accumulated deposits, and lifetimes of hydrodemetallisation catalysts. *Catalysis Today* 7: 531–568.

Toulhoat, H., Hudebine, D., Raybaud, P. et al. (2005). THERMIDOR: a new model for combined simulation of operations and optimization of catalysts in residues hydroprocessing units. *Catalysis Today* 109: 135–153.

Trimm, D.L. (1977). The formation and removal of coke from nickel catalyst. *Catalysis Reviews* 16: 155–189.

Trimm, D.L. (1983). Catalyst design for reduced coking (review). *Applied Catalysis* 5: 263–290.

Van, F., Nevicato, D., Pitault, I. et al. (1996). Fluid catalytic cracking: modelling of an industrial riser. *Applied Catalysis A: General* 138: 381–405.

Verstraete, J.J., Le Lannic, K., and Guibard, I. (2007). Modeling fixed-bed residue hydrotreating processes. *Chemical Engineering Science* 62: 5402–5408.

Voorhies, A. (1945). Carbon formation in catalytic cracking. *Industrial & Engineering Chemistry Research* 37: 318–322.

Ward, J.W. (1993). Hydrocracking processes and catalysts. *Fuel Processing Technology* 35: 55–85.

Weekman, V. (1968). A model of catalytic cracking conversion in fixed, moving, and fluid-bed reactors. *Industrial and Engineering Chemistry Process Design and Development* 7: 90–95.

Wheeler, A. (1951). Reaction rates and selectivity in catalyst pores. *Advances in Catalysis* 3: 249–327.

3

Simulation of the Oxidative Regeneration of Coked Catalysts: Kinetics, Catalyst Pellet, and Bed Levels

Sergey Zazhigalov[1,2], Osman Abdulla[2], and Andrey Zagoruiko[1,2]

[1] *Boreskov Institute of Catalysis, Novosibirsk, Russia*
[2] *Tyumen State University, Tyumen, Russia*

3.1 Introduction

Catalysts are used in most petroleum refining and petrochemical processes and play a key role in their execution. Currently, at least 90% of all chemical products in the world are produced by the catalytic method (Meyers 2016; Speight 2016).

Most hydraulic oil refining processes use catalysts based on molybdenum and tungsten. This is explained by their unique catalytic properties: The ability to accelerate target reactions of the process and low sensitivity to sulfur- and nitrogen-containing compounds, which are strong catalytic poisons for other types of hydrogenation catalysts.

Hydrotreating catalysts are a combination of oxides (or sulfides) of active components (Ni, Co, Mo, W, etc.) supported on a carrier, usually the active aluminum oxide. The carrier in hydrotreating catalysts not only plays the role of an inert diluent but also participates in the formation of active phases and also serves as a structural promoter that creates a specific porous structure, optimal for the processing of specific raw materials.

The main sign of a decrease in catalyst activity is an increase in the sulfur content in the hydrotreating product (Marafi et al. 2017). At the same time, it loses its activity, and the degree of desulfurization of products during its use decreases—normal aging of the catalyst occurs. The most typical process leading to blocking of active centers is the deposition of carbon-containing compounds (coke) on the catalyst. If coking occurs in a reaction medium containing sulfur and oxygen, then these elements partially transform into coke.

An important part of modern oil refining and petrochemical processes in moving and stationary catalyst beds are the processes of regeneration of spent catalysts (Kern and Jess 2005; Zhou et al. 2020; Ferella et al. 2016). Such processes are necessary to remove coke deposited on the catalyst granules during its operation.

Among them, from an energy and environmental point of view, the most effective ones are oxidative regeneration processes based on the oxidation of coke (Furimsky 1991). In such processes, complete removal of coke and other undesirable components is possible; however, such processes are highly exothermic. Considering that most oil refining and petrochemical catalysts have

Mathematical Modeling of Complex Reaction Systems in the Oil and Gas Industry, First Edition.
Edited by Jorge Ancheyta, Andrey Zagoruiko, and Andrey Elyshev.

moderate heat resistance (the maximum permissible temperatures during their processing should not exceed 450–650 °C, depending on the type of catalyst), very stringent requirements are imposed on the thermal conditions of the oxidative regeneration process.

Experimental development of the technological foundations of the process, as well as thermal management strategies, is extremely complex, resource-intensive, and to some extent, risky. This is due to the fact that in laboratory conditions, it is almost impossible to create adiabatic conditions corresponding to the conditions of an industrial process. Accordingly, in laboratory-scale experiments, it is impossible to reproduce the real thermal regime of an industrial process.

To solve this problem, a development strategy was proposed that combines experimental work and mathematical modeling in the following sequence:

- experimental study of the kinetics of reactions occurring during the oxidative regeneration of a spent catalyst;
- experimental study of the oxidative regeneration process in laboratory and pilot-scale reactors;
- preliminary mathematical modeling of the industrial oxidative regeneration process;
- comparison of simulation results with the results of experiments at the pilot reactor;
- verification of the model, making adjustments to the mathematical model and model parameters;
- final mathematical modeling of the industrial process.

The following is a description of experimental studies on the regeneration of a coked catalyst in a bench-scale reactor and mathematical modeling of the process.

3.2 Process Chemistry and Laboratory Experiments

3.2.1 Catalyst and Proposed Reactions

To carry out the experiments, we used a coked cobalt–molybdenum catalyst for the hydrotreatment of the diesel fraction, worked out under real industrial operation conditions at the Omsk Refinery. The characteristics of the catalyst are given in Table 3.1 and its external appearance is demonstrated in Figure 3.1.

The process involves reactions of oxidation of catalyst metal sulfides and oxidation of coke.

Table 3.1 Properties of spent hydrotreating catalyst.

(a)	Appearance	Black extrudate with trefoil cross section, granule length ~ 4 mm, outer diameter ~ 1.3 mm
(b)	Bulk density	~900 g/l
(c)	Chemical composition:	% wt.
	molybdenum (IV) sulfide	16.9
	cobalt sulfide	4.0
	carrier (aluminum oxide)	68.2
	coke, including:	10.9
	carbon	82.2
	sulfur	7.7
	nitrogen	0.9
	hydrogen	9.2

Figure 3.1 Appearance of the original spent catalyst.

Oxidation of sulfides of active metals in the presence of atmospheric oxygen begins already at 200 °C. In this case, metal sulfides are converted into the corresponding oxides and sulfur dioxide (SO_2) is released into the gas phase:

$$2MoS_2 + 7O_2 \rightarrow 2MoO_3 + 4SO_2 \tag{3.1}$$

$$2NiS + 3O_2 \rightarrow 2NiO + 2SO_2 \tag{3.2}$$

$$2CoS + 3O_2 \rightarrow 2CoO + 2SO_2 \tag{3.3}$$

A certain amount of nickel and cobalt oxides in the presence of oxygen reacts with sulfur dioxide to form the corresponding sulfates, which do not decompose at a regeneration temperature, usually not exceeding 550 °C. The interaction of nickel and cobalt oxides with oxygen and sulfur dioxide is described by the following reactions:

$$2CoO + O_2 + 2SO_2 \rightarrow 2CoSO_4 \tag{3.4}$$

$$2NiO + O_2 + 2SO_2 \rightarrow 2NiSO_4 \tag{3.5}$$

Coke oxidation begins at temperatures around 300 °C, with most of the coke burning out in the range of 300–450 °C.

In general, carbon combustion is characterized by reactions that are highly exothermic. The chemical interaction of carbon with oxygen, with the formation of carbon oxides, proceeds according to the following schemes:

$$C + O_2 \rightarrow CO_2 + 395.4\ \text{kJ/mol} \tag{3.6}$$

$$C + 1/2\,O_2 \rightarrow CO + 110.4\ \text{kJ/mol} \tag{3.7}$$

Further transformations of the resulting oxides proceed according to the following schemes:

$$CO + 1/2\,O_2 \rightarrow CO_2 + 285.0\ \text{kJ/mol} \tag{3.8}$$

$$C + CO_2 \rightarrow 2CO - 172.2\ \text{kJ/mol} \tag{3.9}$$

It is also possible that coke carbon interacts with water vapor, since during coke oxidation, water vapor is almost always present in the reaction zone:

$$C + H_2O \rightarrow CO + H_2 + 41\ \text{kJ/mol} \tag{3.10}$$

The sulfur contained in coke is oxidized into sulfur dioxide,

$$S + O_2 \rightarrow SO_2 + 298\ \text{kJ/mol} \tag{3.11}$$

and further into sulfuric anhydride:

$$2SO_2 + O_2 \rightarrow 2SO_3 \tag{3.12}$$

Coke nitrogen is oxidized to the corresponding oxides:

$$2N + O_2 \rightarrow 2\,NO \tag{3.13}$$

$$N + O_2 \rightarrow NO_2 \tag{3.14}$$

$$2NO + O_2 \rightarrow 2NO_2 \tag{3.15}$$

Hydrogen oxidizes to water,

$$2H_2 + O_2 \rightarrow 2H_2O + 136\ \text{kJ/mol} \tag{3.16}$$

Catalysts used in hydrotreating are quite resistant to heat treatment and generally retain their composition and structure. However, at temperatures above 550 °C, sublimation of molybdenum is observed, which forces the process to be carried out below the specified temperature. With a sharp increase (decrease) in temperature, the strength of the catalyst deteriorates. To preserve it, it is not recommended to change the temperature by more than 40 K per hour. A significant temperature difference between the catalyst and the circulating gas can cause destruction to the catalyst structure—the permissible temperature gradient between the gas and catalyst particles should not exceed 150 °C (Froment and Biscoff 1999).

The coke formed on catalysts is not homogeneous in composition. The heterogeneity of composition and structure determines its unequal reactivity to oxidation during the burning process. Thus, when burning coke, two temperature maxima are clearly distinguished. The first corresponds to the combustion of an amorphous coke, which is the most rich in hydrogen; and the second corresponds to the combustion of a slow-burning component of coke, consisting of highly condensed pseudographite structures, the second maximum is observed at a higher temperature in the reaction zone. The ratio of these coke components is determined by the severity of the catalyst operating mode.

3.2.2 Reaction Kinetics

Studies of the kinetics of the ongoing reactions were carried out using the TG-DTA-MS method (thermogravimetry–differential thermal analysis–mass spectroscopy). They were performed on a Netzsch STA 409 PC instrument coupled to an SRS UGA 200 mass spectrometer. The measurements were carried out using corundum crucibles. The DTA sample holder together with the crucibles was preheated to 1000 °C in an air stream.

Before the experiment, the catalyst sample was crushed to a fraction of 0.25–0.5 mm to eliminate the influence of intra-diffusion limitations in the catalyst granules during reactions. The sample weight in each experiment was 76–78 mg.

After loading the sample and installing the crucible on the holder, the temperature was raised to 50 °C and held for 1 hour to extinguish the inertia of the furnace and establish thermal equilibrium. Thermal curves of the samples were recorded in the range from 50 to 450 °C with a heating rate of 3 or 1 K/min in a mixture of argon and argon–oxygen mixture (10% oxygen, 90% argon). The argon–oxygen mixture was supplied to the device furnace at a rate of 10 (or 20) ml/min, and at the same time, a protective gas (argon) was supplied to the scales at a rate of 40 (or 20) ml/min. Correction curves were recorded under similar conditions for each experiment. A typical type of experimental results is shown in Figure 3.2.

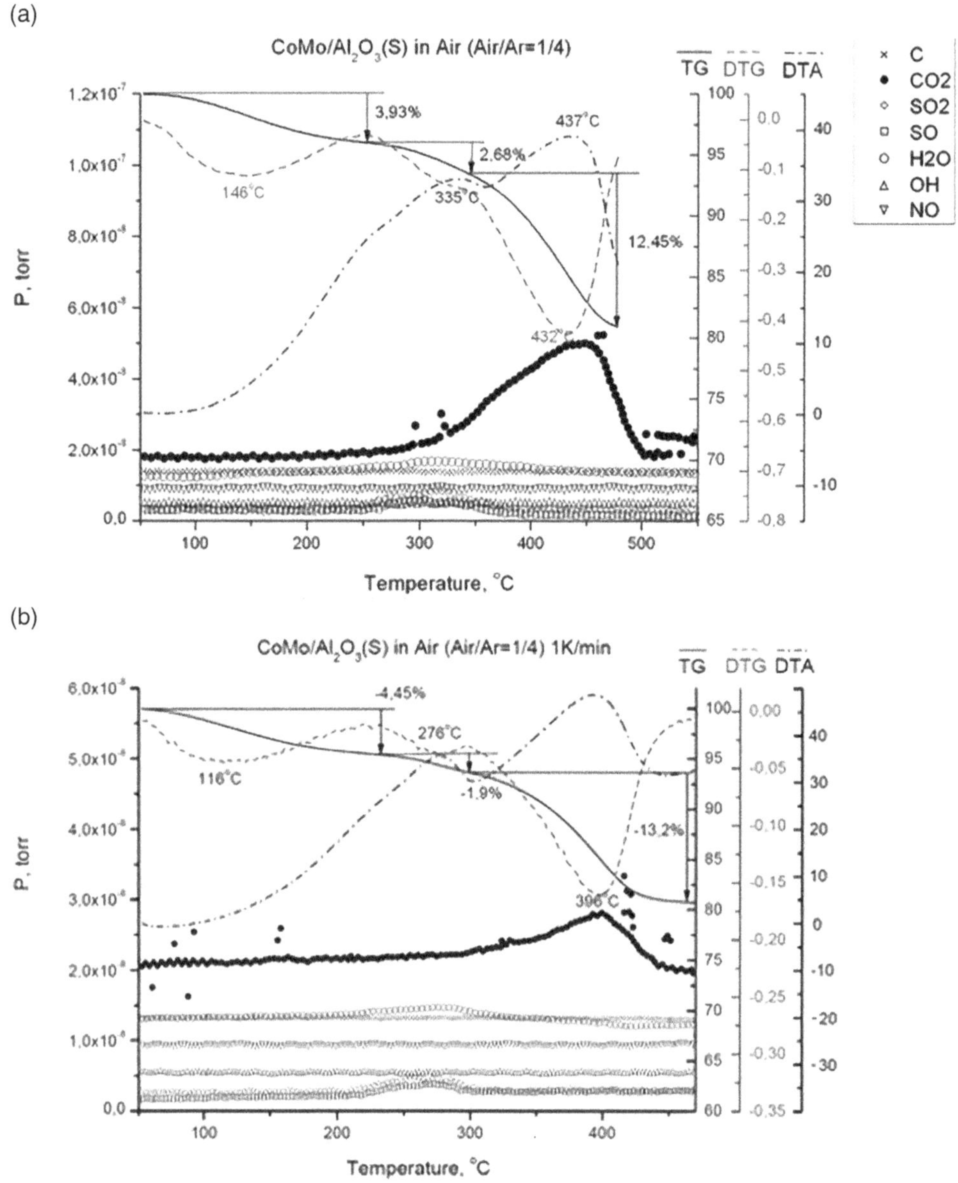

Figure 3.2 Results of TG-DTA-MS analysis of spent catalyst samples. Volume ratio of argon–oxygen mixture and argon flows is 1:4 (calculated oxygen concentration in the mixture is 4% vol.), temperature rise rate is 3 K/min (a) and 1 K/min (b).

According to the experimental data, it is clear that during regeneration, the mass of the catalyst decreases by ~19%, which approximately corresponds to the mass loss calculated on the basis of information obtained from data on the composition of the spent catalyst. In this case, three temperature zones can be distinguished:

a) Low-temperature region (up to 200–220 °C), where the mass loss is ~4%. Judging by the mass spectrometry data of the exhaust gases, it can be assumed that desorption of water, as well as light hydrocarbon components of coke, occurs in this region.

b) Region from 220 to 300–330 °C with mass loss of about 2–3%. In this area, judging by the composition of the exhaust gases, active oxidation of surface sulfides occurs.
c) From 300–330 to 500 °C with a mass loss of about 12–13%. At this range, active oxidation of coke occurs.

When constructing the kinetic model, the low-temperature region (a) was excluded from consideration, since it is not associated with chemical transformations. To describe the rate of oxidation of sulfides and coke, it was proposed to use kinetic equations of the general form,

$$\frac{d\theta}{dt} = -k_0 \exp\left(-\frac{E}{RT}\right)\theta^n C_{O_2}^m \tag{3.17}$$

where θ is the relative mass content of the surface component on the catalyst (kg per kg of catalyst), k_0 is the pre-exponent (kg per kg of catalyst per sec), E is the activation energy, T is temperature (K), C_{O_2} is the oxygen concentration (volume fraction), n and m are the orders of reaction rates for the surface component and oxygen, respectively.

Since, on the basis of experimental information, it is impossible to separate the oxidation rates of cobalt and molybdenum sulfides, and for both sulfides, the rate of their oxidation is described by one summary equation. The initial content θ is taken based on the mass composition of the initial catalyst (0.21 for sulfides and 0.11 for coke).

As calculations using proprietary software have shown, the best description is achieved with the following set of model parameters:

Figure 3.3 shows a comparison of experimental results and data calculated using the proposed model. It can be seen that overall good agreement is achieved.

On the contrary, in the future, the model can be significantly improved, in particular, by taking into account the influence of oxygen concentration in more detail (which will require additional experiments), as well as by taking into account the degree of coke compaction and its dynamic change at different catalyst heating rates.

3.2.3 Experimental Setup

Experiments to study the oxidative regeneration of the catalyst were carried out in a catalytic reactor with a fixed bed of granular catalyst. The regenerating flow is formed by mixing nitrogen taken from the vessel and air passing through filters and a dryer. The feed rate is regulated by mass flow controllers. The general setup diagram is shown in Figure 3.4.

The analysis of the composition of the inlet and outlet gas flows was carried out using an MRU Vario Plus industrial gas analyzer. O_2 and CO_2 were recorded with an accuracy of 0.2%, whereas CO, NO, SO_2, and C_3H_8 with an accuracy of 5 ppm.

Regeneration was carried out until the detection of CO_2, CO, NO, and SO_2 in the outlet reaction mixture ceased.

The experiments were carried out in a reactor equipped with an internal heating system for the catalyst bed, which has good internal thermal insulation and is also designed for catalyst loadings of up to 2.5 liters, which allows it to be operated in conditions close to adiabatic. However, as it turned out later, even in a relatively large reactor, there are very noticeable temperature gradients along the radius of the catalyst bed and the reactor as a whole, that is, its actual operating mode cannot be correctly classified as adiabatic. In this regard, it is more correct to call it quasi-adiabatic.

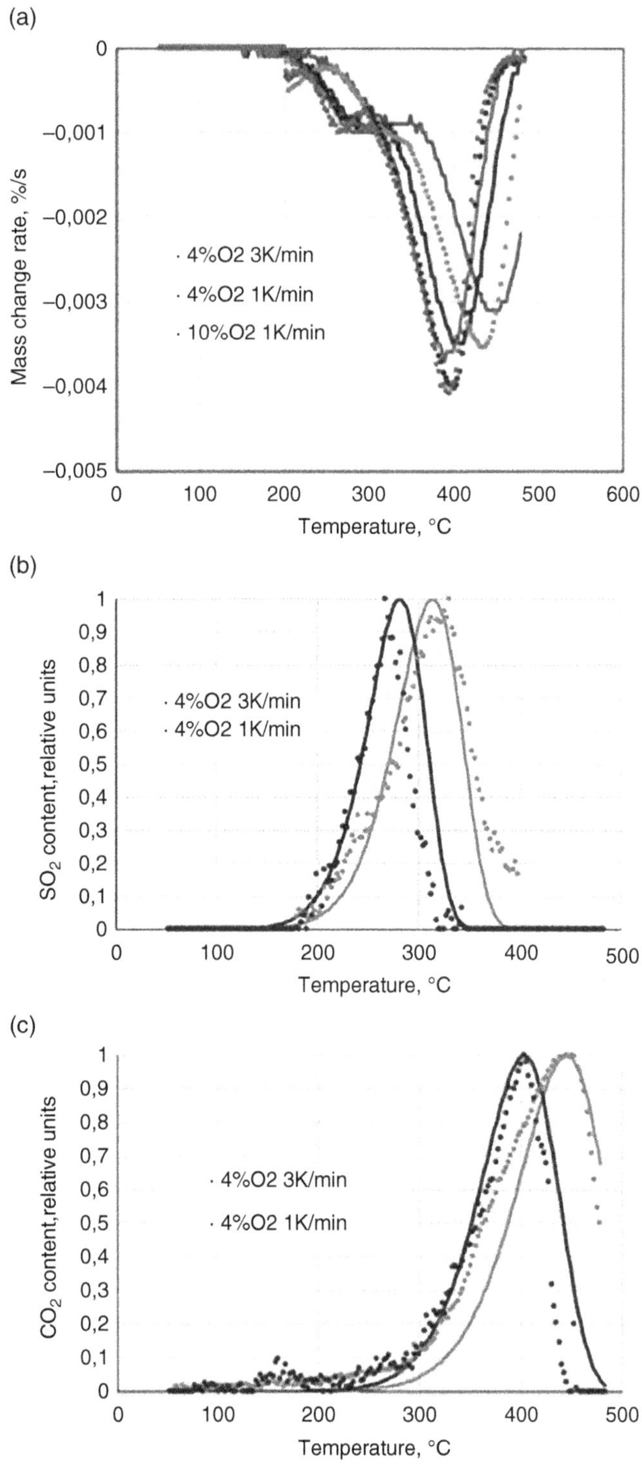

Figure 3.3 The rate of change in the mass of the catalyst (a), the relative content of SO_2 (b), and CO_2 (c) in the outlet gases under various conditions of oxidative regeneration of the catalyst. Points – experiment, lines – calculations.

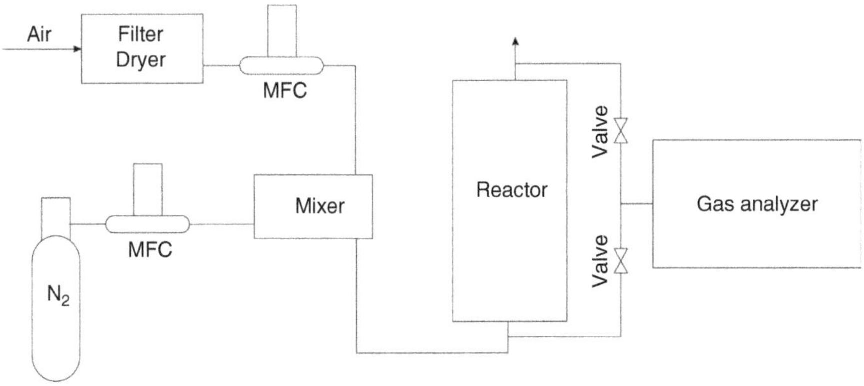

Figure 3.4 Scheme of an experimental setup for oxidative regeneration of catalysts.

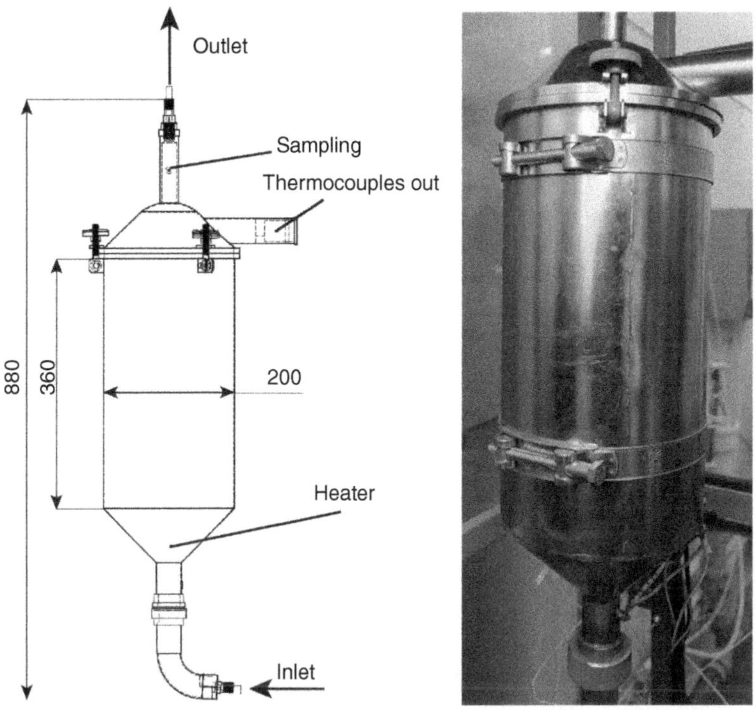

Figure 3.5 The reactor scheme and photo.

The general view of a quasi-adiabatic reactor is shown in Figure 3.5. Along the inner perimeter of the reactor, there is a layer of thermal insulation of 5 cm thick, so the reaction zone does not have thick steel walls, but is formed by a basket of thin steel mesh adjacent to the heat insulator (mineral wool). The heater is made in the spiral form located in the lower part of the reactor, under the catalyst bed, thus the supplied regenerating mixture is heated when passing through the area of the heating spiral. Temperature control in the catalyst bed was carried out using eight thermocouples, some of which were located along the axis of the reactor and some along its wall (Figure 3.6).

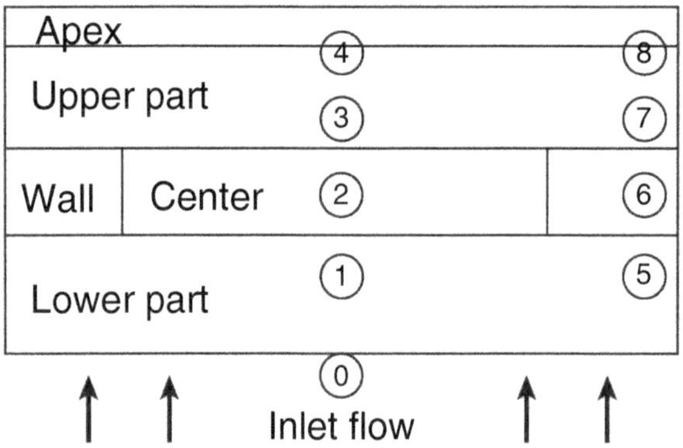

Figure 3.6 Scheme of the catalyst bed and location of thermocouples.

3.2.4 Experiments

250 g of catalyst was loaded into the reactor and the reaction zone was a cylinder with radius and height of 5 cm. The flow rate of the supplied mixture was 38 l/min. The mixture was supplied at ambient temperature; the reaction was initiated by heating a fragment of the catalyst bed in its inlet part (thermocouple 0). After reaching a temperature of more than 300 °C, the heating was turned off, and then the process was carried out in the mode of a moving thermal wave.

Regeneration in quasi-adiabatic reactor was carried out with a flow with variable oxygen content; up to the 18th minute of the process, the oxygen content was 4%, then up to the 30th minute, 6%, and after that 15%. The concentrations of the gas mixture components are shown in Figure 3.7. Figure 3.8 shows the temperature dynamics at various points where thermocouples are located along the bed. The maximum temperature in the bed was 775 °C.

After regeneration, the catalyst was selected for analysis from different parts of the bed. The bed was divided into lower part, middle, upper part, and apex. The middle of the bed was further divided into central and wall regions.

The results of the chemical analysis are presented in Table 3.2, and the textural characteristics of the selected samples are presented in Table 3.3.

It can be seen that, with the exception of the center of the bed middle part, where the highest temperatures were observed, the catalyst was not completely regenerated. The apex corresponds to the top 5 mm of the bed height and has the lowest temperature, so in this part of the bed only about half of the coke contained in the catalyst is burned out.

Figure 3.9 shows an enlarged image of the transverse fracture of catalyst grains from the bottom (inlet) and the top (outlet) of the bed. It is observed that in grains from the lower part of the bed, the black color on the cut, corresponding to coke, is larger than in grains from the top. This corresponds to the carbon content analysis data in Table 3.2. It is also apparent that the transition between the gray and black parts of the slice in Figure 3.9a is more pronounced, while in Figure 3.9b it is more blurred. This suggests worse regeneration of the catalyst in the inlet part of the bed and existence of a greater gradient in the distribution of coke along the grain in this bed section. It may be caused by insufficient development of the heat wave of oxidative regeneration in the very inlet part of the catalyst bed. Subsequent parts of the bed regenerate better due to the increase in temperature at the regeneration front toward the exit of the reactor.

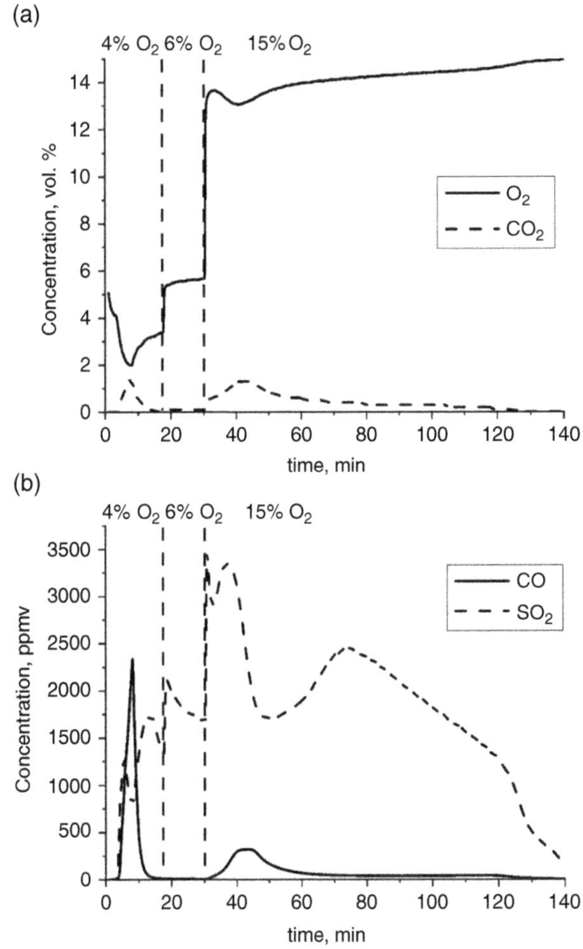

Figure 3.7 Experimentally measured evolution of O_2 and CO_2 (a), CO and SO_2 (b) concentrations at the outlet of the reactor in time.

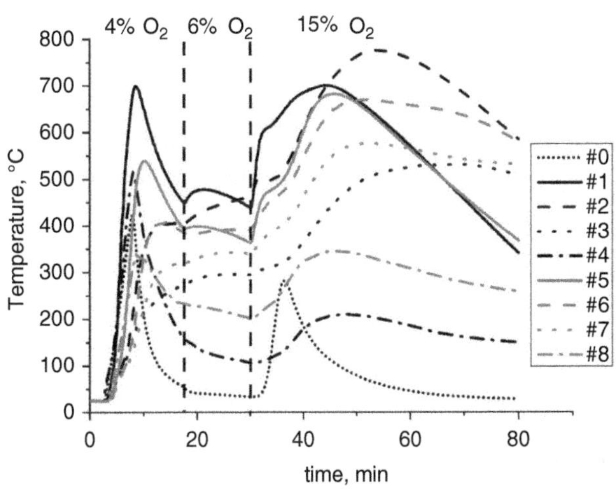

Figure 3.8 The temperature dynamics in central (black) and wall (gray) bed parts.

Table 3.2 Catalyst composition and maximum temperature in different parts of the bed.

Bed part	Mo, wt.%	Co, wt.%	C, wt.%	H, wt.%	N, wt.%	S, wt.%	T_{max}, °C
Lower (inlet) part	10.6	2.91	1.44	1.1	0.036	0.92	700
Middle part – center	11.4	2.89	—	0.98	—	0.16	775
Middle part – wall	11.0	2.74	1.5	1.14	—	0.47	650
Upper (outlet) part	11.4	2.90	1.0	1.17	0.03	0.57	550
Apex	9.75	2.54	5.6	1.3	0.13	1.54	300

Table 3.3 Textural characteristics of the catalyst in different parts of the bed.

Bed part	Specific surface area, m^2/g	Pore volume, cm^3/g	Average pore size, A
Lower part	178	0.50	112
Middle part – center	140	0.56	160
Middle part – wall	189	0.52	110
Upper part	184	0.51	111
Apex	165	0.42	102

3.3 Mathematical Model

To simulate the process of oxidative regeneration in a fixed catalyst bed, a one-dimensional mathematical model was used, which takes into account the course of reaction processes, both heat and mass transfers between the gas flow and the surface of the catalysts and inside the catalyst granules, as well as the thermal conductivity of the bed (Zazhigalov et al. 2012; Reshetnikov et al. 2020).

The regeneration process is described by the oxidation of molybdenum and cobalt sulfides into oxides (18, 19), as well as the oxidation of coke (20):

$$MoS_2 + 3.5O_2 \rightarrow MoO_3 + 2SO_2 \tag{3.18}$$

$$CoS_{0.89} + 1.39O_2 \rightarrow CoO + 0.89SO_2 \tag{3.19}$$

$$C + O_2 \rightarrow CO_2 \tag{3.20}$$

The material balance of the concentrations of substances in the gas phase in the pores of catalyst grains was described by parabolic equations that take into account intra-diffusion limitation:

$$\varepsilon_C \frac{\partial C^p_{SO_2}}{\partial t} = D_{SO_2}\left(\frac{\partial^2 C^p_{SO_2}}{\partial r^2} + \frac{2}{r}\frac{\partial C^p_{SO_2}}{\partial r}\right) + W_{form.SO_2} \tag{3.21}$$

$$\varepsilon_C \frac{\partial C^p_{CO_2}}{\partial t} = D_{CO_2}\left(\frac{\partial^2 C^p_{CO_2}}{\partial r^2} + \frac{2}{r}\frac{\partial C^p_{CO_2}}{\partial r}\right) + W_{form.CO_2} \tag{3.22}$$

$$\varepsilon_C \frac{\partial C^p_{O_2}}{\partial t} = D_{O_2}\left(\frac{\partial^2 C^p_{O_2}}{\partial r^2} + \frac{2}{r}\frac{\partial C^p_{O_2}}{\partial r}\right) - W_{cons.O_2} \tag{3.23}$$

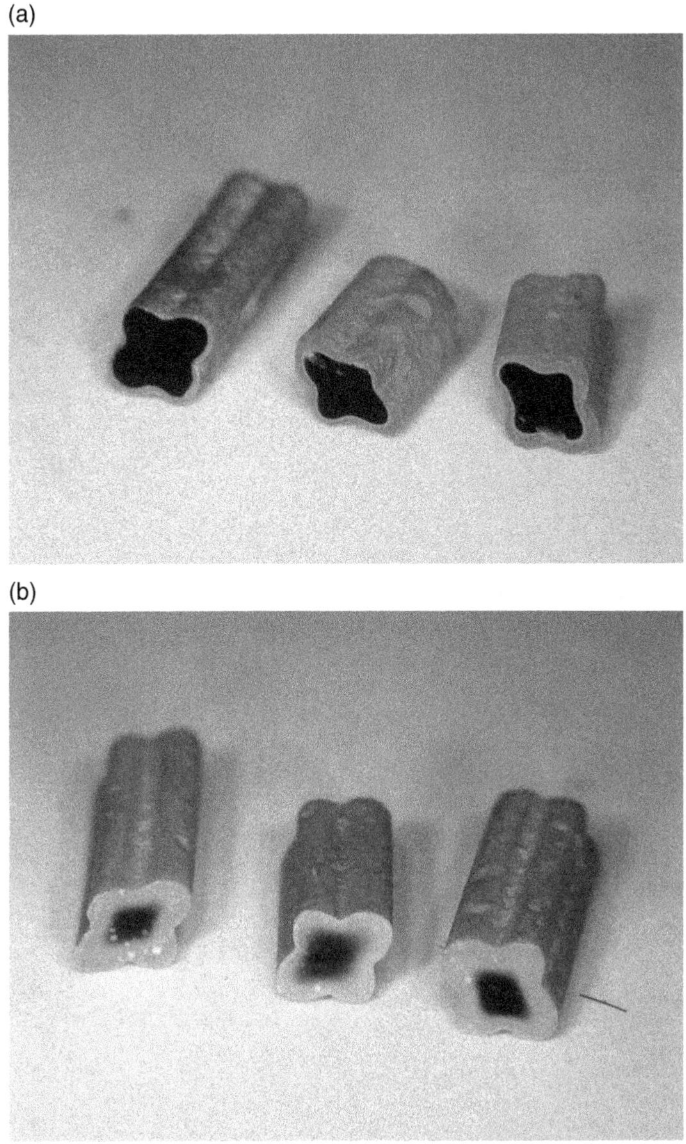

Figure 3.9 Transverse catalyst fracture from the lower (a) and upper (b) parts of the catalyst beds.

$$C^p_{N_2} = 1 - C^p_{SO_2} - C^p_{CO_2} - C^p_{O_2} \tag{3.24}$$

Equations for the rates of consumption and formation of gas components are as follows:

$$W_{\text{form}.SO_2} = 0.27 \cdot 1.667 \cdot \rho_c \cdot k_0^s \exp\left(\frac{-E_s}{RT_p}\right)\theta_S\left(C^p_{O_2}\right)^{0.5} \tag{3.25}$$

$$W_{\text{form}.CO_2} = 1.87 \cdot \rho_c \cdot k_0^C \exp\left(\frac{-E_c}{RT_p}\right)\theta_C\left(C^p_{O_2}\right)^{0.5} \tag{3.26}$$

$$W_{\text{cons}.O_2} = 1.72 \cdot W_{\text{обр}.SO_2} + W_{\text{обр}.CO_2} \tag{3.27}$$

Material balance for sulfide and coke concentrations on the inner surface of grains:

$$\frac{\partial \theta_S}{\partial t} = -k_0^s \exp\left(\frac{-E_s}{RT_p}\right)\theta_S\left(C_{O_2}^p\right)^{0.5} \tag{3.28}$$

$$\frac{\partial \theta_C}{\partial t} = -k_0^C \exp\left(\frac{-E_c}{RT_p}\right)\theta_C\left(C_{O_2}^p\right)^{0.5} \tag{3.29}$$

The change in the concentrations of reagents in the gas mixture along the length of the bed, taking into account external mass transfer processes, was described by the following equations:

$$u\frac{\partial C_{SO_2}}{\partial l} = \beta_{SO_2}S\left(C_{SO_2|r=R}^p - C_{SO_2}\right) \tag{3.30}$$

$$u\frac{\partial C_{CO_2}}{\partial l} = \beta_{CO_2}S\left(C_{CO_2|r=R}^p - C_{CO_2}\right) \tag{3.31}$$

$$u\frac{\partial C_{O_2}}{\partial l} = \beta_{O_2}S\left(C_{O_2|r=R}^p - C_{O_2}\right) \tag{3.32}$$

$$C_{N_2} = 1 - C_{SO_2} - C_{CO_2} - C_{O_2} \tag{3.33}$$

The heat balance equations for the gas flow and grains took into account the heat exchange of the gas phase with the catalyst, thermal conductivity along the bed, and heat release during chemical reactions. It was assumed that the grain volume is isothermal.

$$uc_p\frac{\partial T}{\partial l} = \alpha S(T_p - T) \tag{3.34}$$

$$\gamma_p\frac{\partial T_p}{\partial t} = -\alpha S(T_p - T) + \frac{\lambda}{\rho_c}\frac{\partial^2 T_p}{\partial l^2} + \frac{1}{V}\int_V \left[W_{\text{sulf.ox}}\left(\frac{17}{21}Q_1 + \frac{4}{21}Q_2\right) + W_{\text{coke.ox}}Q_3\right]dV \tag{3.35}$$

For Eqs. (3.21)–(3.34), the corresponding initial and boundary conditions were set:

$t = 0$, $0 < r < R$, $0 < l < L$: (Initial conditions for pellet internals)

$$C_{SO_2}^p(0,r,l) = C_{SO_20}^p(r,l), C_{CO_2}^p(0,r,l) = C_{CO_20}^p(r,l), C_{O_2}^p(0,r,l) = C_{O_20}^p(r,l), \tag{3.36}$$

$$\theta_S(0,r,l) = \theta_{S0}(r,l), \theta_C(0,r,l) = \theta_{C0}(r,l), \tag{3.37}$$

$t = 0$, $r = R$, $0 < l < L$: (Initial conditions for pellet outer surface)

$$T_p(0,R,l) = T_{p0}(l), \tag{3.38}$$

$t > 0$, $r = 0$, $0 < l < L$: (Symmetry conditions in pellet center)

$$\frac{\partial C_{SO_2}^p}{\partial r} = 0, \frac{\partial C_{CO_2}^p}{\partial r} = 0, \frac{\partial C_{O_2}^p}{\partial r} = 0, \tag{3.39}$$

$t > 0$, $r = R$, $0 < l < L$: (Mass exchange conditions on pellet external surface)

$$D_{SO_2}\frac{\partial C_{SO_2}^p}{\partial r} = \beta_{SO_2}\left(C_{SO_2} - C_{SO_2}^p\right), D_{CO_2}\frac{\partial C_{CO_2}^p}{\partial r} = \beta_{CO_2}\left(C_{CO_2} - C_{CO_2}^p\right),$$

$$D_{O_2}\frac{\partial C_{O_2}^p}{\partial r} = \beta_{O_2}\left(C_{O_2} - C_{O_2}^p\right), \tag{3.40}$$

***t* > 0, *r* = *R*, *l* = 0:** (Inlet conditions)

$$C_{SO_2}(t,R,0) = C_{SO_2,in}(t), C_{CO_2}(t,R,0) = C_{CO_2,in}(t),$$

$$C_{O_2}(t,R,0) = C_{O_2,in}(t), T(t,R,0) = T_{in}(t), \tag{3.41}$$

$$\lambda\frac{\partial T_p}{\partial l} = 0, \tag{3.42}$$

***t* > 0, *r* = *R*, *l* = *L*:** (Outlet conditions)

$$\lambda\frac{\partial T_p}{\partial l} = 0. \tag{3.43}$$

In the process of oxidative regeneration under consideration, the catalyst has a shape close to cylindrical. The average grain size was 1.6 mm in diameter and 4 mm in height. To go to the spherical pellet used in the mathematical model under consideration, it is necessary to go to the equivalent pellet diameter.

The equivalent diameter of a cylindrical grain in the spherical approximation is calculated by the following formula:

$$d_{sph} = \frac{6V_{cyl}}{S_{cyl}} = \frac{3dh}{d+2h} \tag{3.44}$$

where *d*—diameter, *h*—cylinder height. For the size of the catalyst used, the equivalent diameter is 2 mm.

The model assumes that the oxidation of molybdenum and cobalt sulfides is described by one summary Eq. (3.28), thus the concentration θ_S is the sum of the weight concentrations of molybdenum and cobalt sulfides.

At the initial time, the catalyst contains 17% wt. MoS_2 and 4% wt. $CoS_{0.89}$. This model assumes that the weight ratio between sulfides will continue to be 17:4.

Equations (3.25)–(3.27) are obtained as follows. Due to the fact that the molar ratio of sulfides is 7:3, on multiplying Eqs. (3.18) and (3.19) by 0.7 and 0.3, we get the total as follows:

$$0.7MoS_2 + 0.3CoS_{0.89} + 2.867O_2 \rightarrow 0.7MoO_3 + 0.3CoO + 1.667SO_2 \tag{3.45}$$

It follows that the rate of formation of SO_2 is equal to the following:

$$W_{form.SO_2} = -1.667\cdot\frac{d\theta_S}{dt}. \tag{3.46}$$

The rate of oxygen consumption can then be expressed as follows:

$$W_{cons.O_2} = \frac{2.867}{1.667}\cdot W_{form.SO_2} + W_{form.CO_2} = 1.72\cdot W_{form.SO_2} + W_{form.CO_2}. \tag{3.47}$$

However, for the correct transition from reaction rates in the solid phase to the rates of formation and consumption of gas components, it is necessary to introduce the corresponding coefficient in Eqs. (3.25) and (3.26). By definition of bulk density, 1 m^3 of catalyst contains $\rho_c\theta_S$ kg of sulfides. In this case, from 1 kg of molybdenum and cobalt sulfides (in a ratio of 17:4) in accordance with Reactions (18) and (19), 0.27 m^3 SO_2 can be obtained. Consequently, from $\rho_c\theta_S$ kg of sulfides,

$0.27\rho_c\theta_S$ m^3 SO_2 is obtained. Thus, per 1 m^3 of catalytic bed, there is $0.27\rho_c\theta_S$ m^3 SO_2. This means that the rate of SO_2 formation can be written as follows:

$$W_{\text{form.SO}_2} = -0.27 \cdot 1.667 \cdot \rho_c \cdot \frac{d\theta_S}{dt}, \tag{3.48}$$

which corresponds to Eq. (3.25).

Carrying out similar reasoning with Eq. (3.20), we obtain Eq. (3.26).

The mathematical model under consideration takes into account changes in the bulk density of the catalyst, heat capacity of the gas, and catalyst. Thus, the coefficients ρ_c, c_p, and γ_p are variables during the regeneration process.

The change in catalyst bulk density ρ_c occurs due to a change in the mass of the catalyst due to coke burnout and the transition of molybdenum and cobalt sulfides into oxides. In this case, the mass of the carrier (Al_2O_3) remains unchanged in a given volume (9000.68 kg/m^3); therefore, the recalculation of bulk density obeys the following equation:

$$\rho_{\text{new}} = 900 \cdot 0.68 + \rho_{\text{old}}\theta_C + \rho_{\text{old}}\theta_S + \rho_{\text{old}}\theta_O, \tag{3.49}$$

here θ_O is the relative mass content of molybdenum and cobalt oxides on the catalyst (kg/kg), which is calculated based on the initial and current content of sulfides (θ_S).

The heat capacity of the gas mixture is calculated based on the current values of the concentrations of the components:

$$c_p = c_{SO_2}c_p^{SO_2} + c_{CO_2}c_p^{CO_2} + c_{O_2}c_p^{O_2} + c_{N_2}c_p^{N_2}, \tag{3.50}$$

here $c_p^{SO_2}$, $c_p^{CO_2}$, $c_p^{O_2}$, and $c_p^{N_2}$ are heat capacities of the corresponding gas components (Table 3.4).

The heat capacities of the catalyst are also calculated based on the concentrations of the corresponding components and their heat capacities:

$$\gamma_p = \theta_S\left(\frac{17}{21}\gamma_{\text{MoS}} + \frac{4}{21}\gamma_{\text{CoS}}\right) + \theta_C\gamma_C + \theta_{\text{MoO}}\gamma_{\text{MoO}} + \theta_{\text{CoO}}\gamma_{\text{CoO}} + \theta_{Al_2O_3}\gamma_{Al_2O_3}, \tag{3.51}$$

where the corresponding terms are determined in Table 3.5.

In the integro-differential Eq. (3.35), the thermal effects of Reactions (18) and (19), Q_1 and Q_2, enter with coefficients 17/21 and 4/21, indicating the fraction occupied by molybdenum and cobalt sulfides in θ_S. $W_{\text{oxid.sulf.}}$ and $W_{\text{oxid.coke.}}$ are the rates of oxidation of sulfides and coke and correspond to the right-hand sides of Eqs. (3.28) and (3.29).

Heat effects of reactions are presented in Table 3.6.

Table 3.4 Heat capacities of gas components.

Component	O_2	N_2	CO_2	SO_2
Heat capacity, kJ/st.m^3/K	1.46	1.36	2.18	2.25

Table 3.5 The heat capacities of solid components.

Component	MoS_2	MoO_3	$CoS_{0.89}$	CoO	Al_2O_3	C
Heat capacity, kJ/kg/K	0.467	0.661	0.579	0.727	1.41	1.73

Table 3.6 Heat effects of reactions.

Reaction	kJ/kg
18	6644
19	4643
20	33,333

Table 3.7 Kinetic parameters of oxidative regeneration reactions.

Reactions	ln k_0 (kg per kg of catalyst per s)	*E/R* (K)	*n*	*m*
18 + 19	12.41	10,108	~1	~0.5
20	10.55	11,279	~1	~0.5

The values of the kinetic parameters for Eqs. (3.28) and (3.29) are obtained from thermodynamic calculations and presented in Table 3.7.

The coefficients of heat and mass transfer and the diffusion coefficient for the components of the gas mixture were equated to the coefficients calculated for oxygen, which were calculated using the following formulae (Aerov et al. 1979):

$$\alpha = \frac{\mathrm{Nu}\lambda_g}{d^{eq}}, \beta_{O_2} = \frac{\mathrm{Sh}_{O_2}\mathrm{D}_{O_2}}{d^{eq}}, \tag{3.52}$$

where

$$\mathrm{Nu} = A\mathrm{Re}^n\mathrm{Pr}^{1/3}, \mathrm{Sh}_{O_2} = A\mathrm{Re}^n\mathrm{Sc}_{O_2}^{1/3}, \mathrm{Sc}_{O_2} = \frac{\mu}{\rho_g D_{O_2}},$$

$$\mathrm{Pr} = 0.69, \mathrm{Re} = \frac{ud^{eq}\rho_g}{\mu\varepsilon}, S = \frac{6(1-\varepsilon)}{d_{sph}}, d^{eq} = \frac{4\varepsilon}{6(1-\varepsilon)}d_{sph}. \tag{3.53}$$

The correspondence between the coefficients A and n and the Reynolds number, Re, is shown in Table 3.8.

Viscosity, thermal conductivity, and gas flow density were calculated using approximations constructed on the basis of reference data in the form of functions of the reference temperature:

$$\mu = 1.551 \cdot 10^{-6}\left(\frac{T_{ref}^{3/2}}{T_{ref} + 124}\right)\mathrm{N} \cdot \mathrm{s/m2},$$

Table 3.8 The correspondence between the coefficients *A* and *n* and the Reynolds number.

Re	*A*	*n*
Re < 2	0.515	0.85
2 ≤ Re ≤ 30	0.725	0.47
Re > 30	0.395	0.64

$$\lambda_g = 5.44 \cdot 10^{-8} T_{ref} + 12.23 \cdot 10^{-6} \text{kJ/m/s/K},$$

$$\rho_g = \frac{353.6}{T_{ref}} \text{kg/m3}.$$

The reference temperature is 400 °C.

The effective thermal conductivity of the bed is calculated using the following formula:

$$\lambda = (1-\varepsilon)\lambda_c + 10\lambda_g \cdot \text{Re} \cdot \text{Pr}, \tag{3.54}$$

where λ_c—thermal conductivity of catalyst supports (Al_2O_3): $\lambda_c = 13$ J/s/m/K.

3.4 Model Solution Method

In view of the complexity of the equations set and the various types of equations included in the model, an analytical solution of the system is not possible. For the numerical solution of the model, a mesh was introduced on the variables r, l, and t. The sequential construction of a model solution at each time step made it possible to obtain a complete model solution at an arbitrary time interval.

To construct a discrete problem, the model equations were approximated by difference schemes. Parabolic-type Eqs. (3.21)–(3.23) and (3.35) were approximated by an implicit difference scheme of the first-order accuracy in the time variable t and the second in the coordinates r and l. As a result of approximation, systems of algebraic equations with tridiagonal matrices were obtained, the solution of which was carried out by the tridiagonal sweep method. In this case, the integral that determines the heat release over the pellet volume, which is part of Eq. (3.35) and defined in its discrete formulation at each time point and coordinate along the length of the bed, was calculated using the trapezoidal rule:

$$\frac{1}{V}\int_V [F(r)]dV = \frac{3}{4\pi R^3}\int_0^R \frac{4}{3}3\pi r^2[F(r)]dr = \frac{3}{R^3}\int_0^R r^2[F(r)]dr$$
$$= \frac{3}{2R^3}\sum_{i=1}^{n_r}\left(r_i^2 F(r_i) + r_{i+1}^2 F(r_{i+1})\right)\Delta r \tag{3.55}$$

Equations (3.28)–(3.32) and (3.34) were approximated by an implicit difference scheme of first-order accuracy.

The resulting discrete problem was solved by the following algorithm. The transition from time layer t to $t+\tau$ was as follows. At the entrance of the bed (at $l = 0$), the concentrations of the components and the temperature of the supplied gas mixture ($C_{SO_2}, C_{CO_2}, C_{O_2}$, and T) are known. Based on these values, solving the algebraic systems obtained from Eqs. (3.21)–(3.23) gives the concentrations of gas components in the pores of the grain ($C^p_{SO_2}, C^p_{CO_2}$, and $C^p_{O_2}$) at each radial point on the $t+\tau$ time layer. In this case, the values of surface concentrations of sulfides and coke (θ_S u θ_C) from the previous time layer were used. Thus, having obtained the value of $C^p_{O_2}$ at the $t+\tau$ layer, the values of θ_S and θ_C are calculated. Consequently, the concentration values of all components are determined at $l = 0$ on the $t+\tau$ layer. When moving to the next section along the length of the layer ($l = \Delta l$), the values of the concentrations of the components and the temperature of the gas flow ($C_{SO_2}, C_{CO_2}, C_{O_2}, T$) are calculated using the values $C^p_{SO_2}, C^p_{CO_2}, C^p_{O_2}$, and T_p from the previous time layer. Thus, according to the algorithm described earlier, the values of

$C^p_{SO_2}, C^p_{CO_2}, C^p_{O_2}, \theta_S$, and θ_C are found for $l = \Delta l$. By performing this procedure, the required number of times, the values of $C^p_{SO_2}, C^p_{CO_2}, C^p_{O_2}, \theta_S, \theta_C, C_{SO_2}, C_{CO_2}, C_{O_2}$, and T are found on the entire time layer $t + \tau$. After determining these values, solving the algebraic system obtained from Eq. (3.35) gives the values of the catalyst temperature T_p over the entire time layer $t + \tau$. Next, the transition to the next time layer is carried out according to the algorithm outlined earlier.

When implementing the algorithm for numerical solution of the model, the step along the bed length was taken equal to $\Delta l = 10^{-3}$ m, and the time step τ was taken equal to 10^{-3} s.

The work verified the mathematical model based on an experiment in quasi-adiabatic reactor. In the model, a mixture with oxygen content and temperature corresponding to the experiment was supplied to a catalyst bed of 5 cm long.

3.5 Modeling Results

For a better description of the experimental data, an effective diffusion coefficient was chosen equal to $D_{O2} = 5 \times 10^{-8}$ m^2/s, which is obviously less than standard value calculated earlier ($D_{O2} = 5 \times 10^{-7}$ m^2/s) for a coked catalyst with pores of radius 12 nm (Reshetnikov et al. 2020). The change in temperature profiles along the length of the bed over time during regeneration is depicted in Figure 3.10 for $D_{O2} = 5 \times 10^{-8}$ m^2/s. The experimental curves (black) are described quite well by the model curves (gray), with the exception of the tail part of the bed, which is not sufficiently heated during the experiments, which is also shown in Table 3.2. Figure 3.11 demonstrates the dynamics of changes in SO_2 and O_2 concentrations at the outlet of the bed in the experiment and model for both of the diffusivities.

It can be seen the case with $D_{O2} = 5 \times 10^{-8}$ m^2/s describes the experiment with better precise than with $D_{O2} = 5 \times 10^{-7}$ m^2/s. The model SO_2 concentration is sharper—the peak value is higher and emission is shorter than the experimental one. This is consistent with the slightly lower oxygen yield in the model. Figure 3.12 shows the coke content averaged by pellet volume in different

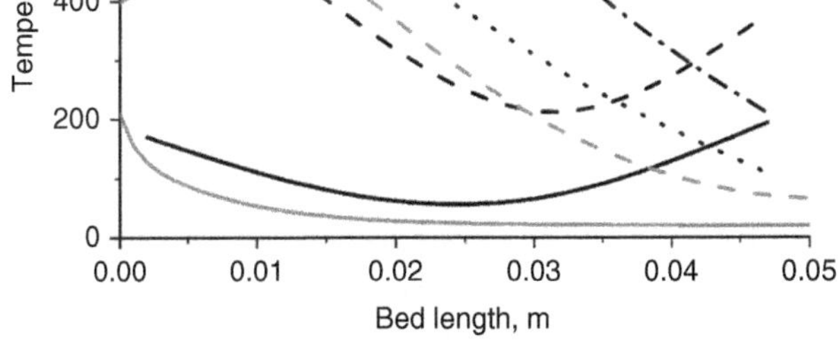

Figure 3.10 Temperature along the catalyst bed in different time points in experiment (black) and model (gray) during the oxidative regeneration.

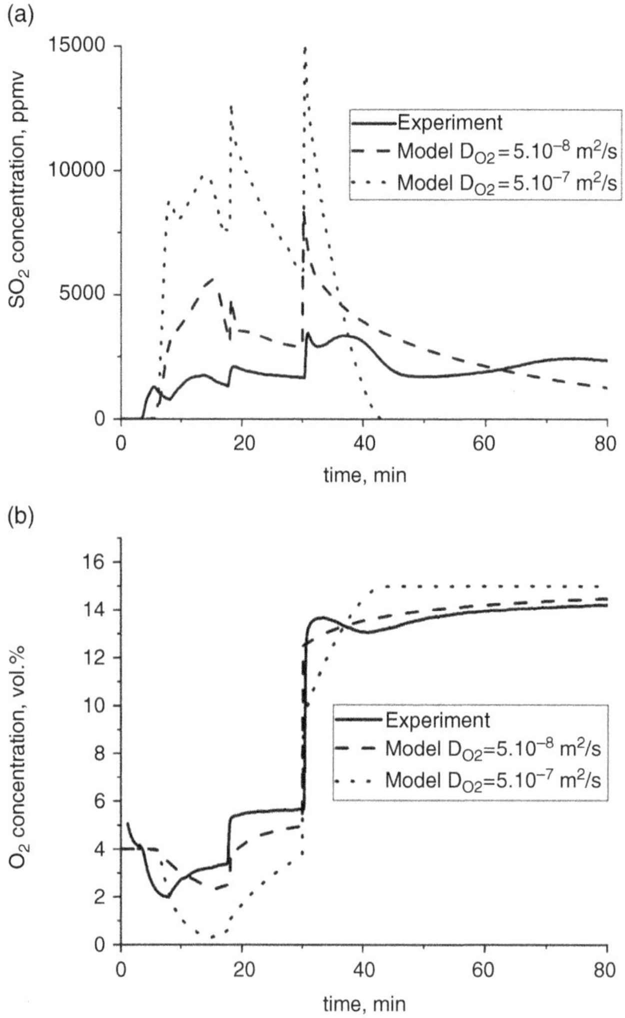

Figure 3.11 Simulated and experimental changing of SO_2 (a) and O_2 (b) concentrations at the reactor outlet in time during the oxidative regeneration.

bed points for $D_{O2} = 5 \times 10^{-8}$ m^2/s. Here, one can observe a good correlation between the experimental and model data, with the exception of the output bed region, which remained unregenerated during the experiment, as mentioned earlier.

3.6 Conclusion

At the moment, the proposed model provides a more or less accurate description of the experimental data and may be considered as essentially verified. At the same time, additional improvement of the model is required to provide better description of experiments. The first improvement may relate to the corrections of the kinetic model to better describe all observed regularities. Most likely,

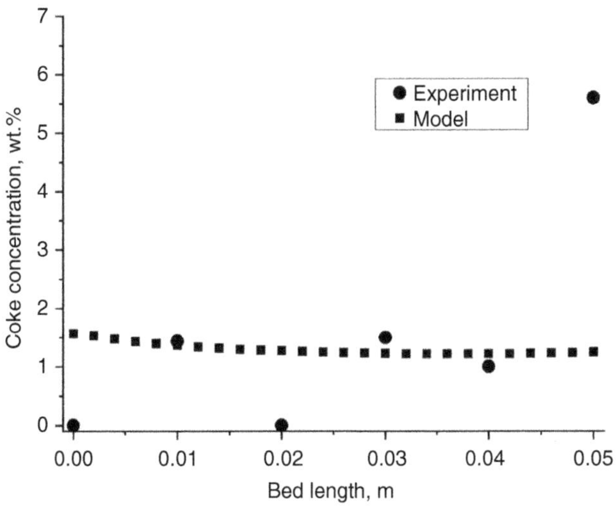

Figure 3.12 Experimentally measured and simulated average mass content of coke in catalyst pellets in the end of regeneration procedure.

it will be necessary to account for at least two types of coke with significantly different oxidation rates. Physically these differences may be related to the different chemical nature of the different coke itself ("crumbly" and "dense" coke).

Another, more serious issue is a more accurate simulation of catalyst pellets under visible influence of mass transfer limitations. The modeling revealed that the value of the effective diffusion coefficient, calculated within the conventional approach, resulted in significant disagreement between calculated and experimental data. A much better description is provided under the decrease of diffusion coefficient by an order of magnitude.

The first possible explanation is based on the hypothesis that the worse diffusion is caused by coke deposition in all pores; in this case, the diffusivity will increase with the decrease of coke content in the catalyst. It was demonstrated that this approach does not provide the appropriate improvement of the description as well.

Another explanation includes the assumption on significant deposition of coke not only in relatively large transport pores but also in micropores, which may become almost or completely blocked by coke. Obviously, the apparent kinetics of coke and sulfide oxidation in these micropores may be significantly different from the bulk one, and the effective diffusion coefficient may be significantly lower than that calculated for large pores. The decrease of diffusivity may be attributed to the decrease in Knudsen diffusion due to the decrease in effective size of pores partially filled with coke as well as to complete blocking of pore mouths, causing the "bottleneck" complication of diffusion at the micropore entrance.

The impact of nonuniform coke distribution among pores of different sizes on the regeneration process was recently theoretically predicted and analyzed (Zhdanov 2022). This assumption is also in good correspondence with the fact that such unusual decelerating of oxidation is not observed at all in kinetic experiments at fine catalyst powder, where the complication of diffusion in micropores does not influence the reaction parameters. In our opinion, this approach deserves further development.

3.7 Notation

C_i, gas flow component concentration, mole fraction;
C_i^p, gas component concentration in the pellet pores, mole fraction;
c_p^i, c_p, gas component and overall heat capacity, J/kg/K;
D_i, effective diffusion coefficients in pellet, m^2/s;
d_{sph}, equivalent diameter of a cylindrical grain in the spherical approximation, m;
E_s, activation energy, J/mol;
k_0, pre-exponential factor, 1/s;
l, bed length coordinate, m;
n_r, number of radial points;
Nu, Nusselt number;
Pr, Prandtl number;
Re, Reynolds number;
Sc, Schmidt number;
Sh, Sherwood number;
Q_i, thermal effects, kJ/kg;
r, pellet radial coordinate, m;
R, pellet radius, m;
S, outer specific surface area, m^2/m^3;
T, T_p, gas and bed temperature, K;
t, time coordinate, s;
u, linear gas velocity, m/s;
V, pellet volume, m^3;
W_i, rate of formation or consumption, 1/s;
α, heat transfer coefficient, $J/m^2/K/s$;
β_i, mass transfer coefficients, m/s;
γ_i, γ_p, surface compounds and catalyst heat capacity, J/kg/K;
Δl, bed length discrete step, m;
ε_c, catalyst porosity;
θ_i, surface concentration, mass fraction;
λ, λ_c, λ_g, effective bed, catalyst, and gas thermal conductivities, W/m/K;
μ, gas dynamic viscosity, kg/m/s;
ρ_c, ρ_g, catalyst bed and gas density, kg/m^3;
τ, time step, s.

Abbreviations

TG, thermal gravimetry
DTA, differential thermal analysis
MS, mass spectrometry

Acknowledgment

This work was partially supported by the Ministry of Science and Higher Education of the Russian Federation within the state assignment for the Boreskov Institute of Catalysis (project FWUR-2024-0037) and by the Tyumen Oblast Government, as part of the West-Siberian Interregional Science and Education Center's project No. 89-DON (3).

References

Aerov, M.E., Todes, O.M., and Narinskii, D.A. (1979). *Vessels with fixed granular bed. Hydraulic and heat operation fundamentals.* Leningrad, Chemistry.

Ferella, F., Innocenzi, V., and Maggiore, F. (2016). Oil refining spent catalysts: a review of possible recycling technologies. *Resources, Conservation and Recycling* 108: 10–20.

Froment, G.I. and Biscoff, K.B. (1999). *Chemical Reactor Analysis Design*, Seconde. New York: Wiley.

Furimsky, E. (1991). Effect of coke and catalyst structure on oxidative regeneration of hydroprocessing catalysts. *Fuel Processing Technology* 27: 131–147.

Kern, C. and Jess, A. (2005). Regeneration of coked catalysts—modelling and verification of coke burn-off in single particles and fixed bed reactors. *Chemical Engineering Science* 60 (15): 4249–4264.

Marafi, M., Stanislaus, A., and Furimsky, E. (2017). *Handbook of Spent Hydroprocessing Catalysts*, Seconde. Elsevier.

Meyers, R. (2016). *Handbook of Petroleum Refining Processes*, Fourthe. New York: McGraw Hill.

Reshetnikov, S.I., Petrov, R.V., Zazhigalov, S.V., and Zagoruiko, A.N. (2020). Mathematical modeling of regeneration of coked Cr-Mg catalyst in fixed bed reactors. *Chemical Engineering Journal* 380: 122374.

Speight, J.G. (2016). *Handbook of Petroleum Refining.* Boca Raton: Taylor&Francis.

Zazhigalov, S.V., Chumakova, N.A., and Zagoruiko, A.N. (2012). Modeling of the multidispersed adsorption-catalytic system for removing organic impurities from waste gases. *Chemical Engineering Science* 76: 81–89.

Zhdanov, V. (2022). Kinetics and percolation: coke in heterogeneous catalysts. *Journal of Physics A: Mathematical and Theoretical* 55 (17): 1–16.

Zhou, J., Zhao, J., Zhang, J. et al. (2020). Regeneration of catalysts deactivated by coke deposition: A review. *Chinese Journal of Catalysis* 41 (7): 1048–1061.

4

Modeling of Unsteady-State Catalytic and Adsorption–Catalytic Processes: Novel Reactor Designs

Sergey Zazhigalov[1,2], *Andrey Elyshev*[2], *and Andrey Zagoruiko*[1,2]

[1] *Boreskov Institute of Catalysis, Novosibirsk, Russia*
[2] *Tyumen State University, Tyumen, Russia*

4.1 Introduction

The protection of atmospheric air from the harmful effects of anthropogenic factors is subject to state supervision worldwide. The existing system of such supervision provides, among other measures, the organization of state, industrial and public control over emissions from industrial enterprises. Development of new scientific and technical approaches to improve the efficiency of purification of exhaust gases from volatile organic compounds (VOCs) (hydrocarbons, alcohols, acids, ethers, aldehydes, ketones, etc.) in industrial enterprises is part of such a system.

The problem of reducing VOC emissions from anthropogenic sources (Garcia et al. 2014; Gonzalez-Velasco et al. 2014; Ojala et al. 2011; Spivey 1987), raised in the 20th century, remains relevant even at present times. The environmental consequences of air pollution with VOC emissions are associated with a destructive effect on the ozone layer and increase in the amount of photochemical smog (Gonzalez-Velasco et al. 2014), as well as with a high level of toxicity of few VOCs. In connection with the increasing production capacities in areas such as energy, transport, chemistry and petrochemistry, mechanical engineering, wood processing, printing, production of building and finishing materials, and so on, research aimed at solving this problem is becoming part of a strategy to ensure the safety of life on earth.

Despite the existence of a large number of developed technologies for the neutralization of VOCs, they cannot be applied always and everywhere. The efficiency and applicability of these technologies should be evaluated considering their VOC abatement efficiency, capital and operation costs, and especially their energy consumption. The latter factor makes not only economic but also environmental sense, as it usually involves the use of additional fuels, which leads to an increase in the carbon footprint of technologies, and sometimes leads to the direct formation of toxic secondary atmospheric air pollutants.

For emissions with high concentrations of VOCs, it is cost-effective to use technologies based on recovery of organic substances from waste gases with their recycling back to the technological process (e.g., absorption, adsorption, condensation, and membrane methods). Despite technological advantages of these methods in application to gaseous wastes with a relatively high VOC content (usually above 1% mass), they become unfeasible in case of lean gases processing, when the cost of the recycled VOCs is significantly lower than the operating expenses for such recovery.

Mathematical Modeling of Complex Reaction Systems in the Oil and Gas Industry, First Edition.
Edited by Jorge Ancheyta, Andrey Zagoruiko, and Andrey Elyshev.

For diluted gases, it is expedient to use thermal and, especially, catalytic oxidation processes of VOCs by atmospheric oxygen to nontoxic products: carbon dioxide and water vapor (Matros et al. 1991). Although this method produces additional emissions of the resulting CO_2 to the atmosphere, the overall environmental effect is strongly positive.

Thermal methods, in absence of catalysts, require extremely high temperatures (often around 1000 °C), leading to limited combustion efficiency, and potentially resulting in the formation of secondary pollutants, such as CO, NO_x, soot, and so on. The concentration of VOC in waste gases is often insufficient for autothermal combustion; so such approaches usually require the consumption of additional fuel, worsening the economic and environmental efficiency.

Catalytic oxidation of VOCs has several advantages over thermal methods: a lower temperature for the oxidation reaction, higher efficiency of oxidation of VOCs, and most importantly, significant decrease of energy consumption in comparison with noncatalytic combustion. Reviews of the results of the neutralization of VOCs on catalysts of various types, such as noble metals or transition metal oxides on various supports, metal-containing zeolites, and so on, are quite widely presented in the scientific literature (Evaraert and Baeyens 2004; Li et al. 2009; Spivey 1987). To date, catalytic methods are considered as the most promising for the neutralization of VOCs in the waste gases from various industrial facilities.

Catalytic oxidation of VOCs occurs at elevated temperatures, typically from 100–150 up to 500–600 °C. However, waste gases very often have ambient or slightly elevated initial temperature, insufficient for complete VOC oxidation. The flow rate of these gases may be significant, and more energy may be required for their preheating. Therefore, the overall energy efficiency of catalytic processes is crucial for the efficiency of heat recuperation through the heat transfer from outlet purified gases to the inlet gaseous stream.

Modern improvement of catalytic processes for the neutralization of VOCs goes both in the direction of creating new effective catalysts for the complete oxidation of organic substances and in the line of creating new options for organizing the oxidation process. Currently, catalytic reverse-flow processes (Matros and Bunimovich 1996), which use the very efficient regenerative heat exchange, are capable of autothermally (i.e. without the supply of additional energy) processing gases with a VOC content of 0.6–0.8 g/m^3 and higher and have the best economic and environmental indicators for processing low-concentration emissions. For gases with lower VOC content, additional energy and capital costs are required. The reverse-flow process implies a periodic change in the direction of supply of the mixture being cleaned in two apparatuses. This tactic allows to save most of the heat released during the oxidation of impurities inside the catalytic bed (Figure 4.1).

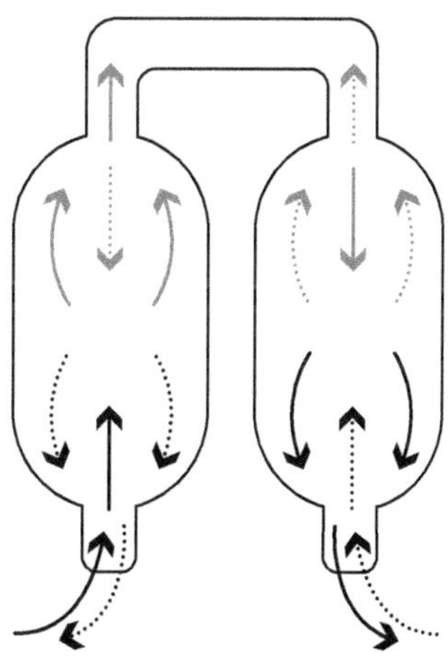

Figure 4.1 Reverse-flow process scheme.

Purification of emissions with extremely low impurity concentrations, that is, below 0.1 g/m^3, remains a serious and widespread practical problem (Garcia et al. 2014). From the viewpoint of compliance with the criteria of energy efficiency, ease, and safety in use and maintenance, adsorption–catalytic processes (ACPs) based on the combination of the principles of adsorption and catalytic technologies are of particular interest (Zagoruiko et al. 1996). Despite wide variety of variants of such processes proposed in the literature, their common essence lies in the fact that during the periodic process, adsorption and regeneration stages alternate (Robinson 1971). There are a significant number of such processes, which are a simple combination of

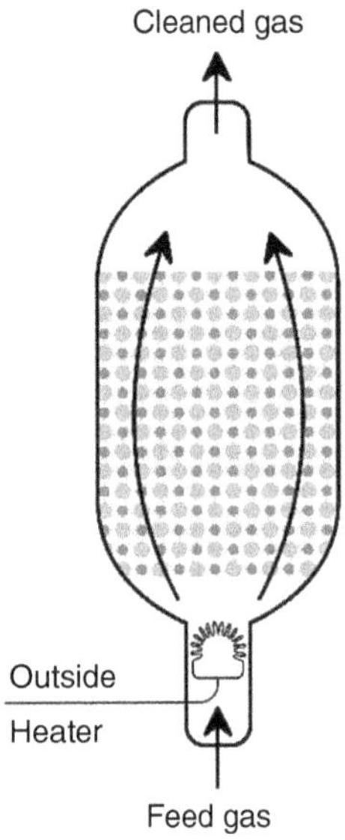

Figure 4.2 Adsorption–catalytic process with external heater scheme.

adsorbers with catalytic reactors. In this case, VOCs are initially adsorbed in the adsorbent bed, and then the sorbent is regenerated (usually by steam or hot air) and the desorbed impurities are catalytically oxidized in a separate apparatus.

Such a multistage scheme with the separation of the adsorbent and catalyst requires relatively large energy and capital costs, especially for the neutralization of large volumes of low-concentration industrial emissions. An interesting alternative to the schemes described above is ACP based on the adsorption of VOC directly on the surface of a deep oxidation catalyst (which also acts as an adsorbent) at ambient temperature. When VOC impurities appear at the exit from the bed, the adsorption stage is completed, and the bed is regenerated. In this case, the gas supplied to the bed is heated using a heater located in front of the reactor inlet (Figure 4.2). Upon completion of regeneration, the adsorbent–catalyst bed is again ready for the first stage, which is adsorption. ACPs at different times and in different versions of their technological organizations were considered and developed by UOP (Stanley and Ritscher 1979) and Haldor Topsoe (Schoubye 1979), as well as by several scientific institutes in the USSR.

Adsorption–catalytic and catalytic reverse-flow processes, with any method of their organization, have significant advantages over traditional catalytic technologies for the deep oxidation of VOCs, especially in cases of processing low-concentration gases. The use of ACP makes it possible to eliminate energy costs for the constant heating of gases to be purified, supplied to the apparatus with a catalyst. Their important advantage is high technological flexibility associated with the possibility of processing gases with varying initial concentrations and types of VOCs (up to the processing of a mixture of various VOCs in exhaust gases).

However, as shown by practical experience on the example of pilot plants and the study of ACPs by means of mathematical modeling (Zagoruiko et al. 1996; Vernikovskaya et al. 1999; Salden and Eigenberger 2001), they also have certain disadvantages:

- Insufficiently low power consumption when heating the reactor from an external heater, associated with inefficient energy consumption for heating the inlet gas duct and the inlet part of the reactor vessel;
- The possibility of partial desorption of nonoxidized VOCs or products of their incomplete oxidation during catalyst heating, leading to a decrease in the degree of gas purification. The release of VOCs during regeneration can also lead to bursts, when for a short period of time the concentration of VOCs in the exhaust gases may even exceed their initial concentration in the treated gases;
- The possibility of ejection of a gas stream with a high temperature (up to 300–500 °C) into the exhaust gas duct during regeneration. Such a hot gas discharge may slat for a relatively short time, but it can cause significant damage to gas communications, especially in ventilation systems where gas ducts are not designed to operate at high temperatures.

At different times, various ways to improve the ACP were proposed; for example, gradual "stepwise" heating of the catalyst during regeneration, regeneration in a filtration combustion wave, optimization of the size of catalyst–adsorbent pellets, application of technological schemes using a postreactor and spiral reactor with a specific system of internal heat exchange (Retallick

1996), the implementation of catalyst regeneration under conditions of a heat wave moving of combustion of adsorbed VOCs, propagating toward the motion of the gas flow due to the thermal conductivity of the catalyst bed.

Next, a proposed way to overcome disadvantages of the ACP is by changing the designs of the reactors used (Zazhigalov et al. 2016, 2017, 2018, 2023b); and the effect of asymmetric reactor designs on the efficiency and stability of the catalytic reverse-flow process (Zazhigalov et al. 2023a) is also estimated. Evaluation of the effectiveness of the proposed modifications of the apparatuses is based on mathematical modeling of processes in the commercial software, COMSOL Multiphysics.

4.2 Novel Reactor Designs for Catalytic Reverse-Flow and Adsorption–Catalytic Processes

4.2.1 Unsteady-State Catalytic Reverse-Flow Process

The stability and efficiency of the reverse-flow process are the two most important indicators that determine it and can vary significantly depending on the conditions of the gas mixture supply. Thus, an asymmetric mode of operation can lead to a significant decrease in stability and arise, for example, as a result of different intervals between switching the direction of the mixture supply or asymmetry of conditions, both at the inlet to the apparatus and at the initial time.

However, one of the most important factors affecting the asymmetry of the operating mode of the process is the asymmetric arrangement of the apparatuses themselves. As a rule, such asymmetry is determined by the nonstandard, nonaxial, location of the inlet or outlet ducts. Such designs can noticeably change the velocity field in the system, the distribution of reagent concentrations, and bed temperature. In this case, some areas of the reactors may remain unused in the process and reduce its overall efficiency.

The following reactor designs (Figure 4.3) will show the different behavior of process parameters in such systems as examples. Thus, reactors with axial, lateral, tangential, tangential–lateral, lateral–axial, lateral–orthogonal gas mixture supplies will be considered. This figure shows one of

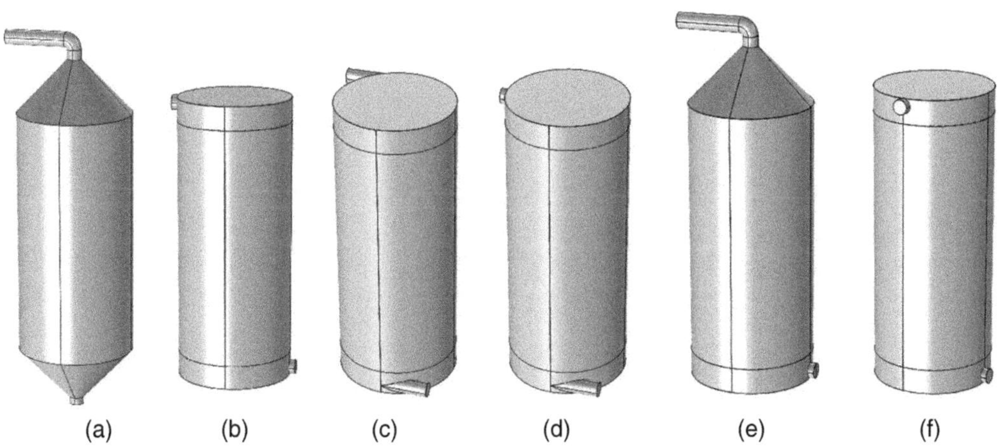

Figure 4.3 Computational geometry (one chamber) for process schemes with axial (a), lateral (b), tangential (c), tangential–lateral (d), lateral–axial (e), and lateral–orthogonal (f) gas inlet/outlet ducts.

the two devices, that is, half of the system being modeled. The catalyst bed has cylindrical shape with 0.5 m in diameter and 1 m in height in all cases.

4.2.2 Adsorption–Catalytic Process

Two important circumstances determine the energy consumption for the ACP—the pumping of the mixture to be purified through the reactor and the periodic regeneration of the bed, during which it is required to heat a certain amount of catalyst to the oxidation reaction temperature (~300 °C). It is shown that heating only the input region of the bed allows for efficient regeneration, as the thermal front moving in the direction of the flow allows the oxidation of impurities in the entire subsequent bed part in an autothermal mode. Due to the fact that the heating element operates for a short time, relative to the entire adsorption cycle, the energy consumption in the process remains quite low. If we talk about systems with large volumes of gases to be purified, then heating such volumes to the required temperature, even for a short period, can lead to quite large energy costs. It should also be noted that in this case both the inlet gas duct and the inlet of the reactor vessel are heated. Of course, such systems require a lot of heater power and must be adapted to peak high power demand.

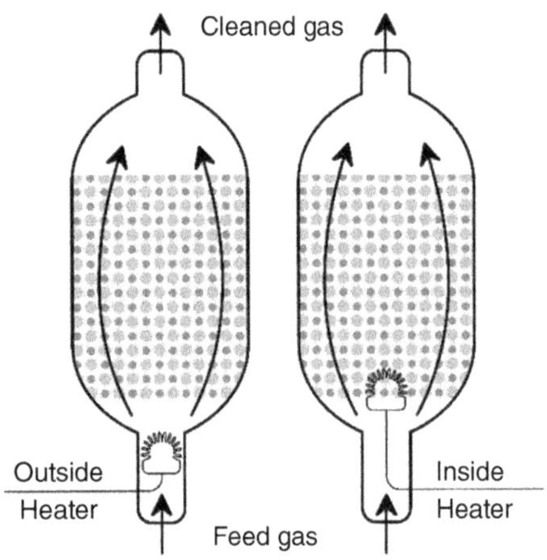

Figure 4.4 Adsorption–catalytic process with external (left) and internal (right) heater schemes.

The problems described above can be solved by few modifications in the system design. The location of the heating system directly in the inlet part of the adsorbent–catalyst bed makes it possible to heat not the supplied gas, but a small volume of the catalyst bed (Figure 4.4).

An evaluation comparison of various purification technologies using toluene as an example is presented in Figure 4.5. Specific energy consumption includes both costs of pumping the mixture and required heating. It can be seen that the catalytic reverse-flow process and ACP have their best performances when their inlet toluene concentrations

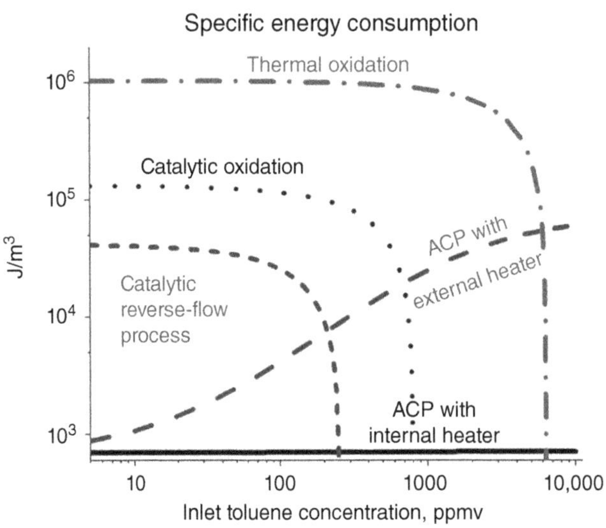

Figure 4.5 Specific energy consumption for different VOC abatement technologies on the toluene example (Adapted from Zagoruiko et al. 2021).

range below 1000 ppmv. Meanwhile, the location of the heating element inside the bed ensures significantly lower energy consumption (up to 1 order in the range of concentrations less than 500 ppmv that are relevant for ACP). Undoubtedly, properties of a particular adsorbent–catalyst, the stream to be purified, the type of impurity, and the organization of the process will affect the quantitative arrangement of the curves on this graph; however, one can qualitatively speak of the undeniable advantage of ACP in the low concentration region, especially, when the heating initiator is located directly in the bed.

Nevertheless, in the case of a large inlet section of the apparatus, it may also be necessary to use a sufficiently powerful internal heater. This problem can be solved using the inlet part of the apparatus, in the form of a truncated cone (Figure 4.6), then the thermal front from a small area at the inlet propagates to the full cross section of the reactor, and it moves not only in axial, but also in radial directions. In such a system, the dependence of energy consumption for regeneration on size of the reactor and volumetric flow rate of the feed mixture is minimized. An analogy of such a process can be drawn with bringing a match to a combustible substance, where the size of the object does not determine the required power of the match, and the process after ignition proceeds in an autothermal mode.

With the described organization of the process, after the passage of a thermal wave, the bed again acquires the temperature of gases to be purified, at which the adsorption of impurities occurs; thus, the process is continuous, that is, there is no need to stop the supply of the mixture during regeneration. In the case of a system with a truncated cone inlet, the question arises which cone angle must be used in the design to ensure stable operation of the system. In addition to the cylindrical reactor, apparatuses with conical angles from 20 to 70° of the same volume (Figure 4.6) are also considered by the method of mathematical modeling.

As described above, one of the problems limiting the application of the ACP is the burst emissions of VOCs and high-temperature stream during the regeneration process. The internal location of the heater, obviously, has almost no effect on these indicators. Meanwhile, the average degree of purification in the process can remain high and the duration of burst emissions can be a low fraction of the process cycle time, but the peak concentration of VOCs can exceed both the concentration in the treated stream and the limit values for a specific substance.

Changing the system design by dividing it into several separated independent adsorption–catalytic systems allows solving the described problem. In this case, it is necessary to provide conditions under which all these systems will have a common inlet and outlet of the gas flow, but separate heating elements for regeneration. Such a system can be implemented not only in separate reactor vessels, but also in one apparatus (Figure 4.7). In this implementation, the gas is supplied to all sections and distributed between them in a natural way. However, as each of the sections has its

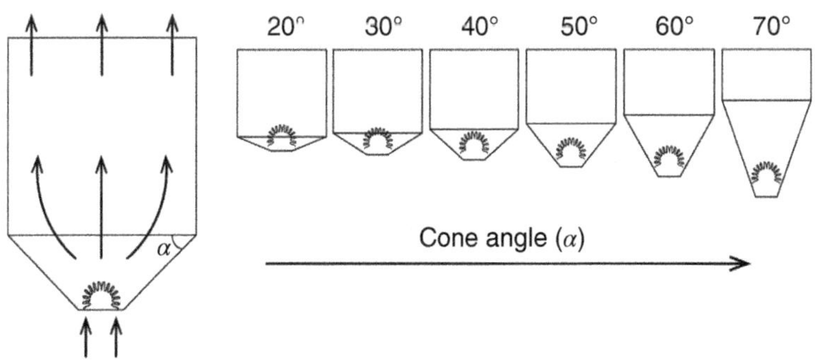

Figure 4.6 Reactors with truncated cones entrance part schemes.

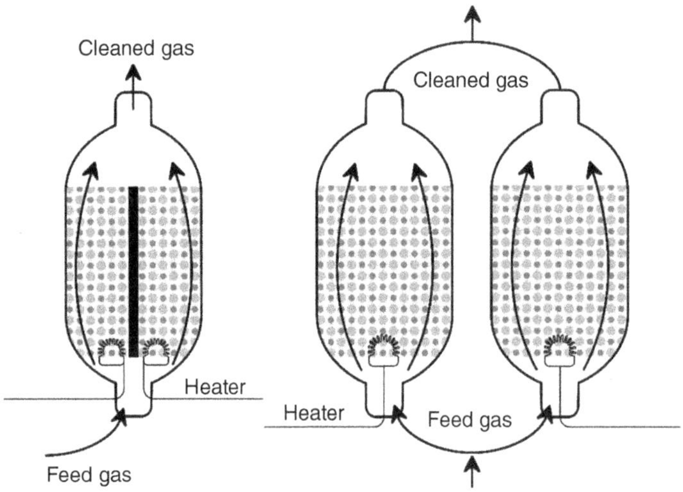

Figure 4.7 Multisectional adsorption–catalytic process schemes with sections in one reactor (left) and in distinct parallel apparatuses (right).

own heating element, regeneration will be carried out in them in turns. The time interval between regenerations is determined individually based on the properties of a specific system.

It is obvious that the concentration of VOCs leaving the regenerated section does not differ much from the concentration in a single-section system. But the concentration of VOCs leaving the remaining sections is close to zero, as these sections operate in adsorption mode. In the total outlet stream, the concentration of VOCs is reduced by mixing these streams, and a decrease in the peak temperature in the effluent gases is also achieved. Such a scheme makes it possible to reduce the peak values of VOC concentration in the outlet stream, without changing the total amount of non-oxidized impurities and, accordingly, the average degree of purification. However, it must be taken into account that the section operating in the regeneration mode has a higher temperature than the others, and therefore, has a higher hydraulic resistance, which limits the flow through it. Such a redistribution of flows probably makes it possible to reduce the total amount of impurities that have not been neutralized. In the case when sections are located in the same apparatus, in addition to the redistribution of flows, each of them has thermal contact with neighboring sections; therefore, the thermophysical properties of the dividing partition will be important here and will also determine the course of the process.

Here, cylindrical apparatuses with one, two, and five sections in one body will be considered and the influence of the number of sections on the main process indicators will be evaluated. Figure 4.8 shows the bed cross sections for these reactors.

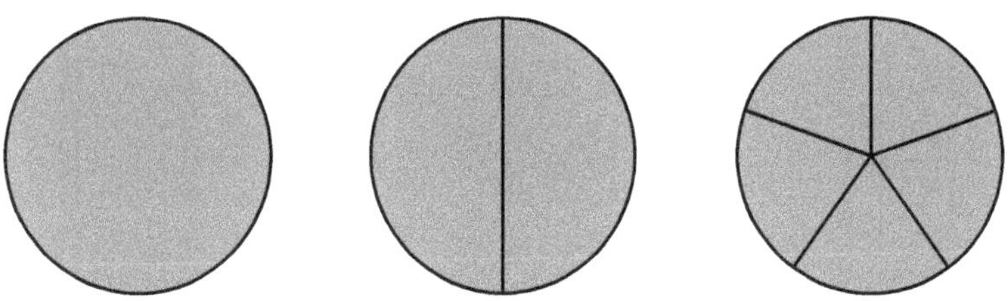

Figure 4.8 Cross section of the bed with one, two, and five sections.

4.3 Mathematical Models of the Processes

Obviously, the considered designs of reactors require the setting of three-dimensional mathematical models, because the ongoing processes have an inhomogeneous radial distribution in the reactor. However, some modifications can also be considered in a two-dimensional axisymmetric formulation, as the geometry and conditions of the problem are axially symmetrical. Such geometry in the model can significantly save computing resources. Basically, in the literature, models of both the reverse-flow process and ACP are found in a simplified one-dimensional formulation. When the radial distribution of process parameters is taken into account, it is also highly desirable for the distribution of the velocity field and pressure in the system to be taken into account. For example, the time-dependent Navier–Stokes Eqs. (4.1) and (4.2) for a free medium and the Brinkman Eqs. (4.3) and (4.4), which are a modification of Darcy's law taking viscous friction into account for a porous medium that presented the catalyst bed.

$$\rho\frac{\partial u}{\partial t}+\rho(u\cdot\nabla)u=\nabla\cdot\left[-pI+\mu\left(\nabla u+(\nabla u)^T\right)-\frac{2\mu}{3}(\nabla\cdot u)I\right], \tag{4.1}$$

$$\frac{\partial\rho}{\partial t}+\nabla\cdot(\rho u)=0. \tag{4.2}$$

$$\frac{1}{\varepsilon_p}\rho\frac{\partial u}{\partial t}+\frac{1}{\varepsilon_p^2}\rho(u\cdot\nabla)u=\nabla\cdot\left[-pI+\mu\frac{1}{\varepsilon_p}\left(\nabla u+(\nabla u)^T\right)-\frac{2\mu}{3\varepsilon_p}(\nabla\cdot u)I\right]-u\mu K^{-1}, \tag{4.3}$$

$$\varepsilon_p\frac{\partial\rho}{\partial t}+\nabla\cdot(\rho u)=0, \tag{4.4}$$

The heat balance can be represented by two equations for the temperature of the gas and catalyst; however, it is quite a reasonable simplification to use a single temperature model. In practice, the phase temperatures are quite close, especially for small grains, and the model includes the following Eq. (4.5), which takes into account the convective heat transfer, effective thermal conductivity of the bed, and heat released during chemical reactions.

$$\left((1-\varepsilon_p)\rho_c C_c+\varepsilon_p\rho C_P\right)\frac{\partial T}{\partial t}+\rho C_P u\cdot\nabla T-\nabla\cdot\left((1-\varepsilon_p)\lambda_c+\varepsilon_p\lambda\right)\nabla T=Q. \tag{4.5}$$

The presented system of equations, of course, should be supplemented by the mass balance, which is represented by slightly different equations for the systems under consideration. In both cases, toluene was used as a model VOC.

4.3.1 Unsteady-State Catalytic Reverse-Flow Process

The mass balance for the nonstationary catalytic reverse-flow process took into account diffusion and convective mass transfer in the flow, as well as changes in the concentrations of the supplied VOC and oxygen during the oxidation reaction. At low concentrations of reacting compounds, the model of diluted species can be applied, which does not take into account the change in the concentration of solvent compounds (4.6).

$$\varepsilon_p\frac{\partial c_i}{\partial t}-\nabla\cdot(D_i\nabla c_i+uc_i)=R_i, \tag{4.6}$$

It was assumed that only the toluene oxidation reaction occurs in the bed

$$C_7H_8+9O_2\rightarrow 7CO_2+4H_2O. \tag{4.7}$$

In each reactor, the catalyst was loaded into a cylindrical mold with a volume of 0.195 m^3, and the residence time was 3.1 seconds. During the entire process, the toluene concentration and the inlet gas mixture temperature were 1000 ppmv and 20 °C, respectively. The process began with the fact that the gas was supplied to a heated reactor, which had a temperature of 400 °C. At a certain point, when the heat wave approached the exit from the second reactor, the direction of the supply changed to the opposite. As a result of such alternating cycles of supply, the system either decays, losing energy, or comes to a stable operating cyclic mode, where the temperature and concentration dynamics in each cycle repeats that of the previous one. The maximum possible cycle times can be chosen as a criterion for assessing the stability and energy efficiency of the process.

Reaction (4.7) was assumed to have first orders in the participating components and the following kinetic parameters.

$$R = kc_{\text{tol}}c_{\text{O2}}, \tag{4.8}$$

where $k = k^0 \exp(-E/(R_g \cdot T))$ and $k^0 = 3 \cdot 10^7$ mol/($m^3 \cdot$ s), $E = 62{,}900$ J/mol.

The void of the bed was taken equal to 0.4, which corresponds to an average spherical filling, and the radius of the catalyst pellets for calculating the hydraulic permeability according to the Kozeny–Carman formula (4.9) was 2 mm. It was supposed to use the commercial copper–chromia–alumina deep oxidation catalyst ICT-12-8 (manufactured by Katalizator Co., Novosibirsk, Russia) in the model. The pellet porosity coefficient was 0.5, and the values of density (3900 kg/m^3), heat capacity (900 J/kg K), and thermal conductivity (27 W/m K) of the catalyst material were taken according to the aluminum oxide data.

$$K = \frac{R^2\, \varepsilon_p^3}{45\left(1-\varepsilon_p\right)^2}\,\text{m}^2. \tag{4.9}$$

4.3.2 Adsorption–Catalytic Process

In ACP, it was assumed that there were no reactions in the flow, and the conversion occurred in the pores of the pellet, on inner surface of the adsorbent–catalyst. The mass balance (4.10) took into account the mass transfer with the surface of the pellets, where the exchange condition (4.11) was set, as well as the diffusion transfer in the pores (4.12) with the corresponding reaction rates on the inner surface (4.13).

$$\varepsilon_p \frac{\partial c_i}{\partial t} - \nabla \cdot (D_i \nabla c_i + u c_i) = h_{D,i} \cdot \left(c_i - c_{pe,i}\right) \cdot S_b, \tag{4.10}$$

$$\left(D_{pe,i} \frac{\partial c_{pe,i}}{\partial r}\right) = h_{D,i} \cdot \left(c_i - c_{pe,i}\right) \text{ at } r = R, \tag{4.11}$$

$$\varepsilon_{pe} \frac{\partial c_{pe,i}}{\partial t} - \frac{1}{r^2}\frac{\partial}{\partial r}\left(-r^2 D_{pe,i} \frac{\partial c_{pe,i}}{\partial r}\right) = R_{pe,i}, \tag{4.12}$$

$$a \frac{\partial c_{surf,i}}{\partial t} = R_{surf,i}. \tag{4.13}$$

The process of VOC adsorption and subsequent surface regeneration are quite complex and consist of various stages, but it can also be described quite accurately by the following simplified scheme.

$$\text{A} + [\text{O}] \leftrightarrow [\text{A}] \tag{4.14}$$

$$[A] \rightarrow [P] \tag{4.15}$$

$$[P] + \nu\, O_2 \rightarrow CO_2 + H_2O + [\,O\,] \tag{4.16}$$

It is assumed that, initially, the gas phase compound A is physically adsorbed on the oxidized sites of the inner surface of the catalyst [O] with the formation of surface compound [A]. Then, the reversibly adsorbed compound [A] passes into the strongly bound irreversible chemisorbed form [P]. And substance [P] can already react with oxygen from the gas phase with the formation of carbon dioxide, water, and surface sites available for the subsequent adsorption of VOCs.

At this reaction rates (4.14)–(4.16) were presented as follows:

$$W_1 = k_1 c_{pe,A} c_{surf,O} - k_{-1} c_{surf,A}, \tag{4.17}$$

$$W_2 = k_2 c_{surf,A}, \tag{4.18}$$

$$W_3 = k_3 c_{pe,O} c_{surf,P}. \tag{4.19}$$

And the reaction rates for the bulk and surface species are as follows:

$$R_{pe,A} = -W_1, \tag{4.20}$$

$$R_{pe,O} = -\nu \cdot W_3, \tag{4.21}$$

$$R_{surf,A} = W_1 - W_2, \tag{4.22}$$

$$R_{surf,P} = W_2 - W_3, \tag{4.23}$$

$$R_{surf,O} = -W_1 + W_3. \tag{4.24}$$

Initially, the adsorbent–catalyst bed does not contain VOC impurities and the entire inner surface is filled with compound [O]. Next, a gas stream of 100 ppmv enters the bed containing compound A, which is gradually adsorbed in the bed, and when 5–10 ppmv VOC appears at the outlet, regeneration is carried out. Regeneration is initiated by turning on the internal heating element in the inlet part of the bed. In this case, for a system with several sections, the heater is turned on in turn at time intervals inversely proportional to the number of sections in the reactor. Thus, the time interval between switching on the heating element in each of the sections is equal to the period between regenerations in a single-section system.

The truncated cone inlet system was also compared with a cylindrical system having a diameter and a height of 50 cm. The conical inlet reactors had the same volume as the cylindrical reactor, so the height of their cylindrical part was changed in such a way as to maintain its diameter. The partitioned system was a cylinder with a diameter of 25 cm, and a height of 1 m, and the thickness of the separating partition was 1 cm.

The heated area was in the bed's entrance and cylindrical in shape with a diameter of 12 cm and height of 10 cm, which corresponded to about 1 kg of catalyst. In a system with several sections, the heating area occupied the entire cross section and had a height of 5 cm. The specific power of the heating element in all systems was 4 kW/l, and the time taken was 300 seconds.

For the cone entrance and multisectional (further in brackets) systems, the pellet radius, weight adsorption capacity, gas flow rate, and residence time were set as 4 (3) mm, 1.5 (2) wt.%, 0.03 (0.005) m^3/s, and 3.3 (10) s, correspondingly.

The unit volume catalyst adsorption capacity was calculated according to its a_{wt} wt.% meaning as follows:

$$a = \frac{a_{wt} \cdot \rho_c}{M_{VOC}} \text{ mol/m}^3. \tag{4.25}$$

Table 4.1 The heat effects, activation energies, and preexponential factors of the reactions (Zazhigalov et al. 2012).

i	1	−1	2	3
Q_i, kJ/mol	49.3×10^3	—	1.4×10^6	2.86×10^6
E_i, kJ/mol	0	37.3	37.3	75.8
k_i^0, mol/m^3/s	10^6 1/s	1.47×10^9	4680	1.53×10^6 1/s

Table 4.2 Thermal properties of wall materials (Pfundstein et al. 2008/Walter de Gruyter GmbH).

	Mineral wool	Expanded clay	Steel
Density, kg/m^3	100	500	7850
Thermal conductivity, W/(m·K)	0.05	0.2	44.5
Heat capacity, J/(kg·K)	1000	840	475

And the kinetic parameters and heat effects for Reactions (4.14)–(4.16) are in Table 4.1. The mass transfer coefficients were calculated in COMSOL using the equation:

$$h_{D,i} = \frac{\text{Sh}_i D_{pe,i}}{R}, \text{Sh}_i = 0.94\,\text{Re}^{1/2}\text{Sc}_i^{1/3}. \tag{4.26}$$

The molecular diffusion coefficients for toluene and oxygen were calculated from the following equations:

$$D_A = 7.7 \cdot 10^{-6}\left(\frac{T}{273}\right)^{3/2} \text{m}^2/\text{s}, D_O = 1.78 \cdot 10^{-5}\left(\frac{T}{273}\right)^{3/2} \text{m}^2/\text{s}. \tag{4.27}$$

The effective diffusion coefficients in pellet pores that are used in Equations (4.11) and (4.12) depend more on pores' properties and may vary in wide range. Based on various estimates, here they are taken to be 50 times smaller than the molecular ones.

As mentioned above, of course, the partition material between the sections in a multisectional system has a direct impact on the process, and so, three different materials are considered here, mineral wool (by default), expanded clay, and steel with thermal properties indicated in Table 4.2. In all corresponding figures, the gas moves from bottom to top.

4.4 Results

4.4.1 Unsteady-State Catalytic Reverse-Flow Process

As stated earlier, depending on the duration of the cycle, the reverse-flow process can either stabilize or die out due to large loss of heat. For each of the considered reactor geometries, the maximum cycle time for stable implementation of the process was calculated. The distribution of the temperature field (top) and toluene concentration (bottom) is presented in Figures 4.9 and 4.10 for

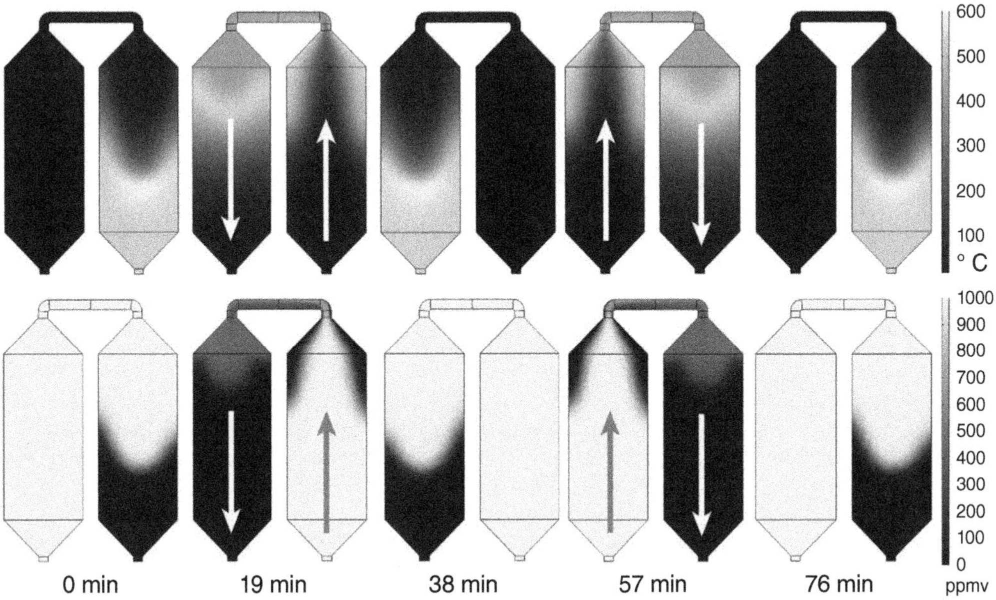

Figure 4.9 The temperature field (top) and toluene concentration (bottom) distribution during the cycle for the system with axial feed.

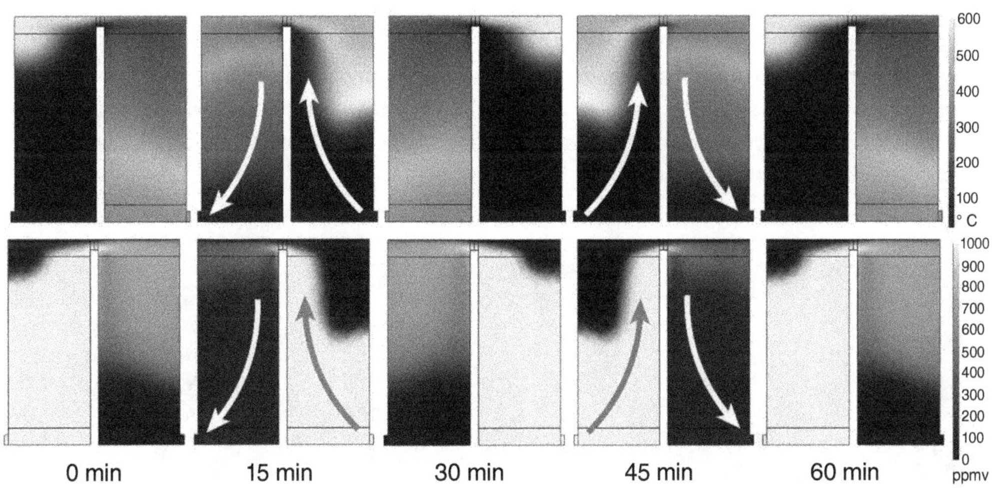

Figure 4.10 The temperature field (top) and toluene concentration (bottom) distribution during the cycle for the system with lateral feed.

axial and lateral feeds, respectively. The presented distributions correspond to the cycle durations of 76 and 60 minutes, and the time point 0 minute is the beginning of the cycle. It can be seen that in the middle of the cycle, the distribution is naturally symmetrical to its beginning and end.

In case of axial feed, small radial gradients are observed, but they are almost symmetrical, as here the reactor is axially symmetrical, and only the connecting tube introduces a slight asymmetry. As for the lateral feed, significant radial gradients are observed. So at the beginning of the cycle, the

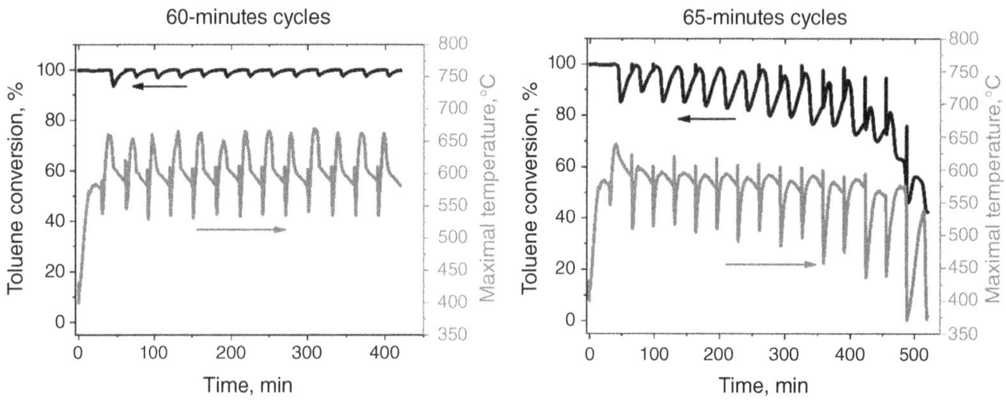

Figure 4.11 Toluene conversion and maximal temperature dynamics for the 60- and 65-minutes cycles in systems with lateral feed.

unevenness is observed mainly in the upper part of the reactors, and as the gas moves along the bed, the unevenness becomes more noticeable in that part of the system into which the gas is supplied.

If you look at the graphs (Figure 4.11) of toluene conversion and maximum bed temperature for cycles of 60 and 65 minutes, you can see that the former ensures stable operation of the system, because successive cycles repeat the previous one, and the second leads to attenuation after 8 cycles. The average conversion of toluene per cycle in the first case is 99.2%.

With regard to the tangential arrangement of feed ducts, the following distribution of temperature and concentration in the central plane (Figure 4.12) can be observed for a cycle of 88 minutes. However, the most interesting is the temperature distribution in the axial planes, which is shown in Figure 4.13 for a half-cycle. Here, one can observe the movement of the heat wave not only along the bed, but also a slight twist around the central axis, which theoretically can have a favorable effect on the efficiency of the process.

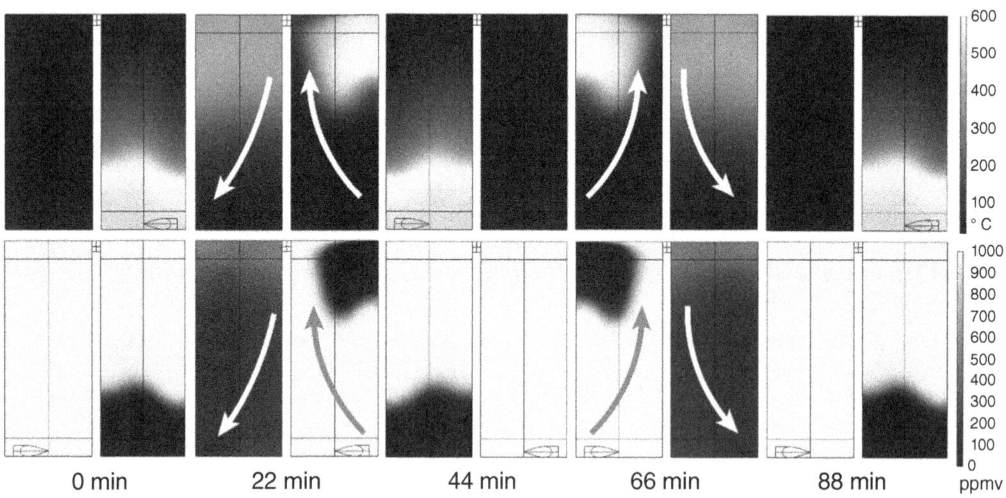

Figure 4.12 The temperature field (top) and toluene concentration (bottom) distribution during the cycle for the system with tangential feed.

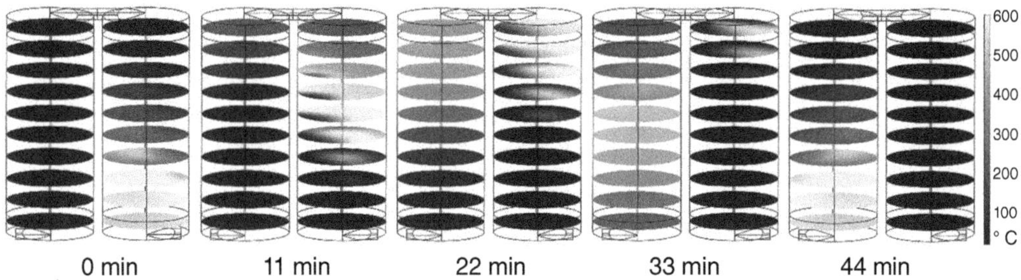

Figure 4.13 The temperature field distribution in axial planes during the half of 88-minutes cycle for the system with tangential feed.

The behavior of the system with a tangential–lateral arrangement of ducts is close enough to the behavior with a tangential arrangement. Similarly, a reactor with lateral inlets and an axial connecting duct gives temperature and concentration distributions close to that of a reactor with lateral ducts. A system with a mutually perpendicular arrangement of lateral ducts forces the gas to move in a not-quite straight direction, thus, showing greater stability than a system with a standard lateral arrangement. This fact is reflected in Table 4.3, which shows the cycle duration at toluene conversion as higher, 99%, including for a one-dimensional model often used to represent the reverse-flow process.

The 1D model does not take radial gradients into account, so it shows longer stable cycles with toluene conversion higher than 99% (97 minutes). The reactor with lateral ducts shows the worst stability among the modifications presented (61 minutes), the axial location of the ducts allows for longer cycles (80 minutes). One of the reasons is more uniform radial distribution of temperatures in the second case. Furthermore, the wider spread of residence time, coupled with uneven temperature distribution in the lateral feed reactor, also results in uneven distribution of toluene oxidation rates. This causes a correspondingly uneven heat release in the bed, and the interaction of these facts leads to a decrease in process efficiency, toluene conversion, and thermal stability.

Table 4.3 The cycle duration at toluene conversion >99% with different connecting ducts.

Type of inlet/outlet ducts connection	Cycle duration at toluene conversion >99%, min
1D	97
Lateral	61
Axial	80
Tangential	80
Tangential–lateral	77
Lateral–axial	63
Lateral–orthogonal	68

The lateral–axial (63 minutes) and lateral–orthogonal (68 minutes) systems are ranked between those described above, and the main inhomogeneities in them arise in the part of the reactor where the gas is supplied. It can be concluded that the design of the region connecting the reactor parts has little effect on the process. Therefore, the behavior of these variants is rather close to the behavior of a system with a lateral arrangement of branch ducts. The lateral–orthogonal system has few advantages, as it causes the gas flow to slightly change its trajectory, twisting slightly.

The systems with tangential feed turned out to be the most stable (77 and 80 minutes). Tangential systems provide stable cycles of up to 88 minutes with a toluene conversion of about 94%, which the axial type of supply cannot provide. The paths of the supplied mixture and the heat wave form a kind of spiral, which also allow us to speak of a more uniform radial distribution in the bed.

Undoubtedly, many factors influence the behavior of the process in all systems considered, including the size of the catalyst granules used. Therefore, for systems with lateral and axial gas supplies, in addition to grains with a radius of 2 mm, variants with radii of 1 and 4 mm were also considered. The size of the particles used affects the hydraulic resistance in the bed and the spatial distribution of process parameters. Figure 4.14 shows the temperature distribution for different particle radii. Large pellets lead to a lower hydraulic resistance, and due to this, the temperature front in the bed is somewhat washed out. The lower permeability at fine particles indicates a more uniform temperature distribution in the bed, with the maximum cycle time being almost independent of the system design (Figure 4.15). However, as particle radius increases, the axial feed system first shows better behavior, and then the lateral feed system shows better stability. This is due to the fact that at low pressure drop, the gas with axial flow moves predominantly along the central axis.

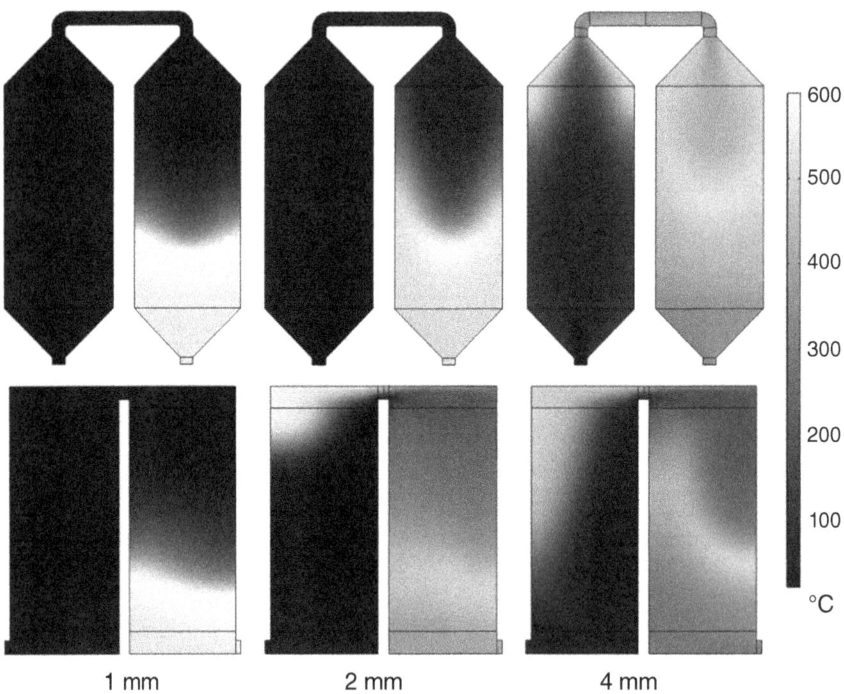

Figure 4.14 Temperature distribution at the end of the stabilized cycle for 1-, 2-, and 4-mm pellet radius in axial and lateral systems.

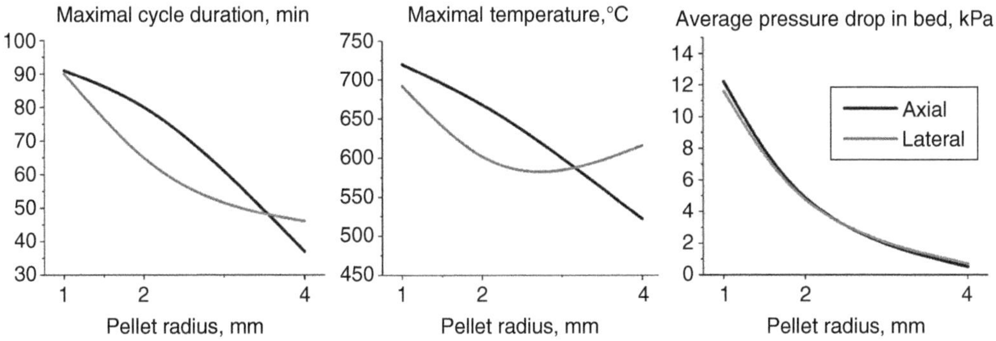

Figure 4.15 The maximal stable cycle duration, maximal temperature, and average pressure drop depending on pellet radius in axial and lateral reactor geometries.

4.4.2 Adsorption–Catalytic Process

4.4.2.1 Reactor with Truncated Cone Entrance

In ACP, the duration of the adsorption stage was determined by the moment of appearance of toluene at the exit from the bed in an amount of 5–10 ppmv. First of all, the simulation of the process in a cylindrical bed was carried out, where gas is supplied to the entire section, and regeneration is carried out by a heating element located in the center of the bed entrance.

The Figure 4.16a shows how toluene moves through the bed, penetrating into the pores of the pellets, adsorbs, and forms a reversibly sorbed compound [A] (Figure 4.16b). Over time, [A] transforms into the chemisorbed form [P] (Figure 4.16c). The temperature at the adsorption stage changes insignificantly and is in the range of 20–23 °C.

After the adsorption cycle is completed, the bed is regenerated by turning on the heating element. When a thermal wave moves along the bed (Figure 4.17a), part of the reversibly adsorbed impurities can be desorbed into the gas phase (in Figure 4.17b, the concentration peak reaches 5000 ppmv) and re-adsorbed further along the bed (Figure 4.17c), and part of it can pass into the compound [P] (Figure 4.17d). Oxidation of adsorbed VOCs leads to a significant release of heat during the reaction, which forms a temperature front that ensures further oxidation of VOCs over the bed and can exist due to the released heat. It can be seen that the thermal wave moves mainly in the axial direction—the direction of the gas flow—and almost does not propagate in the radial direction. After the passage of the heat wave, approximately 85% of the volume of the entire bed remained unregenerated. The oxygen is not consumed completely for oxidation, that is, its concentration in the mixture does not fall below 17 vol.%; this is important to avoid the formation of products of partial oxidation of toluene, possible under oxygen-lean conditions.

It can be assumed that with such an organization, the process cannot be sufficiently efficient and provide a high degree of purification. Because after the first regeneration most of the bed remained unregenerated, the duration of the second adsorption stage was only 10,000 seconds, as toluene appeared quite quickly at the outlet. Such a duration of the adsorption stage cannot provide a sufficient amount of accumulated impurities for carrying out regeneration in an autothermal mode. The duration of the third adsorption stage is 5000 seconds, and in the fourth cycle, the concentration of toluene at the outlet immediately exceeds to 10 ppmv, so the adsorption is inefficient due to the large part of the unregenerated bed. Figure 4.18 shows the distribution of adsorbed toluene in the bed after each of the three cycles.

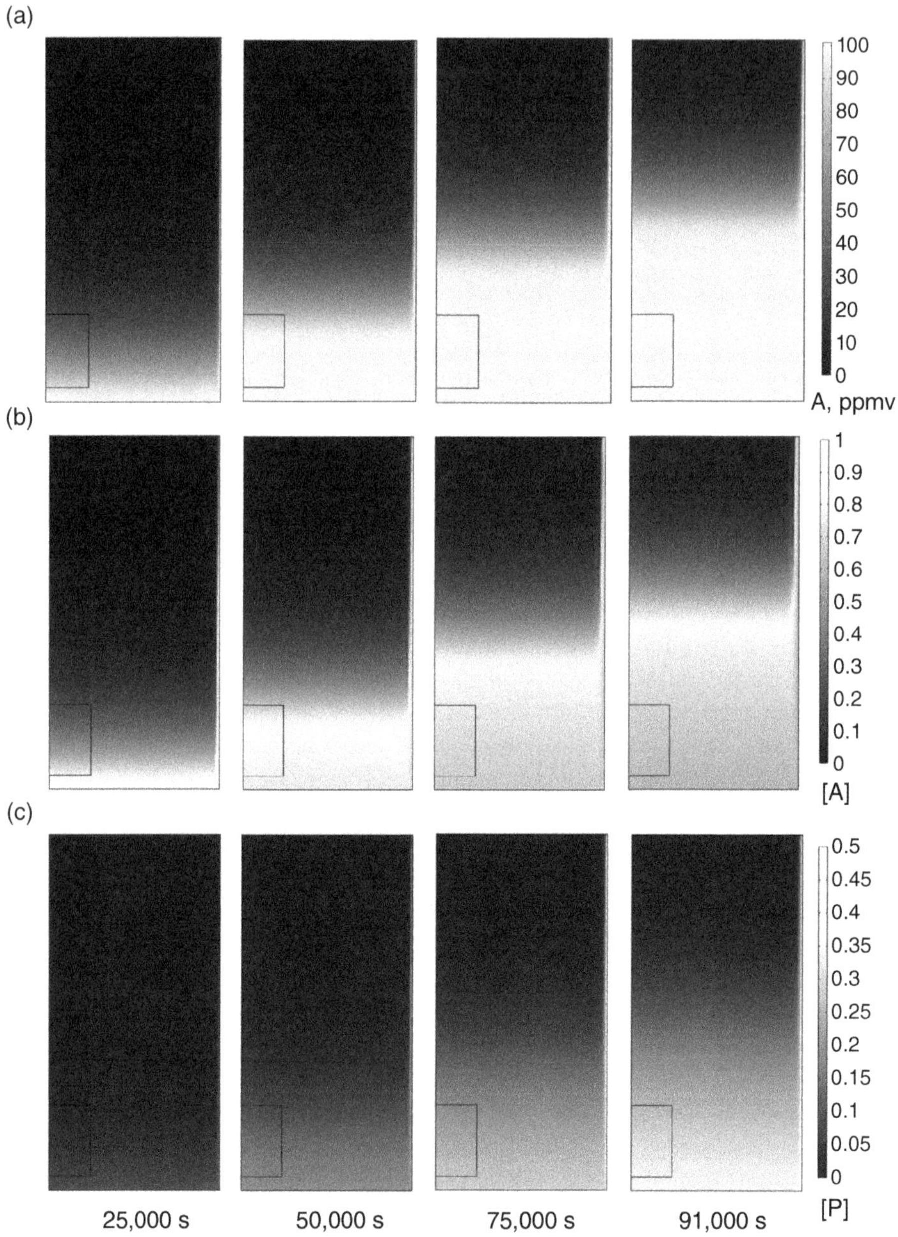

Figure 4.16 Toluene concentration (a), reversibly [A] (b), and irreversibly [P] (c) adsorbed toluene surface fraction averaged on pellet volume in cylindrical bed at the adsorption stage.

The process proceeds in a completely different way in a reactor with an inlet in the form of a truncated cone. The adsorption and regeneration steps are shown in Figures 4.19 and 4.20 using the example of a 70° cone angle reactor. At the adsorption stage (Figure 4.19), the movement of toluene over the bed and the formation of compounds on the inner surface of the adsorbent is fundamentally the same as for the cylindrical bed.

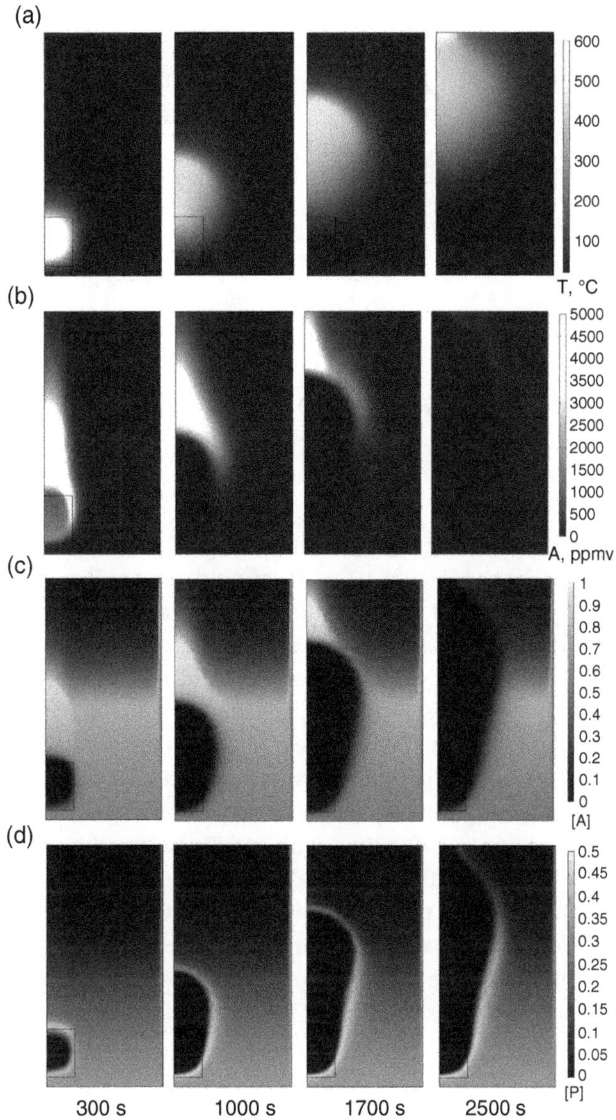

Figure 4.17 The temperature distribution (a), toluene concentration (b), reversibly [A] (c), and irreversibly [P] (d) adsorbed toluene surface fraction averaged on pellet volume in cylindrical bed at the regeneration stage of the first cycle.

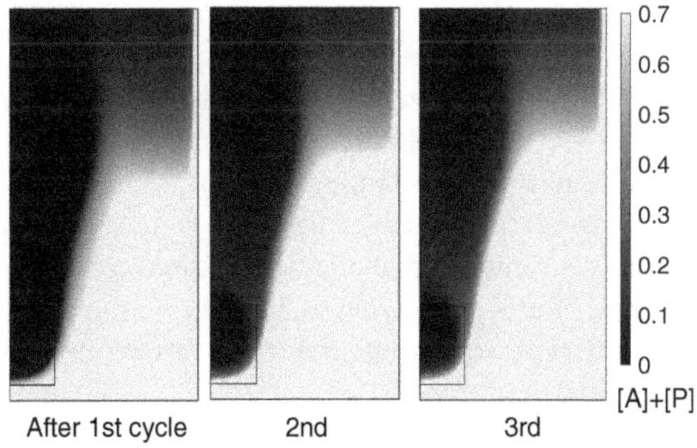

Figure 4.18 Adsorbed toluene distribution [A] + [P] in the end of cycles.

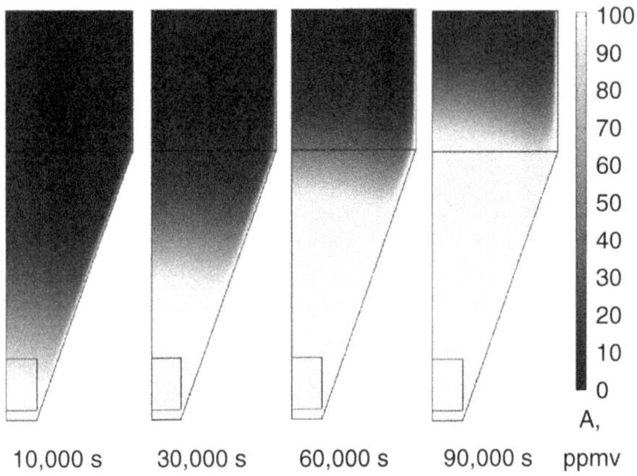

Figure 4.19 Toluene concentration in the bed with truncated cone entrance with 70° angle at the adsorption stage of the first cycle.

However, during the regeneration stage, the heat wave propagates from a narrow inlet, covering most of the bed (Figure 4.20a). It can be seen from Figure 4.20b–d that after the passage of the heat wave, only parts of the bed at the inlet remained unregenerated (because the region is cooled by the supplied cold air) and along the walls of the conical part.

With repeated cycles of adsorption and regeneration, the area of the inactive zone does not expand, it occupies 13% of the bed volume, and the process stabilizes (Figure 4.21).

Undoubtedly, the value of the cone angle required for a cyclically repeating ACP depends on the parameters of the bed and the supplied mixture. For example, with the parameters under consideration, an angle of 60° cannot provide stable repeating cycles (Figure 4.22). The nonregenerated area expands and after the eighth cycle, adsorption cannot be carried out, because the concentration of toluene at the outlet of the bed exceeds 10 ppmv. Increasing the angle to 80° makes it possible to reduce the portion of the unregenerated area to 5%. Changing the process parameters affects the optimal value of the cone angle; however, it is qualitatively shown that the presented geometric modification of the process makes it possible to efficiently conduct ACP when heating at the stage of regeneration of a small area at the entrance to the bed.

4.4.2.2 Multisectional Reactor

Modeling of a multisectional system was carried out on the example of reactors with one, two, and five sections. For a single-section system, the duration of the adsorption stage (until the appearance of toluene at the outlet in an amount of 10 ppmv) was 200,000 seconds. Accordingly, for a system with several sections, the interval between regenerations was $2 \times 10^5/N$, where N is the number of sections.

The Figure 4.23a shows the concentration of toluene in the gas phase at the adsorption stage for a one-section system. Similar to the previously considered process, toluene moves through the bed and adsorbed on the inner surface of the pellet (Figure 4.23b and c). Concentrations of surface substances averaged on pellet volume give a general idea of their content in the bed; as for the distribution in the pellet, Figure 4.24 gives a more detailed picture. Here, using the example of a particle lying at the bed entrance, it is shown how initially a region with the reversibly adsorbed compound [A] is formed near the outer surface of the pellet, which transforms into the form [P].

After completion of the adsorption stage, as in cases described earlier, the bed is regenerated. As a result, a heat wave (Figure 4.25) is formed in the bed, which ensures the oxidation of adsorbed

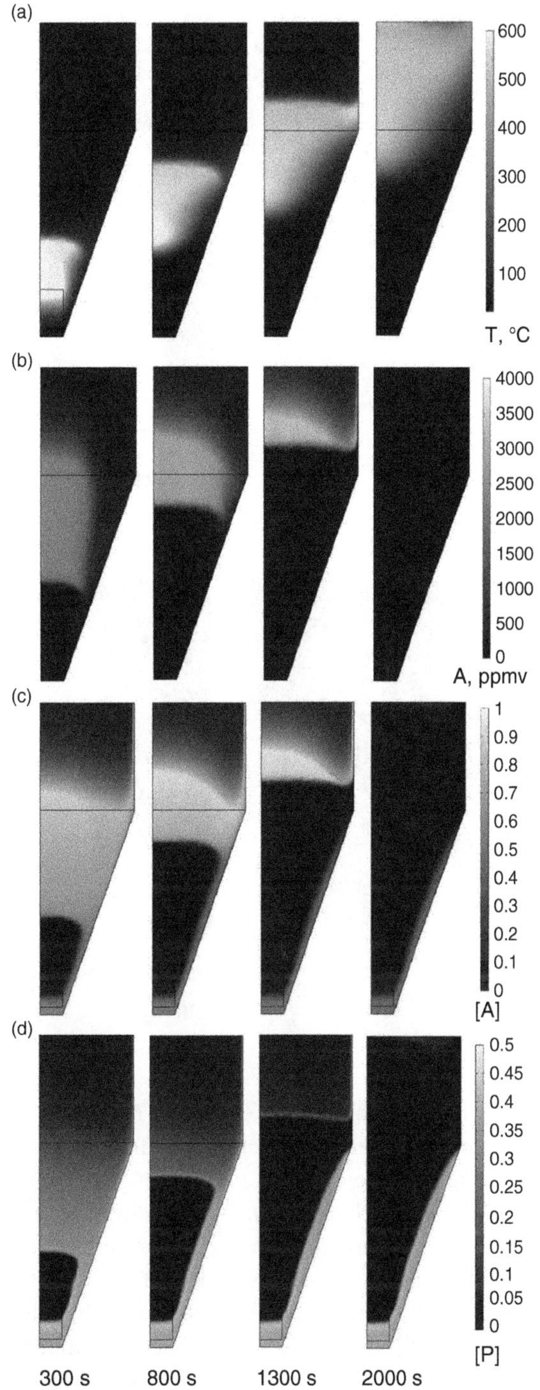

Figure 4.20 The temperature distribution (a), toluene concentration (b), reversibly [A] (c), and irreversibly [P] (d) adsorbed toluene surface fraction averaged on pellet volume in the bed with truncated cone entrance with 70° angle at the regeneration stage of the first cycle.

impurities. As before, part of the reversibly adsorbed impurities can be desorbed into the gas flow and reduce the overall degree of purification, which is calculated as the ratio of the amount of neutralized impurities to the total amount that entered the bed. Qualitatively, the distribution and dynamics of the concentrations of substances and temperature repeat those described for the system in Section 4.4.2.1.

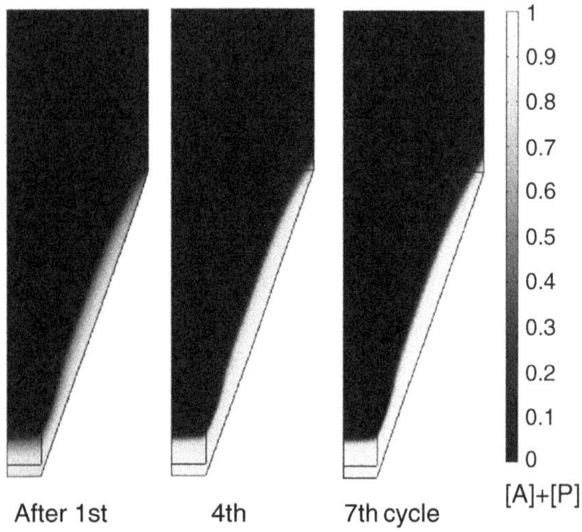

Figure 4.21 Adsorbed toluene distribution [A] + [P] in the end of cycles for the bed with truncated cone entrance with 70° angle.

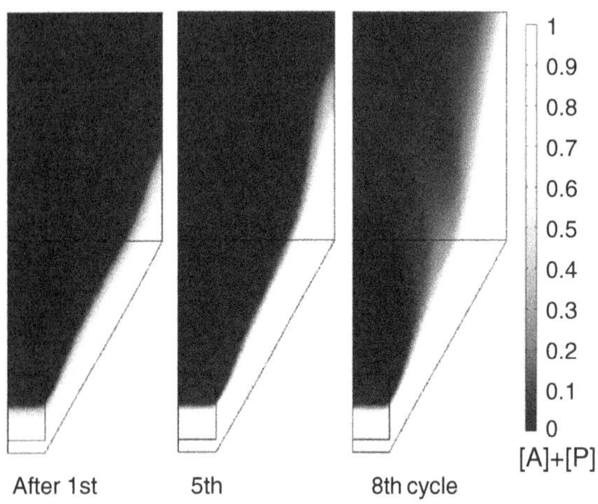

Figure 4.22 Adsorbed toluene distribution [A] + [P] in the end of cycles for the bed with truncated cone entrance with 60° angle.

Surface concentrations [A] and [P] in the pellet during heating change according to the Figure 4.26. After few minutes, the pellet surface becomes regenerated, and adsorption can proceed once again as before.

In a multisection system, when performing regenerations in turn, there is obviously some start-up mode of operation, because a short adsorption stage may not allow autothermal regeneration after the heating element is turned off. But after two complete cycles of regeneration of all sections, the system comes to a stable mode and each cycle repeats the previous one. An example of the distribution of concentrations of surface compounds for stabilized operation of a two-section system is shown in Figure 4.27a and b. Here, regeneration is carried out in the right section, and at the end of the half-cycle, the distribution of concentrations is symmetrical with respect to its beginning.

In a multisection system, the sections have thermal contact through a separating wall, and Figure 4.27c shows the temperature distribution in such a system. The maximum temperature in the nonregenerated section is relatively low and about 80 °C, because the wall material, mineral wool, has a low thermal conductivity. However, such an increase in temperature accelerates the

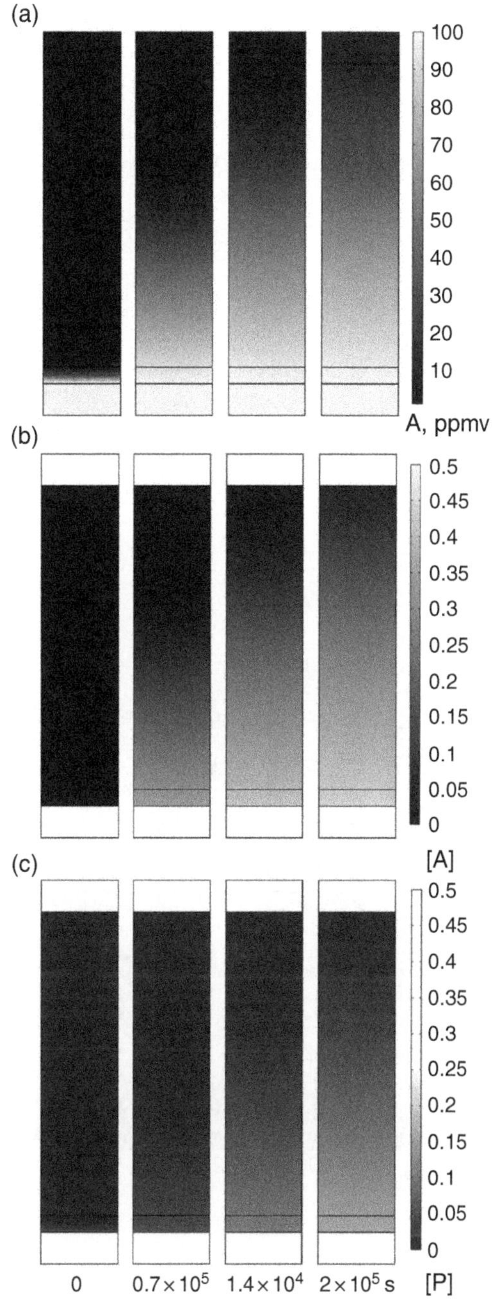

Figure 4.23 Toluene concentration distribution in gas phase (a), reversibly [A] (b), and irreversibly [P] (c) adsorbed toluene surface fraction averaged on pellet volume along the bed at the adsorption stage for one-sectional system.

desorption of toluene, which affects its concentration in the gas phase—it can slightly exceed the supplied 100 ppmv, but the toluene almost does not appear at the outlet, as it is successfully adsorbed further along the bed. At the same time, a rise in temperature also activates the reaction of the transition of toluene to the chemisorbed form, which can increase the final degree of purification.

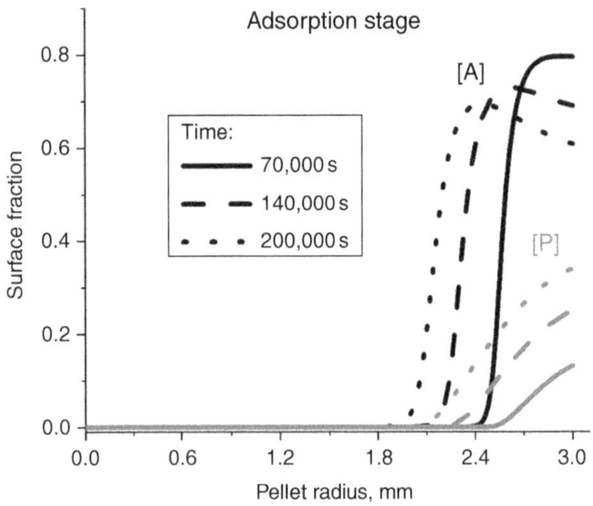

Figure 4.24 [A] and [P] concentrations in pellet at the reactor inlet at the adsorption stage for one-sectional system.

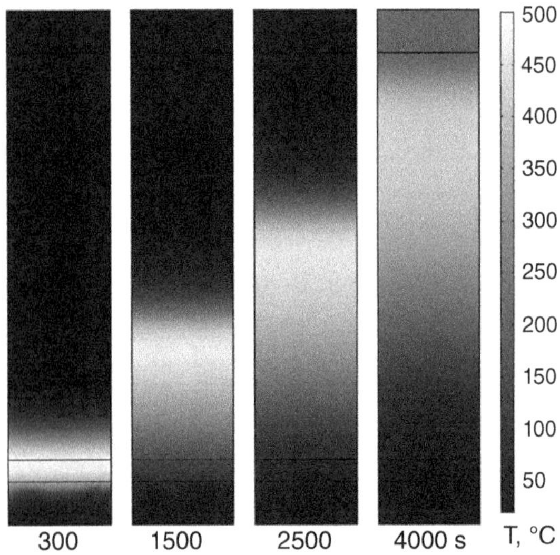

Figure 4.25 Temperature distribution along the bed volume during the regeneration for one-sectional system.

In a system with a larger number of sections, the process proceeds principally in the same way; Figure 4.28 shows the distribution of average concentrations [A] and [P] for a 5-section system. Here, regeneration will be carried out in the area with the highest concentration of adsorbed toluene and then in the remaining sections according to the arrow direction with a period of $2 \times 10^5/5 = 4 \times 10^4$ seconds.

The purpose of dividing the reactor into several sections was to reduce the peak concentrations of toluene and peak outlet gas temperature. The Figure 4.29a shows the concentration of toluene at the outlet of the reactor for a different number of sections. As mentioned earlier, for

Figure 4.26 [A] and [P] concentrations in pellet at the reactor inlet at the regeneration stage for one-sectional system.

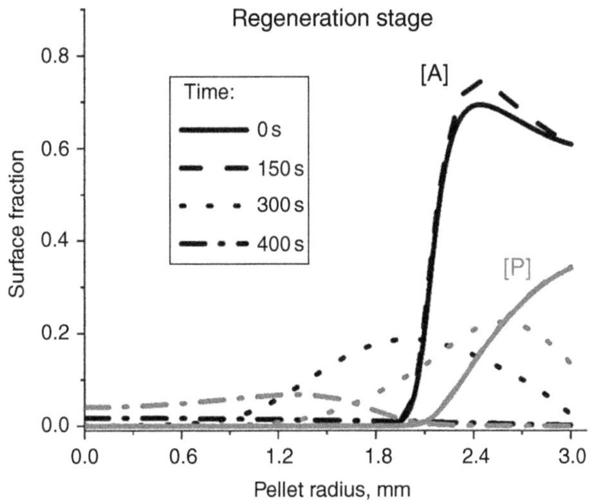

multisection systems, the first two complete cycles are the starting point for reaching stable values. During the first cycle, regeneration may not always take place in an autothermal mode, and in the second cycle, the concentration in the outgoing emissions may be slightly higher than expected in stable mode, as the bed has a larger amount of adsorbed toluene (part of it remained from the first cycle). Similarly, a reduction in peak temperatures in the outlet stream in multisectional systems is achieved (Figure 4.29b).

With the natural separation of the gas flow between sections, its rate in the regenerated section should be lower than in the others, as it has a higher temperature and pressure drop. This reduces the amount of emissions from the regenerated section, and one can observe a decrease in peak concentrations more than proportional to the number of sections. This also confirms the dependence of the degree of purification on the number of sections, presented in Table 4.4. A larger number of sections lead not only to an improvement in the above important indicators but also to a decrease in the peak pressure drop during the regeneration of one of the sections. However, given that too many sections can lead to excessive complexity of the design, we can talk about their optimal number—a compromise between process performance and the cost of manufacturing and operating the reactor.

Another factor limiting the number of sections is the increase in the specific heat transfer area with an increase in the number of sections. Thus, the heat from the regenerated section more actively flows into the adjacent sections and causes redistribution of the concentrations of surface compounds, which can have an ambiguous effect on the average degree of purification. Of course, the material of the partition between the sections, namely, its thermal conductivity, also has an important influence here. The Table 4.5 shows data for three different materials—mineral wool, expanded clay, and steel for a two-section system.

If we compare expanded clay and mineral wool, then the first one has five times higher thermal conductivity, which also affects the performance. Thus, for expanded clay, the maximum temperature in the regenerated section decreased from 440 to 390 °C, and in the neighboring section it increased from 80 to 170 °C. At the same time, the peak concentration of toluene at the outlet and the average degree of purification are almost equal in both cases. Steel baffles have a very high

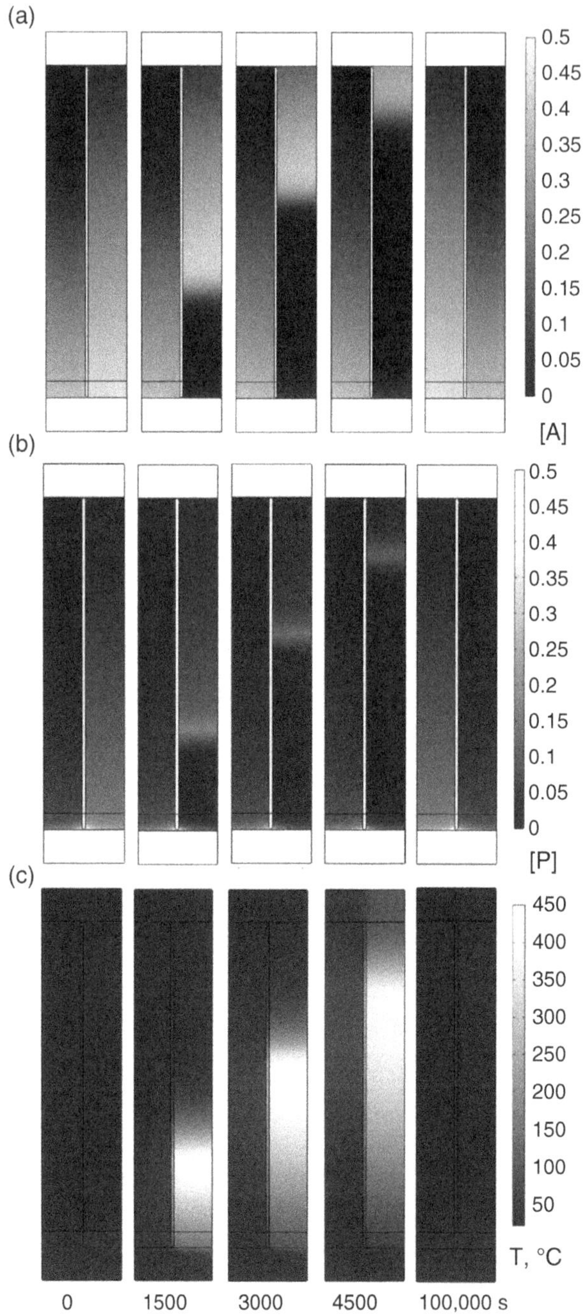

Figure 4.27 The reversibly [A] (a), irreversibly [P] (b) adsorbed toluene surface fraction averaged on pellet volume and temperature (c) during the established half-cycle in two-sectional bed.

thermal conductivity and the standard turn-on time of the heating element of 300 seconds does not allow reaching the ignition temperature due to high heat exchange with the adjacent section. A significantly longer operation of the heater leads to the simultaneous ignition of both sections and the behavior of the process becomes similar to the behavior of a single-section system, which makes the idea of sectioning meaningless.

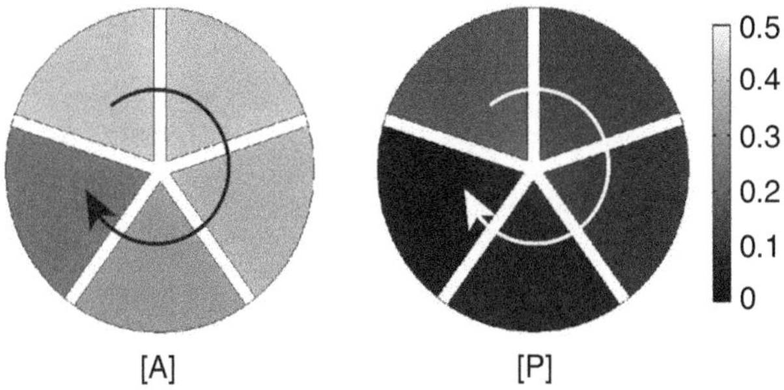

Figure 4.28 Reversibly [A] and irreversibly [P] adsorbed toluene averaged on pellet volume distribution in entrance section. Circular arrow–direction of the regeneration sequence.

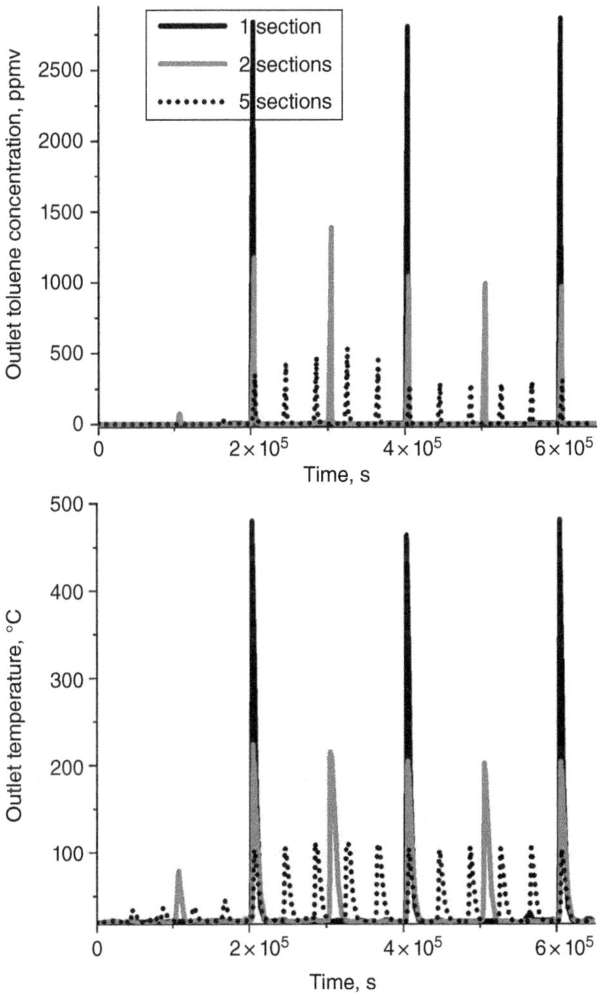

Figure 4.29 Outlet toluene concentration and gas temperature time dependence for different number of sections in the reactor (Zazhigalov et al. 2023b / American Chemical Society).

Table 4.4 The process parameters depending on the number of sections (Zazhigalov et al. 2023b/American Chemical Society).

Number of sections, *N*	1	2	5
Average VOC abatement degree, %	70.9	78.3	82.1
Peak outlet toluene concentration, ppmv	2870	990	290
Peak outlet gas temperature, °C	482	204	103
Average pressure drop, P_a	19.71	19.67	19.68
Maximum pressure drop, P_a	54.15	31.60	23.43

Table 4.5 The process indicators depending on the wall material in a two-sectional reactor (Zazhigalov et al. 2023b/American Chemical Society).

	Mineral wool	Expanded clay	Steel
Average VOC abatement degree, %	78.3	78.1	74.5
Peak outlet toluene concentration, ppmv	990	980	2200
Peak outlet gas temperature, °C	204	269	411

4.5 Conclusion

One of the directions for increasing the efficiency of processes for neutralizing waste gases from impurities of VOCs is, among other things, improving the design of reactors and the organization of the process. The study of the proposed modifications using the method of mathematical modeling allowed us to draw the following conclusions.

In the ACP, placing the heating system directly in the inlet part of the adsorbent–catalyst bed makes it possible to reduce specific energy consumptions by 1–2 orders of magnitude, while organizing the inlet part of the reactor in the form of a truncated cone increases the overall efficiency of the process in this case. Dividing the system into separate sections with an independent heating system allows not only to reduce the peak concentrations of unconverted impurities in the exhaust gases and outflow temperature, but also to increase the process average degree of purification from impurities.

It was shown for the reverse-flow process that both the concentration and temperature radial gradients present in the bed do not permit the use of a classical one-dimensional model for a detailed description of the process. Among the various arrangements of the inlet/outlet ducts of the reactors, the best performance was obtained for their tangential arrangement, where the thermal wave generated in the bed has a slight spiral twist, which ensures more complete bed utilization.

In general, it can be stated that the use of the CFD method for modeling nonstationary catalytic process and ACP opens up new opportunities for optimizing the design of reactors for their implementation with account of complicated dynamic phenomena, especially significant and specific in 3D geometrical representation.

4.6 Notation

[A] – product of reversible adsorption;
[O] – oxidized site at the catalyst surface;
[P] – product of VOC irreversible chemisorption on internal pellet surface;
A – VOC molecule;
a – catalyst adsorption capacity, mol/m^3;
a_{wt} – catalyst weight adsorption capacity, wt.%;
c_i – gas flow component (VOC or oxygen) concentration, mol/m^3;
$c_{pe,i}$ – gas component (VOC or oxygen) concentration in the pellet pores, mol/m^3;
$c_{surf,i}$ – surface compound, surface fraction;
C_c, C_P – catalyst and gas heat capacity, J/kg/K;
D_i, $D_{pe,i}$ – molecular and effective in pellet diffusion coefficients, m^2/s;
E_i – activation energy, kJ/mol;
$h_{D,i}$ – mass transfer coefficients, m/s;
K – bed permeability, m^2;
k_i, k_i^0 – reaction rate constant and pre-exponential factor, mol/m^3/s or 1/s;
M_{VOC} – VOC molar mass, kg/mol;
N – number of sections;
Q – heat source, W/m^3;
Q_i – thermal effects, kJ/mol;
r – pellet radial coordinate, m;
R – pellet radius, m;
$R_{pe,i}$ – reaction source term for gas components, mol/m^3/s;
$R_{surf,i}$ – reaction source term for surface compounds, mol/m^3/s;
S_b – active specific surface area, m^2/m^3;
Sh_i – Sherwood number;
T – bed temperature, K;
u – gas velocity vector, m/s;
ε_p, ε_{pe} – bed and catalyst porosity;
λ_c, λ – thermal conductivity of the catalyst and gas, W/m/K;
μ – gas dynamic viscosity, kg/m/s;
ρ – gas density, kg/m^3
ρ_c – catalyst bed density, kg/m^3

Abbreviations

ACP—adsorption–catalytic process
VOC—volatile organic compound

Acknowledgments

This work was partially supported by the Ministry of Science and Higher Education of the Russian Federation within the state assignment for the Boreskov Institute of Catalysis (project FWUR-2024-0037) and by the Tyumen Oblast Government, as part of the West-Siberian Interregional Science and Education Center's project No. 89-DON (3).

References

Evaraert, K. and Baeyens, J. (2004). Catalytic combustion of volatile organic compounds. *Journal of Hazardous Materials* 109: 113–139.

Garcia, T., Solsona, B., and Taylor, S.H. (2014). The catalytic oxidation of hydrocarbon volatile organic compounds. In: *Handbook of Advanced Methods and Processes in Oxidation Catalysis* (ed. D. Duprez and F. Cavani). Imperial College Press.

Gonzalez-Velasco, J.R., Aranzabal, A., Pereda-Ayo, B. et al. (2014). Catalytic oxidation of volatile organic compounds: chlorinated hydrocarbopns. In: *Handbook of Advanced Methods and Processes in Oxidation Catalysis* (ed. D. Duprez and F. Cavani). Imperial College Press.

Li, W., Wang, J., and Gong, H. (2009). Catalytic combustion of VOCs on non-noble metal catalysts. *Catalysis Today* 148: 81–87.

Matros, Y.S. and Bunimovich, G.A. (1996). Reverse-flow operation in fixed bed catalytic reactors. *Catalysis Reviews – Science and Engineering* 38 (1): 1–68.

Matros, Y.S., Noskov, A.S., and Chumachenko, V.A. (1991). *Catalytic Purification of Industrial Waste Gases*. Novisibirsk: Nauka.

Ojala, S., Pitkaaho, S., Laitinen, T. et al. (2011). Catalysis in VOC abatement. *Topics in Catalysis* 54: 1224–1256.

Pfundstein, M., Gellert, R., Spitzner, M., and Rudolphi, A. (2008). *Insulating Materials: Principles, Materials, Applications*. München: Birkhäuser.

Retallick, W.B., 1996. Air cleaner capable of catalytic regeneration. US Patent No.5487869.

Robinson, E., 1971. Treatment of gaseous effluent. UK Patent No. 1582441.

Salden, A. and Eigenberger, G. (2001). Multifunctional adsorber/reactor concept for waste air purification. *Chemical Engineering Science* 56: 1605–1611.

Schoubye, P., 1979. Method and apparatus for the removal of oxidizable pollutants from gases. UK Patent application 2051761.

Spivey, J. (1987). Complete catalytic oxidation of volatile organics. *Industrial & Engineering Chemistry Research* 26 (11): 2165–2180.

Stanley, P.B., Ritscher, J.S., 1979. Novel combustion process for an organic substrate. USA Patent No. 4234549.

Vernikovskaya, N.V., Zagoruiko, A.N., Chumakova, N.A., and Noskov, A.S. (1999). Mathematical modeling of unsteady-state operation taking into account adsorption and chemisorption processes on the catalyst pellet. *Chemical Engineering Science* 54: 4639.

Zagoruiko, A.N., Kostenko, O.V., and Noskov, A.S. (1996). Development of the adsorption-catalytic reverse-process for incineration of volatile organic compounds in diluted waste gases. *Chemical Engineering Science* 51 (11): 2989.

Zagoruiko, A.N., Bobrova, L., Vernikovskaya, N., and Zazhigalov, S. (2021). Unsteady-state operation of reactors with fixed catalyst beds. *Reviews in Chemical Engineering* 37 (1): 193–225.

Zazhigalov, S.V., Chumakova, N.A., and Zagoruiko, A.N. (2012). Modeling of the multidispersed adsorption-catalytic system for removing organic impurities from waste gases. *Chemical Engineering Science* 76: 81–89.

Zazhigalov, S.V., Mikenin, P.E., Lopatin, S.A. et al. (2016). An improved adsorption–catalytic process for removing volatile organic compounds from exhaust gases. *Catalysis in Industry* 8 (3): 231–241.

Zazhigalov, S., Mikenin, P., Pisarev, D. et al. (2017). Modifications of the adsorption-catalytic system for organic impurities removal. *Chemical Engineering and Processing: Process Intensification* 122: 538–549.

Zazhigalov, S., Chumakova, N., and Zagoruiko, A. (2018). Adsorption-catalytic process for removal of volatile organic compounds from lean waste gases: optimization of the adsorbent-catalyst bed geometry. *Chemical Engineering and Processing: Process Intensification* 132: 1–10.

Zazhigalov, S., Elyshev, A., and Zagoruiko, A. (2023a). Catalytic reverse-flow oxidation process in reactors of various designs: axial, side and tangential gas inlet. *Chemical Engineering Research and Design* 191: 364–374.

Zazhigalov, S., Elyshev, A., and Zagoruiko, A. (2023b). Mathematical modeling of adsorption-catalytic process with internal heater in multi-sectional arrangement. *Industrial & Engineering Chemistry Research.*

5

Molecular Reconstruction of Complex Hydrocarbon Mixtures for Modeling of Heavy Oil Processing

Nikita Glazov and Andrey Zagoruiko

Boreskov Institute of Catalysis SB RAS, Prospekt Akademika Lavrent'yeva, 5, Novosibirsk, Novosibirsk Oblast, Russian Federation

5.1 Introduction

Humanity needs fuel. Energy sources are crucial for today's human society, and the oil refinery is still an essential part of the modern economy. One of the challenging tasks it is facing right now is the efficient processing of heavy petroleum cuts, which becomes even more severe with each passing year. Many technological advancements, both small and big, have been made to solve this problem.

Of course, mathematical modeling is also widely employed in tackling this task on various levels, from the atomic level where the quantum chemistry methods unveil the details of catalytic processes to the level of the whole plant with different planning procedures.

This particular chapter is focused on one narrow application of mathematical modeling for a very common problem that emerges during the building of kinetic models for oil processing: the composition representation problem and a particular set of methods for solving it. These methods are often grouped together under the umbrella term "molecular reconstruction" (Ren et al. 2019).

Unlike the lumping strategy, molecular reconstruction often represents mixtures in terms of molecules/well-studied mixtures, rather than abstract compounds, which has a certain advantage. For example, models that use lumping approach to represent mixture can violate conservation of elements, which is not the case with reconstructed composition. Although, molecular reconstruction methods are often distinguished from lumping strategies, some of them can work with pseudocomponents as well.

5.2 The Problem

The number of chemical species in the heavy petroleum cuts is usually far beyond the capabilities of the current analytical methods. It means that whenever we want to build a kinetic model for an oil refinery process, we cannot simply rely on the composition of the products and feedstock (because it is usually unavailable). There is always a gap between the real composition and the information we have about it. To bridge this gap, we can use math. Notice that in case of the heavy petroleum cuts not only abundance of different molecules is unavailable but also the list of molecules in the mixture. Even the exact number of molecules is unknown.

Mathematical Modeling of Complex Reaction Systems in the Oil and Gas Industry, First Edition.
Edited by Jorge Ancheyta, Andrey Zagoruiko, and Andrey Elyshev.

This task might seem insurmountable or even impossible, however, this challenge could be overcome with the understanding that the composition we need might be just "close" to the real one instead of being the actual composition. The word "close" here is used in the vaguest way possible. We will see that the exact meaning in which the composition is close to the real one varies from method to method, but in general, reconstructed composition is better be seen as a very good model mixture that imitates everything we know about the sample rather than actual composition.

5.3 Illustration

To better understand what different molecular reconstruction methods imply, we will be considering the following toy example: a mixture of normal paraffines from C_{10} to C_{20}, and the only piece of information available would be the average carbon content in the mixture. This example is rather arbitrary and does not have (to the authors' knowledge) any meaningful application. However, it is sufficiently complicated to demonstrate some meaningful aspects of molecular reconstruction methods. Moreover, there is nothing particularly interesting in carbon content, and it all can be easily generalized.

Let α be the average carbon content in the mixture, α_i, M_i, and x_i be the carbon content, molar mass, and the molar fraction of the ith molecule, respectively. It is easy to see that any set of positive molar fractions that satisfies the following two equations represent a mixture with the average carbon content, exactly α.

$$\sum x_i M_i \left(\alpha_i - \alpha\right) = 0 \tag{5.1}$$

$$\sum x_i = 1 \tag{5.2}$$

This system has 2 equations and 11 unknowns, which is an extremely ill-posed problem. Problems in real application are even less well-posed.

5.4 Reconstruction by Entropy Maximization (REM)

From the earlier small analysis, it is clear that the system has numerous degrees of freedom. One way of approaching this problem is to notice that if we consider all possible solutions, there are many solutions with concentrations close to uniform values, than solutions with extreme values. The way to formalize it mathematically is Shannon entropy with Lagrange multiplier.

$$\varepsilon(x, \mu, \lambda) = -\sum x_i \ln x_i + \mu\left(1 - \sum x_i\right) + \lambda\left(0 - \sum x_i M_i \left(\alpha_i - \alpha\right)\right) \tag{5.3}$$

Notice that the framework of entropy maximization is very useful and general for distribution finding. Exponential, normal, beta, gamma distributions, and many more, including even Cauchy distribution, can be obtained through entropy maximization under certain constraints (Lisman and Zuylen 1972).

Application of this approach for molecular reconstruction was performed by Hudebine and Verstraete (2011). The reader is encouraged to read this work for extra details that were left out of this section.

The extremum of (3) will satisfy Equations (5.1) and (5.2) by construction. Moreover, it has a semi-analytical solution.

$$\frac{\partial \varepsilon}{\partial x_i} = -\ln x_i - 1 - \mu - \lambda M_i\,(\alpha_i - \alpha) = 0 \tag{5.4}$$

$$\ln x_i = -1 - \mu - \lambda M_i\,(\alpha_i - \alpha) \tag{5.5}$$

$$x_i = e^{-(1+\mu)} e^{-\lambda M_i\,(\alpha_i - \alpha)} \tag{5.6}$$

$$e^{1+\mu} \sum x_i = \sum e^{-\lambda M_i\,(\alpha_i - \alpha)} \tag{5.7}$$

$$Z = e^{1+\mu} = \sum e^{-\lambda M_i\,(\alpha_i - \alpha)} \tag{5.8}$$

$$x_i = \frac{1}{Z} e^{-\lambda M_i\,(\alpha_i - \alpha)} \tag{5.9}$$

$$\varepsilon(\lambda) = \ln Z \tag{5.10}$$

Equation (5.4) is the partial derivative of Equation (5.3) with respect to x_i. This expression is set to be 0 to find the extremum of the original function. Equations (5.5) and (5.6) are results of algebraic transformation to isolate x_i. Equation (5.7) results from the summation of all concentration and factoring out $\exp(1 + \mu)$ as a common factor. The sum of all components is 1 due to material balance. With the sum of exponents renamed (Equation 5.8), Equation (5.6) can be written as Equation (5.9). The Equation (5.10) is the result of substituting Equation (5.9) into the original entropy Equation (5.3).

For practical application, the extremum for the natural logarithm of Z occurs at the same point as for Z itself. Notice that the problem went from 13 unknowns (11 molar fractions + 2 Lagrange multipliers) to 1 unknown (1 Lagrange multiplier). For this toy example, it might not look that impressive, but it is a huge time-saver when the number of components goes beyond thousands and tens of thousands.

In a more general case, there might be more constraints, but as long as they are linear, the semi-analytical solution exists and reduces the number of unknowns to the number of linear constrains.

$$\varepsilon(x, \mu, \lambda) = -\sum_{i=1}^{N} x_i \ln x_i + \mu \left(1 - \sum_{i=1}^{N} x_i\right) + \sum_{j=1}^{J} \lambda_j \left(f_j - \sum_{i=1}^{N} x_i f_{i,j}\right) \tag{5.11}$$

$$Z = \sum_{i=1}^{N} \exp\left(-\sum_{j=1}^{J} \lambda_j f_{i,j}\right) \tag{5.12}$$

$$\varepsilon(\lambda) = \ln Z + \sum_{j=1}^{J} \lambda_j f_j \tag{5.13}$$

Equation (5.11) is the generalization of Equation (5.3); where $f_{i,j}$ is the weight with which x_i contributes to the property f_j, just like $M_i\,(\alpha_i - \alpha)$ in (1) is the weight with which x_i should sum to 0.

Notice that when at least one of the f_j values is nonzero, the final expression for entropy (13) has an additional sum compared to (10). However, it is always possible to reformulate $f_{i,j}$ in such a way to make all values of f_j zero as shown (14).

$$f_j - \sum_{i=1}^{N} x_i f_{i,j} = f_j \sum_{i=1}^{N} x_i - \sum_{i=1}^{N} x_i f_{i,j} = \sum_{i=1}^{N} x_i f_j - \sum_{i=1}^{N} x_i f_{i,j} = \sum_{i=1}^{N} x_i \left(f_j - f_{i,j}\right) = \sum_{i=1}^{N} x_i f^*_{i,j} \tag{5.14}$$

Entropy maximization procedure can be carried out not only for the exact value but also for the variance, thus making the final distribution closer to normal. The mathematical machinery for deriving the semi-analytical solution is very similar to what was shown before.

$$\sigma^2 = \sum_{i=1}^{N} x_i (f - f_i)^2 \tag{5.15}$$

The simple semi-analytical solution also exists for "uncertain/relax" constraints. The details behind the reason why the suggested method works are beyond the scope of this chapter and curious readers are advised to look for information about a Gaussian prior. Turns out, the way to include "uncertain" constraints for entropy models is by adding logarithm of the Gaussian function, with f_k and σ_k being the mean and standard deviation, respectively.

$$\ln \frac{1}{\sqrt{2\pi\sigma_k^2}} \exp\left(- \frac{\left(f_k - \sum x_i f_{i,k}\right)^2}{2\sigma_k^2} \right) = - \frac{1}{2} \frac{\left(f_k - \sum x_i f_{i,k}\right)^2}{\sigma_k^2} + \text{const} \tag{5.16}$$

The normalization factor does not depend on x_i and for any given value, σ_k is just a constant that does not change the extremum for x_i, so it can be dropped. With this modification, the entropy function has the following form (Equation 5.17):

$$\varepsilon(x, \mu) = - \sum_{i=1}^{N} x_i \ln x_i + \mu \left(1 - \sum_{i=1}^{N} x_i \right) + \sum_{k=1}^{K} - \frac{1}{2\sigma_k^2} \left(f_k - \sum_{i=1}^{N} x_i f_{i,k} \right)^2 \tag{5.17}$$

The similar procedure to Equations (5.4)–(5.10) yields,

$$\varepsilon(\epsilon) = \ln Z + \sum_{k=1}^{K} \left(\frac{1}{2} \epsilon_k^2 - \frac{f_i}{\sigma_k} \epsilon_k \right) \tag{5.18}$$

with

$$Z = \sum_{i=1}^{N} \exp\left(\sum_{k=1}^{K} \epsilon_k \frac{f_{i,j}}{\sigma_k} \right) \tag{5.19}$$

$$\epsilon_k = \frac{f_k - \sum_{i=1}^{N} x_i f_{i,k}}{\sigma_k} \tag{5.20}$$

All mentioned ways of putting constraints can work together, and the most general form of entropy maximization problem (that has a simple semi-analytical solution) is as follows:

$$\begin{aligned} \varepsilon(x, \mu, \lambda, \upsilon) = & - \sum_{i=1}^{N} x_i \ln x_i + \mu \left(1 - \sum_{i=1}^{N} x_i \right) \\ & + \sum_{j=1}^{J} \lambda_i \left(f_j - \sum_{i=1}^{N} x_i f_{i,j} \right) \\ & + \sum_{k=1}^{K} - \frac{1}{2} \frac{\left(f_k - \sum_{i=1}^{N} x_i f_{i,k} \right)^2}{\sigma_k^2} \end{aligned} \tag{5.21}$$

$$\sum_{m=1}^{M} \upsilon_m \left((\sigma_m)^2 - \sum_{i=1}^{N} x_i (f - f_i)^2 \right)$$

which can be simplified to the following:

$$\varepsilon(\lambda, \epsilon, v) = \ln Z + \sum_{j=1}^{J} \lambda_j f_j + \sum_{k=1}^{K} \left(\frac{1}{2} \epsilon_k^2 - \frac{f_i}{\sigma_k} \epsilon_k \right) + \sum_{m=1}^{M} \nu_m \sigma_m^2 \tag{5.22}$$

With

$$Z = \sum_{i=1}^{N} \exp\left(- \sum_{j=1}^{J} \lambda_j f_{i,j} + \sum_{k=1}^{K} \epsilon_k \frac{f_{i,j}}{\sigma_k} - \sum_{m=1}^{M} \nu_m \left(f_{i,m} - f_m \right)^2 \right) \tag{5.23}$$

After finding the maximum of (22), the concentration could be found as follows:

$$x_i = \frac{1}{Z} \exp\left(- \sum_{j=1}^{J} \lambda_j f_{i,j} + \sum_{k=1}^{K} \epsilon_k \frac{f_{i,j}}{\sigma_k} - \sum_{m=1}^{M} \nu_m \left(f_{i,m} - f_m \right)^2 \right) \tag{5.24}$$

Notice an important distinction between exact constraints with standard deviation and uncertain constraints. The first one guarantees that concentration averages to a specific value, and distribution is close to a normal one. The second does not restrict either average or standard deviation; instead, it ensures that the calculated average value is not too far from the desired value. The general formula might be confusing because it uses σ_m as exact constraints on the standard deviation and σ_k as a way to quantify the penalty for discrepancy for inexact constraints. It is important to understand the difference between them.

The derivation of this equation and much more can be found in the works of Hudebine and Verstraete's (2011). This paper also has expressions for $f_{i,j}$ for different analytical methods.

Let us consider our toy example, and see how this method deals with it. We can find the optimal value of the Lagrange multiplier for different values of average carbon content (Figure 5.1).

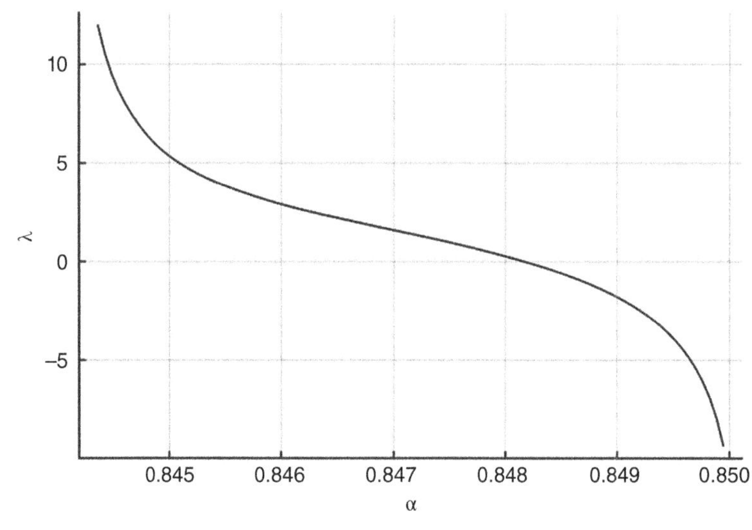

Figure 5.1 Optimal values of Lagrange multiplier for different values of average carbon content.

For comparison, Figure 5.2 shows the distribution for different values of the Lagrange multiplier. It is clear that for $\lambda > 0$; the distribution is skewed toward lighter paraffins; and the bigger the value, the stronger the effect. For $\lambda = 0$; the distribution become uniform and for negative values, it skewed in the other direction.

The quotient of optimal normalized residue and standard deviation for uncertain constraints works very similar to Lagrange multipliers for exact constraints. Figure 5.3 shows it for different values of carbon content and standard deviation. It is easy to see the vertical line that corresponds to the fixed point (uniform distribution), meaning that increasing standard deviation does not affect it. For values

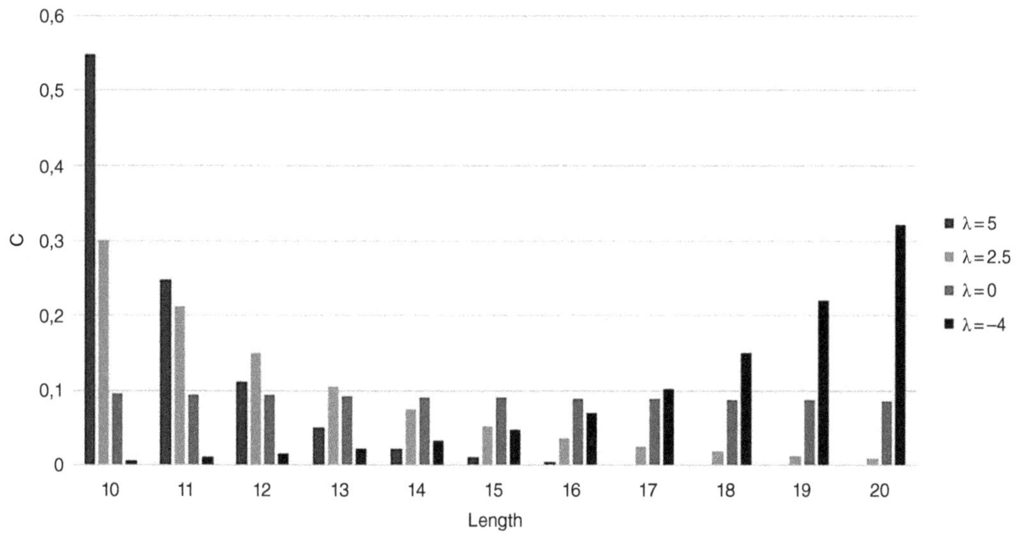

Figure 5.2 Concentration distributions under exact constraints corresponding to λ = 5, 2.5, 0, and −4.

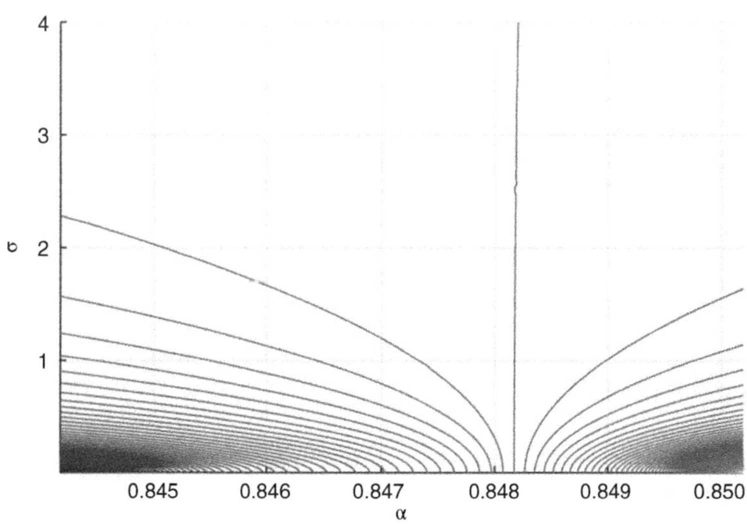

Figure 5.3 Contour of optimal value of quotient of normalized residue for different carbon contents and standard deviations.

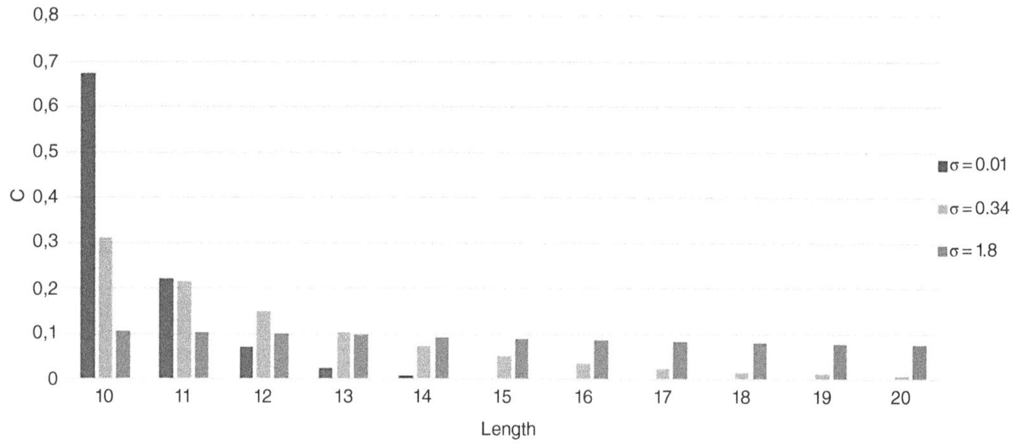

Figure 5.4 Distributions for the same value of carbon content (0.8447) and different standard deviations.

closer to edges, it is clear that high standard deviation cases correspond to low standard deviation cases near the uniform distribution. It just confirms the expected behavior of the system: increasing standard deviation results in a more uniform distribution (Figure 5.4).

Entropy maximization is a powerful technique, but it should be noted that the simplified semi-analytical solution does not exist in case of nonlinear constraints, that is, the same procedure will not exclude molar fraction dependences.

Fortunately, various properties can be encoded exactly or approximately (e.g. density/specific gravity is known for being nonlinear) in linear constraints framework. Although some important properties cannot be effectively approximated with linear constraints, Octan number is a notable example.

Entropy maximization method allows one to find the concentration of a predetermined set of molecules. Each molecule, in this case, can be considered an explicit assumption of the mixture composition. Because even qualitative composition is often unavailable, it is hard to apply this method for heavy petroleum cuts.

That is why the method is often paired with other methods that can produce a detailed composition.

The other issue with this method is that it usually produces exponential distribution (unless variance is also constrained), which can result in the prevalence of few compounds, making the portion of other compounds negligible, which is usually not the case. To ameliorate this problem, it is usually expected that molecules used to reconstruct the mixture are generally very similar to the mixture (Hudebine and Verstraete 2011).

5.5 Stochastic Reconstruction (SR)

One method that can provide us with molecules is the stochastic reconstruction method. It was first introduced by Neurock et al. (1994).

The first insight into the stochastic reconstruction is the realization that we do know something about the sample. It might not seem much, but we know it is a hydrocarbon mixture. We generally have an idea where this mixture came from, what we can expect to see, and what is unlikely to occupy a sizable portion of components. It is possible to greatly reduce complexity by choosing few molecular families that (to our mind) represent the bulk of the mixture/very important part of the mixture for the model we build.

For example, we usually can expect paraffins, iso-paraffins, naphthenes, aromatics, and some heteroatomic compounds in heavy petroleum cuts. If we know that mixture came from hydrotreatment, it might be reasonable to neglect concentrations of heteroatomic compounds. Again, thinking of the reconstructed composition as a very good model mixture helps to understand why other parts are ignored.

Each molecule family can be characterized with several numerical values. They are often referred to as "structure attributes." In a way, structure attributes are very similar to basis vectors that span vector space (chemical compounds). For example, all normal paraffins can be uniquely described by the length of their carbon chain. To represent monosubstituted iso-paraffins, one can use the length of the main chain, length of the branch, and the position of the main chain to which the branch is attached to. It is also possible to use, for example, the total number of carbon atoms in the molecule instead of the length of the main carbon chain/length of the branch. That is to say that the representation is not unique. It is worth mentioning that, at this stage, it is possible to "lump" together different molecules by ignoring certain structure attributes. For example, the location of the branch might not be very important for the model we build and, in this case, we can ignore it, thus making characterization of iso-paraffin much easier. This part of the method is very similar to the structure-oriented lumping (SOL) approach that we are to discuss later.

It is possible to narrow down the possible molecules in the mixture even further by restricting the possible numerical values of the chosen structure attributes. For example, it would be rather unreasonable to assume that methane can occupy a sizable portion of a vacuum gas oil sample.

The second idea on the way to discover the stochastic reconstruction method is the futility of any attempts to describe the composition in an actually detailed way due to the sheer number of molecules that can be found in the mixture with a reasonable number of parameters. For the model to be usable, it must be relatively simple.

Even after greatly reducing the number of possible structures, it is hard to describe their abundance in the mixture accurately without turning their concentration into variables. It is the same problem we faced before: The number of unknowns is far greater than the number of equations. In this approach, however, instead of maximizing the entropy to find the appropriate distribution, a different strategy is used: The structure attributes are assumed to follow a certain type of statistical distribution. Moreover, they are assumed to be independent.

In reality, of course, the distributions of structural attributes in different families are not independent. There is no reason to expect that, for example, the length of an aliphatic chain attached to an aromatic core in general does not depend on the number of aromatic rings in it. However, with this assumption, the problem becomes much more approachable.

The same goes for using specific types of statistical distributions. The distributions of different structural attributes are often "smooth" in petroleum cuts (Wu and Zhang 2010), but there is no rule to guarantee it in general. That is the reason why the distribution for this task is often chosen to be gamma or discrete. Gamma distribution is known for being very flexible (exponential and χ^2 distributions are particular cases of it!) with a relatively small number of parameters (only 2).

The final touch to reach the classical stochastic reconstruction is to generate molecules stochastically with the Monte Carlo approach. The number of molecules generated is often chosen to be 5000–50,000 to balance computational complexity/accuracy (Lopez Abelairas et al. 2016) (the computational complexity increases linearly with the number of molecules, but the standard deviation of the error deceases as a square root). Concentration in this case is just assumed proportional to the number of times the molecule appeared in the mixture.

After all these simplifications, the unknowns in the model are no longer concentrations, but the distribution parameters. Finding them is hindered by the stochastic nature of the problem. Genetic algorithms and simulated annealing are the most common optimization techniques for this task, as they do not rely on derivatives. Lopez Abelairas et al.'s study (2016) cites other source (which we could not find) that claims that genetic algorithms are better suited to this task because they are less dependent on the initial guess compared with simulated annealing. Some works use particle swarm optimization approaches for this problem.

To wrap it up, to reconstruct a mixture with SR, the first thing is to choose structure attributes and corresponding statistical distributions. Generate a mixture with initial (usually random) parameters and evaluate mixture properties. It allows one to use an optimization method to minimize the discrepancy between properties of the real and reconstructed mixtures. After optimization, the parameters can be found to generate a mixture that resembles the real one.

One important question was overlooked in the previous discussion: properties estimation and mixing rules. This approach works well with group contribution methods (e.g. Constantinou and Gani 1994; Skander and Chitour 2006; Ghasemitabar and Movagharnejad 2016) and a great review on this topic in general (Su et al. 2017) for properties estimation (mainly boiling point and density). Unlike the entropy maximization approach, mixing rules can be nonlinear in this method (i.e. the procedure does not spike in difficulty when nonlinear constraints are added). Different mixing rules and group contribution methods can greatly affect the result of reconstruction (Deniz et al. 2018).

For demonstration purposes, we will consider our toy example. We will assume gamma distribution for the length of normal paraffins as it often happens in other works (Lopez Abelairas et al. 2016; De Oliveira et al. 2013; Alvarez-Majmutov and Chen 2017). As we have just two parameters, it is possible to represent the average carbon content visually as a contour for different parameters.

Notice that there are different ways to parametrize the gamma distribution. To make it easier to understand, we used expected value and variance. As shown in the following, the conversion from them to shape and scale/rate parameter is easy.

$$E(X) = k\theta = \frac{\alpha}{\beta} \quad \mathrm{Var}(X) = k\,\theta^2 = \frac{\alpha}{\beta^2} \tag{5.25}$$

$$\alpha = k = \frac{E(X)^2}{\mathrm{Var}(X)} \quad \theta = \frac{1}{\beta} = \frac{\mathrm{Var}(X)}{E(X)} \tag{5.26}$$

The gamma distribution is a continuous distribution, so naturally we need to round off the results to the closest integer. Under normal circumstances, we expect that molecules are in predetermined range and follow their distributions, so the tails outside of the range are normally just rounded to the closest inbound value. For example, in the following plot, the probabilities for values above 20 are added to the probability of exactly 20. It will create some artifacts near the boundaries.

For the first plot (Figure 5.5), we did not use the stochastic approach to generate molecules, and instead used all possible molecules. Their abundance was set to be the probability calculated with cumulative function distribution to show the "ideal" case without distortion caused by the stochasticity of the method. This approach might be harder to perform for a bigger system.

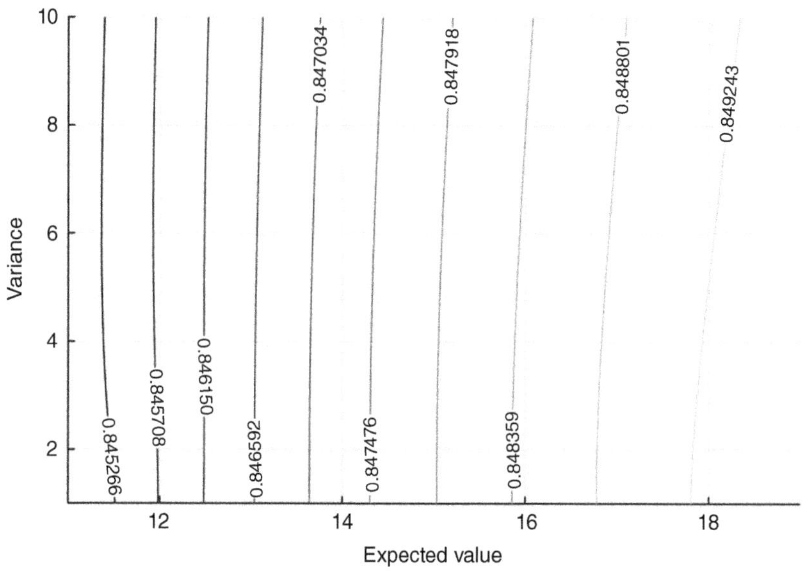

Figure 5.5 The average carbon content in a mixture of normal paraffins from C_{10} to C_{20} distributed with gamma distributions of different variance and expected values.

It is quite evident that the expected value affects the average carbon content much stronger than the variance. We can obtain a similar graph using the Monte Carlo approach with different values of N (number of molecules/mixture). The lines of a constant level in the contours in that case (Figure 5.6) look fuzzy due to the Monte Carlo sampling.

If we were reconstructing a mixture of normal paraffins with a specific value of carbon content, any point on the constant level line would fit. We can see that if the optimization is carried out with low N, the resulting parameters could be much further from the "ideal" case.

This method allows one to represent mixtures in terms of molecules, no matter how complicated they are. On the contrary, constraining structural attribute distributions to a certain type of statistical distributions makes the system much more rigid. The other significant drawback of this method is optimization procedure that usually takes a lot of time. Different approaches, including usage of artificial neural networks, have been employed to solve this problem (Deniz et al. 2017).

Another important concern is connected to the trustworthiness of such compositions. It is important to understand, that some parts of the output might be completely unconstrained by the available analytical data. As a result, some structure attribute values are just remnants of the initial guess for distributions (Glazov et al. 2021), which can potentially affect kinetic model step.

The procedure described here is the basis for other stochastic methods. They can vary in objective functions, optimization procedures (e.g. Dantas et al. (2023) uses particle swarm optimization strategy), and molecule representation methods (e.g. Guan et al. (2022) combine stochastic method with SOL-inspired framework or Zhao et al. (2022) use certain structure descriptors). Some works even attempt to represent mixture with very small (~10 to 100) number of molecules (Campbell and Klein 1997).

The core concept, however, (viewing molecules as a combination of structure attributes which can be described with probability density functions) is mostly unchanged in those works.

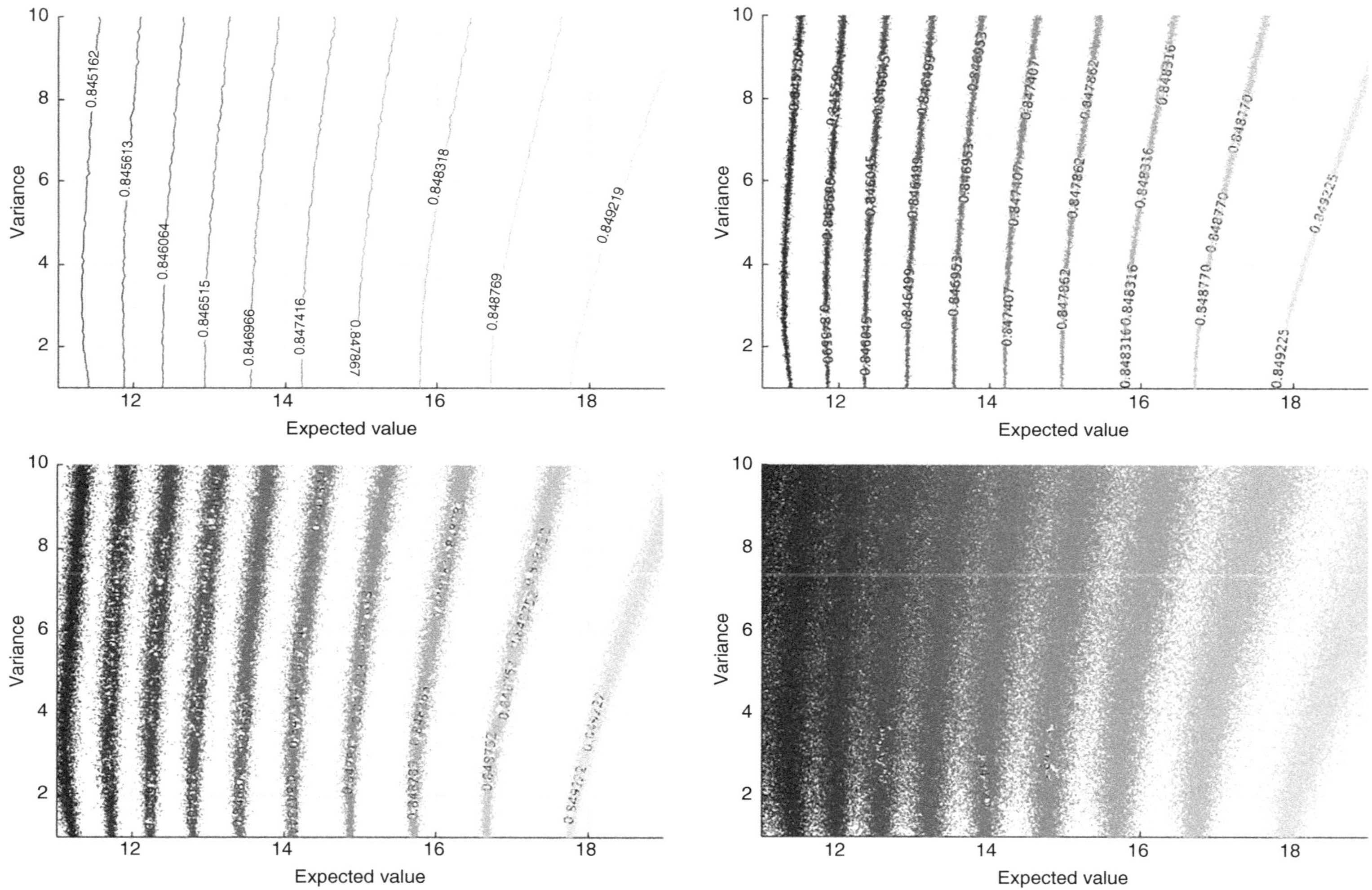

Figure 5.6 The average carbon content in a mixture of normal paraffins from C_{10} to C_{20} distributed with gamma distributions of different variance and expected values obtained by the Monte Carlo method with N = 50,000, 5000, 500, and 50 molecules/mixture (from left to right, from top to bottom).

5.6 SR-EM

One way to solve problems of two previous methods is by combining them together. Stochastic reconstruction method can be used as a basis for molecular library and entropy maximization for fine-tuning the component concentrations. This idea was purposed in (Hudebine and Verstraete 2004).

This combination overcomes problems of individual methods. The entropy maximization approach requires an appropriate initial guess for composition, which can be provided by stochastic reconstruction. On the contrary, stochastic reconstruction is computationally demanding, unlike entropy maximization. The entropy maximization can be used to reconstruct similar samples with relative ease. The downside of this combination is that entropy maximization operates only with linear constraints (or, to be precise, the task became much more complicated when nonlinear constraints are imposed).

Let us test this method with our toy example. It starts the exact same way as stochastic reconstruction. Let us say, at some point of optimization, the following distribution was obtained (Figure 5.7).

This distribution corresponds to $\alpha = 0.8470$. Now, imagine that the actual value we need is 0.8480. We can use REM to alter the distribution (Figure 5.8). The resulting distribution does not follow gamma distribution exactly, but it is still recognizable. For target values even further, the initial distribution might be completely overridden (Figure 5.9). It is the result of the mentioned phenomenon when exponential nature of the exact constraint entropy maximization can turn a mixture into few components. To this end, it might be best to restrict standard deviation (Figure 5.10).

Notice that in this case, entropy maximization process can be simplified even further due to repetition in the molecules. The exponents for all identical molecules are exactly the same, so obviously, the summation of exponents for all components is equivalent to the summation of exponents for unique components multiplied by the number of time that molecule appeared in the mixture. Moreover, dividing by the same factor, all exponents do not affect the position of

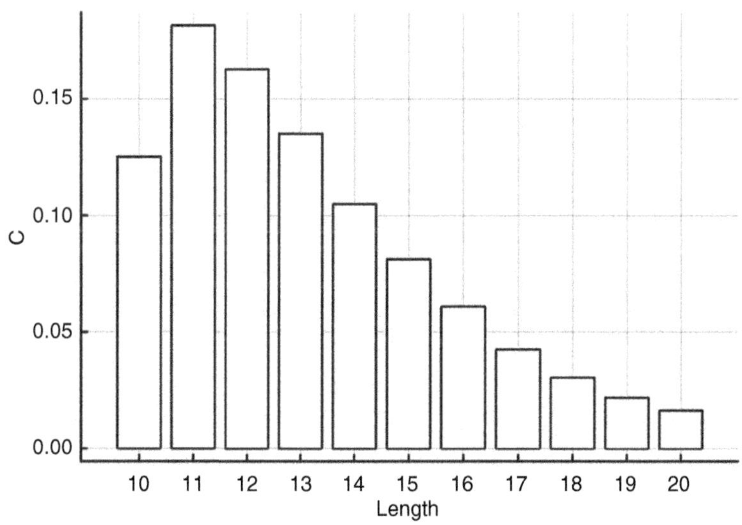

Figure 5.7 A concentration distribution after SR (N = 50,000), α = 0.8470.

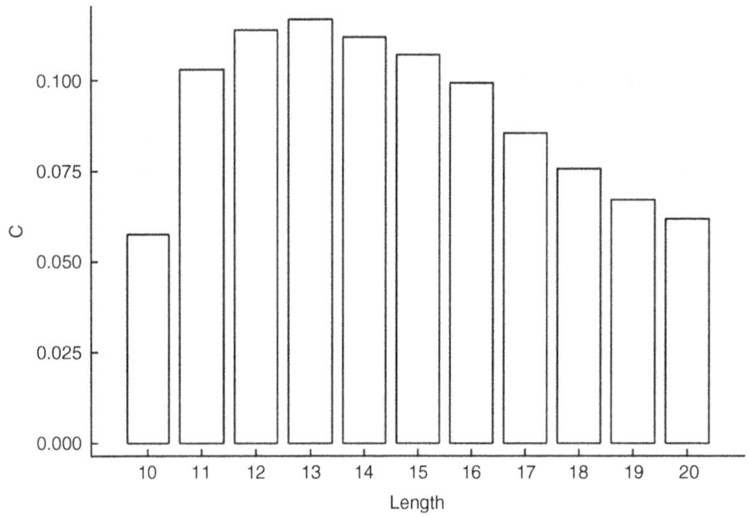

Figure 5.8 The concentration profile after entropy maximization to make $\alpha = 0.8480$.

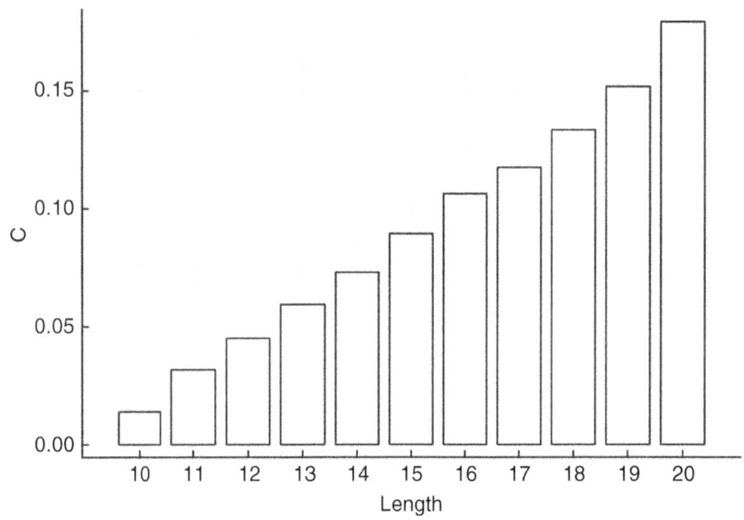

Figure 5.9 The concentration profile after entropy maximization to make $\alpha = 0.8490$.

the extremum, so the number of times a certain molecule appeared in the mixture (n_i) during SR is just N times bigger than the concentration (c_i) assigned by SR.

$$\sum_{i=1}^{N} \exp(-\lambda f_i) = \sum_{i=1}^{N^*} n_i \exp(-\lambda f_i) \sim \sum_{i=1}^{N^*} c_i \exp(-\lambda f_i) \tag{5.27}$$

$$c_i' = \frac{c_i \exp(-\lambda f_i)}{\sum c_j \exp\left(-\lambda f_j\right)} \tag{5.28}$$

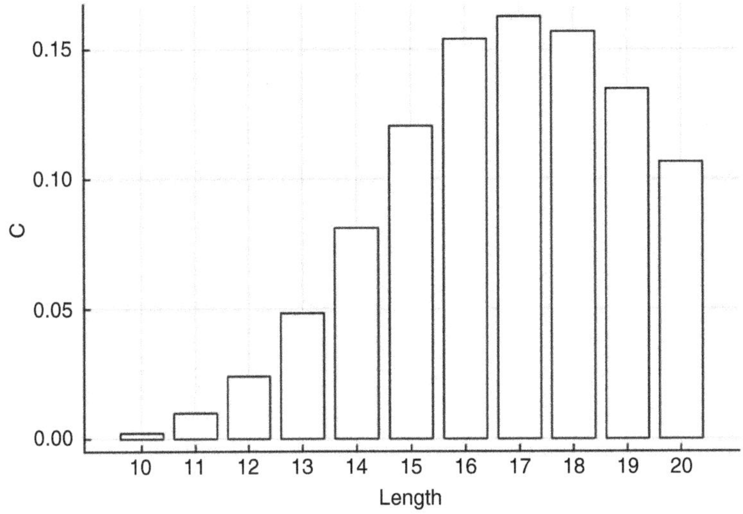

Figure 5.10 The concentration profile after entropy maximization to make $\alpha = 0.8490$ with $\sigma^2 = 0.15$.

where c' is the concentration of a molecule after entropy maximization, and c is the concentration before entropy maximization.

5.7 Structure-Oriented Lumping (SOL) Method

A different approach to molecular reconstruction of hydrocarbon mixtures was developed by Quann et al. (Quann 1998; Quann and Jaffe 1992). The key idea is to lump molecules together on the molecular level. It is very similar to the way molecules in SR could be lumped together. In this framework, molecules are represented with vectors of structural increments. In the original paper, 22 structural increments were proposed (Table 5.1). Later (Jaffe 2005) an extended list of increments was created to describe metal-organic compounds in vacuum residua. This framework can be tailored to other systems too. Tian et al.'s study (2010) used only 7 increments to describe the feedstock, whereas Hou et al.' (2023) used 38 increments.

The basic structure increments are A6, A4, A2 – aromatic rings; N6, N5, N4, N3, N2, and N1 – naphthenic rings; R – total alkyl-group length; IH – incremental hydrogen to specify the

Table 5.1 Structural increments and increment stoichiometry.

	A6	A4	A2	N6	N5	N4	N3	N2	N1	R	br	me	H	A_A	S	RS	AN	NN	RN	O	RO	O=
C	6	4	2	6	5	4	3	2	1	1	0	0	0	0	−1	0	−1	−1	0	−1	0	0
H	6	2	0	12	10	6	4	2	0	2	0	0	2	−2	−2	0	−1	−1	1	−2	0	−2
S	0	0	0	0	0	0	0	0	0	0	0	0	0	0	1	1	0	0	0	0	0	0
N	0	0	0	0	0	0	0	0	0	0	0	0	0	0	0	0	1	1	1	0	0	0
O	0	0	0	0	0	0	0	0	0	0	0	0	0	0	0	0	0	0	0	1	1	1

degree of unsaturation; br – number of branches; me – number of methyl groups in R; AA – biphenyl bridge; NS, NN, and NO – sulfur, nitrogen, and oxygen in naphthenic rings; RS, RN, and RO – mercaptan, amin, and alcohol group, respectively; AN – nitrogen in aromatic ring; and KO – ketone or aldehyde.

The way of determining abundance of different structures in the mixture is not clearly described in the original papers, except "non-negative Lagrange multiplier method, starting with an equimolar composition distribution" (Jaffe 2005). Ren et al. (2019) speculated that the objective function in their work might be negative entropy, sum of squares, or other criteria.

Where SOL method shines the most is the computing of reaction networks. It is possible to describe many important reactions of oil processing (cracking, hydration/dehydration, etc.) in this framework with a set of rules, which is particularly suitable for computers. This approach allows generating huge reaction networks for complex systems. It was shown in the original paper.

5.8 State Space Representation Method

Another approach to this problem, which takes a bit different route to represent mixtures, is the state space representation method (Mei et al. 2017). The basic idea is to represent feedstock as a linear combination of well-known mixtures. The output for kinetic models for the sample is assumed to be the linear combination of known mixtures. This method is not very useful for heavy petroleum cuts as for now, but it might find its use in the future.

Let us start from an n-dimensional vector space with pure components being its orthonormal basis. In that case, they can be represented as follows:

$$\begin{aligned}
&\text{Component 1} \quad && \mathbf{e}_1 = [1 \quad 0 \quad \dots \quad 0]^T \\
&\text{Component 2} \quad && \mathbf{e}_2 = [0 \quad 1 \quad \dots \quad 0]^T \\
&\dots && \\
&\text{Component } n \quad && \mathbf{e}_n = [0 \quad 0 \quad \dots \quad 1]^T
\end{aligned} \tag{5.29}$$

Given these vectors, any mixture composed of these pure components can be represented as a column-vector as well.

$$\nu = (w_1 \quad w_2 \quad \dots \quad w_n)^T \tag{5.30}$$

with w_i being the concentration of ith component. Another (more general) way of looking is to think of w_i as being the projection of the mixture vector on the ith basis vector.

Using L1 norm, any mixture can be represented as a normalized vector:

$$\|\nu\| = \sum_{i=1}^{n} |w_i| \tag{5.31}$$

$$\overline{\nu} = \frac{\nu}{\|\nu\|} \tag{5.32}$$

Pure component basis is not the only viable option. It is also possible to represent mixtures in the basis of other (independent) mixtures $\nu_1, \nu_2, \dots, \nu_n$.

$$\bar{\nu} = \frac{\nu}{\|\nu\|} = \frac{\sum_{j=1}^{n} \nu_j}{\sum_{j=1}^{n} \|\nu_j\|} = \frac{\sum_{j=1}^{n} \|\nu_j\| \bar{\nu}_j}{\sum_{j=1}^{n} \|\nu_j\|} = \sum_{j=1}^{n} h_j \bar{\nu}_j \tag{5.33}$$

where h_j is the blending ratio of the jth mixture ν_j to the final mixture.

Now, if we consider N well-defined mixtures (=vectors), we can rewrite Equation (5.33) in a matrix form as follows:

$$V = W \cdot H \tag{5.34}$$

where V $(n \times N)$ is the sample matrix, W $(n \times n)$ is the basis fraction matrix, and H $(n \times N)$ is the blending matrix.

The simplification comes with the realization that we can approximate V with m (less than n) basis mixtures in a subspace without losing too much information.

The task of finding mixtures representation turns into a matrix decomposition. However, the values of blending ratio and concentration must be positive and therefore "standard" decomposition strategies from linear algebra, such as singular value decomposition, do not work. Fortunately, nonnegative matrix factorization can solve this problem.

This approach was used for the naphtha reconstruction. 59 samples from a plant were collected and represented in terms of 35 lumps. Authors obtained 21 basis fractions through nonnegative factorization. That way, they managed to predict the product yield of pyrolysis of said fractions.

Generally, the feedstock that plants use may vary, but it is not entirely random. This method might be useful to find a (relatively small) set of basic fractions that can span the feedstock subspace with certain accuracy.

5.9 Molecular Type-Homologous Series Matrix

Another framework to model the composition is known as molecular type-homologous series matrix method (MTHS). In this case, the mixture is represented with a matrix. Columns of the matrix correspond to different homologous series, and rows are used for different carbon numbers. The entry of the matrix denotes the abundance of a lump from the given homologous series and carbon number. A modified approach that works with boiling range instead of carbon number (Gomez-Prado et al. 2008) also exists and arguably better suited for heavy petroleum cuts.

From the earlier construction, it is clear that each entry in the matrix could be pure components and pseudo-components alike. Choosing the properties for lumps in this case might be especially challenging. Wu and Zhang (2010) used an interesting approach to this problem: First, they enumerate all isomers which fall into the same entry, and then they estimated their properties with group contribution methods. With this information, they found isomer distribution based on thermodynamic equilibrium assumption, and this equilibrium blend is considered the lump.

As for the methods of determining the abundance of different molecules, there are three main approaches (to the best of our knowledge).

One method assumes that lump distribution within each homologous series can be approximated with a statistical distribution. The distribution of choice is usually gamma distribution. This method is very similar to what we did in stochastic reconstruction to demonstrate an "ideal" case. Except, there are more than one homologous series. Gamma distribution parameters are found by minimizing deviation experimental data and calculated. In this work, the squared relative error was used (Liu et al. 2019).

Second method treats the sample as a blend of well-characterized mixtures. In Aye and Zhang's (2005), they represent mixtures with at least eight other mixtures. The blending proportions are found by minimizing the sum of square of relative errors.

The third method relies on empirical correlations among carbon number and molecular weight, boiling point, and density. This method is hard to use for heavy petroleum cuts.

5.10 Conclusion

Molecular reconstruction is an umbrella term used for different methods, which allows one to represent hydrocarbon mixtures based on limited information. These methods do not normally find the real composition, but they can be useful for kinetic models. In that regard, molecular reconstruction builds a very good model mixture.

There are two main strategies for choosing components that constitute the mixture: deterministic and stochastic. Generally speaking, stochastic strategies require more computations, but the difficulty of using them grows slower than for deterministic one. That is part of the reason why the stochastic approach (at least to the best of authors' knowledge) is used more often for heavy fractions.

As for determining abundance of component in the mixture, it is usually an optimization problem. The properties of individual constituents are found first (though experiments or group contribution methods). The objective function controls how close the blend of component is to the sample. The concentrations are rarely used as the variables in the objective function; instead, an assumption is often made about the connection between them (e.g. concentrations follow a certain distribution, the whole mixture is a linear combination of other mixtures, the distribution maximize entropy, etc.).

Acknowledgment

This work was supported by the Ministry of Science and Higher Education of the Russian Federation within the state assignment for the Boreskov Institute of Catalysis (project FWUR-2024-0037).

References

Alvarez-Majmutov, A. and Chen, J. (2017). Stochastic modeling and simulation approach for industrial fixed-bed hydrocrackers. *Industrial and Engineering Chemistry Research* 56: 6926–6938.

Aye, M.M.S. and Zhang, N. (2005). A novel methodology in transforming bulk properties of refining streams into molecular information. *Chemical Engineering Science* 60: 6702–6717.

Campbell, D.M. and Klein, M.T. (1997). Construction of a molecular representation of a complex feedstock by Monte Carlo and quadrature methods. *Applied Catalysis A: General* 160: 41–54.

Constantinou, L. and Gani, R. (1994). New group contribution method for estimating properties of pure compounds. *AIChE Journal* 40: 1697–1710.

Dantas, T.S.S., Noriler, D., and Huziwara, K.W. (2023). A multi-population particle swarm optimization algorithm with adaptive patterns of movement for the stochastic reconstruction of petroleum fractions. *Computers & Chemical Engineering* 174: 108221.

De Oliveira, L.P., Trujillo Vazquez, A., Verstraete, J.J., and Kolb, M. (2013). Molecular reconstruction of petroleum fractions: application to various vacuum residues from different origins. *Energy & Fuels* 27: 3622–3641.

Deniz, C.U., Yasar, M., and Klein, M.T. (2017). Stochastic reconstruction of complex heavy oil molecules using an artificial neural network. *Energy and Fuels* 31 (11): 11932–11938.

Deniz, C.U., Yasar, S.H.O., Yasar, M., and Klein, M.T. (2018). Effect of boiling point and density prediction methods on stochastic reconstruction. *Energy and Fuels* 32 (3): 3344–3355.

Ghasemitabar, H. and Movagharnejad, K. (2016). Estimation of the normal boiling point of organic compounds via a new group contribution method. *Fluid Phase Equilibria* 411: 13–23.

Glazov, N., Dik, P., and Zagoruiko, A. (2021). Effect of experimental data accuracy on stochastic reconstruction of complex hydrocarbon mixture. *Catalysis Today* 378: 202.

Gomez-Prado, J., Zhang, N., and Theodoropoulos, C. (2008). Characterisation of heavy petroleum fractions using modified molecular-type homologous series (MTHS) representation. *Energy* 33: 974–987.

Guan, Y.M., Guan, D., Zhang, C. et al. (2022). Diesel molecular composition and blending modeling based on SU-BEM framework. *Petroleum Science* 19: 839–847.

Hou, L., Ye, L., Qin, X. et al. (2023). Predicting the physicochemical properties of molecules in petroleum based on structural increments. *Industrial & Engineering Chemistry Research* 62: 7744–7756.

Hudebine, D. and Verstraete, J.J. (2004). Molecular reconstruction of LCO gasoils from overall petroleum analyses. *Chemical Engineering Science* 59: 22–23.

Hudebine, D. and Verstraete, J.J. (2011). Reconstruction of petroleum feedstocks by entropy maximization. application to FCC gasolines. *Oil and Gas Science and Technology* 66: 437–460.

Jaffe, S.B. (2005). Extension of structure oriented lumping to vacuum residual. *AIChE Annual Meeting, Conference Proceedings* 08066: 9578.

Lisman, J.H.C. and Zuylen, M.C.A. (1972). Note on the generation of most probable frequency distributions. *Statistica Neerlandica* 26: 19–23.

Liu, L., Hou, S., and Zhang, N. (2019). Incorporating numerical molecular characterization into pseudo-component representation of light to middle petroleum distillates. *Chemical Engineering Science: X* 3: 100029.

Lopez Abelairas, M., De Oliveira, L.P., and Verstraete, J.J. (2016). Application of Monte Carlo techniques to LCO gas oil hydrotreating: molecular reconstruction and kinetic modelling. *Catalysis Today* 271: 188–198.

Mei, H., Cheng, H., Wang, Z., and Li, J. (2017). Molecular characterization of petroleum fractions using state space representation and its application for predicting naphtha pyrolysis product distributions. *Chemical Engineering Science* 164: 81–89.

Neurock, M., Nigam, A., Trauth, D., and Klein, M.T. (1994). Molecular representation of complex hydrocarbon feedstocks through efficient characterization and stochastic algorithms. *Chemical Engineering Science* 49: 4153–4177.

Quann, R.J. (1998). Modeling the chemistry of complex petroleum mixtures. *Environmental Health Perspectives* 106: 1441–1448.

Quann, R.J. and Jaffe, S.B. (1992). Structure-oriented lumping: describing the chemistry of complex hydrocarbon mixtures. *Industrial and Engineering Chemistry Research* 31: 2483–2497.

Ren, Y., Liao, Z., Sun, J. et al. (2019). Molecular reconstruction: recent progress toward composition modeling of petroleum fractions. *Chemical Engineering Journal* 357: 761–775.

Skander, N. and Chitour, C.E. (2006). A new group-contribution method for the estimation of physical properties of hydrocarbons. *Oil & Gas Science and Technology* 57: 369–376.

Su, W., Zhao, L., and Deng, S. (2017). Group contribution methods in thermodynamic cycles: Physical properties estimation of pure working fluids. *Renewable and Sustainable Energy Reviews* 79: 984–1001.

Tian, L., Wang, J., Shen, B., and Liu, J. (2010). Building a kinetic model for steam cracking by the method of structure-oriented lumping. *Energy and Fuels* 24: 4380–4386.

Wu, Y. and Zhang, N. (2010). Molecular characterization of gasoline and diesel streams. *Industrial and Engineering Chemistry Research* 49: 12773–12782.

Zhao, G., Yang, M., Du, W. et al. (2022). A stochastic reconstruction strategy based on a stratified library of structural descriptors and its application in the molecular reconstruction of naphtha. *Chinese Journal of Chemical Engineering* 51: 153–167.

6

Modeling of Catalytic Hydrotreating Reactor for Production of Green Diesel

Alexis Tirado[1], *Fernando Trejo*[2], *and Jorge Ancheyta*[1,3]

[1] *Department of Petroleum Engineering, Kazan Federal University, Kazan, Russia*
[2] *Instituto Politécnico Nacional, CICATA–Legaria, Ciudad de México, Mexico*
[3] *Instituto Mexicano del Petróleo, Ciudad de México, Mexico*

6.1 Introduction

One of the most crucial worldwide research projects in the energy field is developing alternative fuels to reduce the use of fossil fuels because of the high levels of harmful gases released into the atmosphere. Biofuels are considered a promising alternative because they can be produced from a variety of biomass sources (vegetable oils, animal fats, and even agricultural wastes) and are not harmful to the environment (Kubičková and Kubička 2010; Sotelo-Boyas et al. 2012). These feedstocks are subjected to hydrotreating processes in order to produce high-quality green fuels. Nevertheless, this technology type still presents some challenges, such as high exothermicity, hydrogen consumption, and yields of undesirable by-products (Furimsky 2013; Sivasamy et al. 2009). The use of mathematical models is essential for the design, scaling, and optimization of these processes focused on the conversion of vegetable oils to market-driven transportation renewable fuels.

This chapter presents a study of the kinetic and reactor modeling of hydrotreating vegetable oils for the production of renewable fuels. The different approaches considered in the literature and their results are discussed. Finally, a case study of dynamic reactor modeling is presented in order to provide some insights and considerations on its development, solution, and scaling-up, as well as the interpretation of the results obtained.

6.2 Conversion of Vegetable Oils into Renewable Fuels

Vegetable oils or lignin-derived bio-oils contain compounds such as fatty acid (FA) esters, triglycerides (Tg), phenols, carboxylic acids, and ketones, which are refined to obtain oils rich in Tg capable of being used in the energy industry. Tg present in refined vegetable oils consist of saturated and unsaturated FAs with carbon atom numbers from C_4 to C_{24} and more commonly between C_{16} and C_{18} (Snåre et al. 2006, 2007), which turn them into adequate substitutes for diesel-type fuels. Nevertheless, the high oxygen content in refined vegetable oils (10–12 wt.%) increases thermal and

Mathematical Modeling of Complex Reaction Systems in the Oil and Gas Industry, First Edition.
Edited by Jorge Ancheyta, Andrey Zagoruiko, and Andrey Elyshev.

chemical instability, high viscosity, corrosiveness, immiscibility with fossil fuels, low heating value, and tendency to polymerize. Therefore, removal of oxygen atoms is necessary to obtain high-quality fuels (Noriega et al. 2020).

Catalytic hydrotreating (HDT) has been used for years in the oil industry to increase the hydrogen-to-carbon (H/C) ratio as well as remove sulfur, nitrogen, and metals, among other heteroatoms in diverse petroleum refining streams (atmospheric and vacuum residues, gas oils, and naphtha) through a series of chemical reactions carried out on catalysts under a hydrogen-rich atmosphere (Ancheyta et al. 2010). This process may involve multiple catalytic reaction systems, including fixed-bed, ebullated-bed, and slurry-bed options, and it is crucial due to the environmental restrictions that apply during the combustion of fossil fuels. Such regulations include the emission of sulfur oxides (SO_x), nitrogen oxides (NO_x), particulate matter (PM), and volatile organic compounds (VOCs) like aromatics and olefins. Therefore, this process is appropriate to remove the oxygen atoms in vegetable oils through hydrodeoxygenation (HDO) and decarboxylation/decarbonylation ($deCO_2$/deCO) reactions, obtaining high-quality fuels rich in aliphatic chains called "renewable fuels" or "green fuels." The performance of the hydrotreating process is mainly influenced by the temperature, hydrogen/oil ratio, pressure, catalyst, and reaction time. However, the elimination of oxygen leads to the production of water during HDO reactions, which may impact the catalyst activity, whereby bio-oil conversion into fuels is a processing challenge (Furimsky 2013; Saidi et al. 2014).

Kalnes et al. (2008) summarized some advantages and disadvantages of ultra-low sulfur diesel (ULSD) from petroleum, biodiesel, renewable diesel, and synthetic diesel produced by the Fischer–Tropsch synthesis. Table 6.1 shows that ULSD from petroleum has a considerable number of heteroatoms and aromatic compounds enhancing the formation of greenhouse gases. On the other hand, synthetic diesel and biodiesel are abundant in oxygenated compounds, displaying the undesirable properties previously mentioned. In contrast, renewable diesel derived from vegetable oils is rich in *iso-* and *n*-alkanes and free of aromatic and oxygenated compounds. It is notable that renewable diesel, synthetic diesel, and biodiesel have zero sulfur content and emit fewer greenhouse gases than petroleum diesel. Although the boiling point properties are comparable, diesel has a higher cetane number. It is concluded from these data that the properties of renewable diesel are better than those obtained by other processes. In addition, the investment cost of the process to produce renewable diesel can be reduced by using infrastructure utilized in oil refineries, such as the hydrotreating units (Bezergianni and Dimitriadis 2013).

Table 6.1 Comparison of diesel properties obtained from different processes.

Property	Diesel (ULSD)	Biodiesel	Renewable diesel	Synthetic diesel (Fischer–Tropsch)
Oxygen (wt.%)	0	11	0	0
Specific gravity	0.84	0.88	0.78	0.77
Sulfur content (ppm)	<10	<1	<1	<1
Heating value (MJ/kg)	43	38	44	44
Cloud point (°C)	−5	−5 to 15	−20 to 20	–
Cetane index	51–55	50–55	70–90	>75
H/C	0.162	0.176	0.176	—
Polyaromatics (wt.%)	<8	0	0	—
Stability	Standard	Good	Excellent	Good

6.2.1 Commercial Production of Renewable Diesel

In recent decades, there has been a rapid increase in global energy demand, prompting researchers to address this challenge and decrease greenhouse gas emissions. Currently, catalytic hydrotreating technologies are utilized by industrial refineries to produce renewable fuels, including diesel, kerosene, naphtha, and liquefied petroleum gas (LPG), using various raw materials. Neste Corporation developed the NExBTL technology, through which it has reported biofuel production volumes of 2 million tons per year (Douvartzides et al. 2019; Kubička and Tukač 2013). Nonetheless, the production of renewable fuels is still limited. NExBTL process is a technology compatible with current refinery equipment for producing renewable diesel and jet fuel from feedstocks such as used cooking oil, palm oil, and waste animal fat, resulting in a 40–80% lifecycle reduction in CO_2, depending on the feedstock type (Baldiraghi et al. 2009; Pelemo et al. 2020). On the other hand, the Ecofining process is a technology developed by Honeywell UOP and Eni SpA companies. The design of the process usually varies in the type of catalyst, feedstocks, and operating conditions, although there are similarities in the process configuration, as shown in Figure 6.1a. Through these, high-quality fuels suitable for regular engines are obtained, meeting environmental specifications either directly or mixed with petroleum distillates (Baldiraghi et al. 2009; Kalnes et al. 2008). In addition, companies such as Haldor Topsøe, ConocoPhillips, and Petrobras have developed hydrotreating technologies by co-processing renewable feedstocks and petroleum distillates, as shown in Figure 6.1b (Zhang et al. 2018). Therefore, the process is focused on reducing the heteroatom content while dropping the carbon footprint. However, hydroprocessing of this feedstock type is more exothermic than petroleum-derived streams, causing a high heat release and hydrogen consumption. In industrial practice, the heat loss in these units is negligible compared with the heat released by exothermic reactions. As a result, the behavior of these units is remarkably close to an adiabatic operation. The problem for industrial units is not achieving

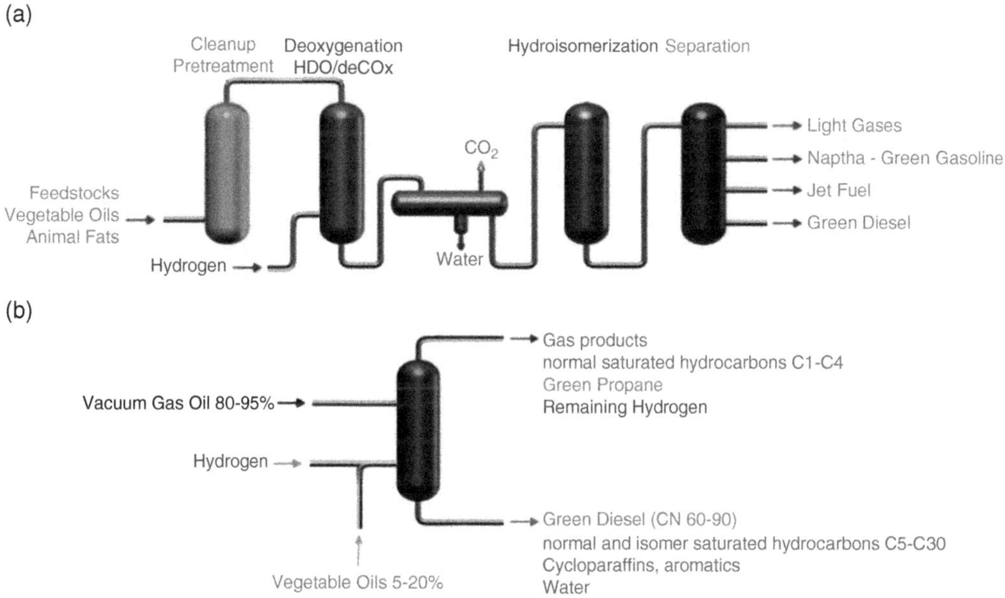

Figure 6.1 Schematic representation of (a) standalone renewable feedstock hydroprocessing plant and (b) co-processing hydroprocessing plant for renewable feedstock and petroleum distillates (Douvartzides et al. 2019/ MDPI /CC BY). *Other hydrocarbons include products from hydrocracking, polymerization, and cyclization.

complete feedstock conversion but rather controlling the high levels of reaction heat released. An increase in reactor temperature beyond permissible limits enhances the generation of hot spots that affect the catalyst activity, change the reaction product yields, and damage the reactor infrastructure. Additionally, keeping the temperature constant in a large-scale trickle-bed reactor is complicated since the removal or addition of heat to the reactor walls is considered economically unfeasible.

Haldor Topsøe revamped a mild hydrocracking unit to carry out the co-processing of light gas oil with 30 vol.% of raw tall diesel (RTD). This unit has a capacity of approximately 10,000 b/d and reports that the reaction heat increased with the increasing concentration of RTD.

In hydrotreating units for petroleum-derived streams, a proper quench system effectively manages the high heat released. The catalytic bed volume is divided into small beds separated by quench zones to mix the hot feedstock streams from the previous catalytic bed with cold streams. This ensures a controlled mixture before it passes to the next bed (Egeberg et al. 2009, 2011). Based on the rate and amount of reaction heat released, the catalytic bed must be configured with multiple catalytic beds. The necessary number of quench zones is adjusted not to exceed a permissible limit temperature because of mechanical restrictions and excessive feedstock hydrocracking.

The design and operation of these processes present notable challenges as they rely on vast experimental data and prior comprehensive knowledge of the design variables acquired in laboratory-scale hydrotreating reactors. In addition, modeling an industrial reactor is a practical tool for ensuring proper start-up and stable operation throughout the process. However, these types of reactor models are scarce because they require the incorporation of aspects such as deactivation of catalysts, hydrogen consumption, pressure drop, and reaction heat released to describe the commercial unit performance (Donnis et al. 2009; Scharff et al. 2013).

6.3 Hydrotreating Kinetic Models and Reaction Pathways

Elucidation of reaction schemes and determination of kinetic parameters are important steps for catalyst screening, as well as designing and scaling-up of processes. Due to the complexity of the hydrotreating process of vegetable oils, which involves a large number of chemical compounds, and chemical and physical phenomena occurring within the reactor, mathematical models are widely utilized in predicting the reactor performance, determining optimal processing conditions, and carrying out the design and scale-up of the process for industrial application. Nevertheless, kinetic and reactor models related to vegetable oils hydrotreating are still scarce in the literature, where most of the reports are focused on experiments involving model compounds (Hickman et al. 2004; Tirado et al. 2018). Table 6.2 summarizes the operating conditions, feedstock, reactor, and catalyst types reported in kinetic studies focused on producing green fuels by the hydrotreating process.

6.3.1 Model Compounds

The development of kinetic models focused on the hydrotreating of vegetable oil is based on accurate and reliable experimental data obtained during studies of reaction mechanisms and catalyst performance. Model compounds such as FAs and Tg serve as representative samples of complex systems such as vegetable oils (Boda et al. 2010; Chen et al. 2013; Kubička et al. 2009; Kubička and Kaluža 2010; Snåre et al. 2006). Thus, kinetic models consider possible reaction pathways and their rates by determination of kinetic parameters under certain operating conditions (Furimsky 2013; Hermida et al. 2015; Mortensen et al. 2011). Figure 6.2 depicts diverse reaction

Table 6.2 Kinetic studies performed for the production of renewable fuels through the hydrotreating process.

Feedstock	Reactor type	Temperature (°C)	Pressure (bar)	Time (h)	Stirring rate (rpm)	Catalyst/solvent	Conversion (%)	Reference
Ethyl stearate	Semi-batch	270–360	17–40 25 ml Ar/min (5% H_2)			$5Pd/C^{(R)}$/*n*-dodecane	41–100	Snåre et al. (2007)
Stearic acid	Semi-batch	260–290	13–14.5 25 ml Ar/min (5% H_2)	6		$Ni/\gamma\text{–}Al_2O_3\ ^{(O)}$/*n*-dodecane	50–95	Kumar et al. (2014)
Oleic acid	Semi-batch	320–380	20 (10% H_2)	1	2000	$FMoOx/Zeol^{(R)}$	40–100	Ayodele et al. (2015)
Waste cooking oil	Batch	300–375	90	8	1000	Unsupported $CoMoS^{(S)}$	91–99	Zhang et al. (2014)
Methyl heptanoate	Batch	250–330	60–100	6	700	$Rh/ZrO_2^{(R)}$/*n*-hexadecane	15–90	Bie et al. (2015)
Methyl palmitic	Batch	250–330	60–100	1	700	$Rh/ZrO_2^{(R)}$/*n*-hexadecane		Bie et al. (2016)
Tripalmitin–tristearin	Batch	280–360	30	6	1200	$15NiAl^{(R)}$/*n*-dodecane	40–100	Yenumala et al. (2017)
Stearic acid TOFA	Semi-batch	300	7–30	6	1200	$5Ni/\gamma\text{–}Al_2O_3^{(R)}$, $5Ni/Y\text{–}H\text{–}80^{(R)}$, $5Ni/SiO_2^{(R)}$, $Pd/C^{(R)}$	99	Jeništová et al. (2017)
Methyl palmitate	Trickle bed	290	30	0.083–3.75 min	—	Ni_2P/SiO_2	34–82	Shamanaev et al. (2019)
Stearic acid	Batch	275–325	40–70	10–180 min	500–1000	$NiMo/Al_2O_3$		Arora et al. (2019)
Rapeseed oil	Fixed bed	300–380	10	2.7–9.8 h^{-1} (LHSV)	—	$NiCu/CeO_2\text{–}ZrO_2^{(R)}$	10–100	Selishcheva et al. (2014)
Solvolytic oil	Batch	300	80	1	1000	$NiMo/\gamma\text{–}Al_2O_3^{(O\text{–}R\text{–}S)}$, Pd/C (R), $MoS_2^{(S)}$, $NiMo/Al_2O_3\text{–}SiO_2^{(R)}$, $Pd/\gamma\text{–}Al_2O_3^{(R)}$/Tetralin	43–94	Grilc et al. (2014b)
Solvolytic oil	Batch	200–350	20–80	1	250–1000	$NiMo/\gamma\text{–}Al_2O_3^{(S)}$/*n*-dodecane	83–85	Grilc et al. (2014a)
Jatropha oil	Fixed bed	300–320	80	2–12 h^{-1} (LHSV)	—	$CoMo/MTS^{(S)}$, H_2/HC =1200–1500 l/l	65–70	Sharma et al. (2012)
Jatropha oil	Fixed bed	320–360	80	1–8 (LHSV)	—	$CoMo/Al_2O_3^{(S)}$, H_2/HC = 1500 Nl/l	89–98	Anand and Sinha (2012)
Jatropha oil	Fixed bed	340–420	80	0.5–12 (WHSV)	—	$NiW/SiO_2\text{–}Al_2O_3^{(O)}$, H_2/HC = 500–2500 Nl/l	29–100	Anand et al. (2016a)
Palm oil	Fixed bed	335–365	30–60	15–45 min	—	$NiMo/\gamma\text{–}\ Al_2O_3^{(S)}$	73–100	Vélez Manco (2014)
Castor oil methyl esters	Batch	395–425	50	2–4	750	—	44–86	Mederos et al. (2021)

State of catalyst: O, oxide form; R, reduce form; S, sulfide form.

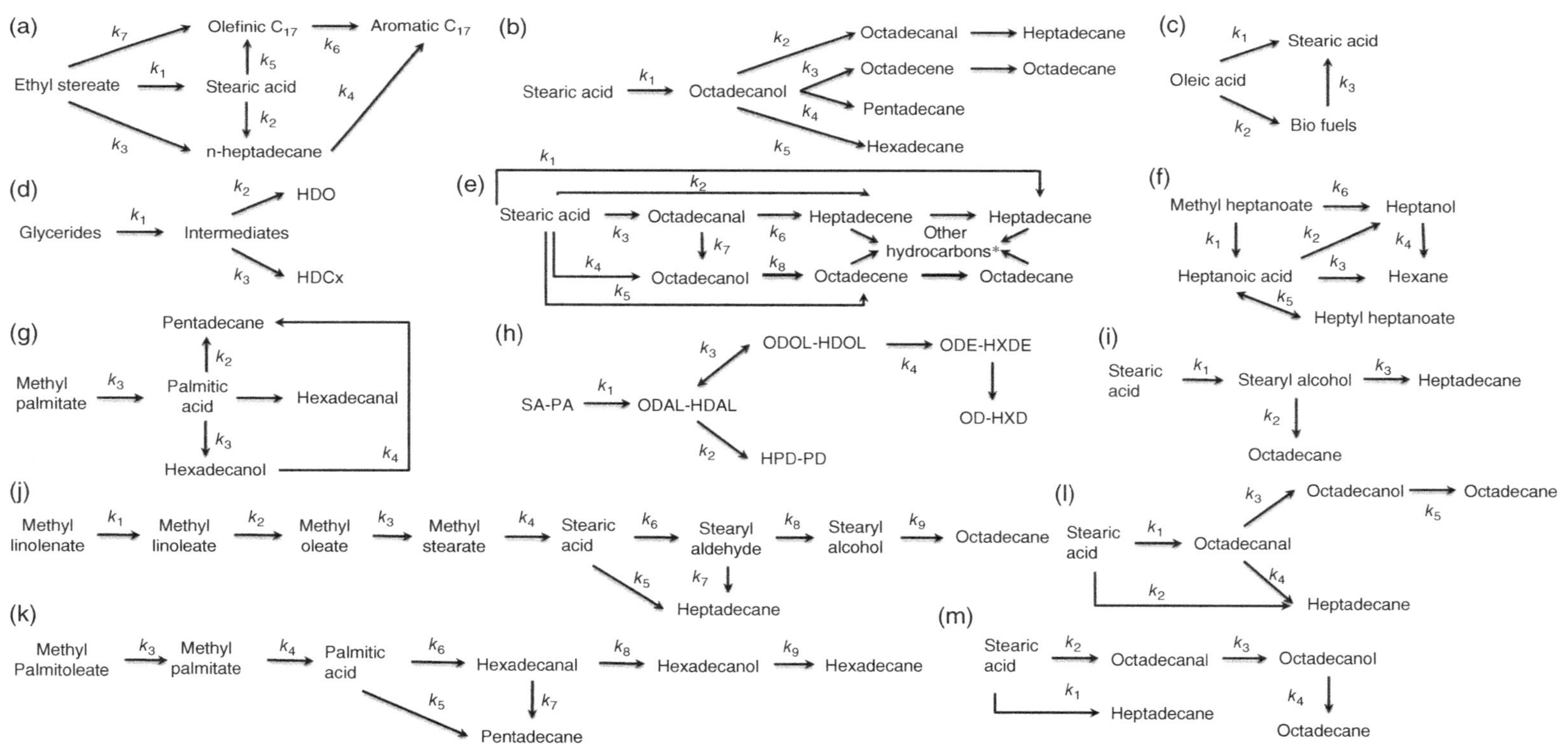

Figure 6.2 Reaction schemes for the hydrotreating of model compounds.

schemes proposed for studying reaction mechanisms as well as the prediction of the conversion and production of compounds.

Snåre et al. (2007) investigated the kinetics of ethyl stearate over a commercial Pd/C catalyst. The reaction scheme, presented in Figure 6.2a, indicates the ethyl stearate (A) conversion to form its corresponding FA (B) and paraffins (C). However, dehydrogenation reactions can form olefins (D) and eventually aromatics by cyclization (E). The Langmuir–Hinshelwood mechanism accurately predicted the experimental concentration profiles for both reactants and products. The adsorption coefficients of some product gases (i.e. CO_2, CO, ethylene, etc.) were neglected. Additionally, the surface reactions are rate-limiting instead the adsorption steps:

$$r_j = \frac{k_j' C_i}{1 + K_A C_A + K_B C_B + K_C C_C + K_D C_D + K_E C_E} \text{ for } j = 1 - 7 \tag{6.1}$$

where k_j' is an apparent reaction coefficient involving the reaction rate and equilibrium adsorption coefficients ($k_j' = k_j \times K_i$) of the diverse reaction pathways j, and C_i is the concentration of reactant species i. Based on the kinetic rate expressions, the mass balances of components are expressed by a set of ordinary differential equations (ODEs) in the function of the catalyst bulk density (ρ_B) and reaction time (t):

$$\frac{1}{\rho_B}\frac{dC_i}{dt} = \sum r_j \tag{6.2}$$

The result obtained with the model confirmed that the adsorption coefficients can be disregarded. Therefore, Eq. (6.1) follows a first-order power-law model.

Kumar et al. (2014) performed a kinetic study for the HDO of stearic acid with a Ni/γ–Al_2O_3 catalyst proposing a reaction pathway (Figure 6.2b) that considers octadecanol as an intermediate product, which undergoes dehydrogenation and decarbonylation reactions to finally produce heptane and one mole of CO. Alternatively, octadecanol is dehydrated and form octadecene, which is hydrogenated to octadecane. The formation of hexadecane and pentadecane is not entirely clear, and it has been speculated that they are the end products of octadecanol. As octadecanal was not detected in the HDO of stearic acid, it was assumed to be rapidly converted into heptadecane. Experiments under different stirring rates (1000–2000 rpm) showed negligible changes in the stearic acid conversion, indicating the reaction follows the intrinsic kinetics (i.e. the transfer at the gas–solid and liquid–solid interfaces is neglected). Nevertheless, resistance due to molecule diffusion inside the pores could not be evaluated. Therefore, kinetic parameters include the internal diffusional resistance effect. The reaction rates were expressed by the power-law model a first reaction order concerning the concentration of reactant species and a hydrogen excess in the reaction system.

$$\frac{dC_i}{dt} = \sum k_j C_i \tag{6.3}$$

The kinetic results indicated a large conversion of alcohols to aldehydes. The highest activation energy was observed for k_4 and k_5, indicating that the formation of pentadecane and hexadecane predominated at higher temperatures. A similar model was used to determine the activity and product yields using a zeolite-based fluoride-ion functionalized molybdenum-oxalate catalyst (FMoOx/zeol) in the oleic acid (OA) HDO (Ayodele et al. 2015). The model considers the hydrogenation of the double bond in the OA to form stearic acid, followed by the subsequent elimination of an oxygen atom, as shown in Figure 6.2c. The production of the C_{18} biofuel is insignificant as its formation occurs with difficulty. The reaction rate and mass balance equations were developed with similar assumptions to determine the kinetic coefficients using the linear Arrhenius equation to calculate the activation energies (E_A, energy quantity required to perform a chemical reaction) and pre-exponential factors (A_0, total number of collisions per time unit leading to reaction).

$$k_j = A_0 e^{-\frac{E_A}{RT}} \tag{6.4}$$

$$\ln k_j = \ln A_0 - \frac{E_A}{RT} \tag{6.5}$$

where R is the universal ideal gas constant. The results indicated that the hydrogenation rate of OA is much higher than the rate of removal of oxygen atoms in stearic acid.

Zhang et al. (2014) proposed a kinetic model for hydrotreating waste cooking oil glycerides over an unsupported CoMoS catalyst. The kinetic parameters calculated using a power-law model showed that glycerides are rapidly transformed into FAs through desterification reactions. FAs are continuously converted into alkanes and other products to form alcohols/aldehydes as intermediates, which are converted via HDO or hydrodecarboxylation/hydrodecarbonylation (HDC_x) reactions (Figure 6.2d). According to the kinetic parameters obtained, the removal of oxygen atoms took place through HDC_x reactions. In addition, the kinetic model includes a catalytic deactivation term, which is discussed in a later section.

Alcohol predominated as the main intermediate, as observed by its large reaction rate coefficient compared with the aldehyde formation in the HDO of FAs (Figure 6.2e). Nevertheless, polymerization and cyclization reactions occur in parallel.

Most of the kinetic models assume an excess of hydrogen relative to the stoichiometric requirement in the reaction system. Thus, the hydrogen concentration does not change significantly and it is held as constant in the reaction rate equations. Nonetheless, some kinetic studies consider the hydrogen pressure effect in order to analyze its effect on the reaction system. Bie et al. (2015) developed a reaction network during HDO of methyl heptanoate on Rh/ZrO_2 catalyst. The simplified route shown in Figure 6.2f was modeled by both power-law and mechanistic models. In the first model, six-reaction steps for reaction j, the concentration of species i, and the hydrogen pressure at its corresponding reaction order were modeled:

$$r_j = k_j C_i P_{\mathrm{H_2}}^{m_j} \tag{6.6}$$

Since the reaction pathway 5 is considered as reversible, its mathematical expression is represented by the van't Hoff equation, considering the temperature-dependent equilibrium constant k_{eq}:

$$r_5 = k_5 C_{\mathrm{B}} C_{\mathrm{C}} - \frac{C_D C_{\mathrm{H_2O}}}{k_{\mathrm{eq}}} \tag{6.7}$$

$$k_{\mathrm{eq}} = 0.338 e^{1136/RT} \tag{6.8}$$

The molar balance of each species n_i in both gas and liquid phases is expressed considering the mass transfer effects in the gas–liquid interphase (a_{GL}) by the following batch reactor model:

$$\frac{dn_i^{\mathrm{G}}}{dt} = -V_{\mathrm{R}} N_{\mathrm{GL},i} a_{\mathrm{GL}} \tag{6.9}$$

$$\frac{dn_i^{\mathrm{L}}}{dt} = W R_i V_{\mathrm{L}} + V_{\mathrm{R}} N_{\mathrm{GL},i} a_{\mathrm{GL}} \tag{6.10}$$

where V_R and V_L represent the effective reactor and the liquid phase volumes, W indicates the mass of the catalyst, and R_i is the reaction rate of each species based on the stoichiometry.

In another study, Bie et al. (2016) used a Rh/ZrO_2 catalyst during HDO of methyl palmitate. The oxygen vacancy sites on ZrO_2 are active for oxygenated species, and the dihydrogen activation through the dissociative adsorption carried out by the active sites on reduced state explains the bifunctionality of noble/transition metal-based catalysts. During the reaction, palmitic acid and methane are produced, as shown in Figure 6.2g, following a similar mechanistic model as previously mentioned. For the reaction rate r_1, hydrogen concentration was considered to better describe the effect of the pressure on the reaction kinetics.

$$r_1 = \frac{k_1 C_A \sqrt{K_{H_2} C_{H_2}}}{\left(1 + \frac{K_B C_B}{\sqrt{K_{H_2} C_{H_2}}} + \frac{K_C C_C}{\sqrt{K_{H_2} C_{H_2}}}\right)\left(1 + \sqrt{K_{H_2} C_{H_2}}\right)} \tag{6.11}$$

Reaction rates r_2–r_3 follow the same mechanism hypothesis reported (Bie et al. 2015), which share a common rate of determination step (the final addition of a hydrogen atom in the reactant species):

$$r_j = \frac{k_j K_B K_{H_2} C_B C_{H_2}}{\left(1 + \frac{K_B C_B}{\sqrt{K_{H_2} C_{H_2}}} + \frac{K_C C_C}{\sqrt{K_{H_2} C_{H_2}}}\right)\left(1 + \sqrt{K_{H_2} C_{H_2}}\right)} \tag{6.12}$$

Finally, the reaction rate r_4 considers the reversible dehydrogenation–decarbonation of hexadecanol in pentadecane:

$$r_4 = \frac{k_4 K_C C_C}{\left(1 + \frac{K_B C_B}{\sqrt{K_{H_2} C_{H_2}}} + \frac{K_C C_C}{\sqrt{K_{H_2} C_{H_2}}}\right)\left(K_{H_2} C_{H_2}\right)} \tag{6.13}$$

It was stated that the kinetic model can be adequately fitted to experimental data on a wide range of reaction conditions and conversions, unlike other models based on power-law kinetics.

Recently, the HDO of Tg was investigated using a mixture of tripalmitin (TP) and tristearin (TS) over Ni/Al_2O_3 (Yenumala et al. 2017). HDO experiments showed that TP and TS were converted instantaneously to FAs and propane, as also observed by Zhang et al. (2014) who reported that Tg were rapidly converted. The scheme presented in Figure 6.2h shows the kinetic model based on the experimental results, where the deoxygenation pathway is mainly observed. Tg formed intermediate aldehydes and alcohols, which were subsequently converted into alkanes and water through reduction/dehydration reactions carried out in unsaturated sites of sulfhydryl groups. In addition, the alkane production is enhanced by decarbonylation reactions because the Ni/Al_2O_3 catalyst is weakly acidic. The reaction rate is modeled using a first-order power-law with respect to compounds in the liquid phase and considering the hydrogen partial pressure (P_{H_2}) during hydrogenation reactions. For stearic or palmitic acid (A) as reactant, the reaction rate equation is:

$$-\frac{dC_A}{dt} = k_1 C_A P_{H_2} \tag{6.14}$$

The conversion of aldehydes (B) to alcohols (D) is a reversible reaction that utilizes the equilibrium constant (k_{eq}) as follows:

$$\frac{dC_B}{dt} = k_1 C_A P_{H_2} - k_2 C_B - k_3 C_B P_{H_2} + \left(k_3/k_{eq}\right) C_B \tag{6.15}$$

$$\frac{dC_D}{dt} = k_3 C_B P_{H_2} - (k_3/k_{eq}) C_D - k_4 C_D \tag{6.16}$$

To precisely determine the hydrogen partial pressure at any given reaction time, it is necessary to subtract the amount of stoichiometric hydrogen consumed by Tg, C, D, and E according to:

$$P_{H_2} = P^0_{H_2} - \frac{V_L(3C_{TG,0} + C_C + 2C_D + 3C_E)RT}{V_G} \tag{6.17}$$

where P^0_i and V_G are the initial pressure of component i and the volume of the gas phase. The decarbonylation reaction (k_2) predominated over the HDO path (k_3), as shown by the kinetic coefficient values. However, similarities in the rate coefficients k_1, k_2, and k_3 were observed for both types of Tg, with only k_4 showing slight variation.

Jeništová et al. (2017) investigated the influence of hydrogen pressure on the HDO of stearic acid and tall oil fatty acids (TOFAs). It was shown that the higher the hydrogen pressure, the higher the FA conversion. The reaction mechanism is similar to previous studies reported elsewhere (Zhang et al. 2014), with some differences in the pathways in which oxygen atoms are subtracted, as observed in Figure 6.2i. The Langmuir–Hinshelwood model was utilized to analyze the effect of hydrogen pressure on the hydrogenolysis reactions (reactions 1 and 3). It is assumed that each reaction rate involves non-competitive adsorption of hydrogen and organic species. For instance, the rate for reaction 1 (r_1) is:

$$-r_1 = \frac{k'_1 C_A P_{H_2}}{(1 + K_A C_A)(1 + K_H P_{H_2})} \tag{6.18}$$

Different catalysts (Ni–H–Y–80, Ni/SiO_2, Pd/C) were used on the HDO of stearic acid, where the kinetic parameters show that decarboxylation/decarbonylation (DC_x) reactions predominated.

Hachemi and Murzin (2018) developed kinetic studies to analyze the FAs methyl esters and Tg HDO using Ni/H–Y–80 and Pd/C catalysts. Reaction schemes for C_{18} and C_{16} are depicted in Figure 6.2j and Figure 6.2k, respectively. The results based on the Langmuir–Hinshelwood mechanism indicated that molecules having different numbers of carbon atoms underwent similar pathways. Reaction rate equations were developed considering that the absorption of acid species is significantly higher than those of the aldehydes, alcohols, alkanes, etc. The Ni-based catalyst promoted the HDO reaction, as indicated by the kinetic parameters ($k_6 > k_5$), while the Pd catalyst enhanced the DC_x reactions ($k_5 > k_6$). It was also found that for both catalysts, the carbon chain length of the feedstock influences the kinetic parameters.

Arora et al. (2019) investigated the kinetics of stearic acid HDO to explore the fundamental chemistry and mechanisms involved in the reactions with a NiMo/Al_2O_3 catalyst (Figure 6.2m). A model based on the Langmuir–Hinshelwood type kinetics was developed, which predicted well trends in variation of selectivities with the reaction conditions, including intermediate compounds such as octadecanol and octadecanal as well as predicting phenomena like inhibiting effects of the FA. The evaluation of the external mass transfer limitations was performed by experimental results varying stirring speeds (500–1000) and the Carberry number. Additionally, the mass transfer resistance in catalyst pores was evaluated by analytical correlations, which presented a moderate influence on the observed rate of reactions for experiments with stearic acid feedstock. However, the results indicated a strong effect of film and pore mass transfer resistances concerning experiments with $C_{18}O$ feedstock.

Recently, Pedroza et al. (2022) performed some research on the kinetics of the HDO of stearic acid over a NiMo/Al_2O_3 catalyst to obtain hydrocarbons in the diesel range. The reaction scheme presented in Figure 6.2n was proposed, and diverse kinetic models were evaluated based on assumptions to the adsorption step, using Eley–Rideal and Langmuir–Hinshelwood mechanisms. The surface reaction-limited model based on competitive adsorption with non-dissociate H_2 adsorbed presented the best results and was used to simulate the process.

6.3.2 Vegetable Oils

Due to the wide variety of chemical species present in real feedstock for the hydrotreating process (such as vegetable oils and petroleum distillates), some assumptions are considered to simplify the calculation of kinetic parameters, i.e. chemical compounds with similar properties are grouped in the so-called lumps or pseudocomponents in order to reduce the number of kinetic parameters to be determined. Kinetic studies of vegetable oil hydrotreating consider various reaction schemes due to differences in feedstocks and methods for grouping their chemical compounds (Zhang et al. 2009). These models have been analyzed and discussed in the literature (Tirado et al. 2018) and mentioned in this section, where the most relevant kinetic models for the simulation of vegetable oil hydrotreating reactors are studied. Figure 6.3 depicts diverse schemes used to analyze the reaction pathways that take place under diverse operating conditions.

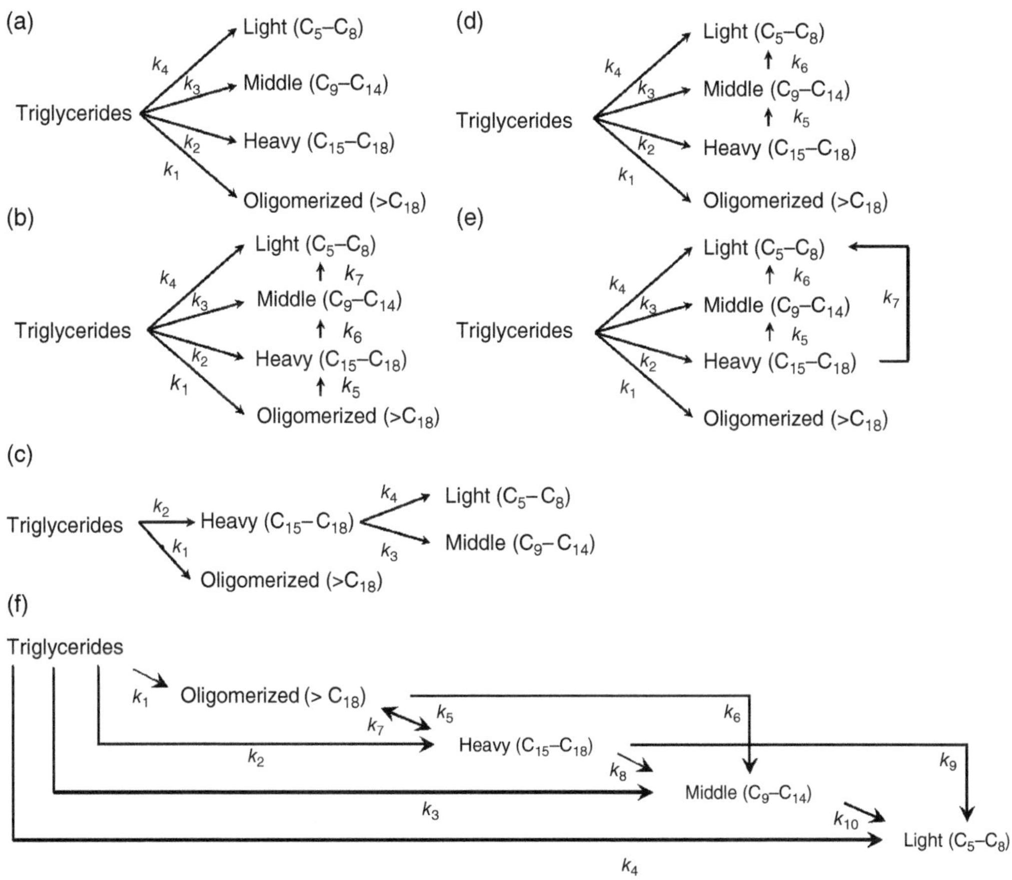

Figure 6.3 Reported kinetic model for hydrotreating of vegetable oils (Anand et al. 2016a; Anand and Sinha 2012; Sharma et al. 2012; Tirado and Ancheyta 2019).

Sharma et al. (2012) developed a kinetic model for Jatropha oil hydrotreating using a CoMo/MTS catalyst. A single lump included the Tg, free fatty acids (FFAs), and some traces of diglycerides and monoglycerides, where different reaction pathways of conversion toward oligomerized compounds ($>C_{18}$), heavy (C_{15}–C_{18}), middle (C_9–C_{14}), and light hydrocarbons (C_5–C_8) were evaluated in the temperature range of 300–320 °C. The model assumed a first-order reaction for the different lumps. The reaction scheme is depicted in Figure 6.3a, and the values of reaction rate coefficients are listed in Table 6.3. However, the model did not account for gas production pathways. The reaction rate coefficients show that the reaction rate of Tg to oligomerized products is much higher than other reactions at 300 °C. The reaction rate coefficients for producing heavy compounds increase by 20 times as the temperature rises from 300 to 320 °C. On the other hand, the high acidity of the CoMo/MTS catalyst increased the formation of light and middle hydrocarbons through cracking reactions. Experimental data indicated only primary reactions, and the rate of reaction of oligomerization is much higher compared with other reaction pathways.

Anand and Sinha (2012) reported a set of kinetic models for the hydrotreating of Jatropha oil using a CoMo/γ–Al_2O_3 catalyst. Diverse reaction pathways were studied in a temperature range of 320–360 °C and pressure of 80 bar. The kinetic parameters obtained for each sequence of reactions are presented in Table 6.3. The results showed that at 320 °C (low temperature), Tg underwent consecutive side reactions with naphtha and kerosene compounds as by-products (C_5–C_8 and C_9–C_{14}, respectively) in small amounts because of lower acidity and metal support interaction compared with the CoMo/MTS catalyst reported by Sharma et al. (2012). At 340 °C (mild temperature), lighter compounds were produced as thermal cracking impacts all pseudocomponents, and even diesel (C_{15}–C_{18}), kerosene (C_9–C_{14}), and non-stable oligomerized compounds underwent cracking reactions. At 360 °C, comparable pathways are pursued, resulting in stable oligomerized products.

Most of the kinetic models reported for vegetable oil hydrotreating assume mass transfer effects to be negligible. Nevertheless, experimental studies are not reported to support this assumption. Anand et al. (2016a) conducted a kinetic and thermodynamic study on the hydrotreating of Jatropha oil to determine the diffusion limitations, kinetic reaction mechanisms, and thermodynamics using a NiW/SiO_2–Al_2O_3 catalyst in a fixed-bed reactor. The results indicated that intraparticle diffusion was negligible since the effectiveness factor was ~1. Additionally, the model results at 340 °C indicated that Tg follow similar kinetic mechanisms, and the reaction rate coefficients have some tendencies as those reported by Anand and Sinha (2012) at 320 °C with a CoMo/Al_2O_3 catalyst. On the other hand, the reaction rate changes slightly at a high temperature (420 °C) since diesel is also converted to naphtha fraction due to the presence of acidic sites on the catalyst and thermal cracking.

Despite the sophistication of these kinetic models, the calculated kinetic parameters obtained do not guarantee to be the optimal ones, i.e. the values that minimize the difference between the experimental and calculated data. When kinetic parameters are inaccurately calculated and not optimized, a model that demonstrates good fit may either disregard reaction pathways that are being carried out or incorporate those that should not be considered. This inappropriate definition of possible reaction schemes has a strong effect on predicting the yields and other reaction parameters that cause poor design of commercial-scale reactors (Alcázar and Ancheyta 2007). Tirado and Ancheyta (2019) recalculated the kinetic parameters using experimental data reported by Anand et al. (2016a) and Anand and Sinha (2012) under different operating conditions and catalyst types. The estimation of the kinetic parameter was performed using a proposed reaction mechanism shown in Figure 6.3f, which involves all reaction pathways reported in the literature for the hydrotreating of vegetable oils as well as a series of statistical analyses based on the algorithm present in Figure 6.4 and ensuring that the kinetic parameters converge to the global minimum of the objective function for each reaction scheme by taking into account the sum of the squares of the errors (SSE) as an objective function to optimize the values. Experimental data obtained during

Table 6.3 Kinetic parameters determined during the HDO of vegetable oils using reaction schemes presented in Figure 6.3.

	Reaction pathway										
Condition	**Tg–Prod.**	**Tg–Ol**	**Tg–Hv**	**Tg–Md**	**Tg–Lh**	**Ol–Hv**	**Ol–Md**	**Hv–Ol**	**Hv–Md**	**Hv–Lh**	**Md–Lh**
CoMo/MTS	Model of Sharma et al. (2012)										
300 (a)	7.39	7.27	0.07	0.02	0.03	—	—	—	—	—	—
320 (a)	14.35	13.25	1.24	0.11	0.04	—	—	—	—	—	—
$CoMo/Al_2O_3$	Model of Anand and Sinha (2012)										
320 (c)	17	7	10	0.3	0.3	—	—	—	—	—	—
340 (b)	17	13	3	0.2	1.1	5	—	—	0.4	—	3
360 (d)	24	2	19	2	1	—	—	—	0.1		0.1
E_a (kJ/mol)	26	87	43	127	47	—	—	—	—	—	—
A_0	31.9×10^{02}	14.5×10^{04}	62.6×10^{05}	4.2×10^{10}	13.7×10^{04}						
$NiW/SiO_2Al_2O_3$	Model of Anand et al. (2016a)										
340 (c)	1.06	0.2	0.85	0.04	0.04	—	—	—	—	—	—
420 (e)	23.4	2.7	18.8	1.2	0.6	—	—	—	0.2	0.1	0.1
$CoMo/Al_2O_3$	Model of Tirado and Ancheyta (2019)										
360 (f)	23.3	0.317	24.85	1.95	0.82	14.37	0.03	2.09	0.1	0	0.268
$NiW/SiO_2Al_2O_3$	Model of Tirado and Ancheyta (2019)										
420 (f)	22.62	2.054	18.494	1.257	0.815	2.662	0	0.553	0.172	0.06	0.094

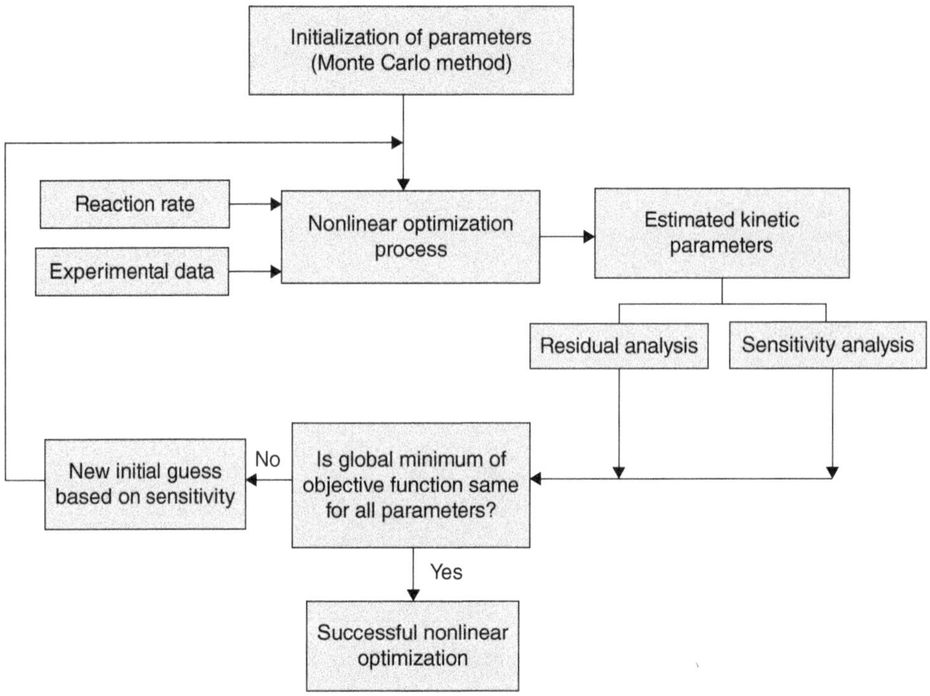

Figure 6.4 General approach for optimization of parameter values of kinetic models (Alcázar and Ancheyta 2007 / with permission of Elsevier).

the hydrotreating of Jatropha oil at 360 °C with a $CoMo/Al_2O_3$ catalyst and 420 °C with a $NiW/SiO_2–Al_2O_3$ catalyst were used to perform the calculation of kinetic parameters based on reaction schemes reported by Anand et al. (2016a) and Anand and Sinha (2012) as well as that proposed (Figure 6.3f). It was reported that at 360 °C with a $CoMo/Al_2O_3$ catalyst, the reactor scheme in Figure 6.3d with six kinetic parameters showed the best fit with experimental data, while at 420 °C with a $NiW/SiO_2–Al_2O_3$ catalyst, the best reaction scheme was the Figure 6.3e with seven parameters (Anand et al. 2016a; Anand and Sinha 2012). The comparison between the obtained kinetic values of the models shows that some reaction pathways are similar. However, certain reaction pathways that were not considered exhibited noteworthy reaction rates. For example, the transformation of oligomerized compounds into large hydrocarbons (k_5) occurs at a rapid reaction rate. This behavior is expected because the oligomerized compounds fraction groups together with oxygenated compounds, such as FAs, aldehydes, and alcohols, undergo deoxygenation reactions and eventually form heavy hydrocarbons (Huber et al. 2007; Zhang et al. 2014). On the other hand, hydrocarbons under severe temperatures can undergo polymerization reactions, resulting in the formation of compounds with high molecular weight (k_7). However, the values of k_6 for the conversion of oligomerized compounds to middle hydrocarbons were negligible, indicating low selectivity toward this reaction pathway. It is corroborated that the yield of light hydrocarbons from heavy hydrocarbons (k_9) is zero at low temperatures with $CoMo/Al_2O_3$ catalyst, but it starts at 420 ° C with $NiW/SiO_2–Al_2O_3$ catalysts. The proposed model gave better fits of the objective function values (SSE) and the determination coefficient (R^2) than those obtained with literature models. Both models predicted higher decomposition of Tg $\left(k_{Tg_0}\right)$ for $CoMo/Al_2O_3$ catalyst compared with the $NiW/SiO_2–Al_2O_3$ catalyst.

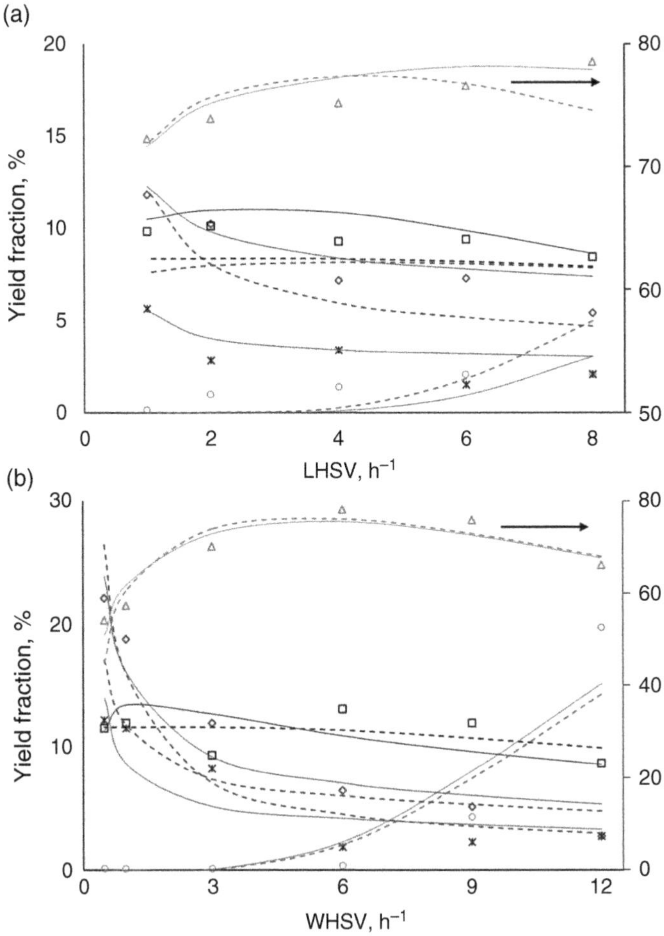

Figure 6.5 Comparison of experimental and calculated data with kinetic parameters of Anand et al. (– –) and with those obtained by Tirado and Ancheyta (2019) / with permission of Elsevier (——) at (a) 360 °C and (b) 420 °C: Tg (○), >C_{18} (□), C_5–C_8 (◇), C_9–C_{14} (∗), C_{15}–C_{18} (Δ).

Figure 6.5 shows the comparison of the experimental data and the calculated profiles of each model with their corresponding kinetic parameters. It is noteworthy that the heavy hydrocarbon yield rises, which is consistent with the LHSV increase as reported by Anand et al. (2016a) and Anand and Sinha (2012). This trend is well-predicted by the proposed model by Tirado and Ancheyta (2019), while the previous model predicts a maximum value around LHSV of 4, and then the yield decreases. This behavior demonstrates that a more detailed kinetic model allows for better prediction of the experimental yields and their trends concerning the reaction time.

In order to ensure that kinetic parameters provide the best fit for their corresponding reaction scheme, sensitivity analysis for the different models is depicted in Figure 6.6 and Figure 6.7. It is observed in Figure 6.6 that the kinetic coefficients reported by Anand et al. (2016a) and Anand and Sinha (2012) do not reach the global minimum, particularly k_1 and k_5 of scheme A7 and k_6 and k_7 of scheme A8, which present the minimum value of the objective function further to 0% disturbance. In contrast, Figure 6.7 demonstrates that the sensitivity analysis of kinetic parameters fits perfectly at 0% disturbance of the objective function, reaching the global minimum.

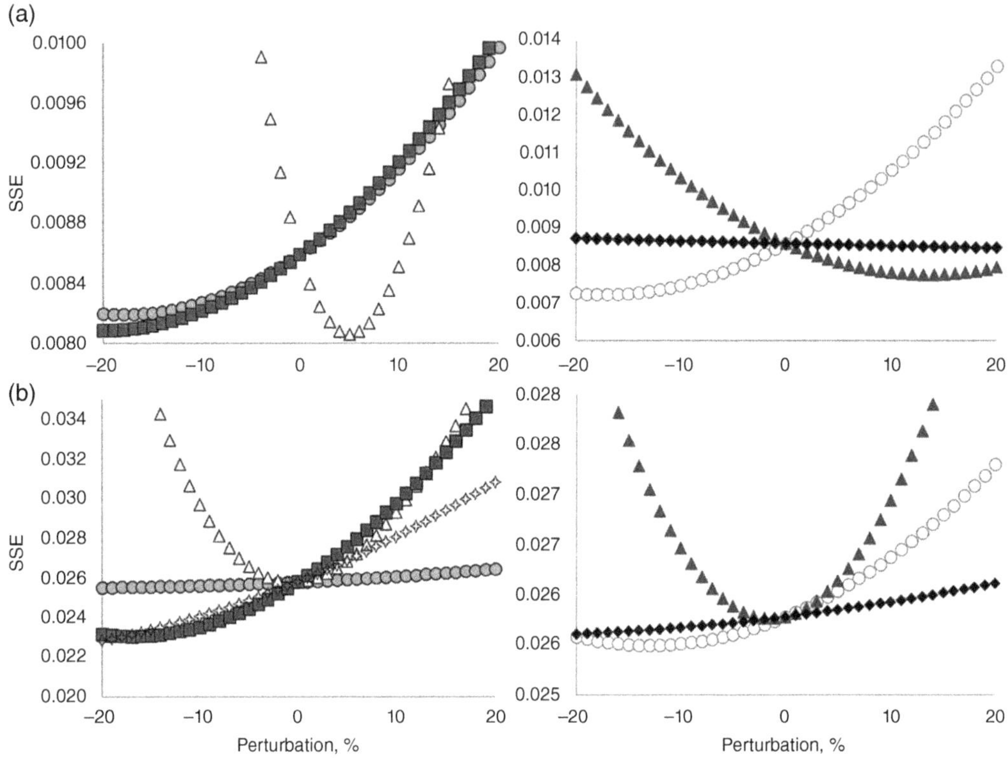

Figure 6.6 Sensitivity analysis for the reaction rate coefficients reported by Anand et al. (Anand et al. 2016a; Anand and Sinha 2012) at (a) 360 °C and (b) 420 °C: k_1 (●), k_2 (○), k_3 (Δ), k_4 (▲), k_5 (■), k_6 (♦), k_7 (✧).

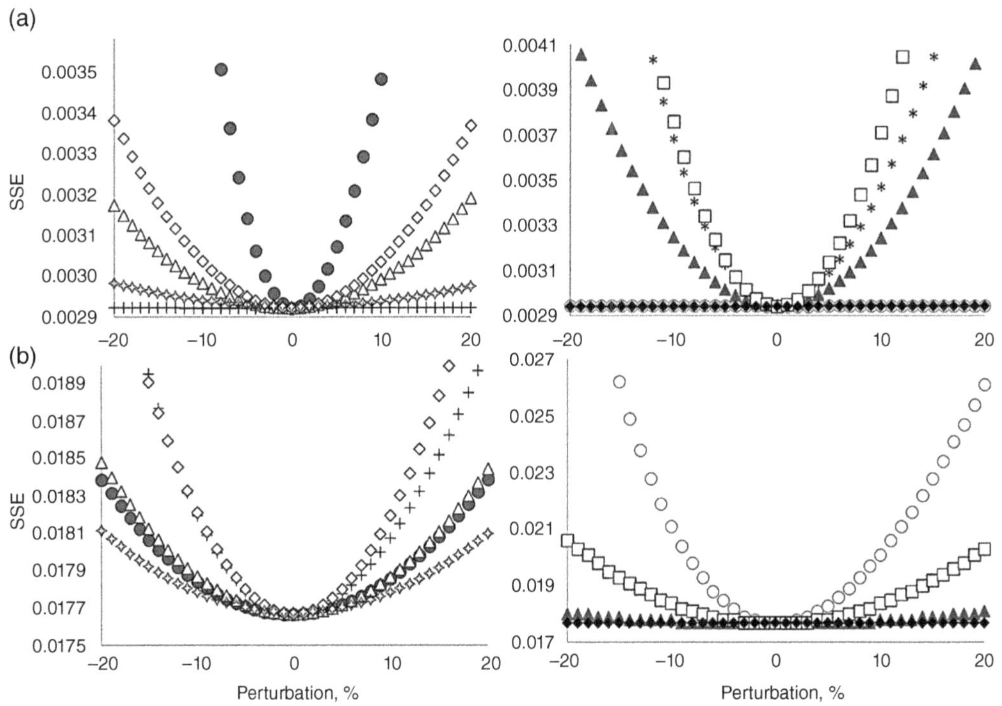

Figure 6.7 Sensitivity analysis for the reaction rate coefficients calculated by Tirado and Ancheyta (2019) / with permission of Elsevier at (a) 360 °C and (b) 420 °C: k_1 (●), k_2 (○), k_3 (Δ), k_4 (▲), k_5 (+), k_6 (∗), k_7 (◇), k_8 (□), k_9 (✧), k_{10} (♦).

Table 6.4 Statistical parameters from regression analysis and degree of confidence.

Catalyst	$CoMo/Al_2O_3$		$NiW/SiO_2Al_2O_3$	
	360 °C		420 °C	
Statistical parameters	Anand and Sinha (2012)	Tirado and Ancheyta (2019)	Anand et al. (2016a)	Tirado and Ancheyta (2019)
	Figure 6.3d	Figure 6.3f	Figure 6.3e	Figure 6.3f
R^2	0.9971	0.9991	0.9901	0.9932
Slope	0.9916	1.0034	0.9861	0.9906
Intercept	0.586	0.3545	0.329	0.2387
Residual (+)	12	8	15	14
Residual (−)	13	17	15	−16
Range	0.0707	0.0337	0.1505	0.0902

Parity plots of the calculated and experimental yield values for both operating conditions indicated that the values obtained by Tirado and Ancheyta (2019) presented a better fit concerning the diagonal line. Regression analysis was used to verify this point, and the results are shown in Table 6.4. The slope value and determination coefficient calculated for the results of the proposed model are closer to 1, while the intercept value is closer to 0 compared with those obtained for the data in the literature.

Residual plots for both operating conditions show no discernible trend among residual values, indicating that the models are not overestimating or underestimating experimental data. The balance of positive and negative residual values further supports this conclusion, as reported in Table 6.4. On the other hand, the range between the lowest negative residual value and the highest positive residual value shows a much lower value for the proposed model at both operating conditions compared with the original models, indicating a greater difference between the calculated and experimental values. Based on the results obtained, it is demonstrated that the proposed kinetic model presents an appropriate estimation of reaction rate coefficients, which show a better fit with experimental data compared with kinetic schemes d and e shown in Figure 6.3. Additionally, the achievement of the global minimum confirmed by the sensitivity analysis makes the proposed model more accurate (Sámano et al. 2020; Tirado and Ancheyta 2019).

Kinetic studies of hydrotreating conducted on model compounds and vegetable oils have resulted in a plethora of reaction schemes owing to the diverse reaction mechanisms that arise from the complex feedstocks, catalysts, and operating conditions employed. Most of them achieve similar conclusions, indicating that Tg and ester molecules of FAs undergo instantaneous hydrogenolysis reactions to produce intermediate acid compounds (Hachemi and Murzin 2018). Depending on these parameters, it is possible to remove oxygen atoms via HDO, decarboxylation, and decarbonylation. The development of kinetic studies requires vital experimental considerations to obtain representative information about the reactive system and its subsequent application in reactor models and simulations. Assessment of external mass transfer and internal diffusional resistance is determined to estimate the intrinsic reaction kinetics. In addition, it has been observed that the mathematical algorithm of methods used to estimate the kinetic parameters significantly influences the prediction of the behavior of the diverse components considered during the kinetic modeling. Table 6.5

Table 6.5 Summary of methodologies used for estimation of kinetic parameters.

Kinetic model	Parameter number	Objective function	Solution method/ software	Optimization algorithm
Snåre et al. (2007)	7	$RSS = \sum_i \sum_t (C_{i,t}^{exp} - C_{i,t}^{cal})^2$	Odessa ODE solver	—
Kumar et al. (2014)	5	$RSS = \sum_i \lvert C_i^{exp} - C_i^{cal} \rvert$	4th Order Runge–Kutta	Modified Levenberg–Marquardt
Zhang et al. (2014)	8	Least squares	Runge–Kutta	Trust-region-reflective
Bie et al. (2015)	6	$RSS = \sum_i \sum_j (w_i^{exp} - w_i^{cal})^2$	Backward difference	Levenberg–Marquardt
Yenumala et al. (2017)	5	$ER = \sum_i \sqrt{(w_i^{exp} - w_i^{cal})^2}$	4th Order Runge–Kutta	Nelder–Mead
Jeništová et al. (2017)	5	—	Backward difference	Simplex and Levenberg–Marquardt
Hachemi and Murzin (2018)	10	—	Backward difference	—
Shamanaev et al. (2019)	—	$OF = \sum_i \sum_j (w_i^{exp} - w_i^{cal})^2$	Gear method	Basin-hopping algorithm
Tirado and Ancheyta (2019)	10	$SSE = \sum_i \sum_j \left(y_{i,j}^{exp} - y_{i,j}^{cal}\right)^2$	4th Order Runge–Kutta	Levenberg–Marquardt
Vélez Manco (2014)	6	$MSE = \sum_i \frac{(C_i^{exp} - C_i^{cal})^2}{2}$	MatLab	—
Mederos-Nieto et al. (2021)	4	$SSE = \sum_i \sum_j \left(y_{i,j}^{exp} - y_{i,j}^{cal}\right)^2$	—	Levenberg–Marquardt

summarizes the characteristics of the algorithms used to estimate the kinetic parameters for renewable fuel production through the hydrotreating process. The diversity in the number of kinetic parameters depends mainly on the required complexity of the model and the capability to analyze and quantify different compounds. Among the different numerical methods for the solution of ODEs and parameter optimization, the Runge–Kutta and the Levenberg–Marquardt are the most used since they have proven accuracy for highly complex systems. In addition, the selection of the objective function plays a key role during nonlinear optimization. The SSE is the most used objective function since it has demonstrated an adequate minimization of the experimental and calculated data. Different objective functions can be used considering weighting factors, absolute differences between calculated and experimental information, and correlation coefficients. Recently, the average absolute error (AAE) has demonstrated a satisfactory balanced weighting to predict components with different magnitude orders in nonlinear optimization algorithms (Félix et al. 2022, 2023).

6.4 Models for Catalytic Deactivation

One of the main challenges to overcome during the hydrotreating of vegetable oils is catalyst deactivation, as it is argued that coke deposition is the main cause of deactivation. Compounds containing oxygen atoms exhibit high reactivity in oligomerization reactions, resulting in coke formation.

Table 6.6 Deactivation model of the hydrotreating process for renewable fuels.

Model	Mathematical expression	Observations
Sebos et al. (2009)	$\alpha = \frac{k_{\text{HDO}_t}}{k_{\text{HDO}_{\text{ref}}}}$	Catalytic activity (α) is defined as the ratio of reaction rate coefficient at any time t concerning a reference time. Activity coefficient is set as reference ($\alpha = 1$) at first 100 h of operation. While catalyst activity declines slowly up to 300 h with a loss of about 13%.
Bykova et al. (2013)	$r_i = \alpha\, k\, C_i\, P_{\text{H}_2}^m$ $\alpha = 1 - \gamma C_B$ $\gamma = c_1 T - c_2$	Catalytic activity decreases linearly with concentration. γ is a temperature-dependent parameter, which is a linear function of temperature, where c_1 and c_2 are constants. The coefficient γ determines the concentration at which the reaction slows down sharply.
Boldrini et al. (2014)	$\varphi_1 = \alpha_1$ $\beta = \beta_0(1 - ct)$ $\varphi_2 = \alpha_2$ $\beta = \beta_0 \exp(-ct)$ $\varphi_3 = \alpha_3$ $\beta = \frac{\beta_0}{(1 + ct)}$	Deactivation coefficients are associated with the number of catalyst reuses, with $\beta = 1$ for the first use of the catalyst. Two independent deactivation phenomena: one associated with the reaction time, and the other one with the batch number. For the model, α_j presented a value of 1.93×10^{26}. For the rest of the models, parameter a showed similar values. The value of parameter b for the different batches (values between 0.83 and 0.13 for Batch 2 and 10. For the low Pd loading, β presented lower values, which could be attributed to the extremely low Pd/oil ratio.
Zhang et al. (2014)	$\frac{d\alpha_j}{dt} = k_{d_j}\alpha_j^2$	The decay rate was assumed to follow a second-order law according to the activity parameter. The sum of both decay coefficients (k_{d1} and k_{d2}) provides the total deactivation rate. A catalytic activity function was applied to calculate the deactivation rate of both HDO and HDCx routes using specific decay constants k_{d1} and k_{d2}, respectively. The decay coefficients indicated a loss of activity of up to 80%. HDCx routes contributed mainly on deactivation ($k_{d2} =$ 0.4475), while the HDO route had no significant effect ($k_{d1} \approx 0$).
Hasanudin et al. (2019)	$\alpha = \frac{r_t}{r_{\text{ref}}}$	Catalytic activity is defined as ratio of reaction rate at particular time (t) to reaction rate of fresh catalyst as expressed. A second-order for deactivation rate was obtained. Deactivation rate constant is highly affected by hydrocracking temperature. Catalyst tends to deactivate rapidly, which shortens its lifetime at high temperatures. Frequency factor (A) and deactivation energy (E_d) are 2.75×10^{12} and 54.27 kJ K^{-1} mol^{-1}, respectively.

The investigation into the decarboxylation of FA in a tubular reactor has shown that high concentrations of stearic acid contribute to catalyst deactivation, which is reflected in a lower value of the corresponding rate constant (Snåre et al. 2007).

Table 6.6 presents the characteristics of the deactivation models reported to produce renewable fuels by the hydrotreating process. These models have presented different approaches for the analysis of the catalytic activity. However, more experimental investigations are required to obtain a better understanding of the phenomena that affect the catalytic activity.

6.5 Reactor Modeling for Vegetable Oil Hydrotreating

Reactor mathematical models are useful and indispensable tools to carry out the scaling-up design and control of reaction units at laboratory and industrial scales. These models consist of mass balance equations and energy balance equations (when the system operates under a non-isothermal state) among different phases in the reacting system (gas, liquid, and solid) to provide a clear idea of the phenomena that take place in the reactor and predict the yield of reactants and reaction products.

Diverse reactor models have been developed to simulate the performance of the hydrotreating process for petroleum distillate streams. Mederos et al. (2009b) reviewed the phenomena observed during the hydrotreating of petroleum distillates in fixed-bed reactors. The authors studied the phenomena of mass and energy transfer between gas–liquid and liquid–solid interfaces. Additionally, they examined the intraparticle diffusion, axial and radial diffusion, accumulation, convection, and generation phenomena in this system type. The consideration of all these phenomena requires some assumptions to simplify the solution of the mathematical models based on experimental data or analytical criteria to discard the influence of these phenomena on the reaction system. Therefore, the complexity of a mathematical equation system of a reactor depends on the prediction requirements and characteristics of the reactive system. Using simple models, convergence with experimental data is not ensured, while overly complex mathematical models may require complicated mathematical solutions and long time-consuming for simulation. For this reason, there are limited models available to illustrate the reactor performance during the hydrotreating of vegetable oils. Furthermore, some of them do not consider the energy balance to predict the thermal profile of the reactor (Forghani and Lewis 2016; Vélez Manco 2014). The first models were focused on the hydrogenation of unsaturated compounds present in bio-oils (Cabrera and Grau 2008; Chen et al. 1981; Fillion et al. 2002; Holser et al. 2002; Takeya et al. 1997). Vélez Manco (2014) established a conceptual industrial reactor design for vegetable oil hydrotreating based on experiments with refined, bleached, and deodorized palm oil (RBDPO) and kinetic studies performed. The reactor model aimed to analyze two cases to avoid hot spots higher than 370 °C, by considering ideal flow and avoiding excessive product cracking (i.e. reactive system is not affected by gradients of concentration or temperature). The molar balance equation for each compound i considers the temperature and liquid residence time based on the catalytic bed length (z) and reactor cross-sectional area (A):

$$\frac{dC_i}{dz} = \frac{A\rho_b}{V_0} r_i \tag{6.19}$$

Energy balance was expressed to consider the exothermal behavior of HDO reaction based on the overall heat-transfer coefficient (U_a) and differential of heat of reaction (ΔH_{ij}) by the equation:

$$\frac{dT}{dW} = \frac{U_a(T_a - T) + \sum_{i=1}^{q} (r_{ij}) \left[\Delta H_{ij}\right]}{\sum_{i=1}^{m} F_i C_{p_i}} \tag{6.20}$$

Additionally, the Ergun equation was used to calculate the pressure drop in packed beds:

$$\frac{dP}{dz} = -\frac{G}{\rho_b d_{pe}} \frac{(1-\varepsilon)}{\varepsilon^3} \left[\frac{150(1-\varepsilon)\mu}{d_{pe}} + 1.75G\right] \tag{6.21}$$

Two situations were analyzed, and in case (a), the feedstock and hydrogen are injected concurrently at the top of the reactor, whereas in case (b), a hydrogen-quenching stream near the reactor inlet is considered. The energy balance depicted a temperature peak placed at 0.55 m due to the energy released by exothermic reactions. The pressure drop, though negligible in both cases, was higher in case (b). However, case (b) was deemed to be the optimal choice as the liquid temperature increased by only 9 °C, resulting in a reduction in the catalyst weight of 800 kg to achieve an alkane/RBDPO ratio comparable to that of case 1. Nevertheless, the reactor model was considered to be free from mass transfer effects.

Forghani et al. (2014) reported a two-dimensional model of a trickle-bed reactor based on the two-film theory to determine the mass and energy transfer coefficients. The reactor model uses the kinetic model reported by Anand and Sinha (2012). For the representation of the reactor mathematical equations, the following assumptions have been considered for the reactor: (a) a plug-flow behavior is assumed; (b) radial concentration gradients are negligible; (c) axial heat dispersion is not present; (d) isothermal behavior is achieved. With these assumptions, the mass balance equations consider evaporation for ideal mixing for each lump (Tg, LC, HC, MC, and OC) as:

$$u_{\mathrm{L}}A\frac{\partial C_i^{\mathrm{L}}}{\partial z} = v_i r_i + \frac{k_i^{\mathrm{L}} a_s A}{R_u T_{\mathrm{G}}}\left(x_i P_i^* - P_i^{\mathrm{G}}\right) \tag{6.22}$$

$$u_{\mathrm{G}}A\frac{\partial P_i^{\mathrm{G}}}{\partial z} = -\frac{k_i^{\mathrm{L}} a_s A}{R_u T_{\mathrm{G}}}\left(x_i P_i^* - P_i^{\mathrm{G}}\right) \tag{6.23}$$

$$\frac{\partial P_{\mathrm{H_2}}^{\mathrm{G}}}{\partial z} = -\frac{k_{\mathrm{H_2}}^{\mathrm{L}} a_s A R_u T_{\mathrm{G}}}{U_{\mathrm{G}}}\left(\frac{P_{\mathrm{H_2}}^{\mathrm{G}}}{H_{\mathrm{H_2}}} - C_{\mathrm{H_2}}^{\mathrm{L}}\right) \tag{6.24}$$

The heat balance was taken into account to evaluate temperature variations along the catalytic bed for both liquid and gas phases as well as for the reactor wall:

Liquid phase:

$$F_{\mathrm{G}}C_p^{\mathrm{L}}\frac{\partial T_{\mathrm{L}}}{\partial z} = \sum r_i \Delta H_i A - h_{\mathrm{G}} a_s A(T_{\mathrm{G}} - T_{\mathrm{L}}) - 2\pi r_{in}\varepsilon h_{\mathrm{L}}(T_{\mathrm{W}} - T_{\mathrm{L}}) \tag{6.25}$$

Gas phase:

$$F_{\mathrm{G}}C_p^{\mathrm{G}}\frac{\partial T_{\mathrm{G}}}{\partial z} = h_{\mathrm{G}} a_s A(T_{\mathrm{G}} - T_{\mathrm{L}}) - 2\pi r_{in}(1-\varepsilon)h_{\mathrm{L}}(T_{\mathrm{W}} - T_{\mathrm{G}}) \tag{6.26}$$

Reactor wall:

$$\lambda_{ax}\frac{\partial^2 T_{\mathrm{W}}}{\partial z^2} + \lambda_{ra}\frac{1}{r}\frac{\partial}{\partial r}\left(\frac{\partial T_{\mathrm{W}}}{\partial r}\right) = 0 \tag{6.27}$$

The system of differential equations was solved with a set of boundary conditions, obtaining the concentration and temperature profiles shown in Figure 6.8. The temperature profile of the liquid phase increases at the reactor inlet, while the temperature of the gas phase, which is fed in countercurrent, increases linearly through the bed due to the constant physical properties of the gas at this temperature range. The main product was diesel (HC), and the evaporation of the liquid products was negligible. In addition, the model fitted well with the experimental data.

Mederos et al. (2020) developed dynamic, plug-flow, one-dimensional, and heterogeneous mathematical models for the catalytic hydrocracking of Jatropha oil using a commercial CoMo catalyst. Multiphase trickle-bed reactor models were used to predict the dynamic behavior of micro-scale and industrial-scale reactors in order to understand the influence on the liquid molar concentration, temperature profiles, and partial pressure. The system of equations is described in detail in Section 6.7. Additionally, the pressure drop in countercurrent flow in a packed bed was modeled by an equation of the Carman–Kozeny type. The simulations fit well with the experimental data displaying significant findings at the industrial scale. The industrial-scale predictions revealed that the most substantial gradients in temperature occur in the first half of the catalytic bed, resulting in the highest Tg conversion to green diesel because of the availability of reactants in this zone. Furthermore, there is an increase in the effectiveness factor beyond the mid part of the catalytic bed caused by the cooling effect of the ascending gas phase through the reactor, which avoids reaching high temperatures.

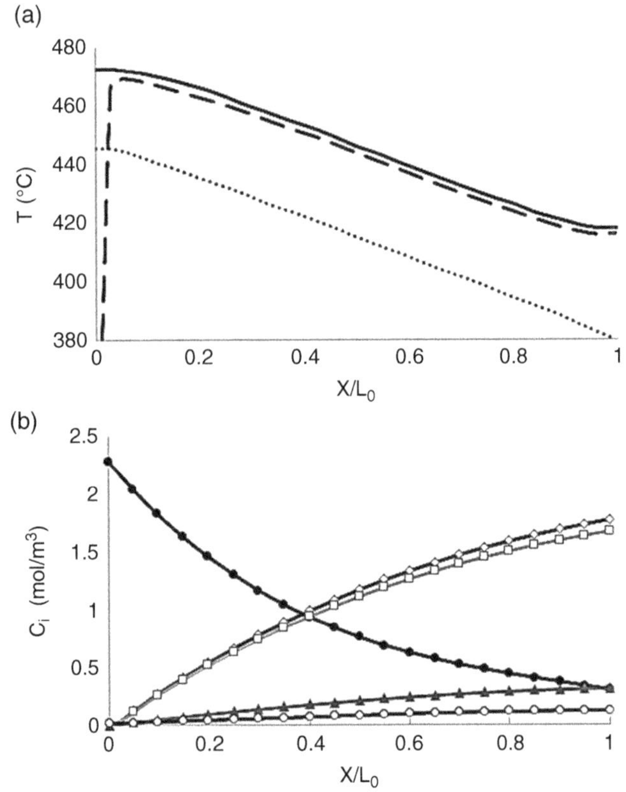

Figure 6.8 (a) Predicted temperature profiles: (——) wall, (- - - -) liquid, (· · ·) gas. (b) Predicted concentration profiles of triglycerides in a hydrocracking reactor: (●) Tg, (◇) HC, (□) OC, (▲) MC, (○) LC (Adapted from Forghani et al. 2014).

Computational fluid dynamics (CFD) has been used to model different reactors to produce renewable fuel by hydrotreating vegetable oils using cooling fluids and considering thermal equilibrium between phases to give a single reactor temperature profile. Muharam et al. (2017) modeled a trickle-bed reactor for the hydrotreating of 5 wt.% triolein in dodecane. The results showed a small increase in the temperature along the catalytic bed (almost 10 °C). Other simulations conducted on slurry bubble column reactors reported minimal issues with heat removal, requiring only a small cooling area (Muharam and Adevia 2018; Muharam and Putri 2018). In the same way, other studies were developed to analyze the effects of operating conditions and heat transfer on the hydrodynamics of the process (Kiran 2015; Mendoza Sépulveda 2013).

6.5.1 Deviation from Ideal Flow Pattern

Adequate knowledge of phenomena occurring along with chemical reactions must be accurate to carry out the design and scaling-up of reactors. During laboratory tests, the collected data can be affected by mass and temperature gradients (deviation of the ideal pattern flow, efficiency of catalytic particle wetting, and wall effects), resistance to mass and heat transfer at gas–liquid and liquid–solid interfaces, as well as in the interior of the catalytic particle. The extensive knowledge of these phenomena on the reactive system allows an appropriate interpretation of experimental data and results from a mathematical model. If these phenomena are unavoidable, their quantification is necessary to obtain reliable and replicable data (Mederos et al. 2009a).

- **Ideal plug-flow**

  One of the main mistakes made during the modeling and scaling-up of trickle-bed reactors is to assume in advance that feedstocks and products present a plug-flow behavior. A deviation from this pattern would deviate the measurement results, causing a misunderstanding of experimental data. Experimental studies have shown that the reactor-to-particle size ratio (L_B/d_{pe}) strongly influences axial and radial dispersion phenomena (Gierman 1988). Therefore, by controlling this ratio, the effect of these phenomena in reactors can be minimized. Mederos et al. (2009a) reviewed a series of analytical criteria to evaluate the effects of deviation from ideal flow patterns in packed beds. A criterion reported by Bischoff and Levenspiel (1962) has been widely used to evaluate the effect of radial dispersion in packed beds as follows:

$$\frac{L_B}{d_{pe}} > 0.04\frac{u_{\mathrm{L}} d_{\mathrm{R}}}{\varepsilon_{\mathrm{L}} D_r^{\mathrm{L}}} \tag{6.28}$$

  On the other hand, axial dispersion effects are present in packed beds. An appropriate selection of dimensions for the catalytic bed can greatly mitigate the impact of axial dispersion. It has been reported that to avoid back-mixing effects in the trickle-bed reactor, the catalyst bed length (L_B) should be at least 100 times the catalyst particle size (d_{pe}) (Perego and Peratello 1999).

- **Catalyst wetting efficiency**

  The distribution of the liquid phase through the catalytic bed influences the performance of reactors since an incomplete wetting of catalytic particles would limit the conversion of the feedstock. In commercial units, the surface velocity of the liquid phase is high enough to completely wet the catalyst particles. However, in laboratory-scale reactors, the superficial velocity is much smaller, causing incomplete wetness (Carruthers and DiCamillo 1988). An even distribution of the liquid phase in trickle flow is observed when the liquid flow is mostly affected by the frictional force rather than the gravity force, causing an even flow through the cross-section of the catalytic bed wetting the interstitial channels available and improving contact with the catalytic surface. Otherwise, a partial wetting of the catalyst is attained and affects the performance of the system as follows:

$$\frac{(dP/dz)_{\mathrm{flow}}}{(dP/dz)_{\mathrm{gravity}}} = \frac{180\mu_{\mathrm{L}}(1-\varepsilon_{\mathrm{L}})^2 u_{\mathrm{L}}}{\rho_{\mathrm{L}} d_{pe}^2 g \varepsilon_{\mathrm{L}}{}^4} > 1 \tag{6.29}$$

- **Wall effects**

  Wall effects are occasioned by the inadequate distribution of the liquid phase through the cross-section of the catalytic bed. The liquid phase displaces toward reactor walls, where the fluid velocities are higher and the conversions are lower due to the smaller amount of catalyst particles available in this part of the reactor. Different researchers have related these phenomena to the reactor-to-particle diameter ratio (d_R/d_{pe}); however, it is not an easy task to find the optimum minimum value to avoid the wall effects because the method of catalyst loading affects the liquid distribution. Therefore, different criteria have been developed based on experimental data reported (Froment et al. 1990; Mederos et al. 2009). To consider that a TBR presents negligible wall effects, the following criterion can be considered since the minimum value of (d_R/d_{pe}) ratio is one of the most rigorous reported in the literature:

$$\frac{d_R}{d_{pe}} > 20 \tag{6.30}$$

6.6 The Importance of Modelling Reactors for Vegetable Oil Hydrotreating

Mathematical models focused on catalytic hydrotreating of vegetable oil are scarce (Tirado et al. 2018). The development of these model types is fundamental for process design and optimization due to:

- It is necessary for scaling up, controlling, and automation of process of catalytic reactors
- Process optimization is only possible if the reactor model adequately predicts the variation of the concentration over time, as in a dynamic model
- Make recommendations for industrial plants and improve the process efficiency based on accurate simulations
- Make modifications in operating conditions by making changes in the feed properties and/or to achieve the product specifications
- Predict the dynamic behavior during the start-up, shutdown, and disturbances in the process and/or establish adequate control strategies
- Describe the influence of operating variables on the catalytic bed length and time-on-stream

6.7 Study Case for the Development of Dynamic Reactor Model

This section describes in detail a methodology for the development of dynamic reactor mathematical models for the hydrotreating of vegetable oil in order to provide some insights and considerations on its development, solution, scaling-up, and interpretation of the results obtained. The reactor models were configured to simulate the process of Jatropha oil hydrotreating with a $NiW/SiO_2–Al_2O_3$ catalyst. Table 6.7 summarizes the physical and chemical properties of the catalyst and vegetable oil.

6.7.1 Equations and Assumptions for Hydrotreating Reactor Modeling

A series of dynamic mathematical models was developed to simulate the performance of a reactor for vegetable oil hydrotreating. The reactor models consist of a set of partial differential equations (PDEs) describing the mass and heat balances in the multiphase system (gas, liquid, and solid) over time-on-stream and spatial variable "z" since the reactants and products follow a concurrent downflow. The mass and heat balance equations are based on a reactor model for vacuum gas oil hydrotreating reported by Korsten and Hoffmann (1996). Mederos and Ancheyta modified this reactor model to analyze the behavior of the reactor in a dynamic regime (Mederos et al. 2006).

Firstly, a reactor model was adjusted with characteristics of an isothermal bench-scale reactor using experimental data obtained during Jatropha oil hydrotreating. Then, the trickle-bed reactor model was modified according to the dimensions of a pilot-scale unit to evaluate the hydrodynamic

Table 6.7 Properties of the feedstock and catalyst for hydrotreating process.

Property	Value
Vegetable oil	Jatropha oil
Density (g/cm^3)	0.9163
Molecular weight (g/mol)	867
Sulfur content (ppm)	4
Composition (wt.%)	
Free fatty acids	1.7
$C_{16:0}$ (Palmitic)	19.5
$C_{18:0}$ (Stearic)	7.9
$C_{18:1}$ (Oleic)	45.4
$C_{18:2}$ (Linoleic)	27.2
Elemental composition (wt.%)	
C	77.0
H	13.6
O	9.4
Distillation curve (vol. %)	
IBP–250 °C	—
250–380 °C	0.20
380 °C–FBP	99.8
Catalyst	4% NiO, 24% WO_3
Support	SiO_2–Al_2O_3
Surface area (m^2/g)	233
Pore volume (ml/g)	0.32
Mean pore diameter (nm)	10.2
Surface acidity (mmol/g)	0.77

IBP, initial boiling point; FBP, final boiling point.

effect by carrying out the scaling-up process. Additionally, the design and simulation of a commercial vegetable oil hydrotreating unit were carried out to examine the dynamic behavior of the process and to establish operational strategies.

The following considerations were made during the development of the reactor model:

1) Reactors operate in a dynamic regime
2) The properties of gas and liquid, such as density, viscosity, superficial velocities, and holdups, are kept constant along the catalytic bed. The same consideration is made for the properties of the catalytic particles and bed (size, porosity, etc.), wetting efficiency, and bed void fraction
3) Organic compounds do not undergo vaporization or condensation
4) Chemical reactions occur when the liquid phase diffuses into the catalyst particles
5) Catalyst activity is constant due to short time-on-stream (<100 h)
6) Liquid phase fills the catalyst pores with capillary forces

7) Analytical criteria are used to quantify the axial and radial dispersion, wetting efficiency, and diffusion limitations within the catalytic particle
8) The heats of reaction are invariant to the reactor temperature

The unsteady-state mass balance equations were formulated for each phase. The flow rate for compound i in the gas phase is influenced by convection and mass transfer phenomena, while the gas–liquid equilibrium is calculated by Henry's law. Therefore, the continuity equation for compounds in the gas phase is defined by:

$$\frac{\varepsilon_G}{RT_G}\frac{\partial p_i^G}{\partial t} = -\frac{u_G}{RT_G}\frac{\partial p_i^G}{\partial z} - K_i^L a_L\left(\frac{p_i^G}{H_i} - C_i^L\right)$$
$$\text{where } i = H_2, C_3H_8, C_4H_{10}, CO, CO_2, H_2O \tag{6.31}$$

Consequently, the solubility of gaseous compounds in the liquid phase is expressed as:

$$\varepsilon_L\frac{\partial C_i^L}{\partial t} = -u_L\frac{\partial C_i^L}{\partial z} + K_i^L a_L\left(\frac{p_i^G}{H_i} - C_i^L\right) - f_w K_i^S a_s\left(C_i^L - C_i^S\right)$$
$$i = H_2, C_3H_8, C_4H_{10}, CO, CO_2, H_2O \tag{6.32}$$

Organic compounds are assumed not to be subject to evaporation/condensation effects based on simulations performed under severe operating conditions (420 °C) (Forghani et al. 2014), by which the mass balance equation for the compounds in the liquid phase is:

$$\varepsilon_L\frac{\partial C_i^L}{\partial t} = -u_L\frac{\partial C_i^L}{\partial z} - f_w K_i^S a_s\left(C_i^L - C_i^S\right)$$
$$i = Tg, Ol, Hv, Md, Lh \tag{6.33}$$

In studies carried out during the hydrotreating of vegetable oil, it was observed that deoxygenation and cracking reaction are negligible in the absence of a catalyst (Anand et al. 2016b). Therefore, it is feasible to assume that the chemical reaction takes place on the wet catalyst surface. In addition, the consumption/generation term is influenced by the effectiveness factor, and the mass and energy transfer terms are affected by hydrodynamic effects (Lange et al. 2004).

$$\epsilon_S(1-\epsilon_B)\frac{\partial C_i^S}{\partial t} = -f_w K_i^S a_s\left(C_i^L - C_i^S\right) + \sum_{j=1}^{N_{RL}} \varphi\, \upsilon_{i,j} r_{i,j}$$
$$i = Tg, Ol, Hv, Md, Lh \tag{6.34}$$

In addition, the catalyst deactivation is considered negligible as the yields of hydrotreating products were constant up to 100 h of continuous operation.

Since most of the chemical reactions occurring during vegetable oil hydrotreating are highly exothermic, the reactor model includes energy balance equations for each phase to analyze the changes in temperature when operating in non-isothermal operating conditions. The unsteady-state heat balance equations for gas, liquid, and solid phases are termed as follows:

$$\varepsilon_G\rho_G Cp_G\frac{\partial T_G}{\partial t} = -u_G\rho_G Cp_G\frac{\partial T_G}{\partial z} - h_{GL}a_L(T_G - T_L) \tag{6.35}$$

$$\varepsilon_L\rho_L Cp_L\frac{\partial T_L}{\partial t} = -u_L\rho_L Cp_L\frac{\partial T_L}{\partial z} - h_{GL}a_L(T_G - T_L) - f_w h_{LS}a_s(T_L - T_S) \tag{6.36}$$

$$\varepsilon_S\rho_S Cp_S\frac{\partial T_S}{\partial t} = f_w h_{LS}a_s(T_L - T_S) + \sum_{j=1}^{N_{RL}} \eta_j v_{i,j} r_{i,j}(-\Delta H_R) \tag{6.37}$$

6.7.2 Kinetic Model for Hydrotreating of Vegetable Oil

Based on the approach reported by Tirado and Ancheyta (2019) for the estimation of optimal kinetic parameters, where the huge number of chemical components inside the HDT system was classified using a single 5-lump reaction scheme (Figure 6.3f), the kinetic model was used under different experimental conditions for the hydrotreating of Jatropha oil with a $NiW/SiO_2–Al_2O_3$ catalyst: 340–420 °C of reaction temperature, a hydrogen pressure of 80 bar, and a WHSV of 0.5–12 h^{-1}. The calculation of the kinetic parameters was performed assuming first-order rate equations represented by the scheme shown in Figure 6.3f:

$$\frac{dy_{Tg}}{d\tau} = -(k_1 + k_2 + k_3 + k_4)y_{Tg} \tag{6.38}$$

$$\frac{dy_{Ol}}{d\tau} = k_1 y_{Tg} + k_7 y_{Hv} - (k_5 + k_6)y_{Ol} \tag{6.39}$$

$$\frac{dy_{Hv}}{d\tau} = k_2 y_{Tg} + k_5 y_{Ol} - (k_7 + k_8 + k_9)y_{Hv} \tag{6.40}$$

$$\frac{dy_{Md}}{d\tau} = k_3 y_{Tg} + k_6 y_{Ol} + k_8 y_{Hv} - k_{10} y_{Md} \tag{6.41}$$

$$\frac{dy_{Lh}}{d\tau} = k_4 y_{Tg} + k_9 y_{Hv} + k_{10} y_{Md} \tag{6.42}$$

The system of ODEs (6.38)–(6.42) was solved by a Runge–Kutta method subjected to some constraints based on the global conversion of compounds. Additionally, each kinetic coefficient must be a positive term ($k_i \geq 0$) and follow the Arrhenius equation. The Levenberg–Marquardt algorithm was used for optimizing the reaction rate coefficients (k_1 to k_{10}). The initial guess values required for the nonlinear optimization were determined using the Monte Carlo method, which evaluates random values for each kinetic parameter at a time and selects those that provide the least value of the sum of squared errors (SSE):

$$\text{SSE} = \sum_{i=1}^{N\ \text{data}} \left(y_{exp} - y_{calc}\right)^2 \tag{6.43}$$

Values of pre-exponential factors (A_0) and activation energies (E_A) were determined by the Arrhenius equation and reported in Table 6.8. In addition, the heats of reaction for each representative reaction pathway were calculated from the heats of formation of the species involved, as shown in Table 6.8. These values are similar to those reported during the hydrotreating of renewable feedstocks (Anand et al. 2016a; Mederos et al. 2019; Vélez Manco 2014).

6.7.3 Hydrogen Consumption and Gas Generation

Hydrogen consumption is a critical parameter to determine because of its large impact on the design and operation of industrial reactors. In addition, it is well known that renewable feedstock often has high hydrogen consumption during the HDT process, so the hydrogen cost is relevant to the feasibility of this technology. The prediction of hydrogen consumption, gases (C_3H_8, C_4H_{10}, CO, and CO_2), and H_2O formation is performed by a method based on reaction pathways reported for the compounds from each lump (Kubička et al. 2009; Mohammad et al. 2013; Sinha et al. 2014). Considering the set of representative reaction pathways shown in Table 6.9 and the reaction rate law, the global reaction rate of hydrogen and gas products is obtained.

Table 6.8 Kinetic parameters and heat of reactions for vegetable oil hydrotreating.

Reaction j	A_0 (h^{-1})	E_A (kJ/mol)	$\Delta H_R^L \left(\frac{MJ}{mol}\right)$
Tg–Ol (1)	9.8723×10^9	128.4918	−1.0235
Tg–Hv (2)	157.5590	12.7951	−1.2620[a] −0.8039[b] −0.9284[c]
Tg–Md (3)	1.1004×10^{21}	277.9029	−1.0574
Tg–Lh (4)	3.4066×10^9	127.7768	−1.0574
Ol–Hv (5)	127.1923	20.7014	−0.0795[a] −0.0732[b] −0.0317[c]
Ol–Md (6)	—	—	—
Hv–Ol (7)	6.1409×10^{12}	173.1821	0.0299
Hv–Md (8)	1.3894×10^{19}	260.8500	−0.0424
Hv–Lh (9)	0.0424	2.1775	−0.0430
Md–Lh (10)	0.2817	6.5759	−0.0202

[a] HDO, hydrodeoxygenation.
[b] HCO, hydrodecarbonylation.
[c] HCO_2, hydrodecarboxylation.

Table 6.9 Representative chemical reactions during the hydrotreating of vegetable oil.

Chemical reaction	Kinetic expression
$C_{57}H_{110}O_6 + 3H_2 \rightarrow 3C_{17}H_{35}COOH + C_3H_8$	$r_1 = k_1 C_{Tg}$
$C_{57}H_{110}O_6 + 12H_2 \rightarrow 3C_{18}H_{38} + C_3H_8 + 6H_2O$ $C_{57}H_{110}O_6 + 6H_2 \rightarrow 3C_{17}H_{36} + C_3H_8 + 3H_2O + 3CO$ $C_{57}H_{110}O_6 + 3H_2 \rightarrow 3C_{17}H_{36} + C_3H_8 + 3CO_2$	$r_2 = k_2 C_{Tg}$
$C_{57}H_{110}O_6 + 6H_2 \rightarrow 3C_8H_{18} + 3C_9H_{20} + C_3H_8 + 3CO_2$	$r_3 = k_3 C_{Tg}$
$C_{57}H_{110}O_6 + 6H_2 \rightarrow 3C_8H_{18} + 3C_9H_{20} + C_3H_8 + 3CO_2$	$r_4 = k_4 C_{Tg}$
$C_{17}H_{35}COOH + 3H_2 \rightarrow C_{18}H_{38} + 2H_2O$ $C_{17}H_{35}COOH + H_2 \rightarrow C_{17}H_{36} + H_2O + CO$ $C_{17}H_{35}COOH \rightarrow C_{17}H_{36} + CO_2$	$r_5 = k_5 C_{Ol}$
$C_{18}H_{38} + C_8H_{18} \rightarrow C_{26}H_{54} + H_2$	$r_7 = k_7 C_{Hv}$
$C_{18}H_{38} + H_2 \rightarrow C_{12}H_{26} + C_6H_{14}$	$r_8 = k_8 C_{Hv}$
$C_{17}H_{36} + H_2 \rightarrow C_8H_{18} + C_9H_{20}$	$r_9 = k_9 C_{Hv}$
$C_{12}H_{26} + H_2 \rightarrow C_8H_{18} + C_4H_{10}$	$r_{10} = k_{10} C_{Md}$

$$r_i = \sum \upsilon_j^i r_j \tag{6.44}$$

where υ is the stoichiometric coefficient for $i = H_2$, C_3H_8, C_4H_{10}, CO, CO_2 and H_2O. Since the amount of H_2 is in excess during the hydrotreating process, each reaction pathway summarized in Table 6.9 does not take it into account, rather hydrogen partial pressure is grouped in the reaction rate coefficient.

6.7.4 Solution of Reactor Models

The method of lines was used to solve each reactor model (Schiesser 1991), where the PDEs were transformed into ODEs by discretization in the axial directions using the backward finite difference method. Thus, the catalyst bed length was divided into N equidistant segments since the gradient in z is equal to the catalyst bed length divided by the number of lines (N). The fourth-order Runge–Kutta method is used to solve the set of ODEs. The initial conditions and boundary constraints to predict the profiles of partial pressure, concentration of the pseudocomponents, and temperature profiles of the phases under non-isothermal conditions are given as follows:

- Initial conditions:

  For $t = 0$, at $z = 0$

$p_i^G = (p_i^G)_0$	$i = H_2$
$p_i^G = 0$	$i = C_3H_8, C_4H_{10}, CO, CO_2, H_2O$
$C_i^L = (C_i^L)_0$	$i = Tg, H_2$
$C_i^L = 0$	$i = C_3H_8, C_4H_{10}, CO, CO_2, H_2O, Ol, Hv, Md, Lh$
$C_i^S = 0$	$i = H_2, C_3H_8, C_4H_{10}, CO, CO_2, H_2O, Tg, Ol, Hv, Md, Lh$

  at $z > 0$

$p_{i,G} = (p_{i,G})_0$	$i = H_2, C_3H_8, C_4H_{10}, CO, CO_2, H_2O$
$C_i^L = 0$	$i = H_2, C_3H_8, C_4H_{10}, CO, CO_2, H_2O, Tg, Ol, Hv, Md, Lh$

  Boundary constraints:

  For $t > 0$, at $z = 0$,

$p_i^G = (p_i^G)_0$	$i = H_2, C_3H_8, C_4H_{10}, CO, CO_2, H_2O$
$C_i^L = (C_i^L)_0$	$i = Tg, H_2$

The initial Tg concentration in the liquid phase was calculated according to the weight fraction of Tg (w_{Tg}) in the vegetable oil:

$$\left(C_{Tg}^L\right)_0 = \frac{(\rho_L)_0}{MW_L}\left(w_{Tg}\right)_0 \tag{6.45}$$

The feedstock gas was considered to be pure hydrogen, and the initial hydrogen concentration in the liquid phase is calculated by Henry's law (Tirado et al. 2018).

$$\left(C_{H_2}^L\right)_0 = \frac{\left(p_{H_2}^G\right)_0}{H_{H_2}} \tag{6.46}$$

where the Henry's law coefficient (H_{H_2}) determines the H_2 solubility in vegetable oil and is based on an Arrhenius-type equation, using a pre-exponential constant (H_0) equal to 8.84 MPa m^3/kmol and an apparent activation energy of absorption (ΔE_A) of −5000 kJ/kmol reported for the solubility of hydrogen in soybean oil (Fillion and Morsi 2000):

$$H_{H_2} = H_0 \exp\left(\frac{-\Delta E_A}{RT}\right) \tag{6.47}$$

On the other hand, the solubility of H_2O and reaction product gases was determined by the Peng–Robinson equation of state (Tirado et al. 2018).

Mass- and heat-transfer parameters in interfaces (gas–liquid and liquid–solid), the properties of vegetable oils as well as gases, were calculated through correlations reported in the literature. Table 6.10 presents a summary of these correlations.

Table 6.10 Correlations to estimate model parameters.

Property	Correlation	References
Liquid–solid specific surface area	$a_S = (6/d_{pe})(1-\epsilon_B)$	Korsten and Hoffmann (1996)
Solid density	$\rho_S = \dfrac{\rho_B}{1-\epsilon_B}$	Tarhan (1983)
Solid holdup	$\varepsilon_S = 1-\epsilon_B$	Tarhan (1983)
Liquid holdup	$\varepsilon_L = 0.185\ \epsilon_B\ {a_S}^{0.333}\chi^{0.22}$	Tarhan (1983)
for bench-unit	$\varepsilon_L = 9.9\mathrm{Re}^{0.333}\mathrm{Ga}^{-0.333}$	Tarhan (1983)
for commercial-unit		
Gas holdup	$\varepsilon_G = \epsilon_B - \varepsilon_L$	Tarhan (1983)
Catalyst porosity	$\in_S = \rho_S V_g$	Tarhan (1983)
Bed void fraction	$\in_B = 0.38 + 0.073\left[1 + \dfrac{(d_R/d_{pe}-2)^2}{(d_R/d_{pe})^2}\right]$	Froment et al. (1990)
Molecular weight	$MW = 3\sum x_i MW_i + 38.0488$	Anand et al. (2010)
Molar diffusivity	$D_{Mi} = 8.93x10^{-8}\left(\dfrac{{v_L}^{0.267}}{{v_L}^{0.433}}\right)\left(\dfrac{T_L}{\mu_L}\right)$	Tyn and Calus (1975a)
Liquid viscosity	$\log(\mu_L)_T = -0.6298 + [273.15/(T+88.81)]$ $\mu_L = (\mu_L)_T 10^{\left[\frac{P}{1000\left(-0.0102+0.04042(\mu_L)_T{}^{0.181}\right)}\right]}$	Rodenbush et al. (1999)
Liquid molar volume	$v_L = 0.285{v_c}^{1.048}$	Tyn and Calus (1975b)
Liquid–solid mass transfer coefficient	$\dfrac{k_i a_S}{D_{Mi}} = \alpha_S\left(\dfrac{G_{mL}}{\mu_L}\right)^{n_S}\left(\dfrac{\mu_L}{\rho_L D_{Mi}}\right)^{\frac{1}{3}};\ \alpha_S = -577.8d_{pe} + 184.3,\ n_S = -0.5886d_{pe} + 0.7018$	Goto and Smith (1975)
for bench-unit *for commercial-unit*	$\dfrac{{k_i}^S}{D_{Mi}a_S} = 1.8\left(\dfrac{G_{mL}}{a_S\,\mu_L}\right)^{\frac{1}{2}}\left(\dfrac{\mu_L}{\rho_L D_{Mi}}\right)^{\frac{1}{3}}$	
Liquid side mass-transfer coefficient at gas–liquid interface	$\dfrac{k_i a_L}{D_{Mi}} = \alpha_1\left(\dfrac{G_{mL}}{\mu_L}\right)^{\alpha_2}\left(\dfrac{\mu_L}{\rho_L D_{Mi}}\right)^{\frac{1}{2}};\ for\ d_{pe} > 2.91cm: \alpha_1 = -26.23d_{pe} + 13.63,\ \alpha_2 = -0.08197d_{pe} + 0.4339;\ for\ d_{pe} \le 2.91cm: \alpha_1 = -7.598d_{pe} + 8.211,\ \alpha_2 = -0.08442d_{pe} + 03854$	Goto and Smith (1975)
Thiele modulus for irreversible reactions	$\Phi_j = \dfrac{1}{\phi_S}\left(\dfrac{V_p}{S_p}\right)\sqrt{\left(\dfrac{n+1}{2}\right)\dfrac{\rho_S k_{in}\left(C_i^S\right)^{n-1}}{D_{ei}^L}}$	Froment et al. (1990)
Effective diffusion	$D_{ei}^L = \dfrac{\epsilon_S}{\tau}\left[\dfrac{1}{(1/D_{Mi}^L) + (1/D_{Ki}^L)}\right]$	Froment et al. (1990)
Effectiveness factor	$for\ \Phi_j < 3\ \eta_j^L = \dfrac{\tanh(\Phi_j)}{\Phi_j}$ $for\ \Phi_j \ge 3\ \eta_j^L = \dfrac{1}{\Phi_j}$	Froment et al. (1990), Mederos et al. (2012)

6.8 Analysis and Discussion of Results

6.8.1 Criteria to Ensure Ideal Behaviors in Trickle-Bed Reactor

The influence of hydrodynamic phenomena on the three-phase system was analyzed and quantified by analytical criteria and correlations developed using experimental information. In this way, the changes in partial pressure, temperature, and concentration of the compounds involved in the process can be accurately predicted and simulated as a function of the operating time and length of the catalytic bed. Operating conditions and dimensions of different reactor scales used to evaluate and analyze the reactor mathematical model are summarized in Table 6.11.

- **Ideal plug-flow**

  Analytical criteria were used to evaluate the effect of axial and radial dispersion on the reaction system at operating conditions. The results of these phenomena for bench-scale and pilot-scale reactors are listed in Table 6.12. The results of the commercial-scale reactor are not disclosed because the effect of hydrodynamic phenomena is negligible due to the high liquid velocities in these units. It is understood that both bench-scale and pilot-scale reactors operate under ideal flow conditions because axial and radial dispersions are neglected. Although the particle size used in the pilot-scale reactor is 10 times larger than the bench-scale, the L_B/d_{pe} ratio for both scales resulted to be of similar magnitude due to the greater length of the catalytic bed in the pilot-scale reactor (12.8 times higher than that in the bench-scale reactor).

Table 6.11 Properties of catalyst and operating conditions for different reactor scales.

Scale	Bench	Pilot	Commercial
Mode	Isothermal	Isothermal	Adiabatic
Catalyst mass (g)	2	100	41,514,783.2
Catalyst shape	Crushed	Extrudates	Extrudates
Equivalent particle diameter, d_{pe} (cm)	0.023	0.254	0.254
Bed length, L_B (cm)	3.5	45	853.44
Reactor diameter, d_R (cm)	1.3	3	304.8
Bed volume (cm^3)	4.64	318	62,272,143.7
Inlet temperature (°C)	420	420	340
Pressure (MPa)	8	8	8
WHSV (h^{-1})	1	1	3
H_2/feed ratio (Nl/l)	1500	2500	1500
Liquid mass flow (g/s)	5.55×10^{-04}	2.77×10^{-02}	34,595.6
u_L ($cm_L^3/cm_R^2\ s$)	1.01×10^{-03}	7.27×10^{-03}	7.71×10^{-01}
K_{Tg}^S (s^{-1})	8.24×10^{-05}	1.51×10^{-03}	1.00×10^{-02}
a_S (cm_S^2/cm_R^3)	126.20	11.86	11.20
ε_L (cm_L^3/cm_R^3)	0.430	0.288	0.390

Table 6.12 Analytical criteria to evaluate hydrodynamic effects in packed-bed reactor.

Criterion	Bench-scale	Pilot-scale
Radial dispersion $\frac{L_B}{d_{pe}} > 0.04\frac{u_L d_R}{\varepsilon_L D_r^L}$	$152.17 > 4.74 \times 10^{-05}$ Fulfilled	$177.16 > 3.26 \times 10^{-02}$ Fulfilled
Axial dispersion $\frac{L_B}{d_{pe}} > 100$	152.17 > 100 Fulfilled	177.16 > 100 Fulfilled
Catalyst wetting $\frac{(dP/dz)_{\text{flow}}}{(dP/dz)_{\text{gravity}}} = \frac{180\mu_L(1-\varepsilon_L)^2 u_L}{\rho_L d_{pe}^2 g \varepsilon_L{}^4} > 1$ f_w	0.0298 <1 Not fulfilled 0.435	0.0933 <1 Not fulfilled 0.539
Wall effects $\frac{d_R}{d_{pe}} > 20$	56.52 > 20 Fulfilled	11.81 > 20 Not fulfilled

- **Catalyst wetting efficiency**

  Evaluation of catalyst wetting efficiency for the different reactor scales was carried out through the friction and gravity forces ratio. Results in Table 6.8 indicate that at the operating conditions, bench-scale and pilot-scale reactors are more influenced by the force of gravity, which indicates an incomplete wetting of the catalytic bed (Mederos et al. 2009a). Therefore, as wetting efficiency is an essential scale-up parameter, it was estimated using the correlation reported by El-Hisnawi et al. (1982).

$$f_w = 1.617\,\mathrm{Re}_{\mathrm{L}}^{0.146}\mathrm{Ga}_{\mathrm{L}}^{-0.071} \tag{6.48}$$

  This correlation was obtained with experimental data reported for superficial velocities of liquid phase in the range of $0.55 \times 10^{-3} \leq u_{\mathrm{L}} \leq 7.5 \times 10^{-3}$. For bench-scale and pilot-scale reactors, values of 1.01×10^{-3} and 7.27×10^{-3} are obtained, respectively, which are adjusted to the established range. $\mathrm{Re_L}$ is the Reynolds number and $\mathrm{Ga_L}$ is the Galileo number. Both of them are calculated as follows:

$$\mathrm{Re}_{\mathrm{L}} = \frac{d_{pe}u_{\mathrm{L}}\rho_{\mathrm{L}}}{\mu_{\mathrm{L}}} \tag{6.49}$$

$$\mathrm{Ga}_{\mathrm{L}} = \frac{d_{pe}^3 g \rho_{\mathrm{L}}^2}{\mu_{\mathrm{L}}^2} \tag{6.50}$$

  The result of the calculation of wetting efficiency for the bench-scale reactor was 0.435, while for the pilot-scale reactor was 0.539. The wetness of the catalytic bed in the pilot-scale reactor is 10% higher than at the bench-scale. These results are consistent with the experimental data reported in the literature, where laboratory-scale reactors give wetting efficiencies in the range of 0.12–0.6 (Bhaskar et al. 2004; Satterfield 1975). This efficiency increases as the surface velocity of the liquid phase does until it reaches values similar to those used in commercial-scale reactors, which are close to 1. Wetting efficiency results were plotted in Figure 6.9 along with data experimentally obtained during hydrotreating of heavy crude oil carried out in a pilot-scale reactor with $L_B = 147.3$ and $d_R = 2.94$ cm ratio and hydrotreating of gas oil in a reactor with $L_B = 66.5$ cm and $d_R = 3$ cm (Alvarez and Ancheyta 2012; Korsten and Hoffmann 1996). It is observed that

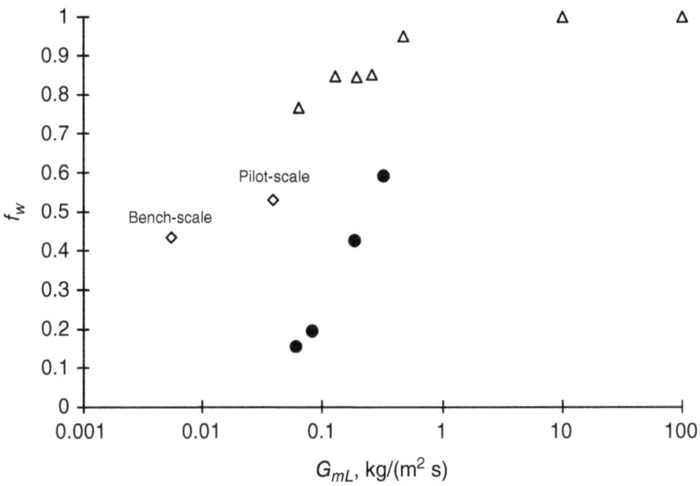

Figure 6.9 Catalyst wetting efficiency as a function of G_{mL}: (Δ) Alvarez and Ancheyta (2012), (●) Korsten and Hoffman (1996), and (◇) dynamic HDT simulation (Tirado and Ancheyta 2020b).

the results of catalyst wetting efficiency obtained in this work are within the reported literature and follow the expected tendency for superficial mass flow velocity of the liquid phase.

- **Wall effects**

  The wall effects were evaluated for both bench-scale and pilot-scale reactors. It was found that the bench-scale reactor fulfilled the criterion ($d_R/d_{pe} = 56.52 > 20$), while the pilot-scale reactor did not meet the criterion ($d_R/d_{pe} = 11.81 < 20$).

  The model considers the contact surface between liquid and solid phases through catalyst wetting efficiency. Therefore, wetting efficiency and wall effects are complementary parameters that help to understand the phenomena occurring in the reactor.

- **Catalyst effectiveness factor**

  The effectiveness factor calculations for the crushed catalyst showed negligible internal diffusion limitations (>0.96) for each reaction pathway at a reaction temperature of 420 °C, hydrogen pressure of 8 MPa, and WHSV of 1 h^{-1}. On the other hand, commercial-size catalysts presented lower values of effectiveness factor. Reaction pathways 2 and 5 have the greater diffusion limitations in catalytic particles, as they showed the lowest values of effectiveness factor (0.814 and 0.953, respectively) because these pathways present the highest reaction coefficients ($k_2 = 18.49$ and $k_5 = 2.66$), which could indicate a large formation of products, preventing the diffusion of reactive species into the catalytic pore. The effectiveness factor of the other reaction pathways shows negligible internal diffusion limitations (>0.96).

6.8.2 Dynamic Profiles of Feedstock and Products of a Bench-Scale Reactor for Catalytic Hydrotreating of Vegetable Oil

The reactor mathematical model considered the characteristics of a bench-scale unit to simulate the hydrotreating process of Jatropha oil and analyze the behavior of each lump involved in the reaction scheme. The heat balance equations were omitted from the model as the reactor operates at isothermal conditions. The yield profiles of each lump along catalyst bed length under steady-state

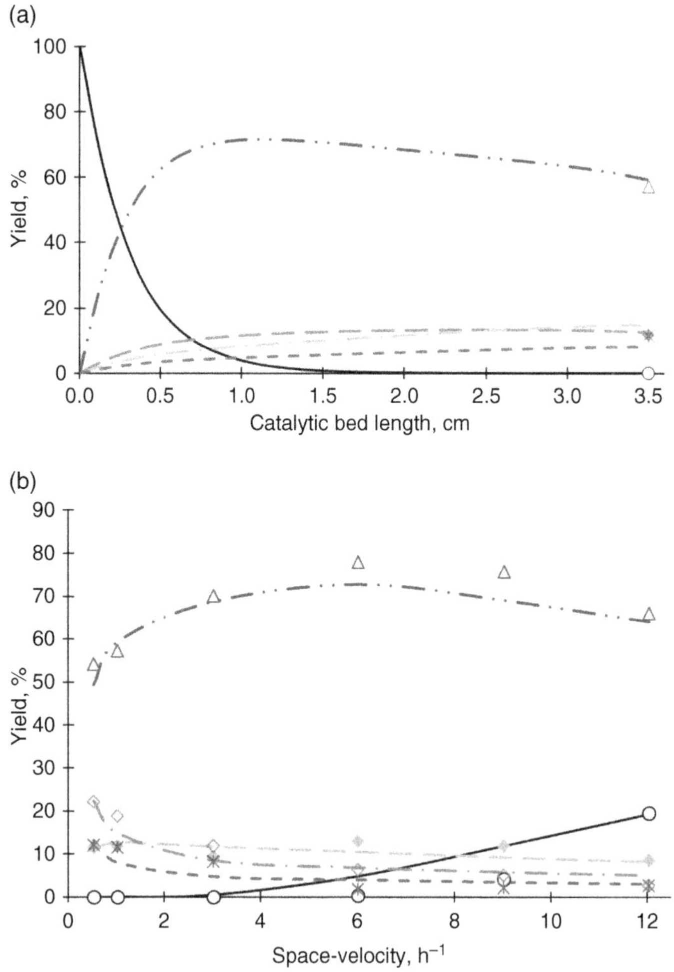

Figure 6.10 Comparison of predicted (lines) and experimental (symbols) yields at (a) WHSV of 1 h^{-1} along the catalytic bed, and (b) varying space-velocity: Tg (–, ○), Ol (– –, ♦), Hv (–··–, Δ), Md (– – –, ∗) and Lh (–·–, ◊) (Tirado and Ancheyta 2020a / with permission of Elsevier).

conditions are shown in Figure 6.10a. It is observed that in the reactor top, the unconverted Tg form mainly heavy hydrocarbons due to the high Tg reaction rate ($k_1 = 2.05$, $k_2 = 18.49$) at the studied operating conditions. The formation of heavy hydrocarbons is enhanced and reaches a maximum yield at one-third of the catalytic bed length until the concentration of Tg is insufficient to continue the reaction. Therefore, heavy hydrocarbon undergoes mostly conversion reactions (k_8 and k_9). The yield profiles of each lump fit well with the experimental data at the end of the catalytic bed. In the same way, Figure 6.10b shows that when varying space–velocity, the predicted values for each lump at the reactor bottom agree well with experimental data obtained at the operating conditions.

Figure 6.11 shows the dynamic profiles of each lump along the catalytic bed length to analyze the behavior of feedstock and products according to chemical reactions advance. Simulations show that the Tg reached half of the catalytic bed in a short time (100 s), whereas they are almost completely converted in 600 s at steady-state conditions. On the other hand, product concentrations increase according to time-on-stream advances. Even after Tg are fully converted, chemical reactions continue to take place mainly due to the conversion of heavy hydrocarbons. This behavior agrees with information obtained during the vegetable oil hydrotreating, which is attributed to the high reaction temperature and acidic nature of the catalyst (Anand and Sinha 2012; Zhang et al. 2014).

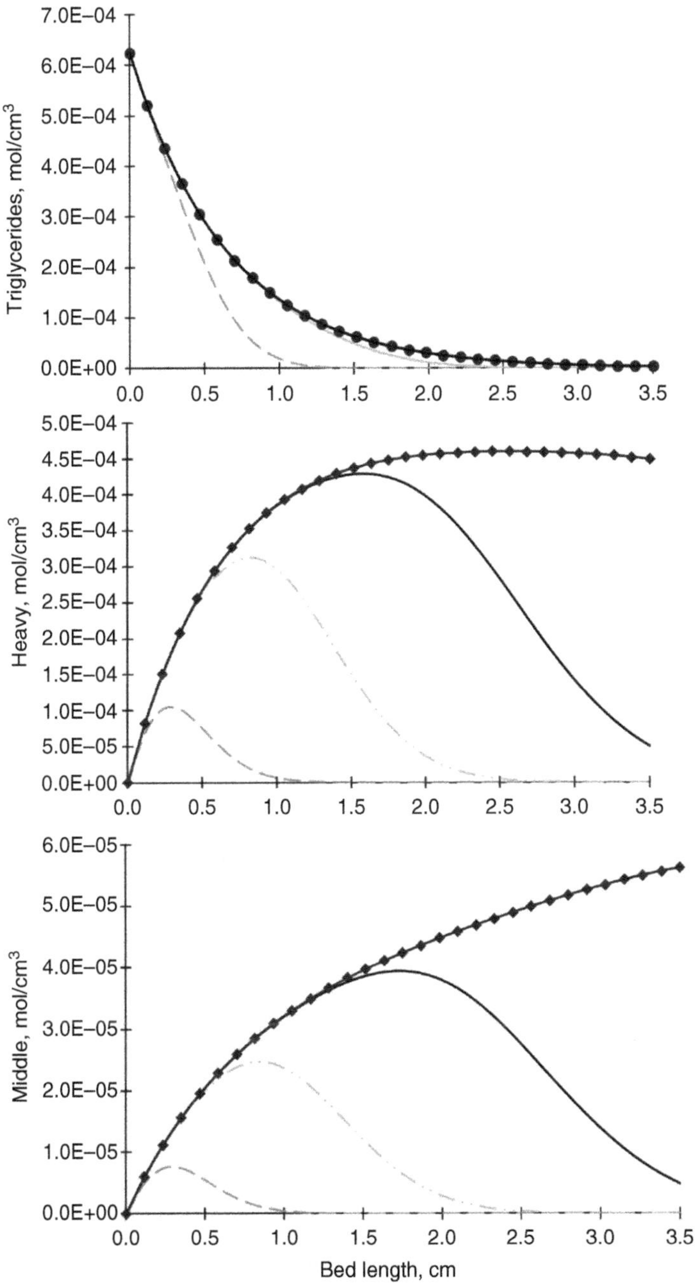

Figure 6.11 Concentration profiles of product pseudocomponents along the catalytic bed at WHSV of 3 h^{-1}: 100 s (– –), 300 s (–··–), 600 s (—) and 1200 s (-◆-) (Tirado and Ancheyta 2020a / with permission of Elsevier).

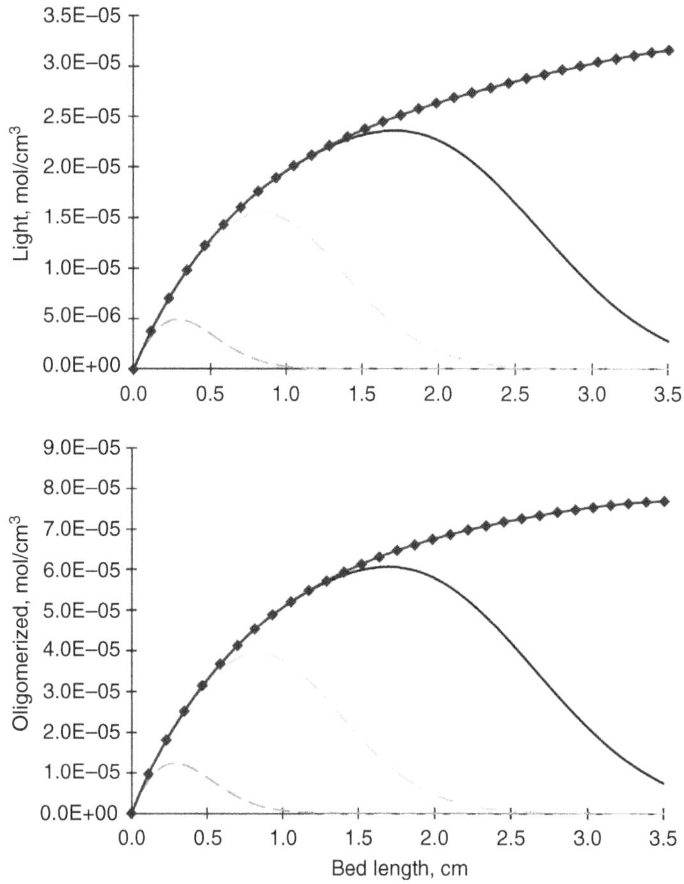

Figure 6.11 (Continued)

The predicted hydrogen pressure and concentration profiles along the catalytic bed are depicted in Figure 6.12, whereas the hydrogen concentration decreases rapidly at the reactor top. When the hydrogen reaction rate decreases, mass transfer at the gas–liquid interface predominates, increasing hydrogen concentration in the liquid phase. This behavior is similar to that reported during catalytic hydrotreating of gas oil due to the balance between mass transfer and reaction rate (Mederos et al. 2012). However, the molecular weight and higher complexity of Tg hinder the hydrogen solubility in vegetable oil. The hydrotreating of vegetable oil consumes large amounts of hydrogen (1640 scf/bbl) compared with that reported during gas oil hydroprocessing (800 scf/bbl) (Ancheyta 2011) because of the high demand for hydrogen in the cracking of Tg molecules (Guzman et al. 2010).

6.8.3 Validation of Hydrotreating Reactor Model with Pilot Plant Data

Once the accuracy of the dynamic model had been verified, the reactor model was scaled up to a pilot-scale unit size to validate the ability of the model by varying the dimensions of the reactor and catalytic particle size. The simulation of the pilot-scale reactor models was conducted at isothermal

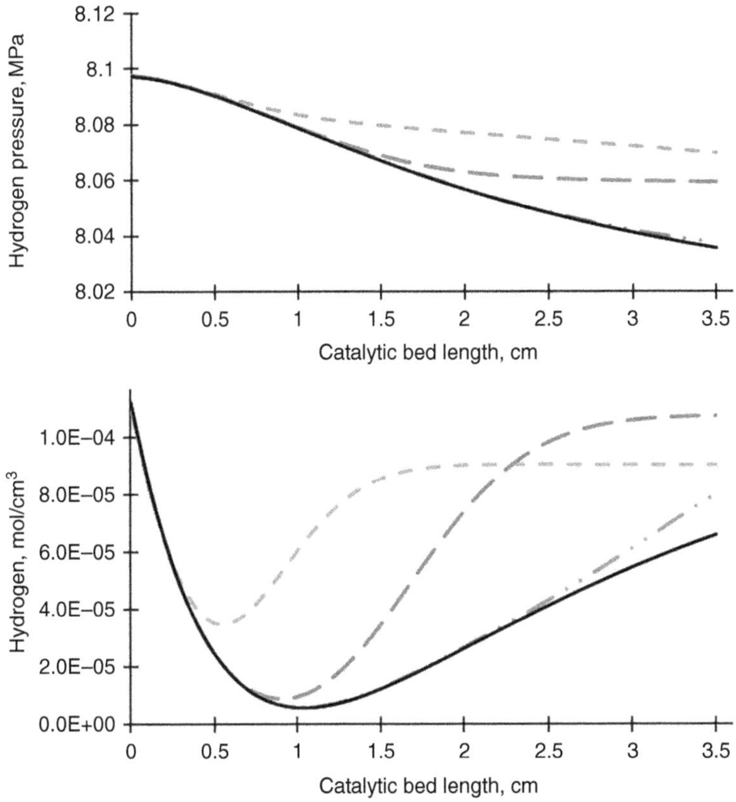

Figure 6.12 Hydrogen pressure and concentration in liquid phase along the catalytic bed at WHSV of 12 h^{-1}: 50 s (- - -), 100 s (– –), 200 s (-··-), 300 s (–).

conditions to compare the predicted yields of each lump. The reactor specifications are summarized in Table 6.11, whereas the predicted yield profiles along the catalytic bed length at steady-state conditions are shown in Figure 6.13. The products obtained in experimental studies were lumped into naphtha ($<C_9$ hydrocarbons), kerosene (C_9–C_{14} hydrocarbons), and diesel (C_{15}–C_{18} + $>C_{18}$). The results show that the simulations are in good agreement with the experimental values. Although the yield profiles seem quite comparable to the results obtained in the bench-scale reactor, there are indeed some important differences:

- Triglyceride conversion is slightly faster in the pilot-scale reactor than in the bench-scale reactor. For instance, for a bench-scale reactor, the unconverted Tg yield is 31.82% at 10% of catalytic bed length, while for a pilot-scale reactor, it is 22.68%.
- The heavy hydrocarbon yield is lower for pilot-scale reactors along the catalytic bed length. This behavior is attributed to the diffusion limitations suffered by the catalytic particles, especially reaction pathways 2 and 5, which showed the lowest effectiveness factor values under the studied operating conditions (0.814 and 0.953, respectively).
- The formation of middle hydrocarbons is preferred in the pilot-scale reactor since the reaction pathways involved during the production of these pseudocomponents are not influenced by internal diffusion limitations (effectiveness factor greater than 0.98 in both scales).

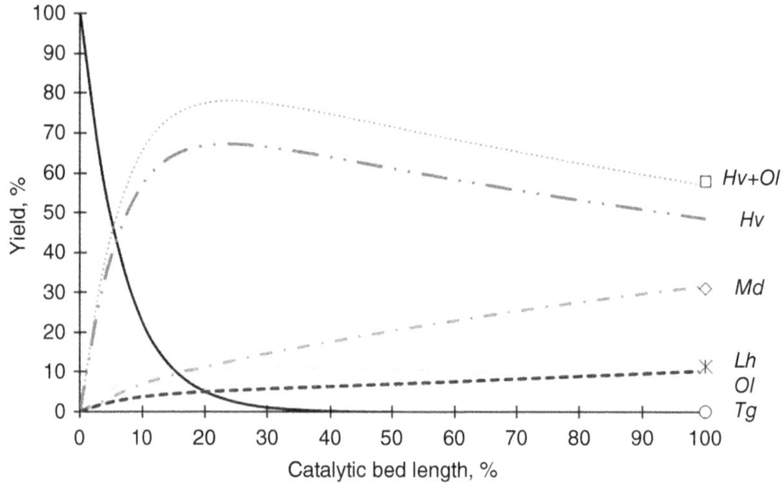

Figure 6.13 Comparison of predicted (lines) and experimental (symbols) yields in isothermal operation of pilot-scale reactor: Tg (–, ○), Ol (– –, ♦), Hv (–··–, Δ), Md (–·–, ◇), Lh (– – –,∗) and Hv+Ol (···, □) (Tirado and Ancheyta 2020b/American Chemical Society).

To measure the deviation among the predicted yields against experimental data, the mean absolute error (MAE) was calculated. The results indicated that for bench-scale and pilot-scale reactors, the MAE values were 2.75 and 1.06, respectively. It validates the ability of the model to correctly predict the behavior of each lump during vegetable oil hydrotreating.

Triglyceride conversion profiles were compared between both reactor scales to compare the differences between them, and the predicted profiles of the bench-scale reactor and pilot-scale reactor are illustrated in the figure at the top and bottom, respectively. In addition, catalytic bed length was normalized for an appropriate comparison of both reactor scales. Simulation of the pilot-scale reactor was developed at higher WHSV to magnify predicted profiles since, at WHSV of 1, the triglyceride conversion profiles practically overlap. Dynamic triglyceride concentration profiles at 420 °C, 8 MPa, and WHSV of 3 h^{-1} are shown in Figure 6.14. The triglyceride concentration increases as the feedstock flows through the catalyst particles until the catalytic bed becomes as wet as possible. Triglyceride concentration profiles shown in the pilot-scale reactor indicate lower values along the catalyst bed than those of the bench-scale reactor, which is caused by a smaller fraction of liquid present in the pilot-scale reactor ($\varepsilon_L = 0.288$) compared with that in bench-scale ($\varepsilon_L = 0.450$). This parameter directly influences the dynamic term in the model, and it is affected by the liquid–solid interfacial area, which has a lower value in the pilot-scale reactor ($a_S = 11.86$) compared with the bench-scale reactor ($a_S = 126.20$), since liquid holdup is inversely proportional to the equivalent particle diameter as follows (Froment et al. 1990; Shah 1979):

$$\varepsilon_L = 0.185\, \epsilon_B\, {a_S}^{0.333} \chi^{0.22} \tag{6.51}$$

$$a_S = \frac{6(1-\epsilon_B)}{d_{pe}} \tag{6.52}$$

On the other hand, the smaller the particle size, the higher the liquid holdup by capillary pressure (Carruthers and DiCamillo 1988; Macías and Ancheyta 2004).

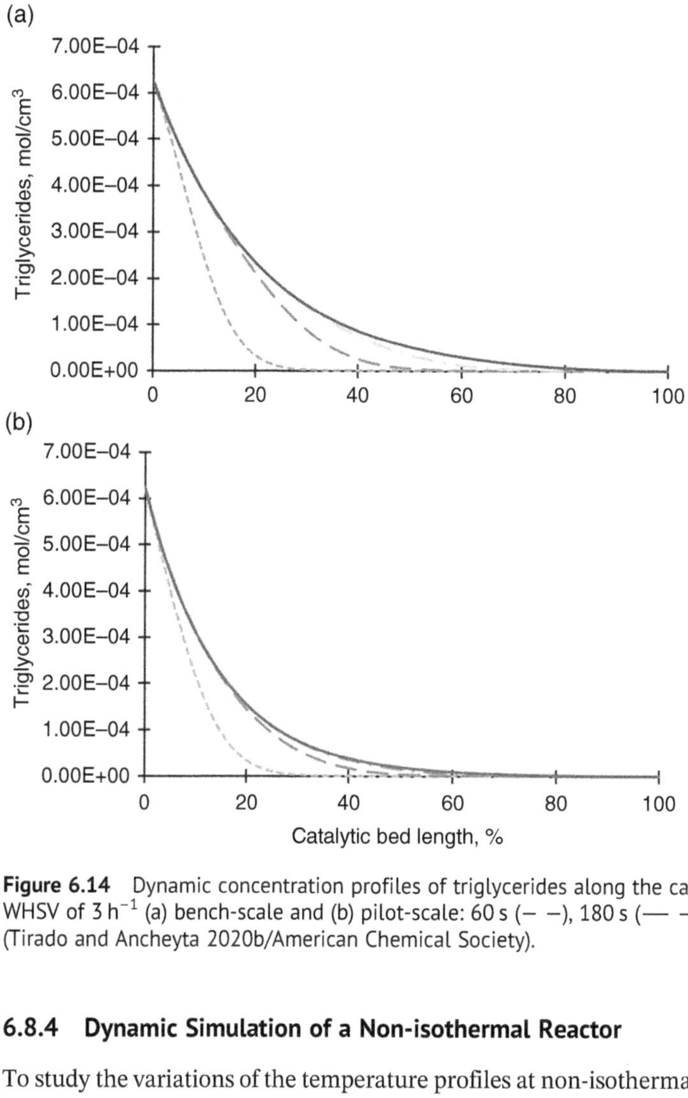

Figure 6.14 Dynamic concentration profiles of triglycerides along the catalytic bed at 420 °C, 8 MPa and WHSV of 3 h^{-1} (a) bench-scale and (b) pilot-scale: 60 s (– –), 180 s (— —), 300 s (-·-), 480 s (——) (Tirado and Ancheyta 2020b/American Chemical Society).

6.8.4 Dynamic Simulation of a Non-isothermal Reactor

To study the variations of the temperature profiles at non-isothermal conditions, the mass and heat balance equations for each phase are solved simultaneously.

6.8.4.1 Comparison of Non-isothermal Model with Experimental Results in Isothermal Reactor

An appropriate comparison of reactor performance at isothermal and non-isothermal conditions was performed by calculating the average temperature (T_{av}) of the liquid phase profile. This parameter is determined by integrating the predicted temperature profile in each simulation step as follows:

$$T_{av} = \frac{\sum_{i=0}^{N} \frac{T_N + T_{N+1}}{2}}{N} \tag{6.53}$$

where T_N is the liquid temperature at point N of the catalytic bed length. The results showed that an inlet temperature of 400 °C was required for WHSV between 0.5 and 3 to achieve an average

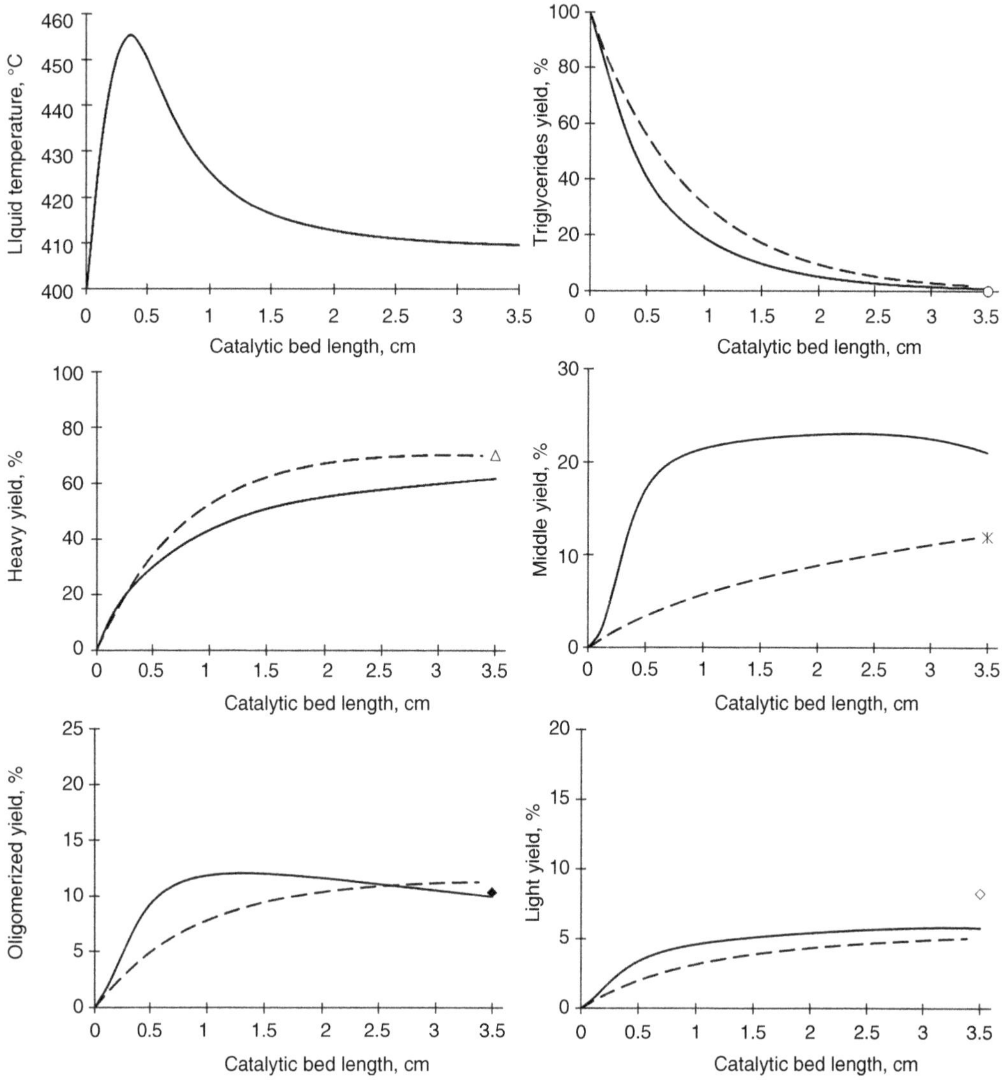

Figure 6.15 Liquid temperature profile and product yields in non-isothermal mode at inlet temperature of 400 °C (solid lines) and isothermal mode at 420 °C (dotted lines). Symbols indicate experimental data at 420 °C, 80 bar, WHSV of 3 h^{-1}, and 1500 Nl/l H_2/feed ratio. (○) Tg, (♦) Ol, (Δ) Hv, (∗) Md, (◇) Lh (Tirado et al. 2020 / with permission of Elsevier).

temperature similar to isothermal conditions (420 °C). Figure 6.15 shows the yield profiles for each lump under both operation modes when reaching steady-state conditions. Based on these behaviors, the following observations arise:

- The temperature profile of the liquid phase increased rapidly at the reactor top due to the fast conversion of triglyceride, which is the most exothermic chemical reaction occurring in the process. This behavior agrees with information reported by Forghani et al. (2014) during the modeling of a hydrocracking reactor for triglyceride conversion into green fuel.

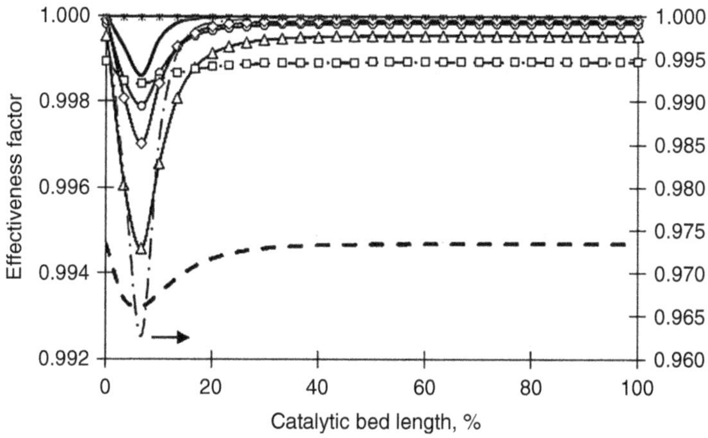

Figure 6.16 Simulated effectiveness factor profiles for each reaction pathway along the catalytic bed at inlet temperature of 400 °C, 80 bar, and WHSV of 3 h^{-1}. (Δ) η_1, (- - -) η_2, (-·-) η_3, (○) η_4, (□) η_5, (◇) η_7, (−) η_8, (+) η_9, (*) η_{10} (Tirado et al. 2020 / with permission of Elsevier).

- Triglyceride presents a lower concentration along the catalytic bed for non-isothermal mode compared with isothermal operation since the increase in temperature in the reactor top enhances the conversion rate of the feedstock.
- Heavy hydrocarbon yield under non-isothermal operating conditions is affected by the slight increase in temperature in the upper part of the reactor, where the hydrocracking reactions mainly occur.
- The formation of middle and light hydrocarbons from Tg (k_3 and k_4) has high activation energy (E_{A3} = 277.90 kJ/mol and E_{A3} = 127.77 kJ/mol), which combined with the higher temperature at the reactor inlet for non-isothermal operation improves the middle hydrocarbon yield
- Production of oligomerized compounds gives a maximum yield of around 31% (1.15 cm) of the catalytic bed length for non-isothermal mode, while increasing linearly during the isothermal behavior. The formation of oligomerized products is enhanced due to the high activation energy of Tg conversion (E_{A1} = 128.49 kJ/mol).

Figure 6.16 depicts the effectiveness factor profiles for each reaction pathway obtained in the bench-scale reactor under non-isothermal operating conditions. It is noted that within the catalytic bed, the impact of internal diffusion limitations is negligible ($\eta > 0.95$) because of using powdered catalysts (Perego and Peratello 1999). The results show that even under a sharp increase in temperature, the internal diffusion limitations are not significant in sprayed catalysts. The lowest value of the effectiveness factor was indicated for the reaction pathway ($\eta_3 = 0.963$) in the upper part of the reactor (7% of the reactor length), while the other reaction pathways present values higher than 0.99. Diminution of effectiveness factors is due to the increase in products that hinder the diffusion of species by the increase in temperature in the reactor top.

6.8.4.2 Comparison of Bench-Scale and Pilot-Scale Reactor Under Non-isothermal Operating Condition

Figures 6.17 and 6.18 show the predicted temperature profiles along the catalytic bed for gas and liquid phases, respectively. The simulations were developed at similar operation conditions (temperature of 420 °C, 8 MPa, and WHSV of 1 h^{-1}) for both reactor sizes. It is observed that the pilot-scale

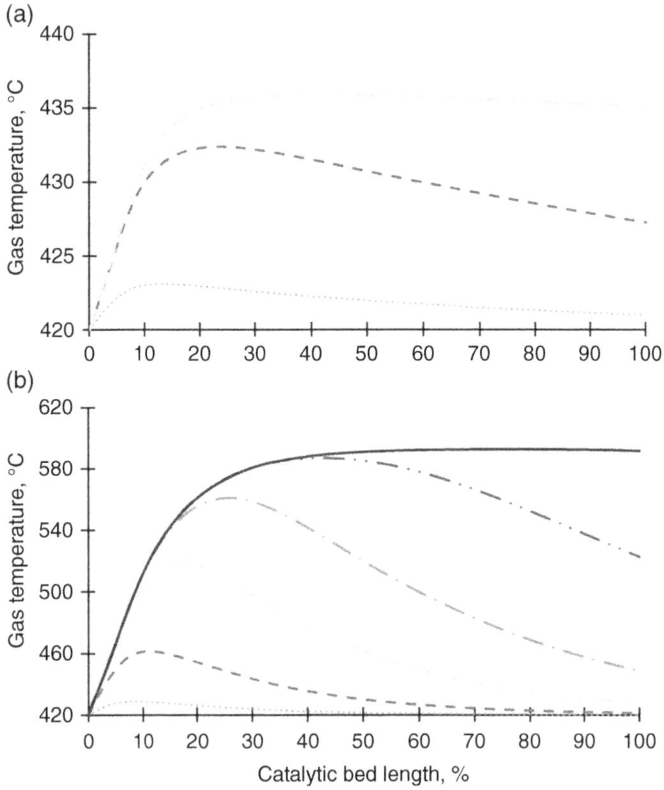

Figure 6.17 Dynamic gas temperature profiles at non-isothermal conditions for (a) bench-scale and (b) pilot-reactor scale: (···) 3 min, (– – –) 10 min, (— —) 30 min, (–·–) 60 min, (–· ·–) 120 min, (–) 240 min.

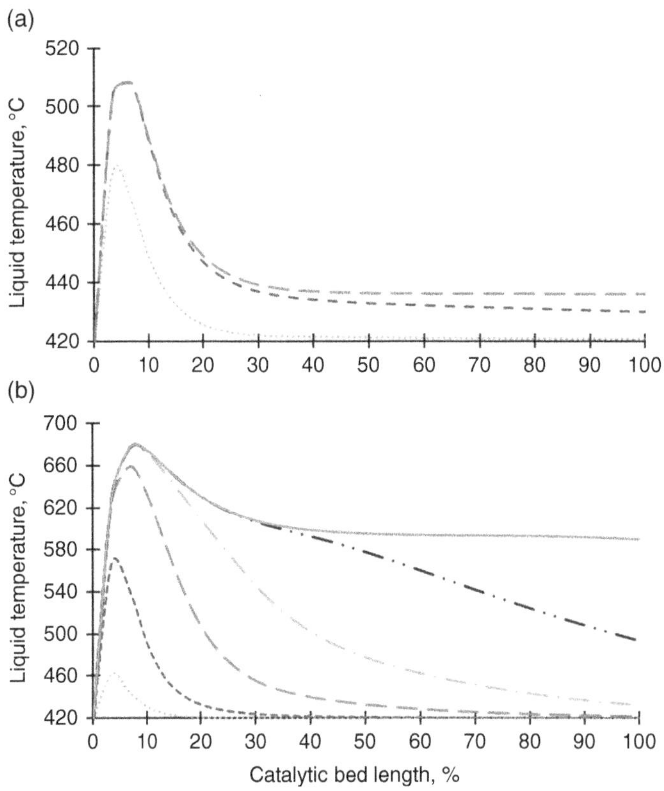

Figure 6.18 Dynamic liquid temperature profiles at non-isothermal conditions for (a) bench-scale and (b) pilot-reactor scale: (···) 3 min, (– – –) 10 min, (— —) 30 min, (–·–) 60 min, (–· ·–) 120 min, (–) 240 min.

reactor exhibits greater temperature increment in the liquid phases at the upper zone of the catalytic bed compared with the bench-scale reactor (85 °C and 260 °C, respectively). This behavior in the temperature profile of the liquid phase corresponds to the profile reported by Forghani et al. (2014), and it is associated with the exothermal decomposition of Tg. This statement agrees with the temperature profile behavior as zero levels of triglyceride have been detected in the latter half of the catalytic bed, resulting in a decrease in temperature and subsequent stabilization.

However, the results are different from those reported for the hydrotreating of straight-run petroleum distillates (Guo et al. 2008; Mederos et al. 2012) and diluted triglyceride (Muharam et al. 2017; Muharam and Putri 2018) feedstocks, obtaining smaller increments in the temperature reactor (15–30 °C). On the other hand, the temperature profile of the gas increases along the catalytic bed by transfer of heat between the gas–liquid interfaces. Solid-phase temperature profiles are not shown for bench-scale reactors; however, they behave similarly to liquid-phase profiles showing negligible resistance at the liquid–solid interface, while for the pilot-scale reactor, a small difference was obtained (2 °C). This resistance to energy transfer is caused by the dimensions of the commercial catalyst used. The temperature profiles of the pilot-scale reactor exhibited noticeable increments in each stage, with the potential for harmful reactions, reactor and particle damages, and difficulties to reach stable temperatures in the latter half of the catalytic bed. Moreover, the abrupt rises in temperature pose challenges in achieving a steady state due to increased feedstock flows and the catalyst amount employed.

The results of the simulations of vegetable oil hydrotreating under non-isothermal conditions provide a clear idea of the temperature behavior of each phase along the catalytic bed. To carry out the process on a commercial scale, it is crucial to establish adequate temperature control. This will help prevent hot spots in the reactor (Alvarez and Ancheyta 2012; Tirado and Ancheyta 2020b).

6.8.5 Dynamic Simulation of an Adiabatic Commercial Reactor

The reactor model for hydrotreating was focused on the design of a commercial-scale unit to examine the effect of the characteristics of the reactor as well as control of temperature and yield of the desired reaction products. The characteristics of the commercial-scale reactor were taken from an existing hydrotreating unit of petroleum-derived streams (Mederos et al. 2012; Rodriguez and Ancheyta 2004). The set of PDEs that constitute the reactor model was modified to provide a realistic industrial environment. Figure 6.19 presents the mass balance equations of the three-phase system, including the thermal effect, hydrogen consumption, product gas formation, and superficial velocity of the gas phase (u_G).

Similar to mass transfer phenomena, the heat balance equations to simulate the adiabatic operation in commercial units and quench stream equations for temperature control are presented in Figure 6.20. The necessary mass flow rate of cold hydrogen quench gas (q) to decrease the temperature of the hot flow of gas (g) and liquid (l) from the previous bed is calculated through an energy balance in the quench zone, where T_q, T_{in}, and T_{out} are the quench, the inlet, and the outlet temperatures of the quench zone, respectively (Alvarez et al. 2007; Alvarez and Ancheyta 2008).

The dynamic reactor model was solved with the kinetic parameters obtained from the bench-scale unit. The high liquid velocities encountered in commercial units provide a complete wetting efficiency and negligible hydrodynamic effects.

Reference data were obtained through a simulation of the commercial hydroprocessing unit without temperature control. The high heat release that was presented in laboratory-scale reactors indicates that the operating conditions must be carefully selected to have an operating range without exceeding a permissible temperature limit. For this reason, mild operating conditions have been

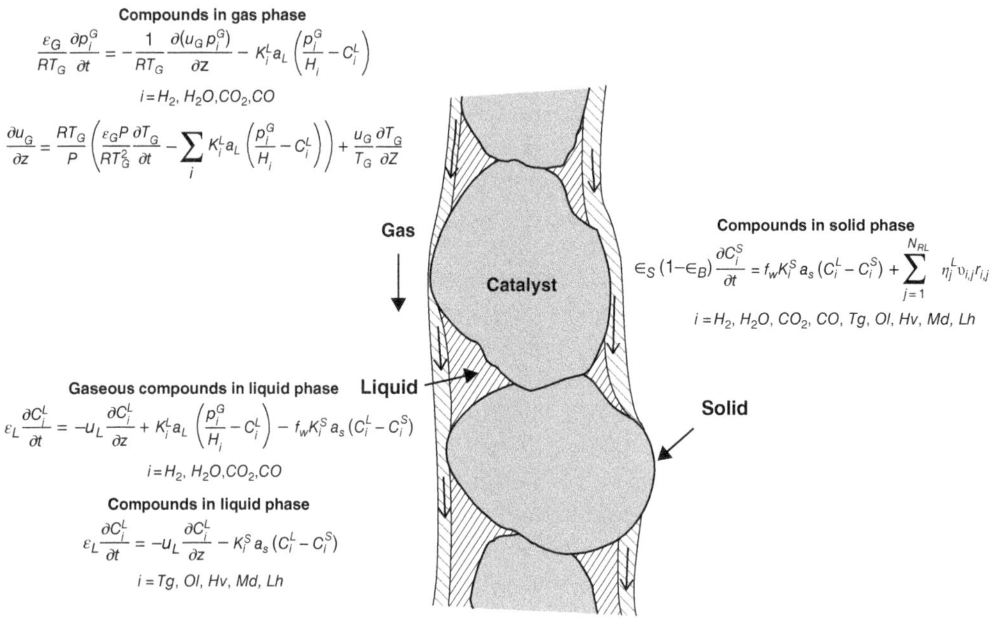

Figure 6.19 Mass balance equations of the commercial-scale fixed-bed reactor model (Tirado et al. 2022 / with permission of Elsevier).

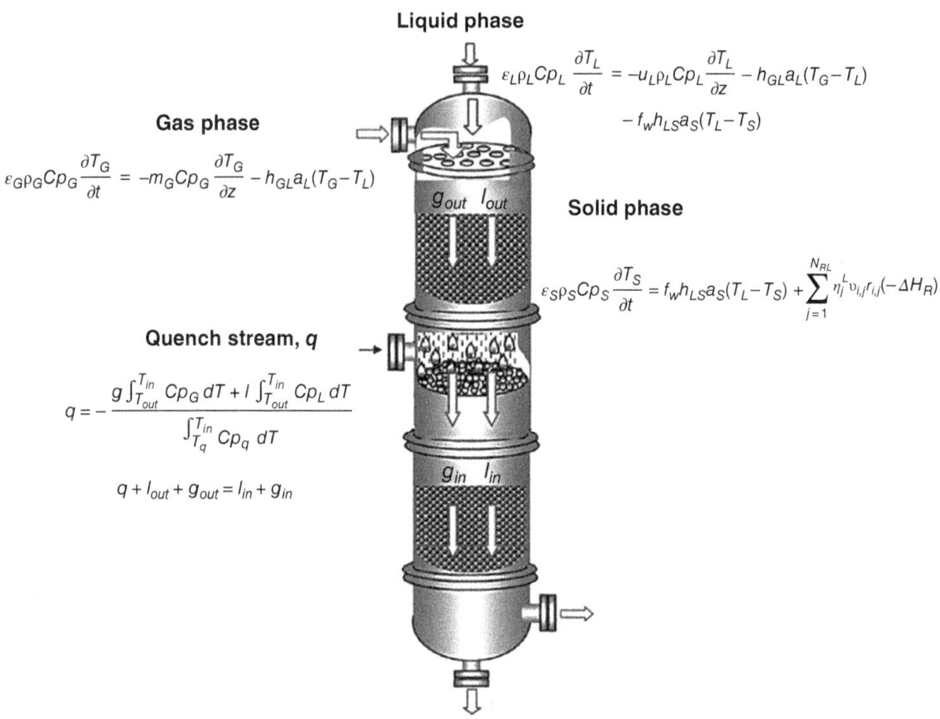

Figure 6.20 Heat balance equations of the commercial-scale fixed-bed reactor model (Tirado et al. 2022 / with permission of Elsevier).

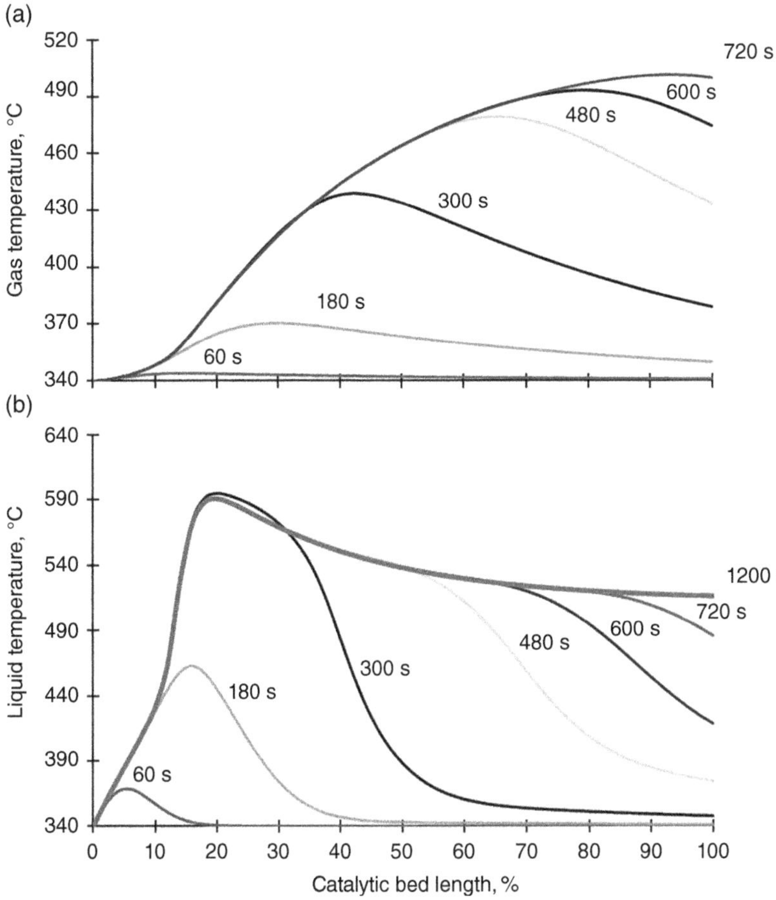

Figure 6.21 Dynamic temperature profiles along the catalytic bed length for (a) gas phase and (b) liquid phase at inlet temperature of 340 °C, 8 MPa, WHSV of 3 h^{-1}, and 1500 Nl/l oil H_2/oil ratio (Adapted from Tirado et al. 2021).

chosen for the commercial unit, as an increase in the feedstock inlet temperature will also increase the reaction temperature and cause the system to overheat (Alvarez and Ancheyta 2012; Fan et al. 2020; Tirado et al. 2020). Based on this information, the simulation was carried out under the following operating conditions: inlet temperature of the feed of 340 °C, WHSV of 3 h^{-1}, 8 MPa, and 1500 Nl/l oil H_2/oil ratio. Figure 6.21 depicts the temperature profiles of the gas and liquid phases along the catalytic bed. A sharp rise in temperature is observed from the start-of-run, especially in the liquid phase, considering that at 180 s a temperature of 470 °C was reached in the upper part of the reactor, while at steady-state conditions (1200 s), the maximum temperature reached was 590 °C. The variation in gas velocity influences the predicted temperature profiles since results obtained when simulating a pilot-scale reactor show higher temperature profiles and longer times to reach a steady state. On the other hand, the temperature difference calculated is higher than those reported during the hydrotreating of oil-derived streams (Alvarez-Majmutov and Chen 2014; Fan et al. 2020; Mederos et al. 2012).

The Preem Gothenburg refinery reported that during the co-processing of straight-run gas oils (SRGO) with 30 vol.% of RTD, the heat released increased; while in the absence of RTD, the Delta-T was about 30 °C, it was around 60 °C when the feedstock contained 30 vol.% of RTD (Egeberg et al. 2011; Egeberg et al. 2009). Therefore, for 100% renewable feedstock, higher temperature differences can be expected as those obtained in this study. Even under non-severe conditions, triglyceride conversion causes a high heat release that raises the temperature of the liquid phase. It increases the feedstock conversion rate even to exceed the temperature under steady-state conditions as observed at 300 s. In addition, the reduction of the triglyceride concentration and the energy transfer at the gas–liquid phase reduces the heat released in the second half of the reactor. Temperature profiles show that the system begins to overheat before the steady-state conditions are reached. It highlights the importance of a dynamic model capable of establishing the appropriate configuration of catalytic beds and the start-up time of quench injections. For industrial practice, a vegetable oil hydrotreating unit should be designed and operated in such a way that the temperature rise is controlled from the start-of-run.

6.8.5.1 Configuration of Hydrogen Quenching

In order to establish strategies for temperature control in a hydrotreating reactor of residual oil, Alvarez and Ancheyta carried out a series of simulations varying the operating conditions and settings for temperature control in the reactor. The simulations indicated that quenching must be started before or as soon as the hydrocarbon front reaches the quench zones to avoid overheating and to keep the temperature under control. The results showed the temperature control was adequate, as well as the hydrogen consumption, and sour gas emission. Thus, control strategies are developed for different hydroprocessing technologies (Alvarez and Ancheyta 2012). Based on information reported by Alvarez and Ancheyta for the temperature control of a residue oil hydrodemetallization unit, a series of strategies are considered for the configuration of hydrogen quench injections along the axial direction of the reactor. The sharp temperature rise and the high velocity of the liquid flow through the catalytic bed were the main issues found in the development of effective configurations, with a mean residence time of 456 s given by the interstitial velocity $\left(u'_L = u_L/\varepsilon_L\right)$. A series of simulations (case 1) where the quenching injections start as soon as the liquid temperature reaches the maximum permissible temperature were developed. Figure 6.22 shows the liquid-phase profiles with quenching injections upon reaching the set limit temperature of 400 °C. The catalyst bed was divided into five catalytic beds with four quench zones to distribute the total heat released below the permissible limit. The location along the catalytic bed length and start time for each quench injection are shown in Table 6.13. It is observed that the liquid-phase temperature gradually increases until cooling occurs, and the steady state is reached. Even after rapid cooling, some profiles show transient temperature variations that do not exceed the allowable limit. However, the heat released at the top of the reactor causes the length of the catalytic beds to become shorter as it approaches the inlet of the reactor.

In a second case study, the cooling injections started simultaneously when the liquid-phase temperature reached the allowable limit (400 °C) at the first quenching zone, to examine the change in the behavior of the temperature of the liquid phase. Figure 6.22 shows that the temperature profiles of the liquid phase are kept under control. This configuration allows reducing the Delta-T in the third and fourth catalytic beds (from 60 to 40 °C) and changes the location of the fourth quench zone, as shown in Table 6.13. However, starting quenching injections when the temperature has not yet risen causes a sharp cooling of the catalytic bed areas and longer times to reach steady state (1200 s).

6.8.5.2 Liquid-Phase Yields and Gas Composition

The yields of products in the liquid phase and gas compositions were analyzed based on the configuration of case 1, since it seems to be the most suitable strategy for temperature control in this type of process. Figure 6.23a depicts the yield profiles of the liquid pseudocomponents along the

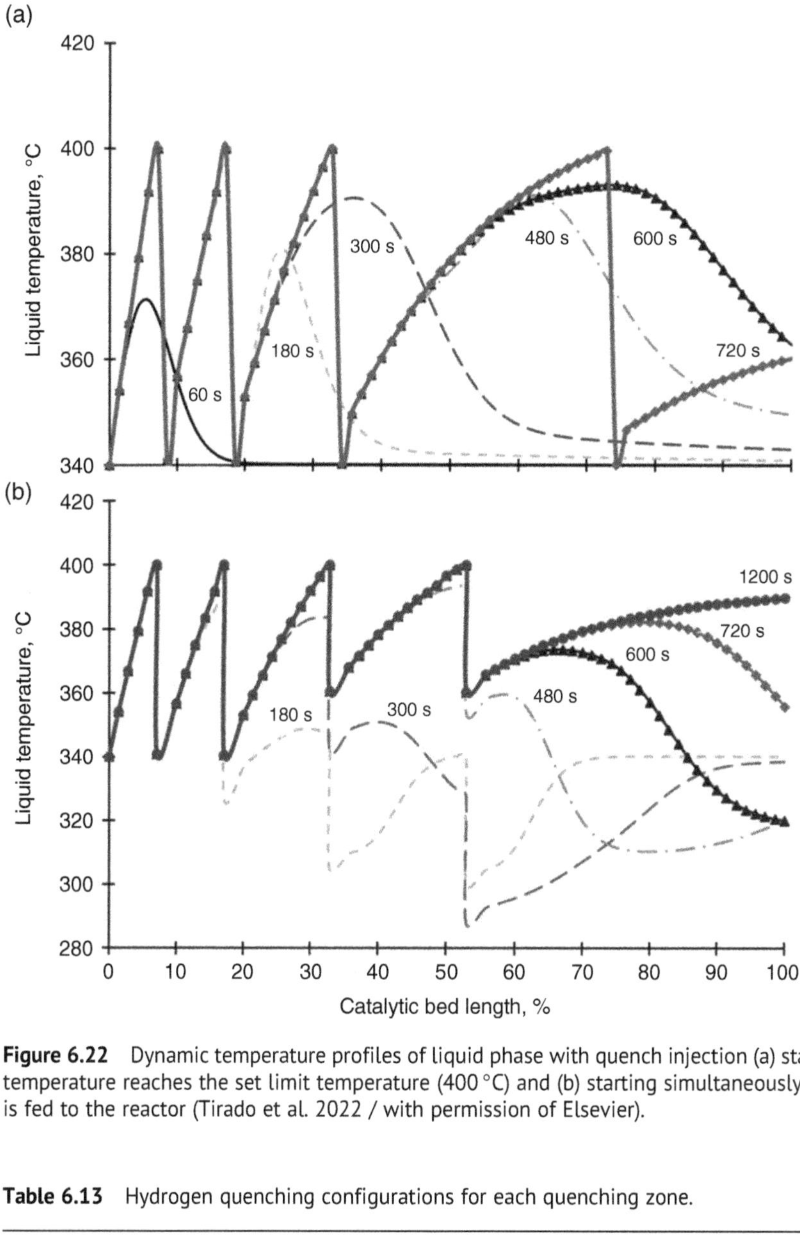

Figure 6.22 Dynamic temperature profiles of liquid phase with quench injection (a) starting when the liquid temperature reaches the set limit temperature (400 °C) and (b) starting simultaneously when the feedstock is fed to the reactor (Tirado et al. 2022 / with permission of Elsevier).

Table 6.13 Hydrogen quenching configurations for each quenching zone.

Configuration	Injection	Localization (cm)	Injection time (s)
Case 1	1	73.15 (8.57%)	95
	2	158.49 (18.57%)	150
	3	292.6 (34.28%)	325
	4	634 (74.28%)	620
Case 2	1	73.15 (8.57%)	95
	2	158.49 (18.57%)	95
	3	292.6 (34.28%)	95
	4	463.3 (54.28%)	95

Numbers in parenthesis indicate the percentage of the catalytic bed from top to bottom of the reactor.

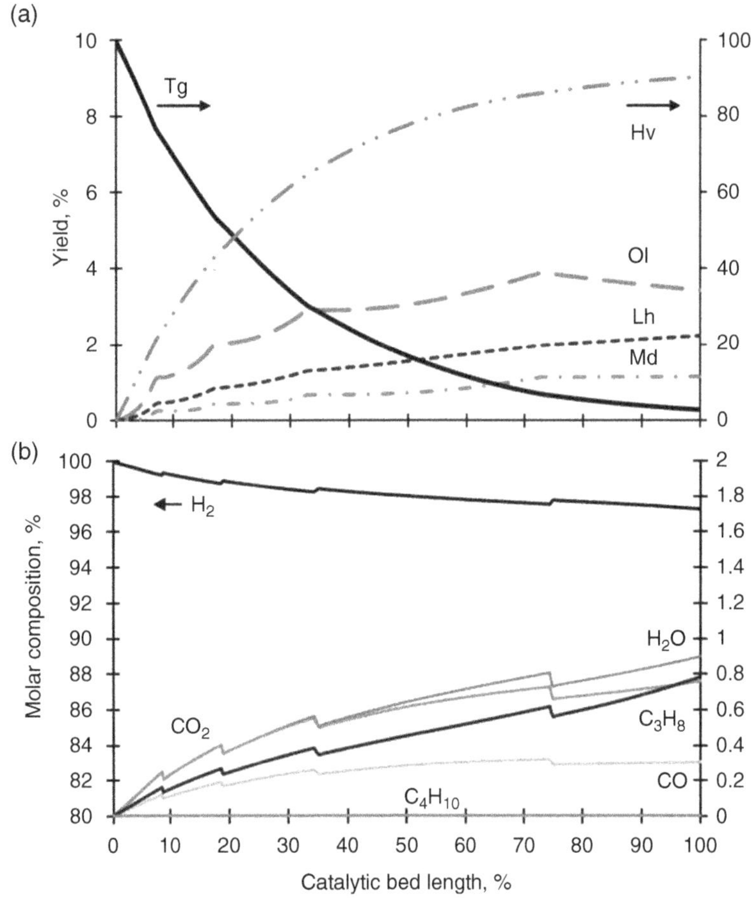

Figure 6.23 Profiles of (a) pseudocomponents yields in the liquid phase and (b) molar composition of each component in the gas phase along the catalytic bed under steady state at inlet temperature of 340 °C, 8 MPa, WHSV of 3 h^{-1}, and 1500 Nl/l oil H_2/oil ratio (Tirado et al. 2022 / with permission of Elsevier).

catalytic bed. The yield of unconverted Tg decreases until a conversion of 98% is reached at the outlet of the reactor, while the yield of heavy hydrocarbons makes it possible to obtain a 90% yield of diesel-type fuel (C_{15}–C_{18}). On the other hand, low-boiling-point fuels (middle and light hydrocarbon) gave lower yields because of the mild severity conditions. The oligomerized compounds are the most temperature-sensitive lumps due to the stepwise increase throughout the reactor. Table 6.14 shows that the yields at the outlet of the reactor follow a trend with the data measured with a catalyst powdered in an isothermal bench-scale reactor, despite the variations in temperature and intraparticle effects at the industrial scale.

Figure 6.23b shows the molar composition profiles of each gaseous compound in a steady state. The gas composition at the inlet of the reactor is pure hydrogen, because the evaporation effects of the vegetable oil are negligible, and the feed gas does not contain lighter hydrocarbons. It tends to decrease along the reactor by the hydrogen consumption and product gas formation. However, staggered fluctuations occur at points where the quench zones are located due to an increase in hydrogen composition and a decrease in liquid phase temperature that diminishes the hydrogen

Table 6.14 Comparison of experimental data with calculated yields.

Pseudocomponent	Experimental 340 °C	Experimental 420 °C	Predicted
Tg unconverted	72.75	0.05	2.85
Ol (>C_{18})	7.65	9.49	3.45
Hv (C_{15}–C_{18})	19.58	70.19	90.32
Md (C_9–C_{14})	0.01	11.96	1.16
Lh (<C_9)	0.01	8.32	2.22

solubility in the vegetable oil. Therefore, the dissolved hydrogen in the liquid phase is transferred to the gas phase. Under the studied configuration, the hydrogen consumption calculation was 353.15 Nm^3/m^3, which agrees with the information reported for the hydroprocessing of renewable feedstock in industrial units with values of 300–400 Nm^3/m^3 (Egeberg et al. 2009; Guzman et al. 2010; Verdier et al. 2019), while the amount of hydrogen for hydroprocessing of petroleum-derived streams is 50–150 Nm^3/m^3 (Ancheyta 2011). Therefore, higher make-up hydrogen and quench gas streams are needed even when co-processing quite small amounts of vegetable oil.

The results show that CO_2 and H_2O are the main product gases formed because HDO and decarboxylation are the predominant reactions, while CO has a lower composition. C_3H_8 continuously increases throughout the catalytic bed due to the triglyceride conversion. On the other hand, C_4H_{10} formation is insignificant because the hydrocracking reaction does not take place for a long extent.

6.9 Conclusions

During the development of kinetic models for the generation of renewable fuels, it is necessary to consider diverse experimental and mathematical assumptions to obtain intrinsic kinetics. Otherwise, it is appropriate to consider that the kinetic parameters are affected by transfer limitations at the interfaces of the reactant system. Mass and heat transfer effects at the gas–liquid and solid–liquid interfaces, as well as inside the catalyst pores, affect the experimental data. Therefore, the influence of these effects on kinetic parameters and the development of reactor models are evaluated. The selection of a reaction scheme is crucial to obtain relevant information from the system. In addition, the optimization of kinetic parameters through complex nonlinear algorithms, mathematical constraints, and verification of the results through statistical analysis is necessary.

Different approaches have been developed to analyze and predict catalytic decay during hydrotreating feedstock capable of producing renewable fuels. However, the analyses of these approaches with feedstocks are still scarce and require further investigation.

The development and scaling of reactor models involve hydrodynamic phenomena related to plug-flow pattern, catalyst wetting efficiency, wall effects, and internal catalyst diffusion. These effects can be determined by analytical criteria in order to achieve adequate prediction of the diverse components in the reaction system during the scale-up of reactor models and their validation with experimental data.

The predicted temperature profiles depict an increase in the upper part of the catalytic bed attributed to triglyceride conversion, while the second half of the catalytic bed undergoes thermal stabilization. This increase in temperature enhances the decomposition of Tg and promotes the hydrocracking reactions. Simulations performed for industrial-scale reactors show high temperature rises along the catalytic bed are unacceptable during the operation of a commercial hydroprocessing unit. Therefore, the operating conditions of a renewable feedstock at a commercial-scale hydroprocessing unit must be carefully selected to control the temperature rise throughout the reactor. These results demonstrate the necessity of using quenching streams in the reactor to control the exothermicity of reactions. Different configurations were simulated to control the reactor temperature and the reaction product yields. From a practical point of view, the results show that during the hydroprocessing of renewable feedstock, the consumption of hydrogen and heat released are much higher than those reported during petroleum distillates HDT. A greater focus on developing mathematical models is required to optimize some process aspects. Hydrogen consumption and quenching configurations are crucial for the choice of high-quality renewable fuels. In addition, it is possible to configure the location and the start-up time of the quenching zones to distribute the high heat of reaction throughout the reactor without exceeding a predetermined temperature control.

References

Alcázar, L.A. and Ancheyta, J. (2007). Sensitivity analysis based methodology to estimate the best set of parameters for heterogeneous kinetic models. *Chemical Engineering Journal* 128: 85–93.

Alvarez, A. and Ancheyta, J. (2008). Simulation and analysis of different quenching alternatives for an industrial vacuum gasoil hydrotreater. *Chemical Engineering Science* 63: 662–673.

Alvarez, A. and Ancheyta, J. (2012). Transient behavior of residual oil front-end hydrodemetallization in a trickle-bed reactor. *Chemical Engineering Journal* 197: 204–214.

Alvarez, A., Ancheyta, J., and Muñoz, J.A.D. (2007). Comparison of quench systems in commercial fixed-bed hydroprocessing reactors. *Energy and Fuels* 21: 1133–1144.

Alvarez-Majmutov, A. and Chen, J. (2014). Modeling and simulation of a multibed industrial hydrotreater with vapor-liquid equilibrium. *Industrial and Engineering Chemistry Research* 53: 10566–10575.

Anand, M. and Sinha, A.K. (2012). Temperature-dependent reaction pathways for the anomalous hydrocracking of triglycerides in the presence of sulfided Co-Mo-catalyst. *Bioresource Technology* 126: 148–155.

Anand, K., Ranjan, A., and Mehta, P.S. (2010). Predicting the density of straight and processed vegetable oils from fatty acid composition. *Energy and Fuels* 24: 3262–3266.

Anand, M., Farooqui, S.A., Kumar, R. et al. (2016a). Kinetics, thermodynamics and mechanisms for hydroprocessing of renewable oils. *Applied Catalysis A: General* 516: 144–152.

Anand, M., Farooqui, S.A., Kumar, R. et al. (2016b). Optimizing renewable oil hydrocracking conditions for aviation bio-kerosene production. *Fuel Processing Technology* 151: 50–58.

Ancheyta, J. (2011). *Modeling and Simulation of Catalytic Reactors for Petroleum Refining*. New Jersey: Wiley.

Ancheyta, J., Trejo, F., and Rana, M.S. (2010). *Chemical Transformation During Hydroprocessing of Heavy Oils*. Boca Raton: CRC Press.

Arora, P., Grennfelt, E.L., Olsson, L., and Creaser, D. (2019). Kinetic study of hydrodeoxygenation of stearic acid as model compound for renewable oils. *Chemical Engineering Journal* 364: 376–389.

Ayodele, O.B., Farouk, H.U., Mohammed, J. et al. (2015). Hydrodeoxygenation of oleic acid into n- and iso-paraffin biofuel using zeolite supported fluoro-oxalate modified molybdenum catalyst: kinetics study. *Journal of the Taiwan Institute of Chemical Engineers* 50: 142–152.

Baldiraghi, F., Di Stanislao, M., Faraci, G. et al. (2009). Ecofining: new process for green diesel production from vegetable oil. *Sustainable Industrial Chemistry* 427–438.

Bezergianni, S. and Dimitriadis, A. (2013). Comparison between different types of renewable diesel. *Renewable and Sustainable Energy Reviews* 21: 110–116.

Bhaskar, M., Valavarasu, G., Sairam, B. et al. (2004). Three-phase reactor model to simulate the performance of pilot-plant and industrial trickle-bed reactors sustaining hydrotreating reactions. *Industrial Engineering Chemistry Research* 43: 6654–6669.

Bie, Y., Kanervo, J.M., and Lehtonen, J. (2015). Hydrodeoxygenation of methyl heptanoate over Rh/ZrO2 catalyst as a model reaction for biofuel production: kinetic modeling based on reaction mechanism. *Industrial and Engineering Chemistry Research* 54: 11986–11996.

Bie, Y., Lehtonen, J., and Kanervo, J. (2016). Hydrodeoxygenation (HDO) of methyl palmitate over bifunctional Rh/ZrO2 catalyst: Insights into reaction mechanism via kinetic modeling. *Applied Catalysis A: General* 526: 183–190.

Bischoff, K.B. and Levenspiel, O. (1962). Fluid dispersion-generalization and comparison of mathematical models-II comparison of models. *Chemical Engineering Science* 17: 257–264.

Boda, L., Onyestyák, G., Solt, H. et al. (2010). Catalytic hydroconversion of tricaprylin and caprylic acid as model reaction for biofuel production from triglycerides. *Applied Catalysis A: General* 374: 158–169.

Boldrini, D.E., Damiani, D.E., and Tonetto, G.M. (2014). Mathematical modeling of the partial hydrogenation of vegetable oil in a monolithic stirrer reactor. *AlChe* 60: 3524–3533.

Bykova, M.V., Zavarukhin, S.G., Trusov, L.I., and Yakovlev, V.A. (2013). Guaiacol hydrodeoxygenation kinetics with catalyst deactivation taken into consideration. *Kinetics and Catalysis* 54: 40–48.

Cabrera, M.I. and Grau, R.J. (2008). Advanced concepts for the kinetic modeling of fatty acid methyl esters hydrogenation. *International Journal of Chemical Reactor Engineering* 6: 1–37.

Carruthers, J.D. and DiCamillo, D.J. (1988). Pilot plant testing of hydrotreating catalysts. Influence of catalyst condition, bed loading and dilution. *Applied Catalysis* 43: 253–276.

Chen, A.H., Mclntire, D.D., and Allen, R.R. (1981). Modeling of reaction rate constants and selectivities in soybean oil hydrogenation. *Journal of the American Oil Chemists' Society* 58: 816–818.

Chen, N., Gong, S., Shirai, H. et al. (2013). Effects of Si/Al ratio and Pt loading on Pt/SAPO-11 catalysts in hydroconversion of Jatropha oil. *Applied Catalysis A: General* 466: 105–115.

Donnis, B., Egeberg, R.G., Blom, P., and Knudsen, K.G. (2009). Hydroprocessing of bio-oils and oxygenates to hydrocarbons. Understanding the reaction routes. *Topics in Catalysis* 52: 229–240.

Douvartzides, S.L., Charisiou, N.D., Papageridis, K.N., and Goula, M.A. (2019). Green diesel: biomass feedstocks, production technologies, catalytic research, fuel properties and performance in compression ignition internal combustion engines. *Energies* 12: 809.

Egeberg, R.G., Michaelsen, N.H., and Skyum, L. (2009). Novel hydrotreating technology for production of green diesel. *Technical Info Haldor Topsøe* 1–20.

Egeberg, R., Knudsen, K., Nyström, S. et al. (2011). Industrial-scale production of renewable diesel. *Petroleum Technology Quarterly* 16: 1–7.

El-Hisnawi, A.A., Dudukovic, M.P., and Mills, P.L. (1982). Trickle-bed reactors: dynamic tracer tests, reaction studies, and modeling of reactor performance. *ACS Symposium Series* 421–440.

Fan, X., Li, D., Feng, X. et al. (2020). Modelling and simulation of industrial trickle bed reactor hydrotreating for whole fraction low-temperature coal tar simultaneous hydrodesulfurisation and hydrodenitrification. *Fuel* 269: 117362.

Félix, G., Ríos, J.J., Tirado, A. et al. (2022). Monte carlo and sensitivity analysis methods for kinetic parameters optimization: application to heavy oil slurry-phase hydrocracking. *Energy and Fuels* 36: 9251–9260.

Félix, G., Tirado, A., Al-Muntaser, A. et al. (2023). *Catalytic mechanism and kinetics*. In: *Catalytic In-situ Upgrading of Heavy and Extra-Heavy Crude Oils* (ed. M.A. Varfolomeev, C. Yuan, and J. Ancheyta), 309–381. New Jersey: Wiley.

Fillion, B. and Morsi, B.I. (2000). Gas-liquid mass-transfer and hydrodynamic parameters in a soybean oil hydrogenation process under industrial conditions. *Industrial and Engineering Chemistry Research* 39: 2157–2168.

Fillion, B., Morsi, B.I., Heier, K.R., and Machado, R.M. (2002). Kinetics, gas–liquid mass transfer, and modeling of the soybean oil hydrogenation process. *Industrial Engineering Chemistry Research* 41: 697–709.

Forghani, A.A. and Lewis, D.M. (2016). Hydro-conversion of oleic acid in bio-oil to liquid hydrocarbons: an experimental and modeling investigation. *Journal of Chemical Technology and Biotechnology* 91: 655–663.

Forghani, A.A., Jafarian, M., Pendleton, P., and Lewis, D.M. (2014). Mathematical modelling of a hydrocracking reactor for triglyceride conversion to biofuel: model establishment and validation. *International Journal of Energy Research* 38: 1624–1634.

Froment, G.F., Bischoff, K.B., and De Wilde, J. (1990). *Chemical Reactor—Analysis and Design*, 3rde. Belgium: Wiley.

Furimsky, E. (2013). Hydroprocessing challenges in biofuels production. *Catalysis Today* 217: 13–56.

Gierman, H. (1988). Design of laboratory hydrotreating reactors scaling down of trickle-flow reactors. *Applied Catalysis* 43: 277–286.

Goto, S. and Smith, J.M. (1975). Trickle-bed reactor performance. *AIChE Journal* 21: 706–713.

Grilc, M., Likozar, B., and Levec, J. (2014a). Hydrodeoxygenation and hydrocracking of solvolysed lignocellulosic biomass by oxide, reduced and sulphide form of NiMo, Ni, Mo and Pd catalysts. *Applied Catalysis B: Environmental* 150–151: 275–287.

Grilc, M., Likozar, B., and Levec, J. (2014b). Hydrotreatment of solvolytically liquefied lignocellulosic biomass over NiMo/Al2O3 catalyst: reaction mechanism, hydrodeoxygenation kinetics and mass transfer model based on FTIR. *Biomass and Bioenergy* 63: 300–312.

Guo, J., Jiang, Y., and Al-Dahhan, M.H. (2008). Modeling of trickle-bed reactors with exothermic reactions using cell network approach. *Chemical Engineering Science* 63: 751–764.

Guzman, A., Torres, J.E., Prada, L.P., and Nuñez, M.L. (2010). Hydroprocessing of crude palm oil at pilot plant scale. *Catalysis Today* 156: 38–43.

Hachemi, I. and Murzin, D.Y. (2018). Kinetic modeling of fatty acid methyl esters and triglycerides hydrodeoxygenation over nickel and palladium catalysts. *Chemical Engineering Journal* 334: 2201–2207.

Hasanudin, H., Rachmat, A., Said, M., and Wijaya, K. (2019). Kinetic model of crude palm oil hydrocracking over Ni/Mo ZrO2–pillared bentonite Catalyst. *Periodica Polytechnica Chemical Engineering* 64: 238–247.

Hermida, L., Abdullah, A.Z., and Mohamed, A.R. (2015). Deoxygenation of fatty acid to produce diesel-like hydrocarbons: a review of process conditions, reaction kinetics and mechanism. *Renewable and Sustainable Energy Reviews* 42: 1223–1233.

Hickman, D.A., Weidenbach, M., and Friedhoff, D.P. (2004). A comparison of a batch recycle reactor and an integral reactor with fines for scale-up of an industrial trickle bed reactor from laboratory data. *Chemical Engineering Science* 59: 5425–5430.

Holser, R.A., List, G.R., King, J.W. et al. (2002). Modeling of hydrogenation kinetics from triglyceride compositional data. *Journal of Agricultural and Food Chemistry* 50: 7111–7113.

Huber, G.W., O'Connor, P., and Corma, A. (2007). Processing biomass in conventional oil refineries: production of high quality diesel by hydrotreating vegetable oils in heavy vacuum oil mixtures. *Applied Catalysis A: General* 329: 120–129.

Jeništová, K., Hachemi, I., Mäki-Arvela, P. et al. (2017). Hydrodeoxygenation of stearic acid and tall oil fatty acids over Ni-alumina catalysts: Influence of reaction parameters and kinetic modelling. *Chemical Engineering Journal* 316: 401–409.

Kalnes, T.N., Marker, T., Shonnard, D.R., and Koers, K.P. (2008). Green diesel production by hydrorefining renewable feedstocks. *Biofuels Technology* 2: 7–11.

Kiran, G.A.R. (2015). *CFD Modeling and Simulations of Catalytic Hydrotreatment of Bio-Oil*. Guwahati.

Korsten, H. and Hoffmann, U. (1996). Three-phase reactor model for hydrotreating in pilot trickle-bed reactors. *AIChE Journal* 42: 1350–1360.

Kubička, D. and Kaluža, L. (2010). Deoxygenation of vegetable oils over sulfided Ni, Mo and NiMo catalysts. *Applied Catalysis A: General* 372: 199–208.

Kubička, D. and Tukač, V. (2013). Hydrotreating of triglyceride-based feedstocks in refineries. *Advances in Chemical Engineering* 42: 141–194.

Kubička, D., Šimáček, P., and Žilkova, N. (2009). Transformation of vegetable oils into hydrocarbons over mesoporous-alumina-supported CoMo catalysts. *Topics in Catalysis* 52: 161–168.

Kubičková, I. and Kubička, D. (2010). Utilization of triglycerides and related feedstocks for production of clean hydrocarbon fuels and petrochemicals: a review. *Waste and Biomass Valorization* 1: 293–308.

Kumar, P., Yenumala, S.R., Maity, S.K., and Shee, D. (2014). Kinetics of hydrodeoxygenation of stearic acid using supported nickel catalysts: effects of supports. *Applied Catalysis A: General* 471: 28–38.

Lange, R., Schubert, M., Dietrich, W., and Grünewald, M. (2004). Unsteady-state operation of trickle-bed reactors. *Chemical Engineering Science* 59: 5355–5361.

Macías, M.J. and Ancheyta, J. (2004). Simulation of an isothermal hydrodesulfurization small reactor with different catalyst particle shapes. *Catalysis Today* 98: 243–252.

Mederos, F.S., Rodriguez, M.A., Ancheyta, J., and Arce, E. (2006). Dynamic modeling and simulation of catalytic hydrotreating reactors. *Energy and Fuels* 20: 936–945.

Mederos, F.S., Ancheyta, J., and Chen, J. (2009a). Review on criteria to ensure ideal behaviors in trickle-bed reactors. *Applied Catalysis A: General* 355: 1–19.

Mederos, F.S., Elizalde, I., and Ancheyta, J. (2009b). Steady-state and dynamic reactor models for hydrotreatment of oil fractions: a review. *Catalysis Reviews* 51: 485–607.

Mederos, F.S., Ancheyta, J., and Elizalde, I. (2012). Dynamic modeling and simulation of hydrotreating of gas oil obtained from heavy crude oil. *Applied Catalysis A: General* 425: 13–27.

Mederos, F.S., Elizalde-Martínez, I., Hernández-Altamirano, R. et al. (2019). Hydrotreating model comparison of raw castor oil and its methyl esters for biofuel production. *Chemical Engineering and Technology* 42: 167–173.

Mederos, F.S., Elizalde-Martínez, I., Trejo-Zárraga, F. et al. (2020). Dynamic modeling and simulation of three-phase reactors for hydrocracking of vegetable oils. *Reaction Kinetics, Mechanisms and Catalysis* 131: 613–644.

Mederos, F.S., Santoyo-López, A.O., Hernández-Altamirano, R. et al. (2021). Renewable fuels production from the hydrotreating over $NiMo/\gamma\text{-}Al_2O_3$ catalyst of castor oil methyl esters obtained by reactive extraction. *Fuel* 285: 119168.

Mendoza Sépulveda, C.C. (2013). *Simulación CFD de la transferencia de calor en un reactor de hidrotratamiento de aceites vegetales de segunda generación*. Medellín.

Mohammad, M., Kandaramath Hari, T., Yaakob, Z. et al. (2013). Overview on the production of paraffin based-biofuels via catalytic hydrodeoxygenation. *Renewable and Sustainable Energy Reviews* 22: 121–132.

Mortensen, P.M., Grunwaldt, J., Jensen, P.A. et al. (2011). Applied catalysis A : general A review of catalytic upgrading of bio-oil to engine fuels. *Applied Catalysis A, General* 407: 1–19.

Muharam, Y. and Adevia, R.T. (2018). Modelling and simulation of a slurry bubble column reactor for green fuel production via hydrocracking of vegetable oil. *Web of Conferences* 67: 1–6.

Muharam, Y. and Putri, A.D. (2018). Simulation of hydrotreating of vegetable Oil in a slurry bubble column reactor for green diesel production. *International Journal of Technology* 9: 1168–1177.

Muharam, Y., Nugraha, O.A., and Leonardi, D. (2017). Modelling of a hydrotreating reactor to produce renewable diesel from non-edible vegetable oils. *Chemical Engineering Transactions* 56: 1561–1566.

Noriega, A.K., Tirado, A., Méndez, C. et al. (2020). Hydrodeoxygenation of vegetable oil in batch reactor: experimental considerations. *Chinese Journal of Chemical Engineering* 28: 1670–1683.

Pedroza, M.A.P., Segtovich, I.S.V., Sermoud, V.d.M., and da Silva, M.A.P. (2022). Hydrodeoxygenation of stearic acid to produce diesel-like hydrocarbons: kinetic modeling, parameter estimation and simulation. *Chemical Engineering Science* 254: 117576.

Pelemo, J., Inambao, F.L., and Onuh, E.I. (2020). Potential of used cooking oil as feedstock for hydroprocessing into hydrogenation derived renewable diesel: A review. *International Journal of Engineering Research and Technology* 13: 500–519.

Perego, C. and Peratello, S. (1999). Experimental methods in catalytic kinetics. *Catalysis Today* 52: 133–145.

Rodenbush, C.M., Hsieh, F.H., and Viswanath, D.S. (1999). Density and viscosity of vegetable oils. *Journal of the American Oil Chemists' Society* 76: 1415–1419.

Rodriguez, M.A. and Ancheyta, J. (2004). Modeling of hydrodesulfurization (HDS), hydrodenitrogenation (HDN), and the hydrogenation of aromatics (HDA) in a vacuum gas oil hydrotreater. *Energy and Fuels* 18: 789–794.

Saidi, M., Samimi, F., Karimipourfard, D. et al. (2014). Upgrading of lignin-derived bio-oils by catalytic hydrodeoxygenation. *Energy and Environmental Science* 7: 103–129.

Sámano, V., Tirado, A., Félix, G., and Ancheyta, J. (2020). Revisiting the importance of appropriate parameter estimation based on sensitivity analysis for developing kinetic models. *Fuel* 267: 117113.

Satterfield, C.N. (1975). Trickle-bed reactors. *AIChE Journal* 21: 209–228.

Scharff, Y., Asteris, D., and Fédou, S. (2013). Catalyst technology for biofuel production: Conversion of renewable lipids into biojet and biodiesel. *OCL—Oilseeds and Fats, Crops and Lipids* 20: 2–5.

Schiesser, W.E. (1991). *The numerical method of lines: integration of partial differential equations*. Texas: Elsevier.

Sebos, I., Matsoukas, A., Apostolopoulos, V., and Papayannakos, N. (2009). Catalytic hydroprocessing of cottonseed oil in petroleum diesel mixtures for production of renewable diesel. *Fuel* 88: 145–149.

Selishcheva, S.A., Lebedev, M.Y., Reshetnikov, S.I. et al. (2014). Kinetics of the hydrotreatment of rapeseed oil fatty acid triglycerides under mild conditions. *Catalysis in Industry* 6: 60–66.

Shah, Y. (1979). *Gas-Fluid-Solid Reactor Design*. New York: McGraw-Hill International Book Co.

Shamanaev, I.V., Deliy, I.V., Aleksandrov, P.V. et al. (2019). Methyl palmitate hydrodeoxygenation over silica-supported nickel phosphide catalysts in flow reactor: experimental and kinetic study. *Journal of Chemical Technology Biotechnology* 94: 3007–3019.

Sharma, R.K., Anand, M., Rana, B.S. et al. (2012). Jatropha-oil conversion to liquid hydrocarbon fuels using mesoporous titanosilicate supported sulfide catalysts. *Catalysis Today* 198: 314–320.

Sinha, A.K., Sibi, M.G., Naidu, N. et al. (2014). Process intensification for hydroprocessing of vegetable oils: experimental study. *Industrial Engineering Chemistry Research* 53: 19062–19070.

Sivasamy, A., Cheah, K.Y., Fornasiero, P. et al. (2009). Catalytic applications in the production of biodiesel from vegetable oils. *ChemSusChem* 2: 278–300.

Snåre, M., Kubičková, I., Mäki-Arvela, P. et al. (2006). Heterogeneous catalytic deoxygenation of stearic acid for production of biodiesel. *Industrial Engineering Chemistry Research* 45: 5708–5715.

Snåre, M., Kubičková, I., Mäki-Arvela, P. et al. (2007). Production of diesel fuel from renewable feeds: kinetics of ethyl stearate decarboxylation. *Chemical Engineering Journal* 134: 29–34.

Sotelo-Boyas, R., Trejo-Zarraga, F., and Hernandez-Loyo, F.d.J. (2012). Hydroconversion of triglycerides into green liquid fuels. *Hydrogenation* 8: 135–152.

Takeya, K., Konishi, H., Kawanarl, M. et al. (1997). Kinetic model for soybean oil hydrogenation using loop reactor. *Food Science and Technology International* 3: 10–16.

Tarhan, M.O. (1983). *Catalytic Reactor Design*. New York: McGraw-Hill.

Tirado, A. and Ancheyta, J. (2019). Defining appropriate reaction scheme for hydrotreating of vegetable oil through proper calculation of kinetic parameters. *Fuel* 242: 167–173.

Tirado, A. and Ancheyta, J. (2020a). Modeling of a bench-scale fixed-bed reactor for catalytic hydrotreating of vegetable oil. *Renewable Energy* 148: 790–797.

Tirado, A. and Ancheyta, J. (2020b). Scaling up the performance of a reactor model for hydrotreating vegetable oil from bench-scale to pilot-scale reactors. *Industrial and Engineering Chemistry Research* 59: 21712–21719.

Tirado, A., Ancheyta, J., and Trejo, F. (2018). Kinetic and reactor modeling of catalytic hydrotreatment of vegetable oils. *Energy Fuels* 32: 7245–7261.

Tirado, A., Trejo, F., and Ancheyta, J. (2020). Simulation of bench-scale hydrotreating of vegetable oil reactor under non-isothermal conditions. *Fuel* 275: 117960.

Tirado, A., Trejo, F., and Ancheyta, J. (2021). Prediction of temperature profiles for catalytic hydrotreating of vegetable oil with a robust dynamic reactor model. *Industrial Engineering Chemistry Research* 60: 13812–13821.

Tirado, A., Alvarez-Majmutov, A., and Ancheyta, J. (2022). Modeling and simulation of a multi-bed industrial reactor for renewable diesel hydroprocessing. *Renewable Energy* 186: 173–182.

Tyn, M.T. and Calus, W.F. (1975a). Diffusion coefficients in dilute binary liquid mixtures. *Journal of Chemical and Engineering Data* 20: 106–109.

Tyn, M.T. and Calus, W.F. (1975b). Estimating liquid molal volume. *Processing* 21: 16–17.

Vélez Manco, J.F. (2014). *Conceptual Design of a Palm Oil Hydrotreatment Reactor for Commercial Diesel Production [Escuela de Química y Petróleos]*. Medellin.

Verdier, S., Alkilde, O.F., Chopra, R. et al. (2019). *Hydroprocessing of Renewable Feedstocks-Challenges and Solutions*. Denmark: *Haldor Topsoe A/S*.

Yenumala, S.R., Maity, S.K., and Shee, D. (2017). Reaction mechanism and kinetic modeling for the hydrodeoxygenation of triglycerides over alumina supported nickel catalyst. *Reaction Kinetics, Mechanisms and Catalysis* 120: 109–128.

Zhang, S., Yan, Y., Li, T., and Ren, Z. (2009). Lumping kinetic model for hydrotreating of bio-oil from the fast pyrolysis of biomass. *Energy Sources, Part A: Recovery, Utilization, and Environmental Effects* 31: 639–645.

Zhang, H., Lin, H., Wang, W. et al. (2014). Hydroprocessing of waste cooking oil over a dispersed nano catalyst: kinetics study and temperature effect. *Applied Catalysis B: Environmental* 150–151: 238–348.

Zhang, B., Wu, J., Yang, C. et al. (2018). Recent developments in commercial processes for refining bio-feedstocks to renewable diesel. *Bioenergy Research* 11: 689–702.

7

Modeling of Slurry-Phase Hydrocracking Reactor

Cristian Calderón[1] and Jorge Ancheyta[2]

[1] *Tecnologico de Monterrey, School of Engineering and Sciences, Mexico State, Mexico*
[2] *Instituto Mexicano del Petróleo, Mexico City, Mexico*

7.1 Introduction

There are different types of reactors commonly used in the petroleum industry where the hydrocracking (HCR) of heavy oils is one of the main processes for converting a heavy carbonaceous feedstock to lower-boiling point products. Those reactors are fixed-bed reactors (FBR), ebullated-bed reactors (EBR), and slurry-phase reactors (SPRs). SPRs are more reliable to achieve high conversions and have shown superiority, especially in the treatment of hydrocarbons containing sulfurous compounds in exceedingly large quantities as well as large amounts of metals, carbon, and asphaltenes (Scheffer et al. 1998; Speight 2004; Bellussi et al. 2013).

SPR involves mixing of the feed oil with dispersed catalysts and hydrogen, whose purpose is the inhibition of coke formation by hydrogenating the coke precursor and removing heteroatoms (Zhang et al. 2007). Also, the catalyst acts as a supporter of coke, which reduces the coking of the reactor wall. At present there are several technologies for slurry-phase HCR processes on a pilot scale and even in industrial applications, which were reviewed by several authors (Bellussi et al. 2013; Zhang et al. 2007; Sahu et al. 2015) and are summarized in Table 7.1. Although SPRs have been used in different applications, in the particular case of HCR, detailed modeling and other aspects of the reactors are scarce because they are owned by manufacturers (Ancheyta 2013).

7.1.1 Characteristics of Slurry-Phase Reactors for Hydrocracking

7.1.1.1 Type of Reactors

SPRs are three-phase reactors that consist of a solid-phase catalyst (known as additive) suspended in a liquid in batch mode or it may move co-currently or counter-currently to the gas flow (Figure 7.1a). Generally, heterogeneous catalysts, typically transition metals (such as Mo, W, Fe, or other elements), are used in the process. The purpose of catalyst and hydrogen is the inhibition of coke formation by hydrogenating the coke precursor and removing heteroatoms. A catalyst with high activity will result in a high yield of light fuel oil and a low yield of coke. In SPR, the catalyst is added to the heavy oil and then the slurry is mixed with hydrogen in the reactor, typically operating at high temperature and pressure. Finally, the products leaving the reactor are separated before they are fractionated (Kriz et al. 2009).

Mathematical Modeling of Complex Reaction Systems in the Oil and Gas Industry, First Edition.
Edited by Jorge Ancheyta, Andrey Zagoruiko, and Andrey Elyshev.

Table 7.1 Main technologies for slurry-phase hydrocracking.

Technology	Pressure (MPa)	Temperature (°C)	Catalyst
VEBA-combi-cracking (VCC) process	15–27	440–485	Red mud (iron-containing material) and a fine coke powder of Bovey coals.
M-coke technology by Exxon Mobil	17	440	Phosphormolybdic acid and molybdenum naphthenate.
HDH technology by INTEVEP	7–14	420–480	An inexpensive natural ore located in Venezuela based on iron or molybdenum.
CANMET process	10–15	440–460	Iron sulfate.
SOC technology by Ashi Industrial	20–22	420–460	A highly dispersed superfine powder and a highly active transition-metal catalyst. With two components: Mo and an ultra-fine particle (black carbon).
Aurabon technology by UOP	20–22	450	Vanadium oxide and Iron sulfate.
EST technology by ENI	13–16	430	Highly dispersed oil-soluble molybdenum catalysts.
HCAT by HTIG	13–15	380–525	Synthesized molecular catalysts from metal–organic precursors such as iron pentacarbonyl or molybdenum 2-ethylhexanoate.
Fushun Research Institute of Petroleum and Petrochemicals	10–15	380–460	Synthesized oil-soluble catalyst.

Another configuration of SPR very common in industry is the slurry bubble column. Slurry bubble column reactors (SBCR) are multiphase systems in which the gas feed stream is continuously bubbled into the slurry phase (Figure 7.1b). In the simplest mode of operation, the liquid phase is stationary while gas is sparged through the vessel. The reactor is an empty vessel placed vertically, with a relationship between length-to-diameter ratio of at least 5 due to large ratios promoting higher conversions but also high pressure drop and low ratios favor higher gas throughputs. There are different internal configurations with devices to promote mass transfer (Zehner and Kraume 2005). SPR operating as a stirred-tank reactor (STR) is shown in Figure 7.1c. This vessel in batch or semi-batch configurations is typically used to perform experiments at a laboratory scale due to the low amount of reactants, small equipment, and easy operation characteristics. Most of these reactors are used to explore the properties of catalysts and to obtain kinetic parameters for a given kinetic model; however, mass transfer limitations, internal configuration of the reactor, or details of operating characteristics are not evaluated (Angeles et al. 2014).

Due to the particular operating conditions of heavy oil HCR, the typical configuration is in the SPR form where the feed oil is mixed with the gas before entering the reactor. Some of the general advantages and disadvantages of SPR for HCR of heavy oil are given below.

The advantages are as follows:

- Nearly isothermal operation.
- Easy temperature control.
- Good interphase contacting.

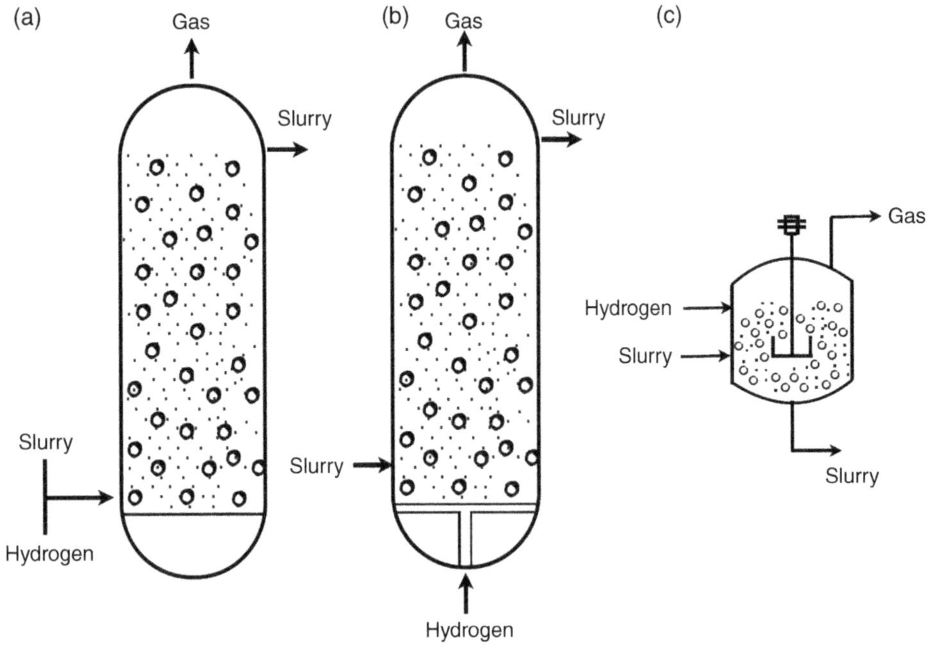

Figure 7.1 Configurations of SPR: (a) Slurry-phase reactor, (b) bubble column reactor, and (c) stirred-tank reactor.

- Large catalyst area.
- Large liquid holdup.
- High conversion rates.
- Operational flexibility.
- Low-pressure drop.
- Low construction and operational costs.
- Easy addition of catalyst.
- Uniform catalyst distribution inside the reactor.
- The catalyst can promote cracking and restrain deposition during the catalytic reactions.

The disadvantages are as follows:

- Difficult separation of solids and liquids.
- Uncertain scale-up.
- Plugging.
- Catalyst sedimentation and agglomeration.
- Formation of coke during reactions.

While this type of reactor offers numerous advantages, particular attention must be paid to potential issues such as plugging, which can arise from catalyst agglomeration or sedimentation at the reactor's bottom. This problem could be avoided if the flow conditions as well as process lines and the equipment assure that the catalyst is completely suspended. Also, using a catalyst of small particle size with similar density to the liquid phase would cause the particles to follow the motion of the liquid avoiding catalyst sedimentation. Another problem that causes plugging in HCR systems is the coke formation during the reactions. While it is indubitable that hydrogen plays an important role during the reaction inhibiting coke formation during the thermal cracking process, the catalyst is also responsible for avoiding the coke formation and prolonging the coke induction period. Therefore, in order to avoid this problem, improving catalyst characteristics is of significant importance.

7.1.1.2 Catalyst Properties

Highly dispersed catalyst particles have different advantages over the supported catalysts commonly used in hydroconversion and hydrotreating of crude oils; for example, they are less susceptible to deactivation during heavy oil upgrading allowing for the elimination of catalyst pore plugging issues, increase the accessibility of highly dispersed active sites by large size reactant molecules, and minimize diffusion control process during the reaction (Angeles et al. 2014; Kennepohl and Sanford 1996). Also, it was recently found that the catalyst deactivation in slurry-phase hydroconversion, in which unsupported catalysts are used, is likely different from the deactivation observed on supported catalysts (Rezair and Smith 2013).

There are two types of catalysts for slurry-phase HCR, heterogeneous solid powder catalysts and homogeneously dispersed catalysts classified as water-soluble catalysts and oil-soluble catalysts (Angeles et al.; Khulbe et al. 1989). In heterogeneous solid powder catalysts, the active catalytic phase is a solid mixed with the feed at the beginning of hydroprocessing. The homogeneously dispersed precursors consist of a catalyst added to the feed in the form of a precursor (water-soluble or oil-soluble non-catalytic compound) that transforms into the catalytic active phase after an intermediate step of activation in situ or during reaction conditions. Catalytic precursors have the advantage that prior being added to the reaction mixture, a dissolution can be made (in water or in oil according to the water-soluble or oil-soluble nature of the precursor). Then, the dissolution is mixed with the feed to form a dispersion where the precursor gets well distributed all over the reaction mixture. The former is no longer used because of the difficulty of separation and equipment wear caused by the high dosage (Noguera et al. 2012). On the other hand, water-soluble catalysts are preferred over oil-soluble catalysts due to their lower cost (Li et al. 1999). Homogeneous dispersed catalysts use transition-metal compounds typically molybdenum, cobalt, iron, and nickel as naphthenates or multi-carbonyl compounds. Molybdenum compounds are preferred to be used as homogeneously dispersed catalysts due to their high hydrogenation activity (Ancheyta et al. 2005a,b; Domokos et al. 2006; Liu et al. 2009; Zhu and Zhou 2011; Todorova et al. 2012).

Panariti et al. (2000) studied the effect of operating conditions over a wide range of catalyst loads, 0–5000 ppm. It was observed that at any level of reaction severity, the coke formation increased at high catalyst concentrations. To avoid this situation, a low catalyst load of around 50–250 ppm is recommended. Ortiz-Moreno et al. (2012) studied the effect of loads from 330 to 1000 ppm in HCR of Maya crude oil at mild conditions in a slurry batch reactor. It was first observed that catalyst load makes no difference in the thermal process; however, the product distribution during the HCR of heavy crude can be directed by changing the amount of catalyst and the operating temperature. On the other hand, with variations in catalyst loads, it was also possible to analyze the thermal and catalytic HCR on the liquid fractions and the different products obtained as shown by Ortiz-Moreno et al. (2012). It was concluded that the catalytic HCR could be divided into two general stages according to the conversion of vacuum residue (VR): below 50%, VR conversion is dominated by the catalytic reactions and the catalyst is able to inhibit the formation of both coke and asphaltenes; and above 50%, VR conversion is dominated by thermal reactions. In this stage the oil phase becomes incompatible to asphaltenes due to the decrease of resins and the increase of light ends, promoting the aggregation of the most dealkylated asphaltenes and its subsequent transformation to coke. The duration of the first stage is expected to be determined by the catalyst load, temperature of operation, and the initial ratio of light/heavy fractions in the feed. These results show that the optimal catalyst concentration of catalyst would depend on different variables such as operating conditions, as well as external factors; use of recycle, catalyst separation, economic balance, pitch specification, and even environmental issues.

7.1.2 SPR Modeling

In the modeling of SPRs due to the small diameter of the catalyst particles ($d_p \leq 150$ μm), it is assumed that there is a very good homogeneity between the solid and liquid phases. As a result, it is considered that the catalyst is part of the liquid forming a pseudo-homogeneous phase (slurry phase), and it is necessary to consider the profile of solid concentration as well as the settling velocity of the particles (Ancheyta 2013; Martínez et al. 2010). The slurry column hydrodynamics is very complex to model; however, if experimental or phenomenological correlations for the hydrodynamic variables are available, the analysis could be simpler. The dispersion and interfacial heat and mass transfer fluxes, which often limit the overall chemical reaction rates, are closely related to the fluid dynamic of the system through the liquid–gas contact area and the turbulence properties of the flow.

The development of mathematical models describing the behavior of catalytic processes involving three phases is complicated because it is necessary to consider various aspects, from the dispersion and interfacial gas–liquid and liquid–solid heat and mass transfer fluxes, which are closely related to the fluid dynamic and turbulence properties of the flow, to pore diffusion and reaction kinetics (Martínez et al. 2010). However, the knowledge of the system behavior in steady-state or dynamic operation is essential in determining proper reactor design and scale-up, and in correctly interpreting data in research and pilot-plant work as well as for the trial and start-up period and optimizing the operating conditions in manufacturing.

7.1.2.1 Classification

Froment et al. (2011) proposed a general classification for adiabatic and non-adiabatic FBR. This classification goes from the simplest model, which considers plug flow, to the most complicated, which involves axial and radial dispersion as well as interfacial and intrafacial gradients. In this classification, the catalyst is big enough to be considered as one phase. It also takes into account the bulk gas-phase temperature and concentration. These variables could be the same at the solid surface (pseudo-homogeneous models) or different (heterogeneous models) as shown in Figure 7.2.

For two-phase bubble columns and SBCR, the most popular classification is based on the $N + 1$ model proposed by Tomiyama (1998) and Tomiyama and Shimada (2001). This model considers $N + 1$ phases, where one phase corresponds to the slurry phase and the N phases correspond to gas bubbles of different sizes. While this classification is similar to the one proposed by Froment et al. (2011), it is based on the system hydrodynamics and not on the state variables T and C. Therefore, the classification presented in Figure 7.3 involves both classifications, in which it is first necessary to define the flow rate of the gas phase, either a simple model with one gas phase or a N gas phases. Once the flow of the gas phase is established, the next step is to determine if the reaction is carried out in the liquid phase (pseudo-homogeneous) or on the surface and inside of the catalyst (heterogeneous). This would depend on the structure and size of the catalyst; for example, with non-porous catalysts with very small particle diameters, the reaction takes place practically in the liquid phase. However, for large particle sizes and porous catalysts, the reaction is carried out in the solid.

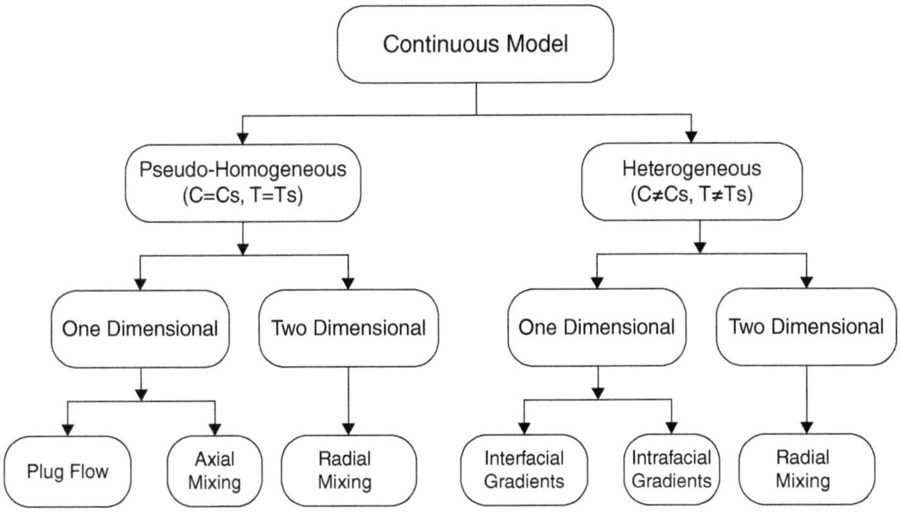

Figure 7.2 Classification of reactor models.

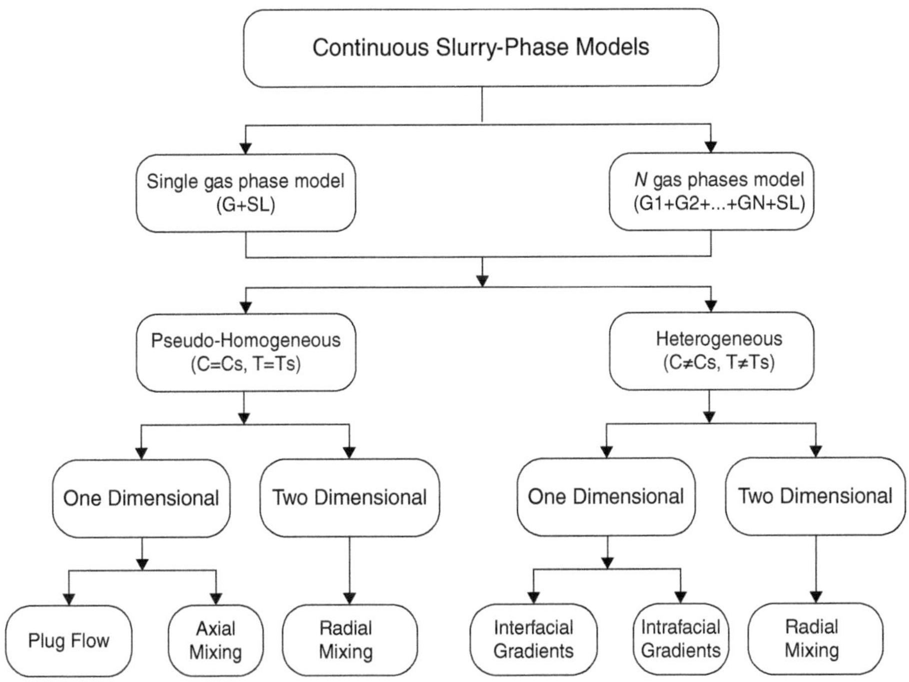

Figure 7.3 Classification of slurry-phase models based on hydrodynamics.

The flow regimes in bubble and slurry bubble columns are classified according to the superficial gas velocity. Two types of flow regimes are commonly observed in slurry and bubble reactors: homogeneous (bubbly flow) regime and heterogeneous (churn turbulent flow) regime. Another regime is called the slug flow regime, which has only been observed in small-diameter laboratory columns at high gas flow rates as can be observed in Figure 7.4.

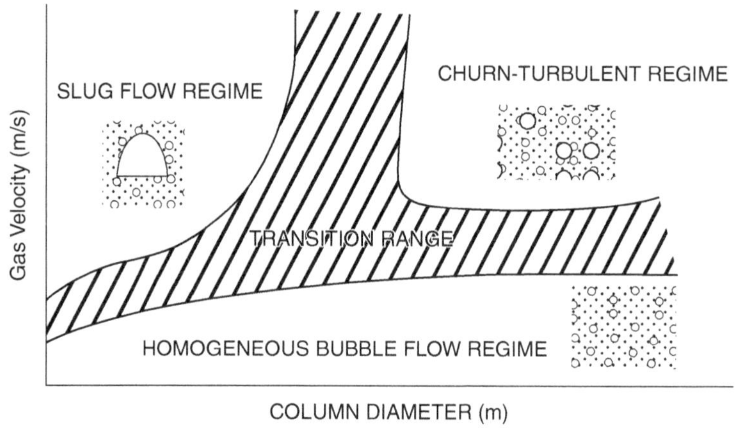

Figure 7.4 Regime flow for bubble and slurry-phase reactors (Adapted from Deckwer, et. al. (1980)).

The homogeneous flow regime is assumed when the high pressure of the system and low gas velocities prevent bubble coalescence. Then the single-phase model can be used. The homogeneous flow is established when the difference between the density of solids and liquid is small or if the liquid viscosity is high. Also operating at low gas flow rates where small bubbles (SB) of gas (1–10 mm) are uniformly distributed into the slurry phase ($L + S$). The bubble size and uniformity depend on the properties of the liquid, the design of the gas distributor, and the column diameter, which are defined for a uniform bubble size distribution in the axial and radial direction of the vessel.

N gas-phase models are used when the churn turbulent flow regime is found inside the reactor. This flow is presented when there is a large gas fraction in a system with a high gas and low liquid velocity and is frequently observed in large-diameter columns at an industrial scale. Typically, at the beginning of this regime, the number of bubbles is low, which is called as ideally separated bubble flow. In this type of flow, the bubbles do not interact with each other directly or indirectly, but as the number of bubbles increases, they start colliding with each other and their size gets reduced. Suddenly a situation comes when they tend to coalesce to form cap bubbles, and the new flow pattern formed is called churn turbulent flow, which has a broad size distribution. This model considers N bubble types plus the slurry phase; therefore, each bubble type would have their respective mass, heat, and momentum equations in order to describe the system dynamics. Nevertheless, previous studies based on the extend of the two-phase ("dilute" and "dense" phases) model proposed by van Deemter (1961) have shown that in a slurry bubble column operating in the churn turbulent flow regime the gas phase can be split up in a "large" bubble population and a "small" bubble population (27–29). In this heterogeneous system, SB combine in clusters to form large bubbles (LB) (20–70 mm). These LB travel up through the column at high velocities (1–2 m/s), in a more or less plug-flow manner. Though this is still a very complicated model, it would lead to a considerable simplification to the fluid dynamic equations of the system (Kantarci et al. 2005; Deckwer et al. 1980).

Due to the high temperature and pressure conditions in heavy oil HCR, large amount of hydrogen, as well as oil and catalyst properties, the typical operating regime is the homogeneous flow. Therefore, when modeling these systems, one of the main considerations is to establish a homogeneous gas flow inside the reactor. Hence in this review, only models concerning homogeneous gas flow will be discussed.

7.1.2.2 Model Complexity

There is no general rule to select the level of sophistication of a reactor model. The complexity of a model is closely related to the purpose of the investigation. In most cases, a typical strategy is selecting the simplest model and adding complexity as the error between experimental and calculated data is reduced. The simplest models are built under ideal considerations neglecting gradients in concentration in all spatial directions (perfect mixing) or recognizing them only in the principal flow direction (plug flow), also a common assumption is a uniform concentration of the solids throughout the reactor. These models can be used in SPR operated as STR. On the other hand, while slurry and bubble column performance often can be fitted with axial dispersion models (ADMs), decades of research have failed to produce a predictive correlation for the axial dispersion coefficient. The so-called computational fluid dynamics (CFD) models solve the fluid dynamics with a numerical solution of the Navier–Stokes equations and a set of partial differential equations (PDEs). These are the most sophisticated models but also the most difficult to solve due to their high nonlinearity, even two-phase transient simulations require high computational cost (Pfleger and Becker 2001). Most CFD models applied to slurry and bubble columns are focused on the hydrodynamics of the system, neglecting mass and heat transfer effects in some cases, which clearly represents a limitation to reactive systems. From the point of view of reaction engineering, a complex model for the field of velocity inside the reactor would not offer advantages if the profile of concentrations of substances involved is described poorly.

The reliability of a model, either ideal, ADM, or CFD, is a function of the validation method such as testing it with independent data and thermodynamic properties at the reaction conditions.

7.1.2.3 Models for Slurry Reactors

Homogeneous Bubble Flow Regime Models Reactor models that feature a practical way to design two/three-phase reactors and that provide a basic understanding of the chemical process on the semi-industrial or even industrial scale have been published only rarely in the usual scientific literature (Jung et al. 2010), and SPRs are not the exception. Unfortunately, only CFD models are often implemented in as much detail as possible for the simulation of slurry HCR reactors, no matter that the complex modeling and computational effort required are extremely time-consuming and costly. However, it has been found that for simple reaction systems such as hydrogenation, three-phase slurry reactor dynamics can be well represented by ADMs where deviation from plug flow is described using an axial dispersion coefficient. For example, de Toledo et al. (2001) proposed two dynamic non-isothermal ADMs applied to describe the dynamic behavior of the reactor during the hydrogenation of O-cresol on Ni/SiO_2 catalyst. The authors conclude that the most complete model that includes internal mass transfer reproduces the dynamic behavior of the reactor in a better way; however, it is possible to use the simplified model if there is not enough experimental information about catalyst properties, in fact this model is the most frequently used in different reactions involving different types of catalyst particles. Both models have been used in different applications of control and optimization (Bahroun et al. 2010; Melo et al. 2005; Costa et al. 2007; Li et al. 2010; Bahroun et al. 2010; Mariano et al. 2011). Shahrzad et al. (2009) presented a dynamical ADM that depicts CO_2 hydrate formation in a slurry bioreactor. The model considers mass transfer phenomena between gas and liquid as well as liquid to solid phases. Also, the particle velocity is considered along with the term of reaction. Chen et al. (2010) proposed a steady state ADM for the direct dimethyl ether (DME) synthesis process from syngas in which the dispersion and velocity of solid particles were introduced and mass transfer resistance was ignored in liquid–solid phase. The authors concluded that the particle and reactor diameters are the two main factors influencing concentration distribution uniformity. On the other hand, for HCR reactions, it has been tested that the field of velocity does not influence on a great way the conversion along the reactor (Matos and

Nunhez 2006). Therefore, it is expected that an ADM with adequate kinetic and hydrodynamic parameters would be enough to describe slurry-phase HCR reactors.

The work by Parulekar and Shah (Parulekar and Shah 1980) attends in a practical way the HCR in SPRs. This model consists of gas, liquid, and solid mass balance equations formulated under the following considerations:

- Isothermal operation.
- Gas phase is assumed to behave according to the ideal gas law.
- All reactions are purely catalytic under the conditions of operation considered.
- The ADM is assumed to be applicable in the case of non-volatile liquid components.
- The gas phase is assumed to move as plug flow.
- The reactor is of concurrent up-flow type.
- All the global reaction rates are expressed in terms of the gas-phase concentration of hydrogen because it is difficult to measure the liquid-phase (oil in the case of HCR) concentration of hydrogen.

They proposed a kinetic model based on three components (AL, heavy oil; BL, light oil; and CG, volatile product) and an HCR reaction, all represented by a simple power-law approach Eqs. (7.1–7.4). Unfortunately, the authors do not give information about catalyst properties.

$$A_L \rightarrow B_L \; r_1 = k_1 C_A C_S \tag{7.1}$$

$$A_L \rightarrow C_G \; r_2 = k_2 C_A C_{\mathrm{H_2}} C_S \tag{7.2}$$

$$B_L \rightarrow C_G \; r_3 = k_3 C_B C_{\mathrm{H_2}} C_S \tag{7.3}$$

$$H_2 + L \rightarrow HL \; r_4 = k_4 C_{\mathrm{H_2}} C_S \tag{7.4}$$

The model equations are as follows:

$$\text{Gas phase H}_2\text{: } \frac{d(U_G C_{\mathrm{H_2}})}{dz} = -k_4 C_{\mathrm{H_2}} C_S \varepsilon_L \tag{7.5}$$

$$\text{Liquid phase } A\text{: } \frac{d}{dz}\left(D_a^L \frac{dC_A}{dz}\right) - \frac{d(U_L C_A)}{dz} - k_1 C_A C_S \varepsilon_L - k_2 C_A C_{H_2} C_S \varepsilon_L MM_C = 0 \tag{7.6}$$

$$\text{Liquid phase } B\text{: } \frac{d}{dz}\left(D_a^L \frac{dC_B}{dz}\right) - \frac{d(U_L C_B)}{dz} + k_1 C_A C_S \varepsilon_L - k_3 C_B C_{H_2} C_S \varepsilon_L MM_C = 0 \tag{7.7}$$

$$\text{Solid distribution } C_S\text{: } \frac{d}{dz}\left(D_a^S \frac{dC_S}{dz}\right) - \frac{d}{dz}\left[U_{SL} - ((1-\varepsilon_G)U_{tS})C_S\right] = 0 \tag{7.8}$$

The major complication in solving these equations is that all hydrodynamic parameters (ε_G, ε_L, ε_S, D_a^S, and D_a^L) and superficial velocities (U_G, U_L, and U_{SL}) vary axially. Therefore, dynamic equations for axial velocities are needed. Thus, it was assumed that superficial velocities vary as a function of the reactive absorption rate into the catalyst and product desorption rate (Eqs. 9 and 10).

$$C_T^G \frac{dU_G}{dz} = C_{\mathrm{H_2}} C_S \varepsilon_L (k_2 C_A + k_3 C_B - k_4) \tag{7.9}$$

$$\rho_L \frac{dU_L}{dz} = C_{\mathrm{H_2}} C_S \varepsilon_L (k_4 MM_{\mathrm{H_2}} - (k_1 C_A + k_2 C_B) MM_C) \tag{7.10}$$

Though this model is limited by particle sizes between 75 and 250 μm with low solid loading ($\varepsilon_S \leq 0.15$) and a simple kinetic model, it is possible to observe the behavior in steady state for the different variables like pressure, temperature, and dimensions of the reactor. How they give an increase in product yield and a maximum in product yield can be found by increasing the rest of the analyzed variables. Thus, the optimum operating conditions in the slurry-phase HCR reactor can be established. Other quantitative results show the behavior of solid concentration along the reactor. In this manner, it is observed that for low particle diameters, the solids distribution along the reactor remains practically constant for a given liquid velocity. On the contrary, when particle diameter is increased, an accumulation of particles is noted at the bottom of the reactor. Although the profile of solids concentration could vary as a function of liquid velocity and column diameter, these variables have no great influence on the particle distribution. Certainly, this is a simple model; however, it could be used as a reference for future works in slurry-phase HCR considering more detail kinetic models as well as transitory state, gas–liquid mass transfer, or even more complete ADMs.

Carbonell and Guirardello (1997) studied an isothermal reactor based on the continuity and momentum balance equations using an Eulerian approach applied in the HCR of heavy oils in a slurry-phase reactor operating at severe conditions of temperature (350–500 °C) and pressure (7–25 MPa). They considered that the thermal cracking reactions do not interfere with the hydrodynamics of the system due to the high mixing of the liquid phase and the low gas absorption rate. Under these conditions, a single-bubble class model could be assumed. Their simulations were performed in two steps: first, fluid dynamic variables were determined as a function of radial position, and then oil cracking conversion was obtained. The power-law kinetics was based on experimental evidence. Feed and products are divided into several boiling range lumps, where each lump is considered a single chemical species with a single cracking rate constant. The model consists of the following assumptions:

- Steady-state operation.
- A symmetric turbulent axial flow.
- Single-bubble model.
- A zero model turbulence in the mixing bed.
- A parabolic radial distribution for gas holdup.
- Dominant convection in the axial dispersion.
- Dominant dispersion in the radial dispersion.
- First-order irreversible reactions.

The hydrodynamic equations for the slurry and the gas phases are:

$$\text{Slurry phase: } \frac{1}{r}\frac{d}{dr}\left(\varepsilon_{SL} r \mu_{SL}^{\text{eff}} \frac{du_{SL}}{dr}\right) - \varepsilon_{SL}\frac{dP}{dz} - g\rho_{SL} + F = 0 \tag{7.11}$$

$$\text{Gas phase: } \frac{1}{r}\frac{d}{dr}\left(\varepsilon_{G} r \mu_{G}^{\text{eff}} \frac{du_{G}}{dr}\right) - \varepsilon_{G}\frac{dP}{dz} - g\varepsilon_{G}(\rho_G - \rho_{SL}) - F = 0 \tag{7.12}$$

where F is the interfacial drag force (N/m^3 reactor) proposed by Torvik and Svendsen (1990), v axial velocity (m/s), r radial position, ε holdup (dimensionless), μ^{eff} effective viscosity (Pa · s), and R_C radius of the reactor. The mass balance is:

$$\varepsilon_L \rho_L u_L \frac{dC_i}{dz} = \alpha_i k \rho_L C_i + \sum \beta_{m,i} k \rho_L C_i \tag{7.13}$$

Despite isolating the hydrodynamic behavior of the system from the chemical reactions, concordance with experimental data was obtained, although a more detailed model, with a more complex study of the kinetics, is required for scaling up and improvements in the operation.

Matos and Guirardello (2002) presented an isothermal CFD model that includes momentum equations, based on the kinetic network of six lumps presented by Krishna and Saxena (1989) for thermal HCR reactions. The lumps are heavy and light aromatics, naphthenes, and paraffins. The HCR is assumed to be thermal, and the catalyst is only used to avoid coke formation and for the removal of heteroatoms. The main assumptions are a pseudo-homogeneous system ($SL + G$) where the solid distribution is not considered due to the very small diameter and the small terminal velocity of the particles, the gas phase is composed mainly of hydrogen, and oil is completely saturated with dissolved hydrogen. Due to hydrogen being found in excess inside the reactor, the reaction is independent of the hydrogen consumption. Because of the lack of experimental information on HCR systems, this model was validated by comparing different experimental data available in the literature for the air–water system. Though this model can describe the behavior of an SPR in a good manner, it is quite complicated to solve because it includes the velocity distribution for gas and slurry phases along the reactor. However, a simple approximation can be obtained by an ADM considering average velocities and the proper parameter values as shown in Figure 7.5.

Matos and Nunhez (2006) presented a fluid dynamic model based on the same considerations as in Matos and Guirardello (2002) using the same kinetic model of Krishna and Saxena (1989), but in this case, the work is focused on how the feed input affects the flow fields and the reactor conversion. Although unsupported by experimental data, their theoretical results show that the fluid dynamic fields inside the bubble column reactor do not affect considerably the reactor conversion.

Recently, Matos et al. (2009) presented a CFD model to estimate operating conditions that minimize sulfur and organometallics concentrations before the petroleum is processed. In the modeling process, it is assumed that the difference between the catalyst and the slurry phase is negligible, preventing catalyst settling. Therefore, a fluid dynamic model capable of suspending the catalyst was deemed sufficient. The model considers the petroleum to be composed of pseudo-components, and HDS and hydrodemetallization (HDM) reactions have a first-order kinetics decomposition controlled by diffusion and occurring on the catalyst pores. As chemical components in the crude mixture present very low reaction rates, there is low gas consumption and the mass transfer between the phases is negligible. However, the main contribution in this work is that the model includes a

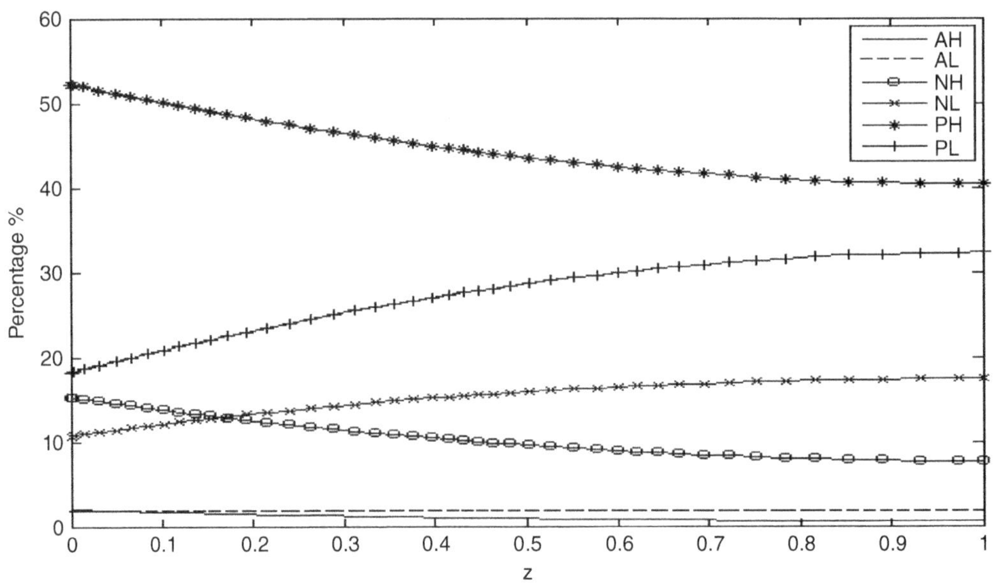

Figure 7.5 Axial dispersion model kinetics reported Adapted from Krishna and Saxena (1989). AH, Heavy aromatics; AL, light aromatics; NH, heavy naphthenes; NL, light naphthenes; PH, heavy paraffins; PL, light paraffins.

kinetic expression for the activity reduction of the catalyst as a function of the catalyst pore characteristics.

Heterogeneous Bubble Flow Regime Models Operation at low gas velocities in a homogeneous bubbly flow regime was considered to be preferred in industrial SPRs. Furthermore, most of the literature models have been focused on operation at relatively low superficial gas velocities (below 0.10 m/s). However, the growth of industrial demand caused the plants to operate at higher production rates, which caused the processes to involve increasing operating conditions such as gas and liquid flows. As a consequence, this generates heterogeneous flow conditions inside the reactor.

So far, the latest study for SBCR modeling is presented by Khadem-Hamedani et al. (2015). They showed a comparison between models in homogeneous and heterogeneous flow regimes for the hydrodesulfurization (HDS) of diesel fuel. The mass and energy balances, as well as the model parameters, are based on the considerations proposed by Mills et al. (1996) and de Swart and Krishna (2002). Both models reproduce the reactor behavior in a good way; however, as it was expected, the more detailed model (heterogeneous model) shows better results in the validation with experimental data. These results are of great importance because they demonstrate that the parameters proposed by Mills et al. (1996) and de Swart and Krishna (1997) generated for *F–T* synthesis for heterogeneous regime can be used in HDS of diesel, which have similar conditions of pressure and temperature to HCR.

7.2 Proposed Generalized Model

7.2.1 Equations for the Generalized Model

Most mass balance equations as well as motion equations are derived from the general equation of continuity and motion, and slurry bubble columns are not the exception. While there are different ways of representing these equations, good approximations of the behavior of this type of reactor have been obtained using an Eulerian approach. The fundamental form of the multi-fluid continuity equation for phase *f* has been presented by several authors (Tomiyama and Shimada 2001; Pfleger and Becker 2001; Jakobsen et al. 2005; Pfleger et al. 1999). In this review, the equations are presented for each component in the *f* phase, thus the reaction rate and convective mass transfer are also considered. The equations are:

$$\frac{\partial\left(\varepsilon_f C_i^f\right)}{\partial t} + \nabla \cdot \left(\varepsilon_f C_i^f \boldsymbol{u}_f\right) = \nabla \cdot \left(\varepsilon_f D_f \nabla C_i^f\right) + \sum_{f=1}^{F-1} N_i^{(f_1, f_{f+1})} + r_i' \tag{7.14}$$

$$\begin{aligned}\frac{\partial\left(\varepsilon_f \rho_f Cp_f T_f\right)}{\partial t} + \nabla \cdot \left(\varepsilon_f \rho_f Cp_f T_f \boldsymbol{u}_f\right) &= \nabla \cdot \left(\varepsilon_f \lambda_f \nabla T_f\right) + \mu\Phi + \Delta H_r + q\left(T_f, T_W\right) \\ &\quad + q\left(T_f, T_I\right) + q\left(T_f, T_{S_{\text{sur}}}\right)\end{aligned} \tag{7.15}$$

This governing set of PDEs consists of the continuity equations for the $N+1$ phases for i components and the energy and momentum equations for $N+1$ phases. It holds that:

$$\frac{\partial\left(\varepsilon_f \rho_f\right)}{\partial t} + \nabla \cdot \left(\varepsilon_f \rho_f \boldsymbol{u}_f\right) = 0 \tag{7.16}$$

$$\sum_{f}^{F} \varepsilon_f = 1 \tag{7.17}$$

$$\varepsilon_f u_f = U_f \tag{7.18}$$

Eq. (7.14) shows the classical terms of transitory state, convective and molecular fluxes, as well as the terms $\sum_{f=1}^{F-1} N_i^{(f_1 f_{f+1})}$ which is related to the convective fluxes between phase resistances for each component, and the reaction rate r_i'. In energy Eq. (7.15), appear the terms related to the transient state temperature, heat exchange caused by convective flow, conduction, viscous dissipation energy (usually negligible, except in systems with large velocity gradients) (Bird et al. 2002), the generation term due to the chemical reaction, the heat fluxes, function of the temperature of each phase, the temperature of fluid exchange, interfaces, and solid temperatures. Being strict, this set of equations consists of three general equations, for the gas, liquid, and solid phases, but for SPRs, it is commonly considered to simplify the gas and liquid phases as a continuous phase in which the temperature is the same due to the large area of contact and favorable transfer coefficients (de Toledo et al. 2001).

As mentioned before, this is a very complex PDE system, particularly for HCR of heavy petroleum where various reactions occur simultaneously, for example, HDS, HDM, hydrodenitrogenation (HDN), and hydrodeasphaltenization (HDA). However, some considerations can be made in order to simplify this model. Taking into account that the model of SPRs is similar to EBR (Martínez et al. 2010) and the reacting system is the same as that presented by Mederos et al. (2009), some of their general criteria can be adapted. For example, though the values of mass and heat transfer coefficients are distinct, the expressions for mass and heat transfer resistances between phases are expected to be the same. Also due to the chemical reactions being merely catalytic, the conversion will be directly associated with the concentration of catalyst particles inside the reactor. However, unlike EBR in SPR, the solid catalyst is found in the wake of the liquid forming an L–S suspension. This unique characteristic generates a greater degree of difficulty in the approach of the mathematical model, since system hydrodynamics becomes more complicated. For the bubble column slurry operation, two suspension states may exist: namely, complete suspension in which all particles are in suspension, commonly described by the axial sedimentation-dispersion model (SDM), and homogeneous suspension, in which the particle concentration is uniform throughout the reactor (Chaudhari and Ramachandran 1980). In the proposed generalized model, the SDM is taken into account. Other considerations are as follows:

- Heterogeneous bubble flow, composed of SB and LB.
- The liquid–solid ($L + S$) suspension behaves as a homogeneous phase due to the small catalyst particle size, whereas the gas phase bubbles up through the suspension.
- Liquid and gas properties (mass and heat dispersion coefficients, specific heats, densities, and viscosities) are constant along the whole catalytic bed.
- Catalyst properties (porosity, size, activity, effectiveness, etc.) are constant along the whole catalytic bed.
- Due to the small particle size, wetting efficiency is considered to be the unity.
- Inside the catalyst particle, mass and heat-effective diffusivity coefficients may also be assumed constant.
- The liquid phase is assumed to be in thermal equilibrium with the gas phases. Then the energy equation can be formulated for the continuous "f" phase defined as ($SB + LB + L$).

A detailed description of each term given in Tables 7.2 and 7.3 is shown by Mederos et al. (2009), except for the solid dispersion terms that are presented below.

Table 7.2 Generalized mass balance equations.

	Accumulation	Convective	Axial dispersion	Radial dispersion	G–L	G–S	SB-LB
(A) Gas phase *SB*	$\frac{\partial \varepsilon_{SB} C_i^{SB}}{\partial t} =$	$\pm \frac{\partial \varepsilon_{SB} u_{SB} C_i^{SB}}{\partial z}$	$+ \frac{\partial}{\partial z}\left(\varepsilon_{SB} D_a^{SB} \frac{\partial C_i^{SB}}{\partial z}\right)$	$+ D_r^{SB}\left(\frac{\partial^2 (\varepsilon_{SB} C_i^{SB})}{\partial r^2} + \frac{1}{r}\frac{\partial (\varepsilon_{SB} C_i^{SB})}{\partial r}\right)$	$-(K_L a)_{SB,i}\left(\frac{C_i^{SB}}{He_i} - C_i^L\right)$	$-(k_{GS} a)_{SB,i}\left(\frac{C_i^{SB}}{He_i}\right) - C_G^{S_{sur}}$	$-(U_{LB} - U_{SB})\frac{K}{H_C}(C_i^{SB} - C_i^{LB})$
(B) Gas phase *LB*	$\frac{\partial \varepsilon_{LB} C_i^{LB}}{\partial t} =$	$\pm \frac{\partial \varepsilon_{LB} u_{LB} C_i^{LB}}{\partial z}$	$+ \frac{\partial}{\partial z}\left(\varepsilon_{LB} D_a^{LB} \frac{\partial C_i^{LB}}{\partial z}\right)$	$+ D_r^{LB}\left(\frac{\partial^2 (\varepsilon_{LB} C_i^{LB})}{\partial r^2} + \frac{1}{r}\frac{\partial (\varepsilon_{LB} C_i^{LB})}{\partial r}\right)$	$-(K_L a)_{LB,i}\left(\frac{C_i^{LB}}{He_i} - C_i^L\right)$	$-(k_{GS} a)_{LB,i}\left(\frac{C_i^{LB}}{He_i}\right) - C_G^{S_{sur}}$	$+(U_{LB} - U_{SB})\frac{K}{H_C}(C_i^{SB} - C_i^{LB})$
	Accumulation	**Convective**	**Axial dispersion**	**Radial dispersion**	***G–L***	***L–S***	
(C) Liquid phase volatile	$\frac{\partial \varepsilon_L C_i^L}{\partial t} =$	$\pm \frac{\partial \varepsilon_L u_L C_i^L}{\partial z}$	$+ \frac{\partial}{\partial z}\left(\varepsilon_L D_a^L \frac{\partial C_i^L}{\partial z}\right)$	$+ D_r^L\left(\frac{\partial^2 (\varepsilon_L C_i^L)}{\partial r^2} + \frac{1}{r}\frac{\partial (\varepsilon_L C_i^L)}{\partial r}\right)$	$-(K_L a)_{SB,i}\left(C_i^L - \frac{C_i^{SB}}{He_i}\right)$ $-(K_L a)_{LB,i}\left(C_i^L - \frac{C_i^{LB}}{He_i}\right)$	$k_i^{LS} a_S (C_i^L - C_i^{S_{sur}})$	
(D) Liquid phase non-volatile	$\frac{\partial \varepsilon_L C_i^L}{\partial t} =$	$\pm \frac{\partial \varepsilon_L u_L C_i^L}{\partial z}$	$+ \frac{\partial}{\partial z}\left(\varepsilon_L D_a^L \frac{\partial C_i^L}{\partial z}\right)$	$+ D_r^L\left(\frac{\partial^2 (\varepsilon_L C_i^L)}{\partial r^2} + \frac{1}{r}\frac{\partial (\varepsilon_L C_i^L)}{\partial r}\right)$		$k_i^{LS} a_S (C_i^L - C_i^{S_{sur}})$	
(E) Solids dispersion	$\frac{\partial \varepsilon_{SL} C_S}{\partial t} =$	$\pm \frac{\partial}{\partial z}((\varepsilon_{SL} u_S - U_{SL}) C_S)$	$+ \frac{\partial}{\partial z}\left(\varepsilon_{SL} D_a^S \frac{\partial C_S}{\partial z}\right)$				
	Accumulation		**Intraparticle diffusion**	***G–S* transfer**	***L–S* transfer**	**Generation**	
(F) Liquid phase in solid surface	$(1-\varepsilon_B)\frac{\partial \varepsilon_S C_{L,i}^{S_{sur}}}{\partial t} =$				$+ k_i^{LS} a_S (C_i^L - C_{L,i}^{S_{sur}})$	$+ \rho_S \sum_{j=1}^{N_{RL}} \eta_j^L \nu_{ij}^L r_j^{\prime L}$	
(G) Gas phase in solid surface	$(1-\varepsilon_B)\frac{\partial \varepsilon_S C_{G,i}^{S_{sur}}}{\partial t} =$			$+ k_i^{GS} a_S (C_i^* - C_{G,i}^{S_{sur}})$		$+ \rho_S \sum_{j=1}^{N_{RG}} \eta_j^G \nu_{ij}^G r_j^{\prime G}$	
(H) Liquid phase in solid inner	$\frac{\partial \varepsilon_S C_{L,i}^{S_{inn}}}{\partial t} =$		$+ \frac{D_{ei}^S}{\xi^2}\frac{\partial}{\partial \xi}\left(\xi^2 \frac{\partial C_{L,i}^{S_{inn}}}{\partial \xi}\right)$			$+ \rho_S \sum_{j=1}^{N_{RL}} \nu_{ij}^L r_j^{\prime L}$	
(E) Gas phase in solid inner	$\frac{\partial \varepsilon_S C_{G,i}^{S_{inn}}}{\partial t} =$		$+ \frac{D_{ei}^S}{\xi^2}\frac{\partial}{\partial \xi}\left(\xi^2 \frac{\partial C_{G,i}^{S_{inn}}}{\partial \xi}\right)$			$+ \rho_S \sum_{j=1}^{N_{RL}} \nu_{ij}^L r_j^{\prime G}$	

Table 7.3 Generalized heat balance equations.

	Accumulation	Convective	Axial dispersion	Radial dispersion	Conductive	f–S transfer	f–W transfer
(A) Gas phase	$Cp_G\rho_G \frac{\partial(\varepsilon_G T)}{\partial t} =$	$\pm Cp_G\rho_G \frac{\partial(\varepsilon_G u_G T)}{\partial z}$	$+ \frac{\partial}{\partial z}\left(\varepsilon_G \lambda_a^G \frac{\partial T}{\partial z}\right)$	$+ \lambda_r^G \frac{\partial^2(\varepsilon_G T)}{\partial r^2} + \frac{1}{r}\frac{\partial(\varepsilon_G T)}{\partial r}$	$-\Delta H_{vi}$	$-h_{GS}a_S(T - T_S^S)$	$-h_{GW}\frac{A_W}{V}(T - T_W)$
(B) Liquid phase	$Cp_L\rho_L \frac{\partial(\varepsilon_L T)}{\partial t} =$	$\pm Cp_L\rho_L \frac{\partial(\varepsilon_L u_L T)}{\partial z}$	$+ \frac{\partial}{\partial z}\left(\varepsilon_L \lambda_a^L \frac{\partial T}{\partial z}\right)$	$+ \lambda_r^L \frac{\partial^2(\varepsilon_L T)}{\partial r^2} + \frac{1}{r}\frac{\partial(\varepsilon_L T)}{\partial r}$	$+\Delta H_{vi}$	$-h_{LS}a_S(T - T_S^S)$	$-h_{LW}\frac{A_W}{V}(T - T_W)$
	Accumulation	**Convective**	**Axial dispersion**	**G–S transfer**	**L–S transfer**	**Generation**	
(C) Isothermal solid phase	$Cp_S \frac{\partial(\varepsilon_S C_S T_S)}{\partial t} =$	$\pm Cp_S \frac{\partial(\varepsilon_S C_S u_{SL} T_S)}{\partial z}$	$+ \frac{\partial}{\partial z}\left(\varepsilon_S \lambda_a^S \frac{\partial T_S}{\partial z}\right)$	$+h_{GS}a_S(T - T_S)$	$+h_{LS}a_S(T - T_S)$	$\rho_B\zeta\left[\sum_{j=1}^{NRL} -\left(\Delta H_{rj}^L\right)\eta_j^L r_j'^L + \sum_{j=1}^{NRG} -\left(\Delta H_{rj}^G\right)\eta_j^G r_j'^G\right]$	
	Accumulation	**Axial dispersion**	**f–W transfer**				
(D) Thermowell	$Cp_W\rho_W \frac{\partial T_W}{\partial t} =$	$+ \lambda_W \frac{\partial^2 T_W}{\partial z^2}$	$+ h_{GW}\frac{A_W}{V}(T - T_W) + h_{LW}\frac{A_W}{V}(T - T_W)$				
	Accumulation	**Intraparticle transfer**		**Generation**			
(E) Non-isothermal solid phase	$Cp_S\rho_S \frac{\partial T_S}{\partial t} =$	$+ \lambda_e^S\left(\frac{\partial^2 T_S}{\partial \xi^2} + \frac{2}{r}\frac{\partial T_S}{\partial \xi}\right)$		$\rho_S\left[\sum_{j=1}^{NRL} -\left(\Delta H_{rj}^L\right) r_j'^L + \sum_{j=1}^{NRG} -\left(\Delta H_{rj}^G\right) r_j'^G\right]$			

7.2.2 Solids Concentration

Eq. (E) in Table 7.4 describes the dynamic behavior of solid concentration in a solid–liquid mixture along the reactor through the SDM. This model considers a solids axial dispersion flux and a solids sedimentation flux superimposed on the average slurry convection flux. One of the main variables in this equation is the particle settling velocity, which is the result between the liquid and particle terminal settling velocities as shown in Figure 7.6. Terminal settling velocity u_{tS} is defined as the highest velocity reach for particles immersed in a fluid; it occurs when the drag and gravity forces are equal. For spherical shapes it typically has the form:

$$u_{tS} = \frac{(\rho_S - \rho_L) g d_p^2}{18\mu_L} \tag{7.19}$$

By definition, the terminal settling velocity could be positive or negative depending on the density difference between solid and liquid phases and on the coordinate system adopted. On the other hand, the prediction of the particles settling velocity is found in the literature given by different correlations; for example, Parulekar and Shah (1980) and Kojima et al. (1989) consider the settling velocity to be the same as the terminal settling velocity. Recently, Sehabiague et al. (2008) applied the correlation proposed by Kato et al. (1972), where the particle settling velocity is proportional to the superficial gas velocity and to the liquid fraction in the slurry phase:

$$u_S = 1.33 u_{tS} \left(\frac{U_G}{u_{tS}}\right)^{0.25} \phi_L^{2.5} \tag{7.20}$$

Most of the models that use the SDM consider a constant average value of the liquid and gas superficial velocities, then an analytical solution of this equation is possible as a function of the Peclet particle number. The mass balance on the external surface of catalyst particles is given in Eqs. (F) and (G) in Table 7.4. The gradients are the product of an effective diffusion of liquid and gas phases through the catalyst surface. It also has the generation terms, which include the catalyst effectiveness factor associated to the reaction in the surface and at the inner of the catalyst particle; however, for SPRs, the catalyst commonly used in FBR is crashed to generate small catalyst diameters, which are known to have a high catalyst effectiveness (Macías and Ancheyta 2004), then it could be assumed to be the unity.

7.2.3 Initial and Boundary Conditions

All transitory mass and energy balance equations presented in Tables 7.2 and 7.3 have initial ($t = 0$) and boundary conditions ($t > 0$), which relate the surface properties to the bulk properties of the reacting system. Considering a symmetric reactor of radius R in which gas, liquid, and solid phases of radius $\xi = d_{pe}/2$ enter at the bottom of the reactor at $z = 0$ and that the reactor outlet is located at $z = H$, all coordinate system has been fixed. The initial conditions for the transitory state are typically set to the inlet properties of the system:

$$C_i^f = C_i^{f0}, C_S = C_S^0, C_{fi}^{S_{\text{sur}}} = C_{fi}^{S_{\text{sur}}0}, C_{fi}^{S_{\text{inn}}} = C_{fi}^{S_{\text{inn}}0}, T_f^0 = T_f^0 \tag{7.21}$$

These conditions could vary depending on the considerations made for the system under study; for example, in some cases, it is considered that the liquid is saturated with the gaseous compound, while in others there is no initial concentration of gas in the liquid phase. Also, the catalyst activity

Table 7.4 Boundary conditions of the generalized model.

Condition	Gas phase	Liquid phase	Solid phase	*i* at solid surface	*i* at solid inner
$z=0,\ 0\le r\le R$	$D_a^G\dfrac{dC_i^G}{dz}=u_G\left(C_i^G-C_i^{G0}\right)$	$D_a^L\dfrac{dC_i^L}{dz}=u_L\left(C_i^L-C_i^{L0}\right)$	$D_a^S\dfrac{dC_S}{dz}=u_{SL}\left(C_S-C_S^0\right)$	$C_{fi}^{S_{\rm sur}}=C_{fi}^{S_{\rm sur}0}$	$C_{fi}^{S_{\rm inn}}=C_{fi}^{S_{\rm inn}0}$
	$\lambda_a^G\dfrac{dT_G}{dz}=u_G\rho_G Cp_G\left(T_G-T_G^0\right)$	$\lambda_a^L\dfrac{dT_L}{dz}=u_L\rho_L Cp_L\left(T_L-T_L^0\right)$	$\lambda_a^S\dfrac{dT_S}{dz}=u_{SL}\rho_S Cp_S\left(T_S-T_S^0\right)$	$\lambda_a^S\dfrac{dT_S}{dz}=u_{SL}\rho_S Cp_S\left(T_S-T_S^0\right)$	$\lambda_a^S\dfrac{dT_S}{dz}=u_{SL}\rho_S Cp_S\left(T_S-T_S^0\right)$
$z=H,\ 0\le r\le R$	$\dfrac{dC_i^G}{dz}=0$	$\dfrac{dC_i^L}{dz}=0$	$U_L C_S=C_S^0$		
	$\dfrac{dT_G}{dz}=0$	$\dfrac{dT_L}{dz}=0$	$\dfrac{dT_S}{dz}=0$	—	—
$0\le z\le H,\ r=R$	$-\lambda_r^G\dfrac{dT_G}{dr}=h_{GW}\left(T_G-T_W\right)$	$-\lambda_r^L\dfrac{dT_L}{dr}=h_{LW}\left(T_L-T_W\right)$		—	—
$\xi=0$	—	—	—	—	$\dfrac{dC_{fi}^{S_{\rm inn}}}{d\xi}=0$
$0\le z\le H,\ 0\le r\le R$	—	—	—	—	$\dfrac{dT_S}{d\xi}=0$
$\xi=\dfrac{d_{\rm pe}}{2}$	—	—	—	—	$-D_{ei}^L\dfrac{\partial C_{fi}^{S_{\rm inn}}}{\partial\xi}=k_i^S a_S\left(C_{Li}^{S_{\rm inn}}-C_i^L\right)=\rho_B\zeta\sum_{j=1}^{\rm NRL}\nu_{ij}^L\eta_j^L r_j^{\prime L}$
$0\le z\le H,\ 0\le r\le R$	—	—	—	—	$-D_{ei}^L\dfrac{\partial C_{fi}^{S_{\rm inn}}}{\partial\xi}=k_i^S a_S\left(C_{Li}^{S_{\rm inn}}-C_i^L\right)=\rho_B\zeta\sum_{j=1}^{\rm NRL}\nu_{ij}^L\eta_j^L r_j^{\prime L}$

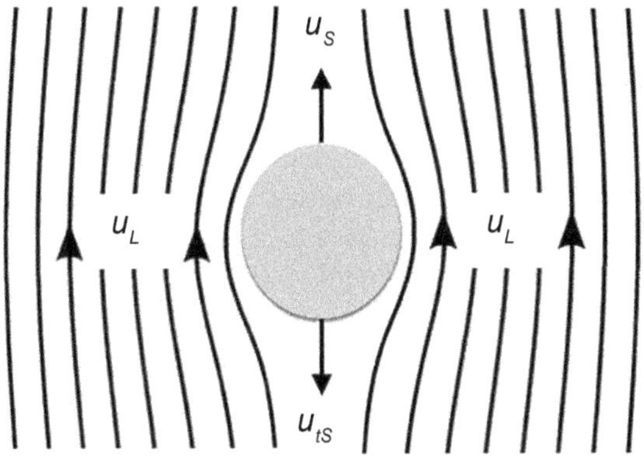

Figure 7.6 Schematic representation of movement of suspended particles.

and initial catalyst concentration may differ if recirculation is used. Moreover, since the system is described by a set of PDE, four boundary conditions are needed in order to describe the longitudinal and radial variation of the system properties ($z = 0, 0 \leq r \leq R, z = H, 0 \leq r \leq R, 0 \leq z \leq H, r = 0, 0 \leq z \leq H, r = R$), which typically have the form of Danckwerts's conditions:

$$\text{at } z = 0, \frac{d\Psi}{dz} = (\Psi - \Psi^0) \tag{7.22}$$

$$\text{at } z = L, \frac{d\Psi}{dz} = 0 \tag{7.23}$$

For the case of the reactor bottom at $z = 0, 0 \leq r \leq R$, there are different ways of setting the inlet conditions; for example, some authors consider that the axial dispersion of mass and temperature is too small compared with the convective term, therefore the gradients at the inlet could be omitted and the conditions are the same as in Eq. (7.21). To describe the dynamics of solid concentration, two conditions are needed because it is a second-order PDE, the fist condition is given at $z = 0$ for all the radial coordinate $0 \leq r \leq R$. Typically Danckwerts's conditions are the most commonly used. In this case, the inlet concentration of the catalyst is obtained as a function of the weight fraction of the solid in the feed slurry through:

$$C_S^0 = \frac{W_S \rho_L}{(1 - [(\rho_S - \rho_L)/\rho_L] W_S)} \tag{7.24}$$

Instead of this equation, simple average concentrations could be also used (Sehabiague et al. 2008). The second condition is given at $z = H$ for all the radial coordinate $0 \leq r \leq R$, the condition is given by a simple balance where the flux of solid entering the reactor must equal that leaving the reactor:

$$U_L C_S = C_S^0 \tag{7.25}$$

On the other hand, though considering internal and external particle diffusion effects in the modeling could lead to a more realistic representation of real reactors, when modeling catalytic slurry-phase reactors, due to reduced dimensions typically found in practice, these effects are commonly ignored, and it is assumed that the reaction takes place on the liquid phase. However, the conditions proposed in Table 7.4 for the concentration of the i component at the solid phase would be ignored.

7.2.4 Estimation of Model Parameters

Even in the most complex models, it is necessary to evaluate several parameters and chemical properties of the system. An adequate description of the system dynamics needs reliable data on parameters that are specific for the SPRs (k_G, k_L, and k_S) and parameters that are not specific for the type of reactor (k' and D_{Mi}), as well as particle and fluid properties (μ and ρ). Those parameters can be estimated with existing correlations, whose accuracy is of great importance for the whole robustness of the model. Density and viscosity of the liquid phase (heavy oil) play an important role in system hydrodynamics. For slurry systems, these properties could be affected by the presence of solid particles. High concentrations of particles in the feed oil could lead to notorious changes in the viscosity of the liquid phase. There are existing correlations that estimate the viscosity and density of the slurry phase as a function of the solids concentration (de Swart and Krishna 2002a,b; Khadem-Hamedani et al. 2015). In general, all liquid and gas phase properties at the operating conditions could be estimated through different correlations reported in the literature as can be seen in the work of Mederos et al. (2009) for heavy oil HCR. On the other hand, there are a few reviews in the literature dealing with SPRs and bubble columns. Chaudhari and Ramachandran (1980), Shah et al. (1982), and Fan (1989) presented a very complete review of previous works published until the 80 decade for parameter estimation for slurry and bubble column reactors. Beenackers and Swaaij (1993) focused only on the mass transfer phenomena in SPRs, and more recently Kantarci et al. (2005) reviewed bubble column reactors with application to bioprocess. However, most of the parameters reported were obtained for two-phase and low-viscosity systems, typically air–water or air–organic liquids. In order to better reproduce the dynamics of the reactor, one should select the parameters that were obtained under similar conditions to those used in the modeling system of interest. Leonard et al. (2015) presented a review of bubble column reactors operating in high pressure and temperature. Their work is focused on overhauling the effect of pressure and temperature on hydrodynamic variables such as gas holdup, as well as mass transfer parameters and axial dispersion coefficients. On the other hand, Mederos et al. (2009) presented a summary of the typical correlations used in HCR systems for diffusion coefficients, viscosities, and densities, which are needed in mass transfer calculations.

7.2.5 Gas Holdup

The gas holdup is one of the most important parameters characterizing the hydrodynamics of BCR and SBCR, which is defined as the percentage of gas volume in the two- or three-phase mixture inside the reactor. The gas holdup in conjunction with the knowledge of mean bubble diameter allows for the determination of interfacial area, thus leading to the mass transfer rates between phases. It depends mainly on the superficial gas velocity, but it is also sensitive to the physical properties of the liquid, gas and solid particles, column dimensions, operating temperature and pressure, and gas distributor design. For practical simplicity, it is usually estimated using the pressure profile method (Li and Prakash 2000), in which the gas holdup is obtained from the static pressure drop as:

$$\varepsilon_G = 1 - \frac{1}{g(\rho_L \phi_L + \rho_S \phi_S)} \frac{\Delta P}{\Delta L} \tag{7.26}$$

where ϕ_L and ϕ_S are the volume fractions of liquid and solid particles in the slurry phase, respectively:

$$\phi_L = \frac{V_L}{V_S + V_L}, \phi_S = \frac{V_S}{V_S + V_L} \tag{7.27}$$

The term $g(\rho_L\phi_L + \rho_S\phi_S)\Delta L$ in Eq. (7.26) has been approximated to the static pressure difference ΔP_0, when no gas is flowing in the column (Tang and Heindel 2006). In these equations, the effect of wall shear stress and liquid acceleration due to void changes are neglected. When more reliable models are required, predictions of radial gas holdup profiles would lead to a better understanding of system hydrodynamics. There are a large number of correlations proposed for the gas holdup, especially for two-phase bubble columns (Kantarci et al. 2005; Shah et al. 1982), although there is no general correlation applicable to any system mainly due to the extreme sensitivity of the holdup to the materials properties and the trace impurities (Bach and Pilhofer 1978). It has been pointed out that for some slurry systems, the presence of non-catalytic solids does not affect the gas holdup significantly (Shah et al. 1982; Ying et al. 1980). However, recent experimental results focused on catalytic systems show that increasing the solid concentration decreases the gas holdup to a significant extent (Vandu and Krishna 2003).

There are several correlations of gas holdup for SPRs (Kito et al. 1976; Begovivh and Watson 1978; Dudukovic and Mills 1984; Reilly et al. 1986; Sauer and Hempel 1987; Schumpe et al. 1987; Chen et al. 1995; Luo et al. 1999; Krishna and Sie 2000; Kumar and Khanna 2014). Unfortunately, most of these correlations have several limitations. On the one hand, these correlations were developed for aqueous systems in small-diameter reactors operating under atmospheric pressure or ambient temperature, and on the other hand, the majority of solids were non-catalytic particles. This leads some authors to choose to solve the dynamic of the holdup along with the state variables of the system introducing turbulence equations, despite the computational effort. Addressing this situation, the correlations proposed by Behkish et al. (2006) predict the total holdup of single bubbles and LB in pseudo-homogeneous and heterogeneous flows in BCR and SBCR, operating under a range of elevated pressures ($P = 0.1$–15 MPa). In their study, several literature data were considered in order to obtain the correlations. Afterward, the agreement between predicted and experimental values was evaluated, presenting acceptable results with a standard deviation between 15% and 20%.

For HCR systems, most of the literature is focused on kinetic or catalyst studies at batch conditions, thus validation of hydrodynamic parameters is complicated. At the pilot scale and based on previous studies focused on the effect of superficial gas velocity, interfacial tension, and column diameter, Parulekar and Shah (1980) proposed that the gas phase holdup in a slurry HCR system operating at 10 MPa and 350 °C was proportional to the variables according to Eq. (7.28).

$$\varepsilon_G = 0.0866\sigma_L^{-0.01} U_G^{0.5} D_C^{-0.08} \tag{7.28}$$

While this correlation was implemented successfully in the model in order to obtain the quantitative behavior of the system, the lack of experimental data does not allow to corroborate its validity or even the model itself. Recently, the model for diesel hydrodesulfurization of Khadem-Hamedani et al. (2015) includes hydrodynamic parameters obtained by Deckwer et al. (1982), Krishna and Ellenberger (1996), Maretto and Krishna (1999), de Swart and Krishna (2002a,b) for *F–T* synthesis. It should be noted that good results were obtained in the model validations using these approximations.

Moreover, in EBR ($dp \geq 200$ μm), wake models allow to reproduce the holdup of the three phases inside the reactor. These models have been tested experimentally for ambient conditions as well as for high temperature and pressure considering homogeneous and heterogeneous bubble phases

(Jiang et al. 1997; Ruiz et al. 2004). Although these models have been implemented efficiently to EBR under HCR conditions, they have not been used for SPR modeling.

7.2.6 Gas–Liquid Mass Transfer Coefficients

The two-film theory is used to describe the mass transfer flux between the gas–liquid interphase:

$$\frac{1}{K_{Li}} = \frac{RT_G Z_{(P,T_G)}}{k_{Gi} H_i} + \frac{1}{k_{Li}} \tag{7.29}$$

where K_{Li}, k_{Li}, and k_{Gi} are the overall, liquid side, and gas side mass transfer coefficients for the i component on each resistance. In slurry columns, the gas–liquid coefficients depend on the gas and liquid velocity, the geometry and operating conditions of the reactor, physicochemical properties of the system, and can be affected by the presence of solids. Gas side resistance to mass transfer becomes important during the absorption of highly soluble gases. In two-phase bubble columns, typically the gas-phase side resistance is negligible, then the overall mass transfer is governed by the liquid side and only the mass transfer coefficient k_{Li} is considered:

$$\frac{1}{K_{Li}} = \frac{1}{k_{Li}} \tag{7.30}$$

In slurry reactors applied to HCR systems, the gas phase is mainly hydrogen (H_2), which has high solubility, then the gas-phase side resistance is also negligible. For a better comprehension of the gas–liquid mass transfer mechanisms, the parameters k_{Li} and a should be considered separately (Kantarci et al. 2005). Nevertheless, only few literature data were found. On the other hand, there are several correlations proposed in the literature when estimating the G–L mass transfer coefficient along with the interfacial area (k_La). For example, the review published by Beenackers and Swaaij (1993) includes different correlations for two-phase systems (BCR) and the discussion of the influence of solid concentration on the G–L mass transfer coefficient for three-phase systems (SPR) showing different adjusting parameters, while Fukuma et al. (1987) and Kim and Kang (1997) presented correlations applied specifically to SBCR. More recently, considering the effect of operating conditions (T and P), the correlations presented in Table 7.5 were developed. However, the only correlation for high operating conditions that considers the effect of reactor geometry is that proposed by Lemoine et al. (2008), which is based on the considerations reported by Behkish et al. (2006) for gas holdup in large-scale systems. These two correlations have been applied successfully to an industrialized F–T reacting system by Sehabiague et al. (2008). Generally, the mass transfer coefficient is obtained through correlations as a function of superficial gas velocity, liquid, and gas phase properties, and in some cases solids concentration in the slurry. Other important hydrodynamic parameters such as bubble size and gas holdup are not considered directly in the correlations. This may be due to the gas–liquid interfacial area having a relation $a = 6\varepsilon_G/d_b$, thus the competing effect of these variables becomes insignificant in mass transfer coefficient. However, if calculations of the mean bubble diameter are required, there are correlations reported in the literature that depend on the sparger specifications, such as the one proposed by Lemoine et al. (2008), or the method proposed by Kim and Kang (1997) in which the bubble diameter is a function of the Reynolds number based on the bubble diameter and the Schmidt number. The gas phase is usually assumed to move as plug flow along the reactor due to the characteristics of slurry-phase HCR, so calculations of mean bubble diameter are not strictly necessary in the modeling.

Table 7.5 Recent G–L mass transfer correlations.

Reference	Experimental conditions	Correlation
Behkish et al. (2006)	$0.17 \le P(\text{MPa}) \le 0.79$ $0.05 \le U_G(\text{m/s}) \le 0.25$ $0 \le C_V(\%\text{Vol}) \le 36$	$k_{Li}a = 0.18Sc^{-0.6}\left(\frac{\rho_L}{C_i}\right)^{-2.84} \times (\rho_G U_G)^{0.49} e^{-2.66C_V}$
Lemoine et al. (2008)	$0.1 \le P(\text{MPa}) \le 19.8$ $0.003 \le U_G(\text{m/s}) \le 0.574$ $275 \le T(\text{K}) \le 538$ $0.0382 \le D_C(\text{m}) \le 7.72$ $0 \le C_V(\%\text{Vol}) \le 36$ $0.16 \le \mu_L(\text{N/m} \times 10^{-3}) \le 75$	$k_{Li}a = 06.14 \times 10^4 \frac{\rho_L{}^{0.26}\mu_L{}^{0.12}}{\sigma_L^{0.52}\rho_G{}^{0.06}} \frac{\varepsilon_G{}^{1.21}}{U_G^{0.12} d_S{}^{0.05}} \times \frac{D_{Mi}{}^{0.5}}{T^{0.68}} \Gamma^{0.11} \left(\frac{D_C}{D_C + 1}\right)^{0.4} e^{-2.66C_V}$ $d_S = 37.19 \times \frac{\rho_L{}^{1.22}\mu_L{}^{0.08}\rho_G{}^{0.02}T^{1.66}}{MM_G^{0.12}\rho_L{}^{1.52}} U_G^{0.14} \times \left(\frac{D_C}{D_C + 1}\right)^{0.3} \times (1 - \varepsilon_G)^{1.56} \Gamma^{-0.2} \varepsilon_G{}^{-2.29X_W + 2.81C_V + 2.77\rho_p d_p}$
Ferreira et al. (2010)	$293 \le T(K) \le 308$ $U_G\left(\frac{m}{s}\right) > 0.0072$	$k_{Li}a = 5.18\sqrt{\frac{D_L}{\pi}}\sqrt{\frac{U_G g}{\mu_L}} 651 U_G^{0.87} \mu^{\text{eff}\,-0.24} \times (1 - d_p)^{0.036} (1 - \varepsilon_S)^{9.66}$
Jin et al. (2014)	$1 \le P(\text{MPa}) \le 3$ $298 \le T(\text{K}) \le 423$ $D_C(m) = 0.10$ $0 \le C_V(\%\text{Vol}) \le 20$	$k_{Li}a = 3.051Sc^{-0.734}\left(\frac{\rho_L}{C_i}\right)^{-1.193} \times (\rho_G U_G)^{0.524}\left(1 - \frac{C_V}{0.85}\right)^{2.303}$

7.2.7 Gas–Liquid Equilibrium

In the proposed generalized model, it is necessary to calculate the gradient for the *i* component between the gas and liquid phases. For that matter, the equilibrium concentration of the *i* component is needed. When modeling heavy oil HCR in presence of excess hydrogen, in the majority of cases the gas phase is considered to be mainly composed by H_2 and in a lower concentration of H_2S, therefore the volatility of other components of the oil is not considered. Thus, there are two ways for performing equilibrium calculation, in one of them the equilibrium concentration is obtained through Henry's law employing solubility coefficients. The common procedure is the calculation of solubility coefficients from Korsten and Hofffman's correlations. Though the correlation for H_2S solubility is not of great accuracy for the high temperatures commonly used in HCR reactors.

The second method consists of obtaining the Henry's coefficients. Either from assuming local equilibrium at the gas–liquid interface or through fugacity coefficients as shown by Mederos et al. 2009. In both cases, the use of an equation of state is necessary. This represents an advantage because volatile compounds can be considered in the calculations. However, this method implies the use of an equation of state that involves the calculation of thermodynamic parameters such as binary interaction parameters, which may lead to considerable errors in the calculations due to the experimental nature of these parameters, as well as increasing the computing time.

7.2.8 Liquid–Solid and Gas–Solid Mass Transfer Coefficients

Usually, the liquid–solid mass transfer coefficient for SBCR and STR has the following form:

$$\mathrm{Sh} = 2 + \alpha \mathrm{Sc}^{\beta}\, \mathrm{Re}_p^{\gamma} \tag{7.31}$$

where Sh and Sc are the Sherwood and Schmidt numbers. The constant parameters α, β, and γ depend mainly on the system hydrodynamics. Several considerations have been taken into account for estimating the particle Reynolds number Re_p. Usually, the particle slip velocity was used in the computing; however, due to the complexity and uncertainty reported in some systems, the approach based on Kolmogoroff's isotropic turbulence theory has been preferred in SBCR (Shah et al. 1982; Beenackers and Swaaij 1993). In the case of STR, Pangarkar et al. (2002) presented a review on different theories including Kolmogoroff's turbulence theory and correlations based on a dimensional approach. In both cases, the authors concluded that the correlation depends on the operating parameters like impeller speed, particle loading, and reactor geometry and the selection of the correlation should be based on similar operation conditions.

Furthermore in order to describe all the mass transfer phenomena occurring from gas to solid phases, two mechanisms are considered. On the one hand, the gas phase is absorbed into the liquid wake with a velocity controlled by K_{Li}, then the gas absorbed in the liquid reaches the solid surface and the intraparticle diffusion process takes place. In SBCR only this mechanism is commonly used in the modeling and is justified by the assumption of equilibrium conditions between gas and liquid phases. On the other hand, the other mechanism considers the gas phase directly reach the solid phase through the agglomeration of solid particles between the gas and liquid interfaces; this leads to the conversion rate of a reaction limited by *G–L* mass transfer enhanced by the presence of small catalyst particles adhering to gas bubbles (van der Zon et al. 1999, 2001). Ruthiya et al. (2004) compared the classical *G–L–S* model with a modified *G–L–S* including *G–S* mass transfer in a STR. Their results confirmed an increase in reaction rate when solid agglomeration was taken into account; however, the estimation of the mass transfer coefficient k^{SG} is obtained only through experimental fitting. General models should contain a priori all the mass transfer resistances.

However, no correlation for the gas–solid mass transfer coefficient in SBCR or STR has been found in the literature, therefore when this type of approximation is applied, the value of k^{SG} should be obtained through experimental data, or introducing models that include the enhanced factor due to particle agglomeration-adhesion.

7.2.9 Dispersion Coefficients

The extent of liquid mixing is essential in the modeling, design, and optimization of BCR and SBCR. The degree of mixing in these reactors is quantified by the axial dispersion coefficient D_a^f. Most of the models applied to SBCR have been formulated under the ADM whereby this parameter becomes overriding. Despite ADM has shown good results simulating the behavior of different systems, there has been no systematic effort made for modeling effectively the combined effect of superficial gas and liquid velocity and column diameter on liquid axial dispersion (Moustiri et al. 2001). Also, there is an open question on how pressure affects the axial dispersion coefficient (Tang and Heindel 2006); however, it has been found that at high pressure in the homogeneous regime, there is no effect of pressure on the liquid axial dispersion coefficient (Yang and Fan 2003), which is generally the case for HCR systems. These results were obtained through comparing experimental values with the correlation of Joshi (1998) formulated for two and three-phase systems:

$$D_a^L = 0.29 D_C (U_L + U_C) \tag{7.32}$$

So far this is considered one of the most complete correlations because it considers liquid and gas velocities as well as variables related to the gas bubble size and bubble rising velocity. Following these considerations, Moustiri et al. (2001) proposed a general correlation for BCR supported by experimental data for different sizes of columns:

$$\mathrm{Bo} = \frac{U_L D_C}{(1-\varepsilon_G) D_a^L} = \frac{1.3\,\mathrm{Re}_L^{0.1}}{(1-\varepsilon_G)\left[\dfrac{1}{((1-\varepsilon_G))} + \dfrac{1}{235}\left(\dfrac{\mathrm{Re}_L}{\mathrm{Fr}_G^{1/3}}\right)^{-2/3} \mathrm{Ga}^{0.6}\right]} \tag{7.33}$$

This correlation has shown good agreement with experimental data while the operating conditions of both are similar. In HCR systems as well as for hydrogenation systems where the homogeneous bubble flow prevails, a common approximation for the liquid axial dispersion coefficient is considered with the correlation proposed by Deckwer et al. (1974).

Recently Khadem-Hamedani et al. (2015) have shown, for an HDS system under sever conditions of temperature and pressure, that the correlation proposed by Deckwer et al. (1982) formulated for a *F–T* system can be satisfactorily included in the modeling regardless of the type of flow regime (homogeneous or heterogeneous). Other correlations are presented extensively in different works starting with Shah et al. (1982), Fan (1989), Degaleesan and Dudokivic (1998), and recently with Sivaiah and Majumder (2013), who present a summary of the typical correlations for three-phase systems and their respective range of parameters.

Less investigations have been found dealing with the radial dispersion phenomena. In general, it has been found that the radial dispersion coefficient of the liquid phase in liquid fluidized beds increases with the particle size and gas velocity. Radial dispersion coefficients are well represented by the ratio of the fluid velocities and of particle-to-column diameters based on the concept of the isotropic turbulence model (Han and Kim 1990). Some of the correlations found in the literature are presented in Table 7.6. On the other hand, Hillmer et al. (1994) observed experimentally that the

Table 7.6 Correlations for the liquid radial dispersion coefficient.

Reference	Correlation
Kang and Kim (1986)	$\mathrm{Pe}_r = 28.3\left(\frac{d_p}{D_C}\right)\left(\frac{U_L}{U_L + U_G}\right)^{1.16}$
Han and Kim (1990)	$\mathrm{Pe}_r = 7.28\left(\frac{d_p}{D_C}\right)^{0.85}\left(\frac{U_L}{U_L + U_G}\right)^{-0.16}\mathrm{Re}_L^{0.21}$
Kang et al. (2003)	$\mathrm{Pe}_r = 0.086\left(\frac{d_p}{D_C}\right)^{0.58}\left(\frac{U_L(\varepsilon_G + \varepsilon_L)}{U_L + U_G}\right)^{0.691}\mathrm{Re}_L^{0.21}$
Cho et al. (2005)	$\mathrm{Pe}_r = 8.47\left(\frac{d_p}{D_C}\right)^{1.303}\left(\frac{d_p U_L \rho_S}{\mu_L}\right)^{-0.056}\left(\frac{G_S}{U_L \rho_S}\right)^{-0.311}$

type of flow inside liquid fluidized bed bubble column reactors satisfies the condition that the Schmidt number is close to unity. Therefore, the dispersion coefficient could be expressed by:

$$D_f = \frac{\mu_f}{\rho_f} \tag{7.34}$$

where μ_f is the sum of turbulent and laminar viscosity. This consideration was applied successfully by Matos and Guirardello (2002) and Matos et al. (2009) in a CFD model for petroleum hydrotreating.

Moreover, the gas phase has been traditionally modeled as a single phase typically in plug flow or in some cases considering axial dispersion. Nevertheless, when a heterogeneous flow regime is considered, usually the two bubble class models are introduced, then the SB are characterized by a completely mixed phase while the LB are modeled either in plug flow or with an ADM (Rados et al. 2005). Rarely both phases are modeled using axial dispersion. It is known that in the gas phase, the axial dispersion is caused by the interactions among bubbles of the same class, therefore the bubble rise velocity is the characteristic variable in the gas dispersion. Yet, the effect of solids on gas dispersion in three-phase systems is not fully known, hence in most of the cases, correlations for solid-free systems are used instead. However, even for *G–L* systems, these correlations are scarce, some of which are presented by Fan (1989). In most cases, only approximations to the gas Peclet number are used. For example, for the heterogeneous flow regime, de Swart and Krishna (2002a,b) assigned the value of 100 to the gas Peclet number due to their plug-flow regime, and the axial dispersion of SB was assumed to be equal to the axial dispersion coefficient of the liquid phase according to the study of Baten and Krishna (2001). This last assumption has been successfully applied in different heterogeneous models, in some cases only the value of the Peclet number for LB is modified introducing correlations for the axial dispersion coefficient, as shown by Khadem-Hamedani et al. (2015) who used the correlation of Wilkinson et al. (1992) for single- and two-bubble class models.

As mentioned before, in slurry systems the suspension of solids in the liquid phase is usually modeled as a single pseudo-homogeneous phase where the solids velocity is assumed to be equal to the liquid velocity. Also, a uniform concentration of the solids throughout the reactor is supposed. Given the small size of the solid particles that are typically used in SBC reactors ($dp \leq$ 50 μm), the pseudo-slurry assumption is probably reasonable. However, some studies have shown that despite the small size of the solids, the axial profile of solids concentration in the slurry may not be flat (Rados et al. 2005). When the system properties are such that uniform concentration cannot be an accurate consideration, the SDM is suitable to solve this issue. There are different formulations of the SDM, but in all of them, the axial dispersion coefficient for the solid phase is needed. Unfortunately, there is not much information on this issue. Some of the most common correlations are also presented in the works of Kojima et al. (1989), Fan (1989), and Kato et al. (1985).

7.2.10 Heat Transfer Coefficients

The heat transfer coefficients in SPR are many times larger than that for single-phase flow or even in liquid–solid fluidization. Due to the large contact area among phases, a common assumption in the modeling is that the gas and liquid phases are in thermal equilibrium. This leads to several simplifications; for example, a global heat transfer coefficient h between the continuous phase ($G + L$) and the wall is considered instead of one for each resistance (*G–W*, *L–W*). Also, in most of the cases due to the small particle diameter, intraparticle temperature gradients are negligible; however, in some cases, the heat transfer coefficient between *L–S* and *G–S* is introduced.

The heat transfer coefficient h in three-phase fluidized beds has been found to be an increasing function of the respective gas and liquid velocities, the size and density of the particles, column diameter, and the thermal conductivity and heat capacity of the liquid phase, but a decreasing function of the dynamic viscosity of the liquid, and probably of the diameter of a heating cylinder immersed inside the bed (Fan 1989). The heat transfer coefficient has been successfully correlated and reviewed by several investigators in terms of experimental variables such as gas and liquid velocities, particle size, and liquid viscosity (Shah et al. 1982; Kim and Kang 1997; Jamialahmadi and Müller-Steinhagen 1999). The range of application of the correlations could vary depending on the system under study, therefore the lecture on those works is recommended. Also, semi-theoretical correlations have been proposed based on the energy dissipation rate; however, for the design and scale-up of multi-phase flow reactors, it is desirable to predict h from dimensionless correlations. Recently the work of Kim et al. (2014) proposed a correlation for viscous systems where the effects of superficial gas velocity (0.002–0.164 m/s) and solid load (0–20%) were examined. It is concluded that the heat transfer coefficient between the immersed heater and the bed increases with increasing gas velocity. On the contrary, it decreases with viscosity of the slurry phase. These are important results because they could be applied to the HCR of heavy oils. Different correlations can also be found in the work of Jhawar and Prakash (2011).

7.2.11 Example of Simplification of the Generalized Model

As mentioned before, a model that considers all the dynamic equations presented in Tables 7.2 and 7.3 along with the initial and boundary conditions in Table 7.4 is way too difficult to solve, therefore considerations are needed to simplify the model equations. However, these considerations would depend on the feed oil properties as well as catalyst characteristics and operating conditions, therefore it would be difficult to establish a general criterion of simplifications. Although some of the general considerations are as follows:

Homogeneous suspension: In order to consider a slurry-phase reactor, it should be considered a good homogeneity between solid and liquid phases as well as small particle size, negligible effects of the terminal and settling velocity and that the solids follow the liquid motion, therefore a constant solids dispersion inside the reactor could be assumed and the phases liquid and solid could be considered as one phase (slurry). On the other hand, for larger particle diameters, this simplification should be avoided, and the model would be approximate to a three-phase reactor.

Homogeneous bubble flow: As explained before, due to the operating conditions as well as the system properties in HCR of heavy oils, the typical flow regime is the homogeneous bubble flow. In that case, the $N + 1$ model equations simplify only to two general mass balance equations, one for the gas phase and another for the slurry phase.

Gas phase: As seen in previous sections, due to hydrogen being found in excess inside the reactor, most models consider the gas phase as a mixture only of H_2 and H_2S. Thus, only two equations for the gas phase are needed, along with the corresponding equations for these components in the liquid phase. Mass transfer between gas and liquid phases for the other lumps is not considered. Also in HCR systems, because the gas phase is mainly hydrogen (H_2) with high solubility, the gas-phase side resistance is negligible. Another consequence of the large amount of hydrogen in the reactor is avoiding coke formation.

Kinetic model: As it is known there are different types of kinetic models with different complexity, and some of them are based either on lumps or continuous mixtures. In any case, both models implied the aggrupation of components of the oil with similar characteristics. The number of lumps is fundamental in the simplification of the model equations. For example, the kinetic model of Krishna and Saxena (1989) includes six lumps; therefore, the dimension of the slurry-phase equations increases to six (in case volatility of any of these lumps is introduced in the modeling, it would be necessary to include the corresponding gas-phase equations for each volatile lump).

Use of correlations: In the modeling of reactors, a common simplification is the use of correlations for the hydrodynamic parameters. This would allow to solve the model equations, avoiding the use of turbulence equations such as in CFD models. In that case, both variables stay out of the differentials and only the variation of the concentration or temperature are computed. However, care must be taken to select the parameters that were obtained under similar conditions to those used in the modeling system of interest.

Isothermal conditions: So far, all the models reported in the literature concerning heavy oil HCR assumed isothermal conditions because of the good mixture and uniform distribution of temperature achieved with these reactors. However, it is not a general consideration.

7.3 Simplified Models

7.3.1 SPR 1D Model

The goal of this section is to formulate a mathematical model for an SPR that can be applicable at a laboratory scale. Some of the main considerations have been reported in the literature (Bahroun et al. 2010; Maretto and Krishna 1999; Rados et al. 2005; Khadem-Hamedani et al. 2015; Calderon and Ancheyta 2016; de Swart and Krishna 2002a,b) and are shown below:

- Gas- and liquid-phase velocities are constant along the reactor because the variations are quite small and do not affect the conversion of the reactor (Matos and Nunhez 2006). Therefore, only mass and energy balances with constant holdup will be included in the model.
- Due to the small size of the reactor, only axial effects will be considered.
- Due to the high solubilities of hydrocarbons and high mixture between liquid and gas phases, there won't be mass transfer limitations and the hydrocarbon will be saturated for each of the gas-phase components.
- The catalyst particles are completely pulverized, therefore no intraparticle diffusion will be considered.

- The catalyst and the liquid phase are considered to be a homogeneous phase (slurry phase).
- Experimental conditions are isothermal.

Based on the above considerations, the reactor model is:

$$\frac{\partial x_i}{\partial t} = -u\frac{\partial x_i}{\partial Z} + \mathrm{Da}\frac{\partial^2 x_i}{\partial Z^2} + \sum_{i,k}^{NC,NR} r_{i,k} \tag{7.35}$$

With the boundary and initial conditions:

$$t = 0, 0 \le Z \le L, x_i = x_i^{\mathrm{in}} \tag{7.36}$$

$$0 \le t \le t_f, Z = 0, \mathrm{Da}\frac{\partial x_i}{\partial Z} = u\left(x_i - x_i^0\right) \tag{7.37}$$

$$0 \le t \le t_f, Z = L, \frac{\partial x_i}{\partial Z} = 0 \tag{7.38}$$

where x_i stands for the mass fraction of component i in the hydrocarbon mixture (hydrocarbon + gas) of feed and product, while x_i^0 represents the mass fraction of component i at the inlet. u is the superficial gas velocity and Da the axial dispersion coefficient. t stands for the time of the run, and Z is the axial coordinate with L being the reactor length.

7.3.2 SPR 2D Model

To describe the operation of an industrial reactor, most of the considerations presented in the previous section are considered; however, since this case is an industrial reactor, the effects of temperature variation as well as radial dispersion are included. Because these reactors have a large diameter and operate at low surface velocities, it is possible that gradients may occur in the radial direction. Hence, the model becomes:

$$\frac{\partial x_i}{\partial t} = -u\frac{\partial x_i}{\partial Z} + \mathrm{Da}\frac{\partial^2 x_i}{\partial Z^2} + \mathrm{Dr}\left(\frac{\partial^2 x_i}{\partial r^2} + \frac{1}{r}\frac{\partial x_i}{\partial r}\right) + \sum_{i,k}^{NC,NR} r_{i,k} \tag{7.39}$$

$$\begin{aligned}\frac{\partial T_L}{\partial t} &= -u\frac{\partial T_L}{\partial Z} + \frac{k_L}{\varepsilon_L \rho_L Cp_L}\frac{\partial^2 T_L}{\partial Z^2} + \frac{k_L}{\varepsilon_L \rho_L Cp_L}\left(\frac{\partial^2 T_L}{\partial r^2} + \frac{1}{r}\frac{\partial T_L}{\partial r}\right)\\ &\quad + \frac{1}{\varepsilon_L V_R \rho_L Cp_L}\sum_{i,k}^{NC,NR} r_{i,k}(-\Delta H_k) - \frac{\mathrm{ha}(T_L - T_G)}{\varepsilon_L V_R \rho_L Cp_L}\end{aligned} \tag{7.40}$$

$$\frac{\partial T_G}{\partial t} = -u\frac{\partial T_G}{\partial Z} + \frac{k_G}{\varepsilon_G \rho_G Cp_G}\frac{\partial^2 T_L}{\partial Z^2} + \frac{k_G}{\varepsilon_G \rho_G Cp_G}\left(\frac{\partial^2 T_G}{\partial r^2} + \frac{1}{r}\frac{\partial T_G}{\partial r}\right) + \frac{ha(T_L - T_G)}{\varepsilon_G V_R \rho_G Cp_G} \tag{7.41}$$

With the boundary and initial conditions:

$$t = 0, 0 \le Z \le L, 0 \le r \le R, x_i = x_i^{\mathrm{in}} \tag{7.42}$$

$$t = 0, 0 \le Z \le L, 0 \le r \le R, T_{Li} = T_L^{\mathrm{in}} \tag{7.43}$$

$$t = 0, 0 \le Z \le L, 0 \le r \le R, T_G = T_G^{\mathrm{in}} \tag{7.44}$$

$$0 \le t \le t_f, Z = 0, 0 \le r \le R, \mathrm{Da}\frac{\partial x_i}{\partial Z} = u\left(x_i - x_i^0\right) \tag{7.45}$$

$$0 \le t \le t_f, Z = 0, 0 \le r \le R, k_L \frac{\partial T_L}{\partial Z} = u\left(T_L - T_L^0\right) \tag{7.46}$$

$$0 \le t \le t_f, Z = 0, 0 \le r \le R, k_G \frac{\partial T_G}{\partial Z} = u\left(T_G - T_G^0\right) \tag{7.47}$$

$$0 \le t \le t_f, Z = L, 0 \le r \le R, \frac{\partial x_i}{\partial Z} = 0 \tag{7.48}$$

$$0 \le t \le t_f, Z = L, 0 \le r \le R, \frac{\partial T_L}{\partial Z} = 0 \tag{7.49}$$

$$0 \le t \le t_f, Z = L, 0 \le r \le R, \frac{\partial T_G}{\partial Z} = 0 \tag{7.50}$$

$$0 \le t \le t_f, 0 \le Z \le L, r = 0, \frac{\partial x_i}{\partial Z} = 0 \tag{7.51}$$

$$0 \le t \le t_f, 0 \le Z \le L, r = 0, \frac{\partial T_L}{\partial Z} = 0 \tag{7.52}$$

$$0 \le t \le t_f, 0 \le Z \le L, r = 0, \frac{\partial T_G}{\partial Z} = 0 \tag{7.53}$$

$$0 \le t \le t_f, 0 \le Z \le L, r = R, \frac{\partial x_i}{\partial Z} = 0 \tag{7.54}$$

$$0 \le t \le t_f, 0 \le Z \le L, r = R, \frac{\partial T_L}{\partial Z} = 0 \tag{7.55}$$

$$0 \le t \le t_f, 0 \le Z \le L, r = R, \frac{\partial T_G}{\partial Z} = 0 \tag{7.56}$$

where T_L and T_G represent the liquid- and gas-phase temperatures, respectively; k_f, ε_f, ρ_f, and Cp_f are the thermal conductivity, phase holdup, density, and heat capacity of f phase; and D_r is the radial dispersion coefficient and ha the heat transfer coefficient.

7.3.3 Continous Stirred Tank Reactor Model

One of the key factors in reactor modeling is the reactions involved. In this case, the main reactions that take place are HDC and HDT. Therefore, to study reaction kinetics is also necessary to provide a continous stirred tank reactor (CSTR) model. The considerations for this model are similar to the ones explained in previous sections, adding the perfect mixing assumption. Then the model simplifies to:

$$\frac{\partial x_i}{\partial t} = -\dot{m}_T\left(x_i^{\text{in}} - x_i\right) + \sum_{i,k}^{NC,NR} r_{i,k} \tag{7.57}$$

where the first term of the left hand represents the changing of mass fraction for component i in time. The first term of the right hand stands for the convective flux gradients, while the last term of the right hand states the reaction rates. The initial conditions have the form:

$$x_i(0) = x_i^{\text{in}} \tag{7.58}$$

7.3.4 Parameters

All models that describe physical and chemical processes, regardless of their complexity, require the knowledge of properties and parameters that characterize the system under study. However, for crude oil HCR systems, experimentally obtaining parameters is impractical and expensive due to the complexity of the hydrocarbon mixture and the conditions under which these systems are operated.

There are different experimental methods that allow to approximate these parameters by correlations. Within this experimental framework, it is common to use tracers, which consist of using fluids at ambient temperature with characteristics like the original fluids at the operating temperature and pressure. These procedures have been effectively applied in Fischer–Tropsch reaction systems, which is why most of the correlations reported in the literature are for these kinds of systems, which operate at temperatures ranging between 250 °C and pressures of 5 MPa. On the other hand, the lack of correlations is evident when it comes to crude oil HCR systems, which operate at temperatures above 350 °C and pressures between 5 and 14 MPa. Due to the difficulty of obtaining experimental information on these systems, an alternative is to use correlations formulated for similar fluids at similar operating conditions. This will be the case for the axial and radial dispersion coefficients, molecular diffusivity, and heat of reaction, as well as for the density and viscosity of the hydrocarbon at the operating conditions. Besides, an equation of state and correlations will also be used to calculate the equilibrium concentrations for the volatile components.

In these types of systems, a temporary solution is to select the correlations that have been formulated for similar fluids in conditions similar to those of the system studied. This will be the case for the axial and radial dispersion coefficients, molecular diffusivity, and heat of reaction, as well as for the density and viscosity of the hydrocarbon at the operating conditions.

Eqs. (7.59)–(7.65) show the correlations to calculate density at operating conditions (Chen et al. 1995; Hillmer et al. 1994):

$$\rho_L(T,P) = \rho_{L,0} + \Delta\rho_P - \Delta\rho_T \tag{7.59}$$

$$\Delta\rho_P = \left(0.167 + 16.181 \times 10^{-0..0425}\right)\left(\frac{P}{1000}\right) - 0.01\left(0.299 + 263 \times 10^{-0.0603\rho_{L,0}}\right)\left(\frac{P}{1000}\right)^2 \tag{7.60}$$

$$\begin{aligned}\Delta\rho_T = {} & \left(0.0133 + 152.4 \times \left(\rho_{L,0} + \Delta\rho_P\right)^{-2.45}\right)(T-520) \\ & - \left(8.1 \times 10^{-6} - 0.0622 \times 10^{-0.764\left(\rho_{L,0} + \Delta\rho_P\right)}\right)(T-520)^2\end{aligned} \tag{7.61}$$

Here, $\rho_L(T, P)$ represents the liquid density as a function of temperature and pressure, $\rho_{L,0}$ is the liquid density at standard conditions (15 °C and 101.3 kPa), P is the pressure in psia, and T is the temperature in R. On the other hand, the liquid viscosity at operating temperature as a function of API is given by:

$$\mu_L(T) = 3.141 \times 10^{10}(T-460)^{-3.444}(\log_{10}(\text{API}))^a \tag{7.62}$$

$$a = 10.313 \times \log_{10}(T-460) - 36.447 \tag{7.63}$$

Finally, the heat capacities are calculated as:

$$Cp_L = (0.35 + 0.55K)(0.6811 - 0.308SG + (0.000815 + 0.000306)T) \tag{7.64}$$

$$K = \frac{T_{\text{MeABP}}^{1/3}}{SG} \tag{7.65}$$

where K is the Watson factor and T_{MeABP} is the average boiling temperature. In addition, an equation of state is used to improve the model and the accuracy of the Henry coefficients. The Peng–Robinson equation has proved to be quite accurate for HCR systems (Mederos et al. 2009)

$$P = \frac{RT}{v - b} - \frac{a}{(v + \delta_1 b)(v + \delta_2 b)} \tag{7.66}$$

T is the temperature (K), P is the pressure (Pa), R is the ideal gas constant (J/mol K), v is the molar volume (m^3/mol), and δ_1 and δ_2 are universal constants for the Peng–Robinson equation. a and b are temperature-dependent parameters that are calculated with:

$$a = a_m = \sum_{i,k}^{\text{NCL, NRL}} x_i x_k a_{ik} \tag{7.67}$$

$$b = b_m = \sum_{i}^{\text{NCL, NRL}} x_i b_i \tag{7.68}$$

With

$$a_{ik} = a_{ki} = (a_{ii} a_{kk})^{1/2}(1 - k_{ik}) \tag{7.69}$$

where k_{ik} is the binary interaction coefficient whose values are taken from the literature. Additionally, the compressibility factors can be obtained with:

$$(Z^L)^3 - (1 - B)(Z^L)^2 + (A - 2B - 3B^2)Z^L = (AB - B^2 - B^3) \tag{7.70}$$

And the fugacity and Henry's constants:

$$\ln \phi_i^L = \frac{b_i}{b}(Z^L - B) - \ln(Z^L - B) - \frac{A}{B(\delta_2 - \delta_1)}\left(\left(\frac{2\sum_k^{\text{NCL}} x_k a_{ik}}{a} - \frac{b_i}{b}\right) \ln \frac{Z^L + \delta_2 B}{Z^L + \delta_1 B}\right) \tag{7.71}$$

$$A = \frac{aP}{RT^2} \tag{7.72}$$

$$B = \frac{bP}{RT} \tag{7.73}$$

The calculation procedure for the thermodynamic parameters is well documented (Mederos et al. 2009; Kesler 1976). Lastly, the critical properties T_c and P_c as well as the accentric factor ω_i and molecular weight MW_i can be estimated based on the boiling point T_b and specific gravity SG with:

$$T_c = 341.7 + 811SG + (0.4244 + 0.1174SG)T_b + (0.4669 - 3.2623)\frac{10^5}{T_b} \tag{7.74}$$

$$\begin{aligned}\ln P_c = 8.3634 - \frac{0.0566}{SG} - \left(0.2444 + \frac{0.2898}{SG} + \frac{0.11857}{SG^2}\right)10^{-3}T_b \\ + \left(1.4685 + \frac{3.648}{SG} + \frac{0.47227}{SG^2}\right)10^{-7}{T_b}^2 - \left(0.42019 + \frac{1.6977}{SG^2}\right)10^{-10}{T_b}^3\end{aligned} \tag{7.75}$$

$$\omega_i = -7.904 + 0.1352\text{K} - 0.007465K^2 + 8.359T_{br} + \frac{(1.408 - 0.0163K)}{T_{br}} \tag{7.76}$$

7.3.5 Reaction Kinetics

Heavy oils are complex mixtures of a large number of components; therefore, the study of all the reactions involved in HCR and hydrotreating is quite complicated. For the case of HCR, there are different types of kinetic models that have been reported to approximate all the reactions (Ancheyta et al. 2005a,b). Despite the dependence of the kinetic parameters on the properties of the feed and the use of an invariant distillation range of products, a pseudo-component-based model will be used to represent the HCR kinetic due to its simplicity and utility in simulation and control applications. The kinetic model is presented in Figure 7.7. This model was proposed by Sanchez et al. (2005) and consists of five pseudo-components: residue (538 °C+), vacuum gas oil (343–538 °C), middle distillates (204–343 °C), naphtha (IBP-204 °C) and gases, and 10 reaction paths for HCR of all lumps, as well as 10 reaction rate coefficients to be calculated.

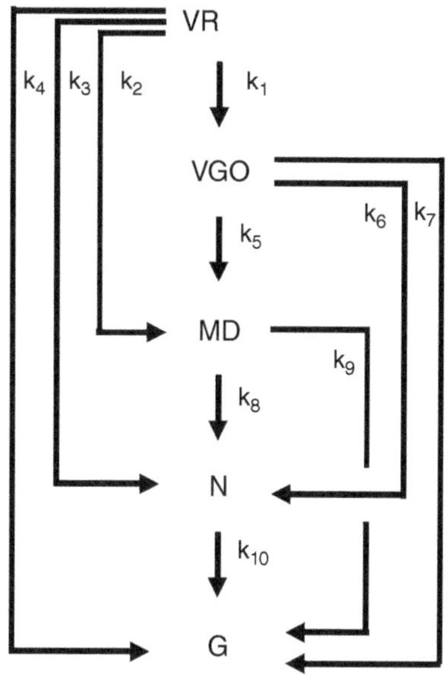

Figure 7.7 Five-lump kinetic model for hydrocracking reported Adapted from Sanchez, et. al. (2005).

Recently Martínez et al. (2010) conducted an experimental study based on the kinetic model presented in Figure 7.7 to obtain the 10 intrinsic reaction rate coefficients in a CSTR. The kinetic expressions for HCR reactions are presented in Eqs. (7.77)–(7.81). Although most literature reports propose that HCR reactions of heavy oils follow a first-order reaction, in this work the HCR reaction of VR was of second order while the rest of the HCR reactions were assumed to be of first order (Martínez and Ancheyta 2012; Ellenberger and Krishna 1994). The second-order reaction is attributed to the broad distribution of reaction rate values of compounds present in the residue fraction, which generates that the reaction rate declines faster than that of a simple first-order reaction because the most-reactive compounds disappear first and fast, leaving the most-refractory compounds in the residue (Sánchez and Ancheyta 2007).

All the reaction rate expressions involve the calculation of effectiveness factors and catalyst deactivation function, due to the large particle size and the deposition of coke on the catalyst surface.

$$r_R = -\varphi_{\text{HDC}}\eta(k_1 + k_2 + k_3 + k_4)y_R^2 \tag{7.77}$$

$$r_{\text{VGO}} = \varphi_{\text{HDC}}\eta\left(k_1 y_R^2 - (k_5 + k_6 + k_7)\right)y_{\text{VGO}} \tag{7.78}$$

$$r_D = \varphi_{\text{HDC}}\eta\left(k_2 y_R^2 + k_5 y_{\text{VGO}} - (k_8 + k_9)y_D\right) \tag{7.79}$$

$$r_N = \varphi_{\text{HDC}}\eta\left(k_3 y_R^2 + k_6 y_{\text{VGO}} + k_8 y_D - k_{10} y_N\right) \tag{7.80}$$

$$r_G = \varphi_{\text{HDC}}\eta\left(k_4 y_R^2 + k_7 y_{\text{VGO}} + k_9 y_D + k_{10} y_N\right) \tag{7.81}$$

On the other hand, as hydrogen is present in the system, hydrotreating reactions also occur simultaneously, so kinetic expressions for all these reactions are needed. It is known that thermal and catalytic reactions act sequentially via thermal scission and hydrogenation. Thus, in order to take into account catalytic and thermal reactions, Martínez and Ancheyta (2014) proposed the expressions presented in Eqs. (7.82)–(7.86). As well as for HCR reactions, all the expressions for hydrotreating reactions require the knowledge of catalyst effectiveness factors and deactivation rates.

$$r_{\mathrm{HDS}} = -\varphi_S \eta_S k_{CS} \frac{x_S^{n_{C_S}} x_{\mathrm{H_2}}^{n_{C\mathrm{H_2}}}}{(1 + K_{\mathrm{H_2S}} x_{\mathrm{H_2S}})^2} + k_{TS} x_S^{\eta_{T_S}} \tag{7.82}$$

$$r_{\mathrm{HDAsph}} = -\varphi_{\mathrm{Asph}} \eta_{\mathrm{Asph}} k_{\mathrm{CAsph}} x_{\mathrm{Asph}}^{n_{\mathrm{Asph}}} + k_{\mathrm{TAsph}} x_{\mathrm{Asph}}^{\eta_{T\mathrm{Asph}}} \tag{7.83}$$

$$r_{\mathrm{HDCCR}} = -\varphi_{\mathrm{CCR}} \eta_{\mathrm{CCR}} k_{\mathrm{CCR}} x_{\mathrm{CCR}}^{n_{\mathrm{CCR}}} + k_{\mathrm{TCCR}} x_{\mathrm{CCR}}^{\eta_{T\mathrm{CCR}}} \tag{7.84}$$

$$r_{\mathrm{HDN_{NBN}}} = -\varphi_{\mathrm{HDN_{NBN}}} \eta_{\mathrm{HDN_{NBN}}} k_{\mathrm{HDN_{NBN}}} x_{\mathrm{HDN_{NBN}}}^{n_{\mathrm{HDNNBN}}} + k_{\mathrm{THDN_{NBN}}} x_{\mathrm{HDN_{NBN}}}^{\eta_{T\mathrm{HDNNBN}}} \tag{7.85}$$

$$\begin{aligned} r_{\mathrm{HDN_{BN}}} = & -\varphi_{\mathrm{HDN_{BN}}} \eta_{\mathrm{HDN_{BN}}} k_{\mathrm{HDN_{BN}}} x_{\mathrm{HDN_{BN}}}^{n_{\mathrm{HDNBN}}} + k_{\mathrm{THDN_{BN}}} x_{\mathrm{HDN_{BN}}}^{\eta_{T\mathrm{HDNBN}}} \\ & -\varphi_{\mathrm{HDN_{NBN}}} \eta_{\mathrm{HDN_{NBN}}} k_{\mathrm{HDN_{NBN}}} x_{\mathrm{HDN_{NBN}}}^{n_{\mathrm{HDNNBN}}} - k_{\mathrm{THDN_{NBN}}} x_{\mathrm{HDN_{NBN}}}^{\eta_{T\mathrm{HDNNBN}}} \end{aligned} \tag{7.86}$$

Catalyst deactivation is originated by coke deposition; however, the estimation of this phenomenon is based on deactivation functions. For HCR reactions, a time-dependent nonselective expression was used, which means that if the selectivity of the HCR reactions is not affected by coke deposition, then the deactivation constant φ is the same for all the reactions. On the contrary, for hydrotreating reactions, a time-dependent selective model was used. In this case, it is necessary to take into account the effect of coke deposition on each hydrotreating reaction, and then there would be a deactivation constant for each reaction, (k_{D_S}, $k_{D_{\mathrm{Asph}}}$, $k_{D_{\mathrm{CCR}}}$, $k_{D_{\mathrm{NBN}}}$, and $k_{D_{\mathrm{BN}}}$). In both cases, the deactivation function is presented in Eq. (7.87), where m_i is the deactivation order for reaction i (HDC, HDS, HDAsph, HDCCR, HDNBN, HDBN).

$$\varphi = \frac{1}{(1 + k_{D_i} t)^{m_i}} \tag{7.87}$$

In the case of the CSTR, Eq. (7.87) can be used directly in the reactor model because the catalyst does not leave the reactor during the operation. However, for the SPR, it is considered that the catalyst does not remain in the reactor; on the contrary, it enters and leaves the reactor along with the feed, then the deactivation function changes to Eq. (7.88).

$$\varphi = \frac{1}{\left(1 + k_{D_i}\left(\frac{Z}{u}\right)\right)^{m_i}} \tag{7.88}$$

where $\frac{Z}{u}$ represents the variation of time that the catalyst remains in the reactor. Also, for the SPR, because the catalyst particles are very small, the effectiveness factor is considered to be the unity for all the reactions ($\eta_i = 1$).

7.3.6 Solution Method

The dynamic models presented in the previous sections for both the CSTR and the SPR are represented by a set of nonlinear differential equations. In the case of the pilot- and industrial-scale SPR, a model is obtained in PDEs since both the composition and the temperature, which are the process variables, change with the axial and radial positions as well as over the course of the process time. These equation systems need to be solved by means of a numerical method. One of the most used methods is the Runge–Kutta method. However, this method can only be applied to systems of ordinary differential equations, so it is necessary to perform a reduction or discretization of the problem in the axial and radial directions. To simulate the behavior of the pilot scale SPR, the orthogonal collocation method will be used to solve the 1D model, which is based on the concept of interpolation of unequally spaced points. Collocation methods have a different perspective from finite difference methods; they belong to the group of so-called spectral methods and are widely used in chemical engineering. One of the advantages of this method is that it does not depend on the geometry of the system and does not need equidistant spacing, unlike finite difference methods. Orthogonal collocation consists of choosing an orthogonal polynomial function (typically Legendre polynomials) to approximate the solution of a differential equation on a given interval. On the other hand, for the 2D model, a finite difference method will be used.

7.4 Numerical Simulations

This section presents the models and simulation results for the different reactors presented in the modeling section. First, the model of a CSTR is presented, which includes mathematical expressions for the kinetics of the HCR and hydrotreating reactions. Subsequently, the case of a tubular SPRs is analyzed for the same reactant system where the reactor model is presented, and the dynamic behavior is analyzed through numerical simulations. Due to certain experimental conditions, it has been determined that the CSTR and the SPR can behave in a similar way. Experimental data obtained in the CSTR were used, and the results were compared with the dynamic behavior of the SPR. Additionally, the dynamic model of an industrial-scale SPR is presented. This reactor has far from ideal behavior, thus the numerical simulations allow to analyze the behavior of this technology at a commercial level. Finally, a parametric sensitivity analysis is presented that will allow exploring the uncertainty of the model as well as possible applications for process control.

7.4.1 Experimental Reactors

To compare the performance of both reactors, the same operating conditions were assumed: An isothermal operation, 891 std m^3/m^3 of H_2–oil ratio, 100 ml of catalyst loaded in the reactors, and a constant pressure of 9.8 MPa. The dimensions of the SPR are 38 cm in length and 7.7 cm in diameter, while the CSTR total volume is 1.7 l. It is also assumed that in both reactors, the heavy oil feed is preheated and mixed with hydrogen and then passed through the reactor.

7.4.1.1 Dynamic Simulations of CSTR and SPR

Martínez and Ancheyta (2012) performed an experimental study to obtain the intrinsic reaction rate coefficients for HCR reactions of heavy oil involving catalyst deactivation and the catalyst effectiveness factor in a continuous STR. In their research, the five-lump 11 kinetic model shown in Figure 7.7 was used. All the reaction rate coefficients and feed properties were taken from that investigation to simulate and compare the performance of both CSTR and SPR reactors. A weight hourly space velocity (WHSV) of 0.98 hours^{-1} and a temperature of 380 °C were considered. The rest of the data can be found in detail in the work of Martínez and Ancheyta (2012).

The dynamic behavior of the small-scale reactors for HCR reactions is presented in Figure 7.8. The continuous line stands for the simulated composition obtained at the exit of the SPR, and the dotted line is the simulated composition obtained in the CSTR. In both simulations, the same volume and weight hourly space velocity were considered. The symbols in each plot represent the experimental data obtained in the CSTR. As can be seen, the simulation of the CSTR is in good agreement with the experimental points. It is also observed that the catalyst starts to deactivate, from the beginning of the reactions, so that the rate of the different HCR reactions reduces until it reaches the steady state near the 150 hours. This situation does not occur in the SPR because the catalyst leaves the reactor along with the hydrocarbon; therefore, a major residue conversion is reached in the SPR in a shorter period, around 5 hours.

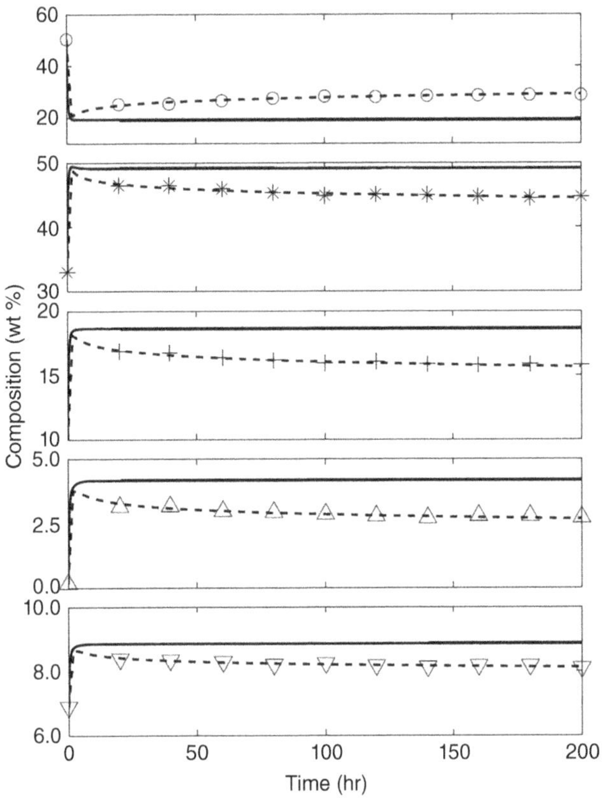

Figure 7.8 Comparison of reactor performance of heavy oil hydrocracking. (–) SPR, (- -) CSTR, (Symbols) Experiment. (o) Residue, (∗) Vacuum gas oil, (+) Middle distillates, (Δ) Naphtha, (∇) Gases.

For thermal and catalytic hydrotreating reactions, the effects of temperature and space velocity were reported in the literature (Martínez and Ancheyta 2012; Martínez and Ancheyta 2014). All reaction rate coefficients were obtained in the kinetic regime; therefore, they are considered to be intrinsic. Since the catalyst used in the experiments is of commercial size, the effectiveness factor is included in the kinetic model for the CSTR. All the parameters needed for the reactor simulations are reported by Martínez and Ancheyta (2012). The dynamic simulation of the small-scale reactors for hydrotreating reactions is presented in Figure 7.9. Similar to HCR reactions, in catalytic hydrotreating, the conversion is favored in the SPR due to the small catalyst size and that the catalyst does not remain inside the reactor, which disfavors catalyst deactivation. On the other hand, the effects of catalyst deactivation and internal diffusion are less severe in hydrotreating reactions compared with HCR reactions, since HCR reactions are considered to occur only on the catalyst surface, and hydrotreating involves thermal and catalytic conversions. In comparison, both reactors have a high conversion at the beginning of the operation because the catalyst has its major activity, as seen in Figures 7.8 and 7.9, for HCR and hydrotreating reactions, respectively. In the case of the CSTR, it is observed that after two hours of operation, the highest conversion is reached, and then the conversion decreases due to catalyst deactivation. Catalyst deactivation is also observed in Figure 7.10 for reactions of HDS and HCR of residue, in which a major effect is presented in HCR reactions due to the heavy molecules present in the liquid phase being easily clogged in the catalyst due to the presence of coke. Evidently, as the reaction

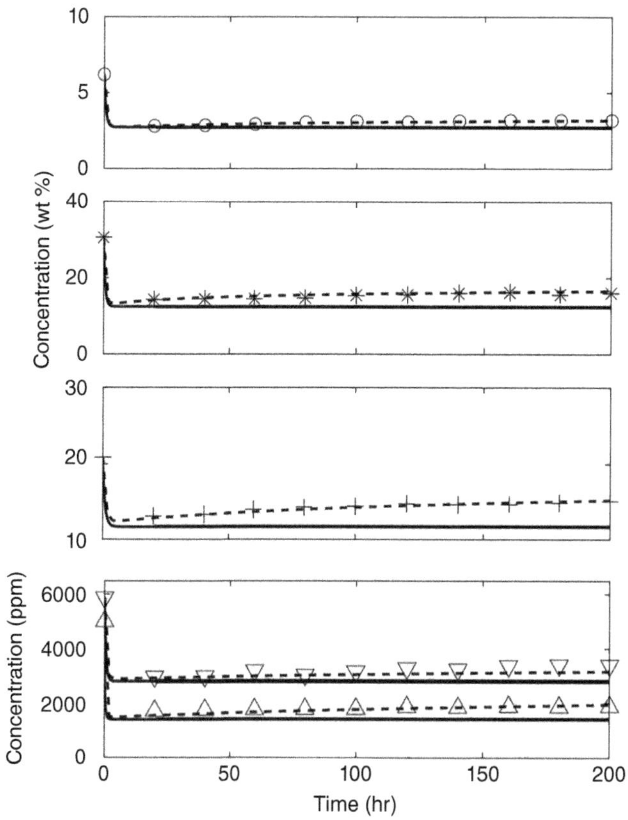

Figure 7.9 Comparison of reactor performance of heavy oil hydrotreating. (–) SPR, (- -) CSTR, (Symbols) Experiment. (o) Sulfur, (∗) Asphaltenes, (+) CCR, (Δ) Basic nitrogen, (∇) Non-basic nitrogen.

time increases, the catalyst deactivation will also increase, generating a negative effect in reactions conversion. In the SPR, the reacting system needs only around two hours to reach the steady state, after that the conversion remains constant while fresh catalyst continuously enters the reactor as the deactivated catalyst leaves it. This behavior is also observed in Figures 7.11–7.13, in which the dynamic axial profile for HCR and hydrotreating reactions, as well as for hydrogen and hydrogen sulfide, is presented. As can be seen, another major conclusion is that in the SPR, a constant axial profile is presented, thus SPR can be modeled as a CSTR.

Figures 7.14 and 7.15 compare the yield of each lump for the conversion of feedstock as a function of the yield of each lump vs the conversion of residue in the SPR and CSTR, respectively. As can be observed in the SPR, a high feed conversion is obtained. The residue can reach a constant 60% of conversion producing lighter fractions after 200 hours of operation, producing mainly vacuum gas oil and distillates that change from 32 and 9 wt.% to 50 and 18 wt.% in the product, respectively. A low production of naphtha and gases is also observed. On the other hand, the CSTR present a similar selectivity; the main difference is observed after a long time of operation, at which the conversion of residue decreases after it reaches its higher value, around 58% of residue conversion. This phenomenon arises from catalyst deactivation, as the catalyst remains inside the CSTR during operation, leading to significant deactivation caused by coke deposition phenomenon is originated due to catalyst deactivation since in the CSTR, the catalyst remains inside the reactor during the operation then there is a great deactivation due to coke deposition.

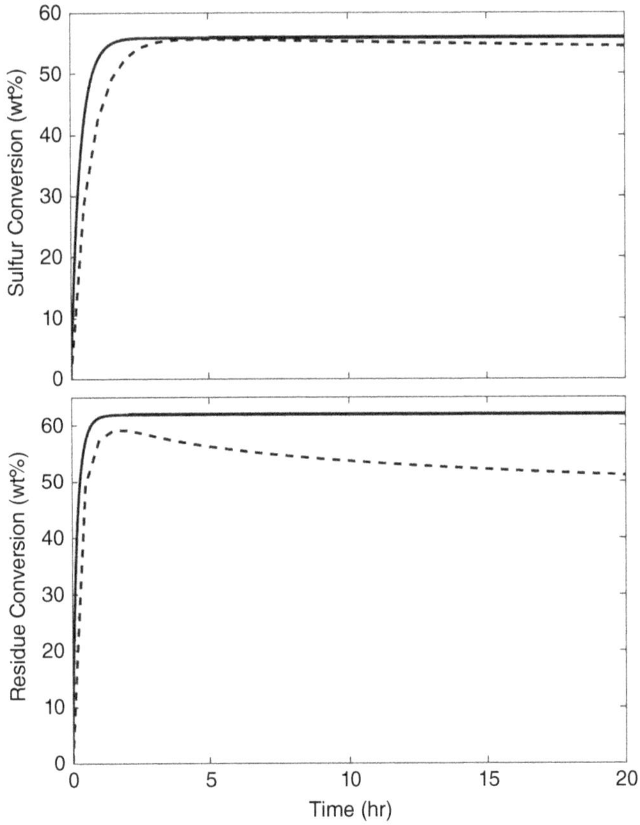

Figure 7.10 Influence of catalyst deactivation in hydrodesulfurization and hydrocracking reactions. (–) SPR, (- -) CSTR.

7.4.1.2 Steady-State Simulations of a SPR

Once the SPR model was validated with the CSTR experimental data, it was used to simulate all the HCR and hydrotreating reactions simultaneously. The feed velocity (WHSV) and temperature vary from 0.98 to 2.56 hours^{-1} and from 380 to 420 °C, respectively, as shown in previous experimental reports (Martínez and Ancheyta 2012; Martínez and Ancheyta 2014). Figures 7.16 and 7.17 show the simulation results of hydrocarbon and impurities compositions at the exit of the reactor once the steady state has been reached. Figure 7.12 shows the steady-state lumps composition. Since residue is only present as a reactant in the kinetic model, a linear effect between the WHSV and temperature variation is observed. In this case, increasing both the residence time of the feed and temperature in the reactor increases the residue conversion; therefore, a low composition of residue would be present in the hydrocarbon product. On the contrary, the vacuum gas oil appears in the kinetic model as both reactant and product; therefore, VGO conversion depends on the selectivity of residue and lighter fractions. For WHSV between 0.98 and 2.56 hours^{-1} and temperatures below 410 °C, a linear function in the VGO composition is observed, which indicates that the rate of HCR of residue is faster than that of the HCR of VGO. For temperatures higher than 410 °C, low composition of VGO is observed, which is sign of a high HCR of this pseudo-component to produce lighter fractions, such as middle distillates, naphtha, and gases. At a steady state, the light fractions

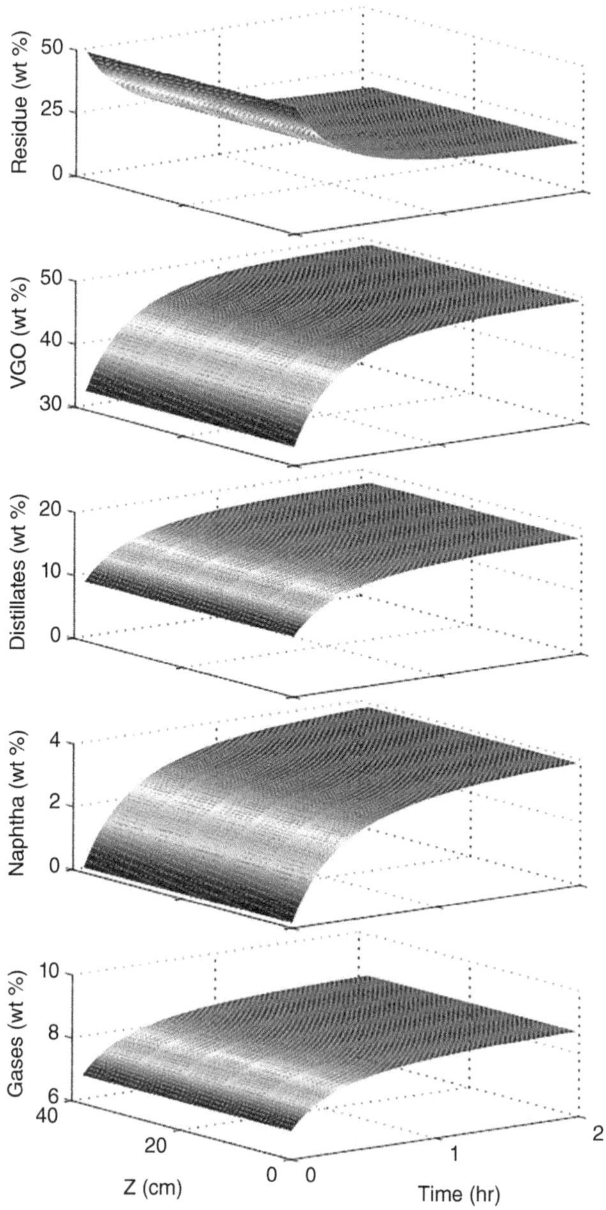

Figure 7.11 Dynamic axial profile in the SPR for hydrocracking reactions.

such as distillates, naphtha, and gases exhibit a constant and linear behavior. In those cases, as well as residence time and temperature increases, the conversion of residue and vacuum gas oil is higher, and a major composition of these fractions is present in the product. A similar effect is shown in Figure 7.17 for impurities contents, in which a high removal of impurities such as sulfur, asphaltenes, Conradson carbon, and nitrogen is observed as temperature and residence time increase.

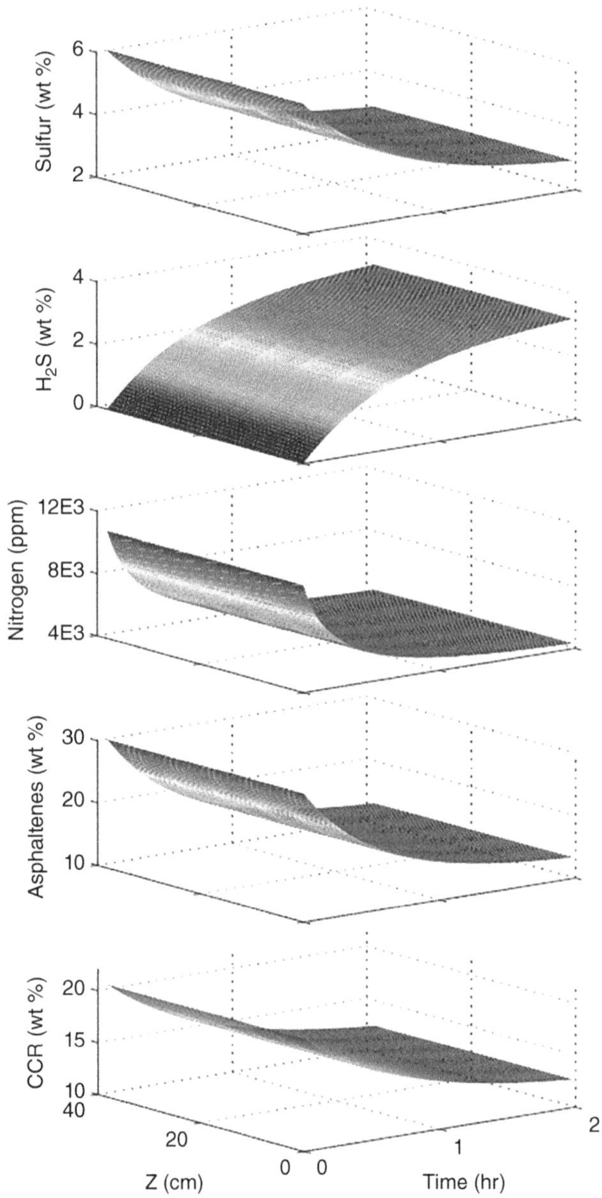

Figure 7.12 Dynamic axial profile in the SPR for hydrotreating reactions.

7.4.2 Industrial-Scale Reactor

Unfortunately, the literature only reports operating temperature and pressure ranges as well as approximate conversions reached by the residue (538 °C+) in a stable operation, as summarized in Table 7.1 for the different technologies documented by Zhang et al. (2007) and Bellussi et al. (2013). Consistent data of a dynamic operation do not exist as far as the knowledge of the authors; thus, an alternative to validate the model is to compare the predictions of the two-directions model

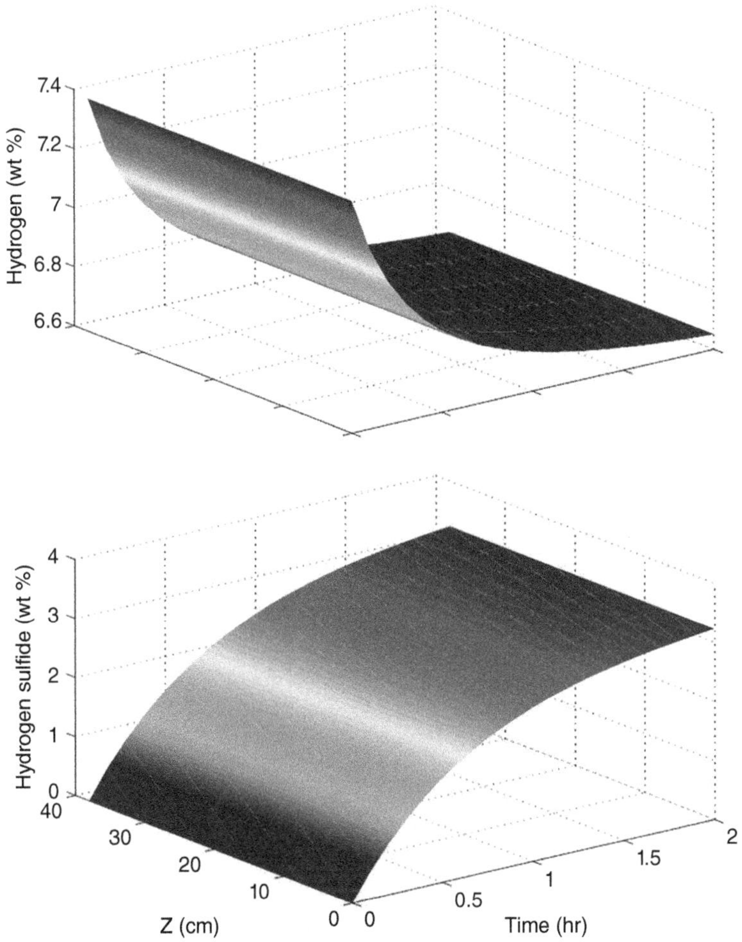

Figure 7.13 Dynamic axial profile of gases in the SPR.

presented above with predictions of a previously validated model used to describe the behavior of an experimental reactor. In such a case, the reactor presented in the previous section will be used as a reference. It is important to mention that this experimental operation was performed isothermally (the energy balance is not required) in an SPR of 38 cm in length and 7.7 cm in diameter at 891 std m^3/m^3 of H_2 – hydrocarbon ratio with 100 ml of catalyst loaded in the reactor, and a constant pressure of 9.8 MPa. The heavy oil feed was preheated, mixed with hydrogen, and fed to the reactor with an LHSV of 0.5. Although the model of the experimental reactor is based only on axial dispersion, it is expected that the predictions of both models are similar, since in the experimental reactor the diffusive effects in the radial direction are minimal due to the high axial velocity and the small diameter size. Such results are shown in Figure 7.18, in which the error variable is defined as the difference between the prediction with the ADM and the prediction with the two directions model. As can be seen, if the order of magnitude of the error is near zero, then both models predict the same response.

Once the model has been verified, what follows is to extrapolate the behavior of the reactor from experimental to industrial scale. Figure 7.19 shows a comparison between the lump composition as

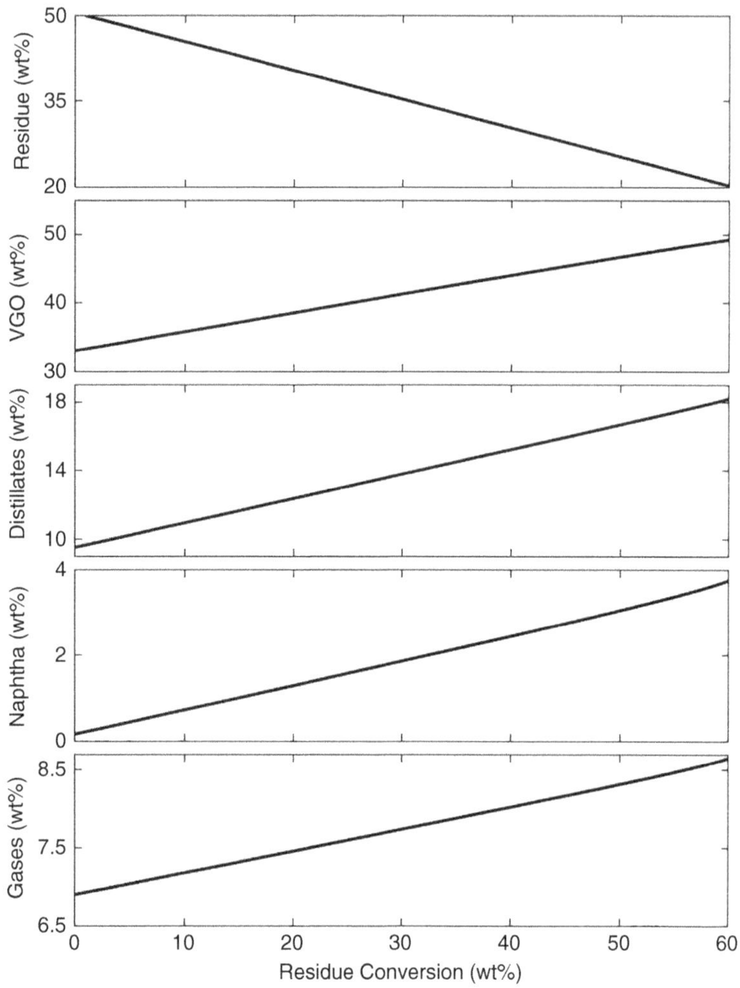

Figure 7.14 Yield of hydrocracked lumps as a function of residue conversion in the SPR.

a function of residue conversion (Figure 7.19a) and as a function of time (Figure 7.19b) for the experimental and industrial reactors. It is expected that the commercial-size reactor exhibits better conversions than the experimental-size reactor, even if they are simulated for the same type of catalyst and operating conditions, since the industrial reactor operates adiabatically and the experimental reactor isothermally. This was indeed corroborated since the industrial reactor reached 70% residue conversion compared with 60% achieved in the experimental reactor, as can be observed in Figure 7.19a. This increase in conversion is mainly because in the industrial-scale reactor, there is a better flow regime, which increases the dispersion of the components in the liquid phase and directly promotes the HCR reactions. Figure 7.19b shows another interesting effect, which is that since the industrial-scale reactor reaches higher residue conversion, the composition of vacuum gas oil increases, which leads to a higher vacuum gas oil HCR that contributes to a higher composition of middle distillates and naphtha during the operation.

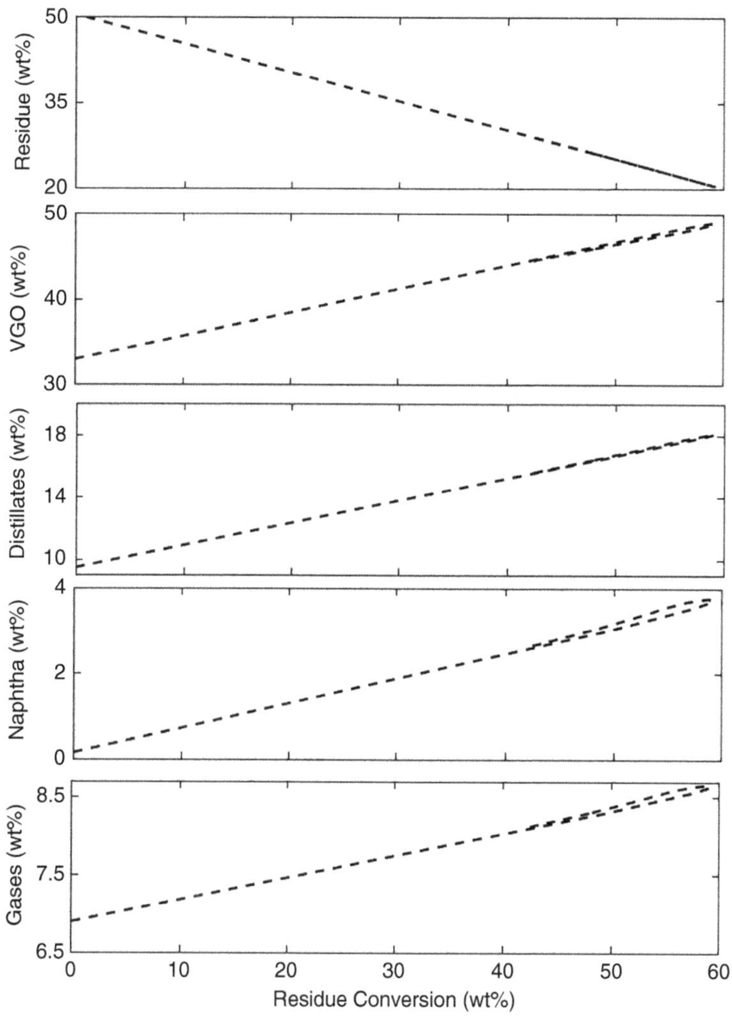

Figure 7.15 Yield of hydrocracked lumps as a function of residue conversion in the CSTR.

7.4.2.1 Dynamic Simulations of the Industrial Slurry-Phase Reactor

In order to simulate the operation of a commercial-size reactor (20 m in length and 2 m in radius), an H_2-to-oil ratio of 2000 std m^3/m^3, 1% of catalyst content in the slurry phase, a constant pressure of 13 MPa in the reactor, an inlet feed and initial temperature of 420 °C, and an LHSV of 0.4 hours^{-1} were assumed as operating conditions. Kinetic parameters for HCR and hydrotreating reactions and feed properties used in the simulations were reported by Martínez and Ancheyta (2012). As the model is proposed to describe a commercial unit, it would be necessary to analyze the temperature profile distribution since this type of reactor operates adiabatically. Figure 7.20 shows the dynamic behavior of reactor temperature in the axial and radial directions. In the upper part of the figure, the temperature surface at the initial time is presented, in which it is observed that the temperature remains constant throughout the reactor. The following surfaces show the temperature profile at 6, 12, 24, 30, and 42 minutes, while the surface at the bottom shows the temperature distribution at steady state after two hours of operation.

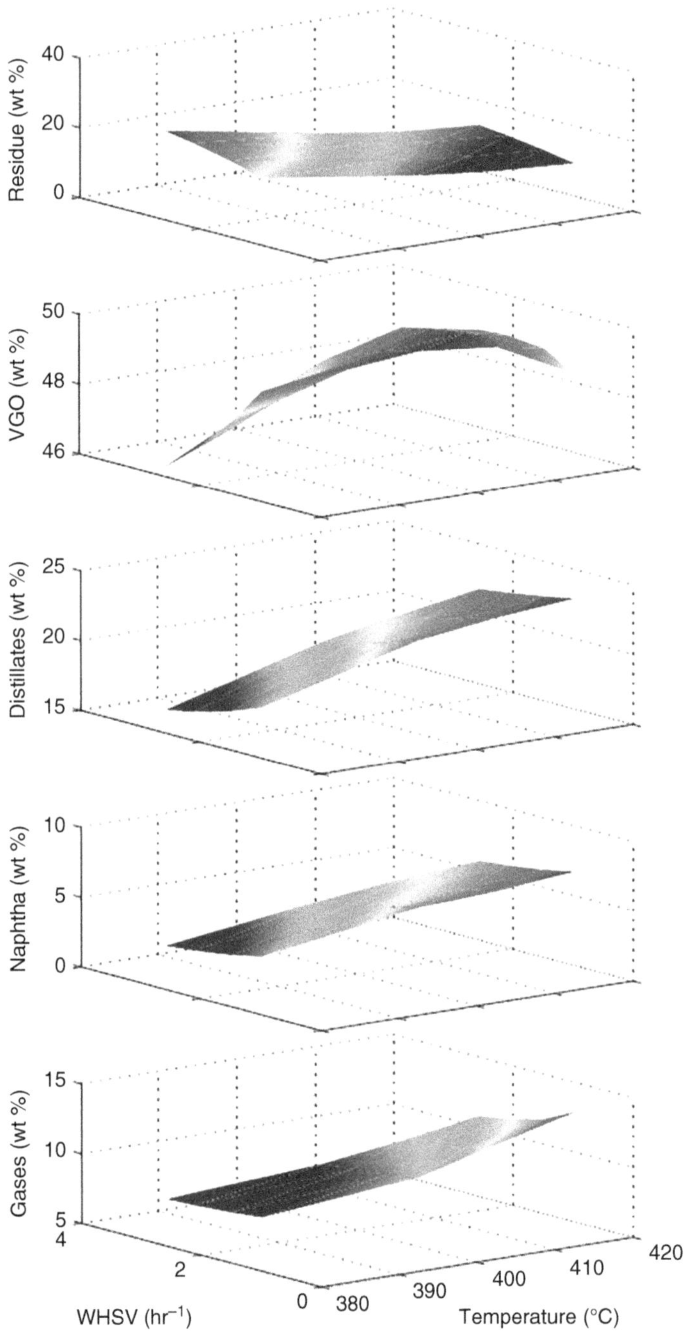

Figure 7.16 Steady-state simulations of hydrocracking reactions.

From this figure, it is possible to appreciate in a remarkable way that, contrary to what would be expected in a large-diameter reactor, the temperature profile remains practically constant in the radial direction during the whole time of operation. This effect is not presented in the axial profile in which a temperature change can be observed from the inlet of the feed through the output of the products. This temperature gradient is approximately 1–2 °C. Nevertheless, for

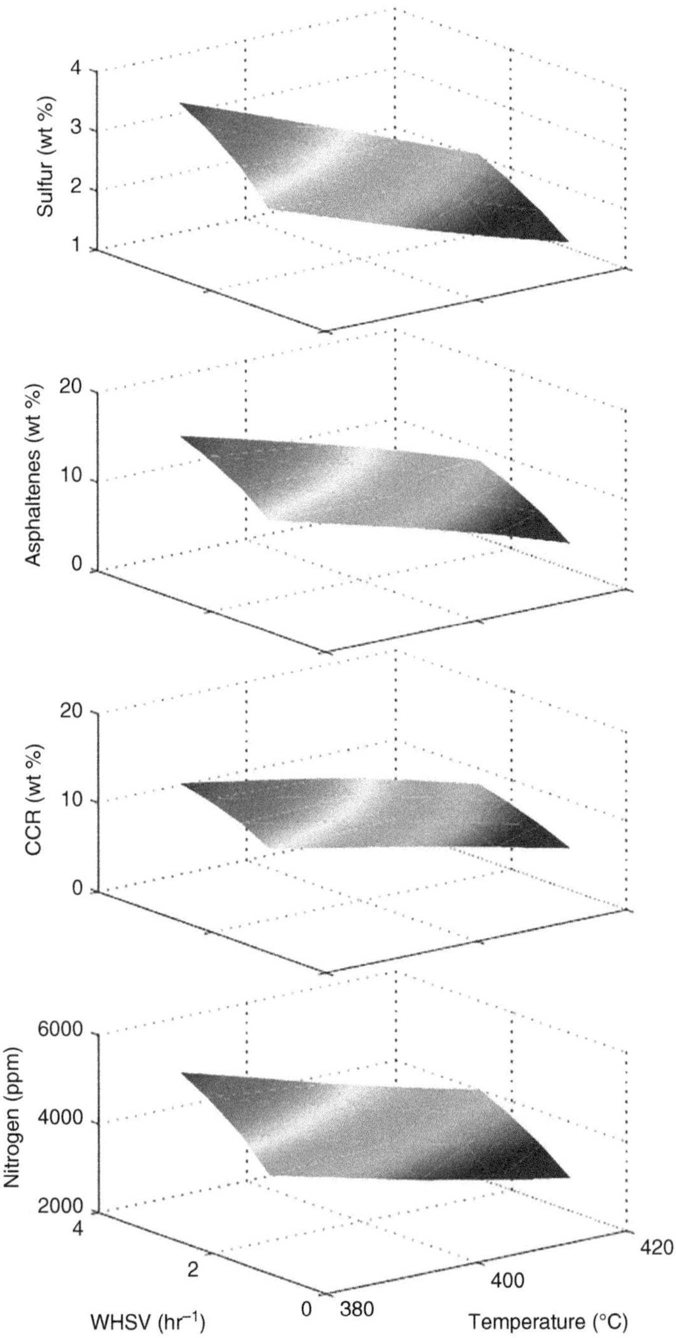

Figure 7.17 Steady-state simulations of hydrotreating reactions.

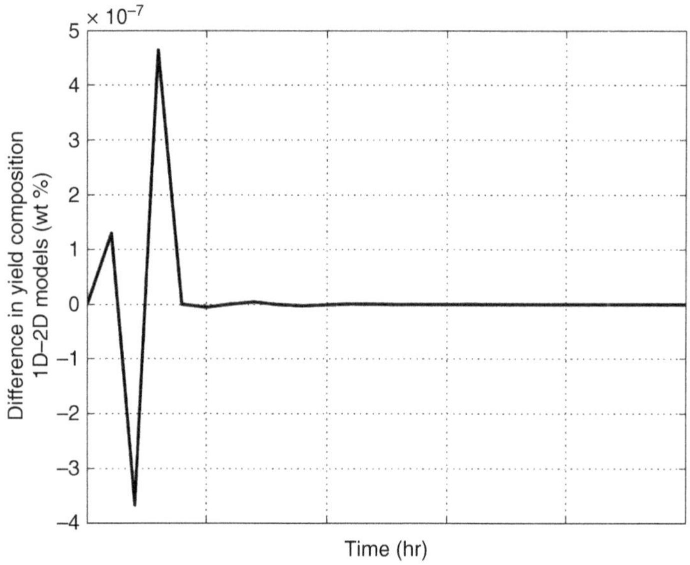

Figure 7.18 Dynamic error of residue composition between the prediction with the axial dispersion model and the axial and radial dispersion model.

simplification purposes, the operation can be considered as almost isothermal, which is one of the main characteristics of SPRs.

Figures 7.21 and 7.22 show the dynamic profiles of HDC and HDT reactions, respectively, at 0, 6, 12, 24, 30, and 42 minutes and at steady state after two hours of operation. In the case of HDC reactions, a decrease in the residue content in the feed of approximately 30% can be observed, while the intermediate fractions of VGO, middle distillates, and naphtha present an increase in their composition until the system reaches the steady state. Naphtha is a product of the HCR of the heavier components, with three direct pathways for its formation with reaction rates (k_3, k_6, and k_8). While its decomposition to light gases occurs with a reaction rate (k_{10}), under the considered operating conditions, the reaction rate of naphtha formation is greater than its decomposition, which is the reason why there is a considerable increase in the composition of naphtha in the steady-state operation. On the other hand, similar to the surface response for temperature, the composition of each of the pseudo-components remains constant in the radial direction, while the axial composition presents variations of around 1 wt.% for each time of operation. This small change in the composition with respect to the entrance and the exit is due to the low feed velocities and high residence times necessary to achieve high conversion rates. Similar behavior can be observed in Figure 7.22, in which the dynamic variations of sulfur, asphaltenes, CCR, and nitrogen are presented. In such cases, there is a decrease in the composition of around 60, 65, 43, and 40%. As well as for the HDC, the composition of HDT components also remains constant in the radial position.

Figure 7.23 shows the surface of residue conversion as a function of pressure and temperature. As expected, an increase in temperature would lead to an increase in residue conversion. The effect of pressure is linear since in the model considerations hydrogen is fed in excess. According to the information reported in Table 7.1 for industrial technologies, for temperatures between 460 and 480 °C, residue conversions of more than 95% can be reached. However, although most of the residue has been hydrocracked, the formation of light gases increases considerably as seen in

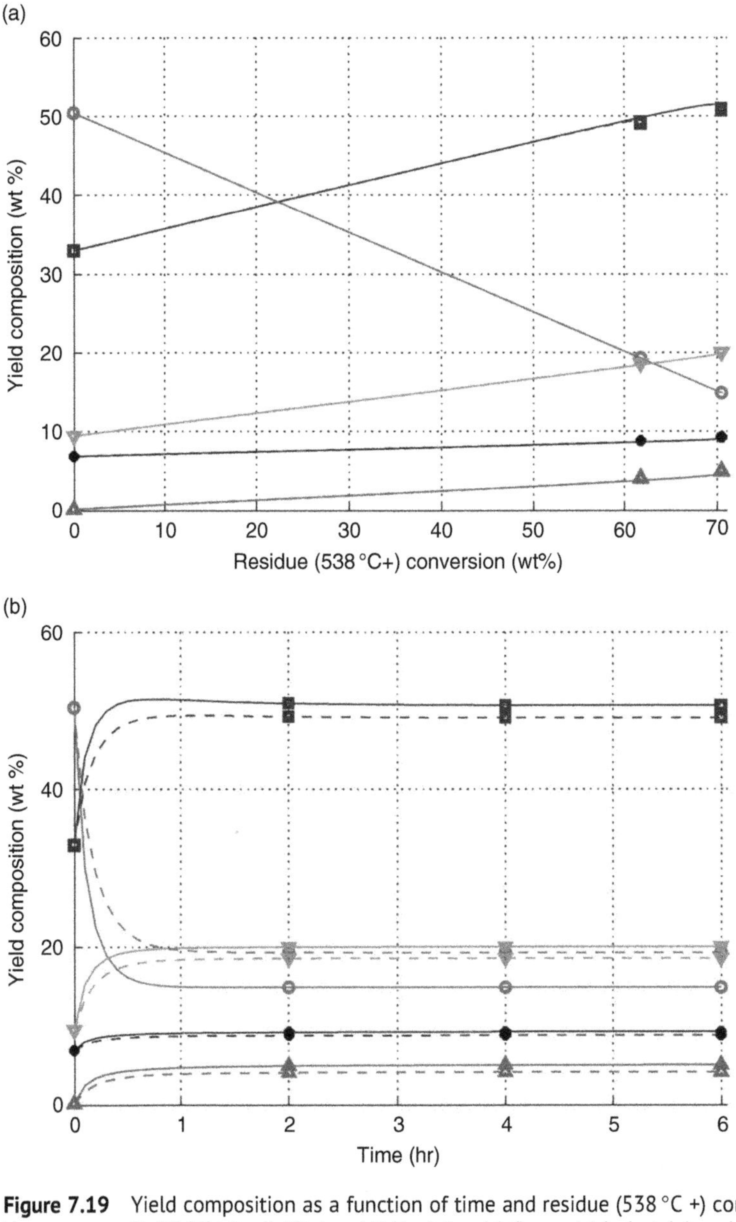

Figure 7.19 Yield composition as a function of time and residue (538 °C +) conversion. (o) Residue, (□) Vacuum gas oil, (∇) Middle distillates, (Δ) Naphtha, (∗) Gases. (–) Industrial-scale reactor, (--) Experimental scale reactor.

Figure 7.24 due to thermal effects. It can also be observed that an optimal operation that promotes high composition of valuable products can be reached at temperatures between 400 and 440 °C.

7.4.2.2 Sensitivity Analysis for the Industrial Slurry-Phase Reactor

Parametric sensitivity is a tool that allows to analyze the effect of the different process variables on the dynamic behavior and steady state of a process and therefore the predictions of the

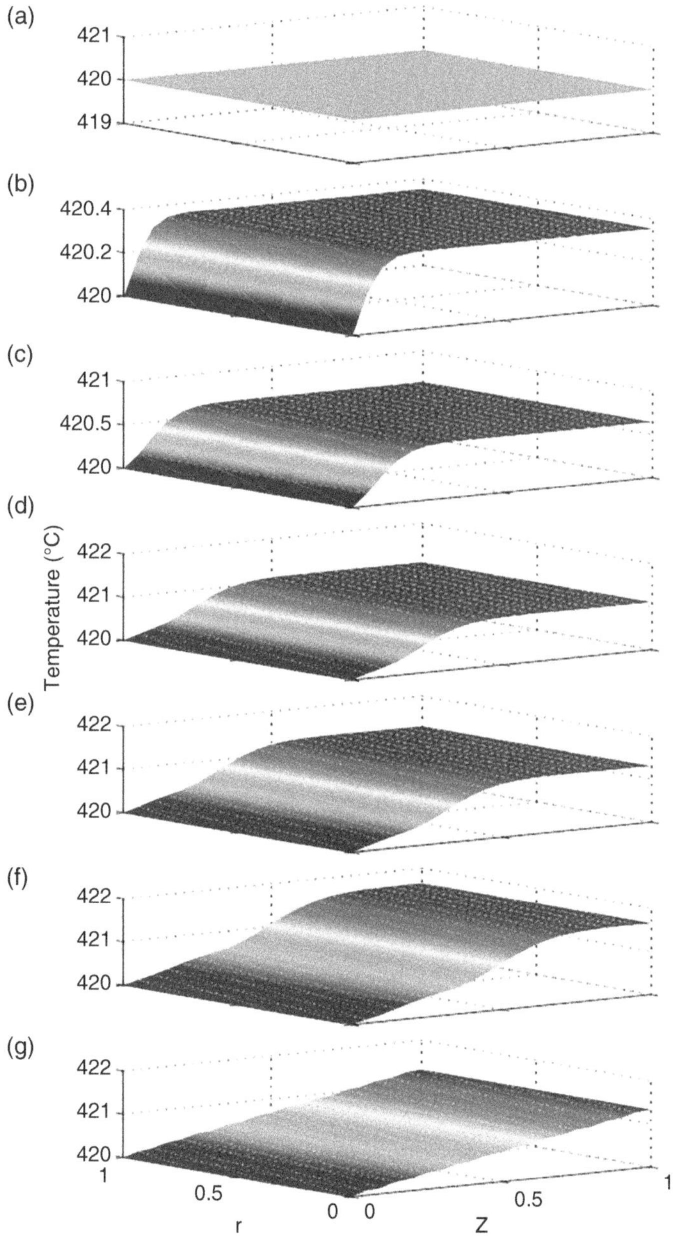

Figure 7.20 Dynamic profile of reactor temperature in axial and radial positions. (a) 0, (b) 6, (c) 12, (d) 24, (e) 30, (f) 42, and (g) 120 minutes.

mathematical model, whether they are process variables (space velocity, catalyst fraction, H_2-to-oil ratio, initial reactor temperature, feed inlet temperature, and reactor operating pressure) or model parameters in which there is a certain degree of uncertainty, for example, the heat of reaction, the axial dispersion coefficient, and the convective coefficient of heat transfer. Figures 7.25–7.27 show the response behavior for the variables and parameters mentioned above. The values that were used as a base case scenario are those proposed in previous sections for the simulation of the commercial unit. From the base case solution, all the parameters of the model were varied by $\pm 50\%$ individually, thus the change percentage is

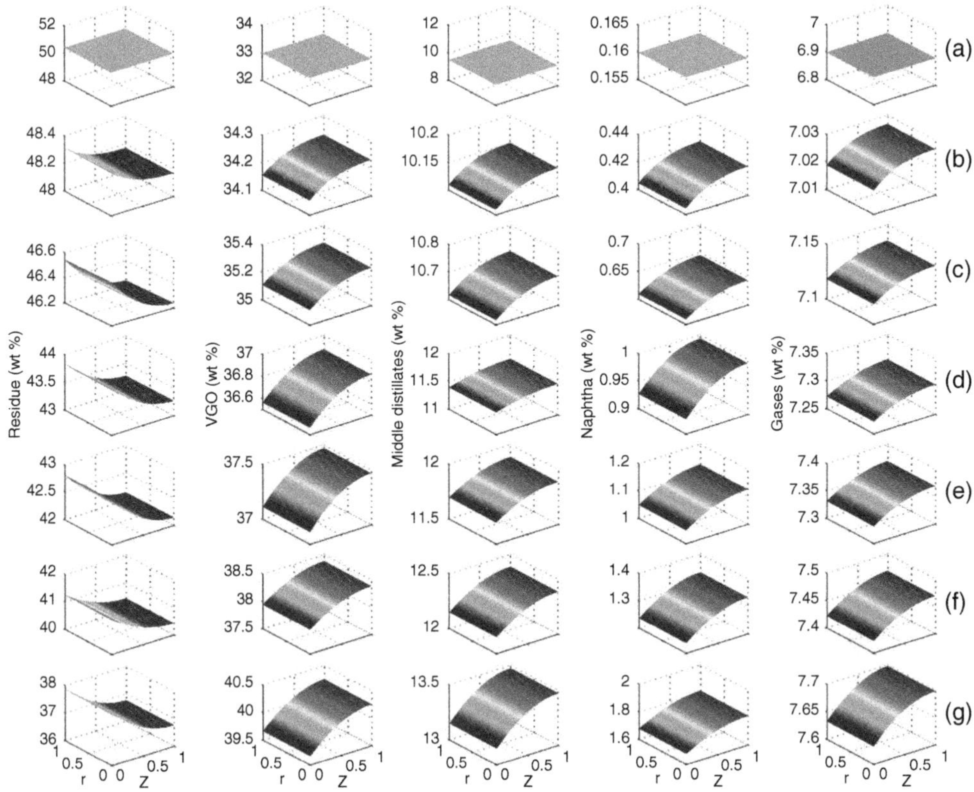

Figure 7.21 Dynamic profile of pseudo-components for HDC reactions in axial and radial positions. (a) 0, (b) 6, (c) 12, (d) 24, (e) 30, (f) 42, and (g) 120 minutes.

defined as $(y_{\text{base case}} - y_{\text{changed}})/y_{\text{base case}}$. In the case of temperatures and operating pressure, simulations were carried out between 400 and 440 °C and from 9 to 17 MPa.

Figure 7.25 shows the variation in the steady-state prediction for changes in the LHSV, catalyst fraction, and H_2-to-oil ratio, from which it was concluded that the HDC and HDT reactions are more sensitive to changes in the solid fraction and H_2-to-oil ratio than LHSV. On the one hand, increasing the catalyst content increases the HCR mainly of the heavier fraction, which finally decreases its composition and gives rise to the increase in the composition of the intermediate fractions. On the other hand, increasing the H_2-to-oil ratio favors the conversion of the intermediate fractions. Another favorable effect on the conversion of residue is the increase in the residence time of the oil in the reactor, which is achieved by decreasing the LHSV. In this case, decreasing LHSV would increase the conversion up to 6% with respect to the previous results. In the case of HDT reactions, a similar effect is observed for LHSV and H_2-to-oil, however with a lower degree of variation. Finally, it can be observed that the effect on the exit temperature is minimum for the three process variables, where the percentages of change are less than 1%. Figure 7.26 shows the changes in the process variables for variations in the temperature and operating pressure, as well as the feed temperature, where it can be observed that these three process variables are the ones that most affect the conversion of the HDC, HDT, and output reaction temperature, generating variations between ±15%. Finally, in Figure 7.27, the variations in the conversion are presented in case of uncertainty in any of the parameters of the model. In general, it can be observed that the errors that occur for any of the parameters are minimal, which means that the accuracy of the predictions of the model is not affected by using correlations that have been formulated for other systems or for

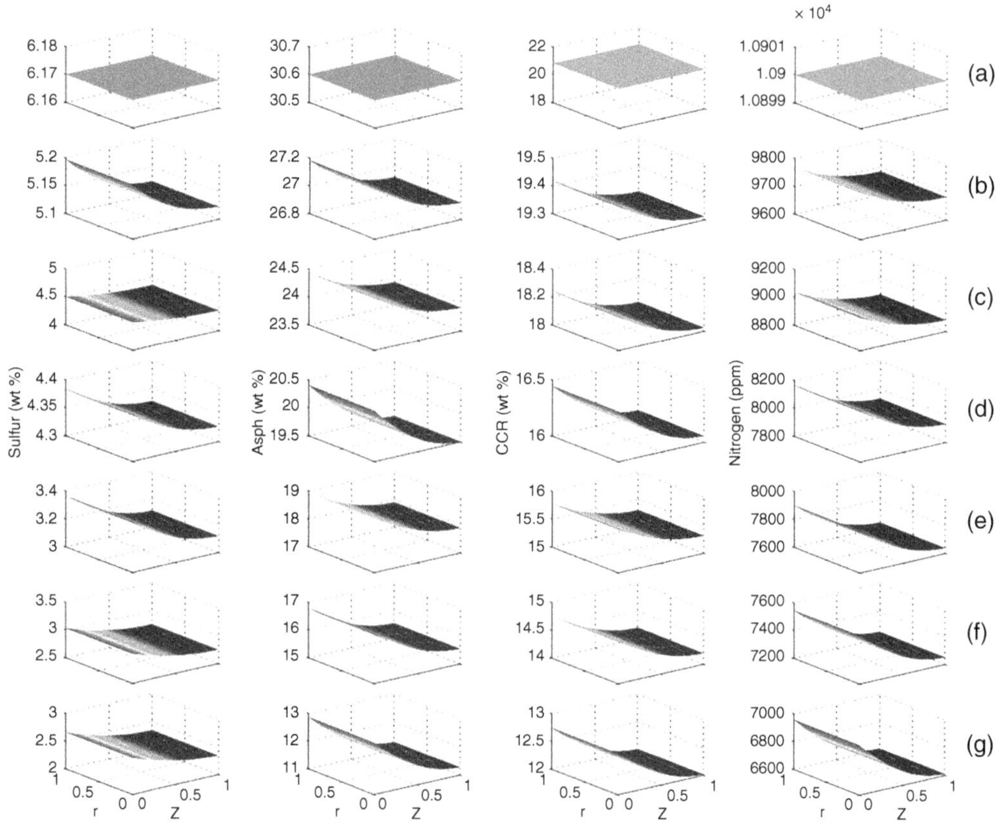

Figure 7.22 Dynamic profile of components for HDT reactions in axial and radial position. (a) 0, (b) 6, (c) 12, (d) 24, (e) 30, (f) 42, and (g) 120 minutes.

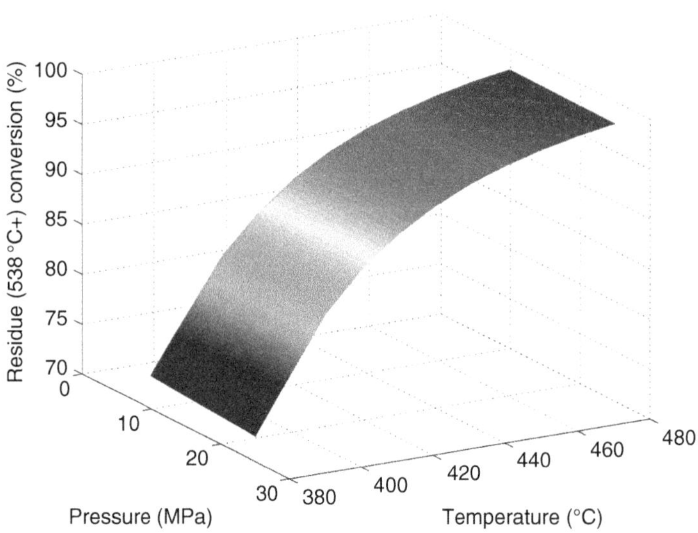

Figure 7.23 Pressure and temperature effects on residue conversion at steady state.

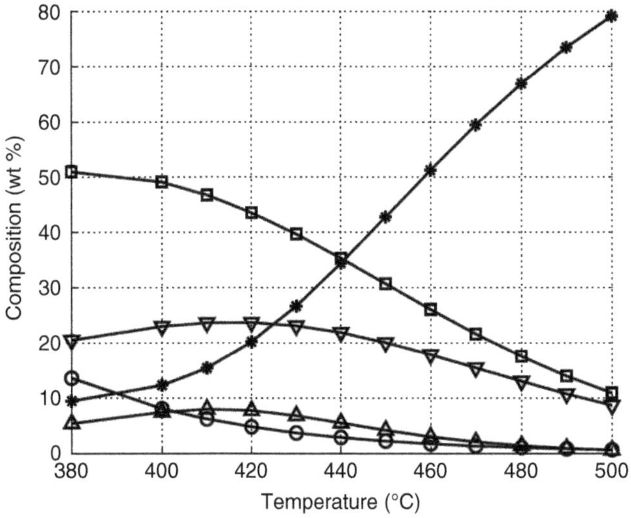

Figure 7.24 Temperature effect on lumps compositions. (o) Residue, (⊡) vacuum gas oil, (∇) middle distillates, (Δ) naphtha, (∗) gases.

utilizing constant values for these parameters. In a general way, the process variable most sensitive to these changes is the residue fraction in the case of HDC and the composition of asphaltenes for HDT. A particular case is the outlet temperature of the reactor, which is greatly affected when changes occur in the initial temperature of the reactor or the inlet feed temperature.

7.5 Conclusions

Mathematical modeling of SPR is a difficult task, mainly due to the complexity of the hydrodynamics involved. It has been found that the main variable is the gas velocity, which has an influence on almost all the parameters involved in the model such as the flow regime, holdup, dispersion and mass transfer coefficients, and even particle settling velocities. The usual modeling configuration is in the form of SBCR in which ADMs are the most common formulation. Although there is much uncertainty in the parameters used in these models, ADMs have been preferred and extensively studied for different types of synthesis as well as for hydrogenation reactions. In addition, under the conditions typically set for these reacting systems, the flow approximation commonly used is the heterogeneous regime for the gas phase, divided as the SB phase moving with constant velocity along the reactor and the LB phase.

Moreover, only a few authors have dealt with HCR reactions in SPR, though in most cases using the CFD modeling approach despite the computational effort. This may be induced by the uncertainty of utilized correlations for hydrodynamic parameters that were obtained under ambient conditions, at severe conditions of temperature and pressure. However, some of the correlations presented in previous sections have been successfully implemented in industrial operations of *F–T* synthesis as well as for HDS systems.

One of the main disadvantages of modeling HCR in continuous operation is the lack of experimental information. Therefore, the existing models have been validated using experimental information obtained for different systems. For example, Matonis et al. (2002) and Matos et al. (2009)

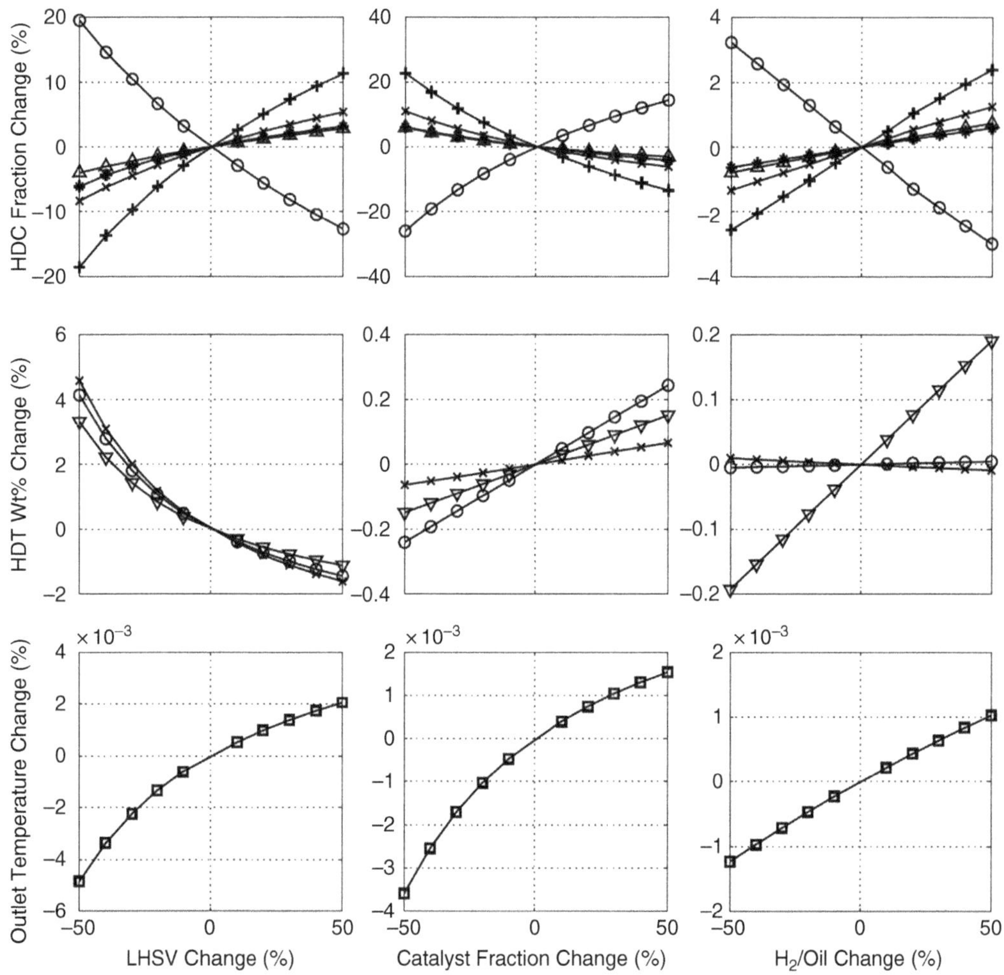

Figure 7.25 Sensitivity analysis of HDC; (o) residue, (Δ) vacuum gas oil, (×) middle distillates, (+) naphtha, (∗) gases, HDT; (o) sulfur, (Δ) asphaltenes, (×) CCR and outlet temperature for LHSV, catalyst fraction and H_2-to-oil ratio changes.

compared their hydrodynamic models with the information previously obtained for air–water systems under different operating conditions, and after that they simulated the behavior including the HCR kinetics. So far, only the ADM of Khadem-Hamedani et al. (2015) for HDS has been validated with experimental data obtained previously by Deng et al. (2010) for this system. In this case, it should be noted that favorable results were obtained with heterogeneous models.

Furthermore, only a few models include an equation for the solid dispersion inside the reactor for HCR systems, mainly because slurry particles, as long as they are small ($d_p \leq 60\,\mu$m), follow the liquid motion except at very high slurry loadings exceeding 20–30%. Due to the operating conditions, all the models developed for HCR in SPR are formulated under the homogeneous bubble flow.

HCR as well as catalytic and thermal hydrotreating were modeled in small-scale isothermal CSTR and SPR based on intrinsic reaction rate coefficients involving catalyst deactivation and catalyst effectiveness factors for each of the reactions. The dynamic simulations have shown similar behavior in the CSTR and SPR. As expected, higher conversion of residue is reached in the SPR (effectiveness factor equals one) and deactivation of catalyst is not taken into account. The results

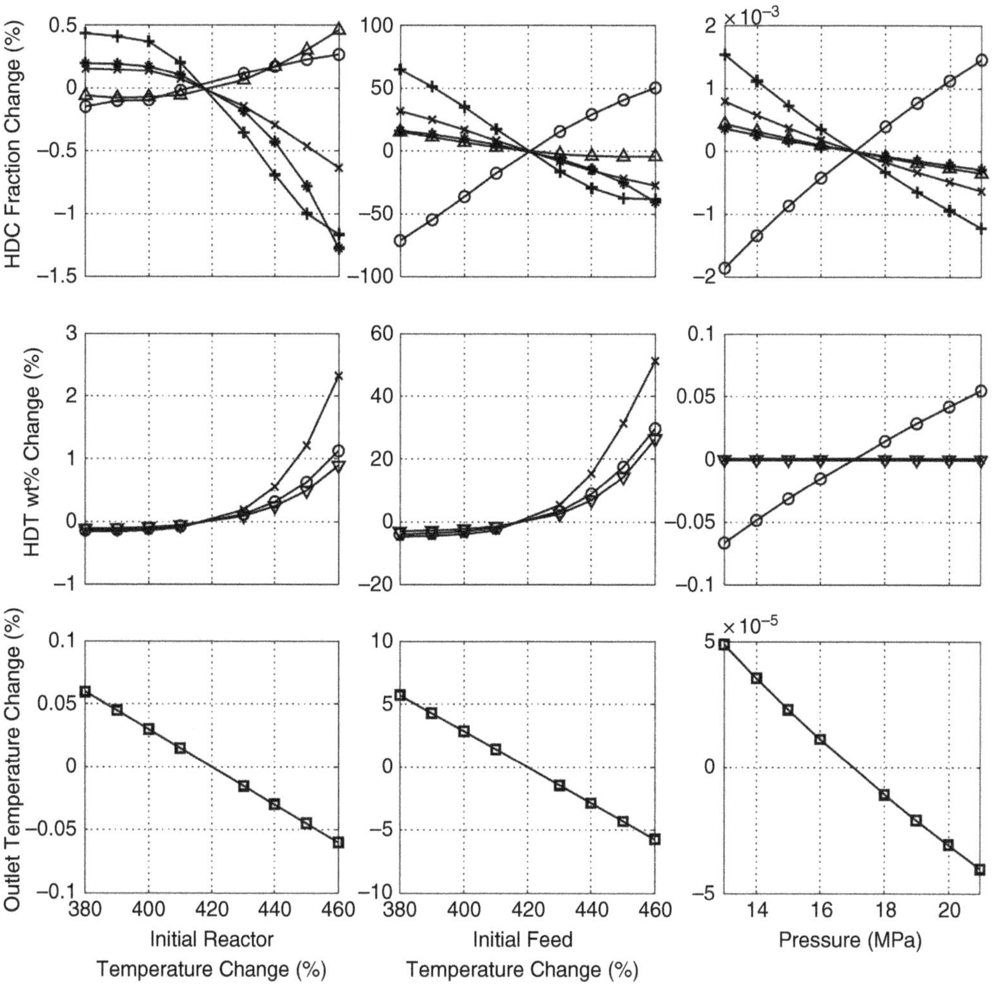

Figure 7.26 Sensitivity analysis of HDC; (o) residue, (Δ) vacuum gas oil, (×) middle distillates, (+) naphtha, (∗) gases, HDT; (o) sulfur, (Δ) asphaltenes, (×) CCR and outlet temperature for inlet reactor temperature, inlet feed temperature, and pressure changes.

of steady-state simulation have shown that it is possible to approximate an SPR for HCR with an ADM with appropriate reaction rate coefficients.

As expected, for experimental purposes, the CSTR has some advantages compared with the SPR. It promises lower cost in the experimental framework due to the amount of catalyst used. It was also found that for modeling purposes, the experimental SPR can be conveniently approximated by means of an ADM; however, to reduce its complexity, the SPR can be approximated to an ideal CSTR model.

A commercial-size SPR for heavy oil HCR and hydrotreating was also modeled, considering axial and radial dispersion effects for the state variables. The model equations were partially discretized with finite central differences in the positional derivatives and adequately solved by a Runge–Kutta method coded in MATLAB in subroutine ode32s and ode45 with a computing time of around 3 minutes for 10 and 5 axial and radial nodes. The operation was assumed to be non-isothermal; however, simulations have shown that the reactor operates nearly isothermal, with a difference between inlet and outlet temperature of 1–2 °C. Also, the dynamic and steady-state simulations have shown that a

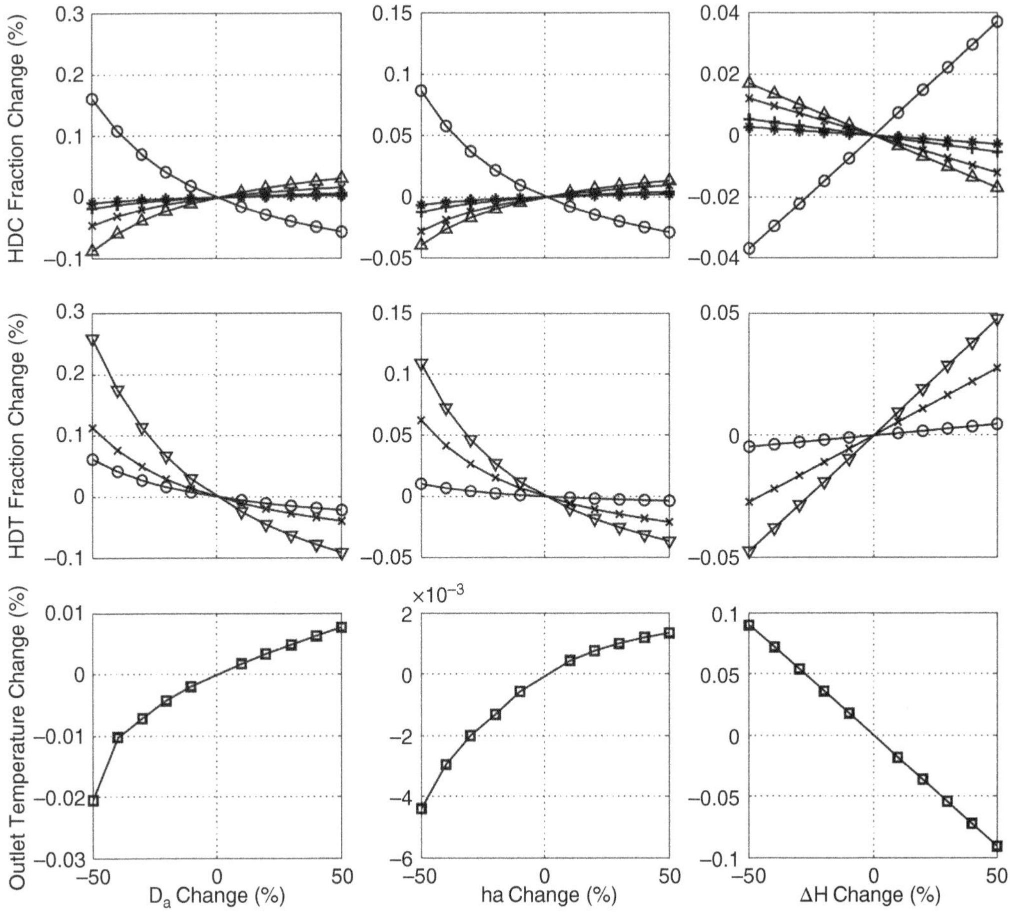

Figure 7.27 Sensitivity analysis of HDC; (o) residue, (Δ) vacuum gas oil, (×) middle distillates, (+) naphtha, (∗) gases, HDT; (o) sulfur, (Δ) asphaltenes, (×) CCR and outlet temperature for axial dispersion coefficient (D_a), convective heat transfer coefficient (h_a), and heat of reaction (ΔH) changes.

constant radial profile can be exhibited due to the large H_2-to-oil ratio and the high residence time of the feed within the reactor. Therefore, an ADM represents a good approximation to describe the system for dynamics and steady-state performances. It was also shown by a sensitivity analysis that the parameters the model requires can be well estimated through existing correlations since variations in the values of these parameters do not produce a high variation in the model predictions, in the case of axial dispersion coefficient, convective heat transfer coefficient, and heat of reaction.

Nomenclature

Symbols

A	Bed-wall reactor heat transfer area, cm^2
a	Interfacial area per unit reactor volume, cm^2/cm_r^2
C_T^f	Total concentration of f phase, g/cm^3
C_i^*	Concentration of compound i in the liquid in equilibrium with the gas phase, mol i/cm_L^3

C_i^f Molar concentration of component i in the f phase, mol i/cm_f^3
$C_{f,i}^{S_{inn}}$ Molar concentration of component i inside the solid filled with f phase, mol i/cm_f^3
$C_{f,i}^{S_{sur}}$ Molar concentration of component i at surface of the solid covered by f phase, mol i/cm_f^3
C_S Solids concentration, g/cm_{SL}^3
C_V Volumetric solids concentration, in the slurry cm_S^3/cm_{SL}^3
Cp_f Specific heat capacity of f phase, $J/g_f \cdot K$
$D_{M,i}^f$ Molecular diffusion coefficient of component i in the f phase, $cm_f^3/cm_S \cdot s$
D_a^f Mass axial dispersion coefficient of f phase, cm_r^2/s
D_r^f Mass radial dispersion coefficient of f phase, cm_r^2/s
D_{ei}^f Effective Fickian diffusivity of component i inside a porous catalyst, $cm_f^3/cm_S \cdot s$
D_C Column diameter, cm
d_p Particle diameter, cm
d_S Bubble mean diameter, cm
F Interfacial drag force, N/m^3
g Gravity constant, cm/s^2
$H_{e,i}$ Henry's law coefficient of component i, $MPa \cdot cm_L^3/mol\ i$
h_{GS} Gas-solid heat transfer coefficient, $J/cm_f^2 \cdot s \cdot K$
h_{GW} Gas-wall heat transfer coefficient, $J/cm_f^2 \cdot s \cdot K$
h_{LW} Liquid-wall heat transfer coefficient, $J/cm_f^2 \cdot s \cdot K$
K Overall LB–SB mass transfer coefficient, $cm_{LB}^3/cm_I^2 \cdot s$
k Intrinsic reaction rate constant, (for first-order reaction) 1/s
k_i^{GS} Gas-solid mass transfer coefficient for i component, $cm_G^2/cm_S^2 \cdot s$
k_i^{LS} Liquid–solid mass transfer coefficient for i component, $cm_L^2/cm_S^2 \cdot s$
k_{Li} Mass transfer coefficient from gas–liquid interface to liquid phase, $cm_L^3/cm_I^2 \cdot s$
K_L Overall gas–liquid mass transfer coefficient, $cm_L^3/cm_I^3 \cdot s$
L Equivalent longitude, cm
LB Large bubbles
MM_f Molecular weight of f phase, g/mol
N Molar flux, mol/cm^2
N_C^f Number of compounds in the f phase
$N_{R,f}$ Number of reactions in the f phase
P Pressure, MPa
P_S Saturation pressure, MPa
P_T Total pressure, MPa
R Gas law constant, $J/mol \cdot K$
r Referring to radial position, cm
$r_j'^f$ Rate of reaction j per unit of catalyst mass in the f phase, mol $i/g_S \cdot s$
SB Small bubbles
t Time, s
T_f Temperature of f phase, °C, K
U_f Superficial velocity of f phase, cm/s
u_f Velocity of f phase, cm/s
u_{tS} Solid terminal settling velocity, cm/s
V Reactor volume, cm_r^3

V_f	Volume of f phase, cm_f^3
W	Weight fraction of solid in the feed slurry, g_S/g_{SL}
X_W	Weight fraction of the primary liquid in the mixture ($1 \leq X_W \leq 0.5$), dimensionless
y^f	Dimensionless concentration of f phase, C_i^f/C_i^{f0}
Z	Compressibility factor, dimensionless
z	Axial position, cm

Subscripts

a	Referring to axial position
G	Referring to gas phase
I	Referring to interphase
i	Referring to i component
L	Referring to liquid phase
p	Referring to solid particles
S	Referring to solid phase
S_{sup}	Referring to solid surface
SL	Referring to slurry phase
W	Referring to cooling fluid

Superscripts

$*$	Equilibrium conditions
f	Referring to f phase
S_{inn}	Solid inner
S_{sur}	Solid surface
0	Inlet conditions
eff	Effective

Dimensionless groups

Bo	Bodenstein number, $U \cdot d_{pe}/D$
Da	Damköler number, $k \cdot \varepsilon_L \cdot H/U_{G0}$
Fr	Froude number, $U/(g \cdot D_C)^{1/2}$
Ga	Galilei number, $g \cdot L^3 \cdot \rho^2/\mu^2$
Pe	Peclet number, $U \cdot Le/D_C$
Re	Reynolds number, $U \cdot Le \cdot \rho/\mu$
Sc	Schmith number, $\mu/\rho \cdot D^f$
Sh	Sherwood number, $k^f \cdot d_p/D^f$
St	Stanton number, $k_L a \cdot H/U_G$

Greek letters

α	Gas-phase contraction factor, dimensionless
α_i, $\beta_{m,i}$	Matrix constant parameter

$\Delta H_{r,j}$ Heat of reaction j, J/mol
$\Delta H_{v,j}$ Heat of vaporization of i component, J/mol i
ΔH Height difference, cm
ΔP Pressure drop, MPa
η_i^f Catalyst effectiveness factor of reaction j in the f phase, dimensionless
Γ Gas distributor parameter
λ_a^f Axial thermal conductivity of f phase, $J/cm_r \cdot s \cdot K$
λ_e^f Catalyst thermal conductivity, $J/cm_S \cdot s \cdot K$
λ^W Thermo-well thermal conductivity, $J/cm_W \cdot s \cdot K$
μ_f Dynamic viscosity of f phase, cP
$\nu_{i,j}^f$ Stoichiometric coefficient of component i in reaction j in the f phase, dimensionless
ϕ Weight fraction of f phase in the slurry, g_f/g_{SL}
Ψ Arbitrary system property
ρ_f Density of f phase, g_f/cm_f^3
σ_L Surface tension of liquid phase, dyn/cm
τ Dimensionless time
θ Dimensionless temperature, T/T^0
ε_B Catalyst bed void fraction or catalyst porosity, cm_{G+L}^3/cm_r^3
ε_f Holdup of f phase, cm_f^3/cm_r^3
ξ Radial coordinate inside spherical catalyst particle, cm_S

References

Ancheyta, J. (2013). *Modeling of Processes and Reactors for Upgrading of Heavy Petroleum.* CRC Press Taylor and Francis Group.

Ancheyta, J., Sanchez, S., and Rodriguez, M. (2005a). Kinetic modeling of hydrocracking of heavy oil fractions: a review. *Catalysis Today* 76–92.

Ancheyta, J., Rana, M.S., and Furimsky, E. (2005b). Hydroprocessing of heavy petroleum feeds: tutorial. *Catalysis Today* 109: 3–15.

Angeles, M.J., Leyva, C., Ancheyta, J., and Ramírez, S. (2014). A review of experimental procedures for heavy oil hydrocracking with dispersed catalyst. *Catalysis Today* 220–222: 274–294.

Bach, H. and Pilhofer, T. (1978). Variation of gas holdup in bubble columns with physical properties of liquids and operating parameters of columns. *German Chemical Engineering* 1: 270–275.

Bahroun, S., Li, S., Jallut, C. et al. (2010). Control and optimization of a three-phase catalytic slurry intensified continuous chemical reactor. *Journal of Process Control* 20: 664–675.

Bahroun, S., Li, S. Jallut, C., Valentin, C., Panthou, F. (2010) Modelling and control of a three-phase catalytic slurry intensified continuous reactor. In European Symposium on Computer-Aided Process Engineering, Ischia, Italy, June 6–9

Baten, J.M. and Krishna, R. (2001). Eulerian simulations for determination of the axial dispersion of liquid and gas phases in bubble columns operating in the churn turbulent regime. *Chemical Engineering Science* 56: 503–512.

Beenackers, A.A.C. and Swaaij, W.P.M.V. (1993). Mass transfer in gas-liquid slurry reactors. *Chemical Engineering Science* 48: 3109–3139.

Begovivh, D. and Watson, J.S. (1978). Hydrodynamics characteristics of three phase fluidized beds. *Fluidization* 190.

Behkish, A., Lemoine, R., Oukaci, R., and Morsi, B.I. (2006). Novel correlations for gas holdup in large scale slurry bubble column reactors operating under elevated pressures and temperatures. *Chemical Engineering Journal* 115: 157–171.

Bellussi, G., Rispoli, G., Landoni, A. et al. (2013). Hydroconversion of heavy residues in slurry reactors: developments and perspectives. *Journal of Catalysis* 308: 189–200.

Bird, R.B., Stewart, W.E., and Lightfoot, E.N. (2002). *Transport phenomena*, Seconde. John Wiley.

Calderon, C. and Ancheyta, J. (2016). Modeling of slurry-phase reactors for hydrocracking of heavy oils. *Energy & Fuels* 2525–2543.

Carbonell, M.M. and Guirardello, R. (1997). Modelling of a slurry bubble column reactor applied to the hydroconversion of heavy oils. *Chemical Engineering Science* 52: 4179–4185.

Chaudhari, R.V. and Ramachandran, P.A. (1980). Three-phase slurry reactors. *AIChE Journal* 26: 177–201.

Chen, Z., Zheng, C., Feng, Y., and Hofmann, H. (1995). Modeling of three phase fluidized beds based on local bubble characteristics measurements. *Chemical Engineering Science* 50: 231–236.

Chen, Z., Zhang, H., Ying, W., and Fang, D. (2010). Mathematical simulation of bubble column slurry reactor for direct dimethyl ether synthesis process from syngas. *World Academy of Science, Engineering and Technology* 4: 1443–1452.

Cho, Y.J., Song, P.S., Lee, C.G. et al. (2005). Liquid radial dispersion in liquid-solid circulating fluidized beds with viscous liquid media. *Chemical Engineering Communications* 192: 257–271.

Costa, C.B., Rivera, E.A., Rezende, M.C.A.F. et al. (2007). Prior detection of genetic algorithm significant parameters: coupling factorial design technique to genetic algorithm. *Chemical Engineering Science* 62: 4780–4801.

Deckwer, W.D., Burckart, R., and Zoll, G. (1974). Mixing and mass transfer in tall bubble columns. *Chemical Engineering Science* 29: 2177–2188.

Deckwer, W.D., Louisi, Y., Zaidi, A., and Ralek, M. (1980). Hydrodynamic properties of the fisher-tropsch slurry process. *Industrial & Engineering Chemistry Process Design and Development* 19: 699–708.

Deckwer, W.D., Serpemen, Y., Ralek, M., and Schmidt, B. (1982). Modeling the Fischer-Tropsch synthesis in the slurry-phase. *Industrial & Engineering Chemistry, Process Design and Development* 21: 231–241.

van Deemter, J. (1961). Mixing and contacting in gas-solid fluidized beds. *Chemical Engineering Science* 13: 143–154.

Degaleesan, S. and Dudokivic, M.P. (1998). Liquid backmixing in bubble columns and the axial dispersion coefficient. *AIChE Journal* 44: 2369–2378.

Deng, Z., Wang, T., and Wang, Z. (2010). Hydrodesulfurization of diesel in a slurry reactor. *Chemical Engineering Science* 65: 480–486.

Domokos, L., Jongkind, H., Rigutto, M. S., van de Voort, E. H. C., Stork, W. H. K. (2006) Hydrocracking catalyst composition. US Patent 0,207,917

Dudukovic, M.P. and Mills, P.L. (1984). *Chemical and Catalytic Reactor Modeling*. American Chemical Society.

Ellenberger, J. and Krishna, R. (1994). A unified approach to the scale-up of gas-solid fluidized bed and gas-liquid bubble column reactors. *Chemical Engineering Science* 49: 5391–5411.

Fan, L.S. (1989). *Gas-liquid-solid fluidization*. Butterworth's Series in Chemical Engineering.

Ferreira, A., Ferreira, C., Teixeira, J.A., and Rocha, F. (2010). Temperature and solid properties effects on gas-liquid mass transfer. *Chemical Engineering Journal* 162: 743–752.

Froment, G.F., Bischoff, K.B., and De Wilde, J. (2011). *Chemical Reactor Analysis and Design*, 3rde. Wiley.

Fukuma, M., Muroyama, K., and Yasunishi, A. (1987). Specific gas-liquid interfacial area and liquid phase mass transfer coefficient in a slurry bubble column. *Journal of Chemical Engineering of Japan* 20: 321–324.

Han, J.H. and Kim, S.D. (1990). Radial dispersion and bubble characteristics in three-phase fluidized beds. *Chemical Engineering Communications* 94: 9–26.

Hillmer, G., Weismantel, L., and Hofmann, H. (1994). Investigations and modelling of slurry bubble columns. *Chemical Engineering Science* 49: 837–843.

Jakobsen, H.A., Lindborg, H., and Dorao, C.A. (2005). Modelling of bubble column reactors: progress and limitations. *Industrial & Engineering Chemistry Research* 44: 5107–5151.

Jamialahmadi, M. and Müller-Steinhagen, H. (1999). Hydrodynamics and heat transfer of liquid fluidized bed systems. *Chemical Engineering Communications* 179: 35–79.

Jhawar, A.K. and Prakash, A. (2011). Heat transfer in a slurry bubble column reactor: a critical overview. *Industrial & Engineering Chemistry Research* 51: 1464–1473.

Jiang, P., Luo, X., Lin, T.J., and Fan, L.S. (1997). High temperature and high pressure three-phase fluidization-bed expansion phenomena. *Powder Technology* 90: 103–113.

Jin, H., Yang, S., He, G. et al. (2014). Gas-liquid mass transfer characteristics in a gas-liquid-solid bubble column under elevated pressure and temperature. *Chinese Journal of Chemical Engineering* 22: 955–961.

Joshi, J.B. (1998). Axial mixing in multiphase contactors: a unified correlation. *Chemical Engineering Research and Design* 58: 155–165.

Jung, S., Becker, M., Agar, D., and Franke, R. (2010). One-dimensional modeling and simulation of bubble column reactors. *Chemical Engineering & Technology* 33: 2037–2043.

Kang, Y. and Kim, S.D. (1986). Radial dispersion characteristics in three-phase fluidized beds. *Industrial & Engineering Chemistry, Process Design and Development* 25: 717–722.

Kang, Y., Cho, Y.J., Lee, C.G. et al. (2003). Radial dispersion and bubble distribution in three-phase circulating fluidized beds. *The Canadian Journal of Chemical Engineering* 81: 1–9.

Kantarci, N., Borak, F., and Ulgen, K. (2005). Bubble column reactors review. *Process Biochemistry. (Oxford, U.K.)* 40: 2263–2283.

Kato, Y., Nishiwaki, A., Fukuda, T., and Tanaka, S. (1972). The behavior of suspended solid particles and liquid in bubble columns. *Journal of Chemical Engineering of Japan* 5: 112–118.

Kato, Y., Morooka, S., Kago, T. et al. (1985). Axial holdup distributions of gas and solid particles in three-phase fluidized bed for gas-liquid (slurry)-solid systems. *Journal of Chemical Engineering of Japan* 18: 308–313.

Kennepohl, D. and Sanford, E. (1996). Conversion of Athabasca bitumen with dispersed and supported mo-based catalysts as a function of dispersed catalyst concentration. *Energy & Fuels* 10: 229–234.

Kesler, M. (1976). Improve prediction of enthalpy of fractions. *Hydrocarb* 153.

Khadem-Hamedani, B., Yaghmaei, B.K.-H., Fattahi, M. et al. (2015). Mathematical modeling of a slurry bubble column reactor for hydrocracking of diesel fuel: single and two bubble configurations. *Chemical Engineering Research and Design* 100: 362–376.

Khulbe, C. P., Belinko, K., Waugh, R. J., Perreault, M. (1989) Process for preparing an iron-coal slurry catalyst for hydrocracking heavy oils. US Patent 4,923,838

Kim, S.D. and Kang, Y. (1997). Heat and mass transfer in three-phase fluidized bed reactor: an overview. *Chemical Engineering Science* 52: 3639–3660.

Kim, H.S., Kim, J.H., Kang, S.H. et al. (2014). Bubble and heat transfer phenomena in viscous slurry bubble column. *Advances in Chemical Engineering and Science* 4: 417–429.

Kito, M., Shimida, M., Sakai, T. et al. (1976). Performance of tubular bed contractor, gas holdup and interfacial area under liquid stagnant flow. *Fluidization* 411.

Kojima, H., Anjyo, H., and Mochizuki, Y. (1989). Axial mixing in bubble column with suspended solid particles. *Journal of Chemical Engineering of Japan* 19: 232–234.

Krishna, R. and Ellenberger, J. (1996). Gas holdup in bubble column reactors operating in the churn turbulent flow regime. *AIChE Journal* 42: 2627–2634.

Krishna, R. and Saxena, A.K. (1989). Use of an axial dispersion model for kinetic description of hydrocracking. *Chemical Engineering Science* 44: 703–712.

Krishna, R. and Sie, S.T. (2000). Design and scale-up of the Fischer-Tropsch bubble column slurry reactor. *Fuel Processing Technology* 64: 73–105.

Kriz, J. F., Ternan, M., Denis, J. M. (2009) Hydrocracking of bitumen and heavy oils. In *5th NCUT Upgrading and refining conference*, synthetic crude oil and heavy oil, Edmonton, Canada, Sept 14–16.

Kumar, S. and Khanna, A. (2014). Experimental analysis and development of correlations for gas holdup in high pressure slurry co-current bubble columns. *Korean Journal of Chemical Engineering* 31: 1964–1972.

Lemoine, R., Behkish, A., Sehabiague, L. et al. (2008). An algorithm for predicting the hydrodynamic and mass transfer parameters in bubble column and slurry bubble column reactors. *Fuel Processing Technology* 89: 322–343.

Leonard, C., Ferrasse, J.-H., Boutin, O. et al. (2015). Bubble column reactors for high pressures and high temperatures operation. *Chemical Engineering Research and Design* 100: 391–421.

Li, H. and Prakash, A. (2000). Influence of slurry concentrations on bubble population and their rise velocities in three-phase slurry bubble column. *Powder Technology* 113: 158–167.

Li, Y., Wang, J., Jiang, L., Zhang, Z., Liu, J., Ren, S., Zhao, B., Jia, Y. (1999) Hydrocracking of heavy oil and residuum with a dispersing-type catalyst. US Patent 6,004,454.

Li, S., Bahroun, S., Valentin, C. et al. (2010). Dynamic model based safety analysis of a three-phase catalytic slurry intensified continuous reactor. *Journal of Loss Prevention in the Process Industries* 23: 437–445.

Liu, D., Kong, X., Li, M., and Que, G. (2009). Study on a water-soluble catalyst for slurry-phase hydrocracking of an atmospheric residue. *Energy & Fuels* 23: 958–961.

Luo, X., Lee, D., Lau, R. et al. (1999). Maximum stable bubble size and gas holdup in high-pressure slurry bubble columns. *AIChE Journal* 45: 665–680.

Macías, M.J. and Ancheyta, J. (2004). Simulation of an isothermal hydrodesulfurization small reactor with different catalyst particle shapes. *Catalysis Today* 98: 243–252.

Maretto, C. and Krishna, R. (1999). Modelling of a bubble column slurry reactor for Fischer-Tropsch synthesis. *Catalysis Today* 52: 279–289.

Mariano, A.P., Costa, C.B.B., deToledo, E.C.V. et al. (2011). Analysis of particle swarm algorithm in the optimization of a three-phase slurry catalytic reactor. *Computers & Chemical Engineering* 35: 2741–2749.

Martínez, J. and Ancheyta, J. (2012). Kinetic model for hydrocracking of heavy oil in a CSTR involving short term catalyst deactivation. *Fuel* 193–199.

Martínez, J. and Ancheyta, J. (2014). Modeling the kinetics of parallel thermal and catalytic hydrotreating of heavy oil. *Fuel* 27–36.

Martínez, J., Sánchez, J.L., Ancheyta, J., and Ruiz, R.S. (2010). A review of process aspects and modeling of ebullated bed reactors for hydrocracking of heavy oils. *Catalysis Reviews* 52: 60–105.

Matonis, D., Gidaspow, D., and Bahary, M. (2002). CFD simulation of flow and turbulence in a slurry bubble column. *AIChE Journal* 48: 1413–1429.

Matos, E. and Guirardello, R. (2002). Modeling and simulation of a pseudo-two-phase gas-liquid column reactor for thermal hydrocracking of petroleum heavy fractions. *Brazilian Journal of Chemical Engineering* 19: 319–334.

Matos, E. and Nunhez, J. (2006). The effect of different feed flow patterns on the conversion of bubble column reactors. *Chemical Engineering Journal* 116: 163–172.

Matos, E., Guirardello, R., Mori, M., and Nunhez, J. (2009). Modeling and simulation of a pseudo-three-phase slurry bubble column reactor applied to the process of petroleum hydrodesulfurization. *Computers & Chemical Engineering* 33: 1115–1122.

Mederos, F., Elizalde, I., and Ancheyta, J. (2009). Steady-state and dynamic reactor model for hydrotreatment of oil fractions: a review. *Catalysis Reviews – Science and Engineering* 51: 485–607.

Melo, D.N., Toledo, E.C.V.D., Santos, M.M. et al. (2005). Off-line optimization and control for real time integration of a three-phase hydrogenation catalytic reactor. *Computers & Chemical Engineering* 29: 2485–2493.

Mills, P.L., Turner, J.R., Ramachandran, P.A., and Dudokivic, M.P. (1996). The Fischer-Tropsch synthesis in slurry bubble column reactors: analysis of reactor performance using the axial dispersion model. *Fuel Energy* 38.

Moustiri, S., Hebrard, G., Thakre, S.S., and Roustan, M. (2001). A unified correlation for predicting liquid axial dispersion coefficient in bubble columns. *Chemical Engineering Science* 56: 1041–1047.

Noguera, G., Araujo, S., Hernández, J. et al. (2012). A comparative activity study of a new ultra-dispersed catalyst system for a hydrocracking/hydrotreating technology using vacuum residue oil: Merey/Mesa. *Chemical Engineering Research and Design* 90: 1979–1988.

Ortiz-Moreno, H., Ramírez, J., Cuevas, R. et al. (2012). Heavy oil upgrading at moderate pressure using dispersed catalyst: effects of temperature, pressure and catalytic precursor. *Fuel* 100: 186–192.

Panariti, N., del Blanco, A., del Piero, G. et al. (2000). Petroleum residue upgrading with dispersed catalysts: part 2. effect of operating conditions. *Applied Catalysis A: General* 204: 215–222.

Pangarkar, V.G., Yawalkar, A.A., Sharma, M.M., and Beenackers, A.A.C. (2002). Particle-liquid mass transfer coefficient in two/three-phase stirred tank reactors. *Industrial & Engineering Chemistry Research* 41: 4141–4167.

Parulekar, S.J. and Shah, Y.T. (1980). Steady-state behavior of gas-liquid-solid fluidized-bed reactors. *Chemical Engineering Journal* 20: 21–33.

Pfleger, D. and Becker, S. (2001). Modelling and simulation of the dynamic flow behaviour in a bubble column. *Chemical Engineering Science* 56: 1737–1747.

Rados, N., Al-Dahhan, M., and Dudokivic, M.P. (2005). Dynamic modeling of slurry bubble column reactors. *Industrial & Engineering Chemistry Research* 44: 6086–6094.

Reilly, I.G., Scott, D.S., Bruijin, T.J.W. et al. (1986). A correlations for gas holdup in turbulent coalescing bubble columns. *The Canadian Journal of Chemical Engineering* 64: 705–717.

Rezair, H. and Smith, K.J. (2013). Catalyst deactivation in slurry-phase residue hydroconversion. *Energy & Fuels* 27: 6087–6097.

Ruiz, R.S., Alonso, F., and Ancheyta, J. (2004). Effect of high-pressure operation on overall phase holdups in ebullated-bed reactors. *Catalysis Today* 9: 265–271.

Ruthiya, K.C., van der Schaaf, J., Kuster, B.F.M., and Schouten, J.C. (2004). Modeling the effect of particle to bubble adhesion on mass transport and reaction rate in a stirred slurry reactor: influence of catalyst support. *Chemical Engineering Science* 59: 5551–5558.

Sahu, R., Song, B.J., Im, J.S. et al. (2015). A review of recent advances in catalytic hydrocracking of heavy residues. *Journal of Industrial and Engineering Chemistry* 27: 12–24.

Sánchez, S. and Ancheyta, J. (2007). Effect of pressure on the kinetics of moderate hydrocracking of maya crude oil. *Energy & Fuels* 653–661.

Sánchez, S., Rodríguez, M., and Ancheyta, J. (2005). Kinetic model for moderate hydrocracking of heavy oils. *Industrial & Engineering Chemistry Research* 9409–9413.

Sauer, T. and Hempel, D.C. (1987). Fluid dynamics and mass transfer in a bubble column with suspended particles. *Chemical Engineering & Technology* 10: 180–189.

Scheffer, B., van Koten, M.A., Röbschlager, K.W., and Boks, F.C. (1998). The shell residue hydroconversion process: development and achievements. *Catalysis Today* 43: 217–224.

Schumpe, A., Saxena, A.K., and Fang, L.K. (1987). Gas-liquid mass transfer in a slurry bubble column. *Chemical Engineering Science* 42: 1797–1796.

Sehabiague, L., Lemoine, R., Behkish, A. et al. (2008). Modeling and optimization of a large-scale slurry bubble column reactor for producing 10,000 bbl/day of Fischer-Tropsch liquid hydrocarbons. *Journal of the Chinese Institute of Chemical Engineers* 39: 169–179.

Shah, Y.T., Kelkar, B.G., Godbole, S., and Deckwer, W.D. (1982). Design parameters estimations for bubble column reactors. *AIChE Journal* 28: 353–379.

Shahrzad, H., Arturo, M., and Phillip, S. (2009). Dynamic simulation of gas hydrate formation in a three-phase slurry reactor. *Industrial & Engineering Chemistry Research* 48: 6983–6991.

Sivaiah, M. and Majumder, S.K. (2013). Dispersion characteristics of liquid in a modified gas-liquid-solid three-phase downflow bubble column. *Particulate Science and Technology* 31: 210–220.

Speight, J.G. (2004). New approaches to hydroprocessing. *Catalysis Today* 98: 55–60.

de Swart, J.W.A. and Krishna, R. (1997). Selection, design and scale-up of the fischer-tropsch slurry reactor. *Studies in Surface Science and Catalysis* 107: 213–218.

de Swart, J. and Krishna, R. (2002a). Simulation of the transient and steady-state behaviour of a bubble column slurry reactor for Fischer-Tropsch synthesis. *Chemical Engineering and Processing* 41: 35–47.

de Swart, J.K.A. and Krishna, R. (2002b). Simulation of the transient and steady-state behavior of a bubble column slurry reactor for Fischer-Tropsch synthesis. *Chemical Engineering and Processing* 35–47.

Tang, C. and Heindel, T. (2006). Estimating gas holdup via pressure measurements in a cocurrent bubble column. *International Journal of Multiphase Flow* 32: 850–863.

Todorova, T., Alexiev, V., and Weber, T. (2012). Energetics and electronic properties of defects at the (100) mos2 surface studied by the perturbed cluster method. *Reaction Kinetics, Mechanisms and Catalysis* 105: 113–133.

de Toledo, E.C.V., de Santana, P.L., Maciel, M.R.W., and Filho, R.M. (2001). Dynamic modelling of a three-phase catalytic slurry reactor. *Chemical Engineering Science* 56: 6055–6061.

Tomiyama, A. (1998). Struggle with computational bubble dynamics. *Multiphase Science and Technology* 10: 369–405.

Tomiyama, A. and Shimada, N. (2001). A numerical method for bubbly flow simulation based on a multi-fluid model. *Journal of Pressure Vessel Technology* 123: 510–516.

Torvik, R. and Svendsen, H.F. (1990). Modeling of slurry reactors. a fundamental approach. *Chemical Engineering Science* 45: 2325–2332.

Vandu, C.O. and Krishna, R. (2003). Gas holdup and volumetric mass transfer coefficient in a slurry bubble column. *Chemical Engineering & Technology* 26: 779–782.

Wilkinson, P.M., Spek, A.P., and van Dierendonck, L.L. (1992). Design parameters estimation for scale-up of high-pressure bubble columns. *AIChE Journal* 38: 544–554.

Yang, G.Q. and Fan, L.S. (2003). Axial liquid mixing in high pressure bubble columns. *AIChE Journal* 49: 1995–2008.

Ying, D.H., Givens, E.N., and Weimer, R.F. (1980). Gas holdup in gas-liquid and gas-liquid solid flow reactors. *Industrial & Engineering Chemistry, Process Design and Development* 19: 635–638.

Zehner, P. and Kraume, M. (2005). *Bubble Columns: Encyclopedia of Industrial Chemistry*. Wiley-VCH.

Zhang, S., Liu, D., Deng, W., and Que, G. (2007). A review of slurry-phase hydrocracking heavy oil technology. *Energy & Fuels* 21: 3057–3062.

Zhu, Y.Q. and Zhou, D.L. (2011). Study on unsupported nano-MOS catalyst for clean fuel. *Advanced Materials Research* 287–290: 1860–1865.

van der Zon, M., Hamersma, P.J., Poels, E.K., and Bliek, A. (1999). Gas-solid adhesion and solid-solid agglomeration of carbon supported catalyst in three-phase slurry reactors. *Catalysis Today* 48: 131–138.

van der Zon, M., Thoolen, H., Hamersma, P.J. et al. (2001). Agglomeration and adhesion of catalyst particles in gas-liquid reactors. *Catalysis Today* 66: 263–270.

8

Modeling of Fischer–Tropsch Synthesis Reactor

César I. Méndez[1] *and Jorge Ancheyta*[2]

[1] *Centro de Investigación en Ciencia Aplicada y Tecnología Avanzada del Instituto Politécnico Nacional, Mexico City, Mexico*
[2] *Instituto Mexicano del Petróleo, Mexico City, Mexico*

8.1 Fundamentals of the Fischer–Tropsch Synthesis to Produce Clean Fuels

Nowadays, the world faces the need to meet high energy demands, for which various fields of study have been opened to explore useful alternatives that are economical and friendly to the environment. Fischer–Tropsch technology has become a viable option for the production of commercial fuels (mainly gasoline, diesel, and airplane fuels), which has attracted the attention of the scientific and industrial community (Jia-Yang et al. 2015). Different reactor technologies exist to carry out the Fischer–Tropsch synthesis (FTS) process. These reactor technologies have been used for design, modeling, and optimization studies and are applied according to the Fischer–Tropsch technology, depending on the type of catalyst and operating temperature used: (i) high-temperature Fischer–Tropsch (HTFT) at a temperature of 290–360 °C with an iron catalyst; and (ii) low-temperature Fischer–Tropsch (LTFT) at a temperature of 180–260 °C with cobalt catalyst. The former used fluidized-bed reactor and circulating fluidized-bed reactor, while the latter used fixed-bed reactor, slurry bubble column, and microchannel reactor (Basha et al. 2015). Despite the advantages of fluidized-bed reactor, slurry bubble column reactor, and the novel microchannel reactor compared with those exhibited by the fixed-bed reactor, the latter has been preferably employed in commercial applications (Chambrey et al. 2006; Maitlis and de Klerk 2013), because it shows superiority over the others in both operational and economic matters, since it can be scaled from a single tube to industrial plants (Saeidi et al. 2014), as can be seen in large commercial scale developed by Sasol (Dry 1996) and Shell (Sie 1998). The Fischer–Tropsch process is quite complex due to the difficulty of predicting the products distribution generated by the synthesis. The use of different catalysts has led to the proposal of various reaction mechanisms that attempt to accurately describe the steps involved in this catalytic reaction (Dry 1996). On the other hand, there is a great diversity of kinetic models that have been developed for the description of the FTS, which arise due to the different reaction mechanisms that have been proposed, the operating conditions, and the type of catalyst.

There is experimental evidence that a complex mixture of hydrocarbons ranging from methane to wax is normally produced in a fixed-bed reactor (Caldwell and Van Vuuren 1986; Wang et al.

Mathematical Modeling of Complex Reaction Systems in the Oil and Gas Industry, First Edition.
Edited by Jorge Ancheyta, Andrey Zagoruiko, and Andrey Elyshev.

2003a). Therefore, it is convenient to provide a comprehensive mechanism that contemplates all the formation steps of the intermediate FTS species, initiators of the growth of the hydrocarbon chain, and the product's distribution, among other aspects related to the behavior of the catalyst, catalytic support, and promoter (Wang et al. 2003a).

8.1.1 Fischer–Tropsch Synthesis Technology

The process of converting syngas into liquid hydrocarbons by using the X-to-liquid technology (XTL) has proven to be an alternative to generate excellent clean-burning fuels, which has increased the interest in studying the FTS to address the widespread demand for liquid fuels for commercial use (Ahón et al. 2005). The XTL process is a viable option for the production of improved clean fuels with a highly attractive business that can be generated from any material containing carbon, natural gas, and biomass, as well as waste (e.g. carbon, natural gas, biomass or waste) (Wilhelm et al. 2001; Mousavi et al. 2015).

The use of gas-to-liquid technology (GTL) for the syngas conversion to commercial value liquid hydrocarbons by the FTS has received much attention due to the following advantages (Forghani et al. 2009; Lee et al. 2015; Luque et al. 2012):

- Increased availability of natural gas located in remote areas and poor access for which there is no nearby market.
- Growing demand for middle distillate fuels used for transportation.
- Improved cost-effectiveness of GTL technology resulting from the development of more active catalysts and more advanced reactor design.
- Involve low costs in transportation of liquid products compared with gaseous products.
- Production of a high cetane number diesel, which is due to the low content of aromatics.
- Wide versatility in terms of different types of raw materials that can be used and different types of products that are obtained, which are expected to be an important alternative for the energy segment soon.
- Products are virtually free of sulfur, nitrogen, and aromatics, providing compatibility and good blending with conventional fuels and the broad possibility to work efficiently with fuel infrastructure nowadays.
- Environment-friendly technology that produces almost free of impurities fuels.

Due to the demand for high-quality fuels to meet the energy needs of the industrial and transport sectors, special attention has been paid to the development of more effective clean fuels production and relatively environmentally friendly technologies. The FTS process, according to the advantages already mentioned, has become a focus of attention around the technological, scientific, and industrial sectors. It is considered a clean option for the environment and is also designated as the heart of the technological processes of gas conversion to liquid (Forghani et al. 2009), which has been widely investigated to improve their aspects of design, implementation, start-up, and optimization at various scales. This technology enables commercial liquid fuel production with low aromaticity and impurities free, which reduces the environmental impact associated with fuel from the conventional refining process of crude oil. Moreover, the effects of the FTS process for fuel production do not cause a strong environmental impact in contrast to conventional processes for fuel production and industrial chemicals. So, the FTS process is a viable technology for the production of synthetic fuels that are eco-friendly (Knottenbelt 2002; van Vliet et al. 2009; Gill et al. 2011).

Syngas is produced with technologies already established in three ways: (i) partial oxidation, (ii) steam reforming, and (iii) autothermal synthesis. Partial oxidation is based on the following exothermic reaction (Vosloo 2001; Todic et al. 2015):

$$CH_4 + \tfrac{1}{2}O_2 \leftrightarrow CO + 2H_2 \tag{8.1}$$

that can be generalized to any hydrocarbon:

$$C_nH_m + \tfrac{1}{2}nO_2 \leftrightarrow nCO + \tfrac{1}{2}mH_2 \tag{8.2}$$

While in the steam reforming the following endothermic reaction is found:

$$CH_4 + H_2O \leftrightarrow CO + 3H_2 \tag{8.3}$$

that can be generalized to any hydrocarbon:

$$C_nH_m + nH_2 \leftrightarrow nCO + (n + \tfrac{m}{2})H_2 \tag{8.4}$$

In autothermal synthesis, steam reforming and partial oxidation are combined. The heat produced by partial oxidation is used to provide energy for steam reforming, making a thermally self-sufficient process. In the typical FTS process, the syngas catalytic conversion occurs mainly in a predominant mixture of linear alkanes and alkenes whose product distribution exhibits a recognizable pattern such as illustrated below:

Paraffins:

$$(2n + 1)H_2 + n\text{CO} \rightarrow \text{C}_n\text{H}_{2n+2} + n\text{H}_2\text{O} \tag{8.5}$$

Olefins:

$$2n\text{H}_2 + n\text{CO} \rightarrow \text{C}_n\text{H}_{2n} + n\text{H}_2\text{O} \tag{8.6}$$

Alcohols:

$$n\text{CO} + 2n\text{H}_2 \rightarrow \text{C}_n\text{H}_{2n+1}\text{OH} + (n-1)\text{H}_2\text{O} \tag{8.7}$$

Until then, a growing number of studies have been significantly reflected in the discussion on the understanding of the integrated nature of surface species and the precise and detailed mechanical sequence, the description by which the FTS reactions proceed over the catalyst. A wide variety of FTS reaction mechanisms have been proposed whose common feature has been the consideration of the polymerization reaction (Nakhaei et al. 2014a) and the stepwise chain growth hydrocarbons process, which is supported by the fact that the carbon number distribution in the FTS products is calculated based only on probabilities for the chain growth matching results observed experimentally obtained from reactors of different sizes and configuration types over a wide range of operating conditions and catalysts (Dry 1996).

According to Wojciechowski (1988), a reaction mechanism proposed for the description of the FTS must comply with the following aspects:

- The complete adsorption of all species on the catalyst surface in a set of active sites resulting from the decomposition of H_2 and CO to hydrogen atoms and the respective adsorption of carbon and oxygen atoms causes interaction between the species located on the surface, leading to the formation of CH_x, OH compounds, etc.
- The CH_2 should be the monomer species to carry out the oligomerization. The formation rate arises from the adsorbed species C and H, and this becomes the rate-determining step for CO hydrogenation kinetics.

- The radical growth on the catalytic surface is immobile except for species $C_1 - C_4$, and the chain grows only through the participation of a nearby monomer. The growing chain may be formed or migrated through the diffusion between the set of appropriate active sites.
- The chain growth surface can spontaneously generate 1–2 accompanying displacements that can result in the formation of branched hydrocarbons.
- The termination stage and therefore the product type are determined by the type of component that occupies the vacant site adjacent to the growth radical.
- Classical distributions involve primary species produced, each having its own chain length distribution of the Anderson–Schulz–Flory (ASF) type. This distribution is typical of a proper collocation grouping of growth, monomers, and the termination stage sites that constitute the growth place location for the molecular species. Such locations have a stable composition and only continue to produce one type of molecular species to a certain set of given reaction conditions.
- Operating parameters, such as temperature, system total pressure, and the H_2/CO ratio, are extremely important factors affecting both the kinetics and the product's distribution.

According to Adesina (1996) and Förtsch et al. (2015), both the rate of the catalytic reaction of Fischer–Tropsch that normally occurs in the catalyst pores and the product's distribution are affected by various physical and chemical phenomena that include the following: (i) the reactants diffusion into a porous particle to active sites, (ii) reactants adsorption on catalytic active sites, (iii) chain initiation, (iv) chain growth (or propagation), (v) chain termination, (vi) product desorption process, (vii) re-adsorption and further reaction, and (viii) products diffusion toward the particle outside. Förtsch et al. (2015) summarized the FTS reaction system's basic stages in simplified schematic reactions involved in the following way:

$$\text{(i) chain initiation: } (\,)^* + CO + 2H_2 \xrightarrow{R_1} (CH_2)^* + H_2O \tag{8.8}$$

$$\text{(ii) chain propagation: } (\mathrm{CH_2})_n^{\;*} + \mathrm{CO} + 2\mathrm{H_2} \xrightarrow{\mathrm{R}_n^{\text{propagation}}} (\mathrm{CH_2})_{n+1}^* + \mathrm{H_2O} \tag{8.9}$$

$$\text{(iii) termination to paraffins: } (\mathrm{CH_2})_n^{\;*} + \mathrm{H_2} \xrightarrow{R_{n,\text{paraffins}}^{\text{termination}}} \mathrm{C}_n\mathrm{H}_{2n+2} \tag{8.10}$$

$$\text{(iv) termination to olefins: } (\mathrm{CH_2})_n^{\;*} \xrightarrow{R_{n,\text{olefins}}^{\text{termination}}} \mathrm{C}_n\mathrm{H}_{2n} \tag{8.11}$$

$$\text{(v) re} - \text{adsorption: } \mathrm{C}_n\mathrm{H}_{2n} \xrightarrow{R_{n,\text{olefins}}^{\text{re-adsorption}}} (\mathrm{CH_2})_n^* \tag{8.12}$$

$$\text{(vi) secondary hydrogenation: } \mathrm{C}_n\mathrm{H}_{2n} + \mathrm{H_2} \xrightarrow{R_n^{\text{hydrogenation}}} \mathrm{C}_n\mathrm{H}_{2n+2} \tag{8.13}$$

It is important to note that for decades, some assumptions have been made about the forms in which the FTS monomer and initiator and its chemical formation occur. The most important monomers identified through the FTS process history are methylene ($-CH_2-$), hydroxyl carbine (CHOH), and CO, which lead to the following three important mechanisms: (i) carbide, (ii) the enol mechanism, and (iii) the CO insertion mechanism, respectively (Mousavi et al. 2015).

Other reaction mechanisms have been proposed that differentiate in the assumptions that arise for monomer, initiator, and their chemical formation such as the alkyl, alkenyl and alkylidene hydride-methylidene, and formate mechanisms (Basha et al. 2015; Frennet and Hubert 2000; Brady and Pretit 1981; Vannice 1976; Satterfielf et al. 1982).

8.1.2 Fischer–Tropsch Synthesis Catalysts

At present, there are some challenges in terms of product selectivity that are of particular interest (such as commercial gasoline and diesel fuels) and improve the FTS reactor performance used under severe operating conditions, for which research on the design of novel catalysts has been an issue of interest. Moreover, it is known that the spectrum of products obtained from the FTS technology is broad and requires a second refining step to improve quality, thus achieving commercial fuel products that are usually of higher quality compared with the fuels generated by conventional refining processes (Zhang et al. 2013). Based on this, for the option of liquid fuels of high quality from syngas by a single-stage FTS without the use of a subsequent refining step, it is necessary to design a novel catalyst with high selectivity toward the desired range of hydrocarbons (Kang et al. 2011; Sun et al. 2012; Valero-Romero et al. 2016). In addition, the FTS process can be very successful by developing novel active catalysts having high selectivity toward generating wax; in other words, the design of a good active catalyst is the key element for excellent operation (Storsaeter et al. 2005).

As for the catalyst types employed in the FTS process, they are based on transition metals such as cobalt, iron, nickel, and ruthenium. It is relevant to mention that one of the biggest problems presented in the design of an FTS process is the apparent complexity exhibited by the reaction mechanism and the large number of species involved (Glasser et al. 2012). Moreover, two main types of products are generated: commercial hydrocarbons (gasoline, diesel, and olefins of low molecular weight) and organic oxygenates (methanol, ethanol, and mixtures of alcohols). That is, the formation of a particular product type is based on the catalyst type employed (Zhang et al. 2013) among other features. The choice of a particular type of catalyst lies mainly in the operating mode (high or low temperature) used and the type of raw material employed (Luque et al. 2012). The FTS process can proceed at low and high temperatures, distinguishing between LTFT synthesis and HTFT synthesis, respectively, and each of these two classifications will typically correspond to two specific catalysts, cobalt-based catalyst and iron-based catalyst respectively.

8.1.2.1 Cobalt-Based Catalysts

The cobalt catalyst has been one of the most used in LTFT technology due to the following advantages:

- Exhibits high activity and selectivity toward the production of long-chain paraffins (Storsaeter et al. 2005; Hinchiranan et al. 2008; Trépanier et al. 20009).
- Provides satisfactory stability in the production of linear hydrocarbons from mixtures of CO/H_2 (Singh and Gu 2010).
- Shows low selectivity toward the production of oxygenates (byproducts) (Hinchiranan et al. 2008).
- The gas–water shift (WGS) reaction has low activity that can be discarded in kinetic models proposed to describe the FTS process (Hinchiranan et al. 2008; Trépanier et al. 2009).
- The water concentration increases with time-on-stream (TOS) in FTS for cocatalysts because of enhanced catalyst deactivation, which reduces the selectivity to water–gas shift (Fazlollahi et al. 2012).
- High resistance to wear catalyst when employed in suspension phases, e.g. slurry reactor operations (Luque et al. 2012).
- Provides higher yields and longer lifetime compared with iron catalysts (Chaumette et al. 1995).
- Not inhibited by water, thus exhibiting higher productivity compared with catalysts involving reaction water–gas shift and higher syngas conversion (van Berge and Everson 1997).

The most important disadvantages of cobalt catalysts are higher prices compared with iron and low performance due to excessive methane production, which can occur when the proper support and promoter are not used and/or when it is not manufactured under the required quality standards. However, despite these disadvantages, it remains the most preferred catalyst (Trépanier et al. 2009).

In addition, the lifetime of the catalysts is influenced by at least two factors: (i) the physical properties such as attrition of the catalyst, wax buildup within the catalyst pellet, pressure drop through the catalytic bed, and sudden variations in temperature and pressure, and (ii) loss of catalyst active sites through poisoning and/or contamination (Luque et al. 2012). Some other causes that lead to catalyst deactivation are (Tsakoumis et al. 2010): (i) poisoning, (ii) re-oxidation of cobalt active sites, (iii) formation of carbon species on the catalyst surface, (iv) carbidization, surface reconstruction, (v) sintering cobalt crystallites, (vi) wear, and (vii) solid-state reactions between metal and support. A mechanism of cobalt catalyst regeneration has been also proposed, which consists of three steps: (i) dewaxing, (ii) oxidation, and (iii) reduction. Luque et al. (2012) recently reported classification of the different types of deactivation mechanisms that affect the cobalt-based catalysts used in the LTFT system: (i) poisoning by sulfur compounds and/or nitrogen in the syngas feed, (ii) oxidation of cobalt metal to active inactive cobalt oxide, (iii) formation of support-cobalt compounds, (iv) sintering cobalt small crystallites, (v) structural reconstruction, and (vi) carbon formation.

8.1.2.2 Iron-Based Catalysts

The main FTS products of iron catalysts are gasoline and diesel components with a relatively higher fraction of olefins (Yan et al. 2014). The iron catalyst can be used in both configurations of FTS at high and low temperatures. The following are the main advantages of iron catalysts:

- Much lower cost compared with cobalt catalysts (Rao et al. 1992; Jager and Spinoza 1995).
- Exhibit high activity in the water–gas shift reaction and high selectivity toward olefins that appear to be stable when converting syngas with high H_2/CO ratio (Rao et al. 1992; Jager and Spinoza 1995).
- Can operate under a wide range of temperatures and ratios of H_2/CO showing low selectivity to methane (Fu and Li 2015).
- Improve the syngas conversion from raw materials such as carbon and biomass with low H_2/CO ratios due to the high activity of the water–gas shift reaction (Fun and Li 2015; Mabry 2014).
- Quite small iron catalyst particles (diameter approximately 100 μm) such that carbon deposition on the catalyst surface does not adversely affect reactor performance (Mabry 2014).
- Heavy paraffinic waxes as the main product that can be converted into usable products such as commercial waxes or high-quality diesel (Mabry 2014).
- Production of hydrocarbons such as linear alkanes, alkenes, and oxygenates (Fu and Li 2015).

Some disadvantages of iron catalysts reported in the literature are as follows: (i) due to the high activity shown by the iron catalysts toward the gas–water shift reaction, hydrothermal sintering and oxidation can occur if the partial pressure of H_2O increases progressively as the FTS reaction proceeds through the catalyst bed in fixed-bed reactors (Schanke et al. 1995); (ii) for the commercial FTS fixed-bed reactors, the iron-catalyst bed becomes progressively deactivated toward the reactor exit as a result of the oxidation of iron when considerable amounts of H_2O and CO_2 are produced by the water–gas shift reaction (Dry 1991), which depends on the operating conditions; and (iii) it is susceptible to produce aromatics when operating at high temperatures in the FTS, which would have a negative effect on the environment (van Dijk 2001).

8.1.2.3 Catalyst Support

The design of catalytic supports is extremely essential to significantly increase the FTS catalyst activity, and product's selectivity of major interest, among others. The most common supports are metal oxides such as aluminum, silicon, and titanium (Storsaeter 2005; Hinchiranan et al. 2008; Girardon et al. 2005; Fischer et al. 2013). The following are some advantages of the supports that significantly improve the performance of FTS heterogeneous catalysts (Fu and Li 2015; Surisetty et al. 2010; Gardezi et al. 2012):

- Positive effect of the FTS process performance due to interactions that arise between metal-support, the acidity of the support, the porosity, and the mass transfer limitations.
- Dispersion of particles of cobalt on the catalyst surface after the reduction process and then to maintain stable cobalt particles during the reaction.
- Porosity of the support, i.e. the average pore diameter, volume, and surface area positively influence the dispersion and reducibility property, as well as activity–selectivity of the FTS.
- Improved diffusion properties of both reactants and products within the granules of the catalyst support.
- Dissipation of heat reaction and reduction of temperature gradients in a fixed-bed reactor.

Catalyst activity increases according to the following order of supports: $MgO > MnO > C > SiO_2 > Al_2O_3$ (Reuel and Bartholomev 1984).

Some catalysts supported on carbon have been widely suggested by the high potential shown to increase the catalytic performance of the FTS process, within which the nanofibers, nanotubes, spheres, and mesoporous carbon exhibit unique properties such as high mechanical strength, good electrical conductivity, high thermal stability, and large surface area (De Jong and Geus 2000; Tavasoli et al. 2010). It has been indicated that silica is one of the catalytic supports suitable for the design of cobalt Fischer–Tropsch catalysts for its implementation in fixed-bed reactors due to its greater surface area, porosity, stability, and weak metal support interaction (Oukaci et al. 1999).

8.1.3 Fischer–Tropsch Synthesis Kinetic Models

One of the high difficulties in developing accurate kinetic models for describing the behavior of FTS over a range of carbon numbers obtained experimentally has been the detailed and improved construction of FTS mechanisms (Glasser et al. 2012; Dry 2001). The adequate kinetic formulation that describes the reactions involved in the FTS system has been considered a task of high complexity for use in industrial practice, becoming a prerequisite for the design of industrial-scale processes, optimization, and numerical simulation (Atashi et al. 2010; Visconti et al. 2011a). Different approaches for kinetic modeling have been reported to describe the reactions occurring during the FTS, such as kinetic rate designed from a semi-empirical and/or empirical mode, models derived from the use of certain catalytic reaction mechanisms, as well as kinetic expressions represented by a type of power-law equation (Zimmerman and Bukur 1990; van der Laan and Beenackers 1999). The kinetic models depend on the type of catalyst and operating conditions used in the FTS process (Basha et al. 2015).

Kinetic models are to be developed on the basis of a fairly large set of experimental data distributions under certain process conditions that are varied enough and independently, such as space velocity, temperature, and total and partial pressures of CO and H_2, with heat transfer and mass gradients almost negligible (Van der Laan and Beenackers 1999). The design of kinetic models is quite broad, which arises from the wide variety of catalysts used.

The reported detailed kinetic models have differences due to the assumptions regarding the rate of determining steps (Gideon Botes et al. 2009). Some disadvantages of these models are as follows (van Berge and Everson 1997): high complexity of the catalytic mechanism, statistical uncertainties for many kinetic constants to be estimated, and questionable assumptions to generalize the models developed. Furthermore, the cases of FTS kinetic models that use groups based on the number of carbon atoms in the molecules, as well as those that are simply treated as a group or division set of paraffins and olefins, can give oversimplified information of syngas total conversion, so that they are not able to provide detailed information on the product's distribution (Wang et al. 2003b).

Kinetic models for the FTS process using catalysts based on cobalt have been organized into the following three different classes: (i) kinetic models based on the occupation of a single active site by CO or carbon surface, (ii) kinetic models based on occupancy of two active sites by the CO or a reaction intermediate, and (iii) kinetic models based on nontraditional reaction orders for hydrogen (Gideon Botes et al. 2009). Recently 14 kinetic models were investigated for the FTS process with cobalt catalyst (Keyvanloo et al. 2016). Some correspond to models of the Langmuir–Hinshelwood (LH) type, Eley–Rideal type, and power-law.

The FTS kinetic modeling is a catalytic process that involves a comprehensive approach of the reaction steps involved during the initiation, propagation, and termination processes of the hydrocarbon chain, which is a highly complex task that recognizes the scientific community (Hall et al. 1952; Brötz 1949). There are several studies cited in the literature about various kinetic modeling schemes of the FTS process that derive from a catalytic reaction mechanism developed under certain considerations and assumptions that may be valid for a set of operating conditions that define a specific operational scenario. However, it is important to mention that there is no generalized model that accurately and uniquely describes syngas consumption and the hydrocarbon chain formation and distribution process because it is a highly catalytic reaction.

There is a great diversity of FTS kinetic models for both iron-based and cobalt catalysts and kinetic models for the WGS reaction that is involved in the water formation during the reaction process. In addition, the main difference between the kinetics of cobalt-based catalysts and the iron catalysts lies in the former's inactivity toward the WGS reaction since H_2O is not adsorbed on the catalyst active sites.

8.1.3.1 Kinetic Models Developed with Iron Catalyst

Table 8.1 reports the different kinetic models developed from experiments carried out in fixed-bed reactors with iron catalysts. Table 8.2 shows some reaction rates for water–gas shift with catalysts based on iron. Also, Table 8.3 shows the bases for the kinetic model development and the specifics of the reaches of each kinetic model for the FTS catalyzed with iron.

8.1.3.2 Kinetic Models Developed with Cobalt Catalyst

Table 8.4 shows the different kinetic models applied in fixed-bed reactors with cobalt catalysts, and Table 8.5 shows the bases for the kinetic model development and the specifics of the reaches of each kinetic model for the FTS catalyzed with cobalt.

One of the differences between cobalt catalysts and iron catalysts is that for kinetic modeling, the WGS reaction is not considered, which is due to the very small amount of water produced that can be neglected (Basha et al. 2015). However, there are cases of FTS fixed-bed reactor modeling that

Table 8.1 FTS kinetic models based on iron catalyst.

Type of catalyst/ promoter	Operation conditions P (MPa)	T (°C)	$^{H_2}/_{CO}$	Kinetic model	Reference
Fe	—	—	—	$-R_{FT} = k_{FT}\frac{p_{H_2}^2}{p_{CO}}$	Hall et al. (1952)
Reduced fused	2.2.4.2	250–320	2.0	$R_{FT} = k_{FT}P_T$	Brötz and Systematik (1949)
Fe	2.2	225–240	0.25–2.0	$-R_{FT} = ap_{H_2}^{0.6}p_{CO}^{0.4} - fp_{H_2O}^{0.5}R_{FT}^{0.5}$	Anderson and Karn (1960)
$Fe/K_2O/Al_2O_3/SiO_2$	1.0–2.8	225–265	1.2–7.2	$R_{FT} = k_{FT}p_{H_2}$	Dry (1972)
Fe	0.5–4.0	200–340	1.0–7.3	$R_{FT} = k_{FT}\frac{p_{H_2}p_{CO}}{p_{CO} + a_c p_{H_2O}}$ $a_c = {}^{K_{H_2O}}/_{K_{CO}}$	Dry (1976)
Reduced fused $Fe/K_2O/Al_2O_3/SiO_2$	2.0	250–315	2.0	$R_{FT} = k_{FT}\frac{p_{H_2}p_{CO}}{p_{CO} + b_c p_{H_2O}}$	Atwood and Bennet (1979)
Precipitated copper promoted by potassium/iron	1.0–2.0	220–270	1.0–6.0	$R_{FT} = k_{FT}\frac{p_{H_2}}{p_{CO}^{0.25}}$	Feimer et al. (1981)
Reduced precipitated Fe/Cu/K	1.0–3.2	220–300	1.1–2.8	$R_{FT} = k_{FT}\frac{p_{H_2}p_{CO}}{p_{CO} + C_c p_{H_2O}}$ C_c ratio of the adsorption equilibrium constants of the two kinetic models	Liu et al. (1995)
Reduced precipitated $Fe/CuO/K_2O/Na_2O/SiO_2$	0.6–2.1	250–350	3.0–6.0	$R_{C_nH_{2n+2}} = k_5\frac{p_{H_2}\left(\frac{k_1p_{CO}}{k_1p_{CO} + k_5p_{H_2}}\right)\alpha^{n-1}}{1 + \left(\frac{k_1p_{CO}}{k_1p_{CO} + k_5p_{H_2}}\right)\frac{1}{\alpha - 1}}$	Lox and Foment (1993a)

(*Continued*)

Table 8.1 (Continued)

Type of catalyst/ promoter	Operation conditions P (MPa)	T (°C)	H_2/CO	Kinetic model	Reference
				$R_{C_nH_{2n}} = k_6 \dfrac{p_{H_2}\left(\dfrac{k_1 p_{CO}}{k_1 p_{CO} + k_5 p_{H_2}}\right)\alpha^{n-1}}{1 + \left(\dfrac{k_1 p_{CO}}{k_1 p_{CO} + k_5 p_{H_2}}\right)\dfrac{1}{\alpha - 1}}$ $\alpha = \dfrac{k_1 p_{CO}}{k_1 p_{CO} + k_5 p_{H_2} + k_6}$	
Fe	2.4	220–260	1.87–2.0	$R_{FT} = k_{FT} p_{H_2} \dfrac{1}{1 + 1.6\dfrac{p_{H_2O}}{p_{CO}}}$	Jess et al. (1999)
Fe/Cu/K	1.1–3.1	220–269	1.0–3.0	$R_{CH_4} = \dfrac{k_{5M} p_{H_2} \alpha_1}{1 + \left(1 + \dfrac{1}{k_2 k_3 k_4} + \dfrac{p_{H_2O}}{p_{H_2}^2} + \dfrac{1}{k_2 k_3 p_{H_2}} + \dfrac{1}{k_4}\right)\sum_{i=1}^{N}\left(\prod_{j=1}^{j}\alpha_j\right)},\ (n = 1)$ $R_{C_nH_{2n+2}} = \dfrac{k_5 p_{H_2} \prod_{j=1}^{j}\alpha_j}{1 + \left(1 + \dfrac{1}{k_2 k_3 k_4} + \dfrac{p_{H_2O}}{p_{H_2}^2} + \dfrac{1}{k_2 k_3 p_{H_2}} + \dfrac{1}{k_4}\right)\sum_{i=1}^{N}\left(\prod_{j=1}^{j}\alpha_j\right)},\ (n \geq 2)$ $R_{C_nH_{2n}} = \dfrac{k_6 (1 - \beta_n)\prod_{j=1}^{j}\alpha_j}{1 + \left(1 + \frac{1}{k_2 k_3 k_4} + \dfrac{p_{H_2O}}{p_{H_2}^2} + \dfrac{1}{k_2 k_3 p_{H_2}} + \dfrac{1}{k_4}\right)\sum_{i=1}^{N}\left(\prod_{j=1}^{j}\alpha_j\right)},\ (n \geq 2)$ $R_{CO_2} = \dfrac{k_V\left(\dfrac{p_{CO} p_{H_2O}}{p_{H_2}^{0.5}}\ \dfrac{p_{CO_2} p_{H_2}^{0.5}}{K_P}\right)}{1 + k_V \dfrac{p_{CO} p_{H_2O}}{p_{H_2}^{0.5}}}$ $\alpha_1 = \dfrac{k_1 p_{CO}}{k_1 p_{CO} + k_{5M} p_{H_2}};\ n = 1$ $\alpha_n = \dfrac{k_1 p_{CO}}{k_1 p_{CO} + k_5 p_{H_2} + k_6};\ (n \geq 2)$	Wang et al. (2001)

				$\beta_n = \dfrac{\frac{k_{-6}}{k_6} p_{C_nH_{2n}}}{\left[\alpha_A^{n-1} \dfrac{k_1 p_{CO}}{k_1 p_{CO} + k_5 p_{H_2}} + \dfrac{k_1 p_{CO}}{k_1 p_{CO} + k_5 p_{H_2} + k_6} \sum_{i=2}^{n} \alpha_A^{j-1} p_{C_{(n-i+2)}H_{(n-1+2)}}\right]}; (n \geq 2)$	
				$\alpha_n = \dfrac{k_1 p_{CO}}{k_1 p_{CO} + k_5 p_{H_2} + k_6}; (n \geq 2)$	
				k_1, k_2, k_3, k_4, and k_{5M}: constants given by the Arrhenius equation	
				K_P: equilibrium constant that is determined by a polynomial.	
				k_V: rate constant of CO_2 formation	
Reduced Fe/Mn	1.0–3.0	267–327	1.0–3.0	$R_{CH_4} = \dfrac{A_k}{(B_k + C_k + D_k)^2}$	Yang et al. (2003)
				$R_{C_nH_{2n+2}} = \dfrac{k_7}{k_{7M}(B_k + C_k + D_k)^2}$	
				$R_{C_nH_{2n}} = \dfrac{\dfrac{k_8^+ (1-\beta_n) K_3' p_{CO}}{p_{H_2O}} \prod_{j=1}^{j} \alpha_j}{(A_k + B_k + C_k)^2}$	
				$A_k = \dfrac{k_{7M} K_4 K_6 K_3' p_{CO} p_{H_2}^3}{p_{H_2O}}$	
				$B_k = 1 + \sqrt{K_4 p_{H_2}} + K_1 p_{CO} + \dfrac{K_3' p_{CO} p_{H_2}^2}{p_{H_2O}}$	
				$C_k = K_1 K_2 p_{CO} p_{H_2} + \dfrac{K_6 K_4^{0.5} K_3' p_{CO} p_{H_2}^{0.5}}{p_{H_2O}}$	
				$D_k = \dfrac{K_3' p_{CO} p_{H_2}^2}{p_{H_2O}} \left(1 + K_6 \sqrt{K_4 p_{H_2}}\right) \sum_{i=1}^{N} \left(\prod_{j=1}^{i} \alpha_j\right)$	
				whose kinetic rate constants involved in the model are established in the form given by Arrhenius	

(Continued)

Table 8.1 (Continued)

Type of catalyst/ promoter	Operation conditions			Kinetic model	Reference
	P (MPa)	T (°C)	H_2/CO		
Fe/Co/Ni	0.1–0.7	250–270	1.0–2.5	$-R_{FT} = \frac{k_{FT} b_{CO} p_{CO} b_{H_2} p_{H_2}}{(1 + b_{CO} p_{CO} + b_{H_2} p_{H_2})^2}$ b_{CO} and b_{H_2}: adsorption coefficients of CO and H_2, respectively	Mirzaei et al. (2015)
Fe	1.1–2.61	260–300	0.67–2.05	$\alpha_{C_1} = \frac{k_{HC1(1)} p_{CO}}{k_{HC1(2)} p_{CO} + k_{HC5(l)} p_{H_2} + k_{HC7(1)} \frac{p_{H_2O}}{K_{HC3} K_{HC4}}}$ $\alpha_{C_n} = \frac{k_{HC1(n)} p_{CO}}{k_{HC1(n+1)} p_{CO} + k_{HC5(n)} p_{H_2} + k_{HC6(n)} + k_{HC7(n)} \frac{p_{H_2O}}{K_{HC3} K_{HC4}}}; (n \geq 2)$ $\beta_{C_n} = \frac{k_{HC6re(n)}}{k_{HC1(n+1)} p_{CO} + k_{HC5(n)} p_{H_2} + k_{HC6(n)} + k_{HC7(n)} \frac{p_{H_2O}}{K_{HC3} K_{HC4}}}; (n \geq 2)$ $\varphi = \left\{ 1 + \alpha C_1 + \sum_{i=2}^{n} \left(\prod_{j=i+1}^{n} \alpha C_j p_{C_i H_{2i}} \right) + \beta_{C_n} p_{C_n H_{2n}} \right\}$ $R_{CH_4} = \frac{k_{HC5(1)} \alpha_{C_1} p_{H_2}}{\varphi}$ $R_{C_n H_{2n+2}} = \frac{k_{HC5(n)} \left[\prod_{i=1}^{n} \alpha_{C_i} + \sum_{i=2}^{n} \left(\prod_{j=i+1}^{n} \alpha_{C_j} \beta_{C_i} p_{C_i} H_{2i} \right) + \beta_{C_n} + p_{C_n H_{2n}} \right] P_{H_2}}{\varphi}; (n \geq 2)$ $R_{CH_3 H_{2n}} = \frac{k_{HC6(n)} \left\{ \prod_{i=1}^{n} \alpha_{C_i} + \sum_{i=2}^{n} \left(\prod_{j=i+1}^{n} \alpha_{C_j} \beta_{C_i} p_{C_i} H_{2i} \right) + \beta_{C_n} + p_{C_n H_{2n}} \right\} - k_{HC6re(n)} p_{C_n H_{2n}}}{\varphi};$ $(n \geq 2)$ $R_{CH_3 OH} = \frac{k_{HC7(1)} \alpha_{C_1} p_{H_2O}}{K_{HC3} K_{HC4} \varphi}$ $R_{C_n H_{2n+1} OH} = \frac{k_{HC7(n)} \left\{ \prod_{i=i}^{n} \alpha_{C_i} + \sum_{i=2}^{n} \left(\prod_{j=i+1}^{n} \alpha_{C_j} \beta_{C_i} p_{C_i H_{2i}} \right) + \beta_{C_n} p_{C_n H_{2n}} \right\} p_{H_2O}}{K_{HC3} K_{HC4} \varphi}; (n \geq 2)$	Zhou et al. (2016)

Table 8.2 Kinetics models for water–gas shift based on iron catalyst.

Type of catalyst/ promoter	Operation conditions			Kinetic model	Reference
	P (MPa)	T (°C)	H_2/CO		
Fe	0.5–4.0	200–340	1.0–7.3	$R_{WGS} = R_{WGS}p_{CO}$	Dry (1976)
Reduced precipitated $Fe/Cu/K_2O$	1.0–2.0	220–270	1.0–6.0	$R_{WGS} = k_{WGS}p_{CO}$	Feimer et al. (1981)
Reduced precipitated $Fe/CuO/K_2O/Na_2O/SiO_2$	0.6–2.1	250–350	3.0–6.0	$R_{WGS} = k_{WGS}\dfrac{\left(p_{CO}p_{H_2O} - \dfrac{p_{CO_2}p_{H_2}^{0.5}}{K_{eq}}\right)}{\left(1 + a\dfrac{p_{H_2O}}{p_{H_2}^{0.5}}\right)^2}$	Lox and Froment (1993b)
Fe	2.4	220–260	1.87–2.0	$R_{WGS} = k_{WGS}p_{H_2O}$	Wang et al. (2001)
Reduced precipitated Co/MnO	0.6–2.6	210–250	1.6–4.1	$R_{WGS} = k_{WGS}p_f\left(p_{CO} - \dfrac{p_{CO_2}p_{H_2}}{K_{eq}p_{H_2O}}\right)$	Kayser et al. (2000), Yang et al. (1979)
Reduced Fe/Cu/K					
Fe/Cu/K	1.1–3.1	220–269	1.0–3.0	$R_{WGS} = k_V\dfrac{\left(\dfrac{p_{CO}p_{H_2O}}{p_{H_2}^{0.5}} - \dfrac{p_{CO_2}p_{H_2}^{0.5}}{K_P}\right)}{\left(1 + k_V\dfrac{p_{CO}p_{H_2O}}{p_{H_2}^{0.5}}\right)}$	Wang et al. (2003b), Yang et al. (2003)
Reduced Fe/Mn	1.0–3.0	267–327	1.0–3.0	$R_{WGS} = a_c\dfrac{\left(\dfrac{p_{CO}p_{H_2O}}{p_{H_2}^{0.5}} - \dfrac{p_{CO_2}p_{H_2}^{0.5}}{K_{eq}}\right)}{\left(1 + b_c\dfrac{p_{CO}p_{H_2O}}{p_{H_2}^{0.5}}\right)}$	Yang et al. (2003)

consider the WGS reaction, given that the amount of water experimentally produced is considerable (Dai et al. 2014; Moazami et al. 2015a).

8.1.4 General Aspects of Fischer–Tropsch Catalytic Mechanisms

Generally, not only taken as a valid assumption but also accepted as necessary is the fact that there are different ways or reaction systems that depend on the type of catalyst and operating conditions used, among other features that directly affect the approach of a single generalized reaction surface mechanism during the FTS, thus giving rise to the existence of several alternate paths parallel operation also becoming an object of scientific research.

Currently, the scientific literature has reported a wide variety of studies on the reaction mechanisms involved in the FTS process, which today has resulted in the existence of a dispute about which chemical reactions occur in this process. It is also important to keep in mind that for decades various chemical formation pathways of the monomer and the initiator of hydrocarbon chain growth have been proposed. However, the most important monomers identified throughout the history of

Table 8.3 Bases for the kinetic model development and reaches of the FTS process description based on iron catalyst.

Reference	Bases for the kinetic model development	Scope
Hall et al. (1952)	The differential method was used for the development of the kinetic model, which is a function of the partial pressures of the syngas (CO and H_2).	The model describes the synthesis gas consumption well, but for a narrow range of operating conditions. Because the kinetics was not based on a reaction mechanism, i.e., the phenomena of formation of the hydrocarbon chain, the olefins readsorption, and others are not contemplated, their use is limited for cases where operating conditions of industrial interest are explored.
Brötz and Systematik (1949)	The kinetic model of the FTS reaction is directly proportional to the total system pressure.	The kinetic model was not based on any reaction mechanism, so it is limited for use in case studies under industrially relevant conditions.
Anderson and Karn (1960)	Use of the fractional coverages of the surface by reactants and products. These surface coverages are approximated by Freundlich isotherms since the coverage of each component is proportional to its partial pressure in a positive exponent smaller than one.	Methane production and the effects of CO_2 inhibition are ignored. The postulate of methylene formation as the complex species involved in the hydrocarbon growth chain requires a broad reconsideration of recent concepts on the synthesis mechanism development, especially in the formation of oxygenated molecules.
Dry (1972)	The Fischer–Tropsch synthesis mechanism involves the reaction of the adsorbed hydrogen with the adsorbed CO, such that the first-order dependence of the hydrogen pressure means that the adsorption of hydrogen or the reaction of the hydrogen molecules adsorbed with the CO is the rate-controlling step.	At different low values of the H_2O/CO_2 ratios, the kinetic model can predict that H_2O is the predominant primary product rather than CO_2 and that at a fixed carbon monoxide partial pressure the activity increases with increased hydrogen partial pressure.
Dry (1976)	The CO_2 formation occurs easily under the process conditions used so that the water–gas shift reaction is considered in the kinetics development. Langmuir's adsorption theory was used to represent a system in which the CO molecule competes with the H_2O, CO_2, and H_2 molecules to occupy the adsorption sites.	At higher temperatures, the H_2/CO ratio does not manifest negative consequences in the description of hydrocarbon selectivity, and the selectivity seems to be predominantly controlled by the CO_2 partial pressure. Because Langmuir's adsorption theory was simplified according to the operating scenario used, the model is limited to certain operating conditions and therefore to a good prediction of a broad spectrum of hydrocarbons.
Atwood et al. (1979)	The kinetic equation is based on the CO consumption rate. The diffusional effects on the catalyst particle were neglected, so the derived kinetics is intrinsic.	The kinetic model is limited to the description of a short range of liquid hydrocarbons. However, it can predict the H_2O and CO_2 production.
Feimer et al. (1981)	The model was derived from the power law to represent the formation of C_1 to C_5 hydrocarbons and alcohols based on the partial pressures of CO and H_2.	The model can predict the formation of hydrocarbons with a low number of carbon atoms and alcohol production. The reaction mechanism does not contemplate the CO readsorption or the

Table 8.3 (Continued)

Reference	Bases for the kinetic model development	Scope
		olefins readsorption/hydrogenation, so the reaction kinetics is limited to the description of long-chain hydrocarbons.
Liu et al. 1995	The intrinsic kinetic model was developed based on the concept of passivation and using the lumping method.	Intrinsic kinetics predicts the inhibition of the FTS reaction by the effect of water produced but not its inhibition by CO_2.
Lox and Froment (1993a)	The reaction kinetics was developed using the LHHW theory under the assumption that the elementary step that gives formation to the monomer is the rate-determining step.	The initial rates of carbon dioxide formation and various hydrocarbons depend on the partial pressures of CO and H_2. The model can predict the influence of CO and H_2 partial pressures on the formation of olefins and n-paraffins with more than two carbon atoms. The model is limited to the use of an ideal hydrocarbon growth probability model.
Lox and Froment (1993b), Jess et al. (1999)	The kinetic model assumes that different active sites are arranged on one side for the gas–water shift reaction to occur and other sites on the other side for the formation of *n*-paraffins and 1-olefins to occur. The rate kinetic expressions for the hydrocarbon formation were developed from the elementary reactions dictated in the carbide mechanism that are based on CO adsorption and products hydrocarbon desorption, which are not in equilibrium. For purposes of simplifying the kinetic model, it is assumed that the active sites are occupied by surface hydrocarbon intermediates. The kinetics of the gas–water shift reaction implies a surface intermediate formate on elementary reactions. The intermediate formate formation is the rate-determining step.	The hydrocarbon distribution is limited to the ideal description given by the Anderson–Schulz–Flory model. The model only contemplates the elementary reactions of CO adsorption and hydrocarbon desorption as those of greater relevance in the kinetic model development, leaving aside the re-absorption of short-chain olefins truncating the long-chain paraffins formation. All the elementary reactions leading to hydrocarbon products formation are single-site reactions.
Wang et al. (2001)	The kinetic expression is based on the overall consumption rate of CO, the hydrocarbon distribution model, and assuming the olefin readsorption process.	The kinetic model can accurately describe the paraffins and olefins formation. However, it is limited to the description of oxygenated compounds.
Yang et al. (2003)	The alkylidene mechanism was used in which readsorption and secondary reactions of olefins are considered and deviations of hydrocarbon distribution from the use of the conventional ASF model are described. Both the product desorption and the insertion of methylene into the metal-alkylidene bond are considered the rate-determining steps. Only five surface steps are contemplated for the derivation of the kinetic model: (i) adsorption of reactants (CO and H_2);	The model good fits with the experimental data on the hydrocarbon distribution under the use of industrially relevant conditions. The olefin-to-paraffin ratio was not intrinsically described.

(Continued)

Table 8.3 (Continued)

Reference	Bases for the kinetic model development	Scope
	(ii) chain initiation; (iii) chain propagation; (iv) chain termination and product desorption; (v) readsorption and secondary reaction of olefins. The dissociative adsorption of hydrogen is considered to occur both in the molecular state and in the dissociative state.	
Mirzaei et al. (2015)	The LHHW theory was used and associative adsorption of CO and H_2 were considered.	The diffusion limitations in the catalyst pore play an important role in the catalytic syngas conversion that must be considered for the derivation of a more precise kinetic model.
Zhou et al. (2016)	The elementary steps are based on the carbide mechanism with hydrogen-assisted CO insertion and CH_2 as the growth monomer. The readsorption of olefins was considered, observing a negligible effect on the hydrocarbon product distribution. The oxygenation reation is considered and the reaction scheme used was expressed in terms of a mechanistic carbide model with hydrogen-assisted CO insertion.	The model shows a deviation from the linear distribution of ASF, which is because the structure of the reactants and the transition states intermediates were not considered.

the FTS process are methylene (—CH_2—), hydroxyl carbine (CHOH), and CO, which lead to approach the following three important mechanisms FTS: (i) carbide, (ii) the enol mechanism, and (iii) CO insertion mechanism, respectively. However, today different reaction mechanisms have been developed with the aim of describing in a precise way the FTS process, which have been proposed based on differences between the proposed mechanisms because of the assumptions that arise for monomer, initiator, and their chemical formation. Such mechanisms are the alkyl, alkenyl and alkylidene hydride-methylidene, and more recently the formate mechanism (Frennet and Hubert 2000).

In the case of cobalt-based catalysts, there are two main mechanisms that describe the overall scheme of FTS reactions: (a) the direct CO dissociation (carbide mechanism) and (b) hydrogen-assisted CO dissociation. Although there are several studies that indicate that the mechanism of hydrogen-assisted CO dissociation is the one that describes the FTS product's distribution more precisely, it is not able to predict the presence of atomic carbon on the catalytic surface (Hall et al. 1952). Recently, a mechanism based on a parallel hydrogen-assisted mechanistic pathway was used to develop an LH-type kinetic model to predict the presence of atomic carbon on the catalyst surface. This mechanism is characterized by including the direct dissociation of CO, hydrogen-assisted CO dissociation, and parallel hydrogen-assisted CO dissociation (Keyvanloo et al. 2016). Recently, a work proposed a comprehensive mechanism-derived FTS kinetic model whose results were compared with the most recent findings reported in the literature (eight kinetic models). It was shown that the kinetic model developed based on a combination of the alkyl/alkenyl mechanisms for FTS reactions (to produce the n-paraffins and α-olefins) together with the formate mechanism for the WGS reaction can provide more accurate predictions of the syngas

Table 8.4 FTS kinetic models based on cobalt catalyst.

Type of catalyst/promoter	Operation Conditions			Kinetic model	Reference
	P (MPa)	*T* (°C)	H_2/CO		
Co/MgO/ ThO_2/ Kieselguhr	0.1	185–200	2.0	$k_{FT} = \frac{p_{H_2}^2}{p_{CO}}$	Hall et al. (1952)
Co/CuO/ Al_2O_3	1.7–5.5	235–270	1.0–3.0	$R_{FT} = k_{FT} p_{H_2} p_{CO}^{0.5}$	Yang et al. (1979)
Reduced precipitated	0.015–0.1	250	0.25–5.0	$R_{FT} = k_{FT} \frac{p_{H_2} p_{CO}^{0.5}}{1 + \sqrt{K_{CO} p_{CO}} + \left(\sqrt{K_{H_2} p_{H_2}}\right)^3}$ *K*: pseudo-adsorption equilibrium constant for CO and H_2 species	Outi et al. (1981)
Co/ Kieselguhr	0.2–1.5	190	0.5–8.3	Model 1: $R_{FT} = k_{FT} \frac{p_{H_2}^{0.5} p_{CO}^{0.5}}{\left(1 + a_c p_{CO}^{0.5} + b_c p_{H_2}^{0.5}\right)^2}$; $a_c = 0.446$, $b_c = 0.914$ Model 2: $R_{FT} = k_{FT} \frac{p_{H_2}^{0.5} p_{CO}}{\left(1 + a_c p_{CO} + b_c p_{H_2}^{0.5}\right)^2}$; $a_c = 0.0116$, $b_c = 0.0354$	
Co/TiO_2	2.0265	180–240	1.0–3.5	$R_{FT} = k_{FT} p_{CO}^{-0.24} p_{H_2}^{0.74}$ $R_{FT} = k_{FT} \frac{a_c p_{CO} p_{H_2}^{0.74}}{(1 + b_c p_{CO})^2}$	Zennaro et al. (2000)
Co – Mn/ TiO_2	0.1–1.0	190–280	1.0–3.0	$R_{FT} = \frac{k_{FT} K_{CO} p_{CO} p_{H_2}}{1 + b_c p_{CO}}$ $R_{FT} = \frac{k_{FT} K_{CO} p_{CO} p_{H_2}^2}{\left[1 + 2(b_c p_{CO})^2\right]}$ K_{CO}: equilibrium constant of CO adsorption step	Visconti et al. (2011a)
Co – Ru/ Al_2O_3	0.1	186–207	0.8–3.5	$R_{FT} = \frac{a_c p_{H_2}^{0.5} p_{CO}^{0.5}}{\left(1 + b_c p_{H_2} + c_c p_{CO}^{0.5} + d_c p_{CO}\right)^2}$	Irankhah et al. (2007)
Co – Ce/ SiO_2	At atmospheric pressure	200–300	1.0–3.0	$R_{FT} = \frac{p_{H_2}}{(1 + a_c p_{CO})}$ a_c: constant lumped	Mansouri et al. (2013)
Co/Al_2O_3	1.0–2.5	190–220	1.5–3.0	$R_{FT} = k_{FT} \frac{p_{CO} p_{H_2}}{(1 + a_c p_{CO})}$ a_c: determined by the Arrhenius equation	Kaiser et al. (2014)

Table 8.5 Bases for the kinetic model development and reaches of the FTS process description based on cobalt catalyst.

Reference	Bases for the kinetic model development	Scope
Visconti et al. (2011a)	The two kinetic models were based on the Langmuir–Hinshelwood theory.	Both models predict the synthesis of synthesis gas (CO and H_2) and the selectivity of hydrocarbon products. However, it was only validated for a narrow range of operating conditions. A rigorous kinetic mechanism for the development of kinetics is not contemplated.
Yang et al. (1979)	Kinetics is based on the power rate law. The conventional ASF model was used to describe the hydrocarbon chain growth.	Kinetics indicated that hydrocarbon formation increases linearly with syngas conversion, limiting its use with a reactor model where high linearities are present.
Outi et al. (1981)	FTS kinetics consider that the process initiation is carried out by the CO dissociation and the formation of the CH_2 surface intermediate, where its formation is a rate-determining step. It is assumed that the hydrocarbon chain growth proceeds through the addition of the same CH_2 groups to the growing molecule.	The kinetics can explain the partial pressure effect of the syngas on the olefin/paraffin ratio and on the methane selectivity. The kinetics illustrates that the chain growth probability constant is a complex function of the CO and H_2 partial pressures.
Sarup and Wojciechowski (1989)	Two kinetic models are of LHHW type in which CO conversion to hydrocarbons and methane rate formation are used. The first model is based on the hydrogenation of surface carbon and the second on a hydrogenation-assisted dissociation of CO both considered as the rate-determining steps of the FTS process.	The step of CH formation as a building block of the hydrocarbon chain is assumed in the reaction mechanism which allows the kinetics to predict CO conversion and methane formation good. Both models fit satisfactorily to CO conversion and methane formation data for a limited range of operating conditions.
Zennaro et al. (2000)	Two kinetics (i) of the power law type and (ii) a simplified version of Langmuir–Hinshelwood are developed.	Both kinetics are limited to their application to a wide range of pressures, gas space velocities, and feed syngas ratio. They do not integrate a hydrocarbon chain distribution model.
Irankhah et al. (2007)	LHHW model was used to obtain kinetics and parameters to describe CO conversion to CH_4 formation under supercritical conditions. Kinetics formulation was based on the following steps: (i) adsorption and dissociation of CO and H_2 occur at the active sites of the cocatalyst; (ii) removal of oxygen from the catalytic surface through water and CO_2; (iii) hydrogenation of adsorbed carbon, oligomer production and formation of monomer ($CH_2{*}S$) and alkyl species ($CH_3{*}S$); (iv) methane formation.	The kinetic modeling is validated through a linear relationship between the reaction rate of the CO consumption measured with respect to that estimated in supercritical conditions. However, it lacks validation against the description of hydrocarbon selectivity and estimation of the hydrocarbon chain. The range of operating conditions was narrow.
Mansouri et al. (2013)	Kinetics is based on the LHHW approach. Kinetics assumes the following: (i) intrinsic reaction rates are proportional to the surface coverage of reactants; (ii) CO	The activation energy value is too low compared to the values reported in the literature for a set of similar operating conditions. The kinetics well describes the

Table 8.5 (Continued)

Reference	Bases for the kinetic model development	Scope
	consumption and CH_2 intermediary formation are the irreversible rate-determining step while the other reactions are in quasi-equilibrium; (iii) there is no accumulation in the catalytic surface at steady-state conditions; (iv) operation is considered under isothermal regime; (v) the catalyst is fixed and uniformly distributed; (vi) elementary CO and H_2 adsorption are in quasi-equilibrium within the concentration of gas phase; (vii) water is removed after irreversible CO decomposition; (viii) CO and carbon concentrations are assumed to be the dominant concentrations.	hydrogenation of CO toward light olefins. The LHHW approach is relatively simple. The application of kinetics is limited to the use of operating conditions of industrial relevance.
Kaiser et al. (2014)	Intrinsic kinetics for CO consumption and its parameters were evaluated using the Langmuir–Hinshelwood (LH) approach. The influence of pores diffusion was assumed.	CO_2 and H_2O formation show negligible effect on FTS reaction rate. Water produced has minor effect on product's distribution as predicted by kinetics. Simplified LH approach despises the adsorption of H_2. The model predicts well FTS and CH_4 reactions for medium range of commercially relevant operating conditions.

consumption, hydrocarbons formation, and a better description of the FTS reaction network (Moazami et al. 2017a). An important finding of this study given by Moazami et al. (2017a) was that the combination of the alkyl and alkenyl mechanisms for the formulation of a comprehensive kinetic model that allows to describe well the initiation, propagation, and termination steps of the hydrocarbon chain, since it is a fact that the alkenyl mechanism largely represents the formation of α-olefins, while the alkyl mechanism allows to describe well the formation of n-paraffins. In addition, it was found the alkyl mechanism favors the paraffins formation rather than olefins; hence it is a better choice for this case compared to other mechanisms, e.g. alkenyl, CO insertion, and/or enol mechanisms. However, care should be taken only for the use of the alkyl mechanism since erroneous underestimations of olefin selectivities can be obtained. On the other hand, it was observed that considering the WGS reaction kinetics, the formate mechanism, in which the formate species that were formed from the reaction between the adsorbed CO intermediate and a hydroxyl surface species (—OH), was considered the most relevant route for the construction of a kinetic model, since it is able to describe in a precise way the CO consumption and the higher hydrocarbon distribution (Moazami et al. 2017a).

8.1.5 The Fischer–Tropsch Synthesis Product Distribution Models

The FTS product distribution is usually predicted by the classical model of ASF, which is represented by the following equation:

$$f_C = n(1-\alpha)^2\alpha^{n-1} \tag{8.14}$$

where f_C is the fraction of carbon atoms that make up the length of the hydrocarbon chain of size n and α is the chain growth probability factor that is determined by the following expression (Todic et al. 2015):

$$\alpha = \frac{r_p}{r_p + r_t} \tag{8.15}$$

where r_p and r_t are the propagation and chain termination velocities, respectively. On the other hand, van Dijk (2001) reported the following equation to determine the probability of chain termination:

$$\alpha_t = \frac{1-\alpha}{\alpha} \tag{8.16}$$

The growth of hydrocarbons considering a polymerization FTS reaction scheme is illustrated in Figure 8.1, in which the probability of growth is characterized by C that depends on the type of catalyst and operating conditions. The chain termination leads to the formation of products (olefins and paraffins), where the hydrogenation reaction of paraffins is irreversible whereas for olefins it is reversible, which implies readsorption of these in catalytic surface and again the hydrogenation reaction to continue with the formation of paraffins.

However, it is noteworthy that the idealized ASF model has certain deviations from the precise description of the hydrocarbon spectrum generated in the FTS process. It is known that the selectivity of the carbon number of the FTS products does not fully obey the classical ASF model due to three types of deviations that have been observed and have been reported previously: a production

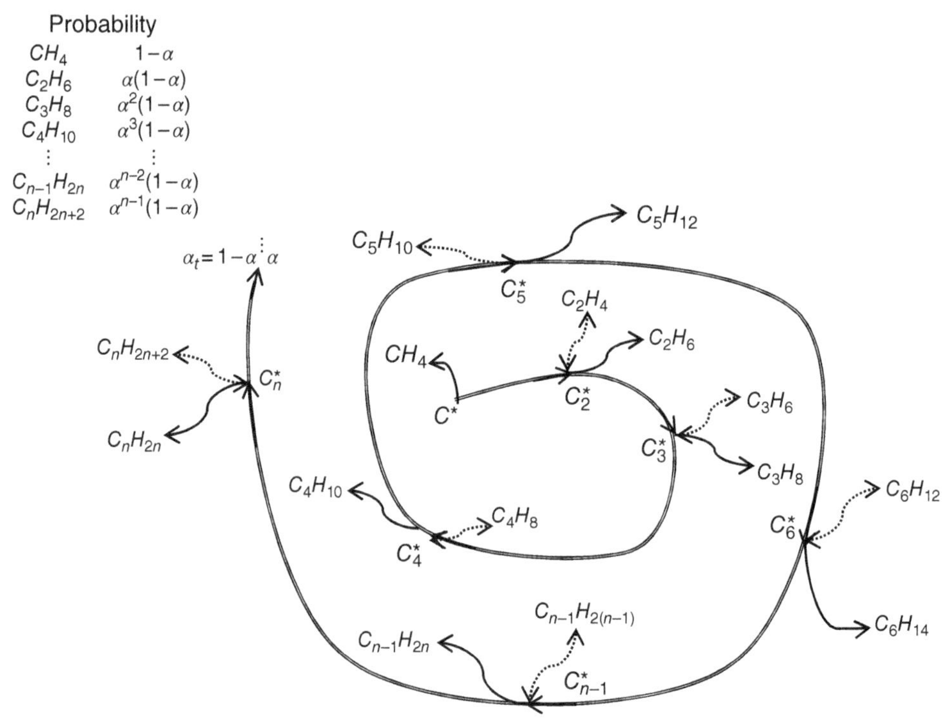

Figure 8.1 FTS hydrocarbons growth simplified scheme.

of methane higher than expected; a lower production of ethene than expected; and higher selectivity to heavy products, which is due to increase in the chain growth probability with the number of carbon (Lozano-Blanco et al. 2009; Todic et al. 2014; Novak et al. 1982). Some authors identified several factors that may be responsible for such deviations, for example, Masuku et al. (2011) and van der Laan and Beenackers (1999) attribute these deviations to phenomena such as diffusion limitations, vapor–liquid equilibrium (VLE), and readsorption. It is also believed that deviations can come from analytical difficulties (Puskas et al. 1993) and/or due to handling and the reactor under unsteady-state conditions (Huff and Saterfield 1985).

However, others suggest that deviations from the ideal ASF model may be due to higher production of methane than expected, low selectivity to ethane and heavier products, as well as a lot of FTS products showing a strong deviation, which may be due to mainly secondary reactions of the primary hydrocarbon product, such as reinsertion into the chain growth process, hydrogenation, and hydrogenolysis (Kuipers et al. 1996; Schulz and Claeys 1999). The secondary reactions exhibit a strong influence on the FTS products spectrum, which is critical to optimizing the selectivity to the desired product range (Kuipers et al. 1996).

Other cases of the current study aim at improving the ASF model by describing in more detail the FTS product's distribution and may be able to explain the deviations that may happen with the classical ASF model (Todic et al. 2015).

Nowadays, the interest in having more accurate predictive models for a better description of the product's distribution has increased due to the improvements in the analysis of all isomers and products involved in the FTS, which cannot be calculated using the classical ASF model distribution (van der Laan and Beenackers 1999).

It is worth noting that product selectivity in the FTS process can not only be affected by the catalyst but also by certain operating conditions such as temperature, syngas feed composition, TOS, and catalyst reduction (Moutsoglou and Sunkara 2011). van der Laan and Beenackers (1999) presented a more detailed discussion about the way the after-mentioned operating conditions affect the performance and overall efficiency of the FTS process.

It has been experimentally found that the water production during the FTS reaction with cobalt catalyst favors the higher hydrocarbons selectivity (C_{5+}). In addition, according to the study conducted by Rytter et al. (2016) for the FTS process with cocatalyst supported on Al_2O_3, SiO_2, and TiO_2, it was found that the selectivity to C_{5+} increases and the selectivity to CH_4 decreases with increasing water partial pressure.

Table 8.6 reports some of the most important products derived from the FTS technology and the corresponding carbon number (Mabry 2014).

Table 8.6 Products of interest in the FTS process and its carbon number range.

Name	Carbon number range	Name	Carbon number range
Fuel gas	$C_1 - C_2$	Diesel	$C_{13} - C_{17}$
LPG	$C_3 - C_4$	Middle distillates	$C_{10} - C_{20}$
Gasoline	$C_5 - C_{12}$	Soft wax	$C_{19} - C_{23}$
Naphtha	$C_8 - C_{12}$	Medium wax	$C_{24} - C_{35}$
Kerosene/jet fuel	$C_{11} - C_{13}$	Hard wax	C_{35+}

8.1.6 Final Remarks

The development of a specific scheme of reactions should involve the full range of FTS products; however, such an approach of a reaction mechanism to accurately describe the synthesis process is a rather complex task. There is a great diversity of works in the literature for the development of FTS kinetic models, which consider all the aspects involved in the catalytic steps that give rise to the formation of intermediary species and initiators of the hydrocarbon chain growth. However, despite the great effort of many researchers, there is still a need for a complete model that describes the range of FTS products under any scheme of operating conditions, types of catalyst, and raw material used.

8.2 Modeling of Catalytic Fixed-Bed Reactors for Fuels Production by Fischer–Tropsch Synthesis

8.2.1 Introduction

A comprehensive review of the various studies reported in the literature up to date on the mathematical modeling of fixed-bed reactors to produce fuels by the FTS was carried out. It is quite clear that most of the proposed models are based on a set of assumptions that allow their wide simplification by reducing the models into forms of low complexity, since in most cases the effects of phase equilibrium are neglected and relatively simple Fischer–Tropsch kinetics of the power-law type are used. In addition, most of the proposed modeling schemes neglect the effects of resistances to gas–liquid and liquid–solid mass transfer. On the other hand, few reports consider the energy effects under the consideration of a non-isothermal operation assuming a plug-flow behavior and a gas–liquid system. A generalized model of a fixed-bed FTS reactor is proposed, which takes into account all the mass and heat transfer phenomena, as well as hydrodynamics and VLE, based on the information given in the literature. It is evident that for fixed-bed reactors for fuel production using Fischer–Tropsch technology, there is little experimental information for validation and a need to explore different types of reactor models, such as reactor models under a trickle-flow regime considering the effects of phase distribution and dispersion under transient state conditions.

8.2.2 Modeling of Fixed-Bed Fischer–Tropsch Reactors

Although different types of reactors have been reported to carry out the FT synthesis for fuels production (diesel and gasoline), the FBR has increased the interest of the scientific and industrial community for the following reasons (Hooshyar et al. 2012; Kaskes et al. 2016):

1) Strong potential over slurry bubble column because it operates with a concentration gradient of plug-flow type.
2) High catalyst holdup compared with other reactor configurations.
3) No requirement for a separation step and recovery of the catalyst, which is a major advantage from the point of view of time and reduction of operating costs.
4) Easy of scale-up.

In addition, it is possible to produce different types of hydrocarbons due to the following reasons:

1) The wide range of catalysts that can operate at low and high temperatures.
2) Different operating conditions.
3) Type of material and geometry employed in the reactor design.

Mathematical modeling is a tool with high potential for an accurate description of the different types of reactors that are used by FTS technology (Moazami et al. 21015b). The importance of developing models is to predict the effects of operating conditions on the thermal behavior of the reactor, especially in industrial application reactors or so-called large-scale reactors (Rados et l. 2013).

In the case of fixed-bed reactors, there are some factors that complicate the modeling (Wang et al. 2001; Wang et al. 2003a):

1) Multidimensional distributed configuration.
2) The nature of the two phases (gas and liquid) involved in the reactor bed.
3) Nonlinear dependence of the reaction rate on the temperature.
4) Uncertainties of both the packaging and the flow through this.
5) The uncertainties of the parameters for heat and mass transfer.
6) Parameters that vary depending on the place within the catalyst bed.
7) The complex nature of the reaction kinetics involving many chemical species.
8) The time-varying reaction rate due to catalyst deactivation.

For modeling of FTS catalytic fixed-bed reactor, the transport parameters of heat and mass, the hydrodynamics, effects of pressure drop, holdups of both phases (gas and liquid), phase equilibrium, chemical kinetics, selectivity, and product distribution must be taken into account to obtain a generalized mathematical model to predict the dynamic behavior under a wide range of operating conditions. In addition, the model must be fitted to the data obtained in experimental runs for validation purposes and scaling-up. To give robustness to the model, further verification is required, which means that the model must accurately describe other experimental data.

8.2.2.1 Classification of Fixed-Bed Fischer–Tropsch Reactor Models

It is well-known that a multi-tubular fixed-bed FTS reactor is like a conventional shell and heat exchanger tube reactor with the FTS reaction that occurs in the tube side. Water is generally used as a coolant, which circulates through the shell side to keep the process under an isothermal regime. The proposed mathematical model for such a system must take into accounted the interactions that arise between the side shell and the catalyst tube where highly exothermic reaction occurs. Therefore, the phenomenon that occurs in the catalytic fixed-bed FTS reactor can be characterized according to the following (Iordanidis 2002):

- Effects of heat and intraparticle mass diffusion.
- Heat and mass convective transport in the fluid phase.
- Heat and mass effects occurring between the fluid and solid phases.
- Effect of heat exchange between the side of the exothermic FTS reaction and tube wall.
- The process of thermal conduction and mass diffusion in the solid phase.

The typical modeling approaches widely used for accurate description of fixed-bed reactors are the continuum models, which are classified according to Figure 8.2 (Jakobsen 2014; Froment et al. 2011):

8.2.2.2 One- and Two-Dimensional Pseudohomogeneous Model

In the pseudohomogeneous model, the presence of the solid is not considered, i.e. the resistances to heat and mass transfer occurring between the solid phase and the fluid phase are negligible, assuming that the catalyst surface is exposed to bulk fluid conditions, and intraparticle diffusion effects are ignored. Such a model is classified into one-dimensional plug-flow, one-dimensional plug-flow with axial dispersion, two-dimensional plug-flow with radial dispersion, and two-dimensional with axial and radial dispersion models (Froment et al. 2011).

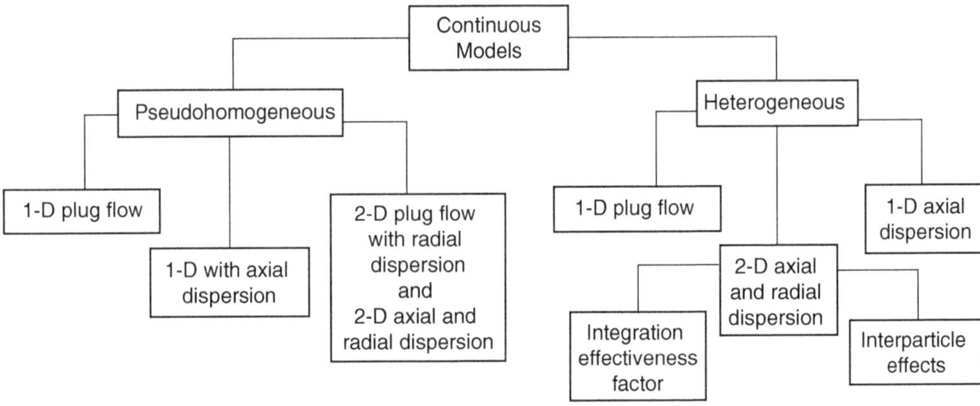

Figure 8.2 Schematic representation of the models commonly used for the description of catalytic fixed-bed reactors (Jakobsen 2014; Froment et al. 2011).

8.2.2.3 One- and Two-Dimensional Heterogeneous Model

In both one-dimensional (axial dispersion) and two-dimensional (axial and radial dispersion) heterogeneous models, the intraparticle diffusion limitations are considered (Froment et al. 2011). In other words, when heat and mass transport within the catalyst particle offer considerable resistance, the reaction rate through the catalyst particle is not uniform (Jakobsen 2014). On the other hand, in the FTS process, the catalysts are completely filled with liquid, for which it is important to take into account the limitations of intraparticle mass transfer for model effects. In addition, the FTS reaction is strongly influenced by the diffusion effects on the catalyst pore and the effective rate (Jess and Kern 2012a). Thus, the modeling of an FTS reactor under a heterogeneous regime is necessary for a more accurate prediction of the FTS process involved in the fixed-bed reactor.

8.2.3 Development of a Generalized Fixed-Bed Fischer–Tropsch Reactor Model

From the literature review, it must be highlighted that it is strictly necessary to develop a general model for FTS fixed-bed reactors to propose a robust mathematical model that can accurately describe certain phenomena naturally occurring in FTS processes for the gasoline or diesel production according to either at high- or low-temperature operation. It is also important to have well-analyzed heat and mass transport phenomena to safely develop a mathematical model that can predict what really happens in commercial fixed-bed reactors.

The formulation of the mathematical model depends on the mass and heat balances for the various phases involved (gas, liquid, and solid), as well as on the balance of momentum and pressure drop through the catalyst bed. It is important to note that during the development of a generalized reactor model, all the terms related to resistances to mass and energy transfer, among other terms, must be incorporated into the mass and energy balance. It is recognized that the solution of the resulting model can be extremely difficult and requires considerable computational time for its numerical solution.

8.2.3.1 General Equations of the Model

To perform proper balances of mass and energy, it is necessary to define the types of phases involved in the FTS process carried out in the fixed-bed system. The species in the gas phase are the syngas mixture (CO and H_2), CO_2 and H_2O (water steam), and light hydrocarbons formed during the FTS process: CH_4, $C_2 - C_4$ (C_2H_4, C_2H_6, C_3H_6, C_3H_8), $i - C_4H_{10}$, $n - C_4H_{10}$, while the liquid phase is constituted by volatile species such as short-chain hydrocarbons (volatile liquid hydrocarbons such as some olefins and paraffins in the gasoline and diesel, respectively, and oxygenated compounds such as alcohols and ketones) and nonvolatile species such as long-chain hydrocarbons (heavy liquid hydrocarbons such as waxes); and solid is the catalyst. Then, the terms involved in the generalized model for the fixed-bed catalytic FTS reactor must be as follows: (i) for mass balance, the terms of accumulation, convective transport, and axial and radial mass dispersion are to be considered, as well as the gas–liquid, liquid–solid, and gas–solid mass transfer effects due to flow regimes; (ii) for the energy balance, the terms of accumulation, convective transport, and axial and radial heat dispersion, as well as heat due to mass transfer effects between the different phases, are to be considered. The mass and energy balances of the generalized model for fixed-bed FTS reactor are summarized in Tables 8.7–8.10. The following assumptions are considered:

- Properties of gas and liquid (mass and heat dispersion coefficients, specific heats, holdups, densities, and viscosities), of the catalyst particle (porosity, size, activity, effectiveness, etc.), and bed void fraction are assumed to be constant throughout the entire catalyst bed.
- Regarding the behavior inside the catalytic particle, the coefficients of mass- and heat-effective diffusivity are considered as constants.
- The gas and liquid superficial velocities are variable.
- For the non-isothermal case, both physicochemical and thermodynamic properties must be evaluated at each point of discretization during the numerical simulation.
- The reactor parameters that are assumed to be constant can be put out of the partial derivatives with respect to the axial and radial spatial coordinates involving them.

Although in the literature some correlations have been reported to predict the variation of some parameters with respect to spatial coordinates, their use in conventional continuum models is quite complex (Mederos et al. 2009).

The proposed generalized model for fixed-bed FTS reactors is based on a set of mass, energy, and momentum balance equations developed for each component in each phase *f*, which are derived from the fundamental form of the multifluid continuity equation considering the terms of reaction rate and convective mass transfer (Jakobsen 2014; Froment et al. 2011; Bird et al. 2001).

Mass Balance The equations involving the mass balances in Table 8.7 assume that each phase in the fixed-bed system is a continuum able to be represented by an Eulerian–Eulerian framework model. Figure 8.3 shows schematically the concentration profiles in a fixed-bed FTS reactor, and the mass transport in different phases is identified.

It is important to note that in the generalized model, the fixed-bed FTS reactor was considered to operate in a trickle-flow regime. According to Caldwell and Van Vuuren (1986), if the operating conditions and the catalyst formulation suitably enhance the FTS process, it is possible to obtain a liquid phase inside the reactor operating in a trickle-flow regime where the syngas and liquid product flow downstream in a concurrent way. On the other hand, it is reasonable to assume that there is a distribution of the gas and liquid phases within the reactor, which varies in time along the axial and radial directions. Therefore, in the proposed generalized reactor model, the behavior of the gas–liquid distribution phases through the bed using dynamic holdups of both phases involved is taken into account. In addition, it is convenient to take into consideration the partial wetting of the catalyst since this effect is common in a trickle-bed reactor. Relatively high surface velocities of the gas and liquid along the reactor allow simplifying the proposed generalized model by considering an axisymmetric flow field of the gas and liquid phases in the whole bed, and dynamic

Table 8.7 Generalized mass balance equations for gas and liquid phases.

	Accumulation	Convective	Axial dispersion	Radial dispersion	G–L transfer	G–S transfer	L–S-transfer
Gas phase	$\epsilon_G \frac{\partial C_i^G}{\partial t} =$	$-\frac{\partial(u_s^G C_i^G)}{\partial z}$	$+\epsilon_G D_{ax}^G \frac{\partial^2 C_i^G}{\partial z^2}$	$+\epsilon_G D_{rad}^G \left[\frac{1}{r}\frac{\partial}{\partial r}\left(r\frac{\partial C_i^G}{\partial r}\right)\right]$	$-k_i^{GL} a_I^{GL}(C_i^G - C_i^L)$	$-(1-f_w)k_i^{GS}a_I^{GS}(C_i^G - C_i^{S_{sur}})$	$+f_w k_i^{LS} a_I^{LS}(C_i^L - C_i^{S_{sur}})$
Liquid phase: volatile species	$(1-f_{St})\epsilon_L \frac{\partial C_i^L}{\partial t} =$	$-\frac{\partial(u_s^L C_i^L)}{\partial z}$	$+\epsilon_L D_{ax}^L \frac{\partial^2 C_i^L}{\partial z^2}$	$+\epsilon_L D_{rad}^L \left[\frac{1}{r}\frac{\partial}{\partial r}\left(r\frac{\partial C_i^L}{\partial r}\right)\right]$	$-k_i^{GL} a_I^{GL}(C_i^G - C_i^L)$	—	$+f_w k_i^{LS} a_I^{LS}(C_i^L - C_i^{S_{sur}})$
Liquid phase: nonvolatile species	$(1-f_{St})\epsilon_L \frac{\partial C_i^L}{\partial t} =$	$-\frac{\partial(u_s^L C_i^L)}{\partial z}$	$+\epsilon_L D_{ax}^L \frac{\partial^2 C_i^L}{\partial z^2}$	$+\epsilon_L D_{rad}^L \left[\frac{1}{r}\frac{\partial}{\partial r}\left(r\frac{\partial C_i^L}{\partial r}\right)\right]$	—	—	$+f_w k_i^{LS} a_I^{LS}(C_i^L - C_i^{S_{sur}})$
Stagnant liquid	$f_{St}\epsilon_L \frac{\partial C_i^L}{\partial t} =$	—	—	—	$-k_i^{GL} a_I^{GL}(C_i^G - C_{L,i}^{St})$	—	$+f_w k_i^{L_{St}S} a_I^{LS}(C_{L,i}^{St} - C_i^{S_{sur}})$

Table 8.8 Generalized mass balance equations for gas–solid phase and liquid–solid phase.

	Accumulation	Intraparticle diffusion	G–S transfer	L–S transfer	Generation
Solid phase/ wet surface	$(1-\varepsilon_B)\epsilon_L^p \frac{\partial C_{L,i}^{S_{sur}}}{\partial t} =$	—	—	$+k_i^{\prime LS} a_I^{LS}(C_{St,i}^L - C_{St,i}^{S_{sur}})$	—
Solid phase/ dry surface	$(1-\varepsilon_B)\epsilon_G^p \frac{\partial C_{G,i}^{S_{sur}}}{\partial t} =$	—	$+(1-f_w)k_i^{GS}a_I^{GS}(C_i^G - C_{G,i}^{S_{sur}})$	—	$+\rho_B \sum_{j=1}^{NS} \nu_{ij}\eta_j R_{FT,j}(C_i^{S_{sur}}, T^{S_{sur}})$
Solid phase/ inner	$\epsilon_G^p \frac{\partial C_{G,i}^{S_{ins}}}{\partial t} =$	**Spherical catalyst** $+\frac{D_{eff,i}^G}{\xi^2}\frac{\partial}{\partial \xi}\left(\xi^2 \frac{\partial C_{G,i}^{S_{ins}}}{\partial \xi}\right)$ **Cylindrical catalyst** $+\frac{D_{eff,i}^G}{\xi}\frac{\partial}{\partial \xi}\left(\xi \frac{\partial C_{G,i}^{S_{ins}}}{\partial \xi}\right)$ **Hollow cylinder catalyst** $+\frac{D_{eff,i}^G}{(\xi\delta_{cyl}+r_{inn})}\frac{\partial}{\partial \xi}\left((\xi\delta_{cyl}+r_{inn})\frac{\partial C_{G,i}^{S_{ins}}}{\partial \xi}\right)$	—	—	$+\rho^S \sum_{j=1}^{NS} \nu_{ij}\eta_j R_{FT,j}(C_i^{S_{ins}}, T^{S_{sur}})$

Table 8.9 Generalized heat balance equations for fluid phases (gas and liquid).

	Accumulation	Convective	Axial dispersion	Radial dispersion	Fluid-interface transfer	Conductive
Gas phase	$\epsilon_G \frac{\partial(\rho^G C_{P,i}^G T^G)}{\partial t} =$	$-\frac{\partial(u_s^G \rho^G C_{P,i}^G T^G)}{\partial z}$	$+\epsilon_G \lambda_{ax}^G \frac{\partial^2 T^G}{\partial z^2}$	$+\epsilon_G \lambda_{rad}^G \left(\frac{\partial^2 T^G}{\partial r^2} + \frac{1}{r}\frac{\partial T^G}{\partial r}\right)$	$-h_c^{GL} a_I^{GL}(T^G - T^I)$	$+\sum_{i=1}^{N_{CG}} \left\{K_{L,i}^{GL} a_I^{GL}\left(\frac{p_i^G}{He_i} - C_i^L\right)[C_{P,i}^G(T^I - T^G) - \Delta H_{vi}]\right\}$
Liquid phase	$\epsilon_L \frac{\partial(\rho^L C_{P,i}^L T^L)}{\partial t} =$	$-\frac{\partial(u_s^L \rho^L C_{P,i}^L T^L)}{\partial z}$	$+\epsilon_L \lambda_{ax}^L \frac{\partial^2 T^L}{\partial z^2}$	$+\epsilon_L \lambda_{rad}^L \left(\frac{\partial^2 T^L}{\partial r^2} + \frac{1}{r}\frac{\partial T^L}{\partial r}\right)$	$-h_c^{GL} a_I^{GL}(T^I - T^L)$	$+\sum_{i=1}^{N_{CL}} \left\{K_{L,i}^{GL} a_I^{GL}\left(\frac{p_i^G}{He_i} - C_i^L\right)[C_{P,i}^L(T^I - T^L) + \Delta H_{vi}]\right\}$

	Fluid–solid transfer	Fluid wall transfer
Gas phase	$-(1-f_w)h_c^{GS} a_I^{GS}(T^G - T^{S_{sur}})$	$-(1-f_w)h_c^{GW}\frac{A_{BW}}{V_r}(T^G - T^W)$
Liquid phase	$-f_w h_c^{LS} a_I^{LS}(T^L - T^{S_{sur}})$	$-f_w h_c^{LW}\frac{A_{BW}}{V_r}(T^L - T^W)$

	Accumulation	Axial dispersion	Generation
Solid phase-non-isothermal	$\rho^S C_P^S \frac{\partial T^{S_{ins}}}{\partial t} =$	$+\lambda_{eff}^S\left(\frac{\partial^2 T^{S_{ins}}}{\partial \xi^2} + \frac{2}{\xi}\frac{\partial T^{S_{ins}}}{\partial \xi}\right)$	$\rho^S\left[\sum_{j=1}^{N_{CG}}(-\Delta H_{FTR,j})R_{FT},j(C_i^{S_{ins}}, T^{S_{ins}})\right]$

Table 8.10 Generalized heat balance equations for the solid phase (catalyst).

	Accumulation	Axial dispersion	Radial dispersion	G–S transfer	L–S-transfer	Generation
Solid phase-isothermal	$\epsilon^S \rho^S C_P^S \frac{\partial T^{S_{sur}}}{\partial t} =$	$+\epsilon^S \lambda_{ax}^S \frac{\partial^2 T^{S_{sur}}}{\partial z^2}$	$+\epsilon^S \lambda_{rad}^S\left(\frac{\partial^2 T^{S_{sur}}}{\partial r^2} + \frac{1}{r}\frac{\partial T^{S_{sur}}}{\partial r}\right)$	$-(1-f_w)h_c^{GS} a_I^{GS}(T^G - T^{S_{sur}})$	$-f_w h_c^{LS} a_I^{LS}(T^G - T^{S_{sur}})$	$\rho_B\left[\sum_{j=1}^{NS}(-\Delta H_{FTR,j})\nu_{ij}\eta_j R_{FT,j}(C_i^{S_{sur}}, T^{S_{sur}})\right]$

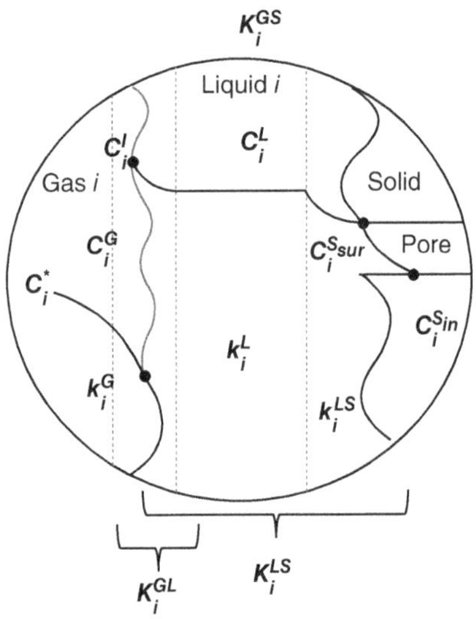

Figure 8.3 Concentration profiles in the fixed-bed FTS reactor.

holdups can be neglected; in consequence, the holdups dynamics in mass balances of gas and liquid phases, the volatile liquid species, and zones of the stagnant liquid are equal to 1. Accordingly, the concept of catalyst wetting factor for modeling purposes can be neglected, since it is very common that in the FTS process the thickness of liquid film on the catalyst surface is negligible, as reported by Jess and Kern (2012a), by which external diffusion limitations can be ignored at operation temperature below 400 °C. The proposed generalized model for a fixed-bed FTS reactor takes into account the wetting effect of the catalyst particle proposed by Mederos et al. (2009) for a trickle-bed reactor applied to the hydrotreating of petroleum fractions, as well as a similar analysis for the mass balance equations of the liquid phase in stagnant zones, on the surface and inside of the catalyst particle.

As for the gaseous species (reactants and volatile liquids produced) present in the fixed-bed FTS reactor, the convective flow term is considered to be plug-flow type, which implies that the concentration and temperature gradients reached only occur in the axial direction, i.e. along the catalytic bed. The mass balance for the gas phase describes the dynamic behavior of the reactor for the generalized non-steady-state model (represented by the accumulation term) and its steady-state behavior (without considering the transient effect).

The effect of axial dispersion of fluid flow in fixed-bed reactors (on a commercial scale) can be neglected in cases where gas and liquid velocities are high enough (Mederos et al. 2009). On the other hand, when the gas-flow pattern exhibits a plug-flow type behavior, the axial dispersion effects can be discarded because the axial dispersion coefficients tend to zero (Salmi et al. 2000).

As in the case of the mass radial dispersion term in a fixed-bed system, this can be neglected when the ratio of diameters d_t/d_{pe} is higher than 25, which means that the variation of the radial porosity within the reactor can be completely omitted (Mederos et al. 2009). In addition, it has been reported that the dispersion coefficient in the axial direction is greater than the radial dispersion coefficient by a factor of 5 when working with Reynolds numbers greater than 10, whereby the radial dispersion of mass in a packed bed may be ignored when operating at high Reynolds number values (Delgado 2006).

The resistance to mass transfer is explained in detail by the theory of the two films as schematized in Figure 8.4. The mass transfer from gas to liquid is determined using a fictitious liquid-phase concentration, which is the concentration that could be in equilibrium with the corresponding bulk partial pressure in the case of light hydrocarbons. Then the following equation is used for its determination:

$$C_i^* = \frac{p_{G,i}}{\mathrm{He}_i} \tag{8.17}$$

where He_i is Henry's constant taken as an equilibrium ratio representing the solubility of the i components in the gas state in the fraction of liquid product. In addition, it is important to considerer that when the liquid phase produced is slightly volatile under the operating conditions of the FTS

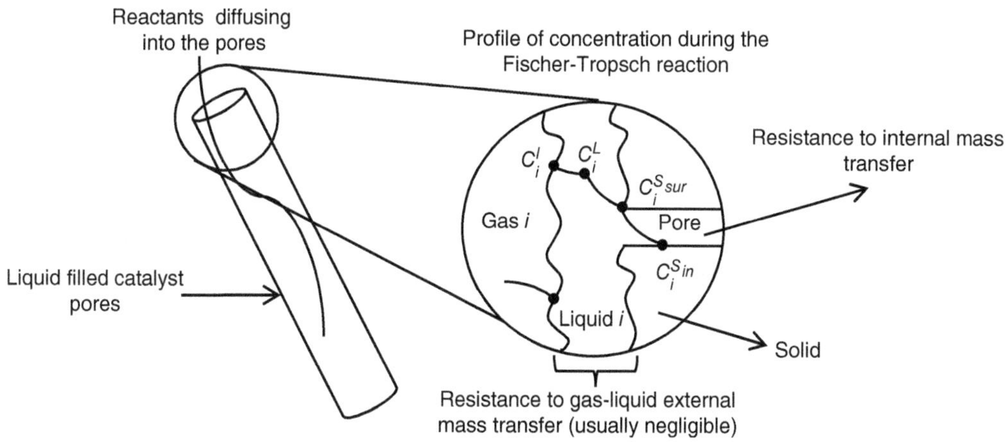

Figure 8.4 Schematic diagram of the Fischer–Tropsch reaction in the fixed-bed system on the catalyst particle.

process, the gas–liquid mass transfer effect on the liquid side can be ignored. However, when the wetting efficiency of the catalytic particle is assumed to be 1 ($f_w = 1$), the direct contact between the gas and the solid catalyst particle is lost, resulting in a null conversion of the reactants CO and H_2, and the reaction finished since it takes place directly into the solid, which is an extreme case to consider.

The dynamic holdup of the liquid phase must be considered similarly to that of the gas phase, because as the FTS reaction occurs along the catalytic bed the gas and liquid fractions inside the reactor change, which is mainly due to the consumption of the syngas (convection) and production of liquid that are a function of time and of the axial and radial positions. In addition, if the liquid produced has a partial volatilization effect, this will reflect a variation of the dynamic holdup along the catalytic bed (Mederos et al. 2009).

It is well-known that inside fixed-bed reactors involving a liquid phase, there may be some areas of liquid stagnation, the so-called dead zones. Many regions of stagnant liquid regularly occur mainly in several regions of the catalytic bed affecting the holdup dynamics of both the liquid and gas. For this reason, in this work, an equation for stagnant liquid is proposed taking into account the gas–liquid transfer effects on the liquid side and mass transfer from stagnant liquid to solid catalyst.

To develop the mass and energy balance in solid catalysts, the following assumptions were proposed based on the schematic diagram depicted in Figure 8.4 (Jess and Kern 2012a, b; Pöhlmann and Jess 2016a):

- The Fischer–Tropsch reaction is carried out inside pores as commonly a heterogeneous catalytic reaction occurs.
- The catalyst pores are considered to be fully filled with liquid.
- Since inner catalyst pores are filled with waxy liquid, the liquid film thickness is commonly neglected.
- Due to the formation of liquid within the catalyst, concentration profile agrees with the two-film theory.
- The wax formed inside the pores is liquid by which solid waxes in pores are discarded.

For the performance of mass balances of the solid catalyst particle, two cases were considered: (i) a dry surface of the catalyst and (ii) a partially wetting. In both cases, the internal mass gradients in the solid phase must be evaluated using the effectiveness factor (Mederos et al. 2009), which is defined as the ratio of the overall reaction rate to the reaction rate that arises when the whole inner surface is exposed to the external condition (Lee and Chung 2012).

The mass gradient that occurs within the catalytic particle results from effective diffusion, which depends mainly on two important factors: (i) the porosity of the catalyst and (ii) the size of the molecule diffusing through the catalytic pores. To achieve maximum FT catalyst effectiveness, it is recommended to operate under conditions where the mass transfer limitations at the liquid–solid interface are neglected (Mederos et al. 2009). There is evidence that reducing the size of the catalyst particle allows for an increase in the effectiveness factor (η_i) by reducing the path lengths inside the particle (Wang et al. 2001). However, the catalyst pellet size plays an important role in reactivity and selectivity during the FTS process because they are size-dependent (Jess and Kern 2012a, b). According to the reactivity and selectivity analysis provided by Wang et al. (2001), a large size of catalyst has a negative effect on the reactivity and selectivity of desired products because strong diffusion effects occur and small-size particles are preferred instead. In industrial-scale applications, catalyst pellets 2–4 mm in diameter are commonly used to keep the pressure drop as low as possible and to remove the generated heat efficiently (Sie and Krishna 1999). However, according to Jess and Kern (2012a) and Jess and Kern (2012b), a particle diameter >3 mm is suitable for obtaining a high production rate of higher hydrocarbons. Moreover, under effective conditions (i.e. with particles of 1 mm as used in fixed-bed reactors), the FTS is affected by internal mass transport limitations, which lead to an increased H_2/CO ratio inside the particle, which has an impact on the local reaction rate and selectivity (Jess and Kern 2012b).

On the other hand, the effect of the particle size of the catalyst on the reaction rate can also be quantified with the effectiveness factor, as well as with the Thiele modulus (Macías and Ancheyta 2004). A simple expression for the effectiveness factor and the Thiele modulus valid for spherical particles and first-order kinetics is (Post et al. 1989):

$$\eta = \frac{1}{\Phi_S}\left(\frac{1}{\tan h(3\Phi_S)} - \frac{1}{3\Phi_S}\right) \tag{8.18}$$

$$\Phi_S = \left(\frac{k_{in}}{D_{eff}}\right)^{0.5} \tag{8.19}$$

In industrial operation, it is impossible to reduce the catalyst pellets to small sizes without having higher pressure drops; however, high yields of products are required (Sie 1998). Furthermore, according to Jess and Kern (2012a) for typical FTS conditions, the particle size suitable to avoid an excessive pressure drop in the catalyst bed must be higher than 1 mm. It is therefore imperative that research on the development of novel FT catalysts strives to design materials that consider the advantages and disadvantages that may arise in their direct application (Macías and Ancheyta 2004). However, it is noteworthy that pore diffusion greatly affects the effective rate constant within a typical temperature range of 200–250 °C for particle diameters above 0.5 mm (Jess and Kern 2012a).

Post et al. (1989) reported significant external diffusion limitations caused by the resistance of the film of gas or liquid around the catalyst particle, which was confirmed by observing no changes in synthesis gas conversion when changing the bed length at the same space-time values.

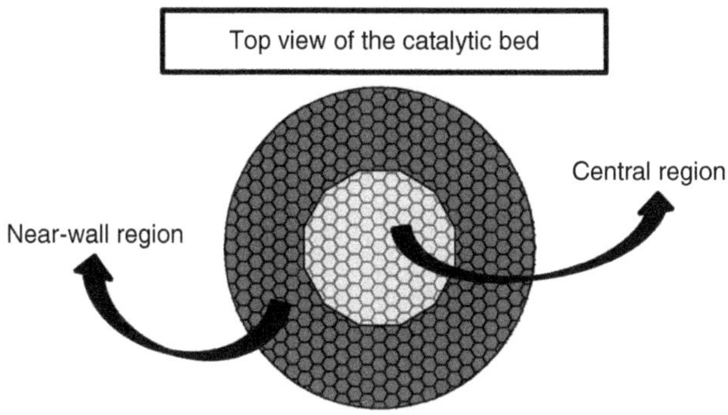

Figure 8.5 Top view of the split in the center and near the wall regions catalyst bed.

To develop the mass balance within the catalyst particle, pores having uniform properties and being completely filled with liquid are to be assumed. A close-up of the catalyst particle is shown in Figure 8.5, in which the concentration profile occurring within the catalyst particle with mass transfer resistance effects can be seen. The equations that model the behavior of the FTS process inside the solid catalyst particle arise from the mass balance on an infinitesimal volume element of the catalyst particle considering different geometric forms (Table 8.8).

Energy Balance It is well-known that the temperature control of a fixed-bed reactor deserves special attention for modeling and simulation. There are several studies of fixed-bed reactor modeling in which an isothermal regime is assumed. However, this assumption could not be reliable since the model could generate erroneous predictions of the actual thermal behavior in a reactor, especially at industrial scale. It may be noted that the FTS reaction is highly exothermic, with a $\Delta H_{FT} = -165\ \text{KJ/mol}_{\text{CO}}$ for the paraffin formation and a $\Delta H_{FT} = -204\ \text{KJ/mol}_{\text{CO}}$ for the olefin formation (Kölbel and Ralek 1980). Then, due to the high exothermicity of the FTS reaction, it is necessary to have an effective temperature control of the fixed-bed reactor, since according to Jager and Espinoza (1995), the following circumstances may occur:

- Generation of hot spots (temperature peaks) along the catalyst bed is common.
- Axial and radial temperature profiles arise inside the tubes.
- To maximize the synthesis gas conversion, the maximum average temperature must not exceed the maximum permissible temperature peak (hot spot) to prevent carbon deposition on the catalyst surface.
- Carbon deposition deactivates the catalyst, which results in activity loss and catalyst replacement.

It is also important to evaluate the temperature profiles as well as the heat transfer rates to ensure control and performance of the reactor (Guardo et al. 2006). As aforementioned, the FTS reaction is highly exothermic and good control of the heat released during the process is required. In addition, the formation of undesirable methane as well as the deactivation of the catalyst usually occurs (Schulz 1999). It is also possible that the handling of high temperatures leads to a considerable loss of selectivity and thermal runaways. Therefore, an efficient alternative is the primary mechanism for removing the generated heat by radial heat transfer, for which it is necessary to match the radial heat transfer rate to the reaction rate of the catalyst to ensure that the catalyst bed temperature can be controlled, according to Zhu et al. (2010).

It has been determined by Pölhmann and Jess (2016b) that, in exothermic reactions such as that of the FTS process, the generation of temperature gradients can occur within the particle, which may lead to an overheating of the particle, which would not be beneficial to reactor performance.

The heat balance equations for the gas and liquid phases considering the isothermal and non-isothermal cases are reported in Table 8.9. In the heat generated by axial and radial dispersion, the fluxes of fluid-interface convection heat are observed where the driving force is produced by a temperature gradient generated among the temperature in the bulk fluid phase (gas and liquid) and temperature at the interface. The transport of energy by conduction of the gas-film side at the gas–liquid interface occurred through the transfer of interfacial mass. The conductive heat flux term for both phases considers the heat generated by evaporation/condensation that occurs between the gas and liquid phases. Convective heat transfers of both the gas and liquid phases on the external catalyst surface are also involved.

In the case of heat balance in the liquid phase, heat evolved by vaporization/condensation is considered. Here ΔH_{vi} represents the latent heat corresponding to the heat that is consumed by vaporization effects only for the products of the FTS reaction.

Wang et al. (2001) investigated the transfer and reaction phenomena in a catalyst pellet for FTS, the interaction between diffusion and reaction in the pellet, and its effects on product selectivity. The results of simulations showed that the temperature differences between the external surface and the center of the pellet were less than 0.02 K due to the excellent thermal conductivity of the Fe–Cu–K catalyst. Accordingly, it is possible to assume, in certain cases of modeling of fixed-bed FTS reactor, an isothermal behavior of the catalyst particle.

In Table 8.10, the heat balance equation for the catalyst is established considering the isothermal case in which concentrations and temperatures on the solid surface must be used. It should be noted that the sign of the heat of reaction ($\Delta H_{FTR,j}$) is negative because the FTS reactions are highly exothermic.

Furthermore, the resistances that exhibit the films (gas and liquid) greatly influence the heat transfer, whereas heat transport within the catalyst particle is normally fast. Thus, the heat transfer phenomena occurring within the particle pores can be described with the Fourier law in terms of the partial differential equation (PDE) with respect to the temperature for the non-isothermal case, as illustrated in Table 8.10.

8.2.3.2 Boundary Conditions of the Proposed Generalized Model

The generalized model for the fixed-bed FTS reactor consists of the mass balances of each species and the heat balance, which imply a system of PDEs, that require a set of initial and boundary conditions to be $z = 0$ and $z = L_B$ at the reactor inlet and outlet, respectively. As indicated by Wärna and Salmi (1996), in a fixed-bed reactor model by an axial dispersion model, appropriate boundary conditions to describe steady-state fluxes are needed (Wang et al. 2001; Danckwerts 1953). It is important to indicate that boundary conditions at the reactor outlet for the case of a dispersion model are needed because back-mixing takes place. The set of initial and boundary conditions that define the proposed generalized model for the FTS reactor is given in Tables 8.11 and 8.12.

At $z = 0$, the concentration of the gas phase is in physical equilibrium with the inlet temperature and pressure, and often the gas concentration at the reactor inlet is expressed as a function of conversion considering the following ideal gas ratio (Dai et al. 2014; Moazami et al. 2015a; Park et al. 2014; Moazami et al. 2015c)

$$C_i^G = p_i^G/(R_u T) \tag{8.20}$$

Table 8.11 Initial conditions ($t = 0$) of generalized mass and heat balance equations of the model.

Condition	Gas phase: $i = H_2$, CO, CO_2, CH_4 and light hydrocarbons: paraffins and olefins ($C_2 - C_4$ and C_{5+})	Liquid phase: i = light and heavy hydrocarbons: paraffins and olefins ($C_2 - C_4$ and C_{5+})	Solid phase: surface (all components)	Solid phase: inside (all components)	Stagnant zone
$z = 0$	$C_i^G = C_{i,0}^G$	$C_i^L = C_{i,0}^L$	$C_{f,i}^{S_{sur}} = 0$	$C_{f,i}^{S_{ins}} = 0$	$C_{f,i}^{S_{sur}} = 0$
$0 \le r \le R$	$T^G = T_0^G = T_0$	$T^L = T_0^L = T_0 = 0$	$T^{S_{sur}} = T_0^{S_{sur}} = T_0$	$T^{S_{ins}} = T_0^{S_{ins}} = T_0$	
$0 \le z \le L_B$	$C_i^G = 0$	$C_i^L = 0$	$C_{f,i}^{S_{sur}} = 0$	$C_{f,i}^{S_{ins}} = 0$	$C_{St,i}^{S} = 0$
$0 \le r \le R$	$T^G = T_0^G = T_0$	$T^L = T_0^L = T_0 = 0$	$T^{S_{sur}} = T_0^{S_{sur}}$	$T^{S_{ins}} = T_0^{S_{ins}} = T_0$	
$0 \le \xi \le R$				$C_{f,i}^{S_{ins}} = 0$	
$0 \le z \le L_B$				$T^{S_{ins}} = T_0^{S_{ins}}$	
$0 \le r \le R$					

$$p_i^G = P_T y_i^G \tag{8.21}$$

Effects of axial and radial dispersion of mass and heat in the dispersion model give second-order differential equations so that the two boundary conditions necessary according to Danckwerts (1953) are as follows:

The Danclwerts' boundary condition at the reactor inlet at $z = 0$ is:

$$-\varepsilon_f D_{ax}^f \left.\frac{\partial C_i^f}{\partial z}\right|_{z=0^+} = u_s^f \left[\left(C_i^f\right)_0 - \left(C_i^f\right)_{z=0^+}\right] \tag{8.22}$$

$$-\varepsilon_f \lambda_{ax}^f \left.\frac{\partial T^f}{\partial z}\right|_{z=0^+} = u_s^f \rho^f C_P^f \left[\left(T^f\right)_0 - \left(T^f\right)_{z=0^+}\right] \tag{8.23}$$

which is simplified as

$$C_i^f = T^f\left(C_i^f\right)_0; T^f = \left(T^f\right)_0 \tag{8.24}$$

- Such boundary conditions are acceptable with reliability since the axial dispersion of mass and heat are relatively small and concentration and temperature gradients at the reactor inlet are usually fairly flat (Chen et al. 2001).

The boundary condition at the exit of the reactor at $z = L_B$ is:

$$\frac{\partial C_i^f}{\partial z} = \frac{\partial T^f}{\partial z} = 0 \tag{8.25}$$

In addition, Pearson (1959) showed a way of manipulating the boundary conditions of Danckwerts to impose the continuity of the reactant concentration in continuous-flow reactors that suggests different boundary conditions can be applied to the inlet and outlet in packed-bed systems, as well as in more complex situations involving problems of diffusion of multicomponent systems with variation in temperature.

Table 8.12 Boundary conditions of generalized mass and heat balance equations of model.

Condition	Gas phase	Liquid phase	Solid phase (surface)	Solid phase (interior)
$z = 0$ $0 \le r \le R$	$C_i^G = C_{i,0}^G$ $T^G = T_0^G = T_0$	$C_i^L = 0$ $T^L = 0$		
$z = L_B$ $0 \le r \le R$	$\frac{\partial C_i^G}{\partial z} = 0$ $\frac{\partial T^G}{\partial z} = 0$	$\frac{\partial C_i^L}{\partial z} = 0$ $\frac{\partial T^L}{\partial z} = 0$		
$r = 0$ $0 \le z \le L_B$	$\frac{\partial C_i^G}{\partial r} = 0$ $\frac{\partial T^G}{\partial r} = 0$	$\frac{\partial C_i^L}{\partial r} = 0$ $\frac{\partial T^L}{\partial r} = 0$		
$r = R$ $0 \le z \le L_B$	$\frac{\partial C_i^G}{\partial r} = 0$ $-\lambda_{rad}^G \frac{\partial T^G}{\partial r} = h_c^{GW}\left(T^G - T^W\right)$	$\frac{\partial C_i^G}{\partial r} = 0$ $+\lambda_{\text{rad}}^L \frac{\partial T^L}{\partial r} = h_c^{LW}\left(T^S - T^L\right)$	$-\lambda_{\text{rad}}^S \frac{\partial T^{S_{\text{sur}}}}{\partial r} = h_c^{GS}\left(T^{S_{\text{sur}}} - T^W\right)$	
$\xi = 0$ $0 \le z \le L_B$	—	—		$\frac{\partial C_{f,i}^{S_{\text{ins}}}}{\partial \xi} = \frac{\partial T^{S_{\text{ins}}}}{\partial \xi} = 0$
$\xi = \frac{d_{pe}}{2}$ $0 \le z \le L_B$	—	—		$-D_{\text{eff}}^L \frac{\partial C_{L,i}^{S_{\text{ins}}}}{\partial \xi} = f_w k_i^{LS} a_I^{LS}\left(C_i^{S_{\text{sur}}} - C_i^L\right)$ $+ k_i^{L_{St}-S}\left(C_i^{S_{\text{sur}}} - C_{L,i}^{St}\right)$ $-D_{\text{eff}}^G \frac{\partial C_{G,i}^{S_{\text{ins}}}}{\partial \xi} = (1 - f_w) k_I^{GS} a_I^{GS}\left(C_i^{S_{\text{sur}}} - C_i^G\right)$ $= \rho_B \varepsilon_B \sum_{j=1}^{N_{NS}} v_{ij} \eta_j R_{FTR}\left(C_i^{S_{\text{sur}}}, T^{S_{\text{sur}}}\right)$ $-\lambda_{\text{eff}}^S \frac{\partial T^S}{\partial \xi} = (1 - f_w) h_c^{GS} a_I^{GS}\left(T^{S_{\text{sur}}} - T^G\right)$ $+ f_w h_c^{LS} a_I^{LS}\left(T^{S_{\text{sur}}} - T^L\right)$

8.2.3.3 Pressure Drop

In designing and modeling of a fixed-bed FTS reactor, it is mandatory to predict the pressure drop through the catalyst bed. In the case of the LTFT process to produce clean fuels, the main products are heavy oil and waxy materials whose molecular weights are relatively high, and in consequence, the effect of flow velocities occurring through the packed bed could cause severe attrition of catalysts.

Among various mathematical models presented in previous sections, there are not many of them that consider the effect of the pressure drop through the bed and many of them make use of the classical Ergun correlation for its prediction. However, the classical model of Ergun to estimate the pressure drop in packed-bed reactors is inefficient when various geometries of catalyst are used. In different fields of engineering, the semiempirical Ergun equation has been employed for predicting pressure drop across packed beds with particles made of various materials (Allen et al. 2013). In the case of FTS, there are several models for fixed-bed FTS reactors that use such a correlation to describe the effect of pressure drop through the packed bed. However, various research works have been conducted in which a comparison between modified correlations of Ergun and contemporaneous correlation has been made, and as a result, a deficiency is obtained in predicting pressure drop by the classical model of Ergun compared to the modified correlations (Allen et al. 2013; Ozahi et al. 2008; Vollmari et al. 2015). It has been established in the literature that deviations from the predicted pressure drop across packed beds that exhibit the classical correlation of Ergun may be due to the high sensitivity of the equation to variations in the bed porosity by which it is necessary to estimate accurately the packed-bed porosity, especially when dealing with non-spherical geometry (Koekemoer and Luckos 2015). It has been found that several factors influence the bed porosity, such as the particle diameter, the particle size distribution, surface roughness, Reynolds number, and others (Vollmari et al. 2015; Bai et al. 2011). In summary, the literature reports give both theoretical and experimental evidence that pressure drop depends on flow velocity and physical properties (viscosity and density), the average bed porosity, shape and surface of the particles packaging, bed height, and contribution of the ratio of particle to the container diameters (Eisfeld and Schnitzlein 2001; Rangel et al. 2001; Montillet et al. 2007).

Flow regimes in catalytic fixed-bed reactors are quite important as they are directly related to the pressure drop. Following four flow regimes in fixed-bed reactors are reported in the literature (Dixon et al. 2006):

- At the flow rate of Darcy ore creeping-flow at $Re_i < 1$, the pressure drop varies linearly with flow rate, and interstitial number range is controlled by viscous forces.
- In a laminar flow regime in a stationary Reynolds interstitial number range of $10 < Re_i < 150$, pressure drop shows a nonlinear relationship with interstitial flow rate.
- In a laminar flow regime in the nonstationary Reynolds interstitial number range of $150 < Re_i < 300$, oscillations of laminar wake are present in the pores and vortices form and the number of $Re_i = 250$.
- In a flow regime considered highly nonstationary at Reynolds interstitial number above 300, eddies arise like turbulent flow in pipes.

The effects of pressure drop involving both the gas phase and liquid phase must strictly be considered separately to get better accuracy and approach to the real pressure drop across the catalyst bed. Bai et al. (2011) proposed an empirical correlation to predict the pressure drop of a two-phase flow through pebble beds in a gas–water system by considering the use of the relative permeability in the gas phase while the bed is subdivided into two regions: (i) one near walls and (ii) another in the central zone of the bed to consider the effects of wall and improvements to low ratios of tube diameter to the particle. The correlation used to calculate the pressure drop in two-phase flow was established in dimensionless terms to cover the whole range of Reynolds and Galileo numbers, assuming that the relative permeability is dependent on the void fraction bed and the subdivision of the bed into two regions to consider the effects of wall. Bai et al. (2011), based on previous studies reported by Eisfeld and Schnitzlein (2001) and Reichelt (1972), proposed the following equations:

$$\Psi_f = \frac{\Delta P_f}{\rho^f \left(u_s^f\right)^2} \frac{d_{pe}}{L} = \kappa_1 \beta_{1w} \frac{\mathrm{Re}_f}{\mathrm{Ga}^f} + \kappa_2 \beta_{2w} \frac{\left(\mathrm{Re}^f\right)^2}{\mathrm{Ga}^f} \tag{8.26}$$

where

$$\mathrm{Re}_f = \frac{u_s^f \rho^f d_{pe}}{\mu^f (1-\varepsilon_B)};\ \mathrm{Ga}^f = \left(\frac{\rho^f}{\mu^f}\right)^2 g_c \left(\frac{d_{pe}\varepsilon_B}{1-\varepsilon_B}\right) \tag{8.27}$$

$$d_{pe} = \frac{6V_p}{A_p} \tag{8.28}$$

κ_1 and κ_2 are Ergun-type constants, and β_{1w} and β_{2w} are the coefficients that consider the effect of wall under the following restrictions:

i) When small packed beds are used ($D_{hyd}/d_{pe} < 10$), the central region of the bed is negligible and the following wall effect coefficients must be considered:

$$\beta_{1w} = \left[a + b\frac{d_{pe}^*}{(1-\varepsilon_w)}\right] \tag{8.29}$$

$$\beta_{2w} = \left(c d_{pe}^*\right)^{-2} \tag{8.30}$$

$$d_{pe}^* = \frac{d_{pe}}{d_h} \tag{8.31}$$

ii) When large packed beds are used (D_{hyd}/d_{pe} <10), the near-wall region is neglected and the coefficients of wall effect will be equal to 1. In Eqs. (8.26)–(8.31), $\kappa_1 = 180$, $\kappa_2 = 1.8$, $a = 0.8$, $b = 2$, $c = 3$, $d = 1$, d_{pe}^* is the dimensionless particle diameter and d_h is the hydraulic diameter of the packed beds.

Figure 8.5 shows a top view of the catalyst bed to specify in detail the effects of pressure drop in the central and near-to-the-wall zones when the ratio of the hydraulic diameter to the equivalent particle diameter is lower than 10.

The proposed generalized model for predicting the pressure drop considers two general Ergun-type approaches: (i) a first Ergun model in which a bed porosity is assumed to be uniform (i.e. a constant bed void fraction); (ii) a second Ergun model in which an oscillatory bed porosity behavior (i.e. bed porosity as a function of radial position) is assumed. Therefore, considering the second Ergun model approach, the variable porosity bed relative to the radial direction (oscillatory behavior of the bed porosity) results in the loss of viscous energy, which is proportional to the linear velocity of the fluid. The second approach for pressure drop is an adapted version from Bey and Eigenberger (1997) and considers the effects of shear stress that are negligible for most cases.

For the proposed model, the pressure drop is calculated by considering the equation of Ergun and the general balance of momentum (Tables 8.13 and 8.14, respectively) recently reported in the literature (Rangel et al. 2001), where the coefficients that determine the viscous effects and kinetics (inertial) correspond to general constants κ_1 and κ_2, respectively. The model incorporates

Table 8.13 General Ergun equation for pressure drop.

General Ergun equation for FTS reactor		
	Viscous energy loss term	**Viscous energy loss term**
$\frac{\partial P_T^f}{\partial z} =$	$\kappa_1 \beta_{1w} \frac{\partial u_s^f}{\partial z}$	$\kappa_2 \beta_{2w} \frac{\partial (u_s^f)^2}{\partial z}$

Model 1 (not oscillatory porosity behavior):

$$\frac{dP_T^f}{dz} = \kappa_1 \beta_{1w} \frac{(1-\varepsilon_B)^2}{\varepsilon_B^3} \frac{\mu^f u_s^f}{(\psi d_p)^2} + \kappa_2 \beta_{2w} \frac{(1-\varepsilon_B)}{\varepsilon_B^3} \frac{\rho^f (u_s^f)^2}{\psi d_p}$$; (Kürten et al. 1966; Hicks 1970)

Model 2 (with oscillatory porosity behavior):

$$\frac{dP_T^f}{dz} = \kappa_1 \beta_{1w} \mu_{eff}^f \frac{[1-\varepsilon_B(r)]^2}{\varepsilon_B^3(r) d_p^2} u_s^f(r) + \kappa_2 \beta_{2w} \rho^f \frac{[1-\varepsilon_B(r)]}{\varepsilon_B(r) d_p} \left[u_s^f(r)\right]^2 + \frac{\mu_{eff}^f}{\varepsilon_B(r)} \frac{d}{dr}\left[\varepsilon_B(r) r \frac{du_s^f(r)}{dr}\right];$$

(Kürten et al. 1966; Hicks 1970; Çarpinlioğlu and Özahi 2008)

Boundary conditions

$P = P_1^f(z=0)$, $P^f = P_2^f(z = L_B)$

$\frac{du_s^f(r=0)}{dr} = 0$, $u_s^f(r=R) = 0$

Table 8.14 Generalized momentum equations for FTS reactor.

	Generalized momentum equation for FTS reactor
Momentum in axial direction	$\frac{\partial[\rho^f u_s^f(z)]}{\partial t} = -\varepsilon_B \frac{\partial P_T}{\partial z} - \frac{1}{r}\frac{\partial}{\partial r}\left[r(\varepsilon_B \rho^f u_s^f(r) u_s^f(z)) - \frac{\partial}{\partial r}\left(\varepsilon_B \mu_{eff}^f \frac{\partial u_s^f(z)}{\partial r}\right)\right] - \frac{\partial}{\partial z}\left(\varepsilon_B \rho^f u_s^f(z) u_s^f(z) - \frac{\partial}{\partial z}\left(\varepsilon_B \mu_{eff}^f \frac{\partial u_s^f(z)}{\partial z}\right)\right) + \frac{1}{r}\frac{\partial}{\partial r}\left(\varepsilon_B \mu_{eff}^f r \frac{\partial u_s^f(r)}{\partial r}\right) + \frac{\partial}{\partial z}\left(\varepsilon_B \mu_{eff}^f \frac{\partial u_s^f(z)}{\partial z}\right) - f_z$
Momentum in radial direction	$\frac{\partial[\rho^f u_s^f(r)]}{\partial t} = -\varepsilon_B \frac{\partial P_T}{\partial z} - \frac{1}{r}\frac{\partial}{\partial r}\left[r(\varepsilon_B \rho^f u_s^f(r) u_s^f(r)) - \frac{1}{r}\frac{\partial}{\partial r}\left(\varepsilon_B \mu_{eff}^f \frac{\partial u_s^f(r)}{\partial r}\right)\right] - \frac{\partial}{\partial z}\left(\varepsilon_B \rho^f u_s^f(r) u_s^f(z) - \frac{\partial}{\partial z}\left(\varepsilon_B \mu_{eff}^f \frac{\partial u_s^f(r)}{\partial z}\right)\right) + \frac{1}{r}\frac{\partial}{\partial r}\left(\varepsilon_B \mu_{eff}^f r \frac{\partial u_s^f(r)}{\partial r}\right) + \frac{\partial}{\partial r}\left(\varepsilon_B \mu_{eff}^f \frac{\partial u_s^f(z)}{\partial z}\right) - \varepsilon_B \mu_{eff}^f \frac{u_s^f(r)}{r^2} - f_r$

the parameters β_{1w} and β_{2w}, which consider the effects of the reactor wall on pressure drop across the bed (Bai et al. 2011).

Table 8.13 displays the widespread Ergun equation corresponding to the prediction of pressure drop caused by both fluids (gas and liquid) involved in the FTS fixed-bed system, while Table 8.14 shows the mathematical terms of the generalized model of momentum.

The effective viscosity ($\mu^f_{eff} = \eta^f_{dyn}$) has been used in the generalized model for pressure drop because this parameter considers the effects of turbulent-flow fluctuations within the package. Therefore, some correlations reported in the literature to estimate the effective viscosity for spherical and cylindrical catalyst particles, respectively, need to be used (Bey 1998; Bey and Eigenberger 1997):

$$\frac{\mu^f_{eff}}{\mu^f} = 1 + \left(7 \times 10^{-6}\frac{d_t}{d_p} + 2 \times 10^{-5}\right)\left(\frac{\rho^f u^f_s d_p}{\mu^f}\right)^2 \tag{8.32}$$

$$\frac{\mu^f_{eff}}{\mu^f} = 1 + \left(1.7 \times 10^{-5}\frac{d_t}{d_p} + 2 \times 10^{-5}\right)\left(\frac{\rho^f u^f_s d_p}{\mu^f}\right)^2 \tag{8.33}$$

8.2.4 Model Parameters

To have a robust mathematical model, it is necessary to have enough information about all parameters involved in the mass and energy balances. Several empirical correlations for estimating the following parameters have been reported: (i) mass transport between phases (G–L, G–S, and L–S); (ii) scattering coefficients for axial and radial mass transfer; (iii) heat transfer such as heat conductivities in the axial and radial directions; (iv) heat transfer in the solid phase; (v) holdups of the gas and liquid phases; (vi) coefficients of convective heat transfer; (vii) wetting factor of the catalytic surface; and (viii) void fraction of the bed, etc. For simulations using the generalized model or simplified forms, it is necessary to evaluate distinct physical and thermodynamic parameters involved to solve the set of PDEs. Consequently, the set of parameters can be estimated by means of existing correlations.

8.2.4.1 Mass Transfer Parameters

Phenomena representing mass transport and dispersion coefficients are two aspects to be considered in the development of mass balances of the FTS reactor model. The mass transfer phenomenon in fixed-bed reactors occurs according to the diffusion of reactants through the liquid that is produced to reach the active sites before the reaction takes place in the solid catalyst. Once the catalyst surface is reached, the reactants diffuse through the inner particle pores from mouth to active sites, and then the chemical reaction occurs under the kinetic conditions.

The mass transfer flux at the gas–liquid interface is described according to the two-film theory valid for fixed-bed configurations:

$$\frac{1}{K_{L,i}} = \frac{R_u T^G Z\left(P_T, T^G\right)}{k^G_{I,i} He_i} + \frac{1}{k^L_{I,i}} \tag{8.34}$$

where the overall external resistance to mass transfer is formed by the mass transfer resistances offered by the gas ($k^G_{I,i}$) and liquid ($k^L_{I,i}$) films.

Delgado (2005) developed a rigorous investigation of variables that directly affect the Newtonian and non-Newtonian fluids axial and radial dispersion coefficients involving fixed-bed packed configurations, as well as the correlations for prediction. The variables that have significant effects on dispersion coefficients were found to be (i) length of the bed, (ii) ratio of bed diameter to particle, (iii) ratio of the bed length to particle diameter, (iv) particle size distribution, (v) particle shape, and

(vi) fluid properties (viscosity, density, velocity, and temperature). The results allowed one to establish two expressions for predicting axial and radial Peclet numbers exhibiting higher accuracy compared with other correlations, which are valid for all values of Peclet and Schmidt numbers (Sc).

Carbonell and Whitaker (1983) proposed a criterion to evaluate the effects of axial dispersion in fixed-bed systems that consists of neglecting the dispersion in the axial direction if the following expression is satisfied:

$$\left(\frac{1-\varepsilon_B}{\varepsilon_B}\right)\frac{D_{ij}^t}{d_p^2} \gg 1 \tag{8.35}$$

On the other hand, Han et al. (1985) showed that the values of the coefficient of axial dispersion when packed beds of uniform size are used and measured at different positions in the bed are a function of the position unless it approximates the following criterion:

$$\frac{L_B}{d_p}\frac{1}{Pe}\left(\frac{1-\varepsilon_B}{\varepsilon_B}\right)^2 \geq 0.3; \text{or } \frac{D_{ij}t}{d_p^2} \geq 0.1 \tag{8.36}$$

8.2.4.2 Heat Transfer Parameters

The heat transfer mechanism is an important aspect used in the design, modeling, simulation, and optimization of catalytic fixed-bed reactors implemented in various chemical processes. According to Dixon (2012), the heat transfer phenomenon that arises in a fixed bed of catalyst particles may be due to different factors such as (i) heat flow by conduction through the fluid phase, (ii) conduction along the particle, (iii) conduction caused by the contact points in existing particle–particle interactions and because of the amounts of gas stagnant around the contact points between the particles, (iv) transfer mechanism radiative heat from particle to particle and intraparticle void fraction to intraparticle void fraction, (v) convective transport of energy due to fluid travels around the catalyst particles along the bed, (vi) heat transfer of solid-particle fluid through the surrounding film, (vii) heat transfer of wall-particle contact, and (viii) heat transport of fluid-wall contact (Barker 1965). Moreover, an important aspect in the modeling of the temperature profile in fixed-bed systems refers to the heat transfer originated near the reactor wall region because in this region a greater amount of catalyst particles is concentrated when a low ratio of the reactor diameter to the catalytic particle diameter is present (Dixon 2011). Therefore, in the region near the wall, a significant increase in resistance to heat transport is found due to some factors (Yagi and Wakao 1959; Tobis and Ziókowski 1988): (i) increasing the void fraction implies a decrease of the thermal conductivity of the bed, (ii) a viscous layer (laminar) that serves as a boundary, (iii) damping of the convective heat transfer, which is mainly due to a decrease of the lateral displacement of the fluid. Furthermore, the oscillatory behavior of the bed porosity along the radial position makes velocity perform an oscillatory behavior directly affecting the heat transfer in packed beds (Mueller 1992; Béttega et al. 2011).

Dixon (2011) reported various factors affecting the effective radial heat transfer solid coefficients. The coefficients of heat transfer by convection of the fluid phase to the wall, as well as several empirical correlations to estimate these two heat transfer coefficients. It is concluded that there are disagreements in estimation of $\lambda_{eff,rad}$ and h_w values; however, the modeling approaches mentioned can be used as complementary tools in studies of the phenomena of fixed-bed packed.

Larachi et al. (2003) collected extensive information on parameters of heat and mass transfer in packed-bed systems, such as parameters for radial effective thermal conductivity and heat transfer on the wall and particle fluid influenced by packing particle size and particle materials. Based on these data, these authors proposed new correlations for calculating these heat transport parameters using artificial neural network, which is a powerful tool to develop more effective correlations in transport equations.

It is known that an important parameter to characterize the thermal resistance between the fluids inside the reactor (syngas) and the cooling shell is the overall heat transfer coefficient. According to Reid et al. (1977), the overall heat transfer coefficient between the cooling medium (commonly boiling water is used in fixed-bed FTS reactors) of the shell side and the bulk of the gas phase inside the reactor is determined by means of the following correlation,

$$\frac{1}{U_r} = \frac{1}{h_c^{GS}} + \frac{A_{inn} In\left(\frac{d_{out}}{d_{inn}}\right)}{2\pi L_r k_w} + \frac{A_{inn}}{A_{out}} \frac{1}{h_c^{LS}} \tag{8.37}$$

where $h_{w,inn}$ is the convection heat transfer coefficient between the gas phase and the reactor wall, and it is calculated as follows (Smith 1981):

$$\frac{h_c^{LW}}{C_{P_{mix,G}} \rho_{mix}^G \mu_{mix,G}} \left(\frac{C_{P_{mix,G}} \mu_{mix,G}}{k_T^G}\right)^{2/3} = \frac{0.458}{\varepsilon_B} \left(\frac{\rho_{mix}^G u_S^G d_p}{\mu_{mix,G}}\right)^{-4.407} \tag{8.38}$$

Moreover, the following Leva correlation is used to calculate the heat transfer coefficient of the cooling medium (boiling water) on the shell side at high pressure:

$$h_c^{LW} = 282.2\left(P_T^{4/3}\right)\Delta T; 0.7 < P_T < 14\,\text{Mpa} \tag{8.39}$$

The Chilton–Colburn analogy (Chilton and Colburn 1934) has proved to be appropriate whose j-factor for the transfer of mass and heat are, respectively, given by

$$j_D = \frac{\text{Sh}}{\text{Re}_f \text{Sc}^{1/3}} = f \tag{8.40}$$

$$j_H = \frac{\text{Nu}}{\text{Re}_f \text{Pr}^{1/3}} = g \tag{8.41}$$

The literature is scarce in showing correlations to estimate the heat transfer coefficients in gas or liquid film interacting with the interfaces (gas–liquid and liquid–solid interfaces). It is often convenient to make use of the Chilton–Colburn analogy ($j_D = j_H$) for calculating these coefficients.

The determination of the heat of reaction is another important issue for modeling the temperature profiles, but not all products could be known in the FTS process. Therefore, it is preferred to use lumping techniques to encompass the FTS products of interest. Furthermore, there are several modeling works in which the formation of CO_2, CH_4, C_2H_4, C_2H_6, C_3H_8, $i - C_4H_{10}$, and $n - C_4H_{10}$ is considered as a net scheme of the FTS reactions involved in a given process (see reaction scheme elsewhere [Bayat et al. 2014; Lozano-Blanco et al. 2009; Park et al. 2015]).

The following expression given by Sandler (2006) can be used to calculate the enthalpy change of any state (T_1, P_1) to (T_2, P_2):

$$\begin{aligned}(T_2, P_2) - H(T_1, P_1) = &\int_{T_1}^{T_2} C_P dT + T_c\left\{\frac{H(T,P) - H^{IG}(T,P)}{T_c}\right\}_{T_{r2},P_{r2}} \\ &- \left\{\frac{H(T,P) - H^{IG}(T,P)}{T_c}\right\}_{T_{r1},P_{r1}}\end{aligned} \tag{8.42}$$

In addition, the database for thermodynamic properties given by Burcat and Ruscic (2005) is used in order to obtain the ideal gas enthalpy of the FTS species as a function of temperature:

$$H^{IG} = R_u T\left(A_1 + A_2\frac{T}{2} + A_3\frac{T^2}{3} + A_4\frac{T^3}{4} + A_5\frac{T^4}{5} + \frac{A_6}{T}\right) \tag{8.43}$$

8.2.4.3 Phase Equilibrium

In most literature reports related to mathematical models of the FTS process in fixed-bed reactors including other reactor technologies, authors have assumed that both the reactants and products remain only in a single phase, either vapor or liquid. Although many of these modeling schemes have served to analyze FTS technology in terms of selectivity, reactivity, and reactor performance, the lack of robustness is common by ignoring the VLE. However, a more complete modeling of the FTS reactor needs to consider the effect of the VLE, because it is necessary to know the gas solubility (reactant and some light hydrocarbons) through the liquid wax produced. In addition, the contribution of VLE in reactor modeling is also because the waxy liquid is partially volatilized at operation conditions of the FTS process (Guardo et al. 2006), whose phase distribution as well as the thermal behavior of the reactor is directly affected. However, according to Pöhlmann et al. (2016) concerning the catalyst pore filling, vaporization of volatile hydrocarbons in the bulk phase of the gas can occur. In summary, The VLE data play an important role in the design, operation, and development of the process (Caldwell and Van Vuuren 1986).

There is evidence in the literature that the cubic state equations widely used to describe the VLE of the FTS system are Soave–Redlich–Kwong (SRK) (Kim et al. 2009; Huang et al. 1988; Wang et al. 1999) and Peng–Robinson (PR) (Marano and Holder 1997; Visconti and Mascellaro 2013; Irani 2014). Wang et al. (1999) developed a modified version of the SRK equation of state to represent the gas–liquid equilibrium (solubility in heavy waxes) in the FTS, considering a wide range of solutes such as CO, H_2, CO_2, CH_4, C_2H_4, C_2H_6, and heavy wax solvents from C_{20} to C_{61} using a wide range of temperatures and pressures. Such an SRK equation has been preferably used (Huang et al. 1988; Wang et al. 1999; Mikhailova et al. 2003; Karimi et al. 2012) because of its high predictive and descriptive ability of the gas–liquid equilibrium behavior in the FTS. Karimi et al. (2012) showed a comparison between the results obtained by the SRK state equation, the PR equation as well as its modified versions, and Henry's empirical equation for the determination of phase equilibrium in the process FTS, where it is strongly shown that the modified SRK equation agrees with the experimental data collected from the literature. Other methods employed including the modified PR equation show high deviations. It should be noted that the modification in the SRK and PR state equations was based on the replacement of the use of the acentric factor by the molecular weight of the FTS products as a characteristic parameter.

The modified SRK state equation for the prediction of VLE equilibrium behavior is:

$$P = \frac{RT}{P - b} - \frac{\alpha_{SRK}a}{\overline{V}\left(\overline{V} + b\right)} \tag{8.44}$$

where P is the pressure, T the absolute temperature, and $\overline{V}$ the molar volume. The constants a and b take the following values:

$$a = 0.42747\frac{R_u^2 T_c^2}{P_c} \tag{8.45}$$

$$b = 0.08664\frac{R_u T_c}{P_c} \tag{8.46}$$

This equation can be rewritten as a function of the compressibility factor as follows:

$$Z^3 - Z^2 + Z\left(A - B - B^2\right) - AB = 0 \tag{8.47}$$

$$A = \frac{aP_T}{R_u^2 T^2} \tag{8.48}$$

$$B = \frac{bP_T}{R_u T} \tag{8.49}$$

The definition of the α parameter by Soave is given as follows:

$$\alpha_{\text{SRK}} = \left(1 - m\left(1 - \sqrt{T_r}\right)\right)^2 \tag{8.50}$$

The parameter m is obtained on the basis that the equation reproduces the vapor pressure of non-polar compounds and is correlated as a function of the acentric factor (ω), whose expression is adopted as follows:

$$m = f(\omega) = 0.48508 + 1.55171\omega - 0.15613\omega^2 \tag{8.51}$$

The acentric factors of heavy n-paraffins that can be used in the SRK equation of state can be obtained by the following correlation proposed by Lee-Kesler (1975):

$$\omega = \frac{-\text{In}(P_r) - 5.97214 + 6.09648\theta^{-1} + 1.28862\text{In}(\theta) - 0.169347\theta^6}{15.2518 - 15.6870\theta^{-1} - 13.472\text{In}(\theta) + 0.43577\theta^6}; \quad \theta = \frac{T_b}{T_c} \tag{8.52}$$

where T_b is the temperature at the normal boiling point and T_c is the critical temperature. However, the parameter α proposed by Soave for the prediction of hydrocarbon vapor pressures implies certain errors, since this form of α increases rapidly at relatively low reduced temperatures. In addition, for a given finite temperature beyond the critical temperature, the parameter α becomes zero and then is raised again with increasing temperature, whose behavior results in obtaining binary interaction parameters quite large and dependent on the temperature in hydrogen–hydrocarbon systems. Such drawbacks clearly indicate that the original version of the SRK equation of state is inadequate for the direct correlation of solubility data of the FTS process (Wang et al. 1999). Wang et al. (1999) improved the term α by replacing the acentric factor with the molecular weight as a characteristic parameter, since the acentric factor of the waxy product is usually inaccurate, whereas the molecular weight of the wax is a property easily calculated. Then the term m was proposed as follows:

$$\begin{aligned} m = f(M_m) &= \sum_{I=i}^{n} A_0 M_m^{i-1} \\ &= 0.45445 + 0.62675 \times 10^{-2} M_m - 0.93697 \times 10^{-5} M_m^2 + 0.37387 \times 10^{-8} M_m^3 \end{aligned} \tag{8.53}$$

The equilibrium between the gases (reactant and products) and the Fischer–Tropsch liquid wax is established as a function of the fugacities of each component in its respective gas or liquid phase as follows:

$$f_i^V = f_i^L \tag{8.54}$$

$$f_i^f = Py_i^f\phi_i^f \tag{8.55}$$

where ϕ_i^f is the fugacity coefficient for component i in f phase, which can be estimated with the SRK equation of state as follows:

$$\text{In}\left(\phi_i^f\right) = \frac{b_i}{b_m}(Z-1) - \text{In}(Z-B) + \frac{A}{B}\left(\frac{b_i}{b_m} - \frac{2}{\alpha_{\text{SRK},i}a_i}\sum_j y_i(\alpha_{\text{SRK}}a)_{ij}\text{In}\left(1+\frac{B}{Z}\right)\right) \tag{8.56}$$

where a_m and b_m are the parameters for mixtures, which are estimated by the following mixing rule:

$$a_m = \sum_i\sum_j y_iy_j\sqrt{a_ia_j}(1-k_{ij}) \tag{8.57}$$

$$b_m = \sum_i y_ib_i \tag{8.58}$$

In addition, the solubility of the gases in the Fischer–Tropsch waxy product can be determined by means of:

$$S_i = \frac{x_i}{\left(1-\sum_i x_i\right)M_m^{\text{wax}}} \tag{8.59}$$

where M_m^{wax} is the molecular weight of the wax and x_i is the molar fraction of each component in the wax.

It should be noted that the critical properties (temperature and critical pressure) with respect to the components of the FTS must be determined for the VLE calculations; however, the information for heavy hydrocarbons is scarce since there are only data reported in the literature for hydrocarbons up to C_{20}. Thus, the following Gasem and Robinson (1985) correlation including data of normal boiling points and number of carbon atoms (N_C) is used:

$$Y = \left[\frac{C_1}{C_4} - \left(\frac{C_1}{C_4} - C_3^{1-C_2}\right)\exp\{-C_4(N_C-1)(1-C_2)\}\right]^{\frac{1}{(1-C_2)}} \tag{8.60}$$

where Y represents either P_c or T_b/T_c, and the constants $C_1 - C_4$ are given in the work by Gasem and Robinson (1985).

Mikhailova (2003) modified the SRK equation, and in this case, the molecular weight distribution of the FTS products was estimated by the ideal ASF polymerization model:

$$M_m = (1-\alpha)^{n-1} \tag{8.61}$$

However, it should be noted that the molecular weight distribution can be estimated by nonidentical ASF models (Förtsch et al. 2015; Kuipers et al. 1996; Donnelly et al. 1988).

8.2.4.4 Catalyst Particles Parameters

In the case of the catalyst particles, it is necessary to determine the different properties involved as well. A review of the different correlations published in the literature was done for the parameter estimation of the catalyst particle used in the FTS process.

Due to the formation of a film or thin layer of waxy liquid product on the catalytic particle surface, the external mass transport effects may be negligible, but the influence of pore diffusion is important (Kaiser and Jess 2014). Thus, the Thiele modulus is an important dimensionless parameter that allows determining the presence of diffusion limitations within the particle, which for the case of a pseudohomogeneous first-order reaction is defined as follows (Jess and Kern 2012a; Thiele 1939):

$$\Phi_S = L_p\sqrt{\frac{k_A}{D_{eff,L}}} = L_p\left(k_{int}\frac{C_{CO}^G}{D_{eff,L}C_{CO}^L}\right)^{0.5} = L_p\left(k_{int}\frac{1}{D_{eff,L}R_u\frac{T}{He_{CO}}}\right)^{0.5} \tag{8.62}$$

where L_p is the characteristic length of the reaction system also known as the form factor, k_A is the kinetic rate constant for the homogeneous first-order reaction, and $D_{eff,L}$ is the effective diffusivity for CO at the liquid phase.

One approach given in the literature to determine the mass transfer limitations of the catalyst consists of solving the mass balance of a differential shell taken into the catalyst particle by considering a spherical geometry of the pellet or catalyst particle, in which a reaction of first order occurs as follows (Iglesia et al. 1993):

$$\frac{C_i^f}{C_{f,i}^{S_{sur}}} = \frac{d_{pe}}{2r}\frac{\sinh\left(3\Phi_S\frac{2r}{d_{pe}}\right)}{\sinh(3\Phi_S)} \tag{8.63}$$

whose modified form for a reaction of order n was established as (Bischoff 1965; Petersen 1965):

$$\Phi_S = L_p\sqrt{(n+1)\frac{k_{int}\left(C_{f,i}^{S_{sur}}\right)^{n-1}}{2D_{eff}^f}} \tag{8.64}$$

The shape factor L_p was used to modify the solution of different geometries, which in the case of a cylinder pellet turns into $d_p/4$ and $d_p/6$ for a spherical pellet. k_{int} is denoted by the intrinsic reaction rate coefficient per volume unit for the nth-order reaction considered, and D_{eff}^f is the effective diffusivity given as follows:

$$D_{eff}^f = D_{\text{comb}}\frac{\varepsilon_P}{\tau} \tag{8.65}$$

$$\varepsilon_P = \frac{V_{\text{pore}}\rho_B}{1-\varepsilon_B} \tag{8.66}$$

On the other hand, for the determination of the effective diffusivity, the following correlations are used in modeling of fixed-bed configuration in which the FTS occurs (Cybulski and Moulijn 1994; Bosanquet 1944):

$$D_{eff,ax,i}^f = D_T\left(1+\frac{1}{192}Pe^2\right) \tag{8.67}$$

where

$$Pe = \frac{d_t u_S^G}{D_{ij}} \text{ and } D_{ij} = \frac{R_u}{6\pi\mu_{\text{mix}}^G d_p}$$

$$D_{eff}^f = \frac{\varepsilon_S}{\tau}\frac{1}{\frac{1}{D_{M_i}^f}+\frac{1}{D_{K,i}^f}}F(\lambda^G) \tag{8.68}$$

Whose restrictive factor $F(\lambda^G)$ is given by Iliuta et al. (2006), in which r_{solute} is the hydrodynamic molecular radius of the solute, r_{pm} is the mean radius of the pore, and $\hat{Z}$ is a dimensionless constant used in the Bosanquet's formula:

$$F(\lambda^G) = (1-\lambda^G)^{\hat{Z}};\lambda^G = \frac{r_{\text{solute}}}{r_{pm}} \tag{8.69}$$

The effectiveness factor of the catalyst can be estimated by the correlations reported in Table 8.15. Mederos et al. (2009) presented some other formulas for the determination of the effectiveness factor and the Thiele modulus for different shapes of catalysts.

It is possible to obtain close approximations if the Thiele modulus is used by making the necessary adjustment for an n-order reaction (Eq. 8.64). However, it is important to mention that there are no analytical methods to determine the effectiveness factor for cases involving nonlinear reaction rates and arbitrary catalyst particle shapes. Thus, its approximation based on the consideration of nonlinear reaction rates and catalysts of an arbitrary shape can be calculated by the equations shown in Table 8.15 using the generalized Thiele modulus (Iordanidis et al. 2002):

$$\Phi_S = L_p \frac{R_{FT}\left(C_{f,i}^{S_{\text{sur}}}, T^{S_{\text{sur}}}\right)}{\sqrt{2d_{pe}}} \left[\int_0^{C_{f,i}^{S_{\text{sur}}}} R_{FT}\left(C_{f,i}^{S_{\text{sur}}}, T^{S_{\text{sur}}}\right) dC_{f,i}^{S_{\text{sur}}}\right]^{-1/2} \tag{8.70}$$

In the case where η tends to 1, the shape adopted by the catalytic particle becomes an important factor that influences the estimated value of the effectiveness factor. Wijngaarden et al. (1998) proposed a novel approach to calculate the effectiveness factor in the case where chemical reactions whose rate is nonlinear and catalytic particles with an arbitrary geometry are contemplated that involved two dimensionless groups:

Zero Aris number:

$$A_{n0} = L_p^2 \frac{R_{FT}\left(C_{f,i}^{S_{\text{sur}}}, T^{S_{\text{sur}}}\right)}{\sqrt{2d_p}} \left[\int_0^{C_{f,i}^{S_{\text{sur}}}} R_{FT}\left(C_{f,i}^{S_{\text{sur}}}, T^{S_{\text{sur}}}\right) dC_{f,i}^{S_{\text{sur}}}\right] \tag{8.71}$$

First Aris number:

$$A_{n1} = L_p \frac{\Gamma}{d_{pe}} \left.\frac{\partial R_{FT}\left(C_{f,i}^{S_{\text{sur}}}, T^{S_{\text{sur}}}\right)}{\partial C_{f,i}^{S_{\text{sur}}}}\right|_{C_{f,i}^{S_{\text{sur}}}} \tag{8.72}$$

Table 8.15 Effectiveness factors for conventional geometry of catalyst particles and their concentration profiles according to analytical solution.

Geometry of the system	L_p	η	Observations
Rectangular	L_B	$\dfrac{\tan h(\Phi_S)}{\Phi_S}$	—
Cylindrical	$\dfrac{d_p}{2}$	$\dfrac{2I_1(\Phi_S)}{\Phi_S I_0(\Phi_S)}$	For pellets of cylindrical catalyst with a particle diameter of less than 200 μm ($d_p < 200$ μm). I_0 and I_1 are the typical Bessel functions of first and second order, respectively.
Spherical	$\dfrac{d_p}{6}$	$\dfrac{3}{\Phi_S}\left[\dfrac{1}{\tan h(\Phi_S)} - \dfrac{1}{\Phi_S}\right]$	Spherical particles with a diameter greater than 200 μm ($d_p > 200$ μm).

where Γ is the geometric factor that depends only on the particle shape. Γ is 2/3, 1, and 6/5 for a slab, an infinite cylinder, and a sphere, respectively. The zero Aris number was introduced to calculate the effectiveness factor in a low region of η, whereas the first Aris number determines η when its value approaches 1. The following expression is a generalized form of the effectiveness factor that implies the two numbers of Aris already defined:

$$\eta = \frac{1}{\sqrt{1 + (1-\eta)A_{n0} + \eta A_{n1}}} \tag{8.73}$$

which can be solved by an iterative procedure.

When k_{int} is unknown, it is not possible to use Eq. (8.64) for the Thiele modulus already described, so if only information on the observed experimental reaction rate is available, the Wagner–Weisz–Wheeler modulus (M_w) must be used (Levenspiel 1999):

$$M_w = \frac{L_p^2}{D_{\text{eff}}^f} \frac{r_v \rho^S}{C_{f,i}^{S_{\text{sur}}}} \frac{n+1}{2} = \Phi_S^2 \eta \tag{8.74}$$

where ρ^S is the density of the pellet and r_v is the observed rate of reaction of species i, which also translates as the CO depletion rate in the FTS reaction per unit mass of the catalyst.

In Eq. (8.75), the diffusivity D_{comb} represents the combination between the bulk diffusivity D_{ij} and Knudsen diffusivity D_K that can be estimated from the correlations reported by Pollard and Present (1949):

$$\frac{1}{D_{\text{comb}}} = \frac{1}{D_{ij}} + \frac{1}{D_K} \tag{8.75}$$

In addition, the diffusivity of each component involved in the gas mixture is given by the following expression (Wilke 1949):

$$D_{i,\text{mix}}^G = \frac{1-y_i}{\sum_{i \neq j} \frac{y_i}{D_{ij}}} \tag{8.76}$$

where D_{ij} (where i diffuses through the liquid wax j) was estimated using the following correlations, which are valid for a wide range of temperatures and pressures (Erkey et al. 1990a; Erkey et al. 1990b; Fuller et al. 1966):

$$D_{ij} = \frac{0.945\sqrt{T}}{M_{m_i}^{0.239} M_{m_j}^{0.781} \left(\sigma_i \sigma_j\right)^{1.134}} \tag{8.77}$$

$$D_{ij} = \frac{10^{-7} T^{1.75} \sqrt{\frac{1}{M_{m_i}} + \frac{1}{M_{m_j}}}}{\rho_{\text{mix}}^G \left(\nu_{ci}^{1/3} + \nu_{vj}^{1/3}\right)^2} \tag{8.78}$$

where M_{m_i} and M_{m_j} are the molecular weight of the solute (either CO or H_2) and the molecular weight of the solvent (waxy liquid product), respectively, both in g/mol; σ_i and σ_j are the hard-sphere diameter for the solute (either CO or H_2) and hard-sphere diameter for the solvent (waxy liquid product), respectively, both in Å; ν_{ci} and ν_{cj} are the critical volumes of components i and j, respectively; and $\overline{V}$ is the specific volume of the different species in m^3/mol. Meanwhile the diffusivity of Knudsen D_K is calculated by the following correlation obtained from the kinetic theory of the gases (Welty et al. 2007):

$$D_K = 4850 d_p \sqrt{\frac{T}{M_{m_i}}} \tag{8.79}$$

Welty et al. (2007) indicate that when D_K exceeds D_{ij} by more than two orders of magnitude under the set of operating conditions used, the effects of diffusion resistance can be neglected.

It should be noted that the diffusivity of CO in the produced wax is usually lower than that of H_2, and it is considered as the limiting diffusion rate. On the other hand, when the synthesis gas feed stream contains low concentrations of H_2, it is convenient to consider that the diffusivity of the H_2 in the waxy product is the limiting diffusivity (Hallac et al. 2015).

In the case of using catalyst pellets with irregular geometry, the equivalent particle diameter (d_{pe}) must be used to utilize the previous correlations.

The equivalent particle diameter is referred to as the diameter of a sphere whose surface area (or volume) is equal to that of the catalyst particle, being a characteristic property of the particle that depends on both its size and shape. One way to determine the equivalent particle diameter for fixed-bed systems is provided by Cooper:

$$d_{pe} = \frac{6\left(\frac{V_p}{S_p}\right)}{\phi_S} \tag{8.80}$$

The geometric forms of the catalyst are ideally spherical, but it would be possible to use any catalyst shape; however, correlations for the effectiveness factor are not possible in all cases mainly for irregular geometry. Therefore, there are other parameters that are involved in calculating the effectiveness factor that can be estimated for other types of catalyst particles such as the particle size also known in the literature as the form factor L_p. It should be noted that the construction of the model for an irregular geometry catalyst particle in the FTS process, would be more complicated.

Nowadays, the design of novel FTS catalysts is of utmost importance for the purposes of improving the synthesis gas conversions, product selectivities, and overall performance of the reactor (Khodakov et al. 2007; Zhang et al. 2010). In the research of catalyst design applied not only to the FTS process but also to the hydrotreatment processes, it is very important to improve the geometries of the catalyst, mainly to diminish the mass and heat transfer intraparticle limitations. The form factor (L_p) defined as the volume ratio to the catalyst particle surface (V_p/S_p) is frequently employed to correlate certain characteristics such as the size and shape of the catalyst with the catalytic activities, as well as to perform studies on the characteristics of the catalyst, packed bed, intraparticle diffusion limitations, catalytic pore size, pressure drop restrictions, etc. Ancheyta et al. (2005) proposed some equations for the determination of the area and the external volume for some porous and nonporous particles of different shapes and sizes. This study is mainly oriented to calculations of V_p and S_p for different types of hydrotreating catalysts; however, its application can be extended to catalysts used in the FTS process. Calculation of V_p and S_p of lobe-shaped catalyst particles are as follows:

$$V_p = n\left(\pi r_c^2 L\right) - A_1 L \tag{8.81}$$

$$S_p = n\left(2\pi r^2 L + 2\pi r_c L\right) \pm 2A_1 - nA_2 \tag{8.82}$$

where n is the number of lobes, A_1 is the lateral area of the geometric shape between the lobes, which represents a rhombus for a bi-lobular geometry, a triangle for a tri-lobular shape, and a frame

Table 8.16 Definition of parameters to determine the volume and surface of particles of different geometries.

Shape	n	θ	r_c	A_1	A_2
Cylinder and pellet	1	—	$d_p/2$		0
Bi-lobular	2	45°	$\frac{d_p}{3.4142}$	$\frac{d_p^2}{8}\left[\frac{\sin(\theta)-\sin^2(\theta)}{(1+\sin(\theta))}\right] = 8.88348\times10^{-3}d_p^2$	$\frac{\pi}{2}r_{cyl}L$
Tri-lobular	3	60°	$\frac{d_p}{4}$	$\frac{d_p^2}{8}\frac{(2\sin(\theta)-1)^2}{\tan(\theta)} = 3.86751\times10^{-2}d_p^2$	$\frac{\pi}{3}r_{cyl}L$
Tetra-lobular	4	45°	$\frac{d_p}{4.8284}$	$d_p^2\left[\frac{2\cos(\theta)-1}{1+2\cos(\theta)}\right]^2 = 2.94373\times10^{-2}d_p^2$	$\frac{\pi}{2}r_{cyl}L$

for tetra-lobular geometry, while A_2 is the common area between each cylinder and between each side of the shape between the lobes. In Eq. (8.82), the term after parenthesis is negative for bi-lobular forms and positive for tri- and tetra-lobular shapes. Table 8.16 shows the parameters involved in Eqs. (8.81) and (8.82) for cylinders, spheres, pellets, bi-lobular, tri-lobular, and tetra-lobular shapes.

An important aspect in the design of catalysts, simulation, and modeling of a fixed-bed reactor is the analysis of transport limitations at the particle level. Mears (1971a) indicated that concentration and temperature gradients can exist in three domains: (i) intraparticle, referring to the gradients that occur within the individual catalytic particles; (ii) interface, wherein the gradients are generated between the external surface of the catalytic particles and the adjacent layer of the fluid; and (iii) interparticle, with the axial and radial gradients within the reactor being considered.

It is of interest to have precise criteria that allow predicting the existence of intraparticle mass transports, since it is of importance in determining suitable pathways for model simplification of the catalyst particle, such as neglecting or not the effects of internal diffusion in the pore. On the other hand, it is stated that it is only possible to obtain information related to the intrinsic kinetics in the absence of effects caused by resistances to mass and heat transport in the catalytic particle (Ibrahim and Idem 2006). There is a criterion given in the literature to determine if the presence of intraparticle diffusion limitations exists; this is the Weisz–Prater criterion given by the dimensionless parameter.

$$C_{WP} = \frac{-R_{FT,obs}\rho^S L_p^2}{D_{eff}^f C_i^G} \tag{8.83}$$

where $R_{FT,obs}$ is the observed reaction rate (mol/(Kg$_{cat}\cdot$s)), ρ^S is the density of the solid catalyst particle, and C_i^G is the limiting reactant concentration in the syngas mixture. This criterion establishes that if the value of C_{WP} becomes much larger than 1, the effects of limiting the internal pore diffusion are significant, i.e. there is a strong resistance to intraparticle diffusion resulting in a concentration gradient from the catalyst surface to its pores (Ibrahim and Idem 2006). Akpan et al. (2007) showed experimentally that the diffusion resistance is null when the particle size (form factor) is less than 0.3 mm. This criterion of Mears (1971a) is often considered for the determination of the onset of external mass transport limitation that is given by

$$\frac{r_{obs}\rho_B L_p n}{k_{FT}} < 0.15 \tag{8.84}$$

To determine that external mass transfer resistance has no significant effect on the reaction rate, the following criterion is used, relating the observed reaction rate to the reaction rate if the resistance to the film controls (Ibrahim and Idem 2006):

$$\frac{\text{observed rate}}{\text{rate if film resistance controls}} = \frac{-R_{FT,obs}}{k_c C_i^f}; i = \text{reactants: CO and H}_2 \tag{8.85}$$

However, Levenspiel (1999) points out that the film resistance to mass transfer should not influence the rate of reaction.

Ibrahim and Idem (2006) established the following two criteria: (1) the heat transfer resistance within the pore and (2) the limitations of heat transfer across the gas film, to be:

$$\Delta T_{\text{max,particle}} = \frac{D_{\text{eff}}^f \left(C_i^{S_{\text{sur}}} - C_i^{cp}\right)(-\Delta H_{\text{RFT}})}{\lambda_{\text{eff}}} \tag{8.86}$$

$$\Delta T_{\text{max,film}} = \frac{L(-R_{FT,obs})(-\Delta H_{\text{RFT}})}{h} \tag{8.87}$$

where $\Delta T_{\text{max,particle}}$ and $\Delta T_{\text{max,film}}$ are the upper limit to temperature variation between the pellet center and its surface and the upper limit to temperature difference between the gas bulk and the pellet surface, respectively; ΔH_{RFT} is the heat of reaction; $C_i^{S_{\text{sur}}}$ and C_i^{cp} are the concentrations at the pellet surface and center, respectively; D_{eff}^f is the effective mass diffusivity; and L is the characteristic length.

One way of assessing the thermal severity resulting from the FTS process in fixed-bed systems is by using the Mears (1971b) rigorous criterion, which allows one to compare the heat generated by the reaction with the capacity of the catalytic bed to transfer heat from the reaction zone toward the wall of the tube. This criterion was used in a pseudohomogeneous modeling of a fixed-bed FTS reactor proposed by Philippe et al. (2009):

$$(1-\varepsilon_B)\frac{d_p}{2}\left|\Delta H_{\text{RFT},j}\right| \cdot \frac{T_a d_t^2}{4\lambda_{\text{eff,rad}} T_w^2} \cdot \left(1 + \frac{8\lambda_{\text{rad}}}{U_r d_t}\right) < 0.4 \tag{8.88}$$

When this criterion is fulfilled, it can be assumed that the radial temperature profile is almost flat, which means that the reaction heat is eliminated in the radial direction. However, this does not ensure that there is no generation of hot spots along the catalytic bed (Philippe et al. 2009). The following criteria can be applied in cases where the resistance to heat transfer at the wall is significant, as well as its impact on catalyst particle:

$$\frac{\left|\Delta H_{\text{RFT},j}\right| R_{FT,B} d_t^2}{4\lambda_{\text{eff,rad}} T_w} \frac{E_a}{R_u T_w} \left(\frac{1}{1 + 8\left(\frac{d_p}{d_t}\right) Bi_w}\right) < \frac{0.4 R_u T_w}{E_a} \tag{8.89}$$

where $R_{FT,\,B}$ is the reaction rate per unit bed volume and Bi_w is the dimensionless number of Biot defined as $h_w d_t/\lambda_{\text{eff,rad}}$. It should be noted that the term $\lambda_{\text{eff,rad}}/8h_w d_t$ represents the heat transfer resistance in the wall that cannot be ignored when low ratios of the diameter of the tubes to the diameter of the particle are used ($d_t/d_p \approx 10$) (Mears 1971a; Mears et al. 1971b). Otherwise, Mears (1971a) pointed out that the magnitudes of heat transfer resistance in experimental reactors behave as follows: interparticle> interphase> intraparticle.

8.2.4.5 Catalytic Bed Parameters

An extremely important parameter in the modeling of fixed-bed systems involves the determination of the bed void fraction (or bed porosity), as this indicates the space within the fixed bed that is filled with catalyst particles, i.e. it states the fraction of the catalyst bed volume that is available for the phases involved in a multiphase multicomponent system. Alternatively, having a reliable estimation of the bed void fraction is important for minimizing the pressure drop across the catalyst bed. Table 8.17 reports some correlations for calculating the void bed fraction applicable to many problems involving fixed-bed configurations.

Antwerpen et al. (2010) found a wide range of correlations to estimate the porosity of the bed of structured packages that can be used in fixed-bed reactor modeling.

Once the bed void fraction is determined, the density of the solid catalyst particle can be calculated according to the following equation:

$$\rho^S = \frac{\rho_B}{1-\varepsilon_B} \tag{8.90}$$

Based on the consideration of volume conservation, the bed void fraction is the sum of the gas and liquid holdups (Whitaker 1973):

$$\varepsilon_B = \epsilon_L + \epsilon_G \tag{8.91}$$

$$\epsilon_L + \epsilon_G + \epsilon_S = 1 \tag{8.92}$$

Relationships between the holdups within the solid catalyst particle are given by

$$\varepsilon_S = \epsilon_L^p + \epsilon_G^p \tag{8.93}$$

$$\epsilon_L^p + \epsilon_G^p + \epsilon_S^p = 1 \tag{8.94}$$

Table 8.17 Correlations for the calculation of the void fraction bed.

Correlation	Reference
$\varepsilon_B = 0.1504 + \frac{0.2024}{\phi_p} + \frac{1.0814}{\left(\frac{d_t}{d_{pe}} + 0.1226\right)^2}$	Benyahia and O'Neill (2005)
(1) $\varepsilon_B(r) = \left[1 + \left(\frac{1-\varepsilon_B}{\varepsilon_B}\right)\sqrt{1-\exp\left(-2\frac{r-r_{\text{inn}}}{d_p}\right)}\right]^{-1}$ for $r_{\text{inn}} \leq r \leq \frac{r_{\text{out}} - r_{\text{inn}}}{2}$ (2) $\varepsilon_B(r) = \left[1 + \left(\frac{1-\varepsilon_B}{\varepsilon_B}\right)\sqrt{1-\exp\left(-2\frac{r_{\text{out}}-r}{d_p}\right)}\right]^{-1}$ for $\frac{r_{\text{out}} - r_{\text{inn}}}{2} \leq r \leq r_{\text{out}}$	White and Tien (1987)
$\varepsilon_B = 0.38 + 0.073 + \left[1 + \frac{\left(\frac{d_t}{d_{pe}} - 2\right)}{\left(\frac{d_t}{d_{pe}}\right)^2}\right]$	Froment et al. (2011), Haughey and Beveridge (1969)

The external surface area of the catalyst particles per unit volume of the fixed-bed reactor can be calculated as

$$a_S = \frac{6(1-\varepsilon_B)}{d_{pe}} \tag{8.95}$$

It is also convenient to estimate the pore size to determine the pore filling of the catalyst particle with water and liquid hydrocarbons using the Kelvin equation (Ermolaev et al. 2015):

$$d_{\text{pore}} = \frac{2\sigma_i}{p_i - p_{\text{sat}} - \dfrac{\rho_i^S R_u T}{M_{m_i}} \ln\left(\dfrac{p_i}{p_{\text{sat},i}}\right)} \tag{8.96}$$

Because it is possible that the formation of waxy liquid is carried out inside the catalytic reactor, and that the velocities reached by the liquid flow are relatively low in the range of a trickle regime, it is necessary to determine the liquid holdup for generalized modeling of the fixed-bed FTS reactor. This parameter is useful in design and operation of a fixed-bed reactor under a trickle-flow regime (Larachi et al. 1991). Liquid holdup has been shown to greatly affect the pressure drop across the bed, the amount of catalyst wetting, and the thickness of the liquid film surrounding the catalyst particle (Ellman et al. 1990). In addition, the holdup knowledge is essential in the case of high exothermicity of the reactions such as the FTS, since it allows avoiding the formation of hot spots in the catalytic bed and the prevention of reactor runaways. On the other hand, it is a parameter that affects the wetting efficiency of the catalyst and hence the selectivity, which depends on whether the reaction is carried out in the wet catalyst area or in the wet and dry catalyst regions alike (Al-Dahhan et al. 1997).

Bazmi et al. (2013) proposed a correlation for the prediction of liquid holdup where the effects of the catalyst shape (extruded trilobe particle) on the pressure drop are considered, as well as the behavior of the liquid holdup to the different methods of loading the catalyst into trickle-bed systems:

$$\epsilon_{L,d} = 0.07\left[\frac{\text{We}_L^{0.5}}{X_L}\left(\frac{\varepsilon_B^3}{1-\varepsilon_B}\right)^{3.5}\left(\frac{\text{Re}_L}{\text{Re}_G}\right)^2\right]^{0.17} \exp\left\{\frac{\text{We}_L^{0.5}}{X_L}\left(\frac{\varepsilon_B^3}{1-\varepsilon_B}\right)^{3.5}\left(\frac{\text{Re}_L}{\text{Re}_G}\right)^2\right\} \tag{8.97}$$

where

$$X_L = \frac{u_s^L(\rho^L)^{0.5}}{u_s^G(\rho^G)^{0.5}} \tag{8.98}$$

$$\text{We}_L = \frac{u_s^L d_{pe}\rho^L}{\sigma_L} \tag{8.99}$$

The equation is valid within the range of gas superficial velocities from 0.01 to 0.1 m/s and the liquid superficial velocities from 0.001 to 0.01 m/s included in the range of the trickle-flow regime. Such a correlation was compared with the correlation proposed by Larachi et al. (1991) as well as with experimental data and showed acceptable improvements in predicting liquid holdups.

Al-Dahhan et al. (1997) also reported a comprehensive review on various correlations given in the open literature for the prediction of the transition from trickle-flow regimes to pulse, for liquid holdups, two-phase pressure drop, and catalyst wetting efficiency (liquid–solid contacting efficiency), among others in high-pressure trickle-bed reactors.

8.2.5 Final Remarks

The search for new and better alternatives to produce improved clean fuels that meet more stringent environmental constraints is a worldwide aim. The scientific and industrial community is increasingly studying the implementation of the FTS as it has been considered as one of the most viable options to replace conventional technologies for producing fuels of commercial value. In addition, the design of more efficient catalysts and novel reactor technologies (microchannel reactor and monolith reactor) has been a field of research currently active to greatly improve the overall FTS process. However, the fixed-bed catalytic reactor has been preferably used because of its marked advantages over other reactor technologies, which include the relatively easy scaling up from a single tube. On the other hand, the FTS process is considerably complex since the distribution of FTS products has not been accurately known because the products are quite broad and difficult to accurately predict. FTS behavior is attributed to the effect of (i) the catalyst type and/or support promoter and (ii) the type of reactor used, among others.

Despite the large number of mathematical models published in literature until now, it is necessary to propose a generalized modeling scheme of the fixed-bed reactor for the FTS obtained from simplified versions reported in the literature. However, solving the generalized model is a rather complex task, since it requires an in-depth study of hydrodynamics, the analysis of the phases distribution in the catalytic bed (vapor–liquid equilibrium), mass and heat transport phenomena in the bulk and particle phases, a robust model for pressure drop, and the search for precise correlations that estimate the several parameters of the model. Therefore, it becomes evident that there is a need to study different modeling schemes that can be derived from this proposed generalized model and that must be validated through comparisons between simulated and experimental data.

8.3 Importance of Proper Hydrodynamics Modeling in Fixed-Bed Fischer–Tropsch Synthesis Reactor

8.3.1 Introduction

The importance of hydrodynamics, particularly gas density, superficial gas velocity, and total pressure in axial and radial directions, was analyzed for the modeling of catalytic reactors using a non-isothermal pseudohomogeneous approach. The modeling of a fixed-bed reactor in one and two stages for the CO conversion by FTS was taken as a study case. For the validation of the proposed model, the results of the simulations for the CO conversion and temperature profiles were compared with experimental data reported in the literature. Simulations for CO conversion and reactor temperature profiles confirmed the model's ability to predict the selectivity of the liquid products in the FTS reactor in one and two stages. The proposed model predicts more suitable profiles of CO conversion and temperature along the reactor, which makes it a more robust and efficient tool for design, optimization, and control purposes.

8.3.2 Mathematical Modeling of the Fixed-Bed Fischer–Tropsch Synthesis Reactor

To predict the syngas conversion and reactor temperature profiles of a fixed-bed multitubular FTS reactor, a two-dimensional pseudohomogeneous model is proposed, which is formulated using cylindrical coordinates. The radial dispersive plug flow is described by the mass and heat balances, and the following assumptions were considered:

- Steady-state operation
- Plug-flow regime with no channeling effects along the catalyst bed
- Negligible axial dispersion effects of mass and heat, as well as concentration gradients in the radial direction
- Variable density, total pressure, and gas velocity along the axial direction
- Radial heat transfer within the catalytic bed as a function of the constant radial effective thermal conductivity
- Constant temperature of the cooling medium
- Negligible formation of a wax layer on the catalyst surface of mass and heat external transfer limitations
- The liquid hydrocarbon diffusion out of the catalyst pores occurs fast enough to neglect any (internal and external) mass transfer resistances.

It is worth mentioning that the liquid hydrocarbon accumulation (mainly wax) on the catalyst particle surface is important for diffusional effects and external mass transport (Pöhlmann et al. 2016). It has been shown experimentally and by means of numerical simulations how the filling of the pores catalyst with liquid hydrocarbons changes as a function of time. Also, the effectiveness factor varies with the degree of catalytic pores filling (Pöhlmann and Jess 2016b). On the other hand, for low-temperature FTS (<250 °C), the catalyst pores are filled with liquid hydrocarbons (waxes), which leads, inevitably, to pore diffusion effects within the catalyst particles and a decrease in optimum catalyst performance and selectivity (Jess et al. 2013; Calderone et al. 2013). Furthermore, it has been shown that the diameter of the reactor and the catalyst particle play an important role in the diffusional and external mass transfer effects (Jess and Kern 2012a; Jess and Kern 2012b). However, for diameters of catalyst particles of 3–4 mm (Pöhlmann et al. 2016; Jess and Kern 2012b), with a reactor diameter to catalyst particle diameter ratio of $dt/dp = 7.6$, the external diffusional effects are not significant. In addition, the risk of the thermal runaway phenomenon occurring becomes negligible (Pöhlmann et al. 2016). Based on these literature reports, the assumption that the external mass transfer effects that are due to the wax layer or liquid hydrocarbons formed on the catalytic surface are negligible can be justified.

8.3.2.1 Reactor Model

For the two-dimensional axisymmetric plug flow inside the fixed-bed FTS reactor, when the effect of axial mass dispersion is neglected and at steady state, the mass balance equation results in the following expressions:

$$\frac{\partial x_{\mathrm{CO}}}{\partial z} = \frac{1 - x_{\mathrm{CO}}}{u_s}\frac{\partial u_s}{\partial z} + \frac{D_{er}}{u_s}\left(\frac{\partial^2 x_{\mathrm{CO}}}{\partial r^2} + \frac{1}{r}\frac{\partial x_{\mathrm{CO}}}{\partial r}\right) + \frac{\rho_B R_{\mathrm{FTS}}}{u_s C_{0,\mathrm{CO}}} \tag{8.100}$$

$$\frac{\partial x_{\mathrm{H_2}}}{\partial z} = \frac{1 - x_{\mathrm{H_2}}}{u_s}\frac{\partial u_s}{\partial z} + \frac{D_{er}}{u_s}\left(\frac{\partial^2 x_{\mathrm{H_2}}}{\partial r^2} + \frac{1}{r}\frac{\partial x_{\mathrm{H_2}}}{\partial r}\right) + \frac{\rho_B \upsilon_j R_{\mathrm{FTS}}}{u_s C_{0,\mathrm{H_2}}} \tag{8.101}$$

While the heat balance equation in the fixed-bed FTS reactor is:

$$\frac{\partial T}{\partial z} = \frac{\lambda_{er}}{\rho^f u_s C_{P_{\mathrm{mix},G}}}\left(\frac{\partial^2 T}{\partial r^2} + \frac{1}{r}\frac{\partial T}{\partial r}\right) - \frac{\rho_B R_{\mathrm{FTS}}}{\rho^f u_s C_{P_{\mathrm{mix},G}}}(-\Delta H_{\mathrm{FTS}}) \tag{8.102}$$

Eqs. (8.100)–(8.102) represent the syngas conversion (CO and H_2) and the temperature variation in the axial and radial directions, which are typical equations used in two-dimensional pseudohomogeneous models (Philippe et al. 2009; Sharma et al. 2011; Rafiq et al. 2011; Dai et al. 2014).

The equations related to hydrodynamics in the axial variation (density, surface velocity, and gas pressure) are as follows:

$$\frac{\partial \rho^f}{\partial z} = \frac{M_{m_{\text{mix},G}}}{R_c}\left(\frac{1}{T}\frac{\partial P_T}{\partial z} - \frac{P_T}{T^2}\frac{\partial T}{\partial z}\right) \tag{8.103}$$

$$\frac{\partial u_s}{\partial z} = -\frac{u_s}{\rho^f}\frac{\partial \rho^f}{\partial z} \tag{8.104}$$

$$\frac{\partial P_T}{\partial z} = -f_p\frac{u_s^2\rho^f}{d_p} \tag{8.105}$$

The boundary conditions used for solving the proposed model are as follows:

$$\text{at } z = 0 \text{ and } 0 \le r \le R_0;\ x_{CO} = x_{CO,0}, T = T_0, u_s = u_{s,0}, \rho^f = \rho_0^f, P_T = P_{T,0} \tag{8.106}$$

$$\text{at } r = 0 \text{ and } 0 \le z \le L_B;\ \frac{\partial x_{CO}}{\partial r} = \frac{\partial T}{\partial r} = 0 \tag{8.107}$$

$$\text{at } r = R_0 \text{ and } 0 \le z \le L_B;\ \frac{\partial x_{CO}}{\partial r} = 0, \lambda_{er}\frac{\partial T}{\partial r} = -h_w(T - T_w) \tag{8.108}$$

It is important to mention that in the proposed FTS reactor model, the intraparticle effects were neglected for simplicity (two-dimensional pseudohomogeneous model). This was done to compare it with the same model given in the literature that lacks robustness in terms of hydrodynamics, i.e. the variation in the superficial velocity and the density of the gas mixture is not taken into consideration. Another difference is the use of a novel correlation for the friction factor used for the calculation of the total pressure.

8.3.2.2 Kinetics

For the development of kinetic models that describe the FTS reactions, the main approach reported in the literature is based on the power-law rate model, in which the kinetics for reactants conversion is described (syngas: CO and H_2) (Pinna et al. 2002). This model has been successfully used and validated in a series of case studies (Van der Laan and Beenackers 1999; Das et al. 2005). The use of this type of kinetic model that represents a simple description of a reaction system in the FTS process is valid (from a theoretical approach) if the products do not affect or participate in the formation mechanism of initiating monomers of the Fischer–Trospch hydrocarbon chain growth (Visconti et al. 2007).

The kinetic equation of FTS is given by a power-law model as a function of partial pressures of CO and H_2, which represents the global consumption of CO during the synthesis:

$$R_{FTS} = A_p e^{-E_a/(R_c T)} p_{H_2}^m p_{CO}^n \tag{8.109}$$

The following kinetic parameters were taken from the literature (Dai et al. 2014): $A_p = 2.45 \times 10^6$ mol/(Kg$_{cat}$/(MPa$^{0.3}\cdot$ s)), $E_a = 73$ KJ/mol, $m = 1.3$, and $n = -1$. This CO consumption kinetics equation has been previously used by others (Visconti et al. 2011b; Zhen et al. 2009; Zennaro et al. 2000; Yates and Saterfield 1991; Outi et al. 1981).

The molar product selectivity was calculated by the following:

$$S_{CO_2}(\%) = \left(\frac{\text{moles of } CO_2 \text{ produced}}{\text{moles of inlet CO} - \text{moles of outlet CO}}\right) \times 100 \tag{8.110}$$

$$S_{CH_4}(\%) = \left(\frac{\text{moles of CH}_4\text{ produced}}{\text{moles of inlet CO} - \text{moles of outlet CO}}\right) \times 100 \quad (8.111)$$

$$S_{C_2-C_4}(\%) = \left(\frac{\text{moles of C}_2 - C_4\text{ produced}}{\text{moles of inlet CO} - \text{moles of outlet CO}}\right) \times 100 \quad (8.112)$$

$$S_{C_{5+}}(\%) = (1 - S_{CO_2} - S_{CH_4} - S_{C_2-C_4}) \times 100 \quad (8.113)$$

In order to guarantee the proposed model's validity in terms of syngas conversion and product selectivity, it is convenient and necessary to respect the same kinetic modeling scheme considered in this case study (Dai et al. 2014). In addition, it is important to mention that the kinetic parameters of a certain FTS kinetic model depend considerably on the operating conditions used (Gideon Botes et al. 2009).

8.3.2.3 Other Parameters and Correlations

Table 8.18 shows the reactor characteristics and fluid properties used for the simulations,[15] while Table 8.19 presents the correlations used to estimate the parameters involved in the proposed model.

8.3.2.4 Numerical Method

The simulations were performed in MATLAB software, in which the finite difference method was applied for the solution of the PDEs system representing the proposed model. The number of nodes used for the discretization of the first partial derivatives in the axial direction was 1200 with a step size of 0.01 m, while for the first and second partial derivatives in the radial direction, 25 nodes with a step size of 0.0005 m were used. The numerical method was presented in a matrix form in which the set of matrices that arise from the discretization of the system of PDEs exhibits a stable value of the matrix norm of the maximum absolute value of the eigenvalues of the matrix, indicating that the matrices are well conditioned, i.e. the numerical method is stable.

Table 8.18 Characteristics of the FTS fixed-bed reactor and fluid properties (Dai, et. al. (2014) / Elsevier).

Parameter	Value	Parameter	Value
Inner tube radius	13 mm	Heat capacity of wáter	$C_{P_W} = 4220\,\text{J}/(\text{Kg}\cdot\text{K})$
Total catalyst loading	30 l	Initial pressure	$P_T = 1.5$ MPa
Catalyst particle diameter	$d_p = 3.20$ mm	Inlet temperature	$T_0 = 493$ K
Bulk density of catalyst bed	$\rho_B = 630$ Kg/m^3	Cooling temperature	$T_W = 486$ K
Bed void fraction	$\varepsilon_B = 0.42$	Synthesis gas flow	$F_{mix,G} = 0.205$ Kg/m^2/s
Gas viscosity	$\mu_{mix,G} = 4.54 \times 10^{-5}$ Kg/(m · s)	Inlet syngas composition (mol %)	63.11 H_2/30.76 CO/ 2.37CH_4/0.62 CO_2/3.14 N_2
Heat capacity of gas mixture	$C_{P_{mix,G}} = 2700\,\text{J}/(\text{Kg}\cdot\text{K})$	Cooling water flow	$F_W = 780$ Kg/h

Table 8.19 Correlations used for estimation of parameters involved in the proposed model.

Parameter	Correlation	Reference
Radial dispersion coefficient	For $\frac{d_p}{d_t} > 0.1 : \frac{\varepsilon_B D_{er}}{u_s d_p} = \frac{1}{m} + \frac{0.38}{Re_f}$ For $\frac{d_p}{d_t} < 0.1 : \frac{\varepsilon_B D_{er}}{u_s d_p\left[1 + 19.4\left(\frac{d_p}{d_t}\right)\right]} = \frac{1}{m} + \frac{0.38}{\mathrm{Re}_f}$ where $m = \begin{cases} \mathrm{Re}_f > 400 & 11 \\ 20 < \mathrm{Re}_f < 400 & 57.85 - 35.36\log(\mathrm{Re}_f) + 6.68(\log(\mathrm{Re}_f)) \end{cases}$	Rase (1990)
Radial effective thermal conductivity	$\lambda_{er} = 0.199 + 0.015\frac{d_t}{d_p} + \frac{0.331}{1 + 8.54\left(\frac{d_t}{d_p}\right)^2}$	Wilhelm (1962)
Convective heat transfer coefficient on the side wall	$h_w = 1000\ \mathrm{W/m^2K}$	Haarlemmer and Bensabath (2016)
Reaction enthalpy	$\Delta H_{FTS} = \Delta H^0_{FTS} + \int_{298^\circ \mathrm{K}}^{T} C_{P,FTS} dT$	Sandler (2006)
Friction factor	$f_p = \frac{172}{\mathrm{Re}_f} + \frac{4.36}{\mathrm{Re}_f^{0.12}};\ \mathrm{Re}_f = \frac{d_p \rho^f u_s}{\mu_{\mathrm{mix},G}(1 - \varepsilon_B)}$	Allen et al. (2013)

8.3.3 Results and Discussion

The results of the simulations with the proposed model were compared with those reported by Dai et al. (2014). They used a one-stage fixed-bed reactor with 12 m of catalytic bed and a two-stage FBR with 6 m of catalytic bed for each stage. The internal diameter of the reactor for both configurations was 13 mm. The experiments were carried out with a $CoMo/\gamma - Al_2O_3$ catalyst at the following reaction conditions: initial temperature (T_0) of 493 K, initial pressure of 1.5 MPa, gas synthesis flow ($F_{mix,G}$) of 0.205 $Kg/m^2/s$, and H_2/CO molar ratio of 2.05. The model proposed by Dai et al. (2014) was developed based on the following assumptions: (i) steady state; (ii) plug-flow regime assumed with no channeling along the bed; (iii) the axial dispersion of mass and heat as well as radial concentration gradients are ignored; (iv) the build-up of wax layer on the catalyst surface is negligible for mass and heat transfer; (v) described by the mass balance of CO and heat balance; (vi) constant surface velocity and gas density; and (vii) the total gas pressure was estimated by the total sum of the partial pressures of each component of the gas mixture.

8.3.3.1 Simulations for the One-Stage Reactor

Figure 8.6 illustrates the results of the simulations carried out with the proposed model and those reported by Dai et al. (2014) for the CO conversion in the one-stage reactor.

It can be seen that both models quite accurately predict the CO conversion at the reactor outlet. The absolute error between experimental and predicted values of CO conversion, calculated as $\frac{(y_{exp} - y_{cal})}{y_{exp}} \times 100$, was 2.7% for Dai et al. (2014) model and 1.8% for the proposed model.

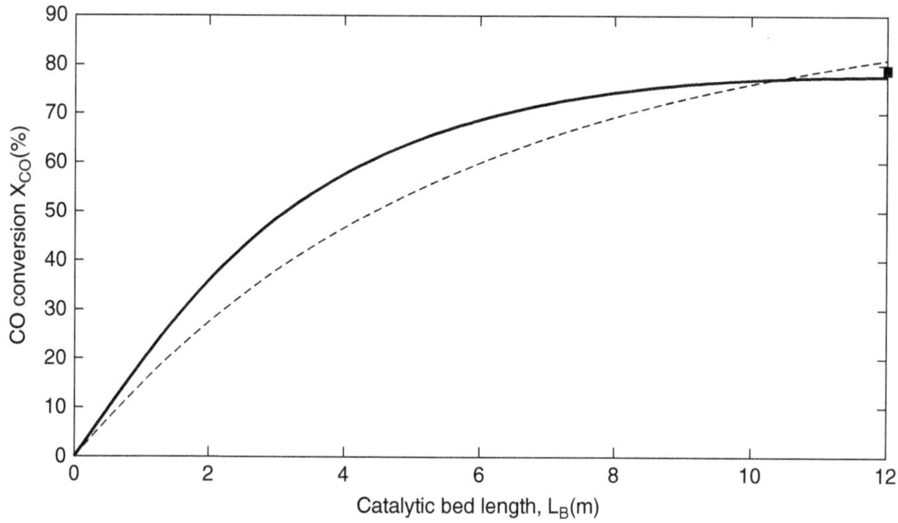

Figure 8.6 CO conversion profile in one-stage fixed-bed FTS reactor. Proposed model (solid line), literature model (Moazami et al. 2015a) (dotted line), experimental data (■) (Dai et al. 2014).

Although the prediction of CO conversion at the reactor outlet is similar, the trend of the simulated CO conversion profiles of both models is different. In majority of the reactors, the proposed model predicts higher values of CO conversion than the literature model. The highest difference is found in the middle of the reactor (~10% CO conversion of difference). It is then anticipated that these dissimilar profiles may cause the prediction of the temperature profile along the axial direction to be different with both reactor models.

In the last third part of the reactor, the proposed model predicts small variations in CO conversion (quasi steady state), while CO conversion with the model presented in the literature continuously increases.

The profiles of CO conversion calculated with the proposed model correlate with those reported by Rafiq et al. (2011), as can be seen in Figure 8.6, who used a similar model with the following assumptions: (i) variable mass and heat radial dispersion; (ii) variable gas density and surface velocity in axial direction; and (iii) variable axial total pressure. It should be noted that the simulations of Rafiq et al. (2011) were carried out in a 2-m long reactor with the use of a spherical unpromoted cobalt catalyst and the following reaction conditions: a wall temperature of 473 K; an initial total pressure of 2 MPa; and a gas hourly space velocity (GHSV) of 37 $NmL/(g_{cat} \cdot h)$.

On the other hand, it is important to mention that the global conversion of hydrogen at the outlet of the reactor in one stage is 89.1% (Figure 8.7). However, neither experimental data nor the simulated value was provided by Dai et al. (2014) for comparison. The hydrogen conversion profile during the FTS reaction has a different tendency than the CO conversion, i.e. higher conversions are achieved at the exit of the reactor (89.1% versus 77.5%). The H_2 conversion is greater than the CO conversion precisely because there is an excessive consumption of hydrogen until it almost reaches its exhaustion, which is due, in principle, to a considerable production of methane and light gases.

When the reaction between H_2 and CO occurs on the surface of the cobalt-based catalyst, it forms methylene groups, (-CH_2-) which, by successively joining these groups, produce long hydrocarbon chains (mainly paraffins). However, it should be noted that each $-CH_2$ formed during the FTS reaction releases a higher average heat than in typical catalytic reactions that are carried out in

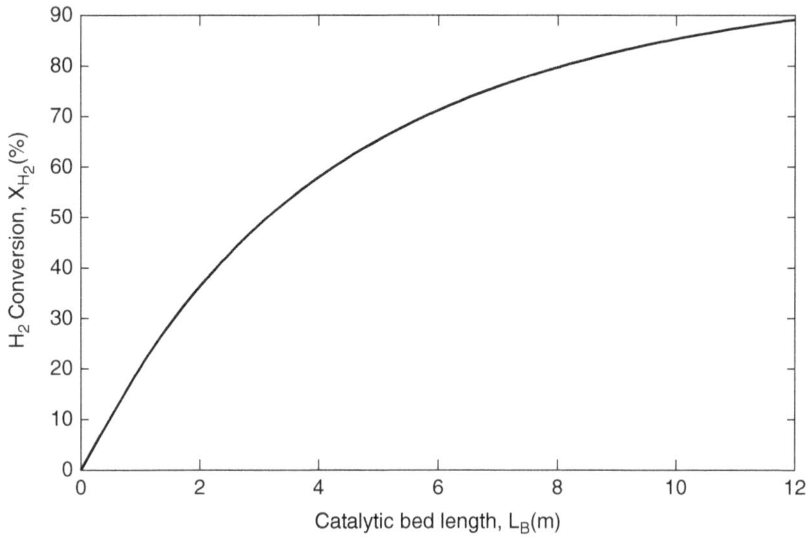

Figure 8.7 H_2 conversion profile in one-stage fixed-bed FTS reactor.

conventional petroleum refining processes (Rafiq et al. 2011). Therefore, it is extremely important to have a safe control reactor temperature due to the high exothermicity of the Fischer–Tropsch reaction, to avoid the runaway phenomenon, and to guarantee a high selectivity and productivity (Rafiq et al. 2011). Thus, the reactor must be modeled in non-isothermal conditions to obtain information about its behavior under this regime and to identify the parameters that have a direct influence on the thermal behavior of the reactor, guaranteeing the control of the temperature.

The literature reports two-dimensional pseudohomogeneous models in which the suitable temperature profile of fixed-bed FTS reactors has been verified; however, that is not the case for unidimensional pseudohomogeneous models (Marvast et al. 2005; Rafiq et al. 20011; Marvast et al. 2005; Lee and Chung 2012; Chabot et al. 2015). It is worth noting that the former model did not take into consideration the variation in the velocity and density of the gas along the axial direction. Assuming they are constant or not, these two hydrodynamic variables may cause differences in the conversion profiles and, hence, in the temperature profile along the reactor.

Figure 8.8 shows the results of the temperature profiles with the proposed model, the model presented in the literature, and experimental data provided by Dai et al. (2014) for one-stage multitubular fixed-bed FTS reactor. The absolute average differences between the experimental and predicted values $\left(\sum_{i=1}^{n}\left|T_i^{exp} - T_i^{cal}\right|\right)$ with the proposed model and the model presented in the literature (Dai et al. 2014) were 4.8% and 4.3%, respectively.

Both models predict a temperature jump (runaway) at the beginning of the reactor, which is slightly lower than the proposed model. This temperature profile behavior is due, in principle, to the gradual increase of the gas mixture velocity in the axial direction of the reactor. This result correlates with Chabot et al. (2015), who studied the effect of the inner diameter of the reactor tube on the performance of the reactor. They confirmed that an increase in the space gas velocity in the axial direction causes an increase in the heat released inside the bed. They also observed that as the diameter decreases, the maximum temperature induced by the heat released by the FTS decreases toward the reference value of the cooling temperature. The temperature profile predicted with the proposed model is like those presented by others with the same type of pseudohomogeneous model in two dimensions (Jess and Kern 2009; Marvast et al. 2005; Rafiq et al. 2011).

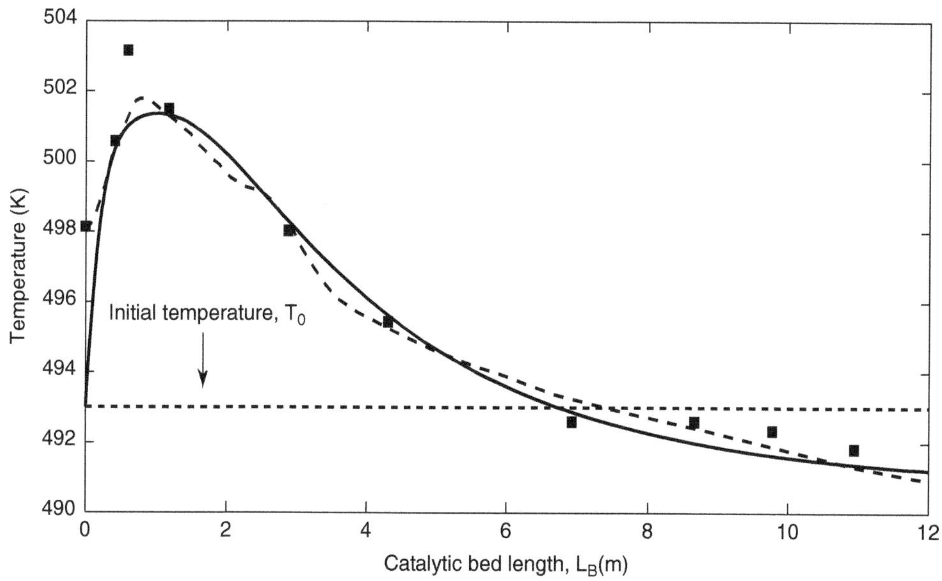

Figure 8.8 Temperature profile in one-stage fixed-bed reactor. Proposed model (solid line), literature model (dotted line) (Moazami et al. 2015a), experimental data (■) (Dai et al. 2014).

It should be noted that the differences in CO conversion profiles shown in Figure 8.6 are due to the high reaction temperatures that are reached at the beginning of the reactor. It should be noted that the model presented in the literature performed simulations with an initial operating temperature of 498 K, 5 degrees higher than that used in our simulations (493 K), which is the correct inlet temperature as indicated by Dai et al. (2014).

Both models predict that the hot spot is slightly lower than the experimental value (1.5-2 K). It has been reported that the presence of a hot spot near the reactor entrance zone is due to two important causes: (i) the composition of syngas in the feed, and (ii) the cooling temperatures that are used (Marvast et al. 2005). The displacement of the hot spot beyond the reactor inlet can be achieved by handling higher cooling temperatures; however, this results in the potential for the reactor to encounter a thermal leakage, leading to the irreversible deactivation of the catalyst (Rafiq et al. 2011).

Figure 8.9 shows the behavior of the total gas pressure in the axial direction of the reactor. The absolute average differences between the experimental values for the proposed model and the literature model $|P_T^{exp} - P_T^{cal}|$ were 0.01 MPa and 0.04 MPa, respectively. Similar to the CO conversion profiles, the total pressure at the reactor outlet is predicted to be more or less equal with both models; however, the variation in the total pressure along the reactor is indeed different since the values calculated with the proposed model were always higher than those found with the model proposed in the literature. This difference is due to the correlation that is used in the proposed model (Allen et al. 2013), which considers the effects of the bed porosity and the Reynolds number of the catalyst particle. Dai et al. (2014) model, however, estimates the pressure drop in the catalyst bed with the sum of the partial pressures of the components of the mixture gas, which may be erroneous since the characteristics of the catalytic bed are neglected.

It is relevant to mention that the correlation proposed by Allen et al. (2013) for the prediction of the pressure drop through the catalytic bed is presumed to be superior to the classical Ergun equation, which is due, in principle, to the fact that the latter tends to over-predict the pressure drop through

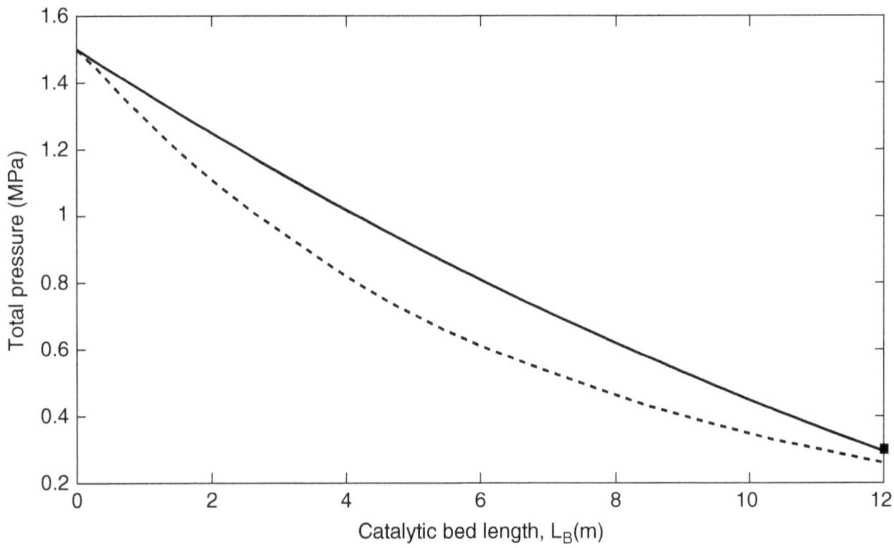

Figure 8.9 Total gas pressure in one-stage fixed-bed reactor. Proposed model (solid line), literature model (dotted line) (Moazami et al. 2015a), experimental data (■) (Dai et al. 2014).

randomly packed or structured beds with smooth spheres at a number >700. In addition, the constant of 1.75 established in the Ergun equation can vary by a factor of up to almost 5, which depends on the alignment and the shape of the obstacle offered by the packing arrangement (Allen et al. 2013). On the other hand, the correlation given by Allen et al. (2013) was derived from an analysis that considers the friction and dragging generated from the tortuosity of a packed bed with smooth or rough spheres and any other geometry such as cylindrical packing of irregular shapes.

Figure 8.10 depicts the gas density profile predicted by the proposed model, in which it can be appreciated that density decreases proportionally to the total gas pressure (Figure 8.9). The

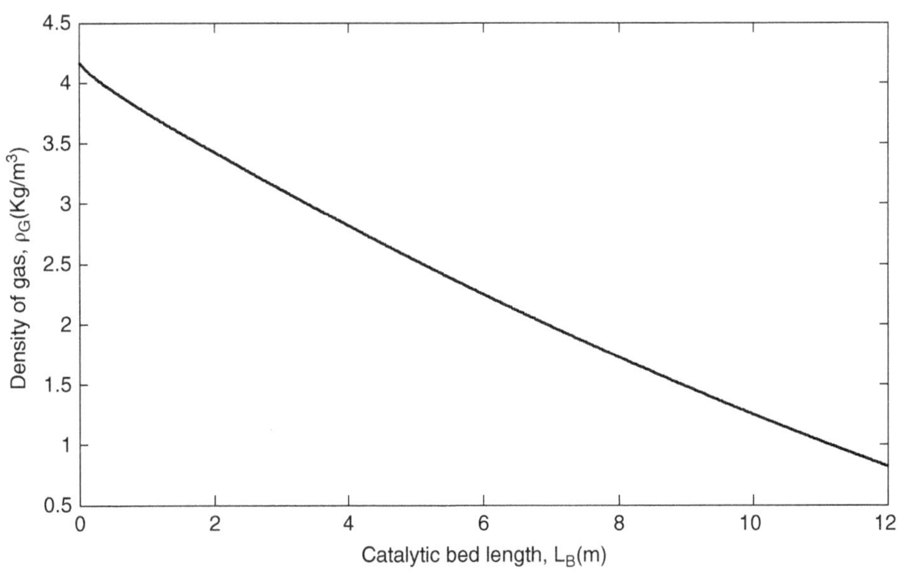

Figure 8.10 Gas density in one-stage fixed-bed reactor according to the proposed model.

reduction in gas density is a result of a decrease in the concentration of the syngas as the FTS reaction proceeds along the axial direction. On the other hand, the calculation of the gas mixture density depends, to a large extent, on the amount of syngas moles and of the light products. As can be seen in Figure 8.10, the gas mixture density tends to decrease because as the syngas is exhausted when the FTS reaction is carried out, a small amount of light gases is generated (CO_2, CH_4, and $C_2 - C_4$), which is denser than the syngas (CO and H_2).

Figure 8.11 shows the gas mixture velocity profile determined using the proposed model. It can be seen that the superficial velocity increases as the FTS reaction occurs. This is because the syngas concentration decreases. On the other hand, it is true that light gases are generated; however, the quantity produced is relatively low, and their appearance is not reflected in a decrease in the superficial velocity of the gas mixture. Another reason for the increase in gas velocity is the decrease in total gas pressure (Rafiq et al. 2011).

Figure 8.12 shows the comparison between the simulation and the experimental value of the product selectivity at the outlet of the one-stage fixed-bed FTS reactor. The light gases produced (CO_2, CH_4, and $C_2 - C_4$) and the generation of liquids with a number of carbon atoms greater than 5 (C_{5+}) are compared. The absolute error between the experimental value given by Dai et al. (2014) and the predicted value acquired using the proposed model of the product selectivity of CH_4, CO_2, $C_2 - C_4$, and C_{5+}, calculated as $\frac{(y_{exp} - y_{cal})}{y_{exp}} \times 100$, was 12.1, 12.1, 20.1, and 0.8 %, respectively. Even though the calculated error for the selectivity of the light gases ($C_2 - C_4$) is ~20%, it can be considered negligible compared with the selectivity of the other products.

It should be noted that a rise in the reaction temperature generates the undesirable production of methane (Eidt et al. 1994). It is known that methane production is undesirable in the FTS process, and a means of inhibiting its production needs to be determined. According to Ranhimpour and Elekaei (2009), if a sufficient amount of hydrogen is added to the FTS process, this can increase the conversion of H_2O to CO_2 by the gas–water shift reaction, which in turn would lead to a decrease in methane production. However, according to Ranhimpour and Elekaei (2009), the decrease in methane production can also be carried out by the two-stage effect. This is because the separation and condensation of water and liquid hydrocarbons that are produced between

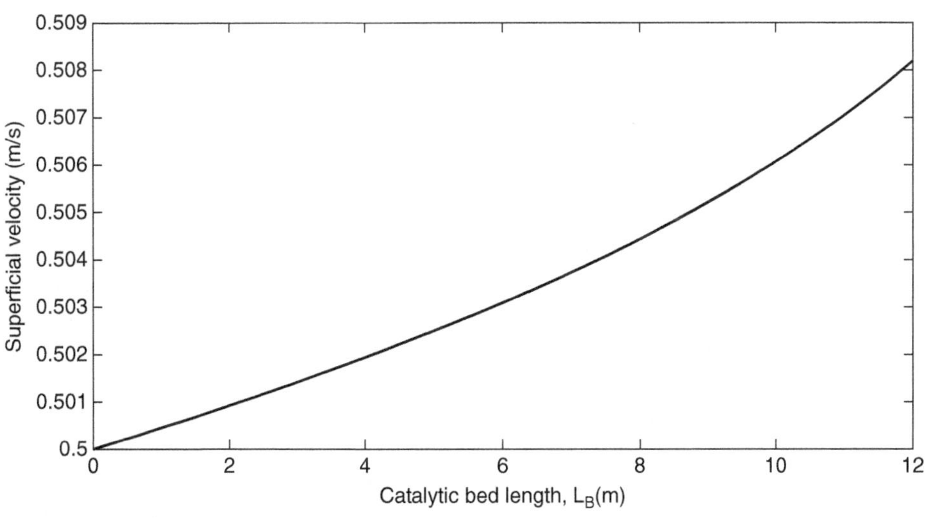

Figure 8.11 Superficial gas velocity in one one-stage fixed-bed reactor according to the proposed model.

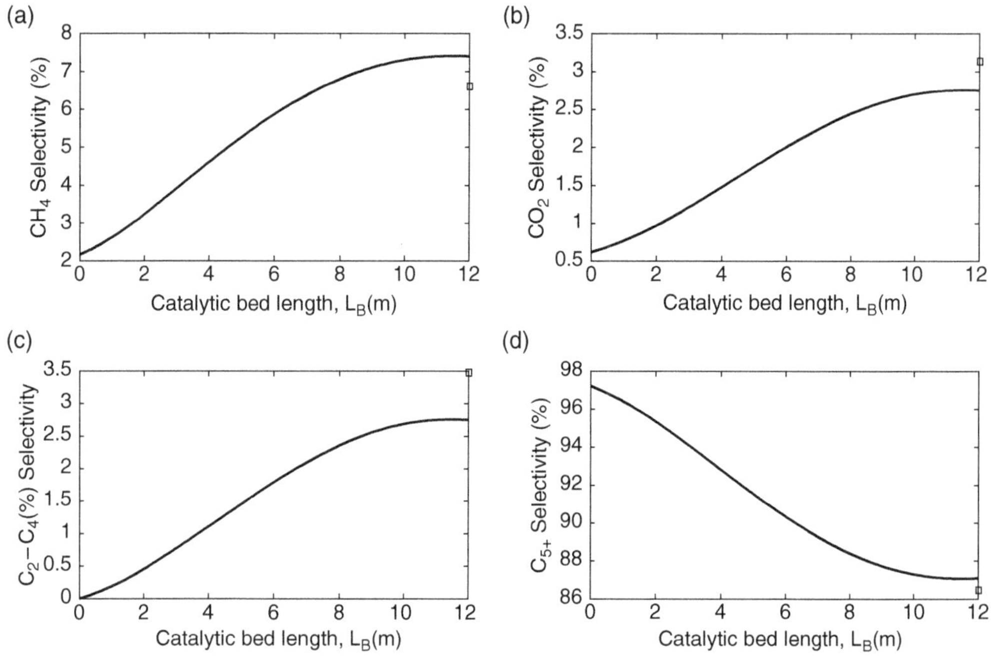

Figure 8.12 Profiles of products selectivities in one stage FTS reactor: (a) CH_4; (b) CO_2; (c) $C_2 - C_4$; (d) C_{5+}. (–) predicted value by proposed model; (□) experimental value by Adapted from Dai et al. 2014.

the two stages of the FTS reactor, and by controlling its temperature, can reduce the generation of methane and, therefore, increase the production of C_{5+} hydrocarbons (Figures 8.12a and 8.12d).

It can be seen in Figure 8.12b that the production of CO_2 is relatively low according to the experimental value given by Dai et al. (2014). However, the value predicted by the proposed model is not far from the experimental value reported, which indicates the efficacy of the model. In addition, low CO_2 production is desirable since it is not a product of interest in the FTS process.

Figure 8.12c shows the light gases selectivity, which is relatively low. This is because a low production of light gases is desired in the use of a multistage reactor, which positively contributes to the production of a significant number of liquid hydrocarbons. However, the low generation of light hydrocarbons also means that the gas mixture density does not increase during the FTS reaction. It can also be seen that the proposed model agrees with the experimental value reported by Dai et al. (2014).

8.3.3.2 Simulations for the Two-Stage Reactor

The predicted and experimental values of CO conversion are presented in Figure 8.13. For the first stage, the absolute error between experimental and predicted values is 7.5% for the proposed model, while it is 6.3% for the model proposed in the literature. For the second stage, the absolute errors are 1.3% and 2.0%, respectively. The proposed model predicts the CO conversion more accurately at the exit of the second stage compared to the model proposed by Dai et al. (2014). In general, the proposed model tends to overestimate the experimental values, while the model from the literature tends to underestimate them.

The H_2 conversion in the two-stage reactor was 71.2% for the first stage and 96.5% for the second stage, and the axial profile is illustrated in Figure 8.14. It can be seen that the achieved

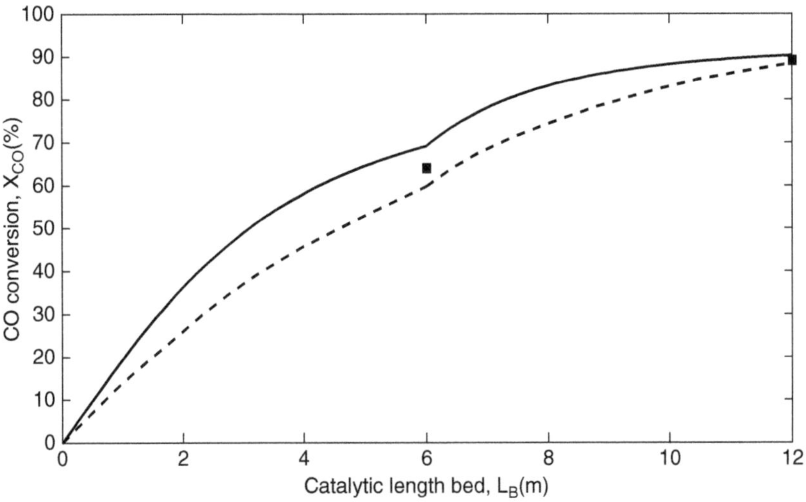

Figure 8.13 CO conversion profile in two-stage fixed-bed FTS reactor. Proposed model (solid line), literature model (dotted line) (Moazami et al. 2015a), experimental data (■) (Dai et al. 2014).

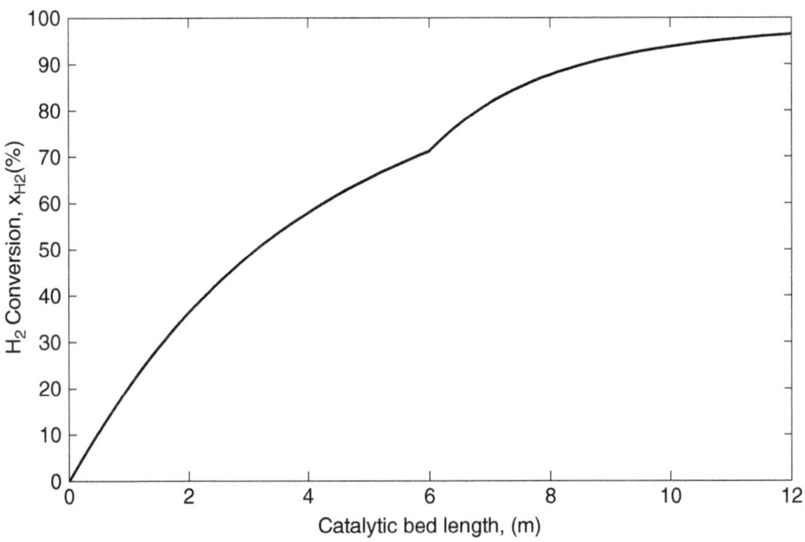

Figure 8.14 H_2 conversion profile for the fixed-bed FTS reactor in two stages according to the proposed model.

H_2 conversions are higher than the CO conversions. This may be because when the surface gas velocity is increased, it produces a low CO conversion but a high H_2 conversion as the residence time is reduced (Rafiq et al. 2011).

Figure 8.15 shows the temperature profile along the axial direction for both stages of the reactor. The absolute error between the experimental values and those predicted by the proposed model were 2.8 K and 2.4 K for the first and second stages, respectively, while for the Dai et al. (2014) model, the errors were 2.0 K and 1.9 K for the first and second stages, respectively. In general,

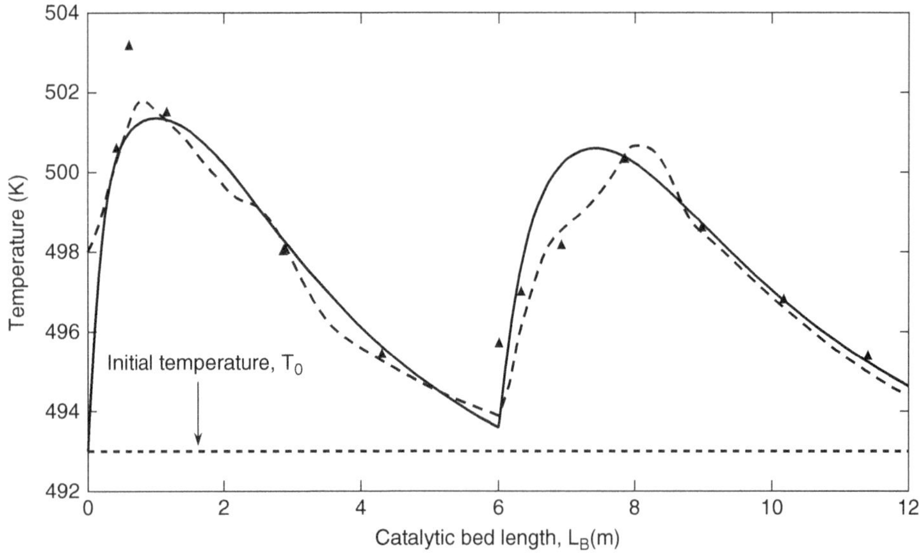

Figure 8.15 Temperature profile in the fixed-bed FTS reactor in two stages. Proposed model (solid line), literature model (dotted line) (Moazami et al. 2015a), experimental data (▲) (Dai et al. 2014).

the error in the prediction of the temperature is negligible, although the Dai et al. (2014) model provides a slightly more accurate prediction than the experimental data. This is because the temperature curve for the two reactor stages shifted to an initial temperature of 498 K. It is important to mention that the proposed model and the Dai et al. (2014) model predict the experimental hot spot with a difference of 2 K and 1 K, respectively.

In recent literature reports (Park et al. 2014; Marvast et al. 2005; Sharma et al. 2011; Rafiq et al. 2011; Chabot et al. 2015), the influence of the variation in the reactor tube diameter on the reactor performance has been analyzed with an aim to reduce the possibility of the formation of high hot spots and temperature runaway. Thus, it is important to optimize the diameter of the reactor tube in order to minimize the heat released by the FTS reaction to eliminate the possibility of a thermal runaway.

It has also been found that the catalytic activity decreases considerably because of the generation of hot spots in the reactor (Chabot et al. 2015). Therefore, temperature control in the reactor is advised to avoid this type of problem; otherwise, the possibility of achieving high yields of liquid products and conversion of syngas is reduced. The proposed model tends to approximate the temperature of the refrigerant (boiling water at high pressure), thus showing its reliability, potential for the prediction of the thermal runaway phenomenon, and application in the design of commercial FTS reactors.

Figure 8.16 shows the results of the comparison of simulation and experimental data for the pressure drop across the catalytic bed for the two-stage reactor. It can be seen that the total pressure at the beginning of the second stage increases from 0.81 to 1.2 MPa with the proposed model, while it increases from 0.6 to 1.2 MPa with the Dai et al. (2014) model. According to Dai et al. (2014), this increase is due to the liquid products and water that are produced and separated by condensation at the end of the first stage until the initial syngas pressure in the second stage of the reactor reaches 1.2 MPa. The increase in the partial pressure of the syngas in the second stage is due to the increase in the syngas concentration. In another report, it was observed that the increase in the syngas

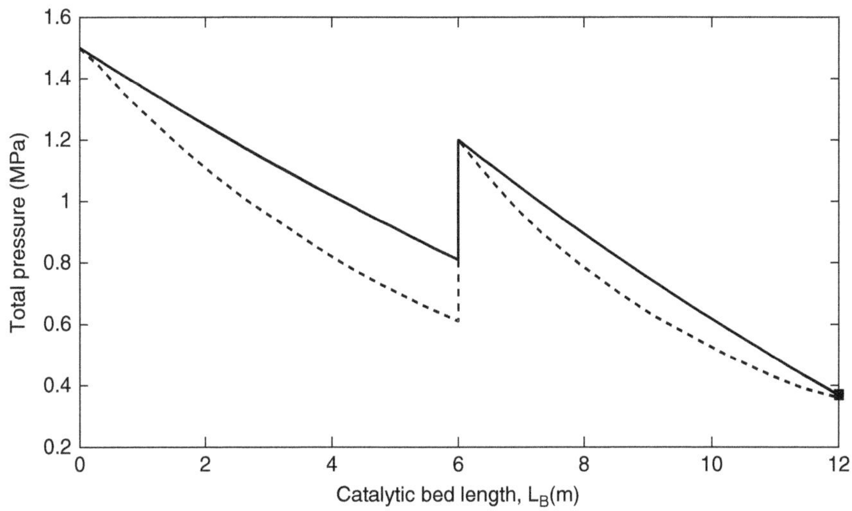

Figure 8.16 Pressure drop profile in the fixed-bed FTS reactor in two stages. Proposed model (solid line), literature model (dotted line) (Moazami et al. 2015a), experimental data (■) (Dai et al. 2014).

pressure contributes favorably to the formation of high-molecular-weight hydrocarbons and the improvement of the FTS reaction rate when using an alumina-supported cobalt-based catalyst (Gideon Botes 2009).

It should be noted that the increase in the operating gas pressure allows higher conversions of CO and H_2 to be obtained. In addition, it has been confirmed in the literature that high pressure and temperature enhance the amount of CO conversion (Rohde et al. 2008; Rahimpour et al. 2011a; Rahimpour et al. 2011b; Wang et al. 2003a). This is the reason why multi-stage fixed-bed configurations improve syngas conversion and the selectivity and productivity of the liquid products.

There is a clear difference in the pressure drop predicted by both models for the two-stage reactor, although the two models accurately predict the experimental value reported at the exit of the reactor. It should be noted that the pressure drop is estimated by Dai et al. (2014) with the sum of the partial pressures of the syngas (CO and H_2), CO_2, and light gases (CH_4 and $C_2 - C_4$) in the axial direction, while in the proposed model it is done by using the correlation recently presented by Allen et al. (2013). This correlation takes into account the packing form of the catalyst particles (including the packing method), since it has been demonstrated that this has a high influence on the estimation of delta-P. This correlation shows the similarity between the calculated pressure drop and the experimental value given by Dai et al. (2014) in both reactors.

Figure 8.17 presents the results of the simulation of the superficial gas velocity according to the proposed model for the two-stage reactor. It is observed that the velocity increases gradually as the FTS reaction occurs along the axial direction in both stages. However, the difference in gas velocity change is not significant in the first stage after 6 m compared with the second stage of the reactor. In general, the increase in the gas velocity observed in the case of both reactor stages allows the reactor temperature to drop so rapidly to approximate the cooling water temperature due to the gradual decrease in the retention time of the reactants (CO and H_2). Such a behavior has been observed by Rafiq et al. (2011), who pointed out that a considerable increase in the synthesis gas velocity would lead to a significant decrease in the conversion of the reactants. The proposed model predicts small velocity gradient.

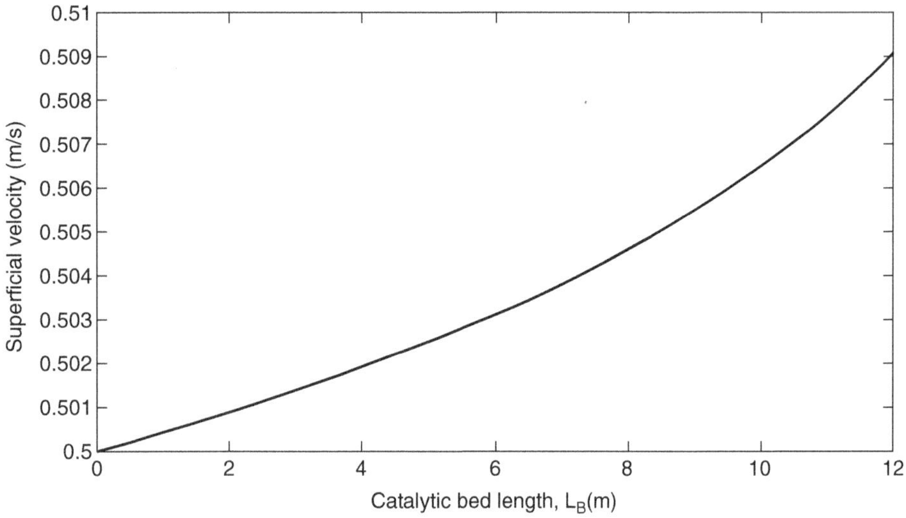

Figure 8.17 Superficial velocity profile in the fixed-bed FTS reactor in two stages according to the proposed model.

On the other hand, Sharma et al. (2011) presented an analysis of the effects of gas superficial velocity on the effective radial thermal conductivity and on the global heat transfer coefficient. In this study, it was observed that an increase in gas velocity produces a significant change in these two thermal parameters, which increases both the released heat of reaction and the temperature of operation, which are beneficial due to the high conversions of CO and H_2 can be achieved.

All these observed behaviors show the importance of considering the effects of the superficial gas velocity for modeling purposes in a single and multi-stage fixed-bed FTS reactor.

In addition, due to the considerable decrease in the pressure drop, the density of the gaseous mixture diminishes proportionally, as can be seen in Figure 8.18. This is due to the decrease in

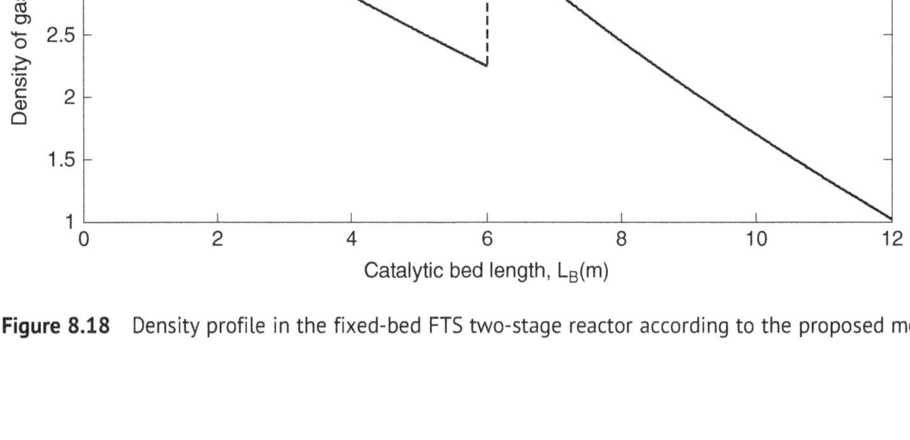

Figure 8.18 Density profile in the fixed-bed FTS two-stage reactor according to the proposed model.

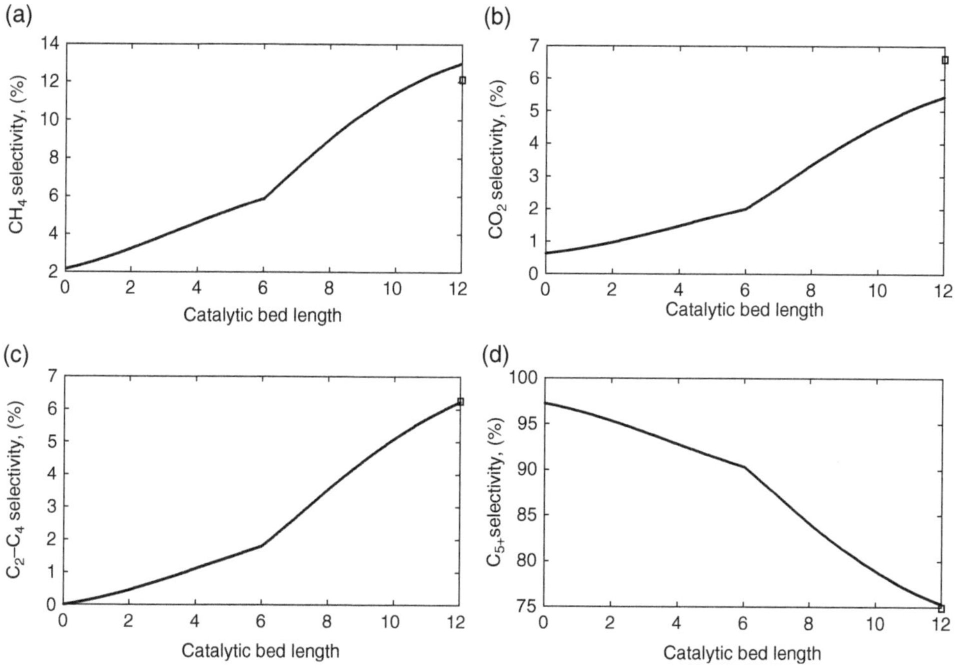

Figure 8.19 Profiles of product selectivities in the two-stage fixed-bed FTS reactor according to the proposed model. (-) predicted value by proposed model; (□) experimental value by Dai et al. (2014). a) CH_4 selectivity, b) CO_2 selectivity, c) $C_2 - C_4$ selectivity, d) C_{5+} selectivity.

the amount of the denser gases ($C_2 - C_4$) as the production of liquid hydrocarbons increases, as reported by Dai et al. (2014).

Figure 8.19 shows the comparison between the simulations and experimental values for the selectivity toward light gases (CO_2, CH_4, and $C_2 - C_4$) and the generation of liquids with a number of carbon atoms greater than 5 (C_{5+}) at the outlet of the two-stage fixed-bed FTS reactor. In general, it can be seen that the production of methane, carbon dioxide, and light gases increases during the operation of the second stage of the FTS reactor. However, the production of C_{5+} hydrocarbons decreases as a result of the effect in multistage operation. The increase observed in gaseous products is also a result of the increase in the pressure system and hence the concentration of syngas. This is because of the condensation of water and the liquid hydrocarbons produced between the two stages of the fixed-bed FTS reactor.

Figures 8.20a and 8.20b show the variations of CO (a) and H_2 (b) conversion, respectively, as a function of the axial and radial directions of the reactor. It can be observed that there is a slight change in the syngas conversion in the radial direction, which could be negligible. The same can be stated for the radial variation of the gas mixture surface velocity, pressure drop, and gas mixture density. However, in the case of radial variation of the reactor temperature, the changes are more significant, which is of vital importance in modeling studies of fixed-bed reactors for FTS.

Figure 8.21 shows profiles of some process variables in axial and radial positions. For instance, from Figure 8.21d, it can be seen that as the tube diameter increases, the temperature of the reactor decreases until the inlet temperature is reached. This phenomenon has already been observed by others who studied the influence of the variation in the reactor tube diameter on the reactor performance (Marvast et al. 2005; Park et al. 2014; Sharma et al. 2011; Rafiq et al. 2011; Chabot et al. 2015).

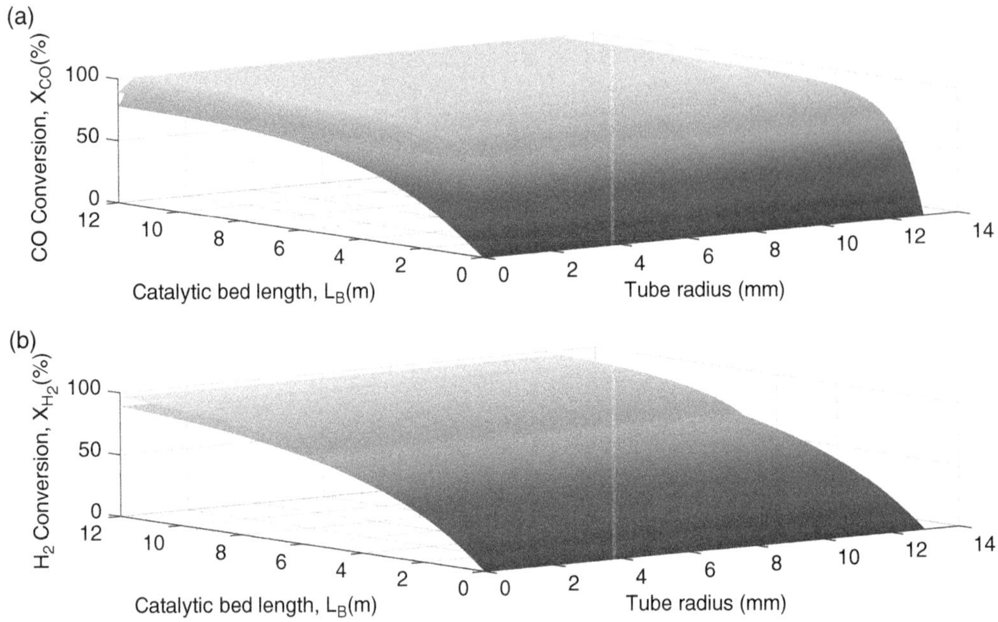

Figure 8.20 Syngas conversion profile in fixed-bed two-stage reactor: (a) CO, (b) H_2.

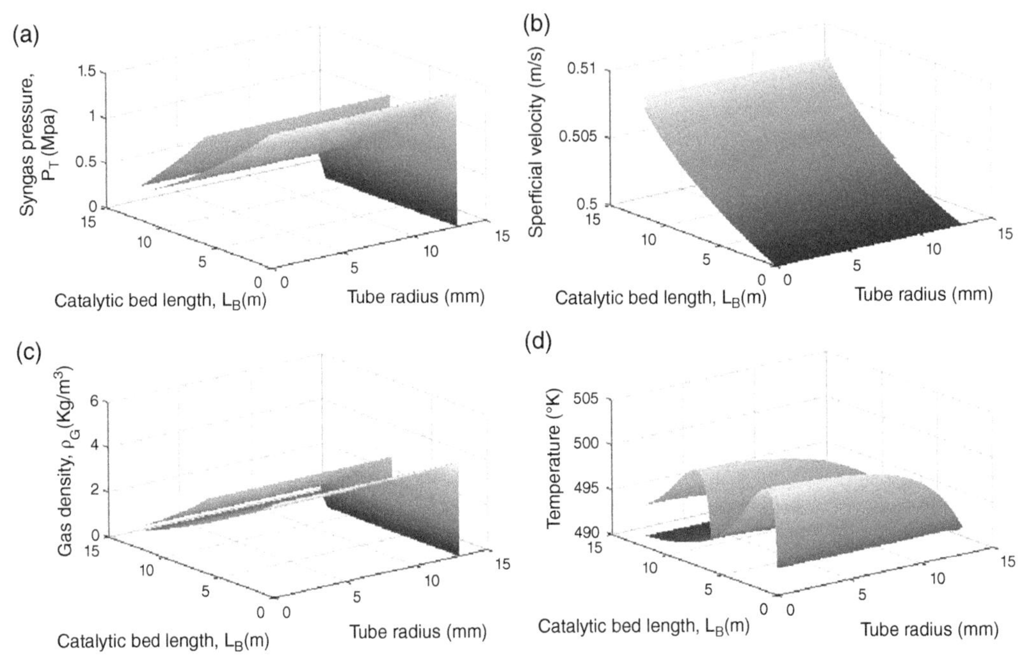

Figure 8.21 (a) Syngas pressure; (b) superficial velocity; (c) gas density; (d) temperature profiles in the fixed-bed FTS two-stage reactor according to the proposed model.

8.3.4 Final Remarks

A more robust model was developed for the simulation of a fixed-bed reactor in one and two stages for the conversion of CO by the FTS. The FTS was used as a study case to demonstrate the importance of considering hydrodynamics in the model formulation that aims for the design, simulation, and optimization of commercial reactors. The developed two-dimensional pseudohomogeneous model was more accurate than the models proposed in the literature for describing the behavior of syngas conversion, product selectivity, and temperature profile in the axial and radial directions. More precise correlations for the diffusivity and thermal conductivity, as well as an equation for the friction factor that allows for a more accurate prediction of the gas pressure in the system, were incorporated in the model. It was proven that the variation of superficial velocity, density, and total pressure of gas allows for a better understanding of the effects of these variables on reactor performance in terms of syngas conversion and temperature profiles. By model simulations, it was confirmed that the FTS can be improved by using multistage fixed-bed reactors since they enhance syngas conversion, product selectivity, and temperature profile.

8.4 Dynamic One-Dimensional Pseudohomogeneous Model for Fischer–Tropsch Reactors

8.4.1 Introduction

A dynamic one-dimensional pseudohomogeneous fixed-bed FTS reactor model in a non-isothermal regime is developed. The proposed dynamic model is validated with experimental data obtained from the literature and is compared with simulations of a steady-state one-dimensional isothermal fixed-bed FTS reactor model recently reported. The best set of initial operating conditions showing highest syngas conversions (CO and H_2) and high heavy hydrocarbons (C_{5+}) selectivities correspond to temperature, pressure, and GHSV of 503 and 518 K, 1.0 and 1.5 MPa, and 1800 (NmL/(gcat·h)), respectively. The proposed model can accurately predict the CO conversion experimental data and product's selectivity. Furthermore, it allows us to evaluate the thermal behavior of the reactor, as well as to analyze the hot spot generation typically observed in highly exothermic reactions such as in the case of the FTS reaction.

8.4.2 Formulation of the Model

8.4.2.1 Model Equations and Solution

The proposed model to describe the dynamic behavior of the fixed-bed FTS reactor was implemented under plug-flow regime at non-isothermal conditions, taking into account the variations of the superficial gas velocity, the gas mixture density, and total pressure of the system in the axial direction, which have been demonstrated to be important parameters to improve predictions of syngas conversion, product's selectivity, and temperature profiles (Méndez et al. 2019).

The proposed dynamic model consists of the following set of equations that describe the behavior of the fixed-bed FTS reactor:

$$\frac{\partial C_i}{\partial t} = -u_s\frac{\partial C_i}{\partial z} + \rho_B\beta\sum_{j=1}^{RN} v_{ij}R_j \tag{8.114}$$

$$\frac{\partial T}{\partial t} = -u_s\frac{\partial T}{\partial z} + \frac{\rho_B\beta}{\rho_{mix,G}Cp_{mix,G}}\sum_{j=1}^{RN}(-\Delta H)_j v_{ij}R_j - \frac{4U_r}{d_t\rho_{mix,G}Cp_{mix,G}}(T-T_w) \tag{8.115}$$

$$\frac{\partial u_s}{\partial z} = -\frac{u_s}{\rho_{mix,G}}\frac{\partial \rho_{mix,G}}{\partial z} \tag{8.116}$$

$$\frac{\partial \rho_{mix,G}}{\partial z} = \frac{M_{mix,G}}{R_c}\left(\frac{1}{T}\frac{\partial P_T}{\partial z} - \frac{P_T}{T^2}\frac{\partial T}{\partial z}\right) \tag{8.117}$$

$$\frac{\partial P_T}{\partial z} = -f_p\frac{u_s^2 \rho_{mix,G}}{d_p} \tag{8.118}$$

subject to the boundary conditions:

$$at\ t = 0, \forall z: C_i = C_{i,0}, T = T_0, u_s = u_{s,0}, \rho_{mix,G} = \rho_{mix,G,0}, P_T = P_{T,0} \tag{8.119}$$

$$at\ z = 0, \forall t: C_i = C_{i,0}, T = T_0, u_s = u_{s,0}, \rho_{mix,G} = \rho_{mix,G,0}, P_T = P_{T,0} \tag{8.120}$$

The proposed model was solved using the finite difference and the Runge–Kutta method using the ode23 routine of MATLAB software. Twenty-five nodes were employed to approximate the first derivatives in the axial direction, and an appropriate step size was also employed to reduce the computational time and cost derived from the dynamic model simulations.

8.4.2.2 Model Parameters, Correlations, and Kinetics

The parameters and correlations used in the proposed model are given in Table 8.20. The global heat transfer coefficient was estimated using the correlation given by Haarlemmer and Bensabath (2016). This parameter is given as a function of the convective heat transfer coefficient on the internal side reactor (h_{int}), the convective heat transfer coefficient of the external side reactor (h_{ext}), and the reactor thermal conductivity (λ_w), which in this case is of stainless-steel material. The pressure drop along the catalytic bed was estimated using a correlation recently reported in the literature for the friction factor. Such a correlation was validated by comparing it with many correlations reported in the literature showing superior accuracy (Allen et al. 2013). The heat conductivity of the steel wall reactor was assumed to be 60 $Wm^{-1}h^{-1}$ (Sharma et al. 2011), and the value of the heat transfer coefficient (film) of the external side corresponding to boiling water coolant was 1000 $Wm^{-1}h^{-1}$ (Haarlemmer and Bensabath 2016).

Table 8.20 Parameters and correlations of the dynamic one-dimensional pseudohomogeneous fixed-bed FTS reactor model.

Parameter	
$\mu_{mix,G} = \sum_{i=1}^{n} y_i \mu_i$	$f_p = \frac{172}{Re} + \frac{4.36}{Re^{0.12}}$
$\mu_i = a + bT + cT^2 + dT^3$	$\varepsilon_B = 0.40;\ \phi = 4/3$
$M_{mix,G} = \sum_{i=1}^{n} y_i \mu_i$	$U_r = \left(\frac{1}{h_{ext}} + \frac{\delta_w}{\lambda_w} + \frac{1}{h_{int}}\right)^{-1}$
$\beta = 0.255$	$h_{int} = \rho_{mix,G}\left(C_{P_{mix,G}}\mu_{mix,G}\right)^{-4.074}\left(\frac{d_p}{\mu_{mix,G}}\right)^{-4.407}\left(\frac{0.458}{\varepsilon_B}\right)$
$R_e = \frac{d_p \rho_{mix,G} u_s}{\mu_{mix,G}}$	$C_{P_i} = a + bT + cT^2 + dT^3$ $C_{P_{mix,G}} = \sum_{i=1}^{SN} y_i C_{P_i}$

Other parameters were considered as constants, such as the hollow fraction of the bed and the active side volume fraction of the catalyst. There are several studies on the kinetics of the FTS reactions focused on the formation of hydrocarbons. Most of the developed kinetic models are of the Langmuir–Hinshelwood–Hougen–Watson (LHHW) type; however, some authors have developed simpler kinetic models based on the power-law type (Marvast et al. 2005; Yang et al. 2003; Wang et al. 2003b), with which good predictions and fit to experimental data have been reported. The kinetic model used in the proposed reactor model is of the following power-law type:

$$R_j = k_j p_{CO}^{m_j} p_{H_2}^{n_j} = A_j \, exp\left(-\frac{E_{a,j}}{RT}\right) p_{CO}^{m_j} p_{H_2}^{n_j} \tag{8.121}$$

The set of FTS reactions considered was:

$$R_1 \; CO + 3H_2 \rightarrow CH4 + H_2O \tag{8.122}$$

$$R_2 \; 2CO + 4H_2 \rightarrow C_2H_4 + 2H_2O \tag{8.123}$$

$$R_3 \; 2CO + 5H_2 \rightarrow C_2H_6 + 2H_2O \tag{8.124}$$

$$R_4 \; 3CO + 7H_2 \rightarrow C_3H_8 + 3H_2O \tag{8.125}$$

$$R_5 \; 4CO + 9H_2 \rightarrow n - C_4H_{10} + 4H_2O \tag{8.126}$$

$$R_6 \; 4CO + 9H_2 \rightarrow i - C_4H_{10} + 4H_2O \tag{8.127}$$

$$R_7 \; 6.05\mathrm{CO} + 12.23H_2 \rightarrow C_{6.05}H_{12.36}(C_{5+}) + 6.05H_2O \tag{8.128}$$

$$R_8 \; CO + H_2O \rightarrow CO_2 + H_2 \tag{8.129}$$

The product's selectivity was calculated using the Eqs. (8.110)–(8.113).

The values of the pre-exponential factors, activation energies, and reaction orders for carbon monoxide and hydrogen were reported by Moazami et al. (2015a) and are summarized in Table 8.21.

8.4.3 Results and Discussion

8.4.3.1 Experimental Data

The simulations with the proposed model are compared with experimental data reported by Moazami et al. (2015a). They used a bench-scale fixed-bed reactor with a length of 52.83 cm, a catalytic bed of 11 cm length, and an internal diameter of 15.7 mm. The experiments were carried out using a Co/SiO_2 catalyst, which was diluted with 2 g of silicon carbide at a mass ratio of 1:12 ($Co/SiO_2/SiC$). A simulated nitrogen-rich syngas was used with a concentration of 17, 33, and 50 vol % of CO, H_2, and N_2 respectively.

8.4.3.2 Conversion of CO and H_2

Table 8.22 reports the operating conditions of 6 experimental cases used for the simulations of the proposed reactor model. Two temperatures (503 and 528 K), four pressures (1.0, 1.5, 2.0, and 2.5 MPa), and four GHSVs (1800, 2400, 3000, and 3600 $\mathrm{NmLg_{cat}^{-1}h^{-1}}$) were used for the validation of the model. Figures 8.22 and 8.23 show the simulated CO conversion profiles for 6 cases as a function of time at the exit of the reactor and along the catalytic fixed-bed at the steady-state conditions, respectively. Figure 8.22 shows the dynamic profile CO conversion for the 6 experimental cases considered. For cases I, II, V, and VI, the conversions are practically 100%. For cases V and VI, the CO conversion reaches the steady state about half the time used for the simulations (180 seg).

Table 8.21 Experimental data of the kinetic parameters of the FTS reaction reported in the literature (Adapted from Moazami, et. al. (2015a)).

Reaction	R_1		R_2		R_3		R_4		R_5		R_6		R_7		R_8	
Kinetic parameters	A_1	E_{a1}	A_2	E_{a2}	A_3	E_{a3}	A_4	E_{a4}	A_5	E_{a5}	A_6	E_{a6}	A_7	E_{a7}	A_8	E_{a8}
	65.56165	83523.9	0.045438	65017	0.001144	49782	1.19E−06	34885.5	1.84E−10	27728.9	6.59E−09	25730.1	1.98E−08	23564.3	3.78E−05	58826.3
Reaction order	m_1	n_1	m_2	n_2	m_3	n_3	m_4	n_4	m_5	n_5	m_6	n_6	m_7	n_7	m_8	n_8
Case 1	0.6095	−0.4679	0.3733	−0.2613	−0.0283	0.1913	0.4354	0.0876	0.0353	1.1163	−0.2515	0.0092	0.8187	0.0342	−0.362	1.2639
Case 2	1.295	−1.0527	−0.4637	0.4007	1.2866	−1.0054	0.6653	−0.0195	0.1816	0.8522	0.1657	0.7574	1.3976	−0.5635	1.4825	−0.6426
Case 3	0.6142	−0.3101	−0.6284	0.2391	1.2625	−0.9602	0.9859	−0.3488	0.2189	0.8523	0.336	0.5542	1.2153	−0.3777	1.4034	0.6075
Case 4	−0.4755	0.8201	0.014	0.0746	−0.0594	0.4128	−0.7131	1.4167	0.1037	−0.4789	−0.2626	1.1713	1.2895	0.4571	0.472	0.7706
Case 5	0.4831	−0.2516	−1.2039	1.395	−0.1382	0.4795	0.1607	0.5001	−0.0166	1.0945	0.3423	0.5573	−0.5105	1.4353	−0.4506	1.4841
Case 6	0.9771	−0.8641	−0.1176	−1.1396	−0.2278	0.5233	1.1815	−0.6838	0.278	0.8884	−0.9539	0.1122	1.0146	−0.1848	−0.4465	1.465

Table 8.22 Experimental conditions for fixed-bed FTS reactor (Adapted from Moazami, et. al. 2015a).

Case	Temperature (K)	Pressure (Mpa)	GHSV (NmL/g_{cat}/h)
I	503	1.0	1800
II	503	1.5	2400
III	503	2.0	3000
IV	503	2.5	3600
V	518	1.0	2400
VI	518	1.5	1800

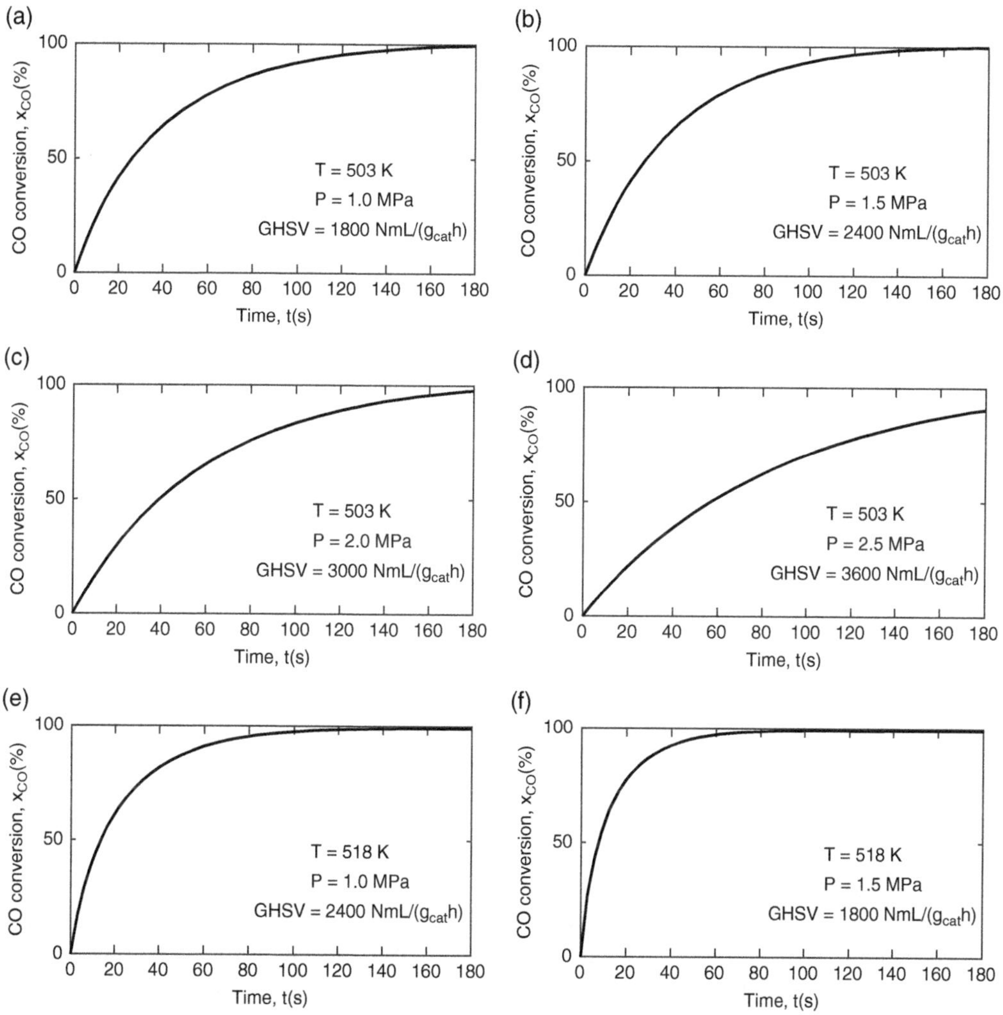

Figure 8.22 Simulated dynamic CO conversion profiles at the reactor outlet. (a) case I; (b) case II; (c) case III; (d) case IV; (e) case V; (f) case VI.

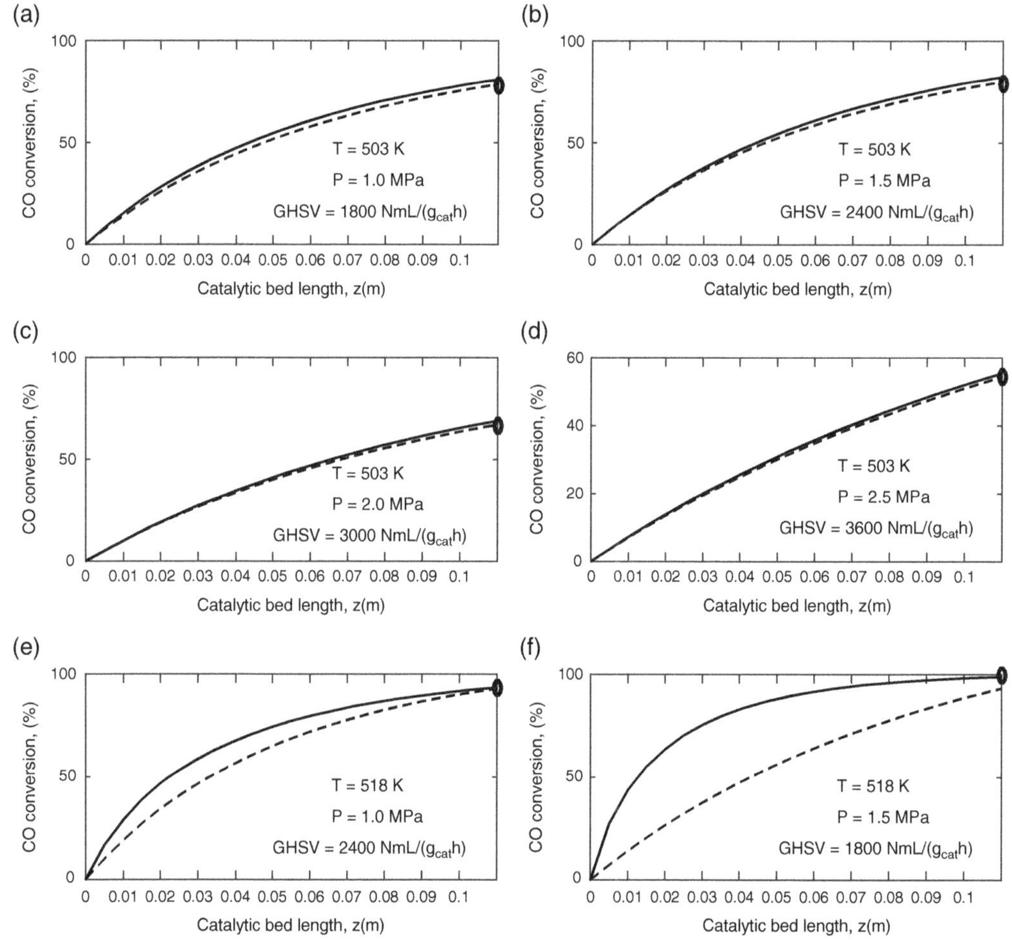

Figure 8.23 Simulated CO conversion profiles in axial position. (○) experimental data; (-) proposed model; (--) literature model [15]; (a) case I; (b) case II; (c) case III; (d) case IV; (e) case V; (f) case VI.

For cases III and IV, the CO conversion is close to 95%. In general, the results of the CO conversion dynamic profile indicate that for relatively short times the conversions are high. The differences in CO conversion shown in all the experimental cases under study are mainly due to changes in total pressure and GHSV. The difference in CO conversion is mostly seen in cases III and IV. In case III, a decrease of about 98% is observed in comparison to the first two experimental cases. This is due to the effect of increasing the gas space velocity. This is expected because an increase in the gas space velocity translates into a decrease in residence time, which results in a decrease in CO conversion (Kwack et al. 2011a). However, the CO conversion in cases V and VI increases considerably until reaching a maximum conversion as the simulation time is extended. Moreover, it is important to note that an increase in total pressure corresponds to a slight increase in CO conversion as reported in the literature (van Berge and Everson 1997; Yan et al. 2009). In addition, the effect of a temperature increase of 503 to 518 K also leads to an improvement in the synthesis gas conversion in terms of an appreciable increase (Dry 1981).

Figure 8.23 shows the simulated CO conversion profiles along the catalytic bed within the reactor at steady-state conditions, their comparison with experimental data, and the results obtained with another model reported in the literature (Moazami et al. 2015a). It is important to mention that the experimental data given for the syngas conversion (CO and H_2) are only reported at the reactor outlet.

The simulated CO conversion profiles with the proposed model and with the literature model (Moazami et al. 2015a) are quite similar for the first 4 cases (Figures 8.23a-8.23d). However, for cases V and VI (Figures 8.23e and 8.23f), the profiles are different, particularly for case VI.

The lower CO conversions predicted by the literature model (Moazami et al. 2015a) are due to the isothermal conditions used by the authors, while in the proposed model the non-isothermal conditions are considered.

The highest difference in predictions with both models is found at the reactor inlet, where the catalytic activity is greater than in any other part of the reactor (van Berge and Everson 1997). On the other hand, the CO conversion is higher for case II; however, for cases III and IV, it is even lower. This is due to the combined effect of the simultaneous changes in the total pressure and GHSV. It has been confirmed experimentally that the increase in total pressure generates an increase in the syngas conversion (van Berge and Everson 1997; Yan et al. 2009), but if the gas space velocity is increased, this causes a decrease in the residence time and decrease in the CO conversion (Kwack et al. 2011a). However, it is important to note that care should be taken with the simultaneous changes in total pressure and gas space velocity because high pressures generate a shift of the FTS process toward the production of heavy hydrocarbons (Dry 1981). Increasing the gas space velocity can cause a slight decrease in CO conversion, but if adequate values are established, this allows for high production rates without altering the quality of the product (Yan et al. 2009).

The errors between the experimental and predicted values at the reactor outlet were calculated with Eq. (8.130). Table 8.23 summarizes the calculated errors for the proposed model and the literature model (Moazami et al. 2015a) for the conversion of CO, H_2, and product's selectivity (CO_2, CH_4, light gases: C_2–C_4, and heavy hydrocarbons: C_{5+}) for the 6 cases studied. In general, lower errors are obtained with the proposed model.

$$error\ (\%) = \left| \frac{y_{i,exp} - y_{i,calc}}{y_{i,exp}} \right| \times 100; \text{where } i = \text{specie} \tag{8.130}$$

Figures 8.24 and 8.25 illustrate the hydrogen conversion profiles as a function of time at the reactor outlet and along the catalytic bed length at steady-state conditions for the 6 cases given in Table 8.22. Both figures present the comparison between the simulations with the proposed and literature models and the experimental data given only at the reactor outlet.

Only for case V (Figure 8.24e), the hydrogen conversion reaches the steady state at 180 seconds. For the rest of the experimental cases, longer time is required, but at 180 seconds the hydrogen conversions are relatively high (above 90%), which is beneficial for the FTS process since the formation of long-chain liquid hydrocarbons is enhanced (Dry 1981). The hydrogen conversion profile along the catalytic bed length of the FTS reactor (axial direction) is illustrated in Figure 8.25. In general, the proposed dynamic model predicts a profile similar to the literature steady-state model.

For case VI (Figure 8.25f), the literature model (Moazami et al. 2015a) tends to predict higher H_2 conversion. It should be noted that no experimental data on hydrogen conversion are available at the reactor outlet, so the accuracy with respect to hydrogen conversion cannot be verified.

Table 8.23 Errors between experimental and predicted values with the proposed model (PM) and the literature model (LM) (Adapted from Moazami, et. al. 2015a).

	Case I		Case II		Case III	
Species	**% error (PM)**	**% error (LM)**	**% error (PM)**	**% error (LM)**	**% error (PM)**	**% error (LM)**
CO	1.9	0.87	1.2	0.92	1.6	0.83
CO_2	0.09	0.11	0.42	0.20	0.21	0.15
CH_4	0.62	0.59	0.56	0.48	0.33	0.47
C_2	0.88	0.69	0.57	0.48	0.35	0.33
C_3H_8	0.23	0.40	0.17	0.21	0.11	0.07
C_4	0.41	0.33	0.09	0.03	0.08	0.04
C_{5+}	0.08	0.06	0.17	0.12	0.09	0.12
	Case IV		**Case V**		**Case VI**	
Species	**% error (PM)**	**% error (LM)**	**% error (PM)**	**% error (LM)**	**% error (PM)**	**% error (LM)**
CO	1.9	0.5	0.05	0.01	0.30	6.0
CO_2	0.05	0	0.04	0	0.23	0.07
CH_4	0.09	0	0.06	0	0.09	0
C_2	1.2	0	0.12	0.04	0.04	0.09
C_3H_8	1.0	0	0.10	0.06	0.03	0
C_4	1.4	0	0.09	0.11	026	0.16
C_{5+}	0.08	0.01	0.21	0.01	0.07	0.02

8.4.3.3 Temperature Profiles

Figures 8.26 and 8.27 illustrate the dynamic and axial temperature profiles at the reactor outlet as a function of time and along the axial direction of the reactor, respectively. No comparison is made with the literature model (Moazami et al. 2015a) since it simulates the FTS reactor in isothermal mode. It is relevant to note that important variations in temperature profiles are observed, thus considering the reactor as isothermal is totally inadequate, which is due to the highly exothermic reactions that occur during FTS. In addition, using fixed-bed FTS reactor under isothermal conditions is: not feasible to be implemented in scales of greater extension (semi-industrial and industrial).

Figures 8.26a, 8.26e, 8.26f and 8.27a, 8.27e, 8.27f show the temperature reactor profiles for experiments I, V, and VI. In these cases, the temperature quickly drops from its initial value to the cooling jacket temperature value. It should be noted that the boiling point of water (cooling agent) was established at a constant value of 498 K for the 6 cases. The temperature was established at this value since the range suggested experimentally is between 5 and 10 K below the initial temperature (Dai et al. 2014) to avoid as much as possible the hot spots formation (Rahimpour and Elekaei 2009). In addition, the use of high temperatures in fixed-bed reactors where a highly exothermic reaction is carried out, such as the FTS, is highly prone to fall into a thermal runaway problem (Kaskes et al. 2016; Yan et al. 2009; Rafiq et al. 2011). Due to this, it is advisable to have good temperature control of the FTS reactor, since poor control of it can cause serious problems in the FTS process, such as catalytic deactivation and poor production of desired products (Lee and Chung 2012). It has been demonstrated that the temperature directly affects the syngas conversion and, therefore, the product's distribution (effects on the hydrocarbon chain length) (Park et al. 2014).

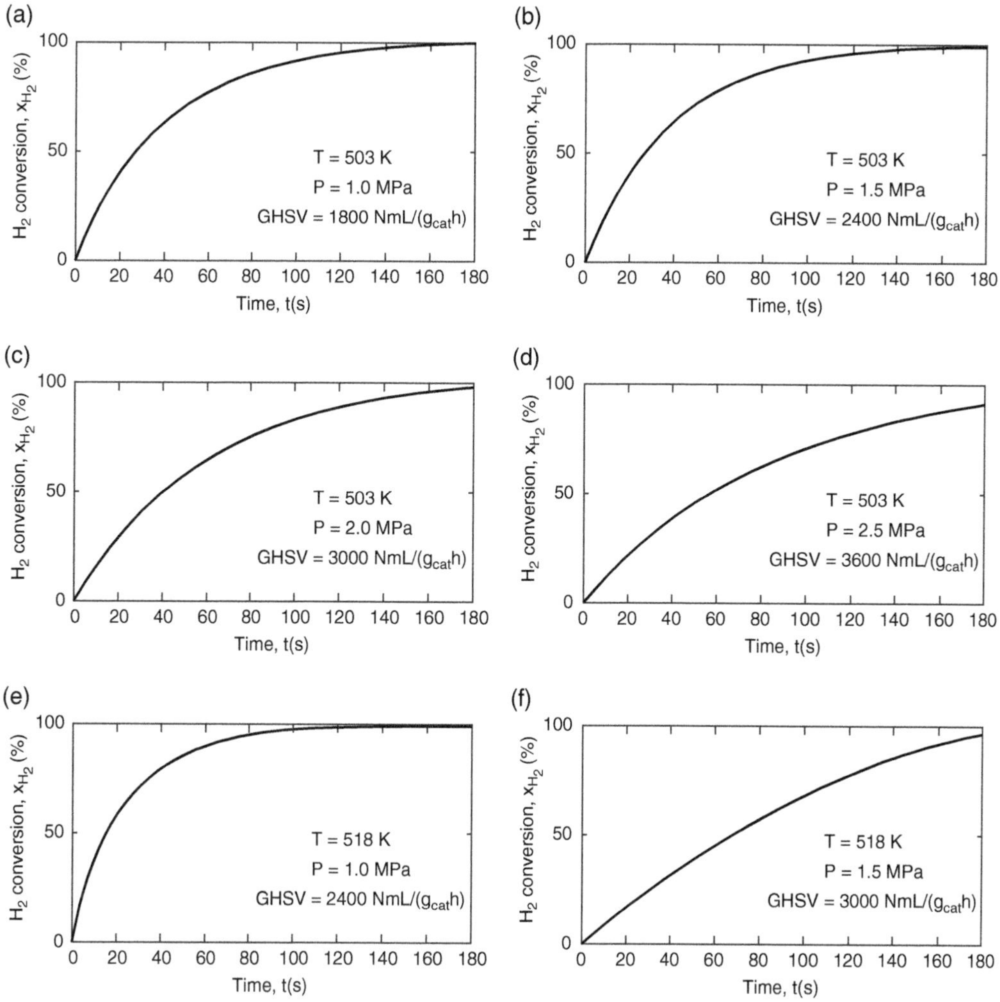

Figure 8.24 Simulated dynamic H_2 conversion profiles at the reactor outlet. (a) case I; (b) case II; (c) case III; (d) case IV; (e) case V; (f) case VI.

On the other hand, if the heat released in a chemical FTS reaction can be adequately removed, then the temperature runaway can be avoided so that a higher hydrogen-to-carbon monoxide feed ratio can be operated safely for a higher conversion. However, too much heat removal makes the reactor stay at relatively lower temperatures, in which the FTS reaction also cannot be activated (Park et al. 2014).

The temperature profile observed in Figures 8.26d and 8.27d, corresponding to case IV, shows a slight hot spot near the reactor inlet. This temperature profile is a good result of the operating conditions combination used for this case and the temperature of the cooling jacket used because it almost eliminates the presence of the hot spot.

However, in Figures 8.26b, 8.26c and 8.27b, 8.27c, corresponding to cases II and III, respectively, the presence of a hot spot near the reactor inlet is quite appreciated. This may be due to the negative effect of the simultaneous increase in the total pressure and gas hourly space velocity. The hot spot increases by 4 and 2 degrees for the operating conditions used in cases II and III, respectively.

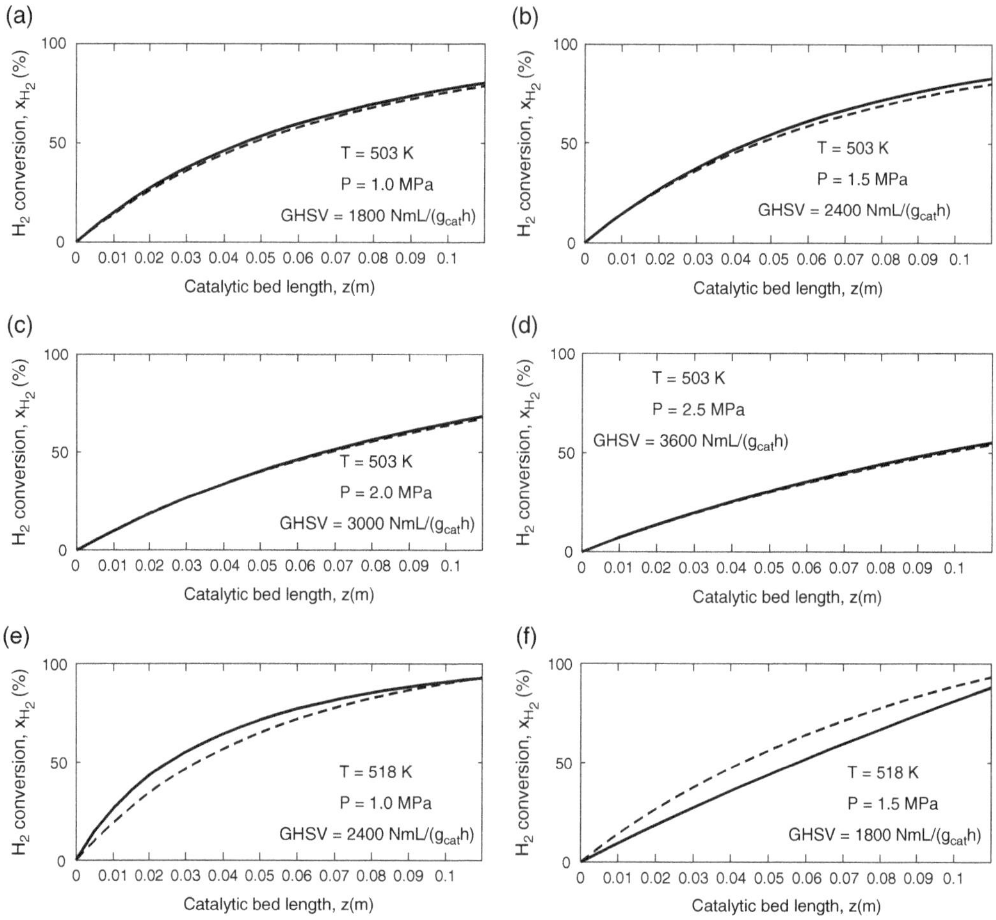

Figure 8.25 Dynamic-state H_2 conversion profile in axial position. (-) proposed model; (--) literature model literature model (dotted line) (Moazami et al. 2015a). (a) case I; (b) case II; (c) case III; (d) case IV; (e) case V; (f) case VI.

However, this does not happen in the case of Figure 8.26d since apparently the considerable increase in gas space velocity appreciably reduces the residence time within the reactor (Kwack et al. 2011a), which favors the FTS reaction in terms of practically negligible hot spot formation.

In fact, the presence of hot spots in the vicinity of the reactor inlet can cause the thermal runaway phenomenon to occur. It is important to mention that the location (distance) of the hot spot with respect to the reactor inlet depends on the feed and cooling temperatures (Rahimpour and Elekaei 2009; Dimitriadis and Pistikopoulos 1995).

Therefore, thermal management is important in an exothermic catalytic reaction to avoid undesirable thermal effects such as the presence of high hot spots and the thermal runaway problem (Lee and Chung 2012).

A solution that is mentioned in the literature to attenuate the presence of hot spots is by eliminating excess heat that is released from reactors in which highly exothermic reactions occur by increasing the heat transfer coefficient or increasing the temperature gradient using lower cooling temperatures (Lee and Chung 2012).

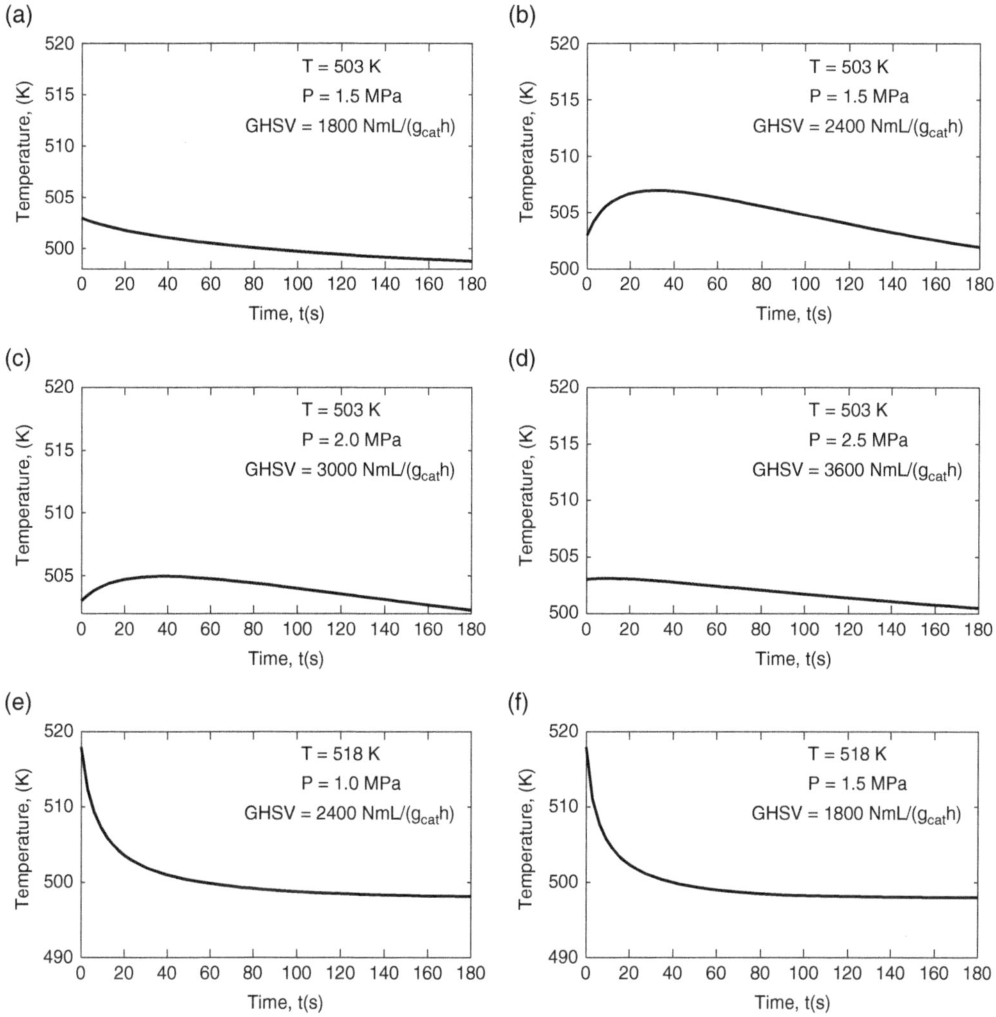

Figure 8.26 Dynamic-state temperature profile, (a) case I; (b) case II; (c) case III; (d) case IV; (e) case V; (f) case VI.

It is important to highlight that the thermal behavior of fixed-bed FTS reactors is only possible to visualize when using a dynamic model. In addition, it is extremely important to improve the design and optimization aspects of fixed-bed reactors in order to verify the feasibility and flexibility of the process under a broad set of operating conditions and uncertainties in the reactor parameters, thus guaranteeing the thermal stability of the process (Dimitriadis and Pistikopoulos 1995; Soroush and Kravaris 1993a; Soroush and Kravaris 1993b).

8.4.4 Product's selectivity

Figure 8.28 shows the CO_2 selectivity profile as a function of time at the reactor outlet, and Figure 8.29 shows the CO_2 selectivity profile along the catalytic bed at unsteady-state conditions. It can be seen that for experimental cases I–IV, corresponding to Figures 8.28a–8.28d and Figure 8.29a–8.29d, the CO_2 selectivity is relatively low, between 4% and 5%.

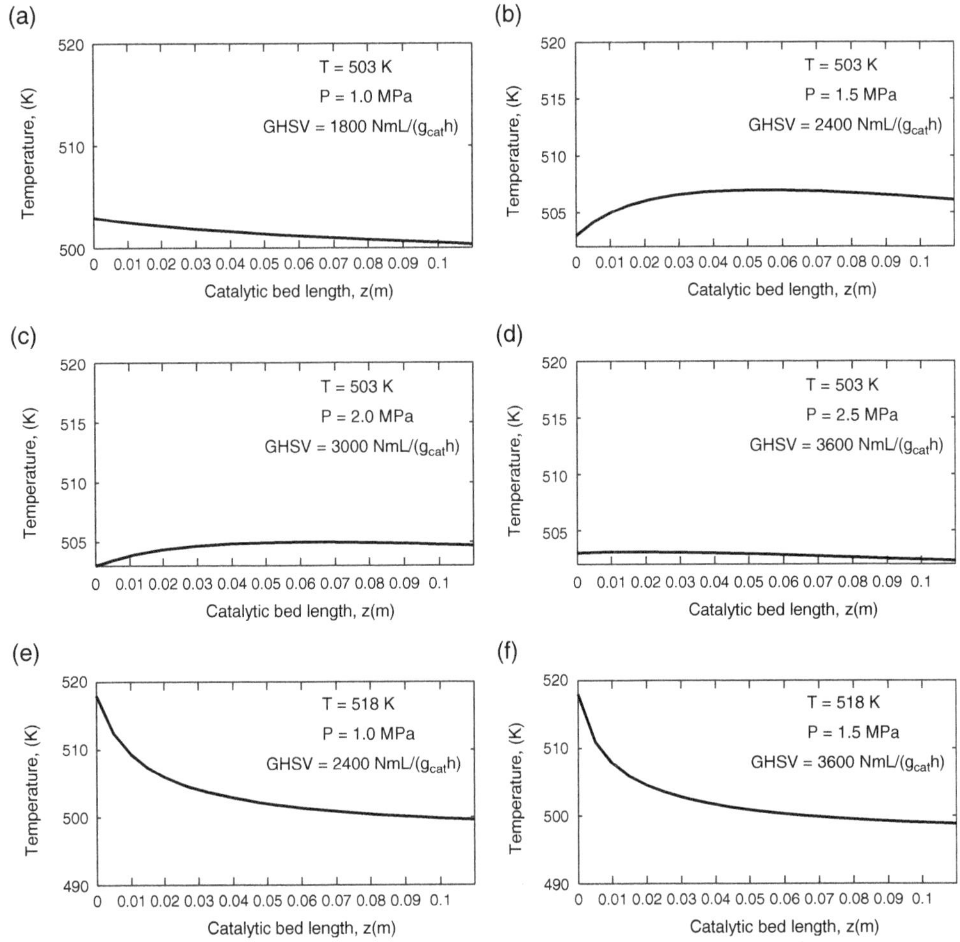

Figure 8.27 Dynamic temperature profile in axial position. (a) case I; (b) case II; (c) case III; (d) case IV; (e) case V; (f) case VI.

However, for cases V and VI (Figures 8.28e, 8.29e and 8.28f, 8.29f, respectively), the CO_2 selectivity variation ranges between 14 and 15%, which is relatively high compared with the results given in cases I–IV. This is due to the simultaneous changes in initial temperature, total pressure, and gas space velocity. The initial temperatures for cases V and VI are high, which favors a higher CO_2 selectivity. It is worth mentioning that CO_2 is an undesirable product for the FTS process so high selectivities should be avoided, which leads to the proper use of feed inlet temperatures. Therefore, the set of operating conditions given in cases I to IV provides an ideal amount of CO_2 produced.

According to the calculated errors given in Table 8.23, a good fit of the proposed model with the experimental data is observed. The error for CO_2 selectivity with the literature model cannot be calculated because the authors reported only the selectivity data of FTS products at the reactor outlet. That is why the simulations of the FTS product selectivity profiles with the literature model are not shown.

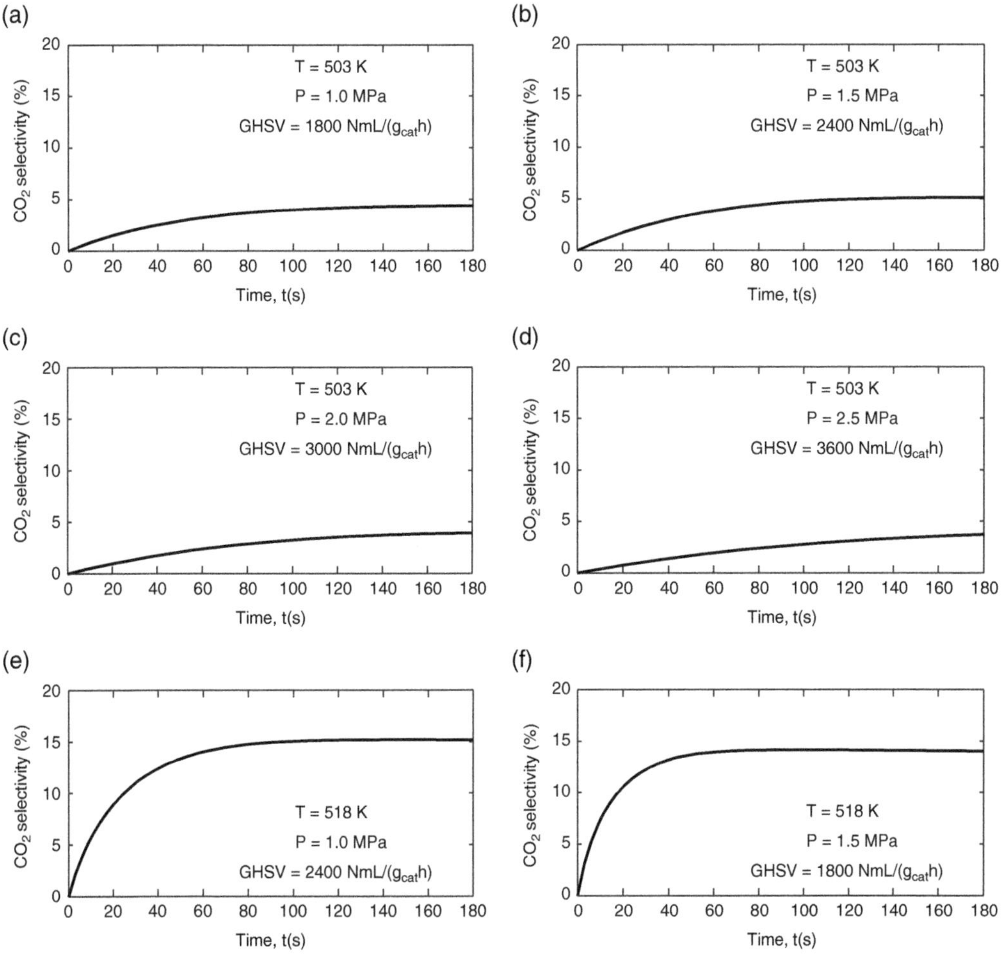

Figure 8.28 Dynamic-state CO_2 selectivity. (a) case I; (b) case II; (c) case III; (d) case IV; (e) case V; (f) case VI.

Figures 8.30 and 8.31 present the results of the proposed model simulations for the methane selectivity profile as a time function and bed length, respectively. For the dynamic case of methane selectivity, Figures 8.30b–8.30e (cases II–V, respectively) show that methane selectivity is high with values of 26, 34, 30, and 31%, respectively. Similar to CO_2, care should be taken in the FTS process with high methane productivity since it is also considered an undesirable product.

The high values of CH_4 selectivity are due to the increase in total pressure and gas space velocity since the latter translates into a decrease in the residence time in the reactor.

The use of high values of gas space velocity leads to an increase in the production of light gases, especially methane, since there is a short time to favor the re-adsorption of olefins for the growth of the hydrocarbon chain, thus giving rise to the formation of high-molecular-weight liquid hydrocarbons (Bechara et al. 2001).

It is observed that an increase in gas space velocity induces a considerable decrease in the production of liquid hydrocarbons. However, the increased total gas pressure in steps of 0.5 MPa does not allow the reduction of the liquid hydrocarbons selectivity to be so drastic. In addition, the light

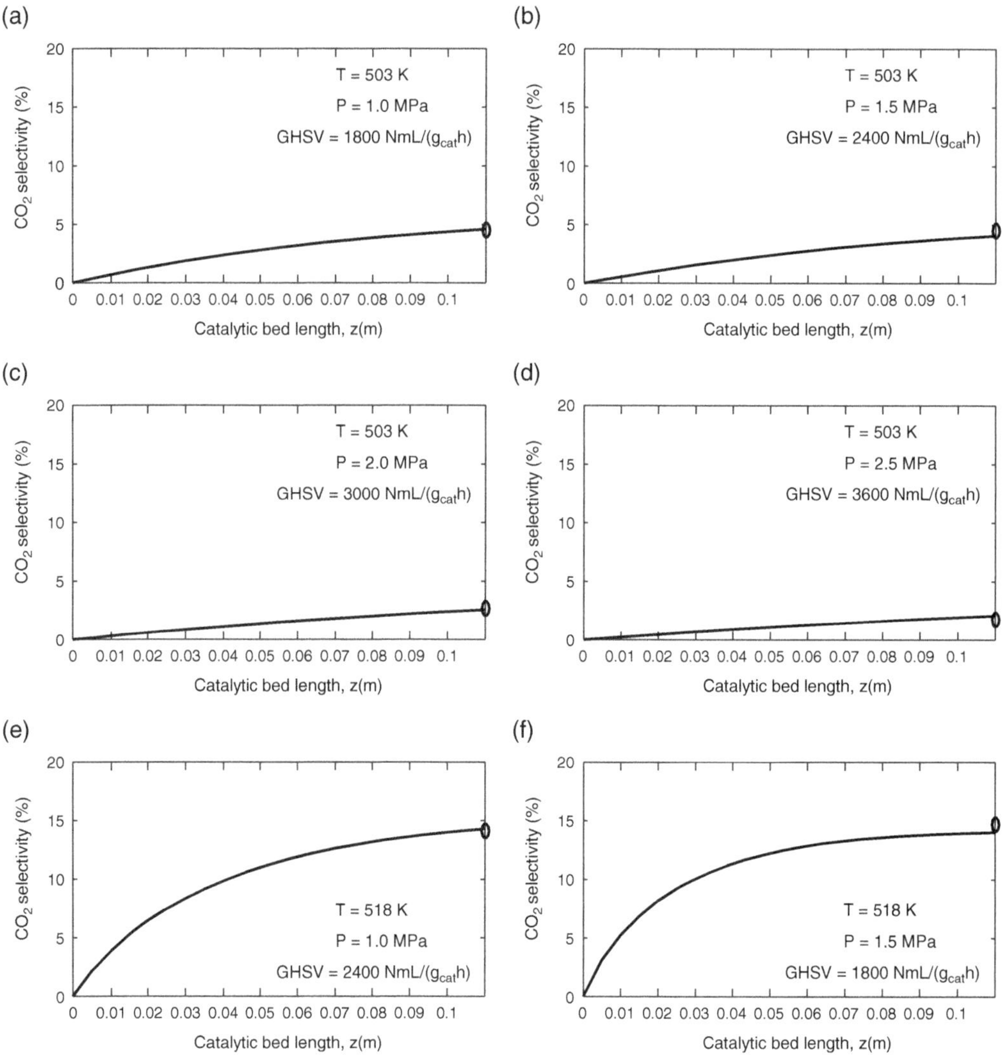

Figure 8.29 Dynamic-state CO_2 selectivity profile in axial position. (O) experimental data; (-) proposed model; (a) case I; (b) case II; (c) case III; (d) case IV; (e) case V; (f) case VI.

gases selectivity does not undergo considerable changes, which is beneficial when an FTS process with high liquid hydrocarbon productivities is desired. Kwack et al. (2011a) corroborated that an increase in total pressure leads to a slight improvement in the production of heavy hydrocarbons. Also, recently Moazami et al. (2017b)found that high pressures favor an increment in the liquid hydrocarbons selectivity, which was not observed in the first four study cases due to the simultaneous increase in gas space velocity (i.e., a decrease in residence time), and there is a low diminish in the concentration of reactants (CO and H_2), which in turn results in a decrease in the long-chain liquid hydrocarbons selectivity.

Under typical reaction conditions, a thin layer of liquid hydrocarbon covers the catalyst particles. As a result, before the reactants reach the catalyst surface, they have to diffuse inside this layer, while reaction products have to do the same in the opposite direction before being desorbed. It

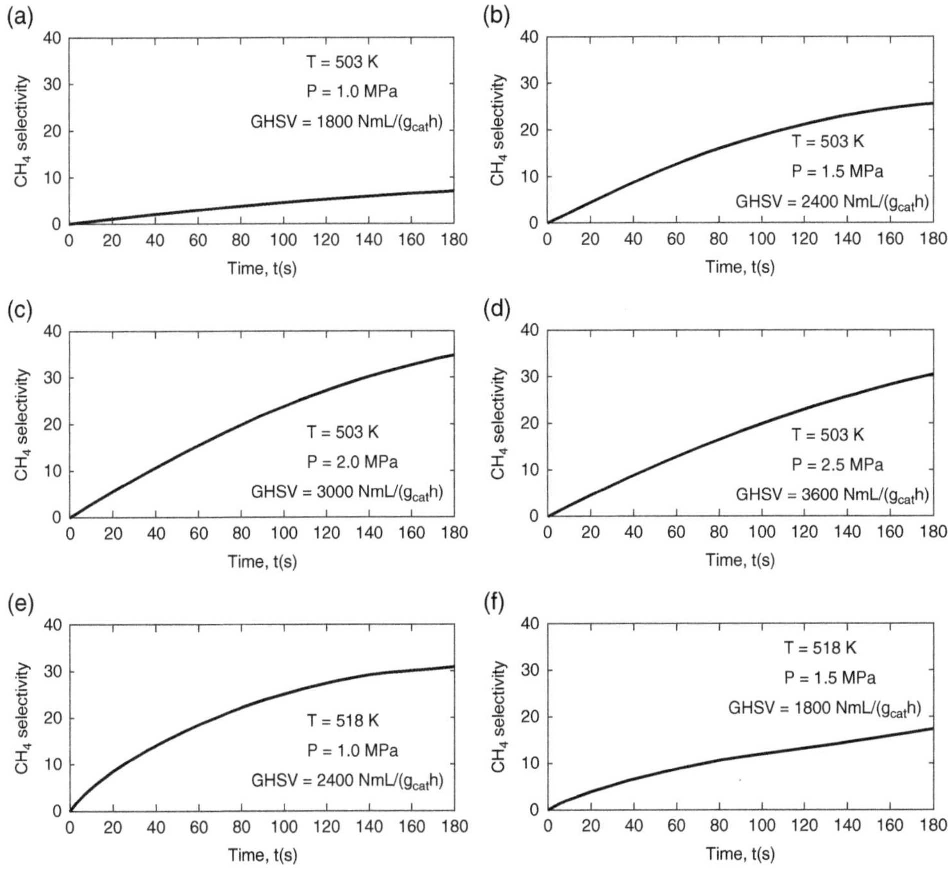

Figure 8.30 Dynamic-state CH_4 selectivity profile. (a) experimental case I; (b) experimental case II; (c) experimental case III; (d) experimental case IV; (e) experimental case V; (f) experimental case VI.

is known that olefins, in contrast to paraffins, can be readsorbed on the active sites, reinserted in the chain growth process, or can be hydrogenated to the corresponding paraffins (Moazami et al. 2017b; Kuipers et al. 1996; van der Laan and Beenackers 1999). However, in experimental case I - (Figure 8.30a), the methane selectivity profile over time is favorable since the selectivity achieved is low, while in experimental case VI (Figure 8.30f), methane selectivity at the reactor outlet is moderate since intermediate selectivity is layered, but even so, it is always desirable to decrease the methane production by establishing an adequate set of operating conditions. In Figure 8.31, the methane selectivity profiles are illustrated along the catalytic bed length. In general, the selectivities are not so high. Cases II, III, IV, and V report selectivities close to 20% (see Figure 8.31b–8.31e). According to Table 8.23, the error between the predictions with the proposed model and the experimental data is relatively small. However, the steady-state literature model presents an error a little bit lower compared with the proposed model. In general, the light gases selectivities (C_2, C_3, and C_4) are relatively low for all the cases, as can be seen in Figures 8.32–8.37.

Kwack et al. (2011a) reported that the variation in the total pressure and gas space velocity considerably affects the light gases selectivity and the product's total distribution. They varied the gas space velocity in an order of 600 units as compared with the experiments carried out by Moazami et al. (2015a). Such a variation in said parameter does not affect the product's selectivity in a

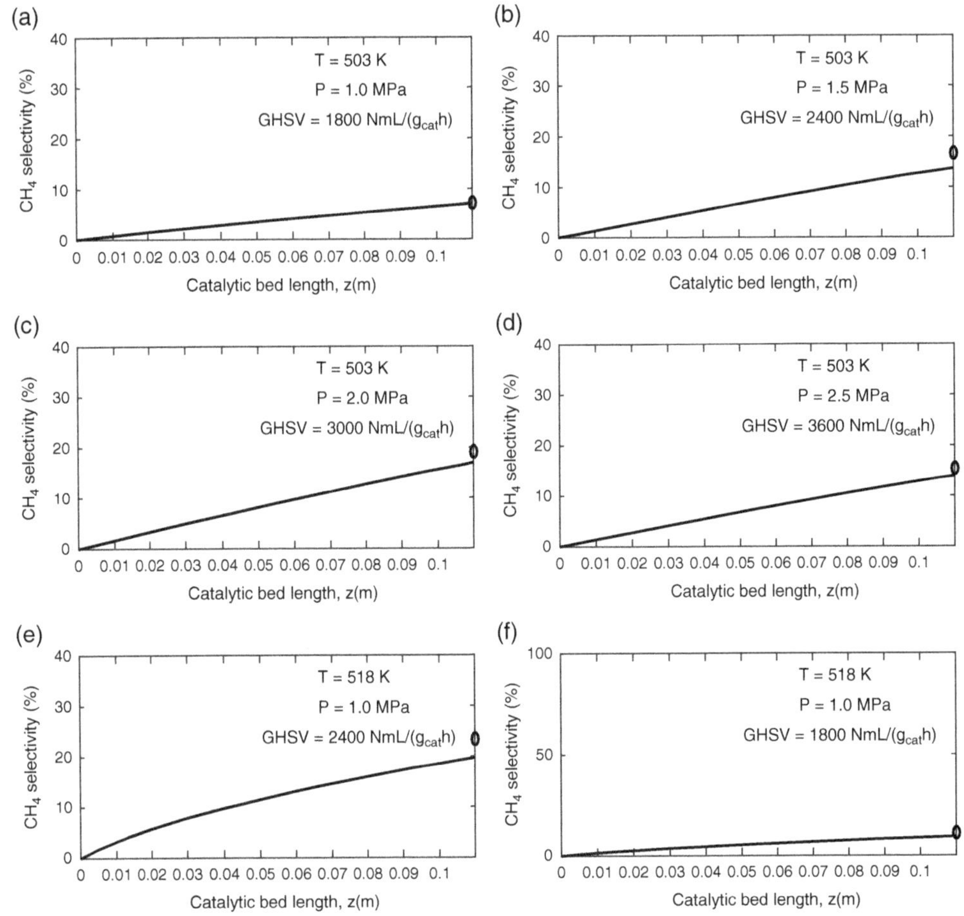

Figure 8.31 Dynamic-state CH_4 selectivity profile in axial position. (O) experimental data; (-) proposed model; (a) case I; (b) case II; (c) case III; (d) case IV; (e) case V; (f) case VI.

relevant manner. For example, the variation of light gases (C_3–C_4), except for C_1 and C_2 (for which it is about 1.3%), represents a change of approximately 2% from one case to another.

The proposed model predicts fairly accurately the experimental data given at the reactor outlet. In the FTS process, it is important to reduce, as much as possible, the light gases production and increase the selectivity of heavy hydrocarbons.

Kwack et al. (2011a) reported that the variation in the total pressure and gas space velocity considerably affects the light gases selectivity and the product's total distribution. They varied the gas space velocity in an order of 600 units as compared with the experiments carried out by Moazami et al. (2015a). Such a variation in said parameter does not affect the product's selectivity in a relevant manner. For example, the variation of light gases (C_3-C_4), except for C_1 and C_2 (for which it is about 1.3%), represents a change of approximately 2% from one case to another.

On the other hand, the liquid hydrocarbons selectivity (C_{5+}) presents a considerable variation due to the simultaneous effects of temperature, total pressure, and the gas space velocity.

Figures 8.38 and 8.39 show the product's selectivity as a function of time and along the axial direction. The lowest productivity is presented in cases II, III, IV, and V (Figures 8.38b–8.38e

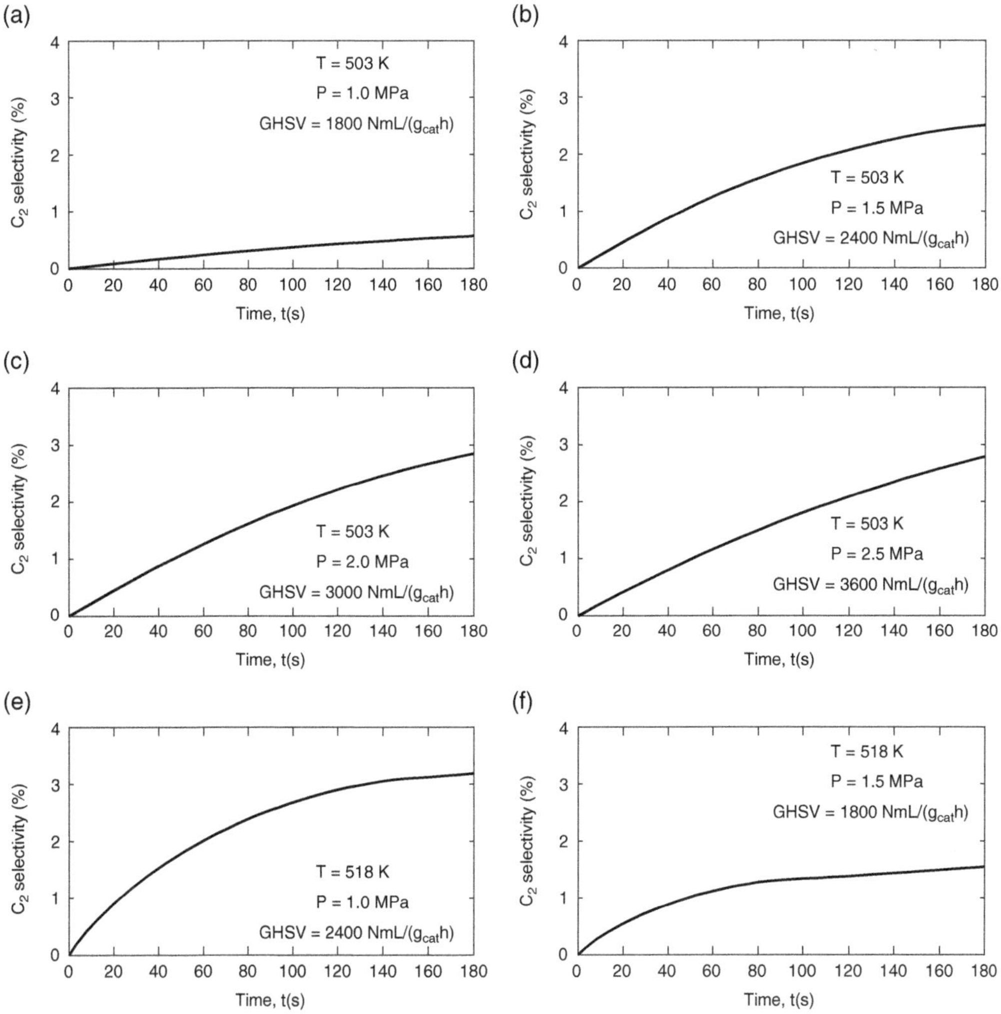

Figure 8.32 Dynamic-state C_2 species selectivity profile. (a) case I; (b) case II; (c) case III; (d) case IV; (e) case V; (f) case VI.

and 8.39b–8.39e), in which the selectivity toward C_{5+} hydrocarbons is around 60–64%. This decrease is mainly because of gas space velocity. The use of high gas space velocities helps maintain an adequate heat transfer due to the short residence time of the syngas inside the reactor, which agrees with the results reported by Rahimpour and Elekaei (2009).

But for the purpose of increasing the CO conversion and improving the product's selectivity, the variations in the gas space velocity do not have a positive impact on the productivity of the FTS process.

In summary, the best set of operating conditions that allows high conversions of the syngas (CO and H_2) and improvements in the product's selectivity in terms of low methane and other light gases (C_2–C_4) production and high production of C_{5+} hydrocarbons are cases I and VI. In these cases, the inlet temperature of the reactor quickly approaches the temperature of the cooling water without giving rise to the formation of hot spots. The syngas conversion is high, around 80 and 100%

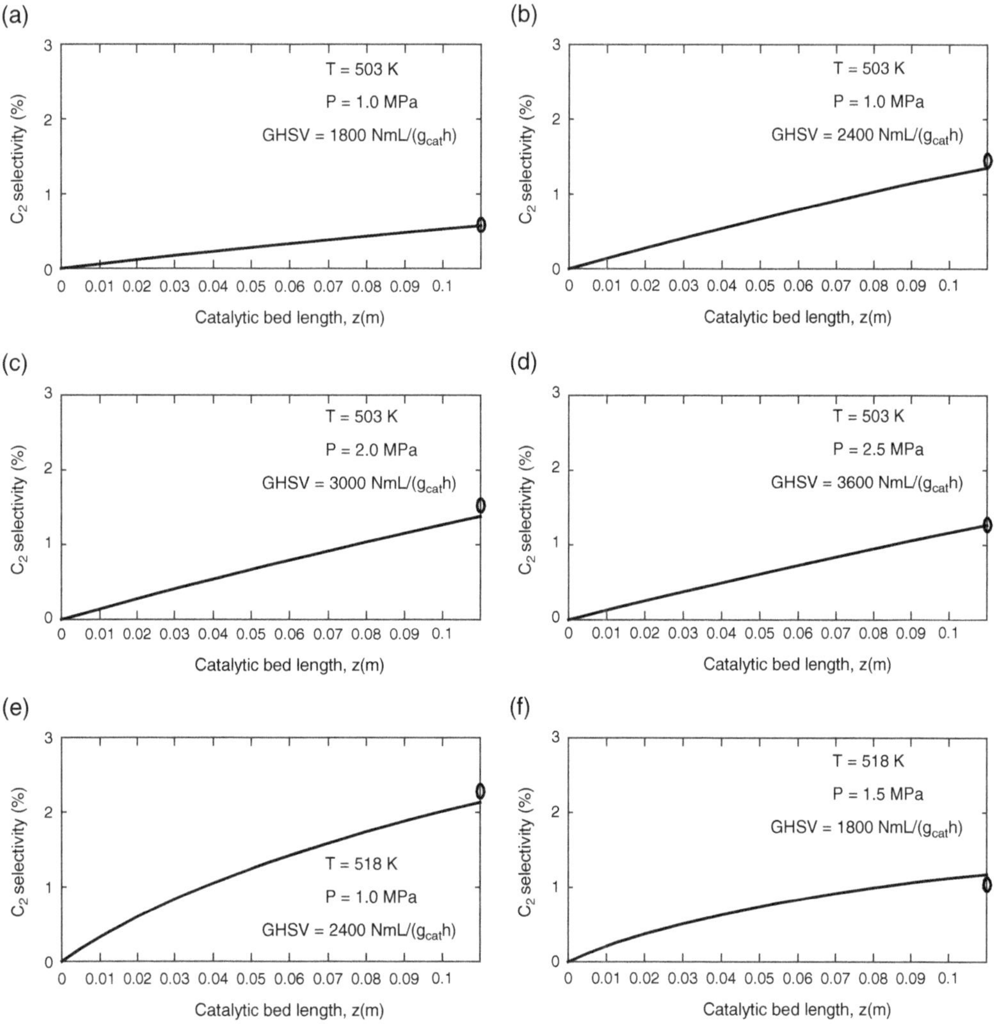

Figure 8.33 Dynamic-state C_2 species selectivity profile in axial position. (O) experimental data; (-) proposed model; (a) case I; (b) case II; (c) case III; (d) case IV; (e) case V; (f) case VI.

for the CO, and around 85 and 90% for the H_2, respectively. In addition, the methane and light gases selectivities are relatively low, while the heavy hydrocarbons selectivity (C5+) is approximately 82 and 80%, respectively, which is favorable.

8.4.5 Final Remarks

The proposed dynamic model for the FTS reactor has been successfully validated with six experimental cases reported in the literature, which considered simultaneous variations in temperature, total pressure, and syngas space velocity at the inlet of the reactor. The reactor dynamics indicates that for most of the operating conditions in each case, the reactor reaches the steady-state condition, high syngas conversions, and improvements in the product's selectivity. The thermal behavior of

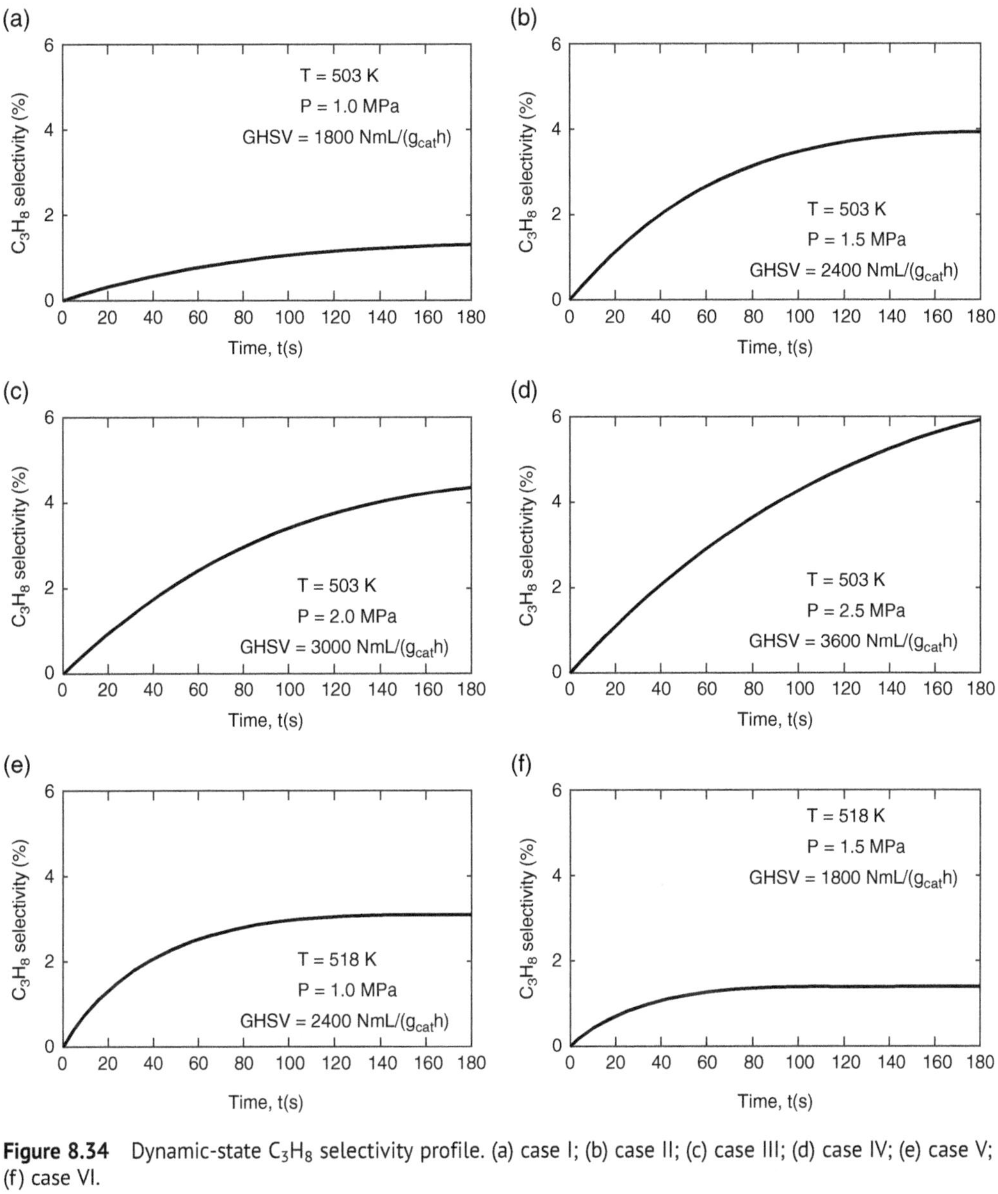

Figure 8.34 Dynamic-state C_3H_8 selectivity profile. (a) case I; (b) case II; (c) case III; (d) case IV; (e) case V; (f) case VI.

the reactor is properly evaluated from the consideration of a non-isothermal regime, whereby the generation of hot spots is analyzed, to prevent a negative effect in terms of the thermal runaway formation or low catalytic activity. With the proposed model, it was confirmed that the best set of operating conditions that allow high syngas conversions (CO and H_2), high C_{5+} selectivity, low production of CH_4, and the absence of hot-spot formation (improvements in the thermal behavior of the FTS reactor) correspond to 503 and 518 K, 1.0 and 1.5 MPa, and 1800 (NmL/(g_{cat}·h)) of temperature, pressure, and gas space velocity, respectively. The proposed dynamic model guarantees the viability of its application for scaling up purposes (semi-industrial and industrial scales), design of control schemes, and dynamic optimization of the FTS process.

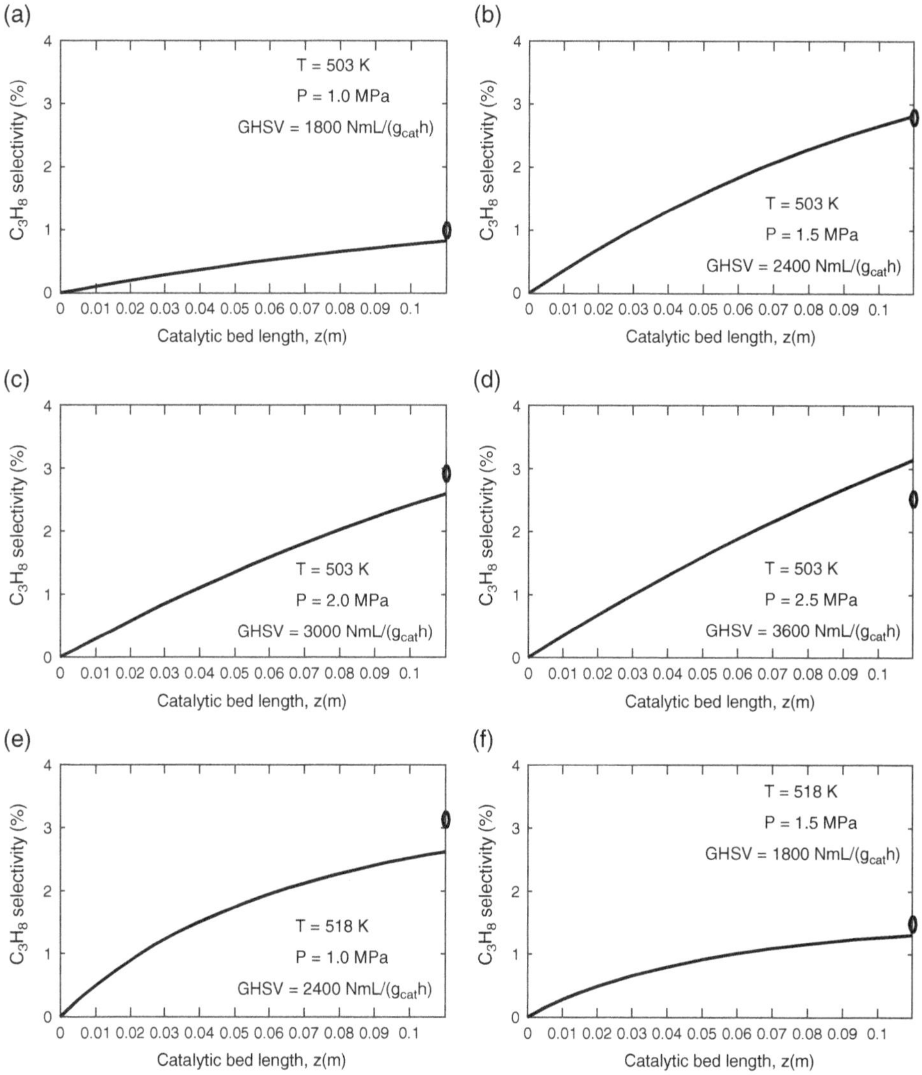

Figure 8.35 Dynamic-state C_3H_8 selectivity profile in axial position. (O) experimental data; (-) proposed model; (a) case I; (b) case II; (c) case III; (d) case IV; (e) case V; (f) case VI.

8.5 Modeling and Control of a Fischer–Tropsch Synthesis Reactor with a Novel Mechanistic Kinetic Approach

8.5.1 Introduction

A modeling and dynamic control study for an FTS fixed-bed reactor under the one-dimensional pseudohomogeneous scheme representation with integration of the cooling jacket dynamic approach has been developed. The proposed reactor model contemplates the use of a novel

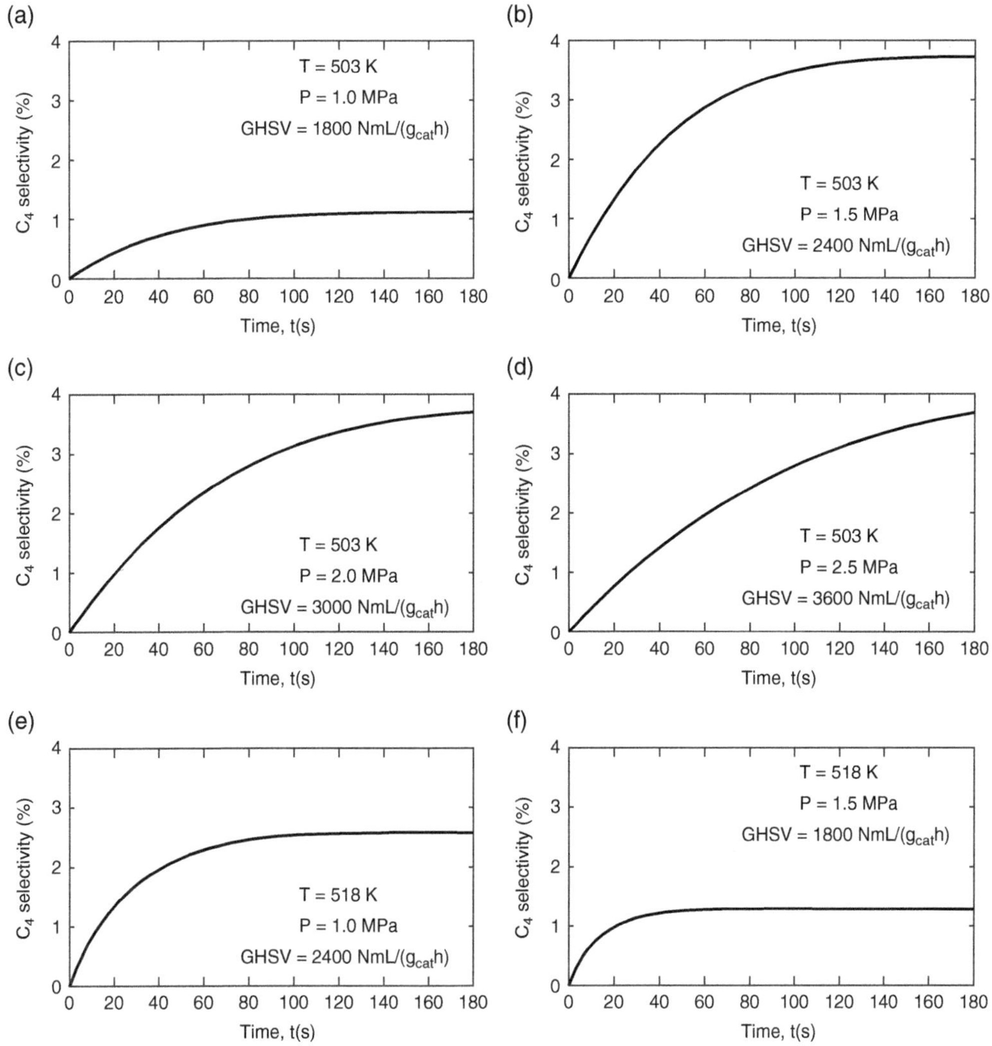

Figure 8.36 Dynamic-state C_4 species selectivity profile. (a) case I; (b) case II; (c) case III; (d) case IV; (e) case V; (f) case VI.

mechanistic kinetic model based on the Langmuir theory recently reported in the literature. To ensure good thermal behavior and improved reactor performance, a proportional–integral controller (PIC) is implemented in the cooling jacket. The robustness of the controller, the dynamic model, and the mechanistic kinetic approach are evaluated and validated against the use of four different sets of initial operating conditions of temperature, pressure, and gas hourly space velocity. The results indicate the reactor model potential under the PIC action since it allows to completely attenuate the typical hot-spot formation in fixed-bed reactors in which highly exothermic reactions are carried out. In addition, the mechanistic kinetic model represents to a great extent the FTS reaction network in terms of the syngas conversion and the light gases and heavy liquid hydrocarbons selectivity.

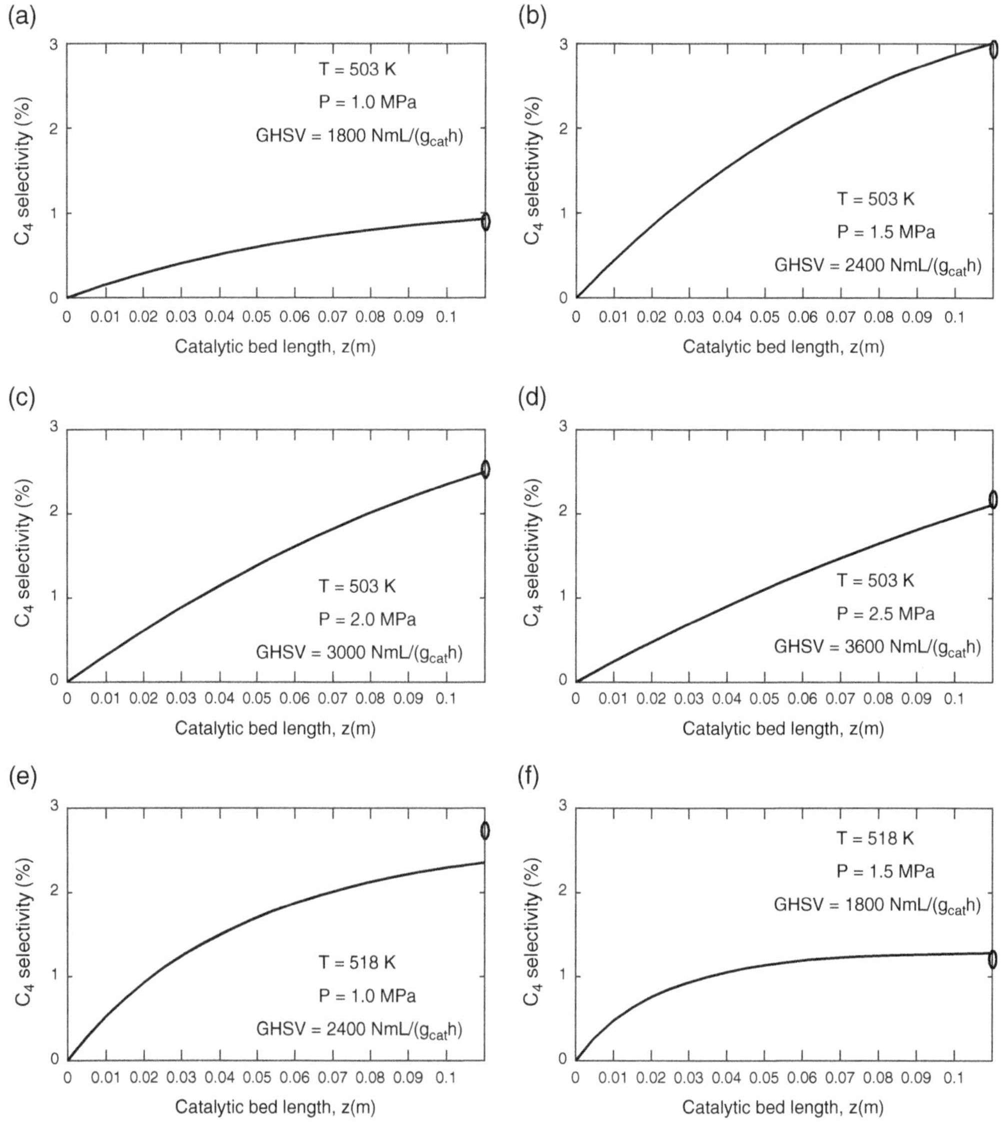

Figure 8.37 Dynamic-state C_4 species selectivity profile in axial position. (○) experimental data; (-) proposed model; (a) case I; (b) case II; (c) case III; (d) case IV; (e) case V; (f) case VI.

8.5.2 Formulation of the Model

8.5.2.1 Model Equations and Solution

The model to describe the dynamic behavior of the fixed-bed FTS reactor was implemented under plug-flow regime at not-isothermal conditions, taking into account the variations of the superficial gas velocity, the gas mixture density, and the total pressure of the system in the axial direction, which have been demonstrated to be important parameters to improve predictions of syngas conversion and temperature profiles. It also incorporates the dynamic model (energy balance) of the cooling jacket (Méndez et al. 2022).

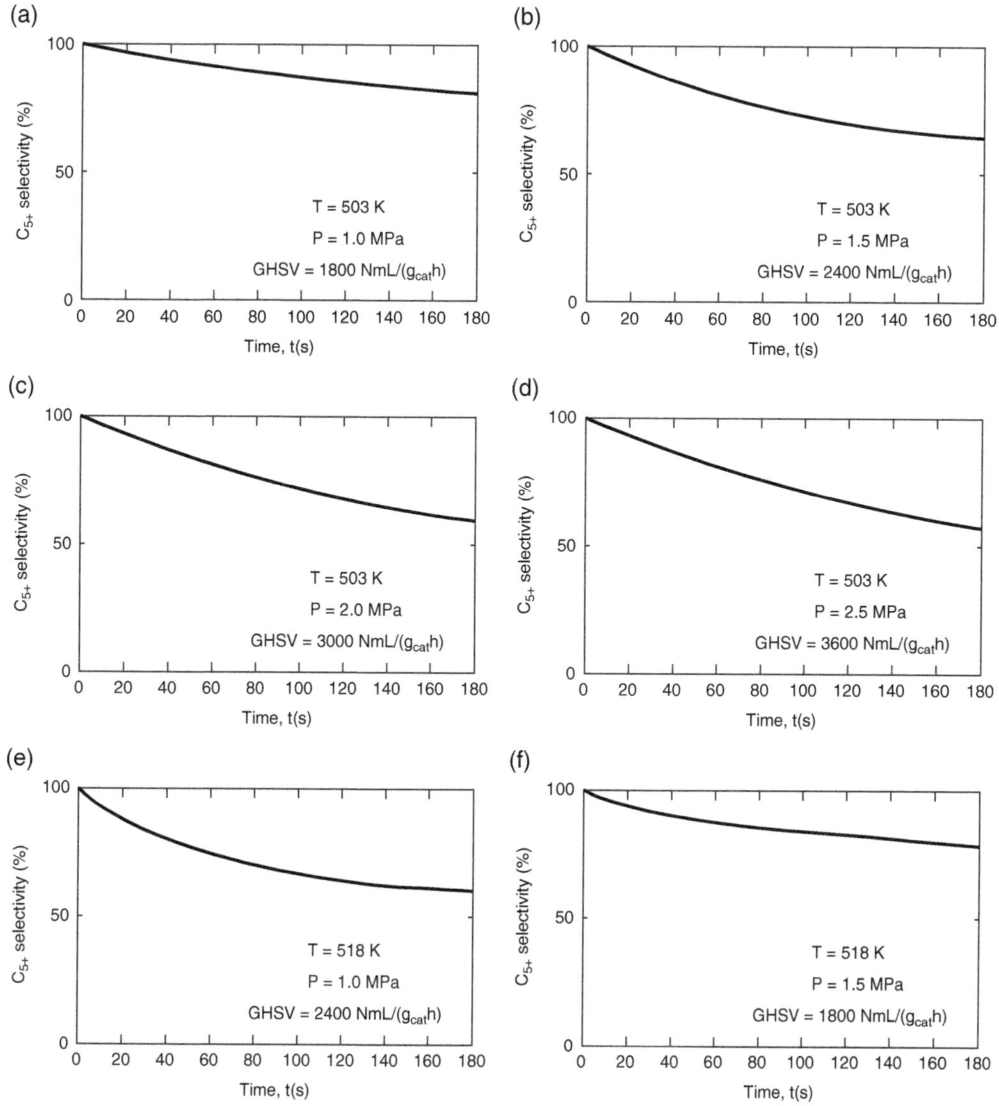

Figure 8.38 Dynamic-state C_{5+} species selectivity profile. (a) case I; (b) case II; (c) case III; (d) case IV; (e) case V; (f) case VI.

The dynamic model is given by the set of Eqs. (8.114)–(8.118) that describe the behavior of the fixed-bed FTS reactor with the boundary conditions given by Eqs. (8.119) and (8.120). The dynamic cooling jacket model is represented by the following equation:

$$C_{P_w}\rho_w V_c \frac{\partial T_w}{\partial t} = -\frac{F_{w,0}}{A_c} C_{P_w}\rho_w V_c \frac{\partial T_w}{\partial z} + F_{w,0} C_w \rho_w (T - T_w) + U_r A_c (T - T_w) \tag{8.131}$$

with the boundary condition:

$$T_w(0, z) = T_w(t, 0) = T_{w,0} \tag{8.132}$$

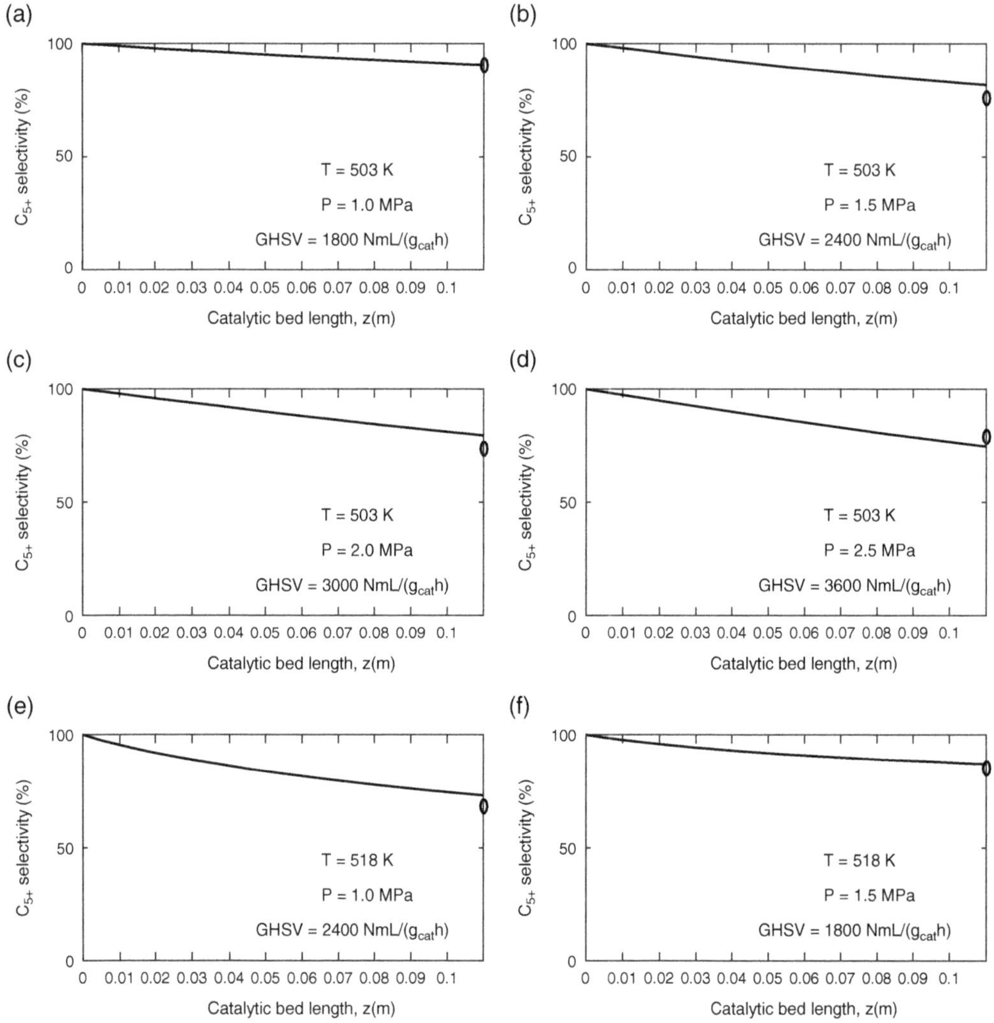

Figure 8.39 Dynamic-state C_{5+} species selectivity profile in axial position. (O) experimental data; (-) proposed model; (a) case I; (b) case II; (c) case III; (d) case IV; (e) case V; (f) case VI.

Cp_w, ρ_w is the heat capacity of water whose value is 4.6 $KJ/(KgK)$ and the ρ_w is the water density. The equations of the proposed model are solved by using the orthogonal collocation numerical method. For approaching to the solution of the model, 7 collocation points were used. The Legendre orthogonal polynomials were used in the estimation of each collocation point (Constantinides and Mostoufi 1999). Python software was used for the development of numerical simulations.

8.5.2.2 Model Parameters and Correlations

The parameters and correlations used in the reactor model are given in Table 8.19. The global heat transfer coefficient was estimated with the correlation proposed by Haarlemmer and Bensabath (2016). This parameter is given as a function of the convective heat transfer coefficient on the internal side of the reactor (h_{int}), the convective heat transfer coefficient of the external side (h_{ext}), and the reactor thermal conductivity (λ_w), which in this case is stainless steel.

The pressure drop along the catalytic bed was estimated using a correlation recently reported in the literature for the friction factor. Such a correlation was validated by comparing it with a large number of correlations showing superior accuracy (Allen et al. 2013).

The heat conductivity of the steel wall reactor was assumed to be 60 $Wm^{-1}h^{-1}$ (Sharma et al. 2011), and the value of the heat transfer coefficient (film) of the external side corresponding to boiling water coolant 1000 $Wm^{-1}h^{-1}$ (Haarlemmer and Bensabath 2016). Other parameters were considered as constants, such as the hollow fraction of the bed and the active side volume fraction of the catalyst. The other parameters of the fixed-bed FTS reactor are given in Table 8.20.

8.5.2.3 The Mechanistic FTS Kinetic Model

The FTS reaction network is classified as a number of lumped reactions by means of the kinetic characteristics of reaction molecules, consisting of 11 reacting components (i.e. CO, H_2, CO_2, H_2O, CH_4, C_2H_4, C_2H_6, C_3H_8, i-C_4H_{10}, n-C_4H_{10}, and C_{5+}). Based on this consideration, the reaction scheme used in the present study is given by Eqs. (8.122)–(8.129).

The kinetic model given by Moazami et al. (2017a) was derived from the LHHW rate theory based on a combination of alkyl/alkenyl mechanism for FTS reactions (for production of n-paraffins and n-olefins), which consist of various elementary reaction step routes and carbon chain distribution pathways (i.e. adsorption, initiation, propagation, and termination steps). The surface elementary reaction steps are given by four categories (van der Laan and Beenackers 199; Moazami et al. 2017a): (i) adsorption of the reactants (molecular CO and H_2 species) on the catalyst surface, (ii) chain initiation step, (iii) chain growth (propagation) step, and (iv) chain termination and desorption of the products step. The following expressions represent the Fischer–Trospch synthesis and water-gas shift (WGS) reaction rates, respectively:

$$R_{FT} = \frac{\sqrt{k_8K_7K_6K_5K_1K_2^{2.5}K_3K_4P_{CO}^{0.5}P_{H_2}^{1.25}}}{\left(1 + K_1P_{CO} + \sqrt{K_2}P_{H_2}^{0.5} + K_1\sqrt{K_2}K_3P_{CO}P_{H_2}^{0.5} + K_1K_2K_3K_4P_{CO}P_{H_2} + \eta_1 + \eta_2\right)^2}$$

where

$$\eta_1 = \frac{K_1K_2^{1.5}K_3K_4K_5K_6}{\sqrt{K_1K_2^{2.5}K_3K_4K_5K_6K_7k_8}}P_{CO}^{0.5}P_{H_2}^{0.25};\ \eta_2 = \sqrt{\frac{K_1K_2^{1.5}K_3K_4K_5K_6K_7}{k_8}}P_{CO}^{0.5}P_{H_2}^{0.75} \tag{8.133}$$

$$R_{WGS} = \frac{\left(k_{WGS_4}K_{W_1}K_{W_2}K_{W_3}K_{W_6}\frac{P_{CO}P_{H_2O}}{P_{H_2}} - k_{WGS_{-4}}\frac{P_{CO_2}}{K_{W_5}}\right)}{\left(1 + K_{W_1}P_{CO} + \sqrt{\frac{P_{H_2}}{K_{W_6}}} + K_{W_2}K_{W_3}K_{W_6}\frac{P_{H_2O}}{P_{H_2}} + K_{W_2}K_{W_6}^{0.5}\frac{P_{H_2O}}{\sqrt{P_{H_2}}} + \frac{P_{CO_2}}{K_{W_5}}\right)^2} \tag{8.134}$$

The reaction rate constants, adsorption coefficients, equilibrium constant, and Gibbs free energy are calculated by using the following equations:

$$k_j = k_{j,0}\,exp\left(-\frac{E_{a,j}}{RT}\right) \tag{8.135}$$

$$K_i = K_{i,0}\, exp\left(-\frac{\Delta C_{ads,i}}{RT}\right) \tag{8.136}$$

$$K_P = \frac{P_{CO_2}P_{H_2}}{P_{CO}P_{H_2O}} \tag{8.137}$$

$$\ln(K_P) = -\frac{\Delta G_R^0}{RT} \tag{8.138}$$

$$\Delta G^0_{R,298.15} = \frac{1}{\alpha_k}\left(\sum_i^{products} \alpha_i \Delta G^0_{f,298.15} - \sum_i^{reactants} \alpha_i \Delta G^0_{f,298.15}\right) \tag{8.139}$$

$$\ln(K_P) = -3.72 + \frac{4861.49}{T} - 6.90\times 10^{-3}T + 1.33\times 10^{-5}T^2 - 8.38\times 10^{-9}T^3 + 1.25\times 10^{-12}T^4 \tag{8.140}$$

The product selectivities were calculated by Eqs. (8.110)–(8.113).

8.5.3 Implementation of the PI Controller

To guarantee a viable operation of the reactor in terms of a safe thermal operation, it is necessary to implement a control scheme in the cooling jacket. It is well-known that the FTS reaction is highly exothermic, which requires good temperature control in any reactor configuration used (Kaskes et al. 2016; Kwack et al. 2011a). The high temperatures of the FTS reaction give rise to the hot-spot formation, which can lead to the presence of the runaway phenomenon (Kaskes et al. 2016a; Rafiq et al. 2011).

The literature indicates that a PI controller is suitable in those cases that involve a highly non-linear process and where it is intended to eliminate the static error (off-set) (Stephanopoulos 1985). On the other hand, it is stated that a PI control is a form of continuous adjustment that originated in the process industries and usually considered part of (automatic) engineering process control (Box and Luceño 1997).

To ensure good temperature control of the fixed-bed FTS reactor, a PIC scheme is implemented in the cooling jacket where the reactor temperature is controlled by adjusting the temperature of the refrigerant inside the cooling jacket by manipulating the flow velocity of the latter. The PI controller is given by the following expression:

$$\zeta(t) = K_C\bar{e} + \frac{K_C}{\tau_I}\int_0^t \bar{e}dt \tag{8.141}$$

where $\zeta(t)$ is the variable control (flow velocity of refrigerant), K_C is the controller gain, τ_I is the reset time, and $\bar{e}$ is the error given by the temperature difference of the set-point (system temperature; 513 K for the first three experimentation cases and 523 K for the fourth experimentation case) and the exit temperature (system response: outlet temperature) as follows:

$$\bar{e} = T_{sp} - T \tag{8.142}$$

whose error is a constant.

The set-point established in the studied control problem becomes the reactor temperature of 503 K, and to achieve a negligible deviation in the process response, the flow velocity of the refrigerant is manipulated. The controller parameters (controller gain K_C and reset time τ_I) are determined by the adjustment rules given by Tyreus and Luyben (1992) whose values were $K_C = 2.6$ and $\tau_I = 0.25$ *min*.

To verify the robustness of the PI controller, changes in total system pressure and gas space velocity were considered as disturbances.

8.5.4 Results and Discussion

8.5.4.1 Experimental Data

The model simulations are compared with experimental data reported by Moazami et al. (2017a). The experiments were carried out using a Co/SiO_2 catalyst, which was diluted with 2 g of silicon carbide at a mass ratio of 1:12 ($Co/SiO_2/SiC$). A simulated nitrogen-rich syngas was used with concentrations of 17, 33, and 50 vol% of CO, H_2, and N_2, respectively. The dimensions of the bench-scale fixed-bed reactor used in this case study are a catalytic bed of 6 m in length and an internal diameter of 13 mm (Dai et al. 2014). An important aspect in the modeling is the extrapolation of laboratory-scale reactor dimensions given by Moazami et al. (2017a) for one-stage reactor dimensions given by Dai et al. (2014), in order to verify the model robustness subject to changes made in the reactor dimensions and in the initial operating conditions. In addition, it is always important to try to scale up the FTS reactor size in order to verify its performance at higher scales (pilot, semi-industrial, and industrial) and its economic feasibility.

8.5.4.2 Simulations of the Syngas Conversion, Light Gases, and Heavy Liquid Selectivity

Although it is known, it is advisable to use kinetic models of reduced complexity to develop mathematical models of FTS reactors (fixed-bed, fluidized-bed, and slurry bubble column) for design and scaling-up purposes (Kwack et al. 2011b). However, it is also essential to have an integrated comprehensive kinetic model that allows us to accurately describe the FTS product's distribution (Kwack et al. 2011b; Mosayebi et al. 2016; Quian et al, 2013).

To validate the proposed model of the fixed-bed FTS reactor with the integration of a novel comprehensive LHHW kinetics, four sets of initial operating conditions of temperature, pressure, and GHSV were used (Table 8.24). From these experimental data, the effect of the main variables (temperature, pressure, and GHSV) can be studied and used for validation of the model.

Figure 8.40 shows the CO conversion profile as a function of time and along the catalytic bed. It is important to point out that the changes in the initial operating conditions were made every 10 min (stage), considering that the syngas initial concentration is the same in each stage.

For the four experimental cases, the simulations show that the maximum CO conversion is reached after 1.3 min. That is, after this period, the reactor reaches a stable steady state. In the first experimental case (Figure 8.40a), it can be seen that the CO conversion is around 99%. In the rest of the cases (II, III, and IV), the CO conversion is close to 90%. The stable state shown for the CO conversion allows to verify that the controller action causes the inside temperature of the FTS reactor to be maintained in almost isothermal conditions. On the other hand, Figure 8.40b shows the

Table 8.24 Experimental cases used in the proposed model simulations.

Case	Temperature (K)	Pressure (Mpa)	GHSV (NmL/g_{cat}/h)
I	518	1.0	2400
II	518	1.5	1800
III	518	2.0	3600
IV	528	1.5	3600

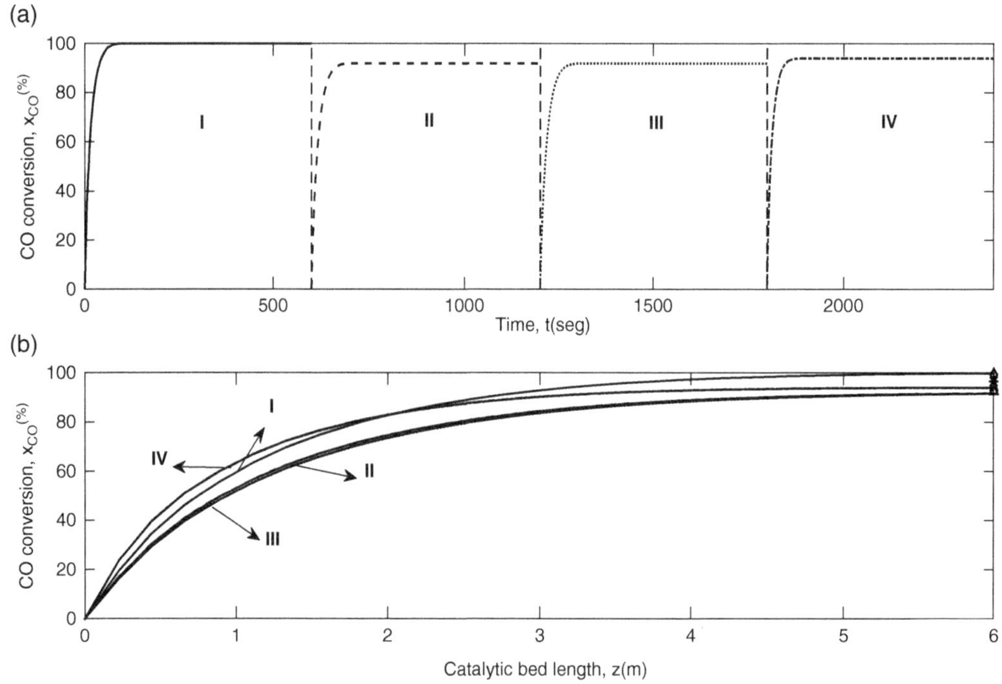

Figure 8.40 CO conversion profile. Proposed model (-), experimental data: case I (○), case II (◊), case III (Δ), and case IV (✱). (a) CO conversion profile as a function of time, (b) CO conversion profile along the catalytic bed.

CO conversion along the catalytic bed (axial direction) for the four case studies. In all the cases, the model manages to predict the experimental data at the exit of the reactor, with a negligible deviation. The maximum error was observed for cases I and III, which is not of considerable attention.

Figure 8.41 illustrates the H_2 conversion profile as a function of time along the catalytic bed length for the four experimental cases. For this profile, no experimental data were reported at the output of the reactor, so there is no comparison with the values predicted by the proposed model.

Figure 8.41a shows the maximum H_2 conversion that is reached at the output of the FTS reactor. In the first three cases, it is slightly above 80%, while for case IV, the H_2 conversion falls below 70%. Figure 8.41b shows how the H_2 conversion profile along the catalytic bed differs for the four cases under study.

The high syngas conversions in terms of CO and H_2 amounts consumed indicate the little influence of the water produced during the FTS reaction. This is because cobalt-based FTS catalysts tend to have little affinity for water production, which in turn results in a non-negative effect on the syngas conversion (CO and H_2), since there is not enough water that reacts with the CO (Alihellal and Chibane 2016). The water–gas shift reaction is spontaneous toward the side in which the hydrogen reacts with the CO_2.

Figure 8.42 shows the CO_2 selectivity for the four experimental cases. All the cases show a low CO_2 production (selectivity < 15% for cases I and II, around 10% for case III, < 20% for case IV). This low CO_2 productivity is mainly due to the little effect of the water–gas shift reaction. That is, the small amount of produced water slightly reacts with the CO to give CO_2. This low CO_2 selectivity is beneficial for the FTS process because it is a greenhouse gas whose production is sought to reduce as much as possible.

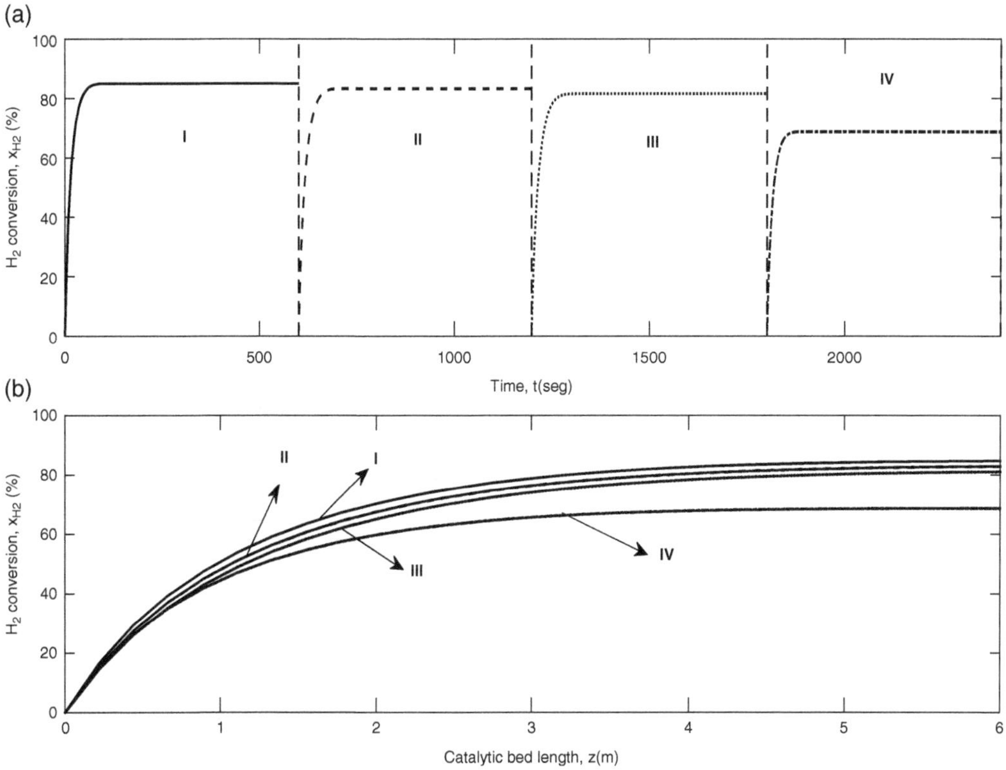

Figure 8.41 H_2 conversion profile according to the proposed model. (a) H_2 conversion profile as a function of time, (b) H_2 conversion profile along the catalytic bed.

The errors between the experimental and predicted values at the output of the reactor were calculated with Eq. (8.130). Table 8.25 summarizes the calculated errors for the proposed model versus the literature model (Moazami et al. 2017a), for the CO conversion and product selectivity (CO_2, CH_4, light gases: C_2, C_4, and hydrocarbons: C_{5+}) for the 4 cases.

The errors in the prediction of CO_2 selectivity in the four cases are relatively low (1.2–5.1%), with case II having the highest error.

It is well known that methane is an undesirable product in the low- or high-temperature FTS process (Gideon Botes 2009; Rahimpour and Elekaei 2009). On the other hand, the FTS reaction is a surface reaction in which the catalytic characteristics are the main cause of rising to high amounts of methane (Nakhaei Pour and Modaresi 2016). This is because the initiating block of the hydrocarbon chain growth process is the methyl building block or monomer (CH_2), which, depending on the catalytic and operating conditions, can give rise to methane production as the main paraffin of the FTS process or chemically re-adsorbed on the catalyst surface to continue with the process of forming short or long-chain olefins or paraffins (Nakhaei et al. 2013; Nakhaei et al. 2014a, 2014b).

Figure 8.43 shows the CH_4 selectivities profile for the four experimental cases. In experimental case II, methane production in terms of selectivity is around 10%, which is considered relatively low. For cases I and III, the selectivity of methane does not account to more than 20%, which is also not of considerable importance. However, for experimental case IV, methane production rises

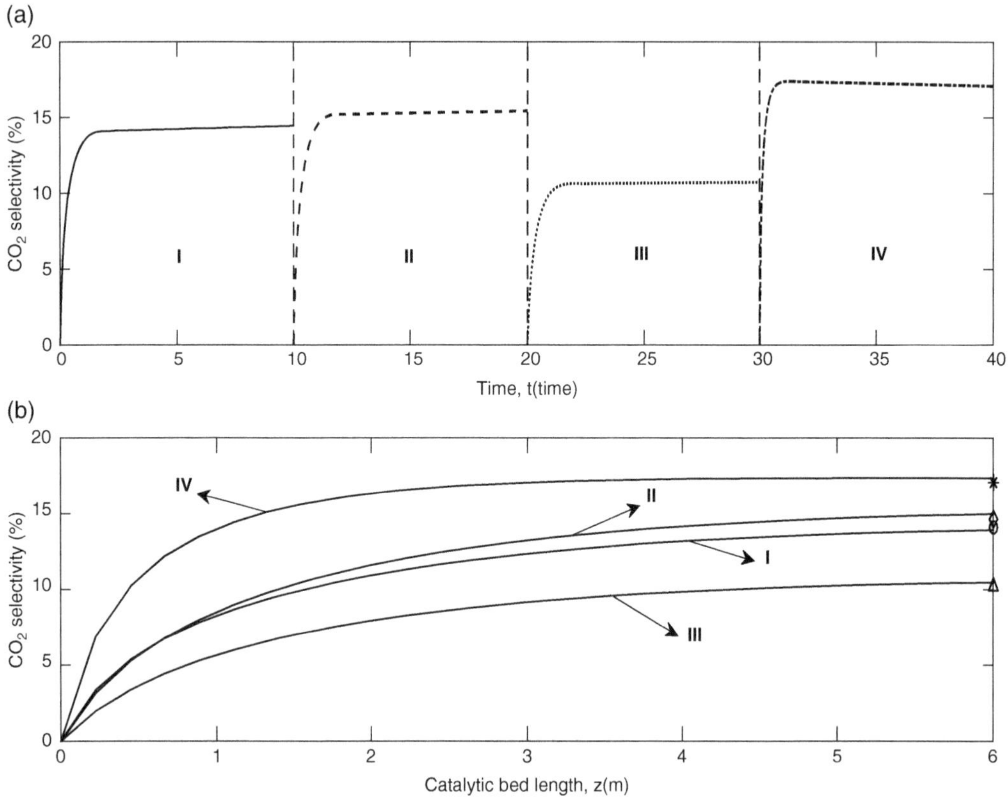

Figure 8.42 CO_2 selectivity profile. Proposed model (-), experimental data: case I (○), case II (◇), case III (Δ) and case IV (✻). (a) CO_2 selectivity profile as a function of time, (b) CO_2 selectivity profile along the catalytic bed.

Table 8.25 Error between model predictions and experimental data given by Adapted from Moazami, et. al. (2015a).

	Error (%)			
Component	**I**	**II**	**III**	**IV**
CO	7.25	0.50	7.58	2.92
H_2	—	—	—	—
CO_2	1.22	5.12	2.21	1.87
CH_4	1.38	10.0	9.10	3.92
C_2H_6	1.21	23.65	7.24	9.34
C_3H_8	10.90	34.30	0.11	6.19
C_4H_{10}	0.27	35.0	5.65	0.59
C_{5+}	0.0003	1.34	3.31	4.45

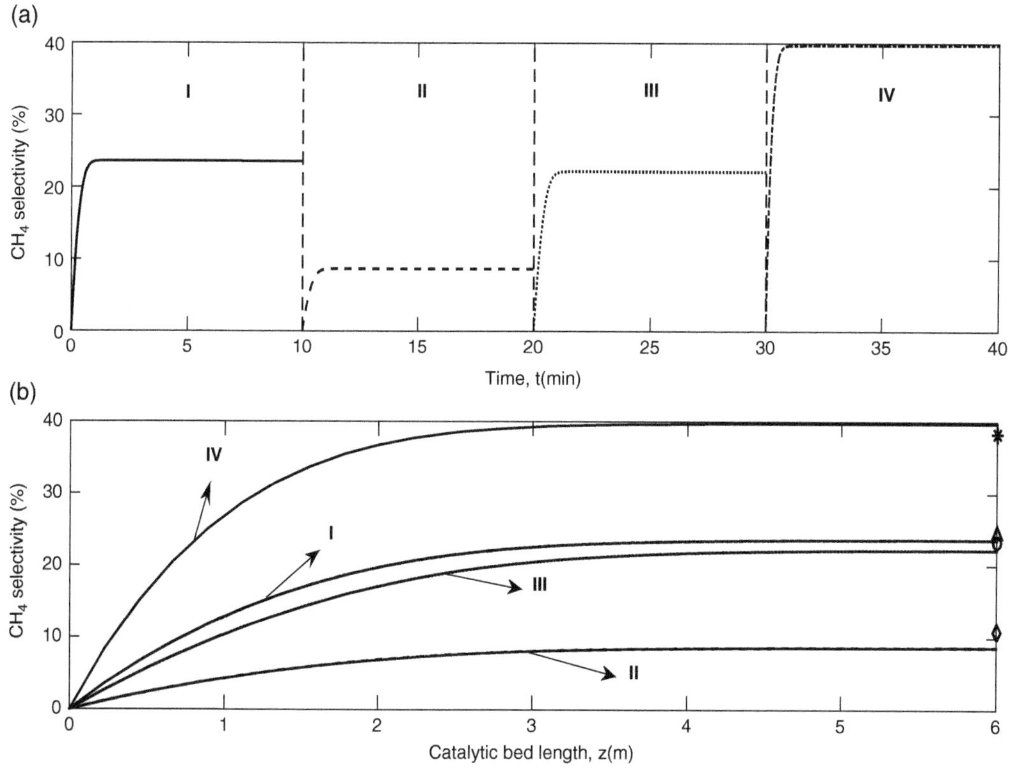

Figure 8.43 CH_4 selectivity profile. Proposed model (-), experimental data: case I (○), case II (◇), case III (Δ), and case IV (✱). (a) CH_4 conversion profile as a function of time, (b) CO conversion profile along the catalytic bed.

to 40%. This is because the reactor operating temperature increases near 10 K, as reported in the literature (Nakhaei Pour and Housaindokht 2014; Rahimpour et al. 2011b).

Figure 8.43b depicts the CH_4 selectivity profile along the catalytic bed, which fits well with the experimental data given at the reactor outlet. The highest error between the calculated and the experimental values is observed for study cases II and III (~10%).

The dynamic profiles of light gases selectivity are shown in Figures 8.44a, 8.45a, and 8.46a, while the profiles of the same gases are illustrated along the axial direction of the FTS reactor in Figures 8.44b, 8.45b, and 8.46b, respectively.

It can be observed that the amount of light gases produced is relatively low (Figures 8.44a and 8.44b) since in the worst case the selectivity reaches at most 5% (case IV). From a point of view of the FTS process, it is advisable to have simultaneously a low selectivity toward light gases and methane, and at the same time a high selectivity toward liquid hydrocarbons. It should be mentioned that case IV reflects the worst scenario in terms of a high selectivity toward light gases (especially CO_2 and CH_4) and a low selectivity toward liquid hydrocarbons (C_{5+}). This effect is due to two main causes: (i) effect of high temperatures: the CO conversion is sometimes favored at the expense of high production of light gases, particularly methane, and more water generation (Dai et al. 2014; Rafiq et al. 2011; Bechara et al. 2001); (ii) effect of space gas velocity: a high value of the space gas velocity translates into a low residence time, which causes the gas to have little reaction time and polymerization of long hydrocarbon (liquid hydrocarbon) chains (Rafiq et al. 2011; Dry 1983; Kuipers et al. 1996).

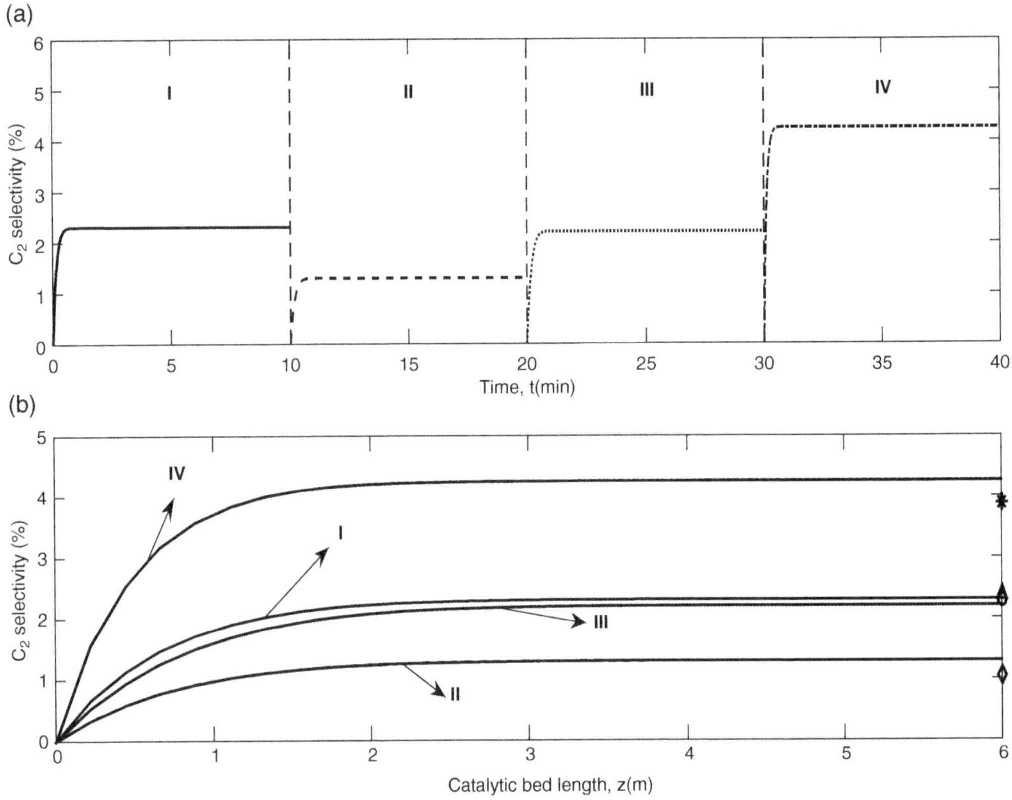

Figure 8.44 C_2H_4–C_2H_6 selectivity profile. Proposed model (-), experimental data: case I (○), case II (◇), case III (Δ), and case IV (✻). (a) C_2H_6 selectivity profile as a function of time, (b) C_2H_6 selectivity profile along the catalytic bed.

Visconti et al. (2016) established that the light gaseous hydrocarbons selectivity decreases due to the variation exerted in the gas pressure, the H_2/CO feed ratio, and the gas space velocity. They emphasized the GHSV effect on the FTS product's selectivity and CO conversion, in which it is verified that at high gas space velocity, the olefin selectivity increases because of the short residence time in the reactor to give time for secondary re-adsorption of olefins to long-chain paraffins (liquid hydrocarbons).

It can be seen in Figures 8.45b, 8.46b, and 8.47b that the light gases selectivity is well adjusted to the experimental data reported at the FTS reactor outlet. The worst experimental case in which a greater error is observed is case II. Such an error may be due to the thermal effect that is considered in the reactor model (non-isothermal regime), the process dynamics, and especially in the reactor parameters uncertainty (global heat transfer coefficient and catalyst active fraction) and kinetic parameters.

Figures 8.47a and 8.47b show the selectivity profiles toward the liquid hydrocarbons (C_{5+}) as a function of the process dynamics and along the axial direction of the FTS reactor, respectively.

High temperatures are found to favor the formation of light hydrocarbons, while low temperatures tend to promote the production of heavy hydrocarbons. That is, the C_{5+} selectivity tends to decrease due to the increase in the inlet temperature. For case IV, the light gases selectivity increases (Figures 8.43–8.46) which is negatively reflected in the liquid hydrocarbons selectivity since it tends to decrease considerably.

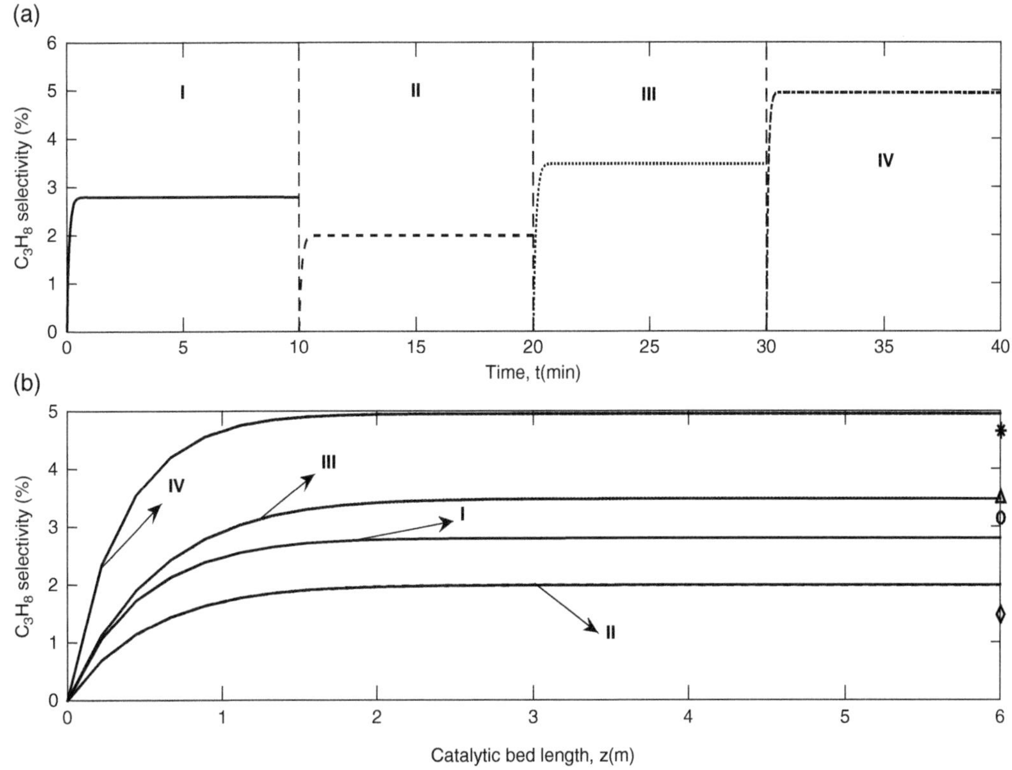

Figure 8.45 C_3H_8 selectivity profile. Proposed model (-), experimental data: case I (○), case II (◇), case III (Δ), and case IV (✱). (a) C_3H_8 selectivity profile as a function of time, (b) C_3H_8 selectivity profile along the catalytic bed.

On the other hand, as mentioned above, the gas space velocity also plays an important role in the FTS hydrocarbons selectivity. For case IV, the reactor temperature and the highest GHSV of the set of initial operating conditions considered results in a low liquid gases productivity that should be avoided in an FTS process when it is sought to make it more efficient.

In cases I and III, the liquid hydrocarbons selectivity is around 70%, where the change in the initial pressure and GHSV conditions cannot be seen to a large extent. Meanwhile, in case II, the highest C_{5+} selectivity is verified, and the CO and H_2 conversions are 93 and 82%, respectively.

For case II, a low light gases productivity is verified, which results in a high C_{5+} selectivity of around 89%. It is important to mention that for this case, a temperature of 518 K favors the FTS process with low productivity in light gases. The kinetic model used in this work shows a satisfactory prediction of the FTS product's distribution (light gases and C_{5+}). The latter is sustained by the fact that the error between the predicted and experimental values is relatively low.

The trend shown in the C_{5+} selectivity profile is according to the predictions made by a two-dimensional steady-state model for a fixed-bed FTS reactor (Gideon Botes 2009) whose data of the dimensions were taken for the simulations development of the model proposed. Therefore, the proposed model predicts with sufficient accuracy the conversion profiles of the syngas and the C_{5+} selectivity, considering the valuable aspect in the extrapolation of the reactor dimensions (from laboratory scale to pilot scale).

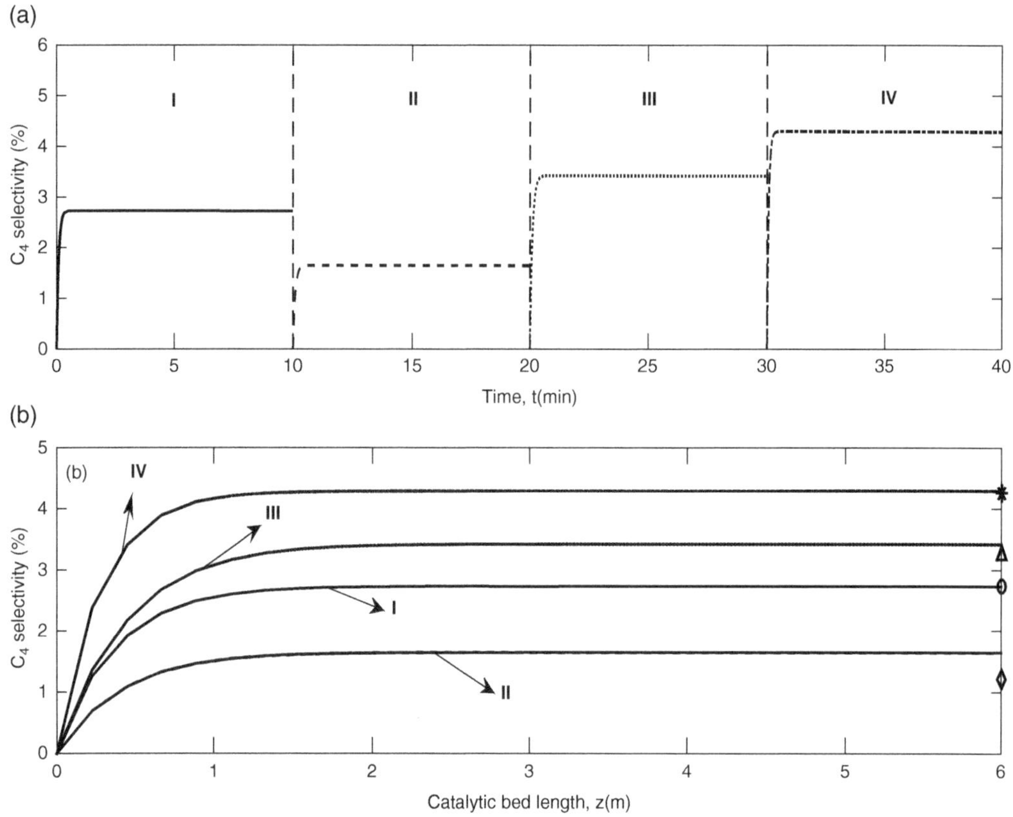

Figure 8.46 n-C_4H_{10}, i-C_4H_{10} selectivity profile. Proposed model (-), experimental data: case I (○), case II (◇), case III (Δ), and case IV (✻). (a) n-C_4H_{10}-i-C_4H_{10} selectivity profile as a function of time, (b) n-C_4H_{10}-i-C_4H_{10} selectivity profile along the catalytic bed.

8.5.4.3 Simulations of the Fischer–Tropsch Fixed-Bed Reactor and the Cooling Jacket Thermal Behavior

Figures 8.48a and 8.48b show the temperature profile for each of the experimental cases under study over time and along the catalytic bed length (axial direction), respectively. It can be appreciated that the PI controller action is fast since it manages to make the FTS temperature reactor be established in the value of the temperature that is set as the set-point. A fast and satisfactory PI controller action is due to a good adjustment of the controller parameters (controller gain and reset time) determined by the adjustment rules given by Tyreus and Luyben (1992).

It is important to point out that to ensure good control of the FTS reactor temperature and to distance itself from the presence of the thermal runaway phenomenon, it is recommended to use a coolant temperature of 5 K below the feed inlet temperature (initial operating temperature) (Jess and Kern 2012a). Therefore, the refrigerant temperature is set at 513 K for the first three cases and at 523 K for case IV.

However, for diameters of catalyst particles of 3 mm (Pöhlmann et al. 2016) to 4 mm (Jess and Kern 2012a), with a diameter ratio of the tube to the catalyst of dt/dp = 7.6, the external diffusional effects are not present in a considerable manner. In addition, the risk of the thermal runaway phenomenon occurring becomes negligible (Pöhlmann et al. 2016).

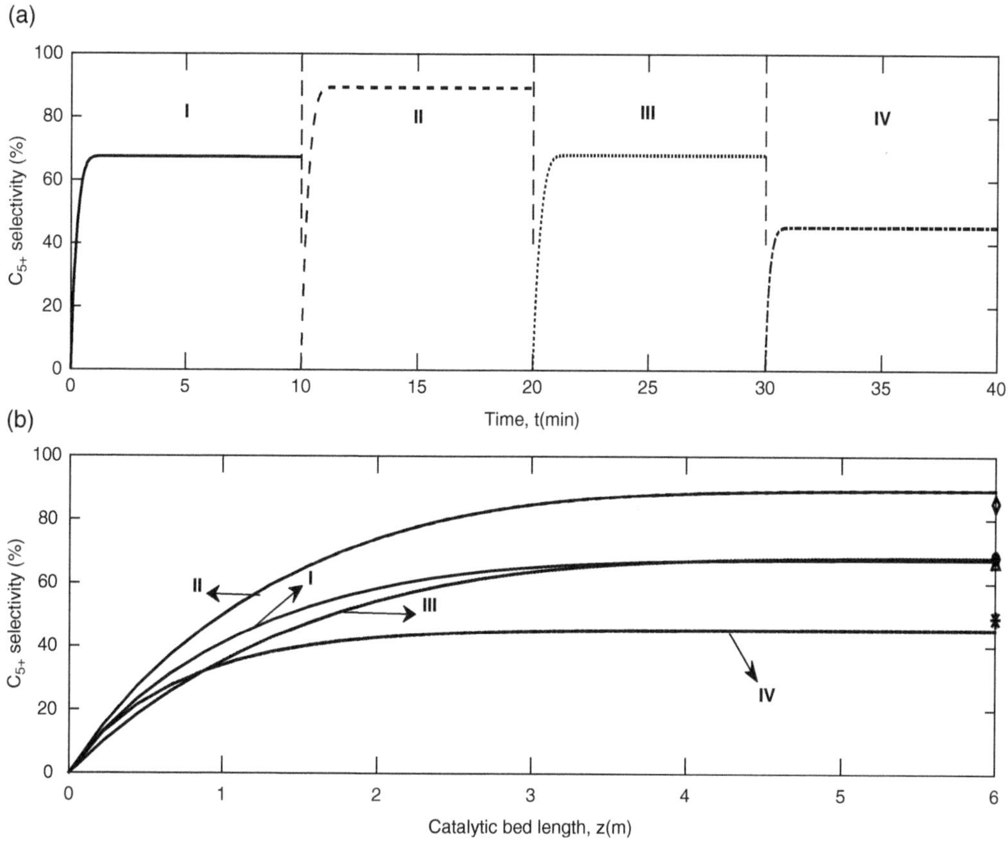

Figure 8.47 C_{5+} selectivity profile. Proposed model (-), experimental data: case I (○), case II (◊), case III (Δ), and case IV (✻). (a) C_{5+} selectivity profile as a function of time, (b) C_{5+} selectivity profile along the catalytic bed.

Figures 8.49a and 8.49b show the temperature profile of the coolant as a function of time and along with the cooling jacket length. It is seen how the refrigerant temperature increases at the jacket inlet because at the reactor inlet the maximum temperature is reached due to the high catalytic activity exhibited by the FTS catalyst. However, the cooling jacket temperature quickly stabilizes at its initial value, which indicates a rapid response from the PI controller.

8.5.4.4 Surfaces of the Syngas Conversion and the Heavy Liquids Selectivity as a Function of the FTS Reactor Temperature

Figures 8.50a, 8.50b, 8.50c, and 8.50d show the surfaces of the change in CO conversion and the liquid hydrocarbons selectivity (C_{5+}) as a function of the FTS reactor temperature for the 4 cases. The objective of the development of these surfaces is to illustrate how the CO conversion and the liquid hydrocarbons selectivity (C_{5+}) are improved simultaneously during the course of the FTS reaction and, at the same time, to show how the C_{5+} hydrocarbons selectivity is favored with the effective CO conversion. It can be seen that case II is the best in terms of the CO conversion and liquid hydrocarbons selectivity.

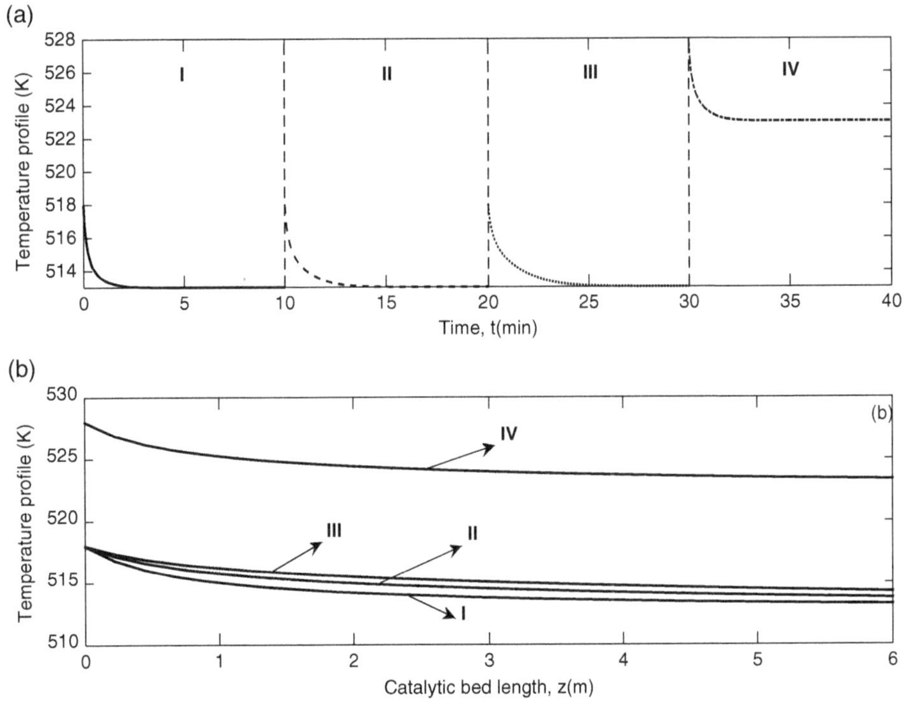

Figure 8.48 Temperature profile of the fixed-bed FTS reactor: (a) temperature profile as a function of time, (b) temperature profile along the catalytic bed length.

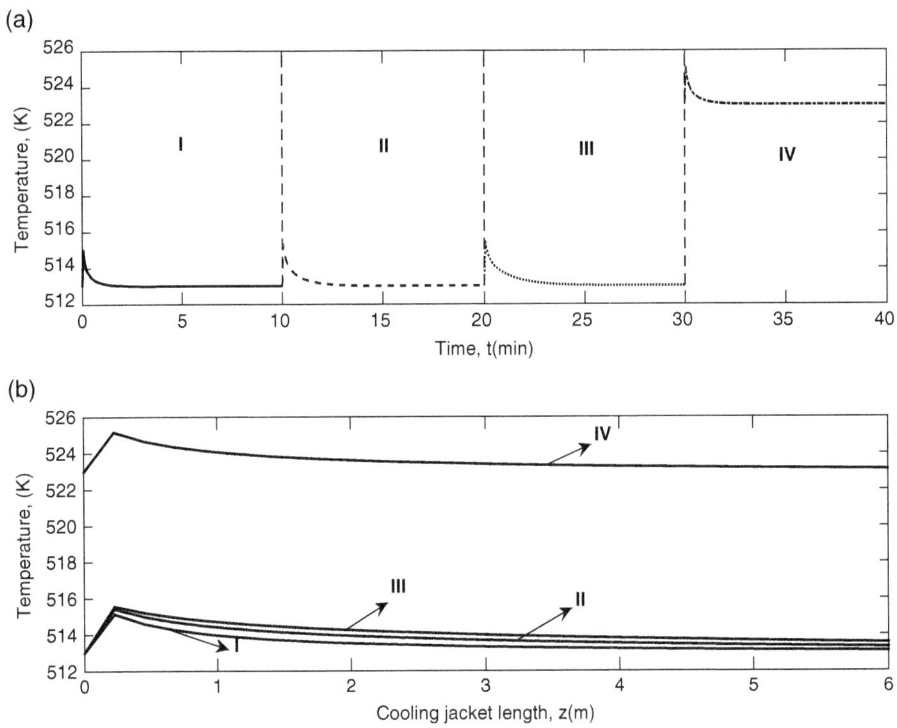

Figure 8.49 Temperature profile of the cooling jacket: (a) temperature profile as a function of time, (b) temperature profile along the catalytic bed length.

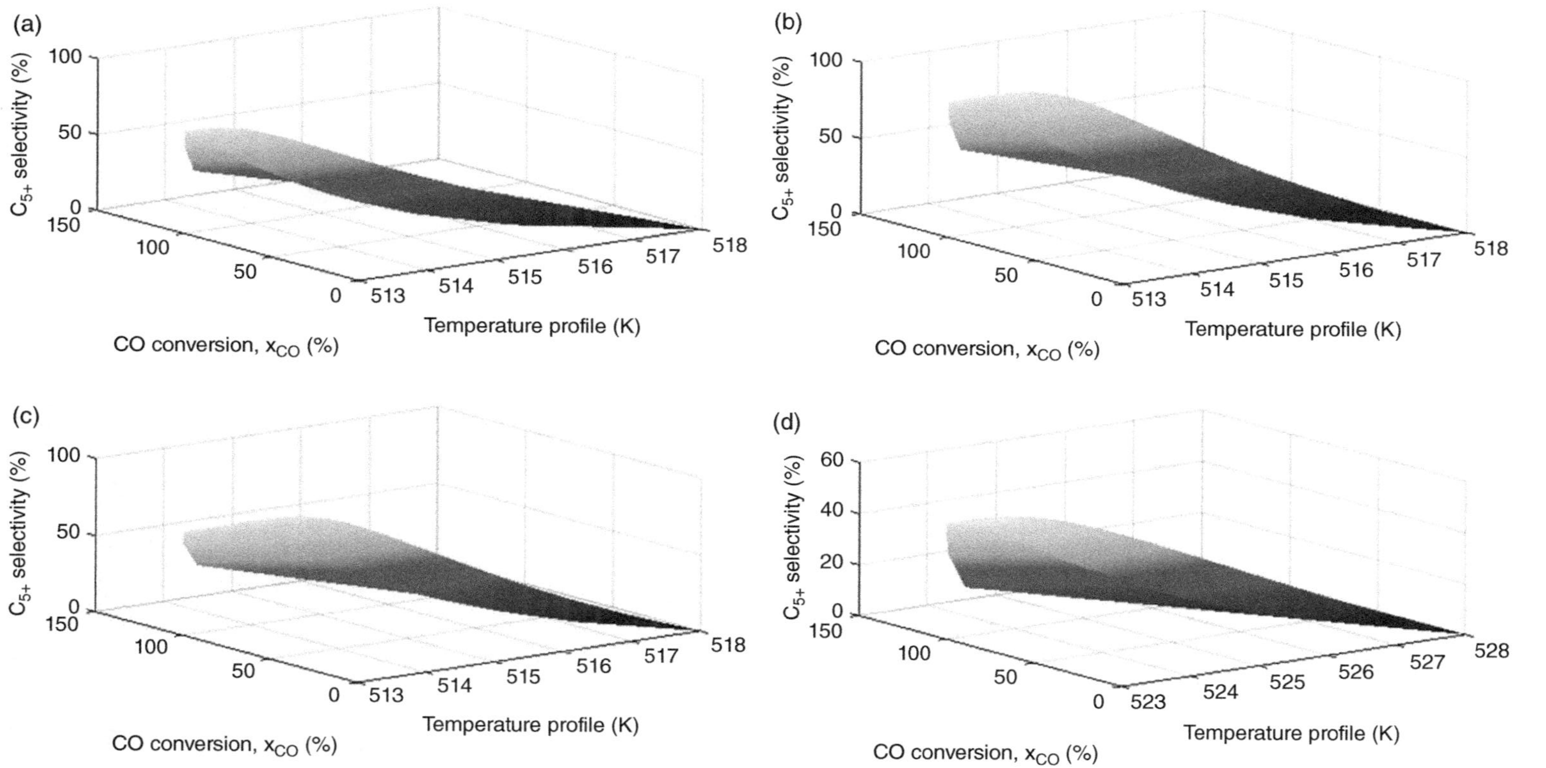

Figure 8.50 Surfaces of the CO conversion and C_{5+} selectivity along the temperature of the fixed-bed FTS reactor. Experimental cases: (a) case I; (b) case II; (c) case III; (d) case IV.

8.5.5 Final Remarks

It can be appreciated that the robustness of the proposed dynamic model, considering a novel kinetic Fischer–Trospch synthesis approach and a simple control scheme, is quite satisfactory. To verify and guarantee the proposed dynamic model with comprehensive FTS kinetics under the PI controller action that was applied to a fixed-bed FTS reactor with pilot-scale dimensions, four sets of operating conditions were used with variable temperature, pressure, and GHSV at the reactor inlet. Case II shows the best result in terms of high CO conversion and liquid hydrocarbons selectivity (C_{5+}), as well as a low amount of unwanted methane production. The PI controller action ensures good control of the reactor temperature allowing the CO conversion and product's selectivity to be favored by the FTS reaction temperature. It is important to highlight that the implementation of a control scheme in the cooling jacket of the fixed-bed FTS reactor is the key element to ensure good thermal behavior and elimination of the runaway phenomenon. This new approach shows great potential to be extended to scales of larger dimensions.

8.6 On the use of Steady-State Optimal Initial Operating Conditions for the Control Scheme Implementation of a Fixed-Bed Fischer–Tropsch Reactor

8.6.1 Introduction

A control scheme implementation for a fixed-bed FTS reactor using the optimal initial operating conditions at steady state is developed. The proposed approach consists of solving a nonlinear constrained optimization problem for a fixed-bed FTS reactor with incorporation of a control scheme in the cooling jacket to find the optimal values of the initial operating conditions of temperature, GHSV, pressure, and coolant flow velocity under a steady-state regime that maximizes CO conversion and C_{5+} product selectivities. Such optimal operating conditions at steady state are used to evaluate the dynamic behavior of the reactor under the PIC scheme action that is implemented in the cooling jacket. An extended dynamic one-dimensional pseudohomogeneous model of a fixed-bed FTS reactor recently reported in the literature is used. The obtained results show the importance of the implementation of dynamic control strategies for the FTS carried out in a fixed-bed reactor, since they allow for ensuring the thermal control of such highly exothermic reactions and assure better reactor performance in terms of high CO conversion and liquid product's selectivity (C_{5+}).

8.6.2 Methodology

In this section, the set of dynamic equations of the reactor model based on mass and energy balances is stated, including the model that describes the dynamic behavior of the cooling jacket. The correlations are cited to calculate the overall heat transfer coefficient and the pressure drop across the catalyst bed. In addition, the optimization problem with non-linear constraints is described.

8.6.2.1 Model Equations and Numerical Solution

The model describing the dynamic behavior of the fixed-bed FTS reactor was implemented under plug-flow regime at non-isothermal conditions, considering the variations of the GHSV, the gas mixture density, and total pressure of the system in the axial direction, which have demonstrated to be important parameters to improve predictions of syngas conversion and temperature profiles

(Méndez et al. 2019). It also incorporates the cooling jacket dynamic equation. The equations of the dynamic reactor model are given by (8.114)–(8.120) and (8.131)–(8.132). The equations are solved by using the orthogonal collocation numerical method. For approaching to the solution of the model, five collocation points were used. The Legendre orthogonal polynomials were used for the estimation of each collocation point (Constantinides and Mostoufi 1999).

8.6.2.2 Model Parameters, Correlations, and Kinetics

The parameters and correlations used in the reactor model are given in Table 8.20. The global heat transfer coefficient was estimated with the correlation proposed by Haarlemmer and Bensabath (2016), which is given as a function of the convective heat transfer coefficient on the internal side of the reactor (h_{int}), the convective heat transfer coefficient of the external side (h_{ext}), and the reactor thermal conductivity (λ_w), being the reactor made of stainless steel for this case. The pressure drop along the catalytic bed was estimated using a correlation recently stated in the literature for the friction factor in which superior accuracy was observed by comparison with several reports (Allen et al. 2013). The heat conductivity of the steel wall reactor was assumed to be 60 $Wm^{-1}h^{-1}$ (Sharma et al. 2011), and the value of the heat transfer coefficient (film) of the external side corresponding to boiling water coolant was 1000 $Wm^{-1}h^{-1}$ (Haarlemmer and Bensabath 2016). Other parameters were considered as constants, such as the hollow fraction of the bed and the active side volume fraction of the catalyst. The kinetic model, reaction scheme, and the corresponding product's selectivity used in the proposed reactor model are given by Eqs. (8.121)–(8.129) and (8.100)–(8.113), respectively. There are several studies on the kinetics of the FTS reactions focused on the formation of hydrocarbons based on the LHHW model; however, some authors have developed simpler kinetic models based on power-law type (Marvast et al. 2005; Yang et al. 2003; Wang et al. 2003) fitting well to experimental data. Table 8.26 details the kinetic parameters (pre-exponential factor, reaction orders for the partial pressures of H_2 and CO, and the activation energy) used in this work and previously reported by Moazami et al. (2015a).

8.6.2.3 Steady-State Nonlinear Constrained Optimization Problem

Satisfactory performance of the FTS process is defined in terms of high syngas conversion and high liquid hydrocarbon productivity ($S_{C_{5+}}$). However, it is important to ensure good thermal behavior of the process by having good control over the FTS reactor, which is temperature-restricted to provide high syngas conversions and liquid hydrocarbons productiveness. For this purpose, the temperature interval allowing a syngas conversion and liquid hydrocarbons productivity above or equal to 80% is 493 K $\leq$ T $\leq$ 513 K.

Then, in order to establish an optimal zone of operation at steady state that faces process disturbances and a fixed value in the uncertain parameters, the following nonlinear constrained optimization problem is defined:

$$\max_{\eta, d, \xi} x_{CO} \tag{8.143}$$

subject to

$$h(x, \eta, d, \xi) = 0 \tag{8.144}$$

$$T^l \leq T \leq T^u \tag{8.145}$$

$$x_{CO} \geq x^l_{CO} \tag{8.146}$$

$$S_{C_{5+}} \geq S^l_{C_{5+}} \tag{8.147}$$

Table 8.26 Kinetic parameters of the FTS reaction (Adapted from Moazami, et. al. 2015a).

Reaction	R_1		R_2		R_3		R_4		R_5		R_6		R_7		R_8	
Kinetics parameters	A_1	E_{a1}	A_2	E_{a2}	A_3	E_{a3}	A_4	E_{a4}	A_5	E_{a5}	A_6	E_{a6}	A_7	E_{a7}	A_8	E_{a8}
	65.56165	83523.9	0.045438	65017	0.001144	49782	1.19E-06	34885.5	1.84E-10	27728.9	6.59E-09	25730.1	1.98E-08	23564.3	3.78E-05	58826.3
Reaction order	m_1	n_1	m_2	n_2	m_3	n_3	m_4	n_4	m_5	n_5	m_6	n_6	m_7	n_7	m_8	n_8
	−0.4755	0.8201	0.014	0.0746	−0.0594	0.4128	−0.7131	1.4167	0.1037	−0.4789	−0.2626	1.1713	1.2895	0.4571	0.472	0.7706

where $h(x, \eta, d, \xi)$ is the proposed model given by Eqs. (8.114)–(8.120) and as a function of state variables (x), process disturbances (d), and parameter uncertainties (ξ). T^u and T^l are the upper and lower bounds of the reactor temperature, and x^l_{CO} and $S^l_{C_{5+}}$ are the lower bounds for the CO conversion and liquid hydrocarbons selectivity ($S_{C_{5+}}$), respectively.

The optimal operating conditions for a steady-state process are often not optimal for a process at unsteady state because the optimality region for a steady-state process differs considerably from the optimal operating region for a dynamic process. Figure 8.51 shows schematically how the optimality region of a process in a steady state tends to move due to process dynamics. In addition, it is important in any dynamic problem to consider the process disturbances effects and the parameters uncertainty to determine a robust solution under different restrictions. On the other hand, the phenomenon of "temperature runaway" is very sensitive to the dynamics of the inlet temperature, the hot spot can appear immediately or later at an initial temperature value. For this reason, the presented optimization method is applied to find the best fixed-bed reactor inlet temperature for steady-state situations.

Figure 8.51a highlights the optimality of the steady-state process considering a fixed value in the uncertain parameters and the disturbances effect. It is seen that certain process constrained separates the unfeasible region. The point located in the constrained line indicates the steady-state optimal design point, and the surrounding region becomes the set of points in an optimal neighborhood in which the process can feasibly operate. However, this point might violate constraints when transient changes are accounted for in the analysis, as schematically illustrated by a black dotted region in Figure 8.51a (Mehta and Ricardez-Sandoval 2016; Rafiei-Shishavan et al. 2017). Nevertheless, for this case study, the effect of the process dynamics does not greatly alter the feasibility.

Therefore, in order to find the optimal values of the inlet operating conditions (temperature, pressure, and coolant flow rate) that determine the optimality region in steady state, the nonlinear constrained optimization problem to solve can be written in the following general form:

$$\max_{\eta, d, \xi} f(x, \eta, d, \xi)$$

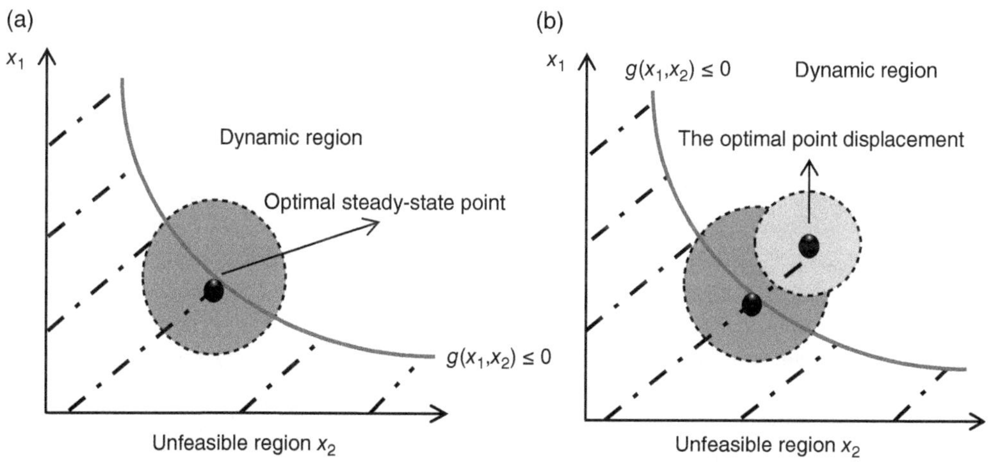

Figure 8.51 Schematic diagram of optimality (Mehta and Ricardez-Sandoval 2016; Rafiei-Shishavan et al. 2017): (a) steady state, (b) dynamic state.

subject to

$$g_j(x) \geq b_{j,} j = 1, \ldots, m \quad (8.148)$$

where $x = (x_1, \ldots, x_n)$ is the vector of decision variables, and each b_j is a constant. The objective f and constraint functions g_j are assumed to have continuous first partial derivatives. Corresponding to each constraint $g_j = b_j$, a Lagrange multiplier λ_j is defined, and $\lambda_j = (\lambda_1, \ldots, \lambda_m)$ is the vector of these multipliers (Tosun 2007). The Lagrangian function for the problem is:

$$L(x, \lambda) = f(x) + \sum_{j=1}^{m} \lambda_j \left[g_j(x) - b_j \right] \quad (8.149)$$

and the first-order necessary conditions for optimality according to Kuhn–Tucker are:

$$\frac{\partial L}{\partial x_i} = \frac{\partial f}{\partial x_i} + \sum_{j=1}^{m} \lambda_j \frac{\partial h_j}{\partial x_i} = 0, i = 1, \ldots, n \quad (8.150)$$

$$g_j(x) \geq b_{j,} j = 1, \ldots, m \quad (8.151)$$

where m are the set of constraints and n are the process variables. In addition, it is essential that the necessary and sufficient second-order Kuhn–Tucker conditions are met to ensure the optimization of the nonlinear constraint optimization problem. The second-order optimality conditions involve the matrix of second partial derivatives with respect to x (the Hessian matrix of the Lagrangian function) and may be written as follows (Biegler 2010; Edgar and Himmelblau 2001):

$$y^T \nabla_x^2 L(x^*, \lambda^*) y > 0 \quad (8.152)$$

for all nonzero vectors y such that:

$$J(x^*) y = 0 \quad (8.153)$$

where $J(x^*)$ is the Jacobian matrix, whose rows are the gradients of the constraints that are active at x^*. The abovementioned equation determines a set of vectors that are orthogonal to the active constraints (Edgar and Himmelblau 2001).

The steady-state nonlinear constrained optimization problem was initialized using an initials temperature, pressure, and flow rate (at the reactor inlet) of 505 K, 3.0 MPa, and 15.6 L/min, respectively, as well as an initial coolant temperature (boiling water) of 498 K. The uncertain parameters considered in the optimization problem were the active catalyst fraction (β) and the overall heat transfer coefficient (U_r). The disturbance that greatly affects the reactor behavior at steady state is the GHSV. The values used for the GHSV were 1800, 2400, 3000, and 3600 $\mathrm{NmLg_{cat}^{-1}h^{-1}}$.

The results of the optimization problem solution indicate optimum initial conditions of temperature, pressure, and flow rate of 503 K, 2.5 MPa, and 13.5 L/min, respectively. On the other hand, the value of the Lagrange multiplier that allows obtaining the best optimal operating conditions was 0.71. Such a set of optimal values defines the solution of the steady-state nonlinear constrained optimization problem accomplishing the established inequality constraints. These optimum values were used to solve the closed-loop reactor model by implementing a PI control scheme in the cooling jacket.

The steady-state nonlinear constrained optimization problem for the FTS fixed-bed reactor was solved using Python software. In Figure 8.52, the algorithm of the nonlinear optimization problem with implementation of the control scheme is shown.

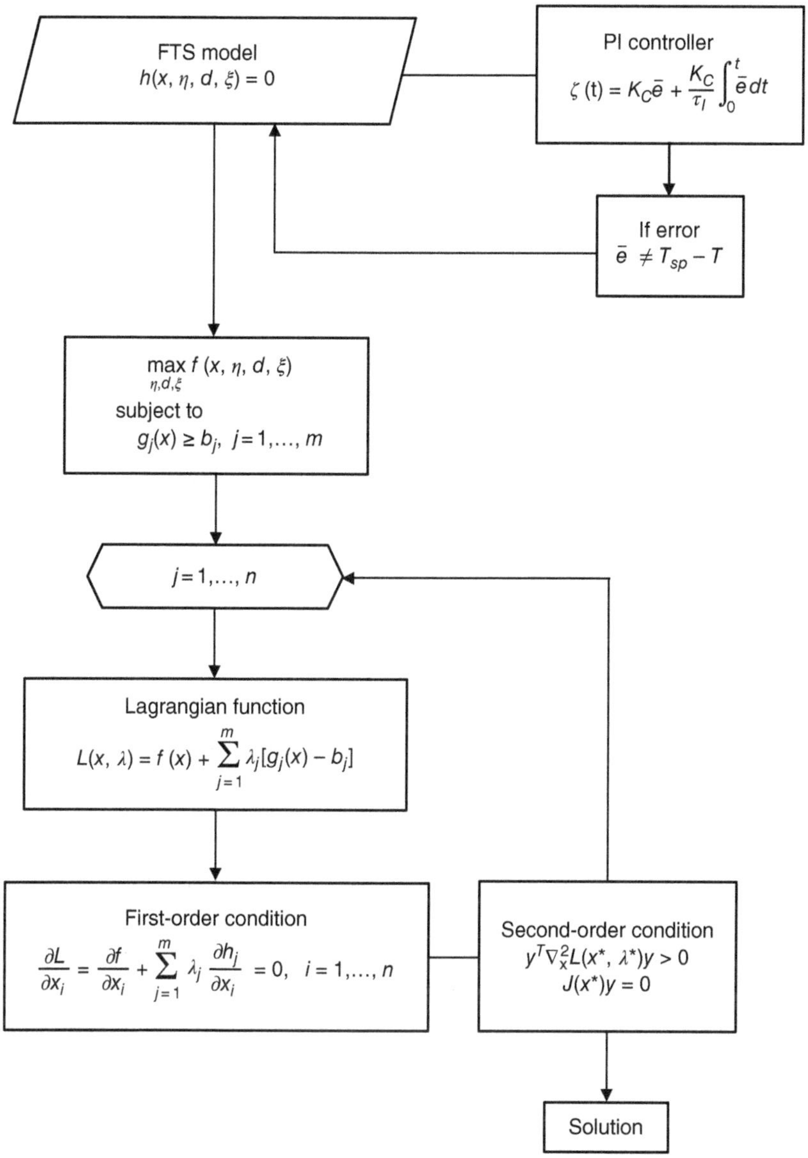

Figure 8.52 Algorithm of the nonlinear optimization problem with implementation of the control scheme.

8.6.2.4 Implementation of the Control Scheme

To ensure a viable operation of the reactor in terms of a safe thermal operation, it is necessary to implement a control scheme in the cooling jacket. It is well known that the FTS reaction is highly exothermic, which requires good temperature control in any reactor configuration used (Kaskes et al. 2016; Kwack et al. 2011a). The high temperatures of the FTS reaction give rise to the hot-spot formation, which can lead to the presence of the runaway phenomenon (Kwack et al. 2011a; Rafiq et al. 2011).

The literature indicates that a PI controller is suitable in those cases that involve a highly non-linear process and where it is intended to eliminate the static error (off-set) (Stephanopoulos 1985). On the other hand, it is stated that PI control is a form of continuous adjustment originating in the process industries and is usually considered part of engineering process control (Box and Luceño 1997).

To ensure adequate temperature control of the fixed-bed FTS reactor, a PIC scheme is implemented in the cooling jacket where the reactor temperature is controlled by adjusting the temperature of the refrigerant inside the cooling jacket by manipulating the flow velocity of the coolant. The PI controller is given by Eq. (8.141), and $\bar{e}$ is the error given by Eq. (8.142).

The set-point established in the control problem becomes the reactor temperature of 503 K, and in order to achieve a negligible deviation in the process response, the flow velocity of the refrigerant is manipulated. The controller parameters (controller gain K_C and reset time τ_I) are determined by the adjustment rules given by Tyreus and Luyben (1992) whose values were $K_C = 2.6$ and $\tau_I = 0.25$ min.

Figure 8.53 illustrates the line diagram of the PI controller implemented in the cooling jacket.

8.6.3 Results and discussion

In this section, the results of simulations are discussed. The conversion of CO and selectivity of liquid hydrocarbons are obtained. The temperature profiles of the FTS reactor and cooling jacket are analyzed. CO conversion and selectivity of liquid hydrocarbons as a function of time under the PI controller action are also shown.

8.6.3.1 Experimental Data

The model simulations are compared with experimental data reported by Moazami et al. (2015a). The experiments were carried out using a Co/SiO_2 catalyst, which was diluted with 2 g of silicon carbide at a mass ratio of 1:12 ($Co/SiO_2/SiC$). A simulated nitrogen-rich syngas was used with a concentration of 17, 33, and 50 vol.% of CO, H_2, and N_2, respectively.

The dimensions of the bench-scale fixed-bed reactor used in this case study are 6 m in length as a catalytic bed with an internal radius (Dai et al. 2014).

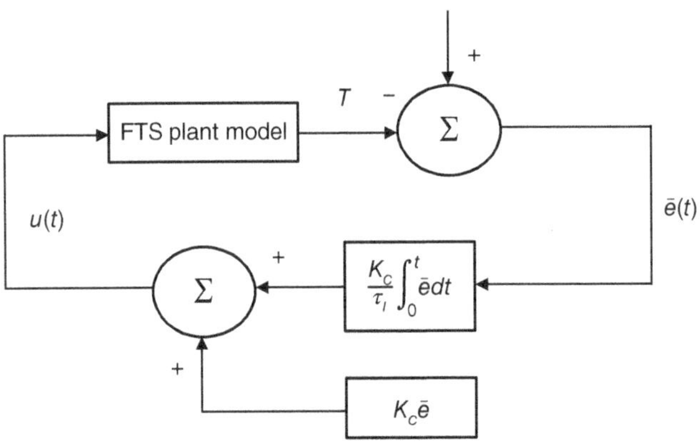

Figure 8.53 Line diagram of the PI controller implemented in the cooling jacket.

8.6.3.2 Simulations of the Steady-State Nonlinear Constrained Optimization Problem: CO Conversion, $S_{C_{5+}}$ Selectivity, and Temperature Profiles

As previously mentioned, the optimal operating conditions at the inlet of the FTS reactor for the nonlinear constrained optimization problem were: $T_{in} = 503$ K, $P_{T,in} = 2.50$ MPa, and $F_{w,in} = 15.6$ L/min. These optimal conditions values were used for the development of the steady-state one-dimensional pseudohomogeneous model simulations.

The CO conversion and selectivity of liquid hydrocarbons are shown in Figures 8.54a and 8.54 b, respectively. The CO conversion reached is close to 100% due to the reactor temperature effect. The optimal initial operating conditions found according to the steady-state nonlinear constrained optimization problem do not affect the restrictions imposed in the CO conversion. On the other hand, the liquid hydrocarbons selectivity is above 81%. It can be verified that the selectivity toward products is enhanced by the effect of the temperature along the catalytic bed. In addition, the low production of light gases also allows the liquid hydrocarbons selectivity to increase because of the effect of GHSV. In such a way that the optimal values for temperature, GHSV, and coolant flow velocity at the reactor inlet do not infringe the constraints imposed in the liquid hydrocarbon products according to the nonlinear problem defined.

Figures 8.55a and 8.55b illustrate the FTS reactor and the cooling jacket temperature profiles along the catalytic bed at steady-state conditions, respectively. It can be clearly seen that the hot spot along the catalytic bed is 3 K, almost formed at the reactor inlet. However, the temperature

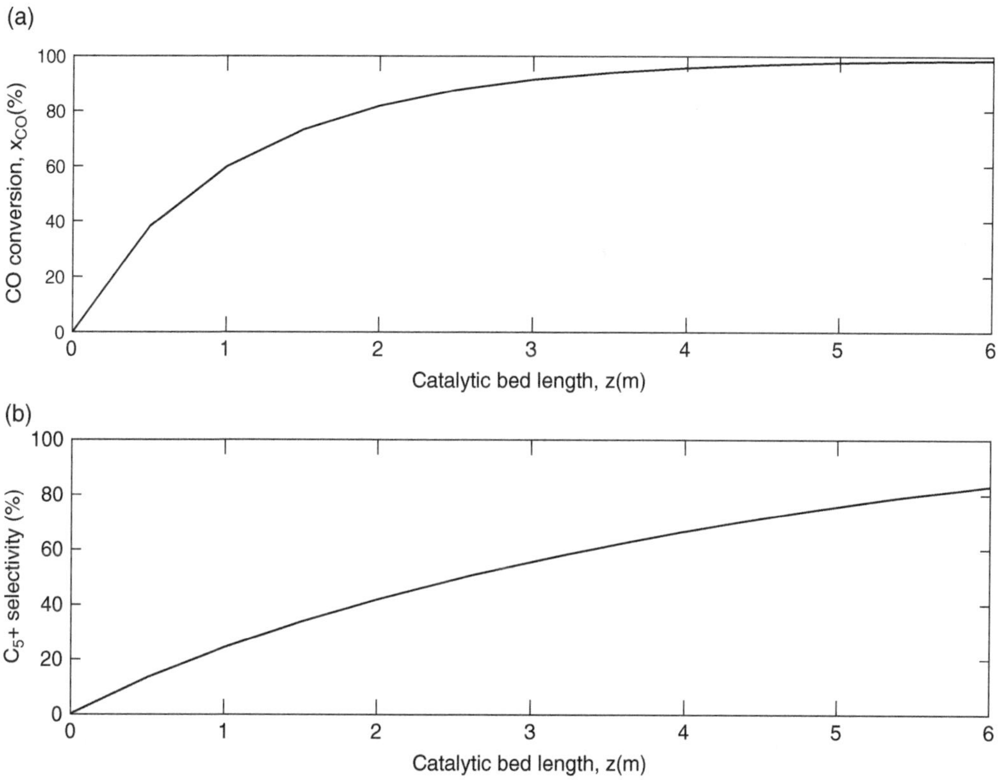

Figure 8.54 CO conversion and liquid hydrocarbons profiles evaluated with the optimal inlet condition values: (a) CO conversion, (b) hydrocarbons liquid ($S_{C_{5+}}$) selectivity.

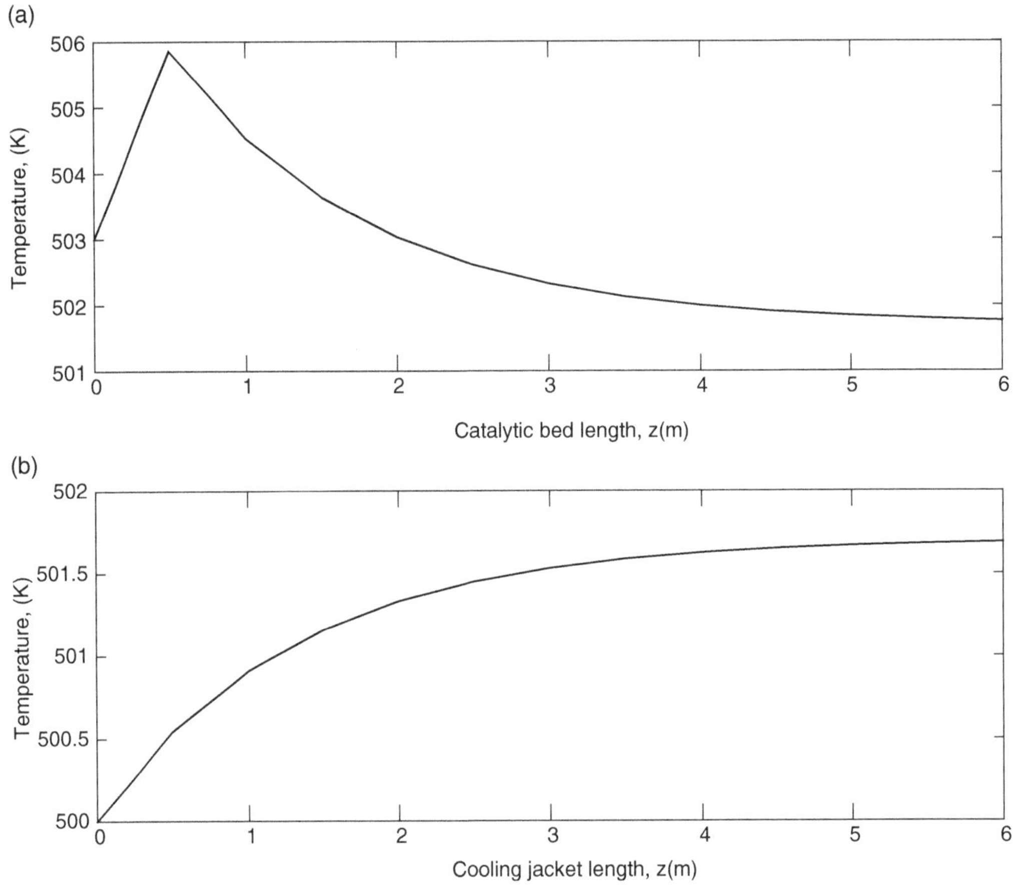

Figure 8.55 Temperature profiles evaluated with the optimal inlet condition values: (a) FTS fixed-bed reactor, (b) cooling jacket.

peak is relatively low such that the runaway phenomenon is not present. The temperature of the cooling jacket rises by about 1.7 K and does not become stable, which is due to the process operating in open loop (without control action on the cooling jacket). It is important to note that the temperature along the catalytic bed of a fixed-bed reactor, in which a highly exothermic reaction is carried out such as in the case of an FTS reaction, can be triggered to cause the presence of the runaway causing damages in the reactor. However, a good control of the FTS reactor temperature is not only guaranteed by the integration of a cooling jacket in the reactor as can be seen in Figure 8.55b. Therefore, it is interesting to verify the integration of a dynamic control scheme in a jacketed fixed-bed reactor involving highly exothermic reaction takes place to ensure good control of temperature, as will be demonstrated below.

8.6.3.3 Simulations of the Control Scheme Implementation: CO Conversion, $S_{C_{5+}}$ Selectivity, and Temperature Profiles

The closed-loop system (FTS reactor with control in the cooling jacket) is studied to evaluate the dynamic behavior of the process under the action of the PI control scheme. To verify the robustness of the PI control implemented in the FTS reactor, variations in the GHSV are made by setting a

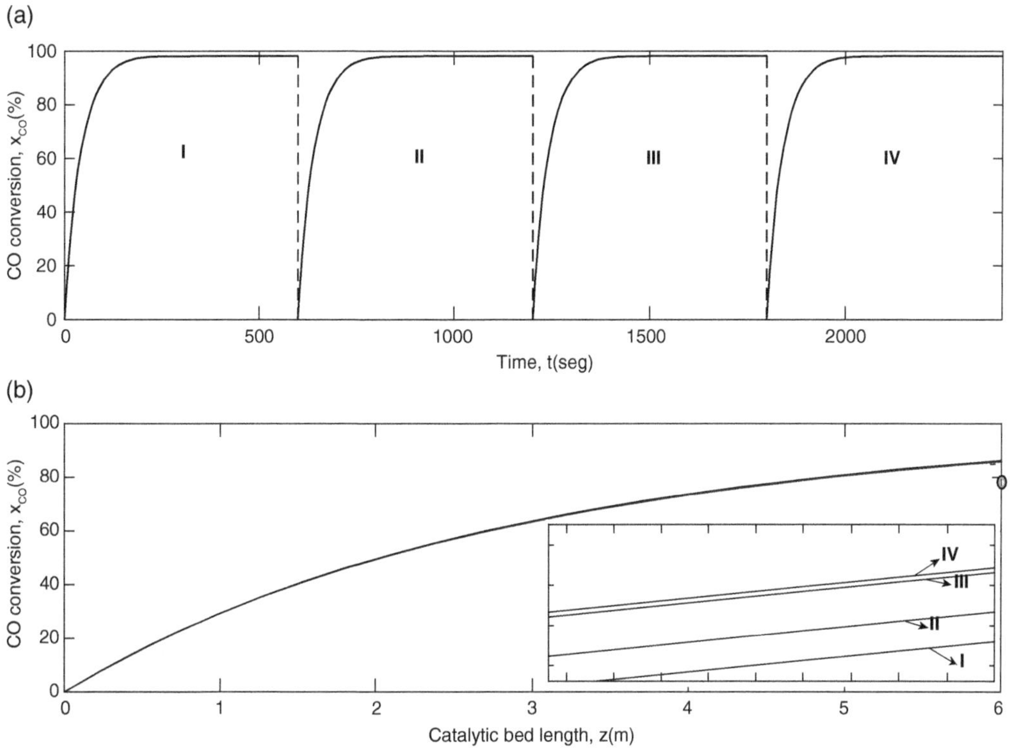

Figure 8.56 Simulated CO conversion profile in the fixed-bed FTS reactor under PI action control (I: GHSV = 1800, II: GHSV = 2400, III: GHSV = 3000, IV: GHSV = 3600 NmL/g_{cat}/h): (a) Dynamic CO conversion, (b) CO conversion along the catalytic bed, (○) Experimental data (Adapted from Kuipers, et. al. (1996)).

value in the uncertain parameters. It should be noted that the continuous processes have been continuously improved to achieve the process optimality around a nominal steady-state operating point (Bansal et al. 2000). In addition, it is important to determine those conditions in which a continuous process can operate in a feasibility region (Pistikopoulos and Mazzuchi 1990; Dimitriadis and Pistikopoulos 1995).

Figure 8.56a shows the CO conversion as a function of time under the PI controller action, for four changes in steps made in the considered disturbance (GHSV). It can be clearly seen that the CO conversion is close to 99% for each experimental run lasting 10 min. On the other hand, the CO conversion along the catalytic bed (Figure 8.56b) is about 88%. The high conversions are due to the effect of the FTS reactor temperature. The higher the pressure, the higher the syngas conversion (van Berge and Everson 1997; Yan et al. 2009), but if the syngas space velocity is increased, a decrease in the residence time is observed, thus diminishing the CO conversion (Lee and Chung 2012). In addition, the error observed between the experimental data reported by Moazami et al. (2015a) at the reactor outlet and the simulated value for the CO conversion is low. However, it is important to note that care should be taken with the simultaneous changes of temperature and space velocity, because high temperatures shift the FTS process toward the production of heavy hydrocarbons (Dry 1981), while increasing the space velocity and decreasing slightly the CO conversion, but if adequate values are established, then high production rates without altering the quality of the product can be obtained (Lee and Chung 2012). The optimal operating conditions

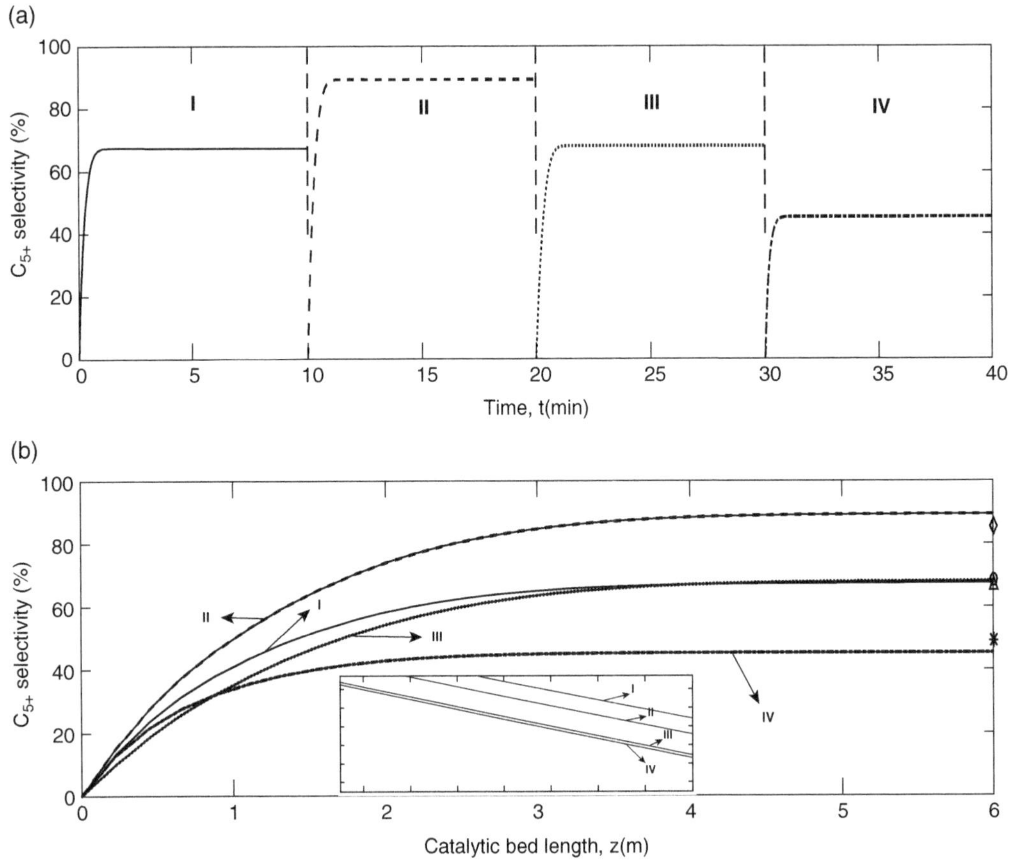

Figure 8.57 Simulated $S_{C_{5+}}$ liquid hydrocarbons selectivity profile in the fixed-bed FTS reactor under PI action control (I: GHSV = 1800, II: GHSV = 2400, III: GHSV = 3000, IV: GHSV = 3600 NmL/g_{cat}/h): (a) Dynamic CO conversion, (b) CO conversion along the catalytic bed; (O) Experimental data (Adapted from Kuipers, et. al. (1996)).

determined for the steady-state nonlinear constrained optimization problem accomplish the constraints imposed on the CO conversion for the dynamic problem under the PI controller action.

Figure 8.57a shows the liquid hydrocarbons selectivity dynamic profile under the PI control action scheme implemented. For the fourth step changes made in the gas space velocity, the liquid product's selectivity is around 44–45%. In this case, the constraints imposed on the liquid hydrocarbons selectivity are not accomplished, which is largely due to the process dynamics (Mehta and Ricardez-Sandoval 2016; Rafiei-Shishavan 2017). However, a value of between 44 and 45% in the liquid product's selectivity is acceptable. Nevertheless, the $S_{C_{5+}}$ selectivity along the axial direction of the reactor remains above 90% (Figure 8.57b). It is important to note that the selectivity of $S_{C_{5+}}$ does not show a significant variation (Figure 8.57b inset), which is caused by the simultaneous effect of temperature, pressure, and GHSV (Kuipers et al. 1996). The error observed between the experimental data reported by Moazami et al. (2015a) at the reactor outlet and the simulated value for $S_{C_{5+}}$ selectivity is practically negligible. Figure 8.58a shows the dynamic temperature profile of the FTS reactor under the PI controller implementation. As GHSV increases, the formed hot spot also increases. However, the action of the PI controller is quite satisfactory since the reactor temperature does not rise considerably and it stabilizes around the established set point. On the

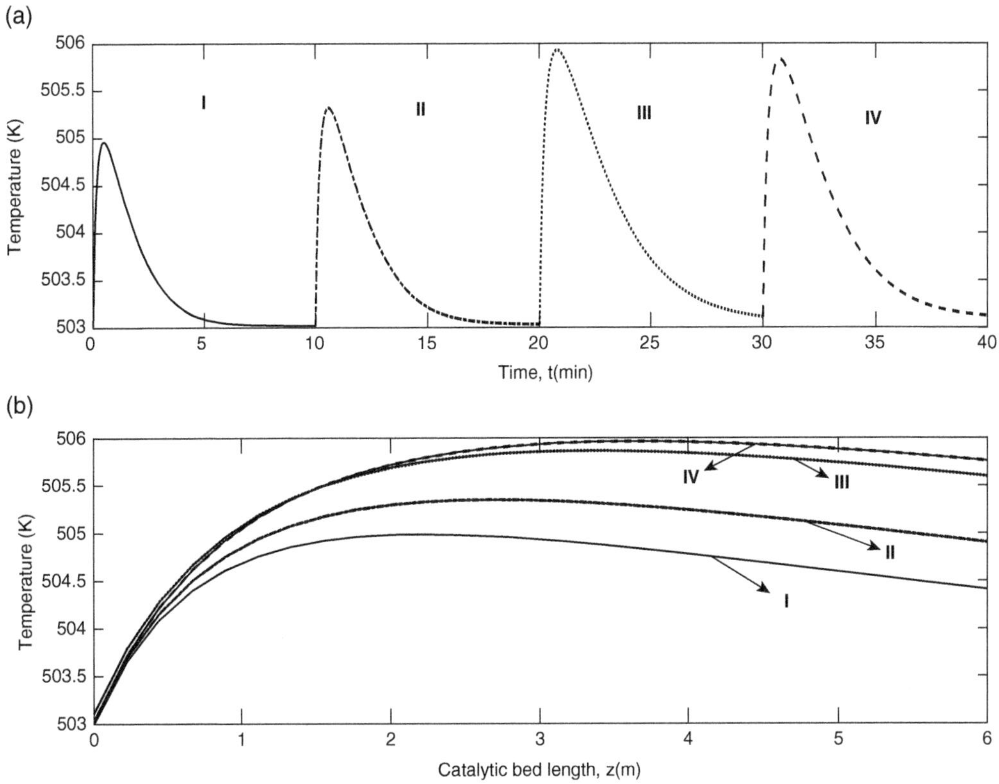

Figure 8.58 Simulated temperature profile in the fixed-bed FTS reactor under PI action control (I: GHSV = 1800, II: GHSV = 2400, III: GHSV = 3000, IV: GHSV = 3600 NmL/g_{cat}/h): (a) Dynamic temperature, (b) temperature along the catalytic bed.

other hand, Figure 8.58b illustrates the temperature profile of the FTS reactor along the axial direction. As the disturbances (GHSV) value increases in each pitch step applied every 10 min, the reactor temperature rises until a hot spot of 3 K is formed above the inlet temperature of the FTS reactor. Thus, the PI controller shows satisfactory robustness because it can stabilize the reactor temperature near the set point with negligible error.

A very important aspect that is verified in the thermal behavior of the FTS reactor is that the hot spot formed undergoes a delay due to the control action. That is, the hot spot that is usually formed has been shown to originate near the reactor inlet. However, the PI controller allows stabilizing the reactor temperature near the set point and achieves the hot spot formed beyond the reactor inlet.

Finally, the temperature profiles along the cooling jacket length and time are shown in Figures 8.59a and 8.59b. Figure 8.59a illustrates the dynamic behavior of the refrigerant temperature (boiling water) under the control action produced by the PI controller. The changes are made in the GHSV (disturbances) with the same initial conditions of concentration and pressure of the syngas. For the temperature at the reactor outlet (which is remarkably close to the input temperature of 503 K) the initial temperature for the next experimental run (simulation) is used.

According to the temperature profile of the coolant, hot spots are formed beyond the cooling jacket inlet, which is neglected (Figure 8.59b). However, in terms of the process response to the controller action, it is verified that the traditional PI control scheme can keep the reactor thermal

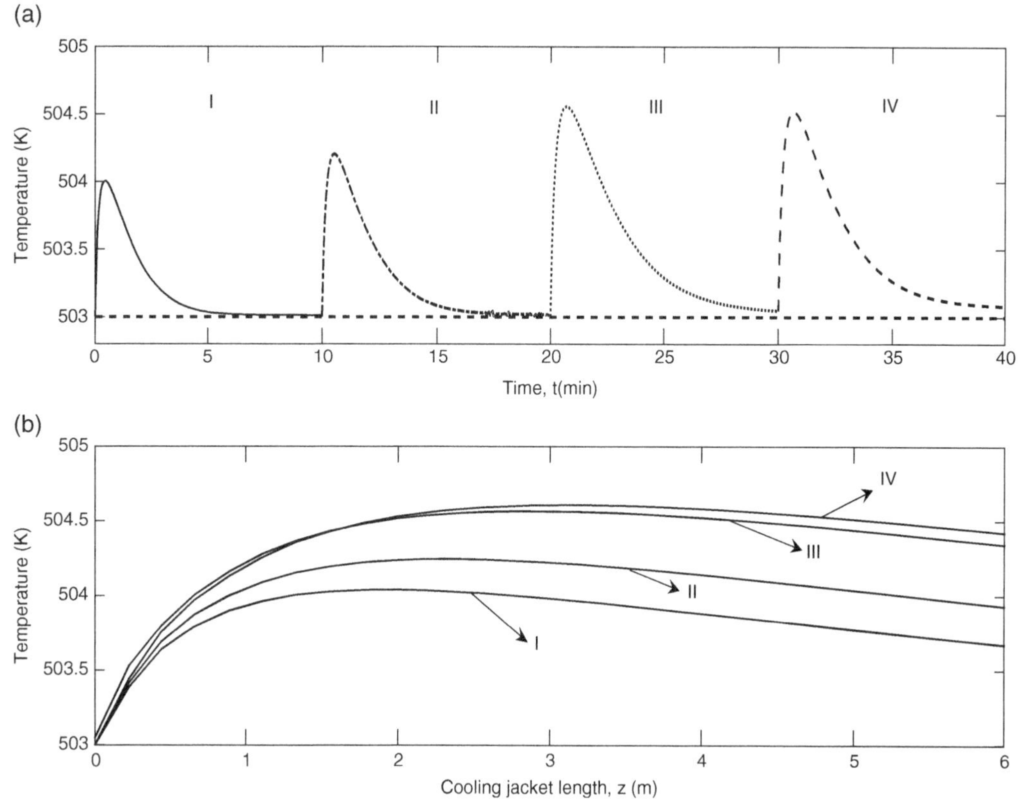

Figure 8.59 Simulated temperature profile in the cooling jacket under PI action control (I: GHSV = 1800, II: GHSV = 2400, III: GHSV = 3000, IV: GHSV = 3600 $\frac{NmL}{g_{cat}}$/h): (a) Dynamic cooling jacket temperature, (b) cooling jacket temperature along the catalytic bed.

operation within an operational feasibility region. On the other hand, the behavior of the controller against the disturbances made in the process (GHSV) and the uncertainty parameters (the active side catalyst and overall heat transfer coefficient) under the optimal operating conditions (according to the solution of the steady-state nonlinear constrained optimization problem) are completely satisfactory.

8.6.4 Final Remarks

The solution of a nonlinear constrained optimization problem at steady-state conditions is: a crucial step in determining the operational feasibility region. In this case study, the process optimality in terms of the inlet operating optimal conditions values is used to evaluate the process dynamics under the traditional PI control scheme implementation. It was possible to demonstrate that the PI controller is robust against disturbances of the GHSV for a fixed value in the uncertain parameters considered. The controller allows keeping the FTS reactor temperature around the established set-point with an insignificant error. In addition, the CO conversion and the liquid hydrocarbons selectivity are satisfactory, verifying a negligible error between experimental and simulated values. On the other hand, the optimal values of the inlet operation conditions (temperature, pressure, and refrigerant flow velocity), although when used in the evaluation of the closed-loop process (on control action), do not accomplish the constraint imposed on the liquid product's selectivity ($S_{C_{5+}}$), the dynamic process behavior is not of considerable attention, since the selectivity reached at the reactor outlet is high. In summary, the implementation of a dynamic control scheme in fixed-bed FTS reactors is extremely important, as demonstrated in this study.

References

Adesina, A.A. (1996). Hydrocarbon synthesis via Fischer-Tropsch reaction: travails and triumphs. *Applied Catalysis A: General* 138: 345–367.

Ahón, V.R., Costa, E.F., Monteagudo, J.E.P. et al. (2005). A comprehensive mathematical model for the Fischer–Tropsch synthesis in well-mixed slurry reactors. *Chemical Engineering Science* 60: 677–694.

Akpan, E., Sun, Y., Kumar, P. et al. (2007). Kinetics, experimental and reactor modeling studies of the carbon dioxide reforming of methane (CDRM) over a new Ni/CeO_2-ZrO_2 catalyst in a packed bed tubular reactor. *Chemical Engineering Science* 62: 4012–4024.

Al-Dahhan, M.H., Larachi, F., Dudukovic, M.P., and Laurent, A. (1997). High-pressure trickle-bed reactors: a review. *Industrial and Engineering Chemistry Research* 36: 3292–3314.

Alihellal, D. and Chibane, L. (2016). Comparative study performance of Fischer-Tropsch synthesis in conventional packed bed and in membrane reactor over iron-and cobalt-based catalysts. *Arabian Journal for Science and Engineering* 41: 357–369.

Allen, K., von Backström, T., and Kröger, D. (2013). Packed bed pressure drop dependence on particle shape, size distribution, packing arrangement and roughness. *Powder Technology* 246: 590–600.

Ancheyta, J., Muñoz, J., and Macías, M. (2005). Experimental and theoretical determination of the particle size of hydrotreating catalysts of different shapes. *Catalysis Today* 109: 120–127.

Anderson, R.B. and Karn, F.S. (1960). Kinetics and reaction mechanism of the Fischer-Tropsch synthesis. *Journal of Physical Chemistry* 64: 805–808.

Atashi, H., Siami, F., Mirzaei, A., and Sarkari, M. (2010). Kinetic study of Fischer–Tropsch process on titania-supported cobalt–manganese catalyst. *Journal of Industrial and Engineering Chemistry* 16: 952–961.

Atwood, H.E. and Bennett, C.O. (1979). Kinetics of the Fischer-Tropsch reaction over iron. *Industrial Engineering Chemistry Process Design Development* 18: 163–170.

Bai, B., Liu, M., Lv, X. et al. (2011). Correlations for predicting single phase and two-phase flow pressure drop in pebble bed flow channels. *Nuclear Engineering and Design* 241: 4767–4774.

Bansal, V., Perkins, J.D., Pistikopoulos, E.N. et al. (2000). Simultaneous design and control optimisation under uncertainty. *Computer and Chemical Engineering* 24: 261–266.

Barker, J.J. (1965). Heat transfer in packed beds. *Industrial and Engineering Chemistry* 57: 43–51.

Basha, O.M., Sehabiague, L., Abdel-Wahab, A., and Morsi, B.I. (2015). Fischer–Tropsch synthesis in slurry bubble column reactors: experimental investigations and modeling – a review. *International Journal of Chemical Reactor Engineering* 13: 201–288.

Bayat, M., Hamidi, M., Dehghani, Z., and Rahimpour, M. (2014). Sorption-enhanced Fischer-Tropsch synthesis with continuous adsorbent regeneration in GTL technology: modeling and optimization. *Journal of Industrial and Engineering Chemistry* 20: 858–869.

Bazmi, M., Hashemabadi, S.H., and Bayat, M. (2013). Extrudate trilobe catalysts and loading effects on pressure drop and dynamic liquid holdup in porous media of trickle bed reactors. *Transport in Porous Media* 99: 535–553.

Bechara, R., Balloy, D., and Vanhove, D. (2001). Catalytic properties of Co/Al_2O_3 system for hydrocarbon synthesis. *Applied Catalysis: A General* 207: 343–353.

Benyahia, F. and O'Neill, K.E. (2005). Enhanced voidage correlations for packed beds of various particle shapes and sizes. *Particulate Science and Technology* 23: 169–177.

Béttega, R., Moreira, M.F.P., Corrêa, R.G., and Freire, J.T. (2011). Mathematical simulation of radial heat transfer in packed beds by pseudohomogeneous modeling. *Particuology* 9: 107–113.

Bey, O. and Eigenberger, G. (1997). Fluid flow through catalyst filled tubes. *Chemical Engineering Science* 52: 1365–1376.

Bey, O. (1998). *Strömungsverteilung und wärmetransport in schüttungen*. VDI-Verlag.

Biegler, L.T. (2010). *Nonlinear programming: concepts, algorithms, and applications to chemical processes*, 1ste. Philadelphia, PA: SIAM.

Bird, R.B., Stewart, W.E., and Lightfoot, E.N. (2001). *Transport Phenomena*, 3rde. New York: Wiley.

Bischoff, K.B. (1965). Effectiveness factors for general reaction rate forms. *American Institute of Chemical Engineers Journal* 11: 351–355.

Bosanquet, C. H. (1944). British TA Report BR-507. *Technical Report.*

Box, G. and Luceño, A. (1997). Discrete proportional-integral adjustment and statistical process control. *Journal of Quality Technology* 29: 248–260.

Brady, R.C. III and Pettit, R. (1981). Mechanism of the Fischer-Tropsch reaction: the chain propagation step. *Journal American Chemical Society* 103: 1287–1289.

Brötz, W. (1949). Zur Systematik der Fischer-Tropsch-Katalyse. *Journal of electrochemistry and applied physical chemistry* 5: 301–306.

Burcat, A. and Ruscic, B. (2005). *Third millenium ideal gas and condensed phase thermochemical database for combustion with updates from active thermochemical tables*. Argonne, IL: Argonne National Laboratory.

Calderone, V.R., Shijn, N.R., Curulla-Ferre, D. et al. (2013). De novo design of nanostructured iron–cobalt Fischer-Tropsch catalysts. *A Journal of the German Chemical Society* 52: 4397–4401.

Caldwell, L. and Van Vuuren, D.S. (1986). On the formation and composition of the liquid phase in Fischer-Tropsch reactor. *Chemical Engineering Science* 41: 89–96.

Carbonell, R.G. and Whitaker, S. (1983). Dispersion in pulsed systems—II: theoretical developments for passive dispersion in porous media. *Chemical Engineering Science* 38: 1795–1802.

Çarpinlioğlu, M.Ö. and Özahi, E. (2008). A simplified correlation for fixed bed pressure drop. *Powder Technology* 187: 94–101.

Chabot, G., Guilet, R., Cognet, P., and Gourdon, C. (2015). A mathematical modeling of catalytic milli-fixed bed reactor for Fischer-Tropsch synthesis: influence of tube diameter on Fischer Tropsch selectivity and thermal behaviour. *Chemical Engineering Science* 127: 72–83.

Chambrey, S., Fongarland, P.V., Karaca, H. et al. (2006). Fischer–Tropsch synthesis in milli-fixed bed reactor: comparison with centimetric fixed bed and slurry stirred tank reactors. *Catalysis Today* 171: 201–206.

Chaumette, P., Courty, P., Kiennemann, A., and Ernst, B. (1995). Alcohol and paraffin synthesis on cobalt based catalysts: comparison of mechanistic aspects. *Topics in Catalysis* 2: 117–126.

Chen, J., Ring, Z., and Dabros, T. (2001). Modeling and simulation of a fixed-bed pilot-plant hydrotreater. *Industrial and Engineering Chemistry Research* 40: 3294–3300.

Chilton, T.H. and Colburn, A.P. (1934). Mass transfer (absorption) coefficients prediction from data on heat transfer and fluid friction. *Industrial and Engineering Chemistry* 26: 1183–1187.

Constantinides, A. and Mostoufi, N. (1999). *Numerical methods for chemical engineers with Matlab applications*, 1ste. New Jersey: Prentice Hall.

Cybulski, A. and Moulijn, J.A. (1994). Monoliths in heterogeneous catalysis. *Catalysis Reviews* 36: 179–270.

Dai, X.P., Liu, P.Z., Shi, Y. et al. (2014). Fischer–Tropsch synthesis in a bench-scale two-stage multitubular fixed-bed reactor: simulation and enhancement in conversion and diesel selectivity. *Chemical Engineering Science* 105: 1–11.

Danckwerts, P. (1953). Continuous flow systems: distribution of residence times. *Chemical Engineering Science* 2: 1–13.

Das, T.K., Conner, K.W., Li, J. et al. (2005). Fischer–Tropsch synthesis: kinetics and effect of water for a Co/SiO_2 catalyst. *Energy and Fuel* 19: 1430–1439.

De Jong, K.P. and Geus, J.W. (2000). Carbon nanofibers: catalytic synthesis and applications. *Catalysis Reviews* 42: 481–510.

Delgado, J.M.P.Q. (2006). A critical review of dispersion in packed beds. *Heat Mass Transfer* 42: 279–310.

Dimitriadis, V.D. and Pistikopoulos, E.N. (1995). Flexibility analysis of dynamic systems. *Industrial and Engineering Chemistry Research* 34: 4451–4462.

Dixon, A.G., Nijemeisland, M., and Stitt, E.H. (2006). Packed tubular reactor modeling and catalyst design using computational fluid dynamics. *Advances in Chemical Engineering* 31: 307–389.

Dixon, A.G. (2012). Fixed bed catalytic reactor modelling-the radial heat transfer problem. *The Canadian Journal of Chemical Engineering* 90: 507–527.

Donnelly, T.J., Yates, I.C., and Satterfield, C.N. (1988). Analysis and prediction of product distributions of the Fischer-Tropsch synthesis. *Energy and Fuels* 2: 734–739.

Dry, M.E., Shingles, T., and Boshoff, L.J. (1972). Rate of the Fischer-Tropsch reaction over iron catalysts. *Journal of Catalysis* 25: 99–104.

Dry, M.E. (1976). Advances in Fischer-Tropsch chemistry. *Industrial and Engineering Chemistry Product Research and Development* 15: 282–286.

Dry, M.E. (1981). The Fischer-Tropsch synthesis. In: *Catalysis. Science and Technology* (ed. J.R. Anderson and M. Boudart). New York: Springer-Verlag.

Dry, M.E. (1983). The Sasol Fischer-Tropsch process. In: *Applied Industrial Catalysis*, vol. 2 (ed. B.E. Leach). New York: Academic Press.

Dry, M.E. (1996). Practical and theoretical aspects of the catalytic Fischer-Tropsch process. *Applied Catalysis A: General* 138: 319–344.

Dry, M.E. (2001). High quality diesel via the Fischer-Tropsch process – a review. *Journal of Chemical Technology and Biotechnology* 77: 43–50.

Edgar, T.F. and Himmelblau, D. (2001). *Optimization chemical processes*, 2nde. New York: McGraw Hill.

Eidt, C.M., Bauman, R.F., Elsenberg, B., and Hochman, J.M. (1994). Current developments in natural gas conversion technology. In: *Proceedings of the 14th World Petroleum Congress*. Chichester, Stavanger, Norway: Wiley.

Eisfeld, B. and Schnitzlein, K. (2001). The influence of confining walls on the pressure drop in packed beds. *Chemical Engineering Science* 56: 4321–4329.

Ellman, M., Midoux, N., Wild, G. et al. (1990). A new, improved liquid hold-up correlation for trickle-bed reactors. *Chemical Engineering Science* 45: 1677–1684.

Erkey, C., Rodden, J.B., and Akgerman, A. (1990a). A correlation for predicting diffusion coefficients in alkanes. *The Canadian Journal of Chemical Engineering* 68: 661–665.

Erkey, C., Rodden, J.B., and Akgerman, A. (1990b). Diffusivities of synthesis gas and n-alkanes in Fischer-Tropsch wax. *Energy and Fuels* 4: 275–276.

Ermolaev, V.S., Gryaznov, K.O., Mitberg, E.B. et al. (2015). Laboratory and pilot plant fixed-bed reactors for Fischer-Tropsch synthesis: mathematical modeling and experimental investigation. *Chemical Engineering Science* 138: 1–8.

Fazlollahi, F., Sarkari, M., Zare, A. et al. (2012). Development of a kinetic model for Fischer–Tropsch synthesis over Co/Ni/Al_2O_3 catalyst. *Journal of Industrial and Engineering Chemistry* 18: 1223–1232.

Feimer, J.L., Silveston, P.L., and Hudgins, R.R. (1981). Steady-state study of the Fischer-Tropsch reaction. *Industrial Engineering Chemistry Product Research and Development* 20: 609–615.

Fischer, N., van Steen, E., and Claeys, M. (2013). Structure sensitivity of the Fischer–Tropsch activity and selectivity on alumina supported cobalt catalysts. *Journal of Catalysis* 299: 67–80.

Forghani, A., Elekaei, H., and Rahimpour, M. (2009). Enhancement of gasoline production in a novel hydrogen-permselective membrane reactor in Fischer–Tropsch synthesis of GTL technology. *International Journal of Hydrogen Energy* 34: 3965–3976.

Förtsch, D., Pabst, K., and GroB-Hardt, E. (2015). The product distribution in Fischer-Tropsch synthesis: an extension of the ASF model to describe common deviations. *Chemical Engineering Science* 138: 333–346.

Frennet, A. and Hubert, C. (2000). Transient kinetics in heterogeneous catalysis by metals. *Journal of Molecular Catalysis A: Chemical* 163: 163–188.

Froment, G.F., Bischoff, K.B., and De Wilde, J. (2011). *Chemical Reactor Analysis and Design*, 3rde. United States: Wiley.

Fu, T. and Li, Z. (2015). Review of recent development in Co-based catalysts supported on carbon materials for Fischer–Tropsch synthesis. *Chemical Engineering Science* 135: 3–20.

Fuller, E.N., Schettler, P.D., and Giddings, J.C. (1966). New method for prediction of binary gas-phase diffusion coefficients. *Industrial and Engineering Chemistry* 58: 18–27.

Gardezi, S.A., Landrigan, L., Joseph, B., and Wolan, J.T. (2012). Synthesis of tailored eggs-hell cobalt catalysts for Fischer–Tropsch synthesis using wet chemistry techniques. *Industrial and Engineering Chemistry Research* 51: 1703–1712.

Gasem, K.A.M. and Robinson, R.L. Jr. (1985). Prediction of phase behavior for CO2 + heavy normal paraffins using generalized parameter Soave and Peng–Robinson equations of state. *American Institute of Chemical Engineering Journal* 24–28.

Gideon Botes, F. (2009). Influences of water and syngas partial pressure on the kinetics of a commercial alumina-supported cobalt Fischer—Tropsch catalyst. *Industrial and Engineering Chemistry Research* 48: 1859–1865.

Gideon Botes, F., Van Dyk, B., and McGregor, C. (2009). The Development of a macro kinetic model for a commercial Co/Pt/Al2O3 Fischer—Tropsch catalyst. *Industrial Engineering Chemistry Research* 48: 10439–10447.

Gill, S., Tsolakis, A., Dearn, K., and Rodríguez-Fernández, J. (2011). Combustion characteristics and emissions of Fischer-Tropsch diesel fuels in IC engines. *Progress in Energy and Combustion Science* 37: 503–523.

Girardon, J., Lermontov, A., Gengembre, L. et al. (2005). Effect of cobalt precursor and pretreatment conditions on the structure and catalytic performance of cobalt silica-supported Fischer–Tropsch catalysts. *Journal of Catalysis* 230: 339–352.

Glasser, D., Hildebrandt, D., Liu, X. et al. (2012). Recent advances in understanding the Fischer–Tropsch synthesis (FTS) reaction. *Current Opinion in Chemical Engineering* 1: 296–302.

Guardo, A., Coussirat, M., Recasens, F. et al. (2006). CFD study on particle-to-fluid heat transfer in fixed bed reactors: convective heat transfer at low and high pressure. *Chemical Engineering Science* 61: 4341–4353.

Haarlemmer, G. and Bensabath, T. (2016). Comprehensive Fischer–Tropsch reactor model with non-ideal plug flow and detailed reaction kinetics. *Computers and Chemical Engineering* 84: 281–289.

Hall, C.C., Gall, D., and Smith, S.L. (1952). A comparison on the fixed-bed, liquid-phase ("slurry") and fluidized-bed techniques in the Fischer-Tropsch synthesis. *Journal of the Institution of Petroleum Technologists* 38: 845–875.

Hallac, B.B., Keyvanloo, K., Hedengren, J.D. et al. (2015). An optimized simulation model for iron-based Fischer-Tropsch catalyst design: transfer limitations as functions of operating and design conditions. *Chemical Engineering Journal* 263: 268–279.

Han, N.W., Bhakta, J., and Carbonell, R.G. (1985). Longitudinal and lateral dispersion in packed beds: effect of column length and particle size distribution. *American Institute of Chemical Engineers Journal* 31: 277–288.

Haughey, D.P. and Beveridge, G.S.G. (1969). Structural properties of packed beds – a review. *The Canadian Journal of Chemical Engineering* 47: 130–140.

Hicks, R.E. (1970). Pressure drop in packed beds of spheres. *Industrial and Engineering Chemistry Fundamentals* 9: 500–502.

Hinchiranan, S., Zhang, Y., Nagamori, S. et al. (2008). TiO2 promoted Co/SiO2 catalysts for Fischer-Tropsch synthesis. *Fuel Processing Technology* 89: 455–459.

Hooshyar, N., Vervloet, D., Kapteijn, F. et al. (2012). Intensifying the Fischer-Tropsch synthesis by reactor structuring – a model study. *Chemical Engineering Journal* 207: 865–870.

Huang, S.H., Lin, H.M., Tsai, F.N., and Chao, K.C. (1988). Solubility of synthesis gases in heavy n-paraffins and Fischer-Tropsch wax. *Industrial and Engineering Chemistry Research* 27: 162–169.

Huff, G.A. and Satterfield, C.N. (1985). Liquid accumulation in catalyst pores in a Fischer-Tropsch fixed-bed reactor. *Industrial and Engineering Chemistry Process Design and Development* 24: 986–995.

Ibrahim, H.H. and Idem, R.O. (2006). Kinetic studies of the partial oxidation of isooctane for hydrogen production over a nickel-alumina catalyst. *Chemical Engineering Science* 61: 5912–5918.

Iglesia, E., Reyes, S.C., and Soled, S.L. (1993). *Computed-aided design of catalysts. Reaction-transport selectivity models and the design of Fischer-Tropsch catalysts*, 1ste. R. Becker and C. J. Pereira.

Iliuta, I., Ring, Z., and Larachi, F. (2006). Simulating simultaneous fines deposition under catalytic hydrodesulfurization in hydrotreating trickle beds-does bed plugging affect HDS performance? *Chemical Engineering Science* 61: 1321–1333.

Iordanidis, A. A. (2002). Mathematical Modeling of Catalytic Fixed Bed Reactors, Ph.D. thesis, Twente University Press.

Irani, M. (2014). Investigating the production of liquid fuels from synthesis gas ($CO + H_2$) in a bench-scale packed-bed reactor based on Fe-Cu-La/SiO_2 catalyst: experimental and CFD modeling. *International Journal of Industrial Chemistry* 5: 1–9.

Irankhah, A., Haghtalab, A., Farahani, E.V., and Sadaghianizadeh, K. (2007). Fischer-Tropsch reaction kinetics of cobalt catalyst in supercritical phase. *Journal of Natural Gas Chemistry* 16: 115–120.

Jakobsen, H.A. (2014). *Chemical Reactor Modeling*, 1ste. New York: Springer Science + Business Media.

Jager, B. and Espinoza, R. (1995). Advances in low temperature Fischer-Tropsch synthesis. *Catalysis Today* 23: 17–28.

Jess, A., Popp, R., and Hedden, K. (1999). Fischer–Tropsch-synthesis with nitrogen-rich syngas: fundamentals and reactor design aspects. *Applied Catalysis A: General* 186: 321–342.

Jess, A. and Kern, C. (2009). Modeling of multi-tubular reactors for Fischer-Tropsch synthesis. *Chemical Engineering and Technology* 32: 1164–1175.

Jess, A. and Kern, C. (2012a). Influence of particle size and single-tube diameter on thermal behavior of Fischer-Tropsch reactors. Part I. Particle size variation for constant tube size and vice versa. *Chemical Engineering Technology* 35: 369–378.

Jess, A. and Kern, C. (2012b). Influence of particle size and single-tube diameter on thermal behavior of Fischer-Tropsch reactors. Part II. Eggshell catalysts and optimal reactor performance. *Chemical Engineering and Technology* 35: 379–386.

Jess, A., Kragl, U., and Wasserscheid, P. (2013). *Chemical Technology: An Integral Textbook*, 1ste. Weinheim: Wiley-VCH.

Jia-Yang, Y.Q., De Chen, J.Y., and Anders, H. (2015). Recent progresses in understanding of Co-based Fischer–Tropsch catalysis by means of transient kinetic studies and theoretical analysis. *Catalysis Letters* 145: 145–161.

Kaiser, P. and Jess, A. (2014). Modeling of multitubular reactors for iron- and cobalt-Catalyzed Fischer-Tropsch syntheses for application in a power-to-liquid process. *Energy Technology* 2: 486–497.

Kaiser, P., Pöhlmann, F., and Jess, A. (2014). Intrinsic and effective kinetics of cobalt-catalyzed Fischer-Tropsch synthesis in view of a power-to-liquid process based on renewable energy. *Chemical and Engineering Technology* 37: 964–972.

Kang, J., Cheng, K., Zhang, L. et al. (2011). Mesoporous zeolite-supported ruthenium nanoparticles as highly selective Fischer–Tropsch catalysts for the production of C_5–C_{11} isoparaffins. *Chemical International Edition* 50: 5200–5203.

Karimi, Z., Rahmani, M., and Moqadam, M. (2012). A study on vapour-liquid equilibria in Fischer-Tropsch synthesis. *Procedia Engineering* 42: 25–33.

Kaskes, B., Vervloet, D., Kapteijn, F., and van Ommen, J. (2016). Numerical optimization of a structured tubular reactor for Fischer-Tropsch synthesis. *Chemical Engineering Journal* 283: 1465–1483.

Keyser, M.J., Everson, R.C., and Espinoza, R.L. (2000). Fischer–Tropsch kinetic studies with cobalt–manganese oxide catalysts. *Industrial and Engineering Chemistry Research* 39: 48–54.

Keyvanloo, K., Lanham, S.J., and Hecker, W.C. (2016). Kinetics of Fischer-Tropsch synthesis on supported cobalt: effect of temperature on CO and H_2 partial pressure dependencies. *Catalysis Today* 270: 9–18.

Khodakov, A.Y., Chu, W., and Fongarland, P. (2007). Advances in the development of novel cobalt Fischer-Tropsch catalysts for synthesis of long-chain hydrocarbons and clean fuels. *Chemical Reviews* 107: 1692–1744.

Kim, Y.H., Jun, K.W., Joo, H. et al. (2009). A simulation study on gas-to-liquid (natural gas to Fischer-Tropsch synthetic fuel) process optimization. *Chemical Engineering Journal* 155: 427–432.

Knottenbelt, C. (2002). Mossgas "gas-to-liquid" diesel fuels-an environmentally friendly option. *Catalysis Today* 71: 437–445.

Koekemoer, A. and Luckos, A. (2015). Effect of material type and particle size distribution on pressure drop in packed beds of large particles: extending the Ergun equation. *Fuel* 158: (8.232–8.238).

Kölbel, H. and Ralek, M. (1980). The Fischer-Tropsch synthesis in the liquid phase. *Catalysis Reviews* 21: 225–274.

Kuipers, E.W., Scheper, C., Wilson, J.H. et al. (1996). Non-ASF product distributions due to secondary reactions during Fischer-Tropsch synthesis. *Journal of Catalysis* 158: 288–300.

Kürten, H., Raasch, J., and Rumpf, H. (1966). Beschleunigung eines kugelförmigen feststoffteilchens im strömungsfeld konstanter geschwindigkeit. *Chemie Ingenieur Technik* 38 (9): 941–948.

Kwack, S.H., Bae, J.W., Park, M.J. et al. (2011a). Reaction modeling on the phosphorous-treated Ru/Co/Zr/SiO2 Fischer–Tropsch catalyst with the estimation of kinetic parameters and hydrocarbon distribution. *Fuel* 90: 1383–1394.

Kwack, S., Park, M., Bae, J.W. et al. (2011b). Development of a kinetic model of the Fischer-Tropsch synthesis reaction with a cobalt-based catalyst. *Reaction Kinetic and Mechanism Catalysis Journal* 104: 483–502.

Larachi, F., Laurent, A., Midoux, N., and Wild, G. (1991). Experimental study of a trickle-bed reactor operating at high pressure: two-phase pressure drop and liquid saturation. *Chemical Engineering Science* 46: 1233–1246.

Larachi, F., Belfares, L., Iliuta, I., and Grandjean, B.P.A. (2003). Heat and mass transfer in cocurrent gas-liquid packed beds. Analysis, recommendations, and new correlations. *Industrial and Engineering Chemistry Research* 42: 222–242.

Lee, B.I. and Kesler, M.G. (1975). A generalized thermodynamic correlation based on three-parameter corresponding states. *American Institute of Chemical Engineers Journal* 21: 510–527.

Lee, H.J., Choi, J.H., Garforth, A., and Hwang, S. (2015). Conceptual design of a Fischer-Tropsch reactor in a gas-to-liquid process. *Industrial and Engineering Chemistry Research* 54: 6749–6760.

Lee, T.S. and Chung, J.N. (2012). Mathematical modeling and numerical simulation of a Fischer-Tropsch packed bed reactor and its thermal management for liquid hydrocarbon fuel production using biomass syngas. *Energy and Fuels* 26: 1363–1379.

Levenspiel, O. (1999). *Chemical Reaction Engineering*, 3rde. New York: Wiley.

Liu, Z.T., Li, Y.W., Zhou, J.L., and Zhang, B.J. (1995). *Journal of the Chemical Society. Faraday Transactions* 91: 3255–3261.

Lox, E.S. and Froment, G.F. (1993a). Kinetics of the Fischer-Tropsch reaction on a precipitated promoted iron catalyst. 1. Experimental procedure and results. *Industrial and Engineering Chemistry Research* 32: 61–70.

Lox, E.S. and Froment, G.F. (1993b). Kinetics of the Fischer-Tropsch reaction on a precipitated promoted iron catalyst. 2. Kinetic modeling. *Industrial and Engineering Chemistry Research* 32: 71–82.

Lozano-Blanco, G., Thybaut, J.W., Surla, K. et al. (2009). Simulation of a slurry-bubble column reactor for Fischer-Tropsch synthesis using single-event microkinetics. *American Institute of Chemical Engineers Journal* 55: 2159–2170.

Luque, R., de la Osa, A.R., Campelo, J.M. et al. (2012). Design and development of catalysts for Biomass-To-Liquid-Fischer–Tropsch (BTL-FT) processes for biofuels production. *Energy and Environmental Science* 5: 5186–5202.

Mabry, J. C. (2014). Utilization of cobalt catalyst for high temperature Fischer-Tropsch synthesis in a fluidized bed reactor, Ph.D. thesis, College of Engineering in the Graduate School Southern Illinois University Carbondale

Macías, M. and Ancheyta, J. (2004). Simulation of an isothermal hydrodesulfurization small reactor with different catalyst particle shapes. *Catalysis Today* 98: 243–252.

Maitlis, P.M. and de Klerk, A. (2013). *Greener Fischer-Tropsch processes for fuels and feedstocks*, 1ste. Weinheim: Wiley.

Mansouri, M., Atashi, H., Mirzaei, A., and Jangi, R. (2013). Kinetics of the Fischer-Tropsch synthesis on silica-supported cobalt-cerium catalyst. *International Journal Industrial Chemistry* 4: 1.

Marano, J.J. and Holder, G.D. (1997). Characterization of Fischer-Tropsch liquids for vapor-liquid equilibria calculations. *Fluid Phase Equilibria* 138: 1–21.

Marvast, M.A., Sohrabi, M., Zarrinpashne, S., and Baghmisheh, G. (2005). Fischer-Tropsch synthesis: modeling and performance study for Fe-HZSM5 bifunctional catalyst. *Chemical Engineering Technology* 28: 78–86.

Masuku, C.M., Hildebrandt, D., and Glasser, D. (2011). The role of vapour–liquid equilibrium in Fischer–Tropsch product distribution. *Chemical Engineering Science* 66: 6254–6263.

Mears, D.E. (1971a). Diagnostic criteria for heat transport limitations in fixed bed reactors. *Journal of Catalysis* 20: 127–131.

Mears, D.E. (1971b). Tests for transport limitations in experimental catalytic reactors. *Industrial and Engineering Chemistry Process Design and Development* 10: 541–547.

Mederos, F., Elizalde, I., and Ancheyta, J. (2009). Steady-state and dynamic reactor models for hydrotreatment of oil fractions: a review. *Catalysis Reviews-Science and Engineering* 51: 485–607.

Mehta, S. and Ricardez-Sandoval, L.A. (2016). Integration of design and control of dynamic systems under uncertainty: a new back-off approach. *Industrial and Engineering Chemistry Research* 55: 485–498.

Méndez, C.I., Ancheyta, J., and Trejo, F. (2019). Importance of proper hydrodynamics modelling in fixed-bed reactors: Fischer-Tropsch synthesis study case. *The Canadian Journal of Chemical Engineering* 97: 2685–2698.

Méndez, C.I., Trejo, F., and Ancheyta, J. (2022). On the use of steady-state optimal initial operating conditions for control scheme implementation of a fixed-bed Fischer–Tropsch reactor. *Arabian Journal for Science and Engineering* 47: 6099–6113.

Mikhailova, I.A., Fadeev, S.I., Khasin, A.A. et al. (2003). Calculating the liquid-gas equilibrium in the Fischer-Tropsch synthesis. *Theoretical Foundations of Chemical Engineering* 37: 167–171.

Mirzaei, A.A., Sarani, R., Azizi, H.R. et al. (2015). Kinetics modeling of Fischer-Tropsch synthesis on the unsupported Fe-Co-Ni (ternary) catalyst prepared using coprecipitation procedure. *Fuel* 140: 701–710.

Moazami, N., Mahmoudi, H., Panahifar, P. et al. (2015a). Mathematical modeling and performance study of Fischer-Tropsch synthesis of liquid fuel over cobalt-silica. *Energy Procedia* 75: 62–71.

Moazami, N., Mahmoudi, H., Rahbar, K. et al. (2015b). Catalytic performance of cobalt–silica catalyst for Fischer–Tropsch synthesis: effects of reaction rates on efficiency of liquid synthesis. *Chemical Engineering Science* 134: 374–384.

Moazami, N., Wyszynski, M.L., Mahmoudi, H. et al. (2015c). Modelling of a fixed bed reactor for Fischer–Tropsch synthesis of simulated N_2-rich syngas over Co/SiO_2: hydrocarbon production. *Fuel* 154: 140–151.

Moazami, N., Wyzynski, M.L., Rahbar, K., and Tsolakis, A. (2017a). Parametric study and multi-objective optimization of fixed-bed Fischer–Tropsch reactor: the improvement of FT synthesis products formation and synthetic conversion. *Industrial and Engineering Chemistry Research* 34: 9446–9466.

Moazami, N., Wyszynski, M.L., Rahbar, K. et al. (2017b). A comprehensive study of kinetics mechanism of Fischer-Tropsch synthesis over cobalt-based catalyst. *Chemical Engineering Science* 171: 32–60.

Montillet, A., Akkari, E., and Comiti, J. (2007). About a correlating equation for predicting pressure drops through packed beds of spheres in a large range of Reynolds numbers. *Chemical Engineering and Processing* 46: 329–333.

Mosayebi, A., Mehrpouya, M.A., and Abedini, R. (2016). The development of new comprehensive kinetic modeling for Fischer-Tropsch synthesis process over Co-Ru/γ-Al_2O_3 nano-catalyst in a fixed-bed reactor. *Chemical Engineering Science Journal* 286: 416–426.

Mousavi, S., Zamaniyan, A., Irani, M., and Rashidzadeh, M. (2015). Generalized kinetic model for iron and cobalt based Fischer–Tropsch synthesis catalysts: review and model evaluation. *Applied Catalysis A: General* 506: 57–66.

Moutsoglou, A. and Sunkara, P.P. (2011). Fischer-Tropsch synthesis in a fixed bed reactor. *Energy and Fuels* 25: 2242–2257.

Mueller, G.E. (1992). Radial void fraction distributions in randomly packed fixed beds of uniformly sized spheres in cylindrical containers. *Powder Technology* 72: 269–275.

Nakhaei Pour, A. and Housaindokht, M.R. (2014). Studies on product distribution of nanostructured iron catalyst in Fischer-Tropsch synthesis: effect of catalyst particle size. *Journal of Industrial and Engineering Chemistry* 20: 591–596.

Nakhaei Pour, A., Housaindokht, M.R., Irani, M., and Kamali Shahri, S.M. (2014a). Size-dependent studies of Fischer-Tropsch synthesis on iron based catalyst: new kinetic model. *Fuel* 116: 787–793.

Nakhaei Pour, A., Khodabandeh, H., Izadyar, M., and Housaindokht, M.R. (2013). Mechanistic double ASF product distribution study of Fischer-Tropsch synthesis on precipitated iron catalyst. *Journal of Natural Gas Science Engineering* 15: 53–58.

Nakhaei Pour, A., Khodabandeh, H., Izadyar, M., and Housaindokht, M.R. (2014b). Detailed kinetics of Fischer–Tropsch synthesis on a precipitated iron catalyst. *Reaction Kinetics, Mechanisms and Catalysis* 111: 29–44.

Nakhaei Pour, A. and Modaresi, S.M. (2016). Methane formation in Fischer-Tropsch synthesis: role of nanosized catalyst particles. *Journal of Nano Research* 35: 39–54.

Novak, S., Madon, R.J., and Suhl, H. (1982). Secondary effects in the Fischer-Tropsch synthesis. *Journal of Catalysis* 77: 141–151.

Oukaci, R., Singleton, A.H., and Goodwin, J.G. (1999). *Applied Catalysis A: General* 186: 129–144.

Outi, A., Rautavuoma, I., and van der Baan, H.S. (1981). Kinetics and mechanism of the Fischer Tropsch hydrocarbon synthesis on a cobalt on alumina catalyst. *Applied Catalysis* 1: 247–272.

Ozahi, E., Gundogdu, M.Y., and Carpinlioglu, M.Ö. (2008). A modification on Ergun's correlation for use in cylindrical packed beds with non-spherical particles. *Advanced Powder Technology* 19: 369–381.

Park, N., Kim, J.R., Yoo, Y. et al. (2014). Modeling of a pilot-scale fixed-bed reactor for iron-based Fischer–Tropsch synthesis: two-dimensional approach for optimal tube diameter. *Fuel* 122: 229–235.

Park, S., Jung, I., Lee, U. et al. (2015). Design and modeling of large-scale cross-current multichannel Fischer-Tropsch reactor using channel decomposition and cell-coupling method. *Chemical Engineering Science* 134: 448–456.

Pearson, J. (1959). A note on the "Danckwerts" boundary conditions for continuous flow reactors. *Chemical Engineering Science* 10: 281–284.

Petersen, E.E. (1965). A general criterion for diffusion influenced chemical reactions in porous solids. *Chemical Engineering Science* 20: 587–591.

Philippe, R., Lacroix, M., Dreibine, L. et al. (2009). Effect of structure and thermal properties of a Fischer-Tropsch catalyst in a fixed bed. *Catalysis Today* 147: S305–S312.

Pinna, D., Tronconi, E., Lietti, L. et al. (2002). Review of kinetic models for the Fischer-Tropcsh synthesis. *La Rivista dei Combustibili* 56: 69–85.

Pistikopoulos, E.N. and Mazzuchi, T.A. (1990). A novel flexibility analysis approach for processes with stochastic parameters. *Computer and Chemical Engineering* 14: 991–1000.

Pöhlmann, F. and Jess, A. (2016a). Influence of syngas composition on the kinetics of Fischer-Tropsch synthesis of using cobalt as catalyst. *Energy Technology* 4: 55–64.

Pöhlmann, F. and Jess, A. (2016b). Interplay of reaction and pore diffusion during cobalt-catalyzed Fischer–Tropsch synthesis with CO2-rich syngas. *Catalysis Today* 275: 172–182.

Pöhlmann, F., Kern, C., Rößler, S., and Jess, A. (2016). Accumulation of liquid hydrocarbons in catalyst pores during cobalt-catalyzed Fischer-Tropsch synthesis. *Catalysis and Science Technology* 6: 6593–6604.

Pollard, W.G. and Present, R.D. (1949). On gaseous self-diffusion in long capillary tubes. *Physical Review* 73: 762–774.

Post, M.F.M., Van't Hoog, A.C., Minderhoud, J.K., and Sie, S.T. (1989). Diffusion limitations in Fischer-Tropsch catalysts. *American Institute of Chemical Engineers Journal* 35: 1107–1114.

Puskas, I., Hurlbut, R.S., and Pauls, R.E. (1993). Telomerization model for cobalt-catalyzed Fischer-Tropsch products. *Journal of Catalysis* 139: 591–601.

Qian, W., Zhang, H., Ying, W., and Fang, D. (2013). The comprehensive kinetics of Fischer-Tropsch synthesis over a Co/AC catalyst on the basis of CO insertion mechanism. *Chemical Engineering Journal* 228: 526–534.

Rados, N., Dahhan, M.H., and Dudukovic, M.P. (2003). Modeling of the Fischer-Tropsch synthesis in slurry bubble column reactors. *Catalysis Today* 79–80: 211–218.

Rafiei-Shishavan, M., Mehta, S., and Ricardez-Sandoval, L.A. (2017). Simultaneous design and control under uncertainty: a back-off approach using power series expansions. *Computer and Chemical Engineering* 99: 66–81.

Rafiq, M., Jakobsen, H., Schmid, R., and Hustad, J. (2011). Experimental studies and modeling of a fixed bed reactor for Fischer-Tropsch synthesis using biosyngas. *Fuel Processing Technology* 92: 893–907.

Rahimpour, M.R. and Elekaei, H. (2009). A comparative study of combination of Fischer–Tropsch synthesis reactors with hydrogen-permselective membrane in GTL technology. *Fuel Processing Technology* 90: 747–761.

Rahimpour, M.R., Mirvakili, A., and Paymooni, K. (2011a). A novel water perm-selective membrane dual-type reactor concept for Fischer–Tropsch synthesis of GTL (gas to liquid) technology. *Energy* 36: 1223–1235.

Rahimpour, M.R., Mirvakili, A., Paymooni, K., and Moghtaderi, B. (2011b). A comparative study between a fluidized-bed and a fixed-bed water perm-selective membrane reactor with in situ H_2O removal for Fischer–Tropsch synthesis of GTL technology. *Journal of Natural Gas Science and Engineering* 3 (3): 484–495.

Rangel, N., Santos, A., and Pinho, C. (2001). Pressure drop in packed shallow beds of cylindrical cork stoppers. *Chemical Engineering Research and Design* 79: 547–552.

Rao, V.U.S., Stiegel, G.J., Cinquegrane, G.J., and Srivastava, R.D. (1992). Iron-based catalysts for slurry-phase fischer-tropsch process: technology review. *Fuel Processing Technology* 30: 83–107.

Rase, H.F. (1990). *Fixed-Bed reactor design and diagnostics*, 1ste. New York: Butterworth Publishers.

Reichelt, W. (1972). Zur berechnung des druckverlustes einphasig durchströmter kugel- und zzylinderschüttungen. *Chemie Ingenieur Technik* 44: 1068–1071.

Reid, R.C., Sherwood, T.K., and Prausnitz, J. (1977). *The Properties of gases and liquids*, 3rde. United states: Mc Graw Hill.

Reuel, C.R. and Bartholomew, C.H. (1984). Effects of support and dispersion on the CO hydrogenation activity/selectivity properties of cobalt. *Journal of Catalysis* 85: 78–88.

Rohde, M.P., Schaub, G., Khajavi, S. et al. (2008). Fischer–Tropsch synthesis with insitu H_2O removal – directions of membrane development. *Microporous and Mesoporous Materials* 115: 123–136.

Rytter, E., Tsakoumis, N.E., and Holmen, A. (2016). On the selectivity to higher hydrocarbons in Co-based Fischer-Tropsch synthesis. *Catalysis Today* 261: 3–16.

Saeidi, S., Talebi Amiri, M., Saidina Amin, N.A., and Rahimpour, M.R. (2014). Progress in reactors for high-temperature Fischer–Tropsch process: determination place of intensifier reactor perspective. *International Journal Chemical Reactor Engineering* 12: 639–664.

Salmi, T., Wärna, J., Toppinen, S. et al. (2000). *Brazilian Journal of Chemical Engineering* 17: 1023–1035.

Sandler, S.I. (2006). *Chemical, Biochemical, and Engineering Thermodynamics*, 5the. United States: Wiley.

Sarup, B. and Wojciechowski, B.W. (1989). Studies of the fischer-tropsch synthesis on a cobalt catalyst II. Kinetics of carbon monoxide conversion to methane and to higher hydrocarbons. *Canadian Journal of Chemical Engineering* 67: 62–74.

Satterfield, C.N., Huff, G.A. Jr., and Longwell, J.P. (1982). Product distribution from iron catalysts in Fischer-Tropsch slurry reactors. *Industrial and Engineering Chemistry Process Design and Development* 21: 465–470.

Schanke, D., Hilmen, A.M., Bergene, E. et al. (1995). Study of the deactivation mechanism of Al2O3-supported cobalt Fischer-Tropsch catalyst. *Catalysis Letters* 34: 269–284.

Schulz, H. (1999). Short history and present trends of Fischer-Tropsch synthesis. *Applied Catalysis A: General* 186: 3–12.

Schulz, H. and Claeys, M. (1999). Reactions of α-olefins of different chain length added during Fischer-Tropsch synthesis on a cobalt catalyst in a slurry reactor. *Applied Catalysis A: General* 186: 71–90.

Sharma, A., Philippe, R., Luck, F., and Schweich, D. (2011). A simple and realistic fixed bed model for investigating Fischer-Tropsch catalyst activity at lab-scale and extrapolating to industrial conditions. *Chemical Engineering Science* 66: 6358–6366.

Sie, S.T. (1998). Process development and scale-up: part 4. case history of the development of a Fischer–Tropsch synthesis. *Chemistry; Reviews in Chemical Engineering* 14: 109–157.

Singh, J. and Gu, S. (2010). Biomass conversion to energy in India – a critique. *Renewable and Sustainable Energy Reviews* 14: 1367–1378.

Smith, J. (1981). *Chemical Engineering Kinetics*, 3rde. United States: McGraw Hill.

Soroush, M. and Kravaris, C. (1993a). Optimal design and operation of batch reactors. 1. Theoretical framework. *Industrial and Engineering Chemistry Research* 32: 866–881.

Soroush, M. and Kravaris, C. (1993b). Optimal design and operation of batch reactors. 2. A case study. *Industrial Engineering Chemistry Research* 32: 882–893.

Stephanopoulos, G. (1985). *Chemical Process Control an introduction to theory and practice*, 1ste. New Jersey: Prentice Hall.

Storsaeter, S., Borg, O., Blekkan, E., and Holmen, A. (2005). Study of the effect of water on Fischer–Tropsch synthesis over supported cobalt catalysts. *Journal of Catalysis* 231 (2): 405–419.

Sun, B., Yu, G., Lin, J. et al. (2012). A highly selective Raney Fe@HZSM-5 Fischer–Tropsch synthesis catalyst for gasoline production: one-pot synthesis and unexpected effect of zeolites. *Catalysis Science and Technology* 2: 1625–1629.

Surisetty, V.R., Dalai, A.K., and Kozinski, J. (2010). Alkali-promoted trimetallic Co–Rh–Mo sulfide catalysts for higher alcohols synthesis from synthesis gas: comparison of MWCNT and activated carbon supports. *Industrial Engineering Chemistry Research* 49: 6956–6963.

Tavasoli, A., Trépanier, M., Dalai, A.K., and Abatzoglou, N. (2010). Effects of confinement in carbon nanotubes on the activity, selectivity, and lifetime of Fischer-Tropsch Co/carbon nanotube catalysts. *Journal of Chemical Engineering Data* 55: 2757–2763.

Thiele, E.W. (1939). Relation between catalytic activity and size of particle. *Industrial and Engineering Chemistry* 31: 916–920.

Tobis, J. and Ziókowski, D. (1988). Modelling of heat transfer at the wall of a packed-bed apparatus. *Chemical Engineering Science* 43: 3031–3036.

Todic, B., Ma, W., Jacobs, G. et al. (2014). Effect of process conditions on the product distribution of Fischer–Tropsch synthesis over a re-promoted cobalt-alumina catalyst using a stirred tank slurry reactor. *Journal of Catalysis* 311: 325–338.

Todic, B., Ordomsky, V.V., Nikacevic, N.M. et al. (2015). Opportunities for intensification of Fischer–Tropsch synthesis through reduced formation of methane over cobalt catalysts in microreactors. *Catalysis Science and Technology* 5: 1400–1411.

Tosun, I. (2007). *Modeling in transport phenomena*, 2nde. Elsevier Science and Technology Books.

Trépanier, M., Tavasoli, A., Dalai, A.K., and Abatzoglou, N. (2009). Co, Ru and K loadings effects on the activity and selectivity of carbon nanotubes supported cobalt catalyst in Fischer–Tropsch synthesis. *Applied Catalysis A: General* 353: 193–202.

Tsakoumis, N.E., Rønning, M., Borg, O. et al. (2010). Deactivation of cobalt based Fischer-Tropsch catalysts: a review. *Catalysis Today* 154: 162–182.

Tyreus, B.D. and Luyben, W.L. (1992). Tuning PI controllers for integrator/dead time processes. *Industrial Engineering Chemistry Research* 31: 2625–2628.

Valero-Romero, M.J., Sartipi, S., Sun, X. et al. (2016). Carbon/H-ZSM-5 composites as supports for bi-functional Fischer–Tropsch synthesis catalysts. *Catalysis Science Technology* 6: 2633–2646.

Van Antwerpen, W.d., Toit, C., and Rousseau, P. (2010). A review of correlations to model the packing structure and effective thermal conductivity in packed beds of mono-sized spherical particles. *Nuclear Engineering and Design* 240: 1803–1818.

Van Berge, P. and Everson, R. (1997). Cobalt as an alternative Fischer-Tropsch catalyst to iron for the production of middle distillates. *Studies in Surface Science and Catalysis* 107: 207–212.

Van Der Laan, G.P. and Beenackers, A.A.C.M. (1999). Kinetics and selectivity of the Fischer-Tropcsch synthesis: a literature review. *Catalysis Reviews* 41: 255–318.

Van Dijk, H. A. (2001). The Fischer-Tropsch synthesis: a mechanistic study using transient isotopic tracing, Ph. D. Thesis, Technische Universiteit Eindhoven, Eihdhoven, the Netherlands

Van Vliet, O.P., Faaij, A.P., and Turkenburg, W.C. (2009). Fischer–Tropsch diesel production in a well-to-wheel perspective: a carbon, energy flow and cost analysis. *Energy Conversion and Management* 50: 855–876.

Vannice, M.A. (1976). The catalytic synthesis of hydrocarbons from carbon monoxide and hydrogen. *Catalysis Reviews* 14: 153–191.

Visconti, C.G., Tronconi, E., Lietti, L. et al. (2007). Development of a complete kinetic model for the Fischer-Tropsch synthesis over Co/Al_2O_3 catalysts. *Chemical Engineering Science* 62: 5338–5343.

Visconti, C.G., Tronconi, E., Groppi, G. et al. (2011a). Monolithic catalysts with high thermal conductivity for the Fischer-Tropsch synthesis in tubular reactors. *Chemical Engineering Journal* 171: 1294–1307.

Visconti, C.G., Tronconi, E., Lietti, L. et al. (2011b). Detailed kinetics of the Fischer–Tropsch synthesis on cobalt catalysts based on H-assisted CO activation. *Topics in Catalysis* 54: 786–800.

Visconti, C.G. and Mascellaro, M. (2013). Calculating the product yields and the vapor-liquid equilibrium in the low-temperature Fischer–Tropsch synthesis. *Catalysis Today* 214: 61–73.

Vollmari, K., Oschmann, T., Wirtz, S., and Kruggel-Emden, H. (2015). Pressure drop investigations in packings of arbitrary shaped particles. *Powder Technology* 271: 109–124.

Vosloo, A.C. (2001). Fischer–Tropsch: a futuristic view. *Fuel processing Technology* 71: 149–155.

Wang, Y.N., Li, Y.W., Bai, L. et al. (1999). Correlation for gas–liquid equilibrium prediction in Fischer-Tropsch synthesis. *Fuel* 78: 911–917.

Wang, Y.N., Xu, Y.Y., Xiang, H.W. et al. (2001). Modeling of catalyst pellets for Fischer-Tropsch synthesis. *Industrial and Engineering Chemistry Research* 40: 4324–4335.

Wang, Y.N., Ma, W.P., Lu, Y.J. et al. (2003a). Kinetics modelling of Fischer-Tropsch synthesis over an industrial Fe-Cu-K catalyst. *Fuel* 82: 195–213.

Wang, Y.N., Xu, Y.Y., Li, Y.W. et al. (2003b). Heterogeneous modeling for fixed-bed Fischer–Tropsch synthesis: reactor model and its applications. *Chemical Engineering Science* 58 (3-6): 867–875.

Wärnå, J. and Salmi, T. (1996). Dynamic modelling of catalytic three phase reactors. *Computers and Chemical Engineering* 20: 39–47.

Welty, J.R., Wicks, C.E., Wilson, R.E., and Rorrer, G.L. (2007). *Fundamentals of momentum, heat, and mass transfer*, 5the. USA: Wiley.

Whitaker, S. (1973). The transport equations for multi-phase systems. *Chemical Engineering Science* 28: 139–147.

White, S.M. and Tien, C.L. (1987). Analysis of flow channeling near the wall in packed beds. *Wärme- und Stoffübertragung* 21: 291–296.

Wijngaarden, R.J., Kronberg, A., and Westerterp, K.R. (1998). *Industrial catalysis: optimizing catalyst and processes*, 1ste. Germany: Wiley.

Wilhelm, R.H. (1962). Progress towards the a priori design of chemical reactors. *Pure and Applied Chemistry* 5: 403–421.

Wilhelm, D.J., Simbeck, D.R., Karp, A.D., and Dickenson, R.L. (2001). Syngas production for gas-to-liquids applications: technologies, issues and outlook. *Fuel Processing Technology* 71: 8.139–8.148.

Wilke, C.R. (1949). Estimation of liquid diffusion coefficients. *Chemical Engineering Progress* 45: 218–224.

Wojciechowski, B.W. (1988). The kinetics of the Fischer-Tropsch synthesis. *Catalysis Reviews* 30: 629–702.

Yagi, S. and Wakao, N. (1959). Heat and mass transfer from wall to fluid in packed beds. *American Institute of Chemical Engineers Journal* 5: 79–85.

Yan, F., Qian, W., Sun, Q. et al. (2014). Product distributions and olefin-to-paraffin ratio over an iron-based catalyst for Fischer–Tropsch synthesis. *Reaction Kinetics, Mechanisms and Catalysis* 113: 471–485.

Yan, Z., Wang, Z., Bukur, D.B., and Goodman, D.W. (2009). Fischer–Tropsch synthesis on a model Co/SiO_2 catalyst. *Journal of Catalysis* 268: 196–200.

Yang, C.H., Massoth, F.E., and Oblad, A.G. (1979). Kinetics of CO + H_2 reaction over Co-Cu-Al_2O_3 catalyst, Hydrocarbon Synthesis from Carbon Monoxide and Hydrogen. *American Chemical Society* 178: 35–46.

Yang, J., Liu, Y., Chang, J. et al. (2003). Detailed kinetics of Fischer-Tropsch synthesis on an industrial Fe-Mn catalyst. *Industrial and Engineering Chemistry Research* 42: 5066–5090.

Yates, I.C. and Satterfield, C.N. (1991). Intrinsic kinetics of the Fischer-Tropsch synthesis on a cobalt catalyst. *Energy and Fuel* 5: 168–173.

Zennaro, R., Tagliabue, M., and Bartholomew, C.H. (2000). Kinetics of Fischer–Tropsch synthesis on titania-supported cobalt. *Catalysis Today* 58: 309–319.

Zhang, Q., Kang, J., and Wang, Y. (2010). Development of novel catalysts for Fischer-Tropsch synthesis: tuning the product selectivity. *ChemCatChem: The European Society Journal for Catalysis* 2: 1030–1058.

Zhang, Q., Deng, W., and Wang, Y. (2013). Recent advances in understanding the key catalyst factors for Fischer-Tropsch synthesis. *Journal of Energy Chemistry* 22: 27–38.

Zhen, Y., Zhou, J.W., Bukur, D.B., and Goodman, D.W. (2009). Fischer-Tropsch synthesis on a model Co/SiO_2 catalyst. *Journal of Catalysis* 268: 196–200.

Zhou, L., Froment, G.F., Yang, Y., and Li, Y. (2016). Advanced fundamental modeling of the kinetics of Fischer–Tropsch synthesis. *American Institute of Chemical Engineers Journal* 62: 1668–1682.

Zhu, X., Lu, X., Liu, X. et al. (2010). Study of radial heat transfer in a tubular Fischer-Tropsch synthesis reactor. *Industrial and Engineering Chemistry Research* 49: 10682–10688.

Zimmerman, W.H. and Bukur, D.B. (1990). Reaction kinetics over iron catalyst used for Fischer-Tropsch synthesis. *The Canadian Journal of Chemical Engineering* 68: 292–301.

9

Computational Fluid Dynamics Modeling of Mass Transfer Processes in Structured Beds of Microfibrous Catalysts

Sergey Lopatin[1,2], *Andrey Elyshev*[2], *and Andrey Zagoruiko*[1,2]

[1] *Boreskov Institute of Catalysis, Novosibirsk, Russia*
[2] *Tyumen State University, Tyumen, Russia*

9.1 Introduction

Different and practically important catalytic processes are significantly influenced by heat and mass transfer limitations, and so the development of new catalysts with improved mass transfer properties is an important task. A possible promising solution in this area is the use of structured catalytic systems using a catalyst based on microfiber carriers made in the form of webs or textiles. Microfibers can be made from glass, minerals, ceramics, carbon, polymers, and other materials. The active components for such catalysts can be selected from a wide range of different noble metals or transition metal oxides depending on the requirements of the particular process.

The most widely used and well-studied microfiber catalysts are glass-fiber catalysts (GFCs), which use glass-fiber cloth as a carrier (Balzhinimaev et al. 2010; Zagoruiko and Lopatin 2019). Such catalysts have demonstrated promising catalytic properties in reactions, such as deep oxidation of hydrocarbons, organic and organohalide compounds (Balzhinimaev et al. 2010; Paukshtis et al. 2010; Lopatin et al. 2015), oxidation of SO_2 and H_2S (Mikenin et al. 2016; Golyashova et al. 2019), low-temperature reduction of NO_x (Li et al. 2014; Golyashova et al. 2023), oxychlorination of light olefins (Shalygin et al. 2013), synthetic gas production (Aldashukurova et al. 2013), selective hydrogenation (Salmi et al. 2000; Gulyaeva et al. 2015), and so on.

The features of microfiber catalysts include their original geometric shape, high mechanical strength and flexibility, as well as the ability to be arranged into structured systems of various geometries and sizes, which make it possible to implement fundamentally new options for catalytic processes.

Various structured GFC packages have been proposed earlier. The most intuitive packaging design for GFC is the multilayer mat or multilayer GFC textiles set. The reaction fluid in these packages flows through the GFC layers. Although it was expected that such a design should be characterized by a fairly high mass transfer, in reality, such packages may have insufficient efficiency due to the suboptimal structure of the internal flow of the reaction mixture (Zagoruiko and Lopatin 2019) and may also have an unreasonably high pressure drop.

GFC-structured cartridges with axial movement of the reaction mixture along the catalytic textile surface are made of parallel layers of GFC alternating with layers of flat and corrugated metal

Mathematical Modeling of Complex Reaction Systems in the Oil and Gas Industry, First Edition.
Edited by Jorge Ancheyta, Andrey Zagoruiko, and Andrey Elyshev.

wire-mesh to ensure the regular channel structure and mechanical strength of the cartridge. Such cartridges may have the outer shape of a circular cylinder with spirally wound layers of glass-fiber/mesh or a prism with square, rectangular, triangular, and other cross-sectional shapes. Spiral shapes are suitable for round reactors with a relatively small diameter, while prismatic cartridges can be used to build catalyst beds of almost any shape and size in large reactors. GFC axial flow cartridges were experimentally studied in our earlier work (Zagoruiko et al. 2017; Zagoruiko and Lopatin 2019). It was shown that such structures are characterized by a high intensity of mass transfer and an extremely low pressure drop. Many peculiar features and properties of such systems have been discovered as well. It has been suggested that the high mass transfer efficiency may be due to the flexibility of the GFC fabric, making possible the mechanical interaction between the textile and the reaction flow, resulting in the variability of the effective area of GFC washed by the flow. The study also revealed a high sensitivity of the cartridge's mass transfer characteristics to the details of the cartridge's internal geometry, which is a possible cause of problems with the reproducibility of experimental results. On one hand, such non-reproducibility and some other complications typical of such catalysts can be considered as a disadvantage of GFCs, but on the other hand, a deep understanding and correct presentation of these details are potential sources of new solutions and improvements in the field of GFC packaging and reactor designs based on them.

Experimental work is an important part of a mass transfer research; however, further optimization of such cartridges and the development of new designs also require the use of modern advanced modeling methods, including those based on computational fluid dynamics (CFD) simulation. This method has already found application in the simulation of catalytic systems of various structures and scales, ranging from traditional catalyst pellets in fixed-beds and tubular reactors (Dixon et al. 2008; Karthik and Buwa 2019; Yang et al. 2021; Bouras et al. 2022) to structured catalysts, including monolithic blocks (You et al. 2019; Klenov et al. 2019), catalysts based on foamed carriers (Bracconi et al. 2018; Dong et al. 2018; Sinn et al. 2019; Wehinger et al. 2019; George et al. 2022), wire-mesh catalytic blocks (Iwaniszyn et al. 2021b; Zazhigalov et al. 2022), and structured catalysts of other forms (Iwaniszyn et al. 2021a; Jurković et al. 2021; Tedesco and Moraes 2021).

The importance of CFD modeling of structured beds of micro-fibrous catalysts lies in its ability to enhance understanding and optimization of the catalytic processes. By accurately simulating the complex flow and reaction phenomena within these structured beds, CFD modeling enables researchers and engineers to analyze the performance of the catalysts, improve their design, and ultimately increase their efficiency and effectiveness in various industrial applications. This modeling approach also facilitates the identification of potential issues, such as pressure drop and heat transfer, and supports the development of innovative solutions to overcome them. Overall, CFD modeling plays a crucial role in the advancement of the development and application of structured beds of micro-fibrous catalysts, leading to significant advancements in catalytic technology.

Until recently, the number of CFD studies of GFC-based catalytic structures was close to zero. CFD has been used to simulate multilayer GFC packaging with a propagating flow of gaseous reaction mixture (Chub et al. 2005; Zagoruiko and Lopatin 2019), but the simulation of GFC cartridges with axial flow has been done for the first time quite recently (Lopatin et al. 2023). This study made it possible to reproduce the main experimentally observed regularities, as well as to provide qualitative confirmation of experimental hypotheses about the unusual hydrodynamic properties of GFCs.

This chapter outlines the most recent achievements in the field of CFD modeling of GFC-based catalytic cartridges.

9.2 Mathematical Model

The object of the simulation was structured catalytic cartridges consisting of glass-fiber catalytic fabrics and structuring metal meshes. Two basic designs of such cartridges were considered: one with plain and corrugated meshes (Figure 9.1a) and the other with only simple meshes (Figure 9.1b).

It is assumed that the gaseous reaction flux moves parallel with the GFC planes in the spaces formed by the structuring grids.

Two main types of glass fabric were used in the modeling: satin (Figure 9.2a), consisting of single, tightly intertwined threads, and openwork (Figure 9.2b), made of bundles of threads with significant holes between them.

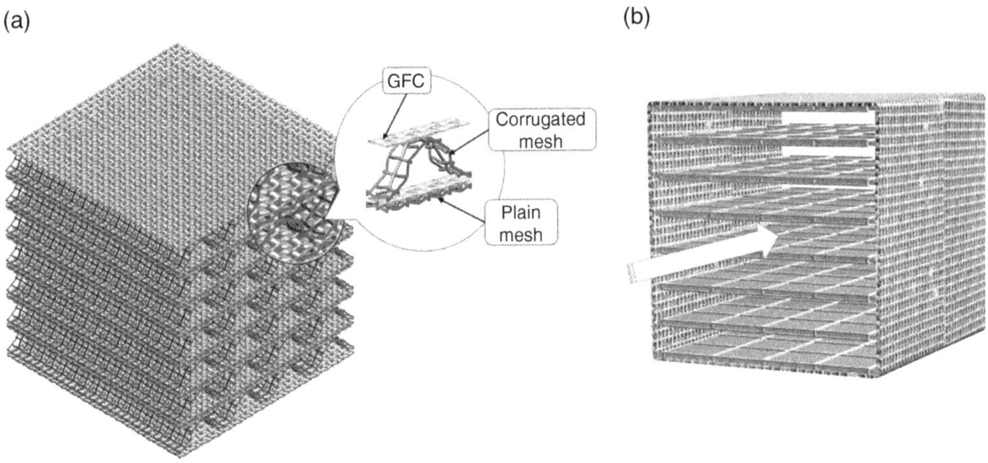

Figure 9.1 Structured cartridges with glass-fiber catalyst: (a) with corrugated and simple structuring meshes and (b) with simple meshes only. The arrows show the direction in which the fluid is moving.

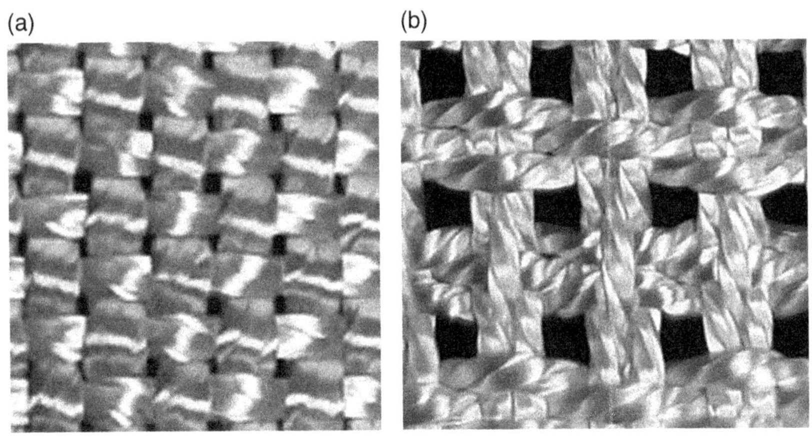

Figure 9.2 The geometric types of the GFC under consideration: (a) satin and (b) openwork.

9.2.1 Model Description

A complete system of equations describing the motion of a viscous fluid in a Cartesian system is discussed.

The coordinate system can be represented in tensor form, as follows:

- Equation of continuity: $\frac{\partial U_j}{\partial x_j} = 0$ (9.1)

- Equations of momentum change (Navier–Stokes equations) averaged by Reynolds:

- $$\rho \frac{\partial U_i}{\partial t} + \rho U_j \frac{\partial U_i}{\partial x_j} = -\frac{\partial P}{\partial x_j} + \mu \frac{\partial^2 U_i}{\partial x_j \partial x_i} - \frac{\partial P\underline{u_i' u_j'}}{\partial x_j} \quad (9.2)$$

where t is time; x_i are Cartesian coordinates ($i = 1, 2, 3$), U_i are the components of the average velocity of fluid flow in the direction x_i, P is the average static pressure, ρ is the density, and μ is the dynamic viscosity coefficient.

The Boussinesq conjecture offers a relationship relating Reynolds stresses to average velocity gradients:

- $$-\rho \underline{u_i' u_j'} = \mu_t \left(\frac{\partial U_i}{\partial x_j} + \frac{\partial U_j}{\partial x_i} \right) - \frac{2}{3} \left(\mu_t \frac{\partial U_i}{\partial x_i} + \rho k \right) \delta_{ij} \quad (9.3)$$

where $k = \frac{\overline{u_i' u_j'}}{2}$ – kinetic energy of turbulent pulsations; μ_t – turbulent (or vortex) viscosity, and δ_{ij} – delta function:

- $$\delta_{ij} = \begin{cases} 1, i = j \\ 0, i \neq j \end{cases} \quad (9.4)$$

The turbulent viscosity coefficient was determined from the selected turbulence model. To close the Reynolds-averaged Navier–Stokes equations, SST transition turbulence models were used.

Uniform distributions of gas flow velocity, temperature, and mass fraction of each component were used as boundary conditions on the input section of the simulated system. In the flow outlet area, the conditions of the "pressure outlet limit" were established, and the gauge pressure value was zero. Temperature dependencies of thermodynamic quantities and viscosity are also considered.

9.2.2 Computing Domain

The commercial software, Ansys Fluent (version 2021 R2, license No. 1148156 for Tyumen State University), with a mesh module was used for CFD simulation of described objects. The parameters of the computational grid were selected in such a manner that the vortices were resolved over the volume of the cartridge, and the resolution of complex geometry was satisfactory. The mesh is built evenly over the volume to allow vortex movements throughout the cartridge volume. A volumetric mesh is created using the Mosaic method with polyhedral elements. The minimum mesh size is 10^{-3} mm. The minimum spelling quality is 0.30, and the maximum aspect ratio is 22.88.

9.2.3 Simulation Object Geometry

The GFC cartridge is made up of similar layers. Assuming that the inlet stream is evenly distributed among the layers, only one layer can be considered for the simulation of the entire cartridge. A part of this GFC cartridge layer is shown in Figure 9.3a. The corrugated mesh forms a wave with a

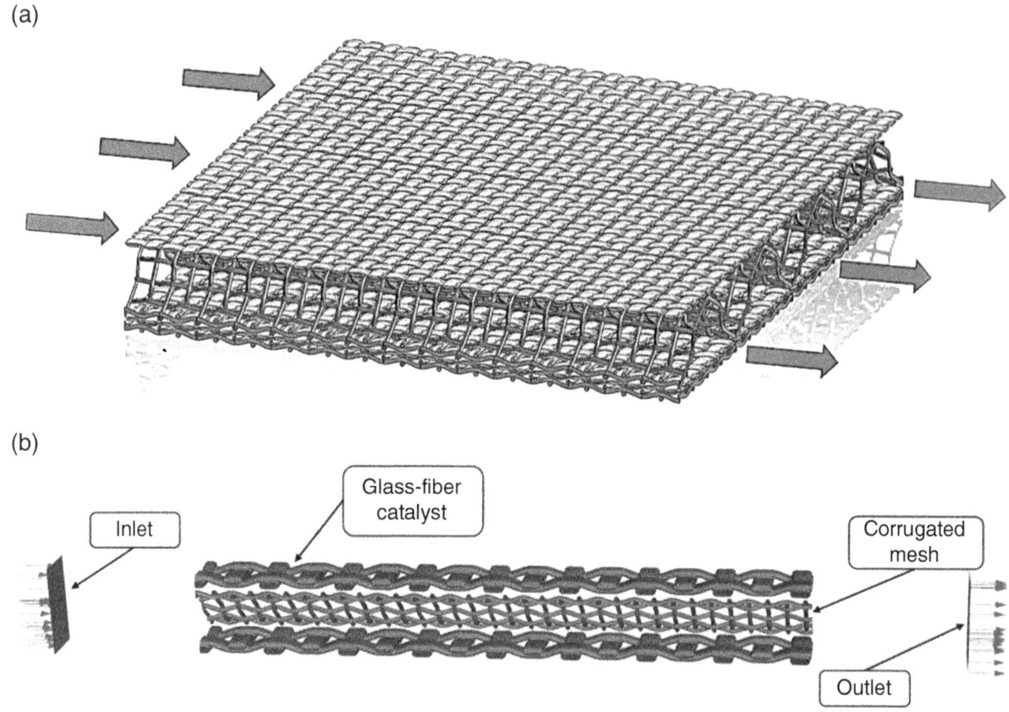

Figure 9.3 Part of the GFC cartridge layer (a) and a reduced simulation segment (b). Arrows show the gas flow movement.

repetitive geometry; so it is possible to minimize the segment for modeling using the symmetry of the channels, thereby reducing the simulated segment to the section of the channel between the upper and lower positions of the corrugated mesh (Figure 9.3b).

To simulate a cartridge with only flat structures, one can use the same geometry after removing the corrugated wire-mesh. The wire-meshes were modeled with a detailed reproduction of their structure as a combination of round wires.

The simulated segment corresponds to the GFC cartridge used in the experiments. The length of the GFC textile was 44 mm, and the total length of the simulated channel with the addition of inlet and internal voids was 66 mm. The width of the channel was taken to be 8.0 mm, and the total height was 5.7 mm.

GFC fabrics have complex geometry. Intuitively, threads in textiles can be thought of as circular in shape in the cross section, but they are actually much closer to flattened ellipsoids. To make it easier to build a geometric model, they can even be considered rectangular. The satin-type fabric model can be further reduced to almost a layer of 0.4 mm thickness, formed by intersecting rectangular bodies of 0.8 mm wide with small holes (0.1×0.1 mm) and small grooves between them (Figure 9.4a). Openwork textiles, which are easier to model, can be represented by a more complex geometry with threads in the form of flattened ellipsoids (1.0 mm wide and 0.45 mm high) with larger holes (1.5×1.5 mm) between them (Figure 9.4b). These values correspond to the geometric parameters of real GFC textiles used in mass transfer experiments.

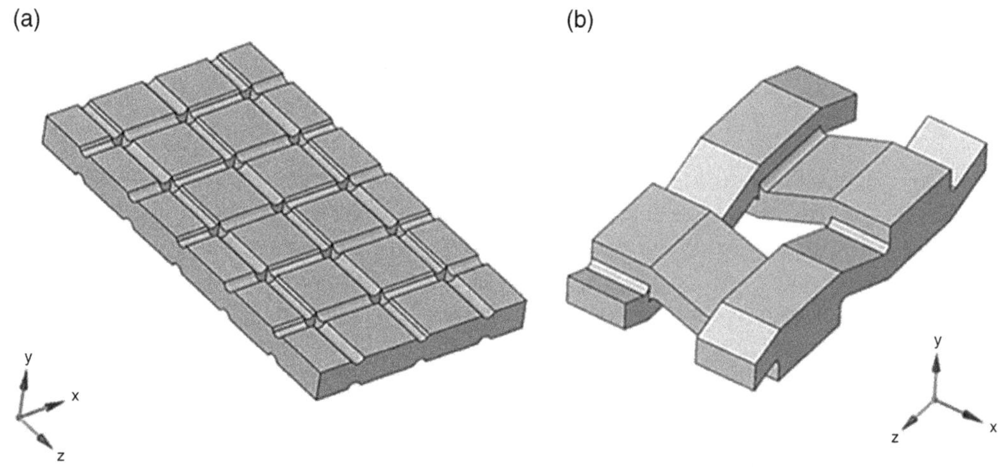

Figure 9.4 Geometric model of satin (a) and openwork and (b) GFC textiles.

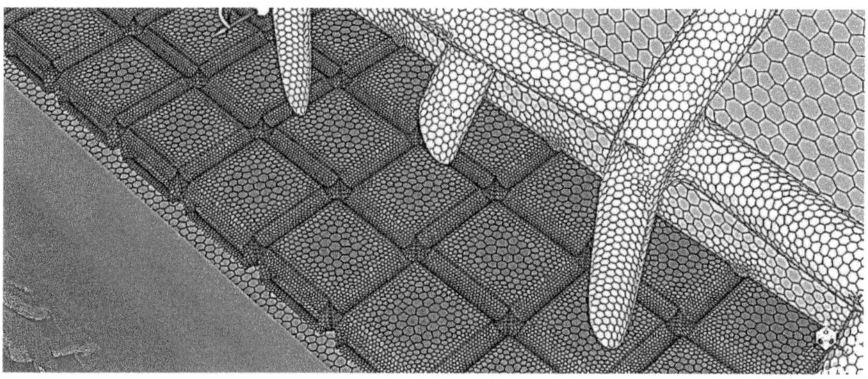

Figure 9.5 An example of an internal cartridge of detailed geometry with a computational mesh.

Figure 9.5 shows an example of the geometric structure of a cartridge showing the use of a computational mesh. The total number of cells in the created mesh has reached 83696753, the face has reached 295431252, and the number of nodes has reached 135098549.

Similar models of cartridges with varying channel heights were built for simulations.

9.2.4 Reaction

The deep oxidation reaction of toluene vapor in air on a Pt-based GFC was taken as a test reaction for CFD simulation:

- $C_7H_8 + 9O_2 \rightarrow 7CO_2 + 4H_2O$ (9.5)

This reaction was previously used for an experimental study of mass transfer in GFC-based cartridges (Zagoruiko et al. 2017; Zagoruiko and Lopatin 2019); so this reaction was chosen to simplify the comparison of simulated and experimental data.

The rate of this reaction can be calculated using the first-order velocity equation with respect to the concentration of toluene:

- $W = k_0 e^{-\frac{E}{RT}} C_{C7H8}$ (9.6)

where W is the reaction rate, k_O is the pre-exponent of rate constant, E is the activation energy, R is the universal gas constant, T is the temperature and, C_{C7H8} is the molar concentration of toluene.

The reaction was assigned to the textile surface GFC (Wall Surface reaction type). The kinetic parameters were taken from our experimental data (Zagoruiko et al., 2017; Zagoruiko and Lopatin 2019) with a corresponding conversion from units related to the mass of GFC to units related to the external surface area of GFC: $k_O = 9.0{*}10^8$ kmol/m^2/s, $E = 125.6$ kJ/mol.

9.2.5 Model Parameters

The simulation was performed for a flow with an inlet temperature of 350 °C. Experimental studies (Zagoruiko et al. 2017; Zagoruiko and Lopatin 2019) have shown that this temperature ensures the achievement of the mass transference regime, when the apparent reaction rate is completely determined by external mass transfer constraints. The basic inlet fluid velocity, recalculated to standard conditions, was assumed to be 0.55 m/s with further variation in some simulation studies. The basic inlet toluene content was set at 100 vppm.

9.3 Simulation Results

9.3.1 Cartridge Channel with Corrugated Structuring Mesh

Figure 9.6 illustrates the distribution of flow parameters in a cartridge channel with satin GFC and a corrugated structuring mesh. The contours are presented in 2D for the middle section of the channel, in a plane that is positioned vertically along the channel cartridge, intersecting the corrugated mesh at half its height.

The static pressure (Figure 9.6a) decreases along the channel from the inlet to the outlet, being almost uniform over the entire height of the channel. This uniformity may be due to the very low pressure drop (below 5 Pa) in the channel. The fluid velocity contours (Figure 9.6b) show intense fluid flows above and below the corrugated mesh with very low flow rates in the mesh and near the GFC surface. A very slight increase in temperature is observed along the channel due to heat emission in exothermic oxidation reaction, though the heat effect is very low due to insignificant concentration of toluene (Figure 9.6c). The concentration of toluene (Figure 9.6d) decreases along the length of the channel due to its consumption in the reaction. Minimum concentrations are observed near the GFC surfaces and maximum concentrations near the channel axis. Obviously, such a distribution is quite expected, and it is formed by the interaction of toluene consumption in the surface reaction to GFC with the mass transfer of toluene from the flow core in the axial region to these surfaces.

The features of the spatial distribution of flow parameters may be better seen in the 3D representation (Figure 9.7). It can be seen (Figure 9.7a) that the corrugated mesh divides the flow into two streams on both sides of the mesh. The maximum flow rate is observed at the cross-sectional corners of the channel furthest from the mesh, and the minimum velocity is observed near the corrugated mesh and GFC surfaces; this deceleration is the result of the surface friction of the flow.

The concentration of toluene (Figure 9.7b), as already shown in Figure 9.6d, is maximized in the fluid core and minimized near the surfaces of the GFC. 3D representation shows that such a

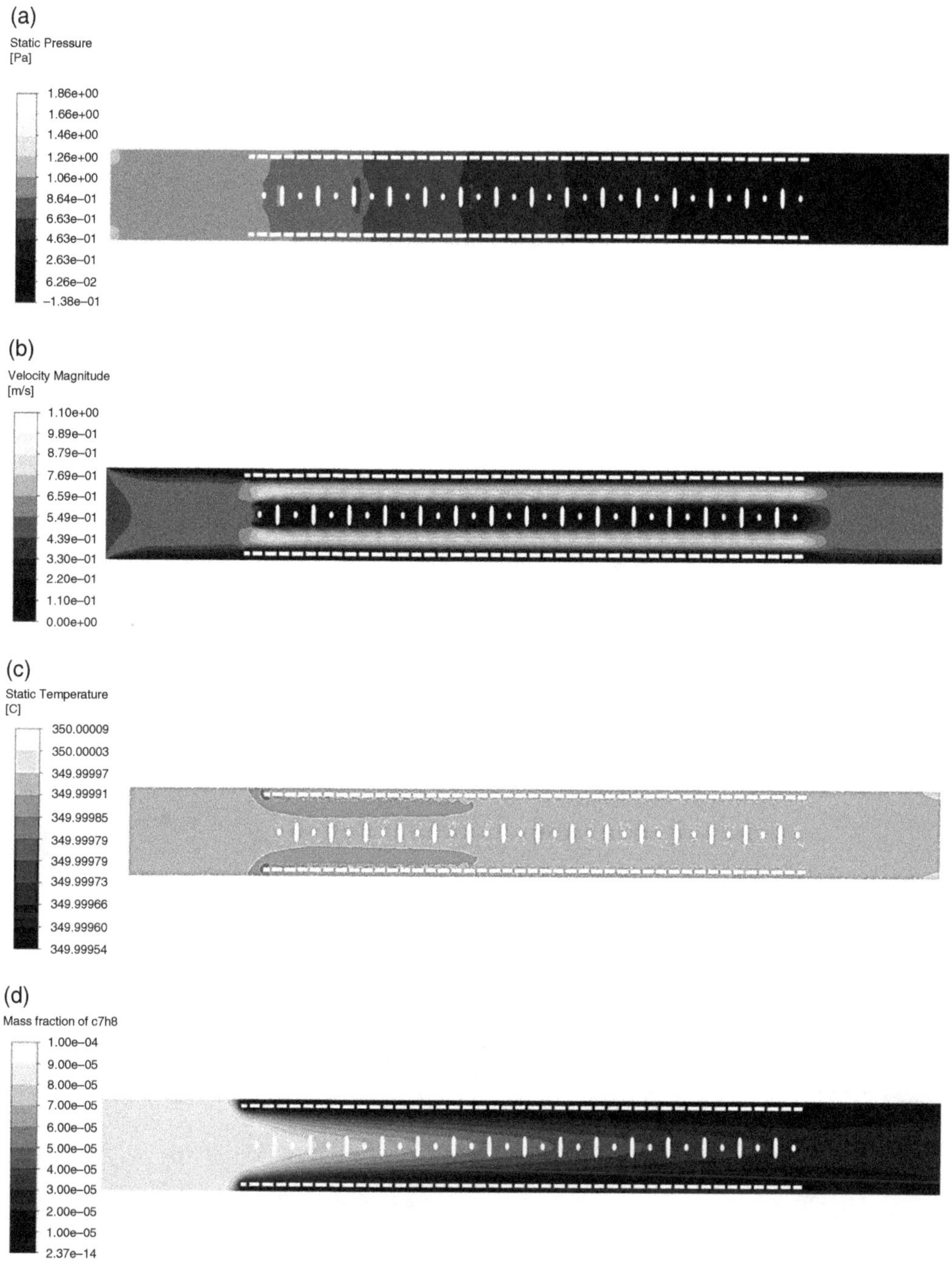

Figure 9.6 Spatial distribution of fluid parameters in the middle sequence of a channel with a corrugated structuring mesh and a satin-type catalyst. (a) Static pressure, (b) velocity of the fluid, (c) static temperature, and (d) mass fraction of the toluene. The dashed lines at the top and bottom of the channel correspond to the GFC layers, the dotted-dotted line in the middle is a cross section of the corrugated mesh.

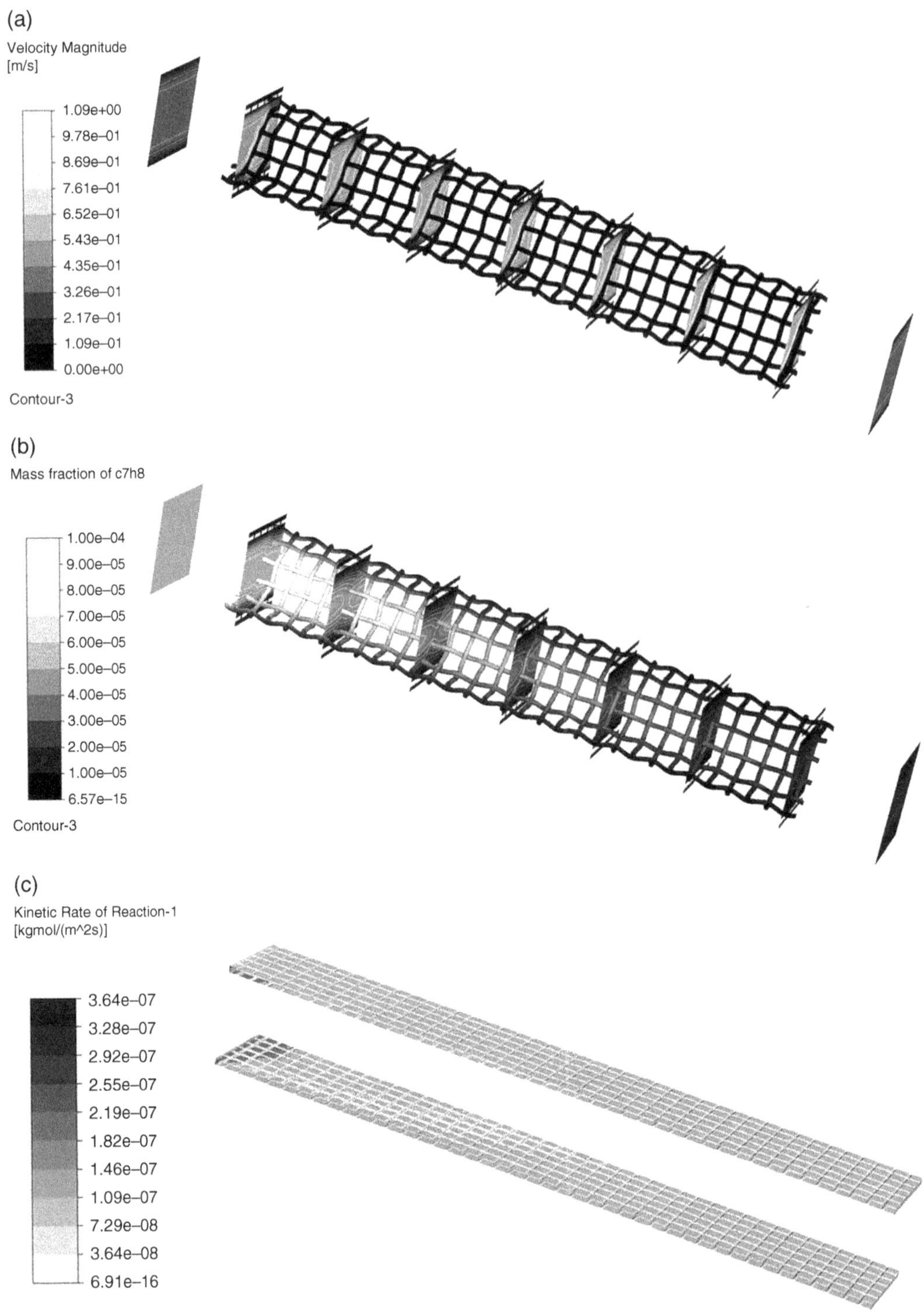

Figure 9.7 Spatial distribution of fluid parameters in a channel with a corrugated structuring mesh and a satin-type catalyst. (a) velocity of the fluid, (b) mass fraction of toluene, (c) reaction rate.

distribution is more complex: In addition to the obvious vertical gradient, it shows some gradient along the width of the channel with a decrease in concentration in the region near the structuring mesh. This pattern is observed in all cross sections along the channel. This is most likely due to the lower fluid velocity in this region, which provides more favorable conditions for mass transfer in its competition with axial mass transport with convective flow.

The maximum reaction rate (Figure 9.7c) is observed on the GFC surface in the inlet part of the channel with a further decrease in the velocity through the channel due to a decrease in the concentration of toluene. A gradient in the reaction rate can also be observed throughout the GFC textile. For example, the lower GFC textile in Figure 9.7c shows the decrease in reaction rate from channel inlet to its outlet. This distribution correlates quite well with the distribution of fluid velocities near the GFC surface (Figure 9.7a): A higher velocity on the left side of the textile results in a shorter residence time of the fluid and thus a lower conversion of toluene, resulting in a higher concentration of toluene (as can be seen in Figure 9.7b) and, in general, a higher reaction rate.

9.3.2 Influence of GFC Textile Shape

Satin and openwork GFCs have different surface shapes, which can cause differences in their mass transfer properties. CFD simulations have shown that the overall pattern observed in the case of the satin-type GFC is approximately the same in the case of the openwork weave catalyst (Figure 9.8).

The most significant difference is observed in the surface distribution of the reaction rate (Figure 9.8c): Although the general trends here (decrease in the reaction rate from the upper left to the lower right corner of the lower GFC textile) are similar to those observed in the previous case of the satin-type catalyst (Figure 9.7c), the overall distribution of the rates in the case of openwork textiles is more uniform. The surface of openwork GFC is much less smooth compared with satin; this can increase the turbulence of the flow near the textile surface and adjacent areas, and thereby improve the overall mass transfer efficiency in the openwork geometry. This hypothesis is supported by the values of the Nusselt number characterizing the intensity of external mass transfer, calculated in CFD studies: For these conditions, they are equal to 7.1 for satin and 9.2 for the openwork catalyst, that is, the estimated increase in apparent activity is $\sim$30%. A similar trend can be traced in experimental data: The apparent reaction rate with an openwork catalyst under such conditions is higher than that of the satin type, and this difference ranges from 7–14% to 25–28%.

9.3.3 Cartridge Channel Without Corrugated Structuring Mesh

Figures 9.9 and 9.10 demonstrate the distribution of fluid parameters in a cartridge channel with openwork GFC and without a corrugated structuring mesh.

Although the pressure distribution over the height of the channel is still uniform in this case (Figures 9.9a and 9.10a), the fluid has a well-defined velocity profile with a maximum velocity of the fluid was along the axis of the channel and minimum velocity was on the GFC surfaces. Compared to the corrugated mesh channel, which breaks down the fluid core into two smaller nuclei, this situation is less favorable from the point of view of external mass transfer. The worst diffusion of toluene from the axial region on the GFC surface is observed in Figures 9.9d and 9.10b. As a result, lower apparent reaction rates lead to reaction propagation along GFC textiles, as can be seen from the temperature distribution (Figure 9.9c) and reaction rate (Figure 9.10c). Unlike the system with a corrugated mesh, where this mesh creates an asymmetry of fluid velocities across the entire width of the textile surface, the distribution of the reaction rate in this case is symmetrical in relation to the channel axis.

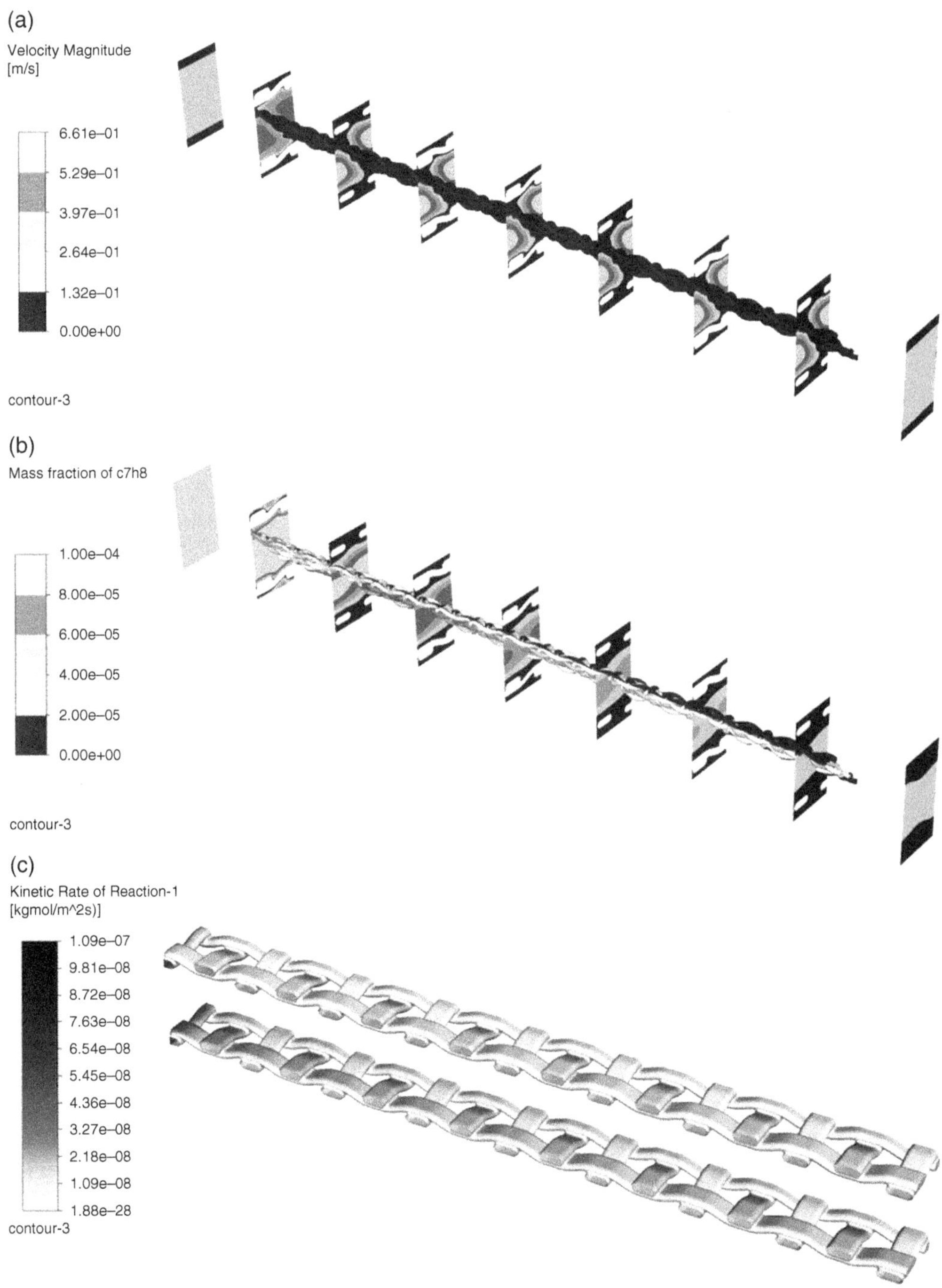

Figure 9.8 Spatial distribution of fluid parameters in a channel with a corrugated structuring mesh and an openwork catalyst. (a) velocity of the fluid, (b) mass fraction of toluene, (c) reaction rate.

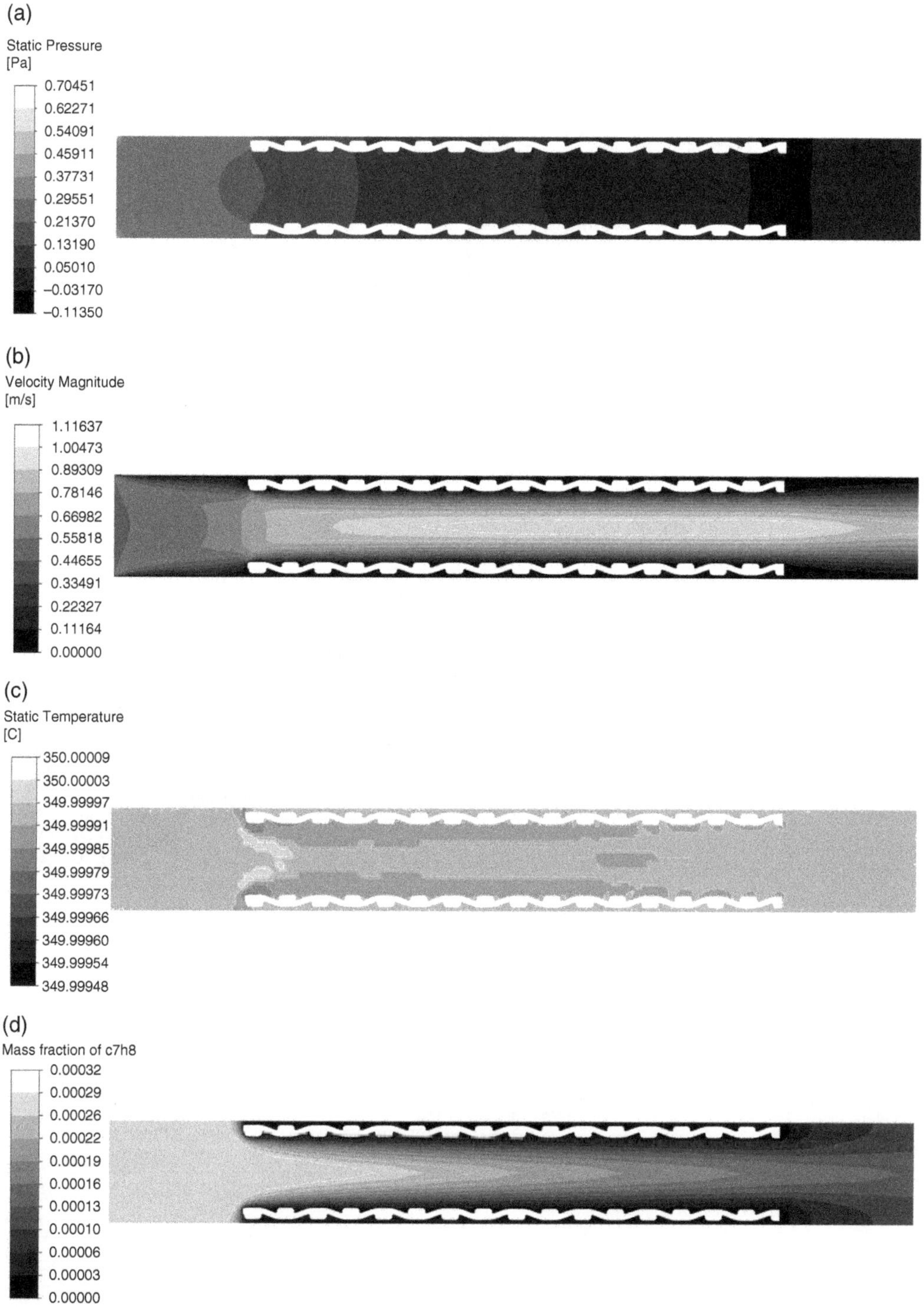

Figure 9.9 Spatial distribution of fluid parameters in a channel without a corrugated structuring mesh and an openwork catalyst. (a) static pressure, (b) velocity magnitude, (c) static temperature, (d) mass fraction of toluene.

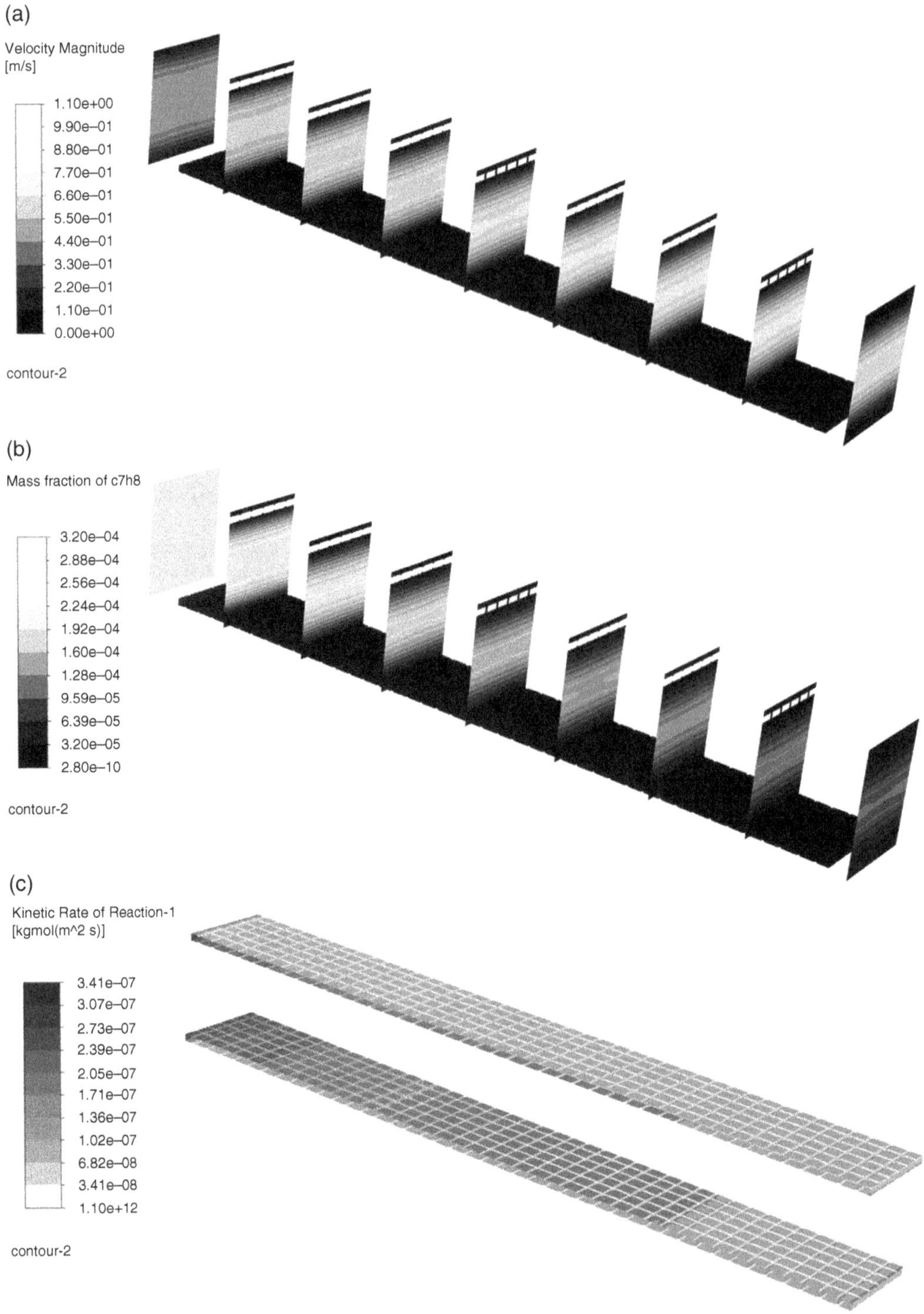

Figure 9.10 Spatial distribution of fluid parameters in a channel without a corrugated structuring mesh and an openwork catalyst. (a) velocity of the fluid, (b) mass fraction of toluene, (c) reaction rate.

Table 9.1 Mass transfer characteristics of GFC-based structured cartridges.

Cartridge/GFC type	Computed Nusselt number	Mass transfer coefficient, m/s	
		Calculated	Experimental
Corrugated mesh + satin-type GFC	7.1	0.025	0.034
Corrugated mesh + openwork GFC type	9.2	0.031	0.036–0.044
Without corrugated mesh + satin-type GFC	6.6	0.023	0.030–0.032
Without corrugated mesh + openwork-type GFC	7.8	0.026	0.035–0.038

According to the quantitative analysis of the calculated mass transfer data, the Nusselt number decreases to 6.6 for satin and 7.8 for the openwork catalyst, respectively, showing 7–14% less-intensive mass transfer than a similar system with a corrugated structuring mesh. The general behavior of the system (better mass transfer in the presence of a corrugated mesh) agrees well with the experimental data, although in experiments, the increase in the Nusselt number under similar conditions can be even more significant (up to 20–30%).

The general quantitative comparison (Table 9.1) revealed that the mass transfer coefficients obtained in CFD calculations are slightly (30%) lower than those obtained in the experiment. On one hand, such a discrepancy may be attributed to the relatively low accuracy of mass transfer experiments. However, this distinction appears to be quite systematic, so it cannot be completely ignored. This issue deserved a further study, including detailed analysis of fluid interactions with GFC, resulting from its flexibility and specific internal structure.

9.3.4 Two-Sided Washing of GFC Textiles

In our previous mass transfer studies (Zagoruiko et al. 2017; Zagoruiko and Lopatin 2019), we discussed possible differences between the behavior of GFC-based structured systems and conventional solid catalysts. It has been suggested that such differences may arise due to the mechanical flexibility of GFC, which is unusual for virtually all other known forms of catalysts. In particular, such flexibility can lead to a specific mechanical interaction between the catalyst and the moving reaction stream, resulting in variability in the external specific surface area of GFC available to the reactants as a function of the fluid velocity. It has been suggested that, in addition to the normal contact between the convective flow of the fluid and the face of the GFC textile oriented to the channel space, part of the flow may also pass through the gap between the GFC and the supporting plain mesh, thus coming into contact with the back side of the textile. This situation is completely different from conventionally structured catalysts; for example, from monolith blocks where the solid washcoat with active component is accessible to reactants from external side only.

This hypothesis has been tested experimentally. For this purpose (Zagoruiko et al. 2017; Zagoruiko and Lopatin 2019), the fluorescent paint was sprayed in the airstream at the inlet of the GFC cartridge using corrugated wire-meshes. It was found that the outer surface of the textile, oriented toward the gas passage and the corrugated nets, contained clearly visible traces of paint applied along the channels between the corrugated elements. Surprisingly, the inner textile surface, oriented toward a flat structural mesh, also contained a significant number of paint particles. The ink particles are relatively large (up to tens of microns in size), and the possibility of its diffusion through the fabric has been excluded. These results made it possible to state that convective gas flow exists in GFC cartridges not only in the cartridge channels but also between the textile and the supporting GFC wire-mesh. This behavior essentially turns the surface area of GFC, which is always assumed to be constant in the traditional approach, into a variable value that depends on the fluid velocity and catalyst cartridge geometry. This phenomenon is solely due to the mechanical flexibility of GFC, which is impossible for the packaging of solid conventional catalysts.

In the current study, an attempt is made to reproduce the described flow structure based on CFD simulations. The first calculation showed that in the case of direct attachment of the GFC textile to the support mesh (or, in terms of the geometry of the model, to the upper and lower boundaries of the modeling area), no flow was observed along the back of the textile. This is reasonable, because the CFD model does not consider the flexibility of the GFC and the ability to change its shape, so the height of the gap between the GFC and the support grid was set to zero and, technically, the gap for the backside flow was simply absent in the model. To explain the existence of this gap, it was assumed that the GFC is located at a short distance from the supporting grid. Figure 9.11 shows the result of a CFD simulation for a system with a GFC plane located 0.3 mm from the supporting smooth mesh.

In this case, there are convective backside flows between the GFC textile and the support meshes in both the upper and lower parts of the channel (Figure 9.11a). These flows provide a response not only on the external side of the GFC but also on its internal one (Figure 9.11b). The calculated fluid structure agrees well with the experimentally observed pattern: While the main frontal lateral flow moves along the GFC surface mainly within the contact lines between the corrugated mesh and the GFC (the "trace" of the corrugated mesh paint is clearly visible in Figure 9.11d), the flow of the reverse side is more uniform along the textile width and mainly observed in the very inlet part of the GFC (Figure 9.11c).

In general, it can be stated that the proposed approach with controlled positioning of GFC textiles is applicable for modeling the complex structure of the fluid arising from the flexibility of the GFC. Of course, the real picture is more complex, and the distance from the reference mesh can be

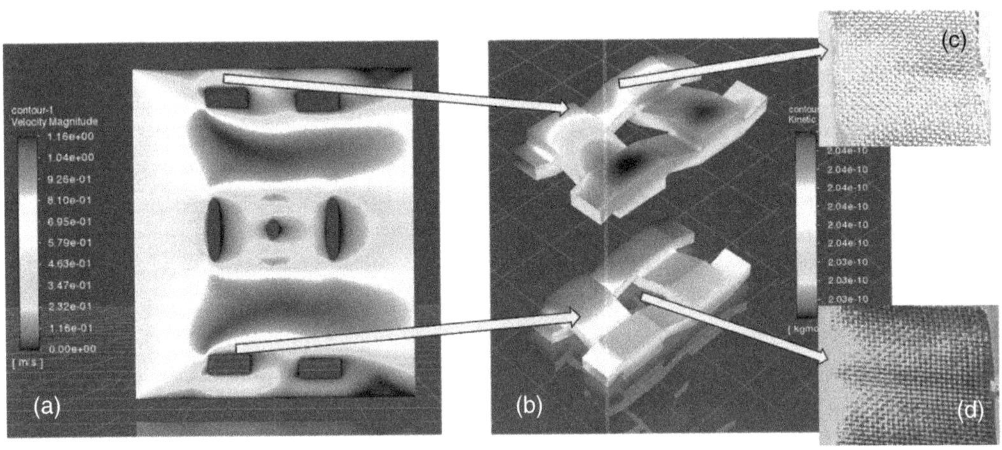

Figure 9.11 The distribution of a calculated values of the fluid velocity (a) and the reaction rate (b) in the inlet part of the channel (Lopatin et al. 2023) and the experimentally measured distribution of paint on the back (c), front, and (d) sides of the GFC textile in the area of the inlet duct (Zagoruiko et al. 2017; Zagoruiko and Lopatin 2019).

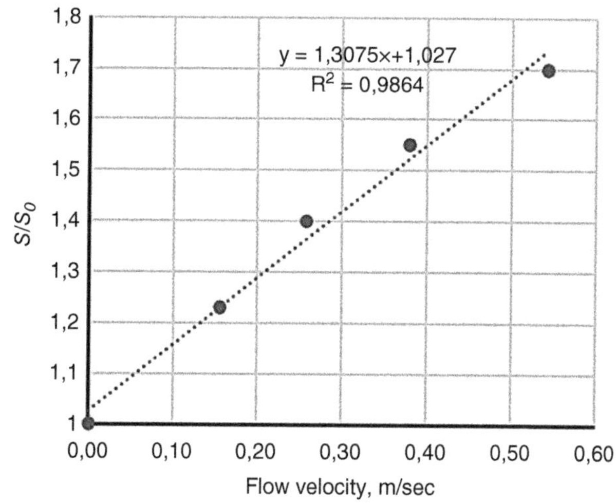

Figure 9.12 Values of S/S_0 ratio, providing the best possible agreement the experimental and calculated data in each experimental point, upon the gas flow velocity in the GFC cartridge channel.

considered varying depending on many factors (flow rate, shape and density of textiles, details of structuring the mesh geometry, etc.). This issue requires more in-depth research.

Another option to reproduce the backside flow was to directly set the value for the washable surface area of the GFC textile. We have performed the variation of the accessible specific GFC surface area S in the range from the surface of the one side of GFC textile (S_0) and up to doubled S_0 value, corresponding to full two-side washing of the textile by the fluid. For each fluid velocity, we varied the values of S/S_0 ratio using the surface area washcoat factor to provide the best possible agreement between the experimental and calculated data in each experimental point. The optimal S/S_0 values are plotted in Figure 9.12.

It is seen that the apparent S/S_0 ratio is ~1 at low gas speed, equivalent to one-sided flow along the textile front side, as in case with the solid catalyst surface; it rises practically linear with the rise of gas flow velocity reaching the value of 1.7. It is equivalent to the two-sided washing of the textile by the flow with the backside flow as high as 70% of the front side one. This phenomenon may be attributed solely to the flexibility of the GFC; it is impossible for all known shapes of conventional solid catalysts.

9.3.5 Convective Flow Inside the GFC Thread

In the beds of conventional catalysts, the transport of reactants in the free layer occurs predominantly or entirely due to the convective mechanism of the movement of the flow, while the transport of substances within the catalyst material is provided by the diffusion of reactants in the pores of the catalyst. Such pores are usually very small in size, in the order of a few tens of nanometers, so convective flow inside them is impossible. The intensity of the diffusion flow is relatively low, and this can be the cause of significant internal diffusion limitation of catalytic reactions.

In the case of GFC, the catalyst microfibers are nonporous, and reactions take place on the surface of these microfibers. In such systems, the internal mass transfer takes place inside the catalytic threads. Given that the distances between the microfibers in the GFC threads are relatively large (up to a few microns), it is possible that the mass transfer by diffusion may be accompanied inside the threads with the convective flows, thus potentially increasing the overall efficiency of internal transport of reactants.

This hypothesis was tested using CFD modeling. To do this, a 2D model of the channel segment in the GFC-based cartridge was built with two catalytic threads located close to each other perpendicular to the observation plane and adjacent to the channel wall. Each thread was represented as a cross section of an elliptical shape, the internal volume of each thread containing 2512 uniformly spaced single microfibers with a diameter of 6 microns, and this corresponds to the real GFC thread parameters. In the simulation, a gas flow with a velocity of 0.5 m/s at the inlet to the channel was considered.

Figure 9.13a shows the calculated distribution of flow velocities in the channel volume. It can be seen that in the core of the flow, the velocity can significantly exceed (up to 1.5 m/s) the initial one due to the narrowing of the channel, while in the lower part of the channel near the filaments and between them the velocity is close to zero, and there are no convective flows inside the filaments. However, when the velocity scale is increased (Figure 9.13b), it becomes clear that there is still a movement of the flow inside the filaments, although the velocities are very small (0.5–1 mm/s).

This velocity is negligible compared with the flow velocity in the channel volume, but its contribution to the overall transport of matter inside the filaments can be noticeable when compared with the diffusion of internal transport of substances. The mechanisms of these types of mass transfer are different and difficult to compare with each other directly, but as a rough estimate, we can use the diffusion Peclet number, which describes the ratio of convective and diffusion mass transfer as follows:

- $$Pe = \frac{ul}{D} \tag{9.7}$$

where u – gas velocity (m/s), l – typical geometrical size (m), and D – effective diffusion coefficient (m^2/s). We may consider the velocity values obtained in CFD calculations $u = 1*10^{-3}$ m/s and use the thread thickness as a characteristic geometric dimension $l = 10^{-3}$ m. The value of the diffusion coefficient of toluene in air was considered at a temperature of 200 °C, at which the internal

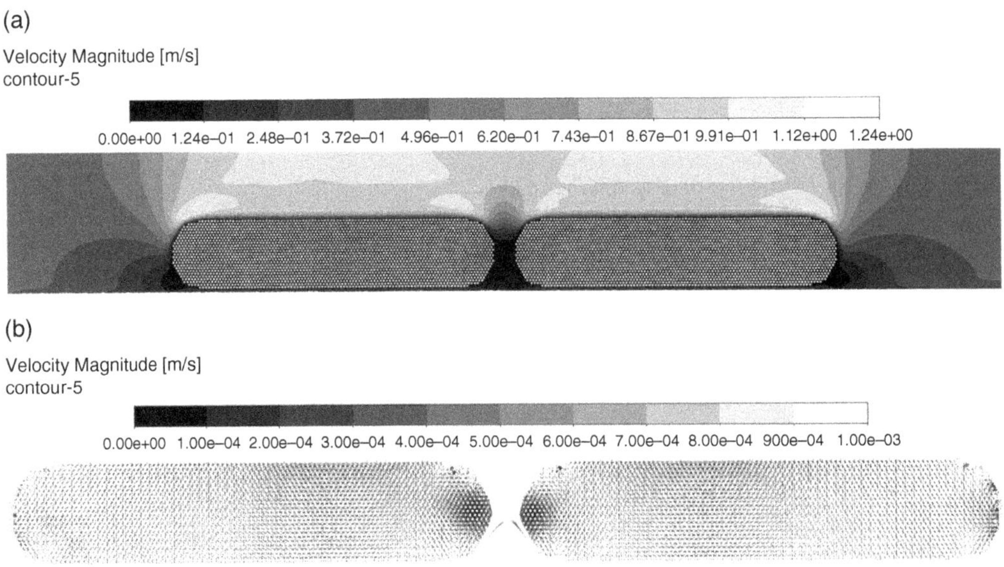

Figure 9.13 Fluid velocity distribution in the void volume of the channel (a) and inside the GFC threads (b).

diffusion limitation in the reaction under consideration becomes noticeable (Zagoruiko and Lopatin 2019), and it is equal to $D = 1.5*10^{-5}$ m^2/s. For these conditions, Pe = 0.07.

On one hand, this value indicates the dominance of diffusion transport of matter within the GFC threads; on the other hand, it indicates that the convective flow within the threads makes a nonzero contribution to improving the overall efficiency of internal mass transfer. This fundamentally distinguishes GFC threads from traditional types of catalysts, where such convective transfer within the catalyst material does not actually exist.

Of course, the aforementioned quantitative estimate is of a particular nature, and the ratio of diffusion and convective transport inside the GFC threads can vary significantly depending on the reaction conditions, types of reactants, characteristics of the filaments, etc. For example, it can be expected that the velocities of internal convective flows will increase with an increase in the total fluid velocity in the channel. These issues require further study of individual cases, however, in any case, the information obtained provides a new understanding of the specifics of mass transfer processes in structured GFC packages, as well as allows for an unexpected approach to optimizing the structural parameters of such packages and increasing the efficiency of catalytic processes based on GFC.

9.3.6 The General Description of Mass Transfer in GFC

Although the proposed CFD-based approach proved to be an efficient tool for simulation of GFC-based cartridges, it is rather complicated and resource-consuming. Sometimes it is necessary to use the simplified approach methods for practical applications. To provide such simplification, it is possible to use the approach based on criterion equations. Within the framework of this approach, the value of the dimensionless criterion that determines the intensity of mass transfer, the mass Nusselt number, may be represented as a general equation:

- $$Nu_m = A\mathrm{Re}^m\mathrm{Sc}^{1/3} \tag{9.8}$$

The Reynolds criterion is a measure of the ratio of the inertial forces acting in the flow to the forces of viscous friction. Physically, it has the meaning of a dimensionless flow velocity or a dimensionless geometric dimension. The Reynolds test is calculated using the following equation:

- $$\mathrm{Re} = \frac{ud_{\mathrm{eq}}\rho}{\mu} \tag{9.9}$$

where d_{eq} is the equivalent size of the cartridge channel, m; Re is the dimensionless Reynolds number; u is the actual fluid in the cartridge, m/s; ρ is the density of the medium, kg/m^3; μ is the dynamic viscosity of the medium, Pa*s.

The mass Nusselt number has the physical meaning of a dimensionless mass transfer coefficient and is calculated using the following formula:

- $$Nu_m = \frac{\beta d_{\mathrm{eq}}}{D} \tag{9.10}$$

where β is the mass transfer coefficient, m/s; D is the coefficient of diffusion of toluene in air, m^2/s.
The Schmidt criterion is a measure of the ratio of viscosity to mass of a substance. The Schmidt criterion is calculated using the following equation:

- $$\mathrm{Sc} = \frac{\mu}{\rho D} \tag{9.11}$$

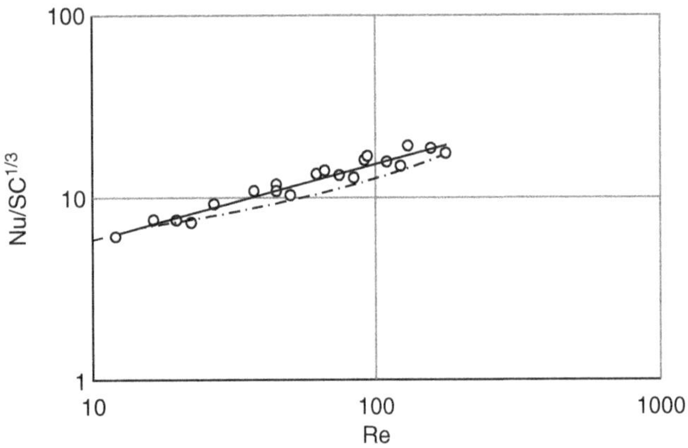

Figure 9.14 Dependence of mass-related Nusselt number upon Reynolds number. Points – values calculated by means of CFD, solid line – approximation of calculated values by power function (9.14), dashed line – approximation of experimental data by power functions (9.12) and (9.13).

Earlier in our work (Zagoruiko and Lopatin 2019), such criterion equations were proposed that generally describe the intensity of mass transfer in ICS cartridges with corrugated mesh structuring elements:

- $\mathrm{Re} < 80, \mathrm{Nu}_m = 2{,}95 \times \mathrm{Re}^{0{,}31} \times \mathrm{Sc}^{1/3}$ (9.12)
- $\mathrm{Re} > 80, \mathrm{Nu}_m = 0{,}84 \times \mathrm{Re}^{0{,}59} \times \mathrm{Sc}^{1/3}$ (9.13)

In this work, a similar dependence was built on the basis of CFD calculations. For this purpose, a series of calculations was carried out with a variation in flow velocity from 0.07 to 0.54 m/s and a channel height in the cartridge in the range of 3–9 mm. Results are shown in Figure 9.14. It is seen that the calculated data obtained with CFD can be approximated with acceptable accuracy over the entire range of Re values by a power equation with a power of 0.42, which is close to the arithmetic mean for the powers in Equations (9.12) and (9.13), which are 0.31 and 0.59, respectively:

- $\mathrm{Nu}_m^{\mathrm{CFD}} = 2{,}20 \times \mathrm{Re}^{0{,}42} \times \mathrm{Sc}^{1/3}$ (9.14)

Besides, the new equation obtained by the CFD method reproduces with acceptable accuracy and experimental results are summarized by Equations (9.12) and (9.13).

9.4 Conclusion

The structure of GFC-based catalytic cartridges is really complex as it contains elements (GFC textiles and structuring meshes) with intricate, fine-shaped parts. At the same time, these details can affect the performance of the cartridge and thus can be important in cartridge modeling. Of course, appropriate simplification procedures were applied, that is, the use of symmetry to reasonably reduce the modeling volume (reducing the entire multilayer cartridge to half a channel) and a simplified representation of the GFC textile structure (especially in the case of satin-type fabric). On the contrary, reducing the channel geometry from 3D to 2D or using a stationary model instead of a transient model proved to be impractical, as such approaches lead to an obvious loss of modeling quality.

As a result, the constructed model turned out to be quite complex, and the modeling procedure turned out to be very resource intensive. For example, some calculations required up to 100 hours of calculations, while modeling a single-dimensional segment of a convection monolithic catalyst with a square channel and a smooth surface took only 6 seconds on the same computer. Perhaps the proposed model could be further simplified to reduce computational time, but this study confirmed that the current model is nevertheless applicable for research purposes in its current form.

The simulation revealed the specificities of fluid mechanics in GFC cartridges. Mass transfer CFD studies have successfully reproduced the qualitative effects of rather complex cartridge design details, such as the GFC fabric structure or the structure of the internal mesh geometry observed earlier in experimental work.

CFD calculations were able to describe the theoretically proposed and experimentally observed regularities of mass transfer processes in GFC structures, including those existing only in GFCs. It was shown that unlike conventional solid catalysts, in case of GFCs, there may exist the flow of reactants along the back side of GFC textile, additional to the main flow along its external surface. This may result in variability of the washable surface area of the GFC textile, which may rise with increase in fluid velocity and reaching up to 170% (and possibly even higher) of the textile external surface area accessible to reactants in absence of flow movement. This is an exceptional feature of GFC, with their flexibility which is not present in the case of solid catalysts.

Another unusual phenomenon is the combined internal mass transfer in GFC threads, which may be provided not only by diffusion, as in usual porous catalysts, but also by convective flow, which is possible only in GFC due to the relatively high permeability of its threads.

The developed model accounts for many sophisticated but important details, and it can be used to optimize the design parameters of existing GFC cartridges as well as to develop new designs of structured systems based on GFC for various practical applications.

Abbreviations

CFD – computational fluid dynamics
GFC – glass-fiber catalyst

Acknowledgement

This work was partially supported by the Ministry of Science and Higher Education of the Russian Federation within the state assignment for the Boreskov Institute of Catalysis (project FWUR-2024-0037) and by the Tyumen Oblast Government, as part of the West-Siberian Interregional Science and Education Center's project No. 89-DON (3).

References

Aldashukurova G.B., Mironenko A.V., Mansurov Z.A., Shikina N.V., Yashnik S.A., Kuznetsov V.V., Ismagilov Z.R. (2013) Synthesis gas production on glass cloth catalysts modified by Ni and Co oxides, *Journal of Energy Chemistry* 22: 811–818. http://dx.doi.org/10.1016/S2095-4956(13)60108-4.

Balzhinimaev, B.S., Paukshtis, E.A., Vanag, S.V. et al. (2010). Glass-fiber catalysts: Novel oxidation catalysts, catalytic technologies for environmental protection. *Catalysis Today* 151: 195–199. http://dx.doi.org/10.1016/j.cattod.2010.01.011.

Bouras, H., Haroun, Y., Bodziony, F.F. et al. (2022). Use of CFD for pressure drop, liquid saturation and wetting predictions in trickle bed reactors for different catalyst particle shapes. *Chemical Engineering Science* 249: 117315. https://doi.org/10.1016/j.ces.2021.117315.

Bracconi, M., Ambrosetti, M., Maestri, M. et al. (2018). A fundamental investigation of gas/solid mass transfer in open-cell foams using a combined experimental and CFD approach. *Chemical Engineering Journal* 352: 558–571. https://doi.org/10.1016/j.cej.2018.07.023.

Chub, O.V., Borisova, E.S., Klenov, O.P. et al. (2005). Research of mass-transfer in fibrous sorption-active materials. *Catalysis Today* 105: 680–688. http://dx.doi.org/10.1016/j.cattod.2005.06.049.

Dixon, A.G., Taskin, M.E., Nijemeisland, M., and Stitt, E.H. (2008). Wall-to-particle heat transfer in steam reformer tubes: CFD comparison of catalyst particles. *Chemical Engineering Science* 63: 2219–2224. https://doi.org/10.1016/j.ces.2008.01.017.

Dong, Y., Korup, O., Gerdts, J. et al. (2018). Microtomography-based CFD modeling of a fixed-bed reactor with an open-cell foam monolith and experimental verification by reactor profile measurements. *Chemical Engineering Journal* 353: 176–188. https://doi.org/10.1016/j.cej.2018.07.075.

George, G., Bockelmann, M., Schmalhorst, L. et al. (2022). Radial heat transport in a fixed-bed reactor made of metallic foam pellets: experiment and particle-resolved computational fluid dynamics. *International Journal of Heat and Mass Transfer* 197: 123376. https://doi.org/10.1016/j.ijheatmasstransfer.2022.123376.

Golyashova, K., Mikenin, P., Elyshev, A. et al. (2019). Structured catalytic cartridges for SO2 oxidation in flue gases of coal-fired powerplants. *Chemical Engineering Journal* 378: 122194. https://doi.org/10.1016/j.cej.2019.112194.

Golyashova, K., Glazov, N., Lopatin, S., and Zagoruiko, A. (2023). Glass-fiber catalysts for selective catalytic reduction of nitrogen oxides by co and hydrocarbons. *Catalysis Communications* 106765: 1–11. https://doi.org/10.1016/j.catcom.2023.106765.

Gulyaeva, Y.K., Kaichev, V.V., Zaikovskii, V.I. et al. (2015). Silicate fiberglass catalysts: from science to technology. *Catalysis in Industry* 7: 267–274. http://dx.doi.org/10.1134/S2070050415040029.

Iwaniszyn, M., Gancarczyk, A., Gąszczak, A., and Kołodziej, A. (2021a). Structured intra-tubular catalyst carrier for highly exothermic processes: modelling and CFD study. *International Journal of Heat and Mass Transfer* 175: 121357. https://doi.org/10.1016/j.ijheatmasstransfer.2021.121357.

Iwaniszyn, M., Sindera, K., Gancarczyk, A. et al. (2021b). Experimental and CFD investigation of heat transfer and flow resistance in woven wire gauzes. *Chemical Engineering and Processing – Process Intensification* 163: 108364. https://doi.org/10.1016/j.cep.2021.108364.

Jurković, D., Jing-Lin, L., Pohar, A., and Likozar, B. (2021). Methane dry reforming over Ni/Al_2O_3 catalyst in spark plasma reactor: linking computational fluid dynamics (CFD) with reaction kinetic modelling. *Catalysis Today* 362: 11–21. https://doi.org/10.1016/j.cattod.2020.05.028.

Karthik, G.M. and Buwa, V.V. (2019). Effect of particle shape on catalyst deactivation using particle-resolved CFD simulations. *Chemical Engineering Journal* 377: 120164. https://doi.org/10.1016/j.cej.2018.10.101.

Klenov, O., Chumakova, N., Pokrovskaya, S., and Noskov, A. (2019). Impact of heat and mass transfer in porous catalytic monolith: CFD modeling of exothermic reaction. *Chemical Engineering Science* 205: 1–13. https://doi.org/10.1016/j.ces.2019.04.010.

Li, L., Diao, Y., and Liu, X. (2014). Low temperature selective catalytic reduction of NOx with NH_3 over Mn-based catalyst: a review. *Journal of Rare Earths* 32: 409–415. http://dx.doi.org/10.3934/environsci.2016.2.261.

Lopatin, S.A., Tsyrul'nikov, P.G., Kotolevich, Y.S. et al. (2015). Structured woven glass-fiber IC-12-S111 catalyst for the deep oxidation of organic compounds. *Catalysis in Industry* 7: 329–334. http://dx.doi.org/10.1134/S2070050415040121.

Lopatin, S., Elyshev, A., and Zagoruiko, A. (2023). CFD modelling of the structured cartridges with glass-fiber catalysts. *Chemical Engineering Research and Design* 190: 255–267. https://doi.org/10.1016/j.cherd.2022.12.027.

Mikenin, P., Zazhigalov, S., Elyshev, A. et al. (2016). Iron oxide catalyst at the modified glass fiber support for selective oxidation of H_2S. *Catalysis Communications* 87: 36–40. https://doi.org/10.1016/j.catcom.2016.08.038.

Paukshtis, E.A., Simonova, L.G., Zagoruiko, A.N., and Balzhinimaev, B.S. (2010). Oxidative destruction of chlorinated hydrocarbons on Pt-containing fiber-glass catalysts. *Chemosphere* 79: 199–204. https://doi.org/10.1016/j.chemosphere.2010.01.050.

Salmi, T., Mäki-Arvela, P., Toukoniitty, E. et al. (2000). Immobile silica fibre catalyst in liquid-phase hydrogenation. *Studies in Surface Science and Catalysis* 130: 2033–2038. https://doi.org/10.1016/S0167-2991(00)80767-3.

Shalygin, A., Paukshtis, E., Kovalyov, E., and Bal'zhinimaev, B. (2013). Light olefins synthesis from C1-C2 paraffins via oxychlorination processes. *Frontiers of Chemical Science and Engineering* 7: 279–288. http://dx.doi.org/10.1007/s11705-013-1338-1.

Sinn, C., Pesch, G., Thöming, J., and Kiewidt, L. (2019). Coupled conjugate heat transfer and heat production in open-cell ceramic foams investigated using CFD. *International Journal of Heat and Mass Transfer* 139: 600–612. https://doi.org/10.1016/j.ijheatmasstransfer.2019.05.042.

Tedesco, G. and Moraes, P. (2021). Innovative design of a continuous flow photoelectrochemical reactor: hydraulic design, CFD simulation and prototyping. *Journal of Environmental Chemical Engineering* 9: 105917. https://doi.org/10.1016/j.jece.2021.105917.

Wehinger, G., Kolaczkowski, S., Schmalhorst, L. et al. (2019). Modeling fixed-bed reactors from metal-foam pellets with detailed CFD. *Chemical Engineering Journal* 373: 709–719. https://doi.org/10.1016/j.cej.2019.05.067.

Yang, X., Wang, S., Zhang, K., and He, Y. (2021). Evaluation of coke deposition in catalyst particles using particle-resolved CFD model. *Chemical Engineering Science* 229: 116122. https://doi.org/10.1016/j.ces.2020.116122.

You, Y., Wu, Z., Zeng, W. et al. (2019). CFD modeling of unsteady SCR deNOx coupled with regenerative heat transfer in honeycomb regenerators partly coated by Vanadium catalysts. *Chemical Engineering Research and Design* 150: 234–245. https://doi.org/10.1016/j.cherd.2019.07.015.

Zagoruiko, A.N. and Lopatin, S.A. (2019). *Structured glass-fiber catalysts*, 158. Francis & Taylor Group, CRC Press https://doi.org/10.1201/9780429317569.

Zagoruiko, A.N., Lopatin, S.A., Mikenin, P.E. et al. (2017). Novel structured catalytic systems - cartridges on the base of fibrous catalysts. *Chemical Engineering and Processing* 122: 460–472. https://doi.org/10.1016/j.cep.2017.05.018.

Zazhigalov, S.V., Shilov, V.A., Rogozhnikov, V.N. et al. (2022). Mathematical modeling of diesel autothermal reformer geometry modifications. *Chemical Engineering Journal* 442 (1): 136160. 1-10. https://doi.org/10.1016/j.cej.2022.136160.

Index

Mathematical Modeling of Complex Reaction Systems in the Oil and Gas Industry, First Edition.
Edited by Jorge Ancheyta, Andrey Zagoruiko, and Andrey Elyshev.

d